T0295385

Thermal Management for Opto-electronics Packaging and Applications

Xiaobing Luo
Professor, IEEE/ASME Fellow
Dean, School of Energy and Power Engineering
Co-Dean, China-EU Institute for Clean and Renewable Energy
Huazhong University of Science and Technology
Wuhan, China

Run Hu
Professor, School of Energy and Power Engineering
Huazhong University of Science and Technology
Wuhan, China

Bin Xie
Assistant Professor, School of Mechanical Science and Engineering
Huazhong University of Science and Technology
Wuhan, China

This edition first published 2024

Registered Offices
John Wiley & Sons, Inc., 111 River Street, Hoboken, NJ 07030, USA
John Wiley & Sons Singapore Pte. Ltd, 1 Fusionopolis Walk, #07-01 Solaris South Tower, Singapore 138628

For details of our global editorial offices, customer services, and more information about Wiley products visit us at www.wiley.com.

Wiley also publishes its books in a variety of electronic formats and by print-on-demand. Some content that appears in standard print versions of this book may not be available in other formats.

Library of Congress Cataloging-in-Publication Data applied for:

Hardback ISBN: 9781119179276

Cover Design: Wiley
Cover Image: © asharkyu/Shutterstock

Set in 9.5/12.5pt STIXTwoText by Straive, Pondicherry, India

SKY8D7B82AF-DC9F-415B-BE8F-0DB92F6678BC_060324

Contents

List of Nomenclatures

Abbreviation

1D	one-dimensional
2D	two-dimensional
3D	three-dimensional
α-GaN	A-gallium nitride
β-Ga_2O_3	B-zirconia
AM	arrangement followed by mixing method
APG	alkyl polyglucoside
AW	atmospheric transparent window
BeO	beryllium oxide
BG	bilayer graphene
BGA	ball grid array
BLT	bond line thickness
BN	boron nitride
BW	bandwidth
CB	conduction band
CCT	correlated color temperature
CFD	computational fluid dynamics
CFs	carbon fibers
CH_2I_2	diiodomethane
CIE	Commission Internationale de L'Eclairage
CLTE	coefficient of linear thermal expansion
CMOS	complementary metal-oxide semiconductor
CMY	Cooper–Mikic–Yovanovich
CNC	computerized numerical control
CNT(s)	carbon nanotube(s)
Com-film	luminescent films containing without hBN
Com-WLEDs	common QDs-WLEDs without hBN
CPCMs	composite phase change materials
CRC	colored radiative cooler
CRI	color-rendering index
CS	crystal structure
CSF	cumulative structure function
CSP	chip-scale package
CTAB	hexadecyl trimethyl ammonium bromide
CTE	coefficients of thermal expansion
CTMS	centralized thermal management system
CVD	chemical vapor deposition
DA	diffusion approximation

DAA	die attach adhesive
DBC	direct bonded copper substrates
DC	dielectric constants
DIP	dual in-line package
DM	directly mixing without arrangement method
DMA	dynamic mechanical analysis
DPC	direct plate copper substrates
DQN	deep Q-learning network
DRL	deep reinforcement learning
DSC	differential scanning calorimetry
DTMS	distributed thermal management system
EBL	electron blocking layer
ECB	electrical conductivity bandwidth
EDS	energy dispersive spectroscopy
EG	expanded graphite
EGE	epsilon greedy exploration
EL	electroluminescence
EMA	effective medium approximation
EQE	external quantum efficiency
ER	electrical resistivity
FCP	flip chip package
FEA	finite-element analysis
FEM	finite-element model
FES	finite-element simulation
FLG	few-layers graphene
FRTE	an extension of RTE integrated with fluorescence
FWHM	full-width-at-half-maximum
GA	genetic algorithm
GNP	graphene nanoplatelets
GNs	graphene nanosheets
GO	graphene oxide
GP	graphene
HAADF-STEM	high-angle-annular-dark-field STEM
HBC	hybrid body cooling
hBN	hexagonal boron nitride
hBNPs	hexagonal boron nitride platelets
hBNS	hexagonal boron nitride sheets
HEC	hydroxyethyl cellulose
HG	Henyey–Greenstein
HP	heat pipe
HRTEM	high-resolution transmission electron microscope
HTHP	high-temperature and high-pressure
$I_{AM1.5}$	the standard AM 1.5 spectrum of solar radiation
IC(s)	integrated circuit(s)
IGBT(s)	insulated gate bipolar transistor(s)
iNEMI	international electronics manufacturing initiative
IQE	internal quantum efficiency
IR	infrared
Iso-film	luminescent films containing with isotropic distributed hBN
Iso-WLEDs	isotropic thermal conductive QDs-WLEDs with isotropic arranging hBN
JAICIPM	jet array impingement cooling system with integrated piezoelectric micropump
JIBC	jet impingement body cooling

JISC	jet impingement surface cooling
KM	Kubelka–Munk
LDs	laser diodes
LE	luminous efficiency
LED(s)	light-emitting diode(s)
LEE	light extraction efficiency
LERP	laser-excited remote phosphor
LFA	laser flash analysis
$LiAlO_2$	lithium aluminate
LM	lattice mismatch
MA	mixing followed by arrangement
MBAM	modified Bruggeman asymmetric model
MCE	mixed cellulose esters
MCM	multi-chip module
MCPCB	metal core-printed circuit board
MD	molecular dynamics
MDM	metal–dielectric–metal
mhBN	magnetically responsive hBN
mhBN-silicone	Imhbn-silicone
m-MPMF	micro multiple piezoelectric magnetic fan
mohBN	modified hBN
MOSFET	metal-oxide-semiconductor field-effect transistor
MPP	mesophase pitch
MQW	multiple quantum well
MWT	maximal working temperature
NICFs	nickel-coated carbon fibers
PAN	polyacrylonitrile
PCB	printed circuit board
pc-LD	phosphor-converted laser diode
pc-LED	phosphor-converted light-emitting diode
PCM(s)	phase change material(s)
PDMS	polydimethylsiloxane
PEEK	poly-ether–ether–ketone
PET	photo-electro-thermal
PG	propylene glycol
PID	proportional–integral–derivative (controller)
PL	photoluminescence
PMMA	polymethyl methacrylate
PTMS	passive thermal management system
QDs	quantum dots
QDs-WLEDs	QDs-converted WLEDs
QE	quantum efficiency
QFP	quad flap package
QSNs	QDs-silica coated nanoparticles
QY	quantum yield
RC	radiative cooling
RDL	redistribution layer
RE	relative error
RMSE	root mean square error
RTDs	resistance temperature detectors
RTE	radiative transfer equation
SAM	self-assembled monolayer

SAM-CH3	alkanethiol type SAM
SAM-NH2	11-amino-1-undecanethiol hydrochloride
SCMC	sodium carboxymethyl cellulose
SEM	scanning electron microscope
SiC	carborundum
SILAR	successive ionic layer adsorption and reaction
SLA	stereolithography
SLG	single-layer graphene
SMD	Sauter mean diameter
SNTP	solid heat conduction, natural air convection, thermal radiation, and phase change processes
SOI	silicon-on-insulator
SP	solid heat conduction and phase change processes
SPD	spectral power distribution
SR	selective radiative cooling radiator
SRH	Shockley–Read–Hall
SSH	the sustainability of the image horizontal direction
SSV	the sustainability of the image in the vertical direction
STEM	scanning transmission electron microscope
TC	thermal camouflage
TCEE	thermal conductivity enhancement efficiency
TD	thermal diffusion coefficient
TEOS	tetraethyl orthosilicate
TES	thermal energy storage
TFFC	thin-film flip-chip
TG	thermal gravimetric analysis
TIM(s)	thermal interface material(s)
TIR	total internal reflection
TMA	thermal mechanical analysis
TmhBN-silicone	through-plane-aligned mhBN-silicone
TMM	transfer matrix method
TMR	thermal mismatch rate
TMS(s)	thermal management system(s)
TO	transistor outline
TOP	tri-*n*-octylphosphine
TRPL	time-resolved PL
TSOP	thin small outline package
TSV	through silicon via
UHMR	ultrahigh magnetic response
UV	ultraviolet
VB	valance band
VC	vapor chamber
VCHP	variable conductance heat pipe
Ver-film	luminescent films containing with vertical hBNSCMC templates
Ver-WLEDs	vertical thermal conductive QDs-WLEDs
V-ORI	orientation in the vertical direction
WLEDs	white light-emitting diodes
WRC	white-color radiative cooler
XPS	X-ray photoelectron spectroscopy
XRD	X-ray diffraction
XRT	X-ray tomography
ZnO	zinc oxide

Nomenclature (Units of Measure)

Rate_{SRH}	recombination rate (s^{-1})
Eu	Euler number (–)
Gr	Grashof number (–)
Nu	Nusselt number (–)
Pe	Berkeley number (–)
Pr	Prandtl number (–)
Re	Reynolds number (–)
k_{SRH}	SRH recombination coefficient ($cm^6 \cdot s^{-1}$)
k_{auger}	the Auger recombination coefficient ($cm^6 \cdot s^{-1}$)
k_{rad}	the radiative recombination coefficient ($cm^6 \cdot s^{-1}$)
A	surface area (m^2)
A_c	cross-sectional area (m^2)
A_r	the radiative surface area (m^2)
n	carrier concentration (cm^{-3})
n_c	phosphor concentration ($g \cdot cm^{-3}$)
k_{f2}	constant coefficient 2 (–)
$\text{Rate}_{\text{auger}}$	the rate of Auger recombination (s^{-1})
k_{f1}	constant coefficient 1 (–)
P_h	heat transfer rate (W)
Q_f	flow rate ($ml \cdot min^{-1}$)
P_{rad}	the rate of thermal radiation energy (W)
e	electron charge (C)
q	heat flux density ($W \cdot m^{-2}$)
$\partial T/\partial x$	temperature gradient ($K \cdot W^{-1}$)
F	driving force (N)
R	resistance of the process ($K \cdot W^{-1}$)
R_{thermal}	thermal resistance ($K \cdot W^{-1}$)
R_c	contact resistance ($K \cdot W^{-1}$)
m_t	transfer amount (W)
m	mass (kg)
k_{f3}	constant coefficient 3 (–)
T_j	junction temperature (K)
T_a	ambient temperature (K)
$R_{\text{c-TIMs}}$	thermal contacted resistance with TIMs ($K \cdot W^{-1}$)
R_{TIMs}	the bulk resistance of the TIMs ($K \cdot W^{-1}$)
R_s	thermal spreading resistance ($K \cdot W^{-1}$)
R_{cu}	total thermal resistance of the copper substrate ($K \cdot W^{-1}$)
C_p	constant pressure heat capacity ($J \cdot kg^{-1} \cdot K^{-1}$)
k_{wl}	coefficient determined by the materiel of surface and liquid
D_h	hydraulic diameter (m)
F_x	external force in x-direction (N)
F_y	external force in y-direction (N)
g	gravity ($m \cdot s^{-2}$)
h	convection heat transfer coefficient ($W \cdot m^{-2} \cdot K^{-1}$)
k_P	Planck's constant ($J \cdot s$)
h_c	stable heat transfer coefficient ($W \cdot m^{-2} \cdot K^{-1}$)
$h_{\text{non-rad}}$	a non-radiative heat transfer coefficient that combines the effective conductive and convective heat exchange ($W \cdot m^{-2} \cdot K^{-1}$)
$h_{\text{microchannel}}$	the heat transfer coefficient of the microchannel liquid cooling ($W \cdot m^{-2} \cdot K^{-1}$)

l	length (m)
l_n	nozzle length (m)
l_p	length of plate (m)
l_y	length in y-direction (m)
l_z	length in z-direction (m)
p	pressure (Pa)
k_{f4}	constant coefficient 4 (–)
k_{f5}	constant coefficient 5 (–)
k_{asp}	aspect ratio of a rectangular duct (–)
p_{in}	pressure of the entrance of the flow (Pa)
p_{out}	pressure of the exit of the flow (Pa)
p_{sw}	shear stress of the partial wall (Pa)
P_{in}	the total output optical power from LED
q_c	convection heat transfer density ($W \cdot m^{-2}$)
q_w	heat transfer density between fluid and contacting solid ($W \cdot m^{-2}$)
s_v	horizontal pipe spacing in the vertical flow direction (m)
s_l	horizontal pipe spacing in the longitudinal pipe spacing along the flow (m)
T_f	qualitative temperature of the fluid (K or °C)
T'_f	average temperature of the fluid at inlet (K or °C)
T''_f	average temperature of the fluid at outlet (K or °C)
T_m	qualitative temperature of the vapor (K or °C)
T_s	saturation temperature (K or °C)
T_w	temperature of the wall and the flow (K or °C)
T_∞	temperature of the flow (K or °C)
u_x	velocity in x-direction ($m \cdot s^{-1}$)
u_{xdim}	dimensionless velocity in x-direction (–)
u_y	velocity in y-direction ($m \cdot s^{-1}$)
u_{ydim}	dimensionless velocity ratio in y-direction (–)
x	coordinate x (m)
x_{dim}	dimensionless length in x-direction (–)
y	coordinate y (m)
y_{dim}	dimensionless length in y-direction (–)
p_{dim}	dimensionless pressure (–)
T_{Edim}	dimensionless excess temperature (–)
∇T	temperature difference (K or °C)
E_λ	the energy radiated per unit volume by a cavity of a blackbody in the wavelength interval λ to $\lambda + \Delta\lambda$ ($\Delta\lambda$ denotes an increment of wavelength) ($W \cdot m^{-3}$)
c	the speed of light ($m \cdot s^{-1}$)
k_B	the Boltzmann constant ($J \cdot K^{-1}$)
T	temperature (K or °C)
rew_t	the reward in Bellman equation (–)
r_{log}	radius of the logging tool (m)
wei_{tar}	the weights of the target network (–)
wei_{main}	the weights of the main network (–)
$P_{cooling}$	cooling power of the emitter (W)
P_{rad}	the power emitted from the emitter (W)
P_{atm}	the input power from the atmosphere absorbed by the emitter (W)
P_{solar}	the incident solar power (W)
$P_{cond+conv}$	the power of the non-radiative heat transfer due to the conductive and convective (W)
T_{amb}	the temperature of the ambient air (K)
I_{BB}	the spectral radiance of a blackbody at temperature T and wavelength λ ($W \cdot m^{-1} \cdot sr^{-1}$)

G	the total solar irradiance ($W \cdot m^{-2}$)
T_{ste}	the steady temperature when the $P_{cooling}$ is zero (K)
H	nozzle-to-surface distance (m)
k_{sro}	the spearman rank order correlation coefficients (–)
k_{sro_L}	the spearman rank order correlation coefficients of *Lightness* (–)
k_{sro_C}	the spearman rank order correlation coefficients of *Chroma* (–)
k_{sro_H}	the spearman rank order correlation coefficients of *Hue* (–)
J	the LED injection current density ($A \cdot m^{-2}$)
q_{eff}	effective radiation ($W \cdot m^{-2}$)
t	time (s)
BE_c	the effective density of states at conduction band edge (m^{-3})
BE_V	the effective density of states at valence band edge (m^{-3})
P_{opt}	the optical power (W)
P_{ele}	a constant input electrical power (W)
P_{heat}	the heat generation power within the LED (W)
k_{heat}	the heat-dissipation coefficient (–)
R_{hs}	the heat sink to ambient thermal resistance ($K \cdot W^{-1}$)
R_{jc}	the junction-to-case thermal resistance ($K \cdot W^{-1}$)
k_{rs}	the relative slope (–)
$\boldsymbol{n}_w$	the unit inward normal vector at the substrate boundary ($z = 0$) (–)
n_p	the refractive index of the phosphor layer (–)
n_0	the refractive index of the air (–)
$\boldsymbol{\Omega}^m$	a specific angular direction with the superscript m (sr)
M	the total angular number (–)
I_j	the radiative intensity at solution node j ($W \cdot sr^{-1}$)
I_g	the global approximation intensity ($W \cdot sr^{-1}$)
N_{sol}	the total number of solution nodes (–)
P_{ph}	the phosphor heating power (W)
R_{ph}	the conductive thermal resistance of phosphor layer ($K \cdot W^{-1}$)
R_{mir}	the conductive thermal resistance of mirror layer ($K \cdot W^{-1}$)
R_{bond}	the conductive thermal resistance of bonding layer ($K \cdot W^{-1}$)
R_{conv}	the convective thermal resistance between heat sink with the ambient ($K \cdot W^{-1}$)
D_{ph}	the diameter of a circular phosphor plate (m)
D_{spot}	the diameter of a circular pump spot (m)
P_{limit}	the critical incident power (W)
R_j	thermal resistance at the joint ($K \cdot W^{-1}$)
R_b	thermal boundary resistance ($K \cdot W^{-1}$)
R_{bulk}	bulk resistance of TIM ($K \cdot W^{-1}$)
G_{int}	interfacial thermal conductance ($K \cdot W^{-1}$)
D/t	aspect ratio (–)
D	diameter (m)
D_n	nozzle diameter (m)
f	volume fraction (%)
T_{ave}	average temperature of heat source (K)
T_{max}	maximum temperature (K)
k_{Fit}	fitting parameters (–)
I	relative PL initial intensity (–)
$P_{heat\text{-}chip}$	heat power of LED chip (W)
$P_{QDs/hBN\text{-}phosphor}$	heat power of QDs/hBN-phosphor (W)
$P_{QDs\text{-}phosphor}$	heat power of QDs-phosphor (W)

P_{el}	input electrical power of WLEDs package (W)
$P_{\mathrm{op.ref}}$	optical power from WLEDs with only silicone (W)
$P_{\mathrm{op\text{-}Qs}}$	optical power from WLEDs with QSNs and silicone (W)
$P_{\mathrm{op\text{-}to}}$	optical power from the type I WLEDs (W)
$P_{\mathrm{op\text{-}tt}}$	optical power from the type II WLEDs (W)
P_{QSNs}	heat power of QSNs (W)
P_{phosphor}	heat power of phosphor silicone gel (W)
$P_{\mathrm{QSNs/phosphor}}$	heat power of QSNs/phosphor-silicone gel (W)
$P_{\mathrm{op\text{-}wq}}$	optical power from WLEDs with QSNs and silicone
r	radius (m)
T_{b}	bottom temperature (K or °C)
T_{t}	top temperature (K or °C)
C_{eff}	equivalent heat capacity of PCM ($\mathrm{J \cdot kg^{-1} \cdot {}^{\circ}C^{-1}}$)
h_{L}	average convective heat transfer coefficient ($\mathrm{W \cdot m^{-2} \cdot K^{-1}}$)
m_{PCM}	mass of PCMs (kg)
LH	total latent heat (J)
SH	total sensible heat (J)
r_{h}	radius of the wellbore wall (m)
T_0	initial temperature (K or °C)
T_{e}	final temperature (K or °C)
T_{l}	final phase change end temperature of PCM (K or °C)
$T_{\mathrm{s\text{-}ini}}$	initial phase change temperature of PCM (K or °C)
D_{s}	Sauter mean diameter (m)
T_{sub}	substrate temperature (K or °C)
T_{sur}	surface temperature (K or °C)
ΔT_{uni}	temperature non-uniformity (K or °C)
$\overline{Q}''$	mean volumetric flow rate ($\mathrm{m^3 \cdot s^{-1} \cdot m^{-2}}$)
Q_{v}	total volume flow rate ($\mathrm{m^3 \cdot s^{-1}}$)
q_{r}	heat flux ratio (%)
w	frequency (Hz)
w_{c}	critical frequency (Hz)
dir^{m}	the direction cosine along z-direction (–)
f_{PCM}	volume fraction of liquid PCM (–)
Abs	absorptivity (–)
$\mathrm{Abs_{emi}}$	the absorptivity of the emitter
DF	the discount factor (–)
$\mathrm{LH_{liq}}$	latent heat of liquefaction ($\mathrm{J \cdot kg^{-1}}$)
k_z	correction factor determined by the number of pipe rows (–)
k_{β}	correction factor determined by the impact angle (–)
Tran	transmissivity (–)
$\mathrm{Tran_{atm}}$	the transmissivity of the atmosphere in the zenith direction
T_{E}	excess temperature (K or °C)
k_{a}	the absorption coefficient ($\mathrm{mm^{-1}}$)
k_{s}	the scattering coefficient ($\mathrm{mm^{-1}}$)
ORI	degrees of orientation (–)
u_{m}	average velocity of the pipe section ($\mathrm{m \cdot s^{-1}}$)
u_{∞}	velocity of the flow ($\mathrm{m \cdot s^{-1}}$)
u_{i}	in-plane-diffused velocity ($\mathrm{m \cdot s^{-1}}$)
T_{hs}	the heat sink temperature (K or °C)
T_{ph}	the phosphor temperature (K or °C)
T_{c}	the critical temperature (K or °C)

L_{PCM}	the latent heat of PCM ($\mathrm{kJ \cdot kg^{-1}}$)
$\boldsymbol{u}$	the air velocity ($\mathrm{m \cdot s^{-1}}$)
C_{air}	specific heat capacity of air($\mathrm{J \cdot kg^{-1} \cdot K^{-1}}$)
Fg	the gravity (N)
J_i	the effective radiation of the surface i ($\mathrm{W \cdot m^{-2}}$)
G_i	the input radiation of the surface i ($\mathrm{W \cdot m^{-2}}$)
T_i	the temperature of the surface i (K)
q_i	the radiation heat transfer of surface i (W)
$k_{F_{ij}}$	the radiation angle coefficient from surface i to surface j (–)
A_i	surface area of surface i ($\mathrm{m^2}$)
A_j	surface area of surface j ($\mathrm{m^2}$)
d_{ij}	the distance between the surface i and the surface j (m)
$C_{\mathrm{PCM\text{-}S}}$	heat capacity of solid PCM ($\mathrm{J \cdot kg^{-1} \cdot {}^{\circ}C^{-1}}$)
$C_{\mathrm{PCM\text{-}L}}$	heat capacity of liquid PCM ($\mathrm{J \cdot kg^{-1} \cdot {}^{\circ}C^{-1}}$)
$V_{\mathrm{PCM\text{-}S}}$	the volume of the solid PCMs ($\mathrm{m^{-3}}$)
$V_{\mathrm{PCM\text{-}L}}$	the volume of the liquid PCMs ($\mathrm{m^{-3}}$)
u_{mud}	the velocity of the logging tool movement ($\mathrm{m \cdot s^{-1}}$)
l_{log}	the length of the logging tool (m)
$l_{\mathrm{insulator\ 1}}$	the length of insulator 1 (mm)
$l_{\mathrm{PCMs\ 1}}$	the length of PCMs 1 (mm)
$l_{\mathrm{PCMs\ 2}}$	the length of PCMs 2 (mm)
$l_{\mathrm{insulator\ 2}}$	the length of insulator 2 (mm)
$T_{\mathrm{heat\ source\ 1}}$	the temperature of heat source 1 (°C)
$T_{\mathrm{heat\ source\ 2}}$	the temperature of heat source 2 (°C)
N_{v}	the number of vertexes in the simplex (–)
X_i	the ith vertex (–)
X_{l}	the optimal vertex (–)
K_{VT}	the temperature rise rate ($\mathrm{{}^{\circ}C \cdot min^{-1}}$)
L_{comp}	the latent heat of CPCMs ($\mathrm{kJ \cdot kg^{-1}}$)
L_{para}	the latent heat of paraffin ($\mathrm{kJ \cdot kg^{-1}}$)
ΔM	weight loss ratio (–)
r_{β}	the Kapitza radius (m)
W_{SL}	thermodynamic parameter ($\mathrm{J \cdot m^2}$)
P_{thermal}	thermal power (W)
r_{p}	the radius of the circular microcontact point (m)
r_{t}	the radius of the heat flux tube (m)
R_{a}	the arithmetic mean deviation (m)
k_{sl}	the slope (–)
H_{a}	heights of asperity (m)
wid	width (m)
S	spacing between nozzles (m)
U	uncertainty (–)
T_{in}	the inlet temperature (K or °C)
T_{out}	the outlet temperature (K or °C)
T_{loc}	the local temperature of the surface (K or °C)
N_{top}	the number of nozzles of the top surface (–)
N_{front}	the number of nozzles of the front surface (–)
N_{right}	the number of nozzles of the right surface (–)
H_{ch}	channel height (m)
R_{fth}	effective fluid thermal resistance ($\mathrm{K \cdot W^{-1}}$)
R_{fconv}	fluid convection thermal resistance ($\mathrm{K \cdot W^{-1}}$)

R_{wcond} conduction thermal resistance to the wall ($\mathrm{K \cdot W^{-1}}$)
R_{wconv} wall conduction thermal resistance ($\mathrm{K \cdot W^{-1}}$)
$l_{\mathrm{chip}\text{-}x}$ the dimensions of the chip in x-direction (m)
$l_{\mathrm{chip}\text{-}y}$ the dimensions of the chip in y-direction (m)
$l_{\mathrm{chip}\text{-}z}$ the dimensions of the chip in z-direction (m)
$\mathrm{wid_{ch}}$ channel width (m)
$\mathrm{wid_{w}}$ channel wall width (m)
$P_{\mathrm{thermal\_sur}}$ thermal power of the surface (W)
N_x the number of discrete mesh in x-direction (–)
N_y the number of discrete mesh in y-direction (–)

Greek Letters

δ thickness (m)
δ_{LED} the thickness of LED active region (m)
δ_{ph} the thickness of the phosphor plate (m)
δ_{b} the thickness of boundary layer (m)
δ_{tb} the thickness of thermal boundary layer (m)
δ_{r} thickness ratio (–)
τ time constant (s)
τ_1 stable PL lifetime (s)
Ω the solid angle (sr)
θ fiber orientation angle (deg)
θ_{c} critical incident angle (deg)
θ' emission angle (deg)
θ'_i emission angle of the surface i (deg)
θ'_j emission angle of the surface j (deg)
θ_{t} the transmitted angle (deg)
θ_{c} contact angle (deg)
α thermal diffusivity ($\mathrm{m^2 \cdot s^{-1}}$)
α_{mud} thermal diffusivity of the mud ($\mathrm{m^2 \cdot s^{-1}}$)
α_{T} through-plane thermal diffusivity ($\mathrm{m^2 \cdot s^{-1}}$)
β coefficient of cubical expansion (–)
Γ surface energy (J)
μ dynamic viscosity ($\mathrm{N \cdot s \cdot m^{-2}}$)
μ_{c} dynamic viscosity of the coolant ($\mathrm{N \cdot s \cdot m^{-2}}$)
μ_{f} dynamic viscosity of average fluid temperature ($\mathrm{N \cdot s \cdot m^{-2}}$)
μ_{l} dynamic viscosity of the saturation liquid ($\mathrm{N \cdot s \cdot m^{-2}}$)
μ_{w} dynamic viscosity of average wall temperature ($\mathrm{N \cdot s \cdot m^{-2}}$)
μ_{v} dynamic viscosity of vapor ($\mathrm{N \cdot s \cdot m^{-2}}$)
μ_{K} kinematic viscosity ($\mathrm{m^2 \cdot s^{-1}}$)
ρ density ($\mathrm{kg \cdot m^{-3}}$)
ρ_{air} density of the air ($\mathrm{kg \cdot m^{-3}}$)
ρ_{0} density of the air initial density ($\mathrm{kg \cdot m^{-3}}$)
ρ_{PCM} density of the PCM ($\mathrm{kg \cdot m^{-3}}$)
$\rho_{\mathrm{PCM\text{-}S}}$ density of the solid PCM ($\mathrm{kg \cdot m^{-3}}$)
$\rho_{\mathrm{PCM\text{-}L}}$ density of the liquid PCM ($\mathrm{kg \cdot m^{-3}}$)
ρ_{l} density of the liquid ($\mathrm{kg \cdot m^{-3}}$)
ρ_{v} density of the vapor ($\mathrm{kg \cdot m^{-3}}$)

ρ_{md}	microcontacts density ($\mathrm{m^{-2}}$)
σ	the Stefan–Boltzmann constant ($\mathrm{W \cdot m^{-2} \cdot K^{-4}}$)
σ_{g}	Gaussian distribution (–)
ζ_{s}	surface tension of the liquid ($\mathrm{N \cdot m^{-1}}$)
ζ_{AW}	air–liquid surface tension ($\mathrm{J \cdot m^{2}}$)
ε	emissivity (–)
$\varepsilon_{\mathrm{atm}}$	the emissivity of the atmosphere (–)
η_{qm}	the phosphor quantum efficiency (%)
η_{lum}	the luminous efficiency (%)
η_{0}	the rated efficiency at the temperature T_0 (%)
η_{EQ}	the external quantum efficiency (%)
η_{con}	the conversion efficiency (%)
η_{ex}	light extraction efficiency (%)
η_{inj}	electrical injection efficiency (%)
η_{IQ}	internal quantum efficiency (%)
η_{wp}	wall-plug efficiency (%)
γ	reflectivity (–)
γ_{m}	mirror reflectivity (–)
γ_{B}	the diffuse reflectivity at substrate surface for blue light (–)
γ_{Y}	the diffuse reflectivity at substrate surface for yellow light (–)
ω^{m}	the angular weight corresponding to the incident direction $\boldsymbol{\Omega}^{m}$ (%)
ω_{GN}	the mass fraction of GNs (%)
λ	wavelength (m)
$\Delta\lambda$	an increment of wavelength (m)
λ_{B}	the wavelength for excitation blue light (m)
λ_{Y}	the wavelength for emission yellow light (m)
φ	distribution function (–)
ξ	the pre-set value (–)
κ_{p}	reduced thermal conductivity (–)
κ	thermal conductivity ($\mathrm{W \cdot m^{-1} \cdot K^{-1}}$)
κ_{air}	air thermal conductivity ($\mathrm{W \cdot m^{-1} \cdot K^{-1}}$)
κ_{PCM}	thermal conductivity of PCM ($\mathrm{W \cdot m^{-1} \cdot K^{-1}}$)
$\kappa_{\mathrm{PCM\text{-}S}}$	thermal conductivity of solid PCM ($\mathrm{W \cdot m^{-1} \cdot K^{-1}}$)
$\kappa_{\mathrm{PCM\text{-}L}}$	thermal conductivity of liquid PCM ($\mathrm{W \cdot m^{-1} \cdot K^{-1}}$)
κ_{mud}	thermal conductivity of the mud ($\mathrm{W \cdot m^{-1} \cdot K^{-1}}$)
κ_{f}	fluid thermal conductivity ($\mathrm{W \cdot m^{-1} \cdot K^{-1}}$)
κ_{m}	thermal conductivity of matrix ($\mathrm{W \cdot m^{-1} \cdot K^{-1}}$)
κ_{T}	through-plane thermal conductivity ($\mathrm{W \cdot m^{-1} \cdot K^{-1}}$)
κ_{TIM}	thermal conductivity of TIM ($\mathrm{W \cdot m^{-1} \cdot K^{-1}}$)
κ_{v}	thermal conductivity of vapor ($\mathrm{W \cdot m^{-1} \cdot K^{-1}}$)
q_{vol}	volumetric heat generation in the solid ($\mathrm{W \cdot m^{-3}}$)

About the Authors

Xiaobing Luo (Date of birth: April 3, 1974) is a professor, doctoral supervisor, recipient of the National Outstanding Youth Fund, IEEE Fellow, ASME Fellow, and leading talent in scientific and technological innovation under the National Ten Thousand Talents Plan. He enjoys special allowances from the State Council and is the Dean of the School of Energy and Power Engineering and the Chinese Dean of the China Europe Energy Institute at Huazhong University of Science and Technology (HUST). He received his PhD in 2002 from Tsinghua University, China. From 2002 to 2005, he worked at the Samsung Advanced Institute of Technology (SAIT) in Korea as a senior engineer and obtained the SAIT Best Researcher Award in 2003. In September 2005, he returned to China and became an associate professor. In November 2007, he became a full professor after an exceptional promotion. He obtained the 2020 Baosteel Excellent Teacher Special Award, the second prize of the 2018 National Teaching Achievement Award (ranked second), the 2016 IEEE CPMT Outstanding Technical Achievement Award, the second prize of the 2016 National Technology Invention Award (ranked second), and the first prize of the 2015 Hubei Provincial Natural Science Award. He served as the Associate Editor for IEEE Packaging and ASME Electronic Packaging, respectively. In 2019 and 2018, he served as the Vice General Chair for the ASME InterPACK Conference in Silicon Valley, USA. As the first or corresponding author, he has published 176 SCI-indexed papers and authorized 56 Chinese invention patents and 5 US patents, of which 20 have been transferred in cash.

Run Hu (Date of birth: September 6, 1987) is a professor and doctoral supervisor at the School of Energy and Power Engineering, Huazhong University of Science and Technology. He received his bachelor and PhD degrees in 2010 and 2015, respectively, from HUST. He then joined the faculty in the School of Energy and Power Engineering. From November 2014 to March 2015, he worked at Purdue University, United States, as a visiting scholar. From November 2016 to November 2017, he worked at the University of Tokyo, Japan, as a JSPS postdoctoral fellow. In 2017 and 2023, he got promoted to associate professor and full professor, respectively. He was the recipient of Outstanding Youth Scholar and Chutian Scholar in Hubei Province. He served as the youth editor for several journals like Research. As the first or corresponding author, he has published more than 90 SCI-index papers and granted 20 Chinese invention patents.

Bin Xie (Date of birth: February 20, 1993) is an assistant professor at the School of Mechanical Science and Engineering, HUST. He received his bachelor and PhD degrees in 2014 and 2019, respectively, from HUST. From 2019 to 2020, he worked at Huawei Technologies Co., Ltd. in Wuhan as a senior engineer. From 2020 to 2022, he worked at HUST as a postdoctoral researcher. He then joined the faculty at the School of Mechanical Science and Engineering, HUST. He was the recipient of the Natural Science Prize of Guangdong Province (second class) and the Outstanding Paper Award of the International Conference on Electronic Packaging Technology (ICEPT). He served as the guest editor in several journals such as *Electronics*. As the first or corresponding author, he has published more than 20 SCI-index papers and granted 14 Chinese invention patents.

Preface

Date back to 2002 when the first author Xiaobing Luo graduated from Tsinghua University and began to work at Samsung Electronics Co. Ltd. in Suwon, Korea, he realized the importance of thermal management for electronics and devoted himself to the development of advanced thermal management solutions for integrated circuits (ICs), power electronics, microelectromechanical systems, etc. Three years later, although he had gotten promoted to senior engineer at Samsung, he resolutely and determinedly returned to China and joined the faculty at the School of Energy and Power Engineering at Huazhong University of Science and Technology, where he had spent his undergraduate and postgraduate periods from 1991 to 1998. Since then, he has established the Thermal Packaging Laboratory, which aims at developing advanced active/passive thermal management solutions for optoelectronics and electronics. He built a strong and fruitful collaboration with Prof. Sheng Liu from the School of Mechanical Engineering and Science and Wuhan National Laboratory for Optoelectronics at Huazhong University of Science and Technology and developed high-performance, high-power light-emitting diode (LED) packaging technologies, which perfectly solved the state-of-the-art optical, thermal, and mechanical challenges of high-power LED packages. Besides, he firmly believed that liquid cooling is the trend for active thermal management of power electronics, and the core of liquid cooling is the pump. Since 2008, he began to develop the miniature mechanical pump and successfully developed the prototypes of the hydraulic suspension pump, super-thin vortex pump, and piezoelectric pump, which have been granted many patents in China and the United States and transferred to companies for industrial commercialization. In 2010 and 2014, Run Hu and Bin Xie joined his group as PhD students and began to devote themselves to developing different kinds of thermal management solutions for optoelectronics and power electronics, such as high-power white-light LEDs, quantum-dot LEDs, insulated gate bipolar transistor (IGBT), IC packaging, and so on. In particular, they realized the importance of package-inside thermal management, which provides unique solutions for high-performance photoluminescent materials and optoelectronics. More importantly, due to increasing miniaturization of chips, devices, and power electronics, thermal management has become the bottleneck and attracts booming attention from both academic and industrial aspects, which is almost the common consensus across the world. So it is quite the right time to summarize what we did in the field of thermal management for optoelectronics and electronics. This book intends to assemble what we learned in the past years into a useful reference book for both the LED community and the IC packaging community, in the hope that the results to be presented will benefit engineers, researchers, and young students.

Therefore, this book's subject matter is thermal management for electronics, which can be a reference book for thermal engineers, mechanical engineers, packaging engineers, reliability engineers, and graduate students. This book covers the three basic ways for thermal management (i.e. thermal conduction, thermal convection, and thermal radiation) and the specific applications of these three ways for advanced thermal management in optoelectronics and power electronics. Moreover, this book will also demonstrate some specific applications, such as opto-thermal modeling, thermal interface materials, liquid cooling, packaging-in thermal dissipation, phase-change materials in downhole electronics, and so on. This book will not only present the specific applications but also provide fundamental research to satisfy the interests of active researchers.

There have already been five nice books about thermal management of electronics available to readers in English. They are *Thermal Management Handbook: For Electronic Assemblies* by Jerry Sergent and Al Krum in 1998; *Heat Transfer: Thermal Management of Electronics* by Younes Shabany in 2009; *Thermal Management for LED Applications* by Clemens J.M. Lasance and Andras Poppe in 2013; *The Art of Software Thermal Management for Embedded Systems* by Mark Benson in 2014; and *Thermal Management of Microelectronic Equipment* by Lian-Tuu Yeh in 2016. Moreover, there are some books on specific applications, such as lithium-ion batteries, electric vehicles, aircrafts, engines, heat pipes, LEDs, telecommunications equipment, power plants, data centers, and military applications. However, there is not such a book focusing on the thermal management of opto-electronics from both fundamental analysis and application design aspects. In addition, there have been no books dedicated to package-inside thermal management and its corresponding applications.

All authors feel obligated to explore these subjects and contribute to the whole community through our recent findings to promote the healthy development of thermal management technologies for optoelectronics and power electronics. Chapter 1 provides an introduction of packaging, with an emphasis on thermal issues and challenges. Chapters 2–4 provide the fundamentals and development trends of thermal management solutions with respect to thermal conduction, convection, and radiation, respectively. Specifically, Chapter 2 will cover the introduction of different thermal materials, ranging from thermal conduction materials, thermal interfacial materials, heat pipes, phase-change materials, and thermal metamaterials. Chapter 3 will introduce the air and liquid cooling technologies. Chapter 4 will introduce spectral and directional radiative cooling materials for electronics cooling. The upcoming chapters will begin to provide some specific examples, including the problems and their solutions. Chapter 5 focuses on the opto-thermal modeling of photoluminescent materials such as phosphors in silicone for light converters in white-light LEDs. Chapter 6 emphasizes on the thermally enhanced thermal interfacial materials, covering the modeling of thermal interfacial resistance and modulation of thermal conductivity. Chapter 7 is devoted to liquid cooling for high-heat-flux electronics, including jet impingement cooling, spray cooling, direct body cooling, and microchannel cooling. Chapter 8 tends to introduce the concept of package-inside thermal management for phosphors and quantum dots. Chapter 9 will provide a unique example of high-temperature thermal management for downhole electronics with distributed phase-change materials. We hope this book will be a valuable source of reference for all those who have been facing the challenging thermal problems created by the ever-expanding application. We also sincerely hope it will aid in stimulating further research and development on new thermal materials, analytical methods, testing and measurement methods, and even newer standards, with the goal of achieving a green environment and an eco-friendly energy-saving industry.

The organizations that know how to develop thermal management have the potential to make major advances in developing their own intellectual properties (IPs) in packaging and applications to achieve benefits in performance, cost, quality, and size/weight. It is our hope that the information presented in this book may assist in removing some of the barriers, avoiding unnecessary false starts, and accelerating the applications of these techniques. Developing thermal management solutions for opto-electronics packaging and applications may be limited only by the ingenuity and imagination of engineers, managers, and researchers.

Xiaobing Luo, PhD,
IEEE/ASME Fellow, Professor
School of Energy and Power Engineering,
Huazhong University of Science and Technology,
Wuhan, Hubei, China

Run Hu, PhD,
Professor
School of Energy and Power Engineering,
Huazhong University of Science and Technology,
Wuhan, Hubei, China

Bin Xie, PhD,
Assistant Professor
School of Mechanical Science and Engineering,
Huazhong University of Science and Technology,
Wuhan, Hubei, China
March 1st, 2024

1

Introduction

1.1 Development History of Packaging

Since the invention of the first semiconductor transistor in 1947, the electronic industry has experienced rapid development following Moore's law in the past decades. Packaging plays a key role in the electronic industry: it integrates numerous packaging components (i.e., chip, solder, printed circuit board [PCB], encapsulant, and cooling device) together to form a full-featured electronic system. The rapid development of the electronic industry raises numerous requirements for packaging. On the one hand, packaging is moving toward high integration, high frequency, high power, and low cost. On the other hand, packaging is expected to be more energy efficient, environment-friendly, and sustainable. These requirements pose several challenges for packaging including material development, process innovation, electrical design, and thermal management.

Nowadays, a single-chip package contains most functions of an electronic system and chips are the core component of the package. However, bare chips are far from application due to several problems. First, chips should be connected with the external circuit by pin, bonding wire, or solder ball. Second, chips are very sensitive to external factors including external mechanical force, moisture, and dust. Therefore, the chips should be isolated from these external factors by encapsulation. Third, chips generate heat during operation, which raises the chip junction temperature. If, without excellent thermal management, the chip junction temperature reaches an extremely high value, this would cause serious efficiency drop, lifetime reduction, and even chip failure. These problems are becoming more serious due to the increasing requirement for high integration, high frequency, and high power. For electronic connection, high integration of chips results in dense pin, bonding wire, and solder joint. For chip isolation, to place more chips on a fixed-size circuit, the encapsulation should be more compact. For thermal management, the high integration and high power of chips result in extremely high heat flux, which requires advanced cooling technologies such as vapor chamber, microchannel, microjet cooling, and spraying cooling. The main function of packaging is to solve the three problems listed earlier. Therefore, packaging is also regarded as chip packaging. It not only plays an important role in placing, fixing, sealing, and protecting the chips but also connects the chips with the external circuits and provides thermal management for chips.

In the past decades, packaging has undergone rapid development. In general, the development of packaging can be divided into three stages. The first stage is the through-hole insertion technology before the 1980s. This technology inserts the chip pins directly into the through holes of PCB. Because the technology requires extremely high alignment of the pins and holes, it presents low packaging density and frequency. The second stage is the surface-mount technology that emerged in the mid-1980s. It mounts the chips on the PCB through tiny pins. Compared to the through-hole insertion technology, it enhances the electrical characteristics of chips and improves the automation degree of production significantly. Although this technology has advantages of high density, small pin spacing, low cost, and suitability for surface mounting, it still fails to meet the packaging requirements of some advanced electrical systems. The third stage is the ball grid array (BGA) and chip scale package (CSP) technologies after the 1990s. During this period, the electronic industry developed rapidly, so the previous packaging technologies no longer met the packaging requirements. In this situation, the BGA and CSP were developed. They utilize the solder balls as input/output (I/O) pins, which greatly increases the package density. The emergence of BGA and CSP led to the explosive growth of the electronic industry.

Thermal Management for Opto-electronics Packaging and Applications, First Edition. Xiaobing Luo, Run Hu, and Bin Xie.

During the development of packaging, a lot of packaging technologies have been developed, such as transistor outline (TO) package, dual in-line package (DIP), quad flat package (QFP), thin small outline package (TSOP), BGA, CSP, flip-chip package (FCP), multichip module (MCM), and 3D packaging. From TO to 3D packaging, the technical indicators of packaging have greatly improved, such as closer chip area to package area ratio, higher integrating density, better thermal management, denser pins, closer pin spacing, smaller weight, and higher reliability. As the packaging requirements increase, some packaging technologies have been gradually eliminated. However, some advanced packaging technologies have been used until today, such as BGA, CSP, MCM, and 3D packaging, which will be introduced in the following text.

1.1.1 BGA

With the development of electronic industry, the number of I/O pins and power consumption of chips increase dramatically. Therefore, the traditional QFP and TSOP technologies can no longer meet the packaging demand. To solve this problem, the BGA packaging technology was developed in 1998. It synthesizes solder balls at the bottom of the package and uses them as I/O pins to connect with the PCB. Compared to the QFP and TSOP, it has many advantages:

① Dense I/O pins but larger pin pitch, which improves the assembly yield greatly.
② Good electrical and thermal performance.
③ Small size. Compared to the QFP, its thickness is reduced by more than 1/2 and its weight is reduced by more than 3/4. Compared to the TSOP, its packaging size is reduced by more than 2/3.
④ Small signal transmission delay and higher frequency.
⑤ High reliability.

Attributing to these advantages, it became the best choice for high-density, high-performance, multifunction, and high I/O pins packaging as soon as it was invented.

1.1.2 CSP

Although the rise and development of BGA solved the difficulties faced by QFP and TSOP, it still occupies a large substrate area. In order to integrate more chips on a fixed-size PCB, the CSP was invented in 1994. The perimeter of a CSP is no more than 1.2 times the perimeter of the chip it contains, thereby allowing more chips to be arranged in the same area and reducing the overall electrical system significantly. The structure of the CSP is similar to that of the BGA with smaller solder balls and ball spacing, so that more I/O pins can be arranged at the same size package. Compared to the prior packaging technologies, the CSP has several advantages:

① Small size and low weight. The size and weight of the CSP are smaller than any other packages. For packages with the same I/O number, the CSP technology reduces the weight of the package by more than 4/5 and the size of the package by 2/3–9/10, when compared with the QFP.
② Large number of I/O pins. For same size package, the number of I/O pins of the CSP is ~3 times that of QFP and ~1.5 times that of the BGA package.
③ Good electrical performance. The interconnection length of the CSP between the chip and the package shell is much shorter than that of QFP and BGA packages, so it has lower signal transmission delay and higher frequency.
④ Good thermal performance. Because the thickness of the CSP is extremely less, the heat generated by the chips can be easily transferred to the outside of the packaging.

Although the CSP technology was invented more than 20 years ago, it is still in the early development stage and many problems remain to be solved, such as the packaging standard and I/O pins alignment. It is undeniable that the CSP will be one of the mainstream packaging technologies in the future.

1.1.3 MCM

The rapid development of electronic systems raises high demand for multifunction and multichip packaging technologies. However, the traditional packaging technologies only contain one chip in the package and integrate numerous packages on a PCB, which increases the system size and leads to low reliability and high signal transmission delay caused by the long interconnecting wire between the packages. To tackle this issue, the MCM packaging technology was developed.

It integrates two or more large-scale chips together in one packaging shell, so the signal transmission delay between the chips is reduced greatly due to the tiny chip spacing. The MCM has many advantages:

① Low signal transmission delay and high signal transmission speed. Compared to the single-chip package, the MCM increases transmission speed by 4–6 times.

② Compact size and low weight. The chips in the MCM package can be tiled on a single layer or stacked on multiple layers. The multiple-layer structure can decrease the packaging size significantly, resulting in low weight. Compared to the single-chip packaging, it decreases the weight by more than 80%.

③ High reliability. The failure of the electronic systems is mainly caused by failure of circuit interconnections, while the MCM reduces such interconnections, thereby improving the reliability of the electronic system.

④ Multiple functions. The MCM integrates chips with various functions together to form multifunctional electronic systems directly.

Although the MCM has many advantages, it is not as widely applied in industrial production as the BGA and CSP technologies. There are two reasons: the cost of the MCM is much higher than other packaging technologies and the MCM package presents poor thermal performance because the heat cannot be quickly dissipated to the outside due to the high chip integration along the vertical direction. For most commercial electronic systems, the BGA and CSP technologies would be a better choice due to their relatively low cost.

1.1.4 3D Packaging

In recent years, the development of the electronic industry has failed to obey Moore's law, which predicts that the number of transistors on a chip will double every 24 months, due to the physical limitations present in the complementary metal-oxide semiconductor (CMOS) processing technology. The 3D packaging is expected to break this limitation. By stacking more than two chips in the vertical direction through silicon via (TSV), the 3D packaging assembles more chips on the electronic system without increasing the size of the PCB. Compared to the 2D packaging, it has advantages of higher assembly density, lower cost, smaller size, lower power consumption, higher signal transmission speed, and smaller signal transmission delay. So far, the 3D packaging is rarely applied in the industrial production because there are many problems that need to be solved, such as the cost and reliability of the TSV and redistribution layer (RDL) and the severe thermal issue caused by the high chip integration along the vertical direction.

From the TO in the 1970s to the current 3D packaging, packaging technology has undergone tremendous development in materials, processes, and applications. With the further development of the electronic industry, more changes are taking place in the field of packaging.

① High thermal conductivity packaging materials. The low thermal conductivity of some key packaging materials, i.e., thermal interface material (TIM), encapsulant, and chip substrate, has become a serious problem that blocks further development of high integration and power electronic systems. For the TIM and encapsulant, the scholars and industry are trying to enhance their thermal conductivity by adding high thermal conductivity particles, such as graphene nanoplatelets (GNP), carbon nanotube (CNT), hexagonal boron nitride (hBN), and metallic oxide. And the concentration and arrangement of the particles are optimized by considering the material properties of the liquid matrix and particles. For the chip substrate, high thermal conductivity materials, i.e., silicon (Si), silicon carbide (SiC), aluminum nitride (AlN), beryllium oxide (BeO), and diamond, are developed to replace the conventional sapphire (Al_2O_3).

② Advanced cooling technologies. For some electronic systems with high chip power and integration, the local heat flux could reach an extremely high level $>500\ \mathrm{W\ cm^{-2}}$, which requires advanced cooling technologies to dissipate the heat immediately. Therefore, conventional air cooling technologies can no longer meet the cooling requirement. To solve this issue, liquid cooling technologies with microchannel, vapor chamber, and phase change materials have been developed.

③ Environmental packaging material. To protect the environment, it is meaningful to use lead-free, halogen-free, and easy-clean packaging materials. In recent years, the use of lead-free solders has been attracting extensive attention. However, most of the lead-free solders have relatively high melting point and poor wettability, which results in voids in the solder layer and thereby worsens the reliability of the electronic devices. Therefore, it is very important to develop lead-free solders with a low melting point and good wettability. Besides, the conductive adhesive could also be a prior choice.

④ Reliability of the electronic systems. Most chips are sensitive to moisture, oxide, and high temperature environment. Therefore, it is of great importance to investigate the effect of these factors on the reliability of electronic systems and enhance the reliability by developing advanced packaging technologies.

1.2 Heat Generation in Opto-electronic Package

In general, the chips can be regarded as resistors, and the heat power of the chips is proportional to the driven current and equal to the input electrical power in most cases. However, for some electronic packages with opto-electro chips, only part of the input electrical power is converted into heat and the heat generation mechanism can be extremely complex. The accumulation of heat increases the chip junction temperature sharply and induces many thermal problems including thermal stress, performance degradation of the chip, and mechanical and electrical reliability of the package. The opto-electro packages have undergone rapid development in recent decades due to their wide application in solid-state lighting. Therefore, it is of importance to understand the heat generation mechanism of the opto-electro chips.

Light-emitting diode (LED) is a typical opto-electro chip, so we use it as an example in the following text to make the description clearer. The LED was invented by Holonyak and Bevacqua in 1962 [1], and widely applied to general lighting until Nakamura et al. [2] invented blue LED chips with high power and light efficiency in 1991. As a typical type of solid-state lighting, LED converts part of the input electrical power into light power. Compared to the conventional lighting sources (i.e., incandescent lamp and fluorescent lamp), it has the advantages of high light efficiency (>100 $\text{lm} \cdot \text{W}^{-1}$), long lifetime (>50,000 hours), high reliability, compact size, and environmental protection. Therefore, it has become the mainstream light source in the twenty-first century [3, 4].

Figure 1.1 shows the working principle of LED. The core functional structure of the LED chip is the PN junction that is composed of P-type semiconductor and N-type semiconductor. In P-type semiconductor, the carriers that transport electrical energy are holes, while in N-type semiconductors, the carriers that transport electrical energy are electrons. Under the drive of an electric field, the electrons and holes move relatively and recombine in the multiquantum well (MQW) layer to emit light. In addition, part of the electrical power is converted into heat due to the nonradiative combination of electrons and holes, Shockley–Read–Hall (SRH) recombination, Auger recombination, surface recombination, current crowding and overflow, light absorption, etc.

1.2.1 Heat Generation Due to Nonradiative Recombination

Figure 1.2 shows the schematic of the band gap of the semiconductor material, which has an electronic band structure determined by the crystal properties of the material. The discrete energy distribution is affected by the absolute temperature. Above absolute zero temperature, the existing energy levels are filled with electrons according to the Boltzmann distribution. The free electrons range from their bounds to a freely moving state, which is called a conduction band (CB). The valance band (VB) is the highest range of electron energies in which electrons are bounded. The difference between CB and VB is called the band gap or forbidden band, since ideally there is no electron energy state within this region [5].

The freely moving electrons in the meta-stable state exist in the CB until they fall to the VB and recombine with an electron hole. This process is referred to as recombination. There are two types of recombination within the active region of LED chips, i.e., radiative recombination and nonradiative recombination. For radiative recombination, the electron fills a hole in the VB by releasing a photon with energy equal to the band gap energy of the semiconductor material [6]. This process is the

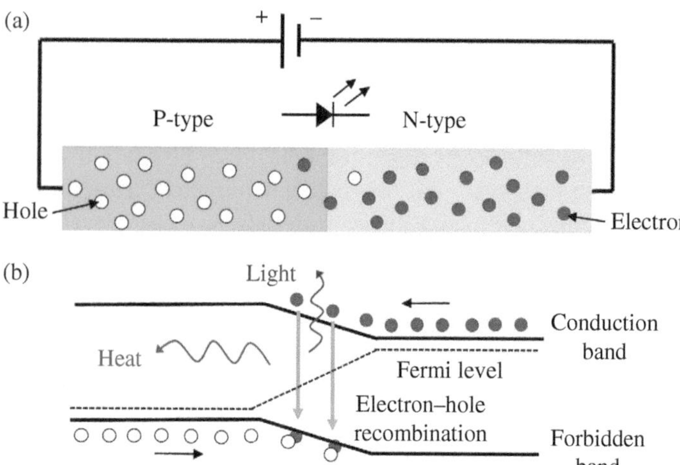

Figure 1.1 (a, b) Working principle of a typical LED chip.

foundation of the LED working mechanism. For nonradiative recombination, the releasing energy exists in the form of atom vibrations within the crystal, such as phonons; if the energy is not collected, it dissipates as heat. Apparently, the nonradiative recombination should be minimized for high device performance and low heat generation [7].

In the active layer of the LED chips, there are two major nonradiative recombination processes, i.e., defect-related SRH recombination and Auger recombination, which will be described in detail next.

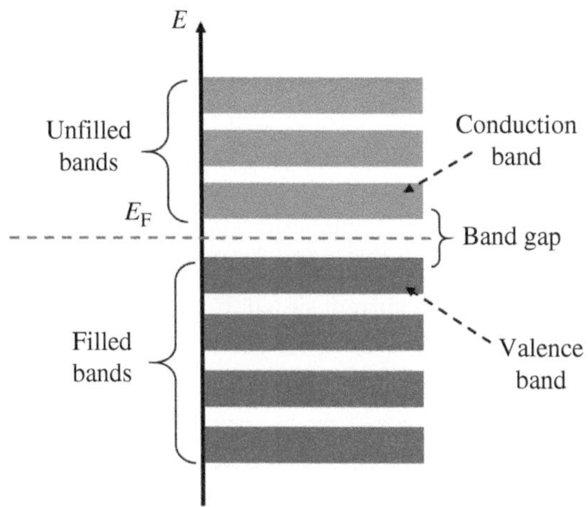

Figure 1.2 Schematic of the band gap of the semiconductor material.

1.2.2 Heat Generation Due to Shockley–Read–Hall (SRH) Recombination

Figure 1.3 shows the band diagram which illustrates the recombination process. In Figure 1.3, the SRH recombination is used to describe the recombination of the electron and hole at the undesired energy level, which is created within the band gap by defects in the lattice. They were first investigated by Shockley, Read, and Hall in 1952 and were used as a model to study the nonradiative recombination caused by defects [8].

Defects in the crystal structure are the main cause of SRH nonradiative recombination. These defects include unwanted foreign atoms and crystallographic defects. All of the defects have energy-level bands that are different from the major semiconductor atoms. Therefore, one or more new energy levels can be generated within the forbidden band gap. Unfortunately, these energy levels within the gap of the semiconductor are efficient recombination centers, especially when the deep level is near the middle of the gap. Detailed analytical expressions are obtained for the lifetime estimation of the SRH recombination [7]. These expressions reveal that when the trap level is at or close to the mid-gap energy, the lifetime is twice the minority lifetime and the probability of SRH recombination is increased. Moreover, the increase in temperature will raise the nonradiative recombination probability. For simplicity, the recombination rate can be estimated by $\text{Rate}_{\text{SRH}} = k_{\text{SRH}}n$, where k_{SRH} is the SRH recombination coefficient and n is the carrier concentration [9].

1.2.3 Heat Generation Due to Auger Recombination

Auger recombination describes the process in which the electron in CB gives off excess energy and recombines with a hole in VB. During this process, the excess energy is obtained by a second electron or hole instead of emitting the energy as a photon. The newly excited electrons or holes release their energy through collision with the crystal lattice and return back to the band edges.

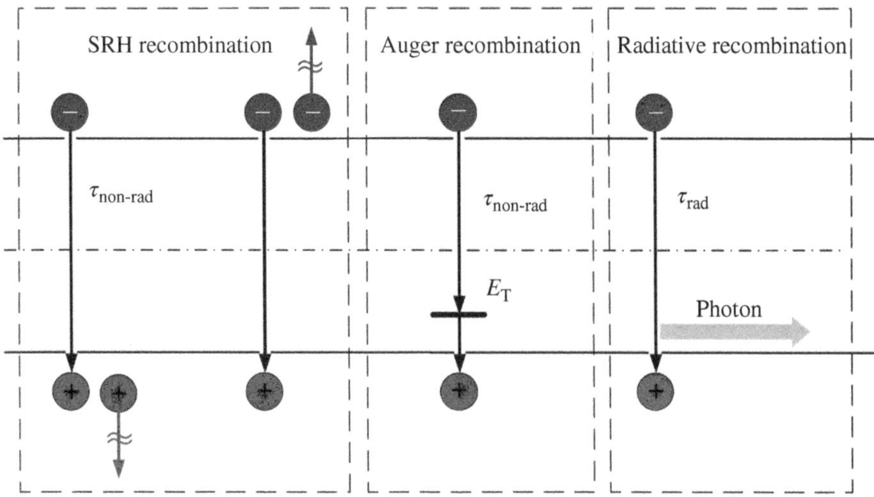

Figure 1.3 Band diagram illustrating: SRH recombination, Auger recombination, and radiative recombination. Adapted from Schubert [7].

The probability of Auger recombination increases with the concentration of charge carriers since this process is based on the ability of the charge carriers to exchange energy. The rate of Auger recombination can be expressed as $Rate_{Auger} = k_{auger}n^3$, where k_{auger} is the Auger recombination coefficient and n represents the carrier concentration. The coefficient k_{auger} on the scale of 10^{-28}–10^{-29} $cm^6 \cdot s^{-1}$ for III–V semiconductors plays an important role [10]. Normally, k_{auger} decreases with increased energy band gap. According to the reported Auger recombination coefficients in Table 1.1, the simulation work adopts 10^{-30} $cm^6 \cdot s^{-1}$ for GaN-based LED chip design, which exhibits good prediction of device performance [15]. More detailed discussions about the Auger recombination coefficients can be found in reports by Cho et al. [16]. At low carrier concentrations, the Auger recombination is neglected for practical reasons. However, at very high excitation intensity or carrier injection in the current situation, n is much higher and Auger recombination should be considered.

Table 1.1 Auger recombination coefficients reported for GaN-based LED chips.

Material	Auger recombination coefficient ($cm^6 \cdot s^{-1}$)	Reference
$In_{0.10}Ga_{0.90}N/GaN$	1.5×10^{-30}	[11]
InGaN/GaN	3.5×10^{-31}	[12]
$In_xGa_{1-x}N$ ($x \sim 9\%$–15%)	1.4–2.0×10^{-30}	[13]
GaInN/GaN	2.5×10^{-31}	[14]

1.2.4 Heat Generation Due to Surface Recombination

Nonradiative recombination also occurs at the semiconductor surface. At surfaces, the periodicity of the crystal lattice ends. Therefore, the band diagram will change at the surface since the strict periodicity of the crystal arrangement is perturbed. Additional electronic states will appear within the forbidden gap of the material [17]. Fortunately, surface recombination can be greatly reduced if the injected carriers are away from the surface. This can be realized by carrier injection under one contact, which is smaller than the LED chip.

1.2.5 Heat Generation Due to Current Crowding and Overflow

Heat generation within the active region of LED chips due to the nonradiative recombination is introduced. Here, the focus will be placed on heat generated outside the LED active region. Particularly, heat generation due to current crowding and overflow will be considered. The solutions are largely attributed to efficient LED chip design at the epitaxial and device levels.

In the realm of semiconductor physics, current crowding is used to describe a nonhomogeneous distribution of current density through the semiconductor at the vicinity of the contacts and over the PN junctions. As shown in Figure 1.4(a), in conventional LED chips, the GaN layer grows on insulating substrates (e.g., sapphire) where electrons laterally spread from N-pad to P-pad. Due to this geometry, the finite resistance of the Ohmic contact and the confinement layer causes the current to "crowd" near the edge of the contact. As shown in Figure 1.4(b), the current density distribution can be undesirably nonuniform. The current density could drop from 99 $A \cdot cm^{-2}$ at N-pad to 11 $A \cdot cm^{-2}$ at P-pad [18].

Due to the nonlinear distribution of the current density, the crowding phenomenon becomes more severe in high-power LEDs operating at high current density. The remarkable current concentration at the edges of P-type and N-type electrodes has a detrimental effect on the device performance. On the one hand, the local increase in carrier density leads to a high recombination rate, causing nonuniformity of light emission in the active region. This will induce localized overheating of the heterostructure at certain points and the formation of hotspots, whereas a large portion of the device remains inactive during operation. On the other hand, the nonhomogeneous distribution of current will increase the electromigration effect, and voids will be formed. Overall, current crowding will induce the local overheating of the heterostructure, lower the device performance, and increase the series resistance.

Various solutions have been proposed to reduce the current crowding problem, such as multifingered chip design [19]. Through simulation, Joshi et al. report that the current crowding problem is finally eliminated by a combination of multi-finger with delta-doping design. Vertical chip configuration can also solve this problem. Figure 1.5 presents the schematic

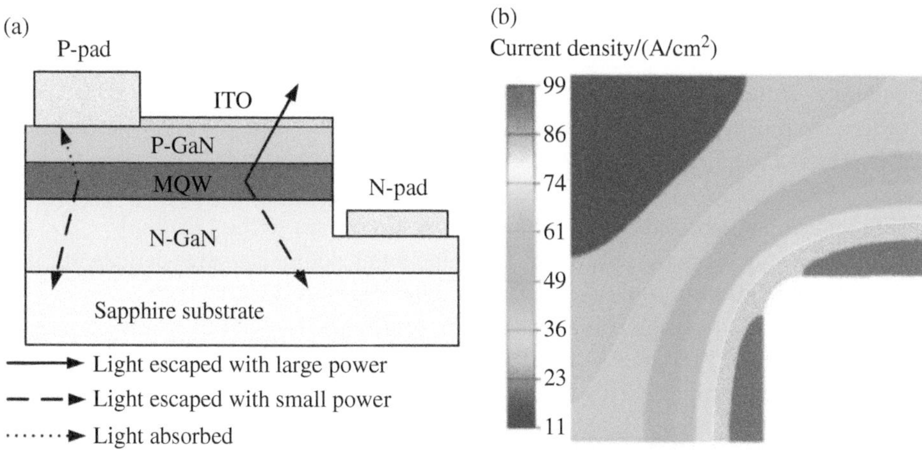

Figure 1.4 (a, b) Current crowding in GaN/InGaN LEDs on insulating substrates. Adapted from Cao et al. [18].

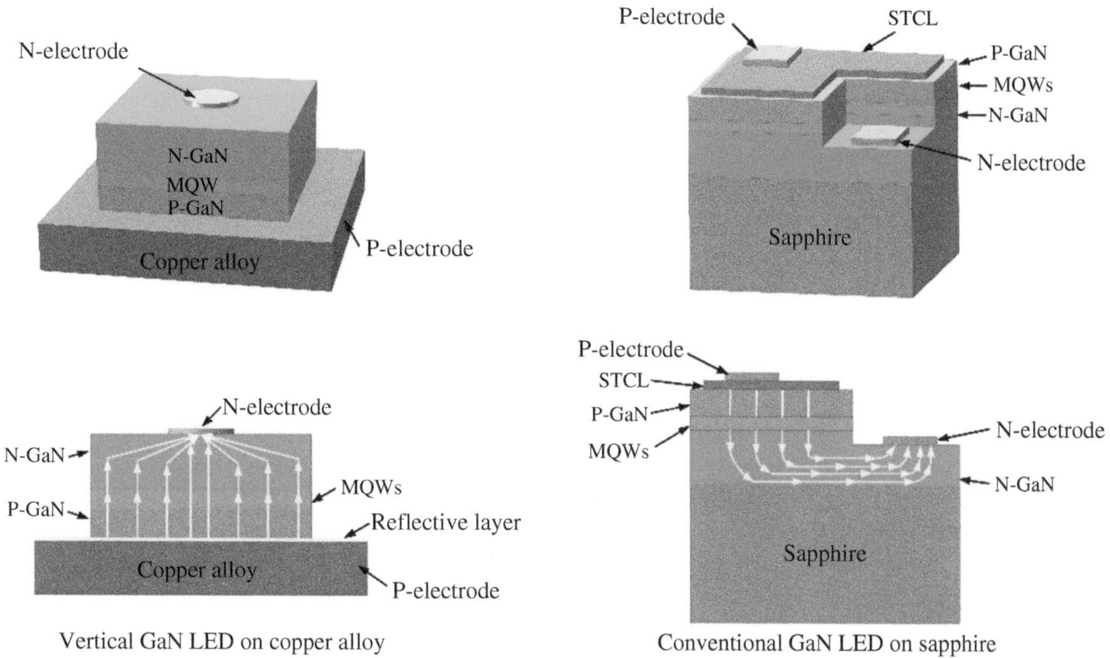

Figure 1.5 Schematic of current injection. Adapted from Joshi et al. [19].

for current injection in conventional chips and vertical chips, respectively. Compared to conventional chips (right side of Figure 1.5), vertical LED chips (left side of Figure 1.5), whose GaN layer is grown on a conducting substrate (e.g., SiC) directly or is transferred to a metal substrate by a laser lift-off process, can also reduce current crowding.

Current overflow, also referred to as electron leakage, describes the process in which the energetic electrons move from the N-type through the active region and recombine with holes in the P-type GaN without being confined in the active region. Since only the carriers confined in the active region are able to participate in the radiative recombination, the recombination caused by electron overflow generates unwanted heat. Current overflow is mainly caused by the higher carrier mobility of the electrons [20] (about 200 $cm^2 \cdot V^{-1} \cdot s^{-1}$) than that of the holes (about 10 $cm^2 \cdot V^{-1} \cdot s^{-1}$) [21]. The longer current spreading length caused by the higher mobility leads to fewer holes than electrons being injected into the active region. This is the reason why the AlGaN electron blocking layer (EBL) is usually adopted on the P-side of the active region.

1.2.6 Heat Generation Due to Light Absorption

Photons generated by radiative recombination may not be able to escape from the LED chip if they are totally internally reflected at the semiconductor/air interface. If the incident angle of the light ray is close to normal, they are effectively extracted outside of the LED chip; otherwise, they will be trapped inside [7] and converted into heat. This occurs because of the large refractive index difference of GaN ($n = 2.5$)/air ($n = 1$) interface. Total internal reflection (TIR) occurs when the incident angle exceeds the critical value $\theta_c = \arcsin(n_1/n_2)$, which can be calculated based on Snell's law. The TIR leads to a narrow escape cone of only 23.5°, with an escape probability of only 4% from the top surface of the LED [22].

Techniques have been developed to improve light extraction efficiency (LEE), including photonic crystal, periodic surface texturing, surface roughing, patterned sapphire substrate, and reflectors using Al or Au electrodes.

Thin-film flip-chip (TFFC) design is widely used in current commercial LEDs, which possess high LEE as compared to that of conventional LED package design. Thin-film LEDs could be realized by removing the sapphire substrate by the laser lift-off technique. On the other hand, flip-chip LEDs are achieved by submounting the P-GaN on a high-reflectance metallic mirror to form the vertical LED configuration. This allows the photons to emit from the thicker N-GaN layer side and enables a more flexible surface texturing and patterning process on NGaN to enhance LEE without potential effect on the InGaN QW's active region. The TFFC LEDs combine these two techniques and therefore show great potential to enhance LEE.

1.3 Thermal Issues and Challenges

The increasing power and integration density of chips bring several thermal issues and challenges, including thermal management and mechanical and electrical reliability.

1.3.1 Thermal Management

Thermal management is an important function of packaging; it transfers the heat from the tiny chips to the large-scale heat sink to maintain the chip temperature below the reliable operating temperature. There are two main issues related to the thermal management; one is the hotspot of packages and the other is the temperature distribution uniformity inside the package. The hotspot mainly takes place inside the chips because chips are the primary heat source of most packages. In addition, the temperature distribution of the package also greatly depends on the chip power and chip arrangement. Therefore, the key issue lies in the thermal management of package: to transfer the heat of the chips to the environment as soon as possible and minimize the temperature difference between the chips and other packaging components.

There are three heat dissipation paths: conduction, convection, and radiation. Typically, the reliable operating temperature of most chips is below 120 °C, so the radiation of the chips as well as other packaging components can be ignored. Therefore, conduction and convection are the main heat dissipation paths. The heat generated in the chips transfers to the heat sink by conduction through several packaging materials and thermal interfaces and then it transfers from the heat sink to the environment by convection. There are four types of thermal resistance that should be taken into consideration: material bulky resistance, thermal interfacial resistance, thermal spreading resistance, and heat sink-to-ambient resistance. Thermal bulky resistance is determined by the dimension and thermal conductivity of the material. Thermal interfacial resistance highly depends on the contact condition of TIMs and their adjacent solid packaging materials, which is determined by the liquid properties of the TIMs, such as the viscosity, surface tension, and wettability, on the adjacent rigid packaging materials. If the TIMs fail to wet their adjacent rigid packaging materials, air/vapor voids form between these two packaging components, which induce extremely high thermal interfacial resistance due to the extremely low thermal conductivity of air/vapor. The thermal interfacial resistance could be the key thermal resistance in a package with multiple-layer packaging structure, such as MCM and 3D packaging. Thermal spreading resistance depends on the size difference between two packaging components and the thermal boundary conditions. For most of the packages, the chip size is much smaller than the PCB as well as other packaging components. The heat transfer from the tiny chips to other large-scale packaging materials could suffer extremely high thermal spreading resistance, which is the main reason for inducing the hotspot on chips and nonuniform temperature distribution in packages.

Heat sink-to-ambient resistance varies greatly with the thermal management technologies. In general, the thermal management technologies could be divided into passive cooling and active cooling. The key difference between these two methods is that active cooling requires an external power source to generate airflow or liquid flow to take the heat away from the heated surface quickly, while passive cooling dissipates the heat by natural convection, heat conduction, and phase change. Compared to active cooling, passive cooling offers the advantages of simple structure, easy fabrication, flexibility, low cost, and high reliability. However, the heat dissipation capacity and efficiency of passive cooling are always limited by its dimensions and structure. Therefore, passive cooling technologies are becoming outdated as they cannot meet the packaging requirements of high-power electronic packages. Active cooling technologies, such as external forced convection, pumped loops, refrigeration, microchannel cooling, microjet cooling, and spray cooling, have higher heat dissipation capacity and efficiency than passive cooling technologies. Although active cooling technologies suffer low reliability and high system complexity, they are expected to solve the bottleneck issue of the thermal management of high-power electronic packages. Moreover, as the chip size decreases and chip power increases, microscale thermal management technologies are attracted extensive attention in recent decades. They achieve extremely low thermal resistance and high heat transfer coefficient at the high heat flux region and, therefore, offer excellent hotspot cooling and significantly improve the temperature distribution uniformity of the packaging. These technologies can also be passive or active, in which the passive cooling could be vapor chamber and microchannel, while the active cooling could be Peltier cycle thermoelectric devices and Stirling refrigeration cycle. However, the miniaturization of thermal management systems brings challenges to packaging processes, material synthesis, and system integration, which still have a long way to go.

1.3.2 Mechanical/Electrical Reliability

As introduced earlier, the electronic packages suffer from hotspots in high heat flux regions and nonuniform temperature distribution inside the packages if without excellent thermal management. These two thermal problems could worsen the mechanical/electrical reliability or even cause catastrophic failure of the packages. The mechanical reliability is caused by two reasons: one is the thermal stress inside the package induced by the temperature difference of various packaging components and the other is the thermal degradation/quenching/carbonization of the chips and other packaging components induced by overheating.

The thermal stress inside the packages can induce cracking of chips, TIMs, electrodes, and bonding wires. Die attaching is the key packaging process for improving the mechanical reliability of packages. In the die attaching process, chips or power modules are bonded on the substrate or PCB through TIMs. As mentioned earlier, the thermal interface resistance that lies between the TIMs and their adjacent rigid packaging materials could be the key thermal resistance in the package with a multiple-layer packaging structure. The high thermal interface resistance makes it difficult for the heat generated by the chip to be transferred to the substrate, PCB, and other packaging components. As a result, the heat accumulates inside the chips and packages, which causes hotspots in chips and nonuniform temperature distribution in packages. The deformation of the high-temperature region is much more serious than that of the low surface temperature region, which causes the thermal stress. Cracking of packaging components happens if the thermal stress is much higher than the fatigue stress of the packaging materials. Increasing the thermal conductivity, reducing the thickness of TIMs, and promoting the contact condition of TIMs with their adjacent rigid packaging materials can reduce the thermal interface resistance, thereby reducing the temperature difference and improving the mechanical reliability of the packages. In addition, developing advanced microscale thermal management technologies to offer higher heat transfer coefficient at the high heat flux region could also reduce the temperature difference inside the packaging, thereby improving the mechanical reliability of the packages.

Thermal degradation/quenching of the chips and other packaging components worsens the electrical/optical/thermal performance and decreases the lifetime of the packages, while carbonization of the packaging components causes irreversible failure to the packages. These issues are particularly prominent in opto-electro packages. In the opto-electro packages, both light and heat are generated from chips/fluorescent materials (phosphors and quantum dots) and then transmitted or conducted through many packaging materials and interfaces. Meanwhile, part of the transmitted light converts into heat along the light propagation. In return, the accumulation of heat leads to rise in temperature and thermal degradation/quenching of the chips/fluorescent materials, thereby generating more heat. Therefore, more severe challenges lie in the research of the thermal reliability of opto-electro packages, such as temperature dependence of the electro-opto conversion of chips and opto-heat conversion of fluorescent materials, light scattering, reflection, and absorption.

In summary, the increasing power and integration density of chips pose significant challenges to the upstream chip manufacturing/designing and material synthesis, midstream packaging processes, and thermal management technologies. Among these challenges, thermal management is thought to be the key issue for supporting Moore's law that is well known in the semiconductor industry. However, the existing thermal management technologies are far from meeting the development requirements of the semiconductor industry. Different from the existing thermal management technologies that aim at lowering the working temperature of the whole package and the electrical system, the fundamental goal of the future thermal management technologies is to cool the hotspot in chip/packaging and minimize the temperature difference inside the packaging and the electrical system. To face the challenges mentioned earlier, great progress should be made in numerous scientific fields, including chip manufacturing/designing, packaging processes, material synthesis, and system integration and reliability testing.

1.4 Organization Arrangement

In this book, we will introduce the thermal management for opto-/electronic packaging and applications in 10 chapters. In this chapter, the development history of opto-/electronic packaging, heat generation in opto-electro package, and related thermal issues and challenges were presented. In Chapter 2, the basic concepts of thermal conduction and thermal resistance are presented and thermal management solutions based on thermal conduction are introduced, including high thermal conductivity materials, tunable interfacial thermal conduction, heat pipe, vapor chamber, phase change materials, and thermal metamaterials. In Chapter 3, the basic concept of thermal convection and solutions are introduced; air cooling and liquid cooling technologies are also discussed. In Chapter 4, the basic concept of thermal conduction radiation and solution are presented and radiative cooling and near-field thermal radiation are introduced. In Chapter 5, the opto-thermal, electro-thermal, and opto-electro-thermal interactions are presented. In Chapter 6, thermal conductivity enhancing principles and solutions of TIMs are presented, including modeling and validation of thermal interface resistance and thermal conductivity manipulation of TIMs. In Chapter 7, the packaging-in thermal management for quantum dots-converted LEDs is presented. In Chapter 8, the application of phase change material in downhole devices is introduced. In Chapter 9, liquid cooling for high heat flux electronic devices is presented. In Chapter 10, we will give a summary of the contents included in this book.

References

1 Holonyak, N. Jr. and Bevacqua, S.F. (1962). Coherent (visible) light emission from Ga $(As_{1-x}P_x)$ junctions. *Appl. Phys. Lett.* 1 (4): 82–83.

2 Nakamura, S., Takashi Mukai, T.M., and Masayuki Senoh, M.S. (1991). High-power GaN P-N junction blue-light-emitting diodes. *Japan. J. Appl. Phys.* 30: L1998–L2001.

3 Luo, X.B., Hu, R., Liu, S. et al. (2016). Heat and fluid flow in high-power LED packaging and applications. *Prog. Energy Combust. Sci.* 56: 1–32.

4 Liu, S. and Luo, X.B. (2011). *LED Packaging for Lighting Applications: Design, Manufacturing, and Testing*. Wiley.

5 Neamen, D.A. (2003). *Semiconductor Physics and Devices Basic Principles*. New York, NY: McGraw-Hill.

6 Kawakami, Y., Omae, K., Kaneta, A. et al. (2001). Radiative and nonradiative recombination processes in GaN-based semiconductors. *Phys. Status Solidi A* 183 (1): 41–50.

7 Schubert, E.F. (2006). *Light-Emitting Diode*. Cambridge University Press.

8 Shockley, W. and Read, W.T. (1952). Statistics of the recombinations of holes and electrons. *Phys. Rev.* 87 (5): 835–842.

9 Dai, Q., Shan, Q., Wang, J. et al. (2010). Carrier recombination mechanisms and efficiency droop in GaInN/GaN light-emitting diodes. *Appl. Phys. Lett.* 97: 133507.

10 Olshansky, R., Su, C.B., Manning, J., and Powazinik, W. (1984). Measurement of radiative and nonradiative recombination rates in InGaAsP and AlGaAs light sources. *IEEE J. Quantum Electron.* 20 (8): 838–854.

11 Zhang, M., Bhattacharya, P., Singh, J., and Hinckley, J. (2009). Direct measurement of auger recombination in $In_{0.1}Ga_{0.9}N$/GaN quantum wells and its impact on the efficiency of $In_{0.1}Ga_{0.9}N$/GaN multiple quantum well light emitting diodes. *Appl. Phys. Lett.* 95 (20): 201108.

12 Laubsch, A., Sabathil, M., Baur, J. et al. (2010). High-power and high-efficiency InGaN-based light emitters. *IEEE Trans. Electron. Dev.* 57 (1): 79–87.

13 Shen, Y.C., Mueller, G.O., Watanabe, S. et al. (2007). Auger recombination in InGaN measured by photoluminescence. *Appl. Phys. Lett.* 91 (14): 141101.

14 Laubsch, A., Sabathil, M., Bergbauer, W. et al. (2009). On the origin of IQE-'droop' in InGaN LEDs. *Phys. Status Solidi C* 6 (S2): S913–S916.

15 Delaney, K.T., Rinke, P., and Van de Walle, C.G. (2009). Auger recombination rates in nitrides from first principles. *Appl. Phys. Lett.* 94 (19): 191109.

16 Cho, J., Schubert, E.F., and Kim, J.K. (2013). Efficiency droop in light-emitting diodes: challenges and countermeasures. *Laser Photon. Rev.* 7 (3): 408–421.

17 Nguyen, H.P.T., Djavid, M., and Mi, Z. (2013). Nonradiative recombination mechanism in phosphor-free GaN-based nanowire white light emitting diodes and the effect of ammonium sulfide surface passivation. *ECS Trans.* 53 (2): 93–100.

18 Cao, B., Zhou, S., and Liu, S. (2013). Effects of ITO pattern on the electrical and optical characteristics of LEDs. *ECS J. Solid State Sci.* 2 (1): R24–R28.

19 Joshi, B.C., Pradhan, N., Mathew, M. et al. (2009). Delta doping: new technique to reduce current crowding problem in III-nitride LEDs. *Optoelectron. Adv. Mater.* 3 (10): 985–988.

20 Götz, W., Johnson, N.M., Chen, C. et al. (1996). Activation energies of Si donors in GaN. *Appl. Phys. Lett.* 68 (22): 3144–3146.

21 Oh, M.S., Kwon, M.K., Park, I.K. et al. (2006). Improvement of green LED by growing p-GaN on $In_{0.25}GaN$/GaN MQWs at low temperature. *J. Cryst. Growth* 289 (1): 107–112.

22 Ee, Y.K., Kumnorkaew, P., Arif, R.A. et al. (2009). Optimization of light extraction efficiency of III-nitride LEDs with self-assembled colloidal-based microlenses. *IEEE J. Quantum Electron.* 15 (4): 1218–1225.

2

Thermal Conduction and Solutions

Thermal management of high-performance integrated circuit (IC) chips has been a challenging issue with the rapid advances of high-power micro-scale electronic devices. According to the International Electronics Manufacturing Initiative (iNEMI) technology roadmap released in 2004, the maximum energy consumption and heat flux of the central processing chips will reach 360 W and 190 $W \cdot cm^{-2}$ by 2020, respectively [1, 2]. Furthermore, the design trend of modern electronics is more integrated, high power, and better performance, which will lead to higher operating heat. If the heat cannot be dissipated out of the packaging timely, the temperature of the device and components will keep rising, leading to an accelerated reduction of device reliability and lifetime [3]. As displayed in Figure 2.1, currently, heat dissipation of most of the electronic devices is realized by attaching them to an external heat sink or a cold plate, which is cooled by forced convection of air or circulated liquid [4]. Therefore, it is essential to dissipate the heat generated from chips to the external heat sink. It is well known that there are three modes of heat transfer: thermal conduction, thermal convection, and thermal radiation. Limited by the temperature range, a working medium, electrical requirement, and thermal requirement, conduction is the major way to dissipate heat from chip to heat sink in a thermal packing system. In this chapter, we will focus on the heat dissipation by conduction in the electronic packaging.

2.1 Concept of Thermal Conduction

The essence of temperature is the intensity of the thermal motion of molecules on microscopic scale. According to molecular motion theory, temperature is the symbol of the average kinetic energy of molecular motion. When a temperature gradient exists in a body or the contacted interfaces of bodies, there will be a heat transfer phenomenon caused by the thermal motion of microscopic particles such as molecules, atoms, free electrons, and phonons [5]. This process is named as thermal conduction. Temperature difference is the driving force of heat transfer. Specially, heat can only be transferred spontaneously from hotter region to colder region. In 1822, French physicist Joseph Fourier used mathematical methods to derive a differential form expression named Fourier's law. It is expressed as

$$q = \frac{P_\mathrm{h}}{A_\mathrm{c}} = -\kappa \frac{\partial T}{\partial x} \tag{2.1}$$

where P_h is the heat transfer rate, q is the heat flux density, A_c is the cross-sectional area which is perpendicular to the direction of heat conduction, κ is the thermal conductivity (TC) of the material, and $\partial T/\partial x$ is the temperature gradient in the direction. The minus sign indicates that heat transfers in a direction opposite to that of the temperature rise as shown in Figure 2.2.

Thermal Management for Opto-electronics Packaging and Applications, First Edition. Xiaobing Luo, Run Hu, and Bin Xie.

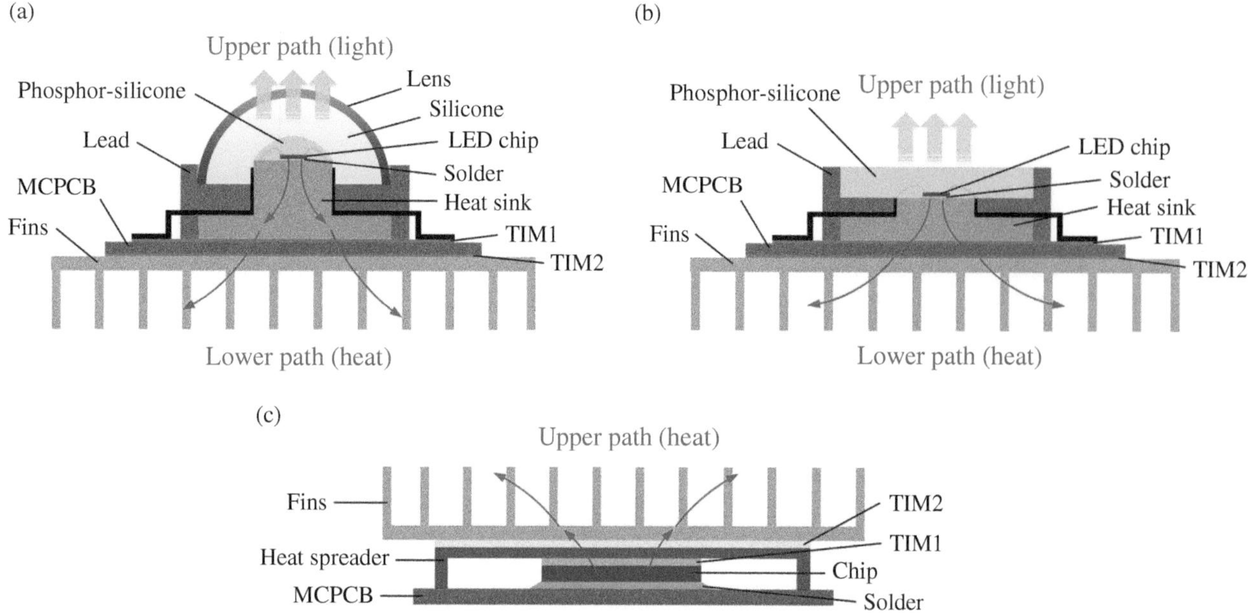

Figure 2.1 Schematic of heat transfer path in (a, b) two typical white light-emitting diodes (WLEDs) packages and (c) IC chips package.

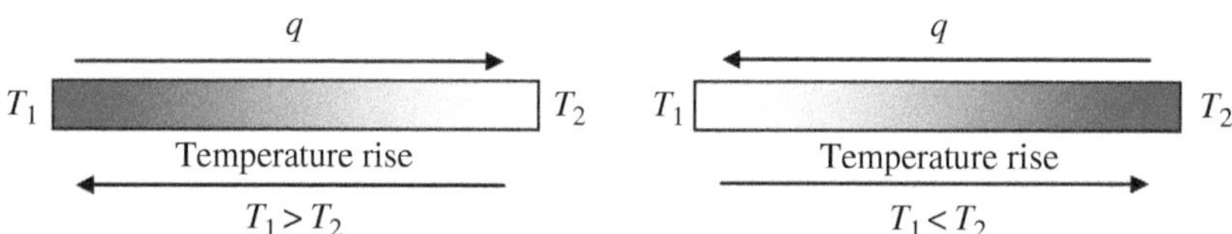

Figure 2.2 Directional relationship between temperature gradient and heat transfer.

2.2 Thermal Resistance

2.2.1 Basic Concept of Thermal Resistance

For a stationary-state heat conduction condition as described in Figure 2.2, the heat flux density q is constant; therefore, Fourier's law in this case can be described as

$$P_{\mathrm{h}} = -\kappa A_{\mathrm{c}} \frac{\partial T}{\partial x} = -\kappa A_{\mathrm{c}} \frac{\mathrm{d}T}{\mathrm{d}x} = \kappa A_{\mathrm{c}} \frac{\Delta T}{\delta} \tag{2.2}$$

where δ is the thickness of the object and ΔT is the temperature difference between T_1 and T_2. This equation indicates that the heat flux is proportional to the temperature difference, while inversely proportional to the thickness.

It is necessary to introduce the concept of thermal resistance, which describes the resistance to heat transfer. It is well known that there are driving force and resistance in every kind of transfer process, where the transfer amount m is closely related to the driving force F and resistance of the process R. Their relationship is shown as

$$m = \frac{F}{R} \tag{2.3}$$

A typical example of Equation (2.3) is Ohm's law in electrics

$$I = \frac{U}{R} \tag{2.4}$$

where I is the current of some critical component, U is the voltage between two terminals of the component, and R is the electrical resistance of the component. Similarly, heat conduction, as a kind of transfer process in nature, follows the same law as Equation (2.3). Temperature difference is the driving force of heat conduction. Heat flux is similar to m in Equation (2.3). As a result, the resistance of heat conduction called thermal resistance is defined as

$$R_{\mathrm{th}} = \frac{\Delta T}{P_{\mathrm{h}}} \tag{2.5}$$

where ΔT is the temperature difference of some critical components and P_{h} is the heat flux in the component at a steady state.

According to Equations (2.2) and (2.5), the thermal resistance of conduction is given by

$$R = \frac{\delta}{\kappa A_{\mathrm{c}}} \tag{2.6}$$

Similar to electric resistance, the thermal resistance in series or in parallel can be calculated by using the method for calculating electric resistances in series or in parallel. After the system thermal resistance is calculated, the temperature difference between the system and the environment can be calculated by Equation (2.5). The junction temperature of the die can be calculated as the ambient temperature is known.

In electronic device packaging, heat generated from chips is dissipated to the outside of the packaging structure through multiple package structures in the way of conduction. It is a typical thermal conduction process of a multi-layer structure. According to Fourier's law, heat flux density is proportional to the temperature gradient

$$q = \frac{P_{\mathrm{h}}}{A_{\mathrm{c}}} = \kappa \frac{T_1 - T_2}{L} = \kappa \frac{\Delta T}{\delta} \tag{2.7}$$

where κ is the TC, A_{c} is the cross-sectional area, δ is the thickness of the heat conduction material, and q is the heat flux density or the dissipation power per unit area. For multi-layer composite materials, the heat flux through each layer is constant. According to the series relationship of thermal resistance, the entire thermal resistance can be expressed as the sum of every part's thermal resistance

$$\frac{\Delta T}{P_{\mathrm{h}}} = \sum_i \frac{\delta_i}{\kappa_i A_{\mathrm{c}}} \tag{2.8}$$

For example, for the double-layer materials shown in Figure 2.3, the entire thermal resistance can be calculated as

$$R = R_1 + R_2 + R_3 = \frac{\delta_1}{\kappa_1 A_{\mathrm{c}}} + \frac{\delta_2}{\kappa_2 A_{\mathrm{c}}} + \frac{\delta_3}{\kappa_3 A_{\mathrm{c}}} \tag{2.9}$$

According to the equation, reducing the thickness of materials and using materials with high TC will minimize thermal resistances.

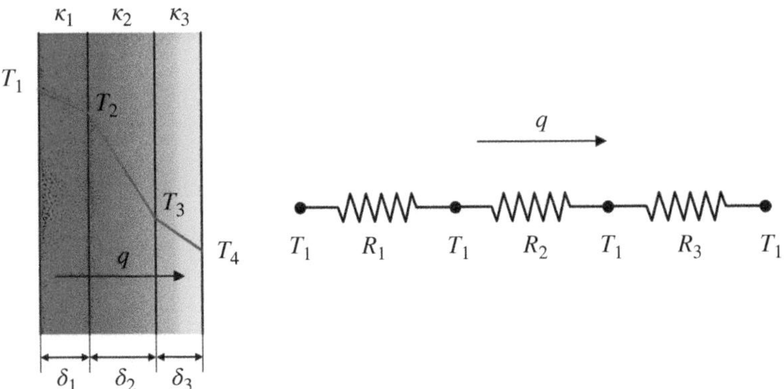

Figure 2.3 Heat transfer process and thermal resistance of double-layer material.

Besides, when temperature difference and heat flux are known, the following formula is usually used to describe the thermal resistance in packaging:

$$R_{j-a} = \frac{T_j - T_a}{P_h} \tag{2.10}$$

where T_j is the junction temperature of the chip, T_a is the ambient temperature, and P_h is the heat transfer rate of the chip. Under the condition of the same heat flux and ambient temperature, the higher the thermal resistance, the higher the junction temperature and the lower the reliability.

2.2.2 Thermal Contact Resistance

As shown in Figure 2.1, heat generated from chips must pass through several solid–solid interfaces, such as die–slug, die–board, slug–board, and board–heat sink. Figure 2.4 shows the mesoscopic contact condition of two solid surfaces. When two solid surfaces are joined, only a few surface asperities are "really" contacted between the two bodies. Consequently, the heat flux across such interfaces tends to be constricted at the micro-contact spots due to the lower TC of air. In addition, for two different materials, when energy transfer encounters the interface, reflection and transmission happen simultaneously. When energy carriers (e.g. phonons and electrons) attempting to traverse the interface (intrinsic interface scattering of energy carriers), the probability of transmission after scattering will depend on the available energy states on both sides of the interface [6]. Due to different physical properties of materials, the microscopic thermal resistance between heterogeneous contact materials is generated. Therefore, it brings in a temperature difference (ΔT) at the interface and then introduces a thermal resistance called thermal contact resistance (R_c):

$$R_c = \frac{\Delta T}{P_h} \tag{2.11}$$

It is seen that thermal contact resistance introduces extra temperature rise for chip cooling. Therefore, to reduce R_c, a conventional solution is to fill the air gap between the surfaces with highly thermally conductive materials, which are called the thermal interface materials (TIMs). Figure 2.5 shows the schematic of TIMs in electronic packaging. The TIMs have a finite bond line thickness (BLT) at the joint and cannot completely fill the air gap due to their incapability of completely wetting the surfaces [7]. Therefore, the thermal contacted resistance with TIMs ($R_{c\text{-TIMs}}$) contains two parts: ① the bulk resistance of the TIMs (R_{TIMs}) due to its finite BLT; ② the R_c between TIMs–solid interfaces due to the incomplete wetting. According to Equation (2.8), $R_{c\text{-TIMs}}$ can be written as

$$R_{c\text{-TIMs}} = \frac{\text{BLT}}{\kappa_{TIMs}} + R_{c1} + R_{c2} \tag{2.12}$$

Thus, there are three parameters that determine $R_{c\text{-TIMs}}$: ① R_c, ② TC of TIMs (κ_{TIMs}), and ③ BLT. To reduce $R_{c\text{-TIMs}}$ and improve heat transfer efficiency of the interface, the possible operation is reducing BLT and R_c, or increasing κ_{TIMs}. There are several kinds of TIMs, such as thermal grease, thermal pad, phase-change materials (PCMs), thermal gel, thermal glue, solder, and liquid metal, these will be expounded in detail in Section 2.4.

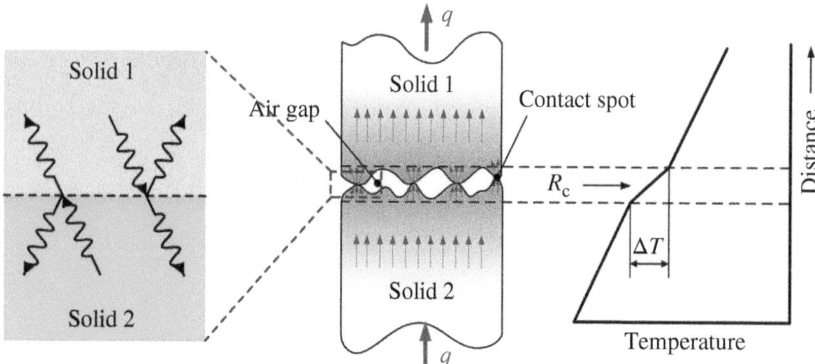

Figure 2.4 Schematic of the R_c between the solid–solid interface.

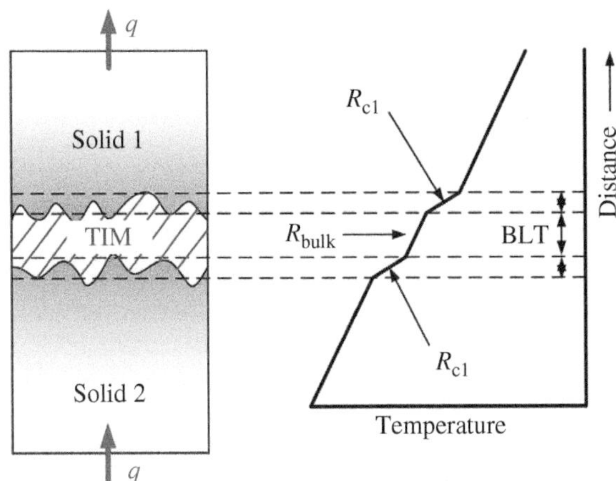

Figure 2.5 Schematic showing the RTIM at the solid–TIM–solid joint.

2.2.3 Thermal Spreading Resistance

Recently, the microelectronic/light-emitting diode (LED) modules have become powerful and compact. Their chips are on the scale of millimeters, while their modules are on the scale of centimeters, and the size of their products (smartphones, lamps, laptops, etc.) may change from several centimeters to several meters. The size difference in the internal heat conduction process will lead to thermal spreading resistance (R_s) [8]. R_s occurs when heat spreads from a smaller area to a bigger area. Normal, multiple chips or LEDs are soldered in a substrate as shown in Figure 2.6(a). A vapor chamber is usually used to spread the heat from the heat source, which is helpful for heat dissipating to the environment via a heat sink. Hence, thermal spreading resistance is a key part of thermal resistance in electronic packaging and applications. Figure 2.6(b) shows a thermal simulation result illustrating the existence of thermal spreading resistance. When the chip area and the substrate are the same ①, the temperature is uniform. But when the chip area is much smaller than the substrate area ②, the temperature of the substrate is nonuniform and the hotspot occurs. In the applications of high heat flux, R_s can occupy 60%–70% of the total thermal resistance.

In order to control the thermal spreading resistance, models should be proposed to evaluate or calculate the resistance. Kennedy [9] proposed an analytical solution for a cylinder with a constant heat flux over a part of one end and a variety of boundary conditions at the other surfaces. Kadambi and Abuaf [10] and Krane [11] proposed an analytical solution for two- and three-dimensional (3D) rectangular bodies with insulated sides and a convective boundary condition on the surface opposite to the heat input. Ellison [12] derived an exact 3D solution for the steady-state heat conduction equation with the source on an adiabatic surface and Newtonian cooling on the opposing surface. Muzychka et al. [13–16] studied the effect of the heat source location, boundary conditions, and types of flux channels on the R_s and provided a series of analytical expressions for calculating the R_s.

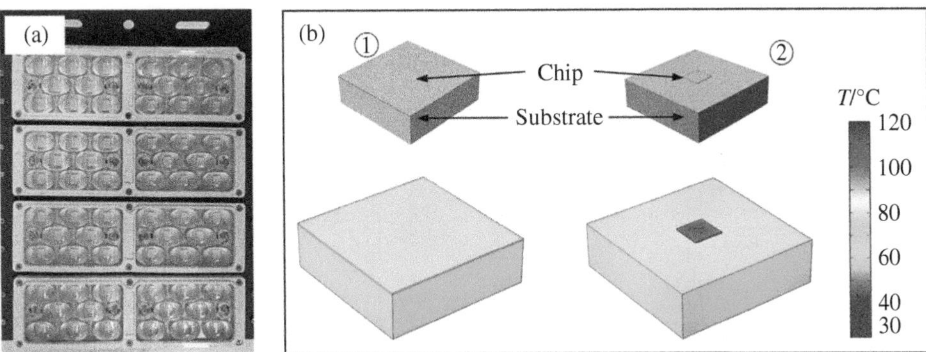

Figure 2.6 (a) LED chips array. (b) Schematic showing the R_s due to size difference in the heat conduction process.

Based on the abovementioned models, many approaches have been proposed to control the thermal spreading resistance. The most convenient and intuitive method is to increase the ratio of the heat source size to the heat sink size. Besides, heat source distribution is extremely important. A centered heat source will lead to larger R_s and hotspot phenomenon, while distributed heat sources with the same power input will show lower R_s, a more uniform temperature distribution, and a lower junction temperature. This provides a critical rule in designing electronic devices. Therefore, some studies optimized the heat source distribution to reduce the junction temperature based on this rule. Yang et al. [17] studied the effect of LED chip size on R_s and found that when the chip size increased from 15 mil to 40 mil, the total thermal resistance decreased sharply, and the R_s had a great effect on the total thermal resistance of LED packages. Yang et al. [18] analyzed the thickness of copper substrate on the R_s of LED packages and found that with the initial increase in copper thickness (up to 0.6 mm), the total thermal resistance of the copper substrate (R_{cu}) decreases with copper thickness due to the decrease of R_s. Then, R_{cu} starts to increase with the copper thickness due to the increase of the bulk thermal resistance of the copper substrate. The minimum R_{cu} value occurs at the copper thickness of about 0.6 mm. For LED lamps containing multiple LED modules, R_s becomes the dominant resistance since the size difference is very large. Therefore, the LED distribution should be optimized to control the R_s. Cheng et al. [19] applied the thermal spreading resistance network to optimize the LED chip arrangement for an 80 W LED lamp. Ha and Graham [20] analyzed the thermal resistance of the LED array as a function of chip distance by the analytical method and compared it with the finite-element analysis (FEA) and found that the pitch is an important factor in determining the R_s. Although the larger pitch can achieve a lower thermal resistance, the pitch is usually determined by considering other factors such as optical performance. For effective thermal management, it is better to minimize the interference of heat spreading in the substrate level by tuning the substrate type, thickness of each layer, etc.

2.2.4 Thermal Resistance Network

Based on all the above thermal resistances, the static thermal resistance network of each packaged device in Figure 2.1 is shown in Figure 2.7. Due to different packaging structures, the static thermal resistance networks change from one another.

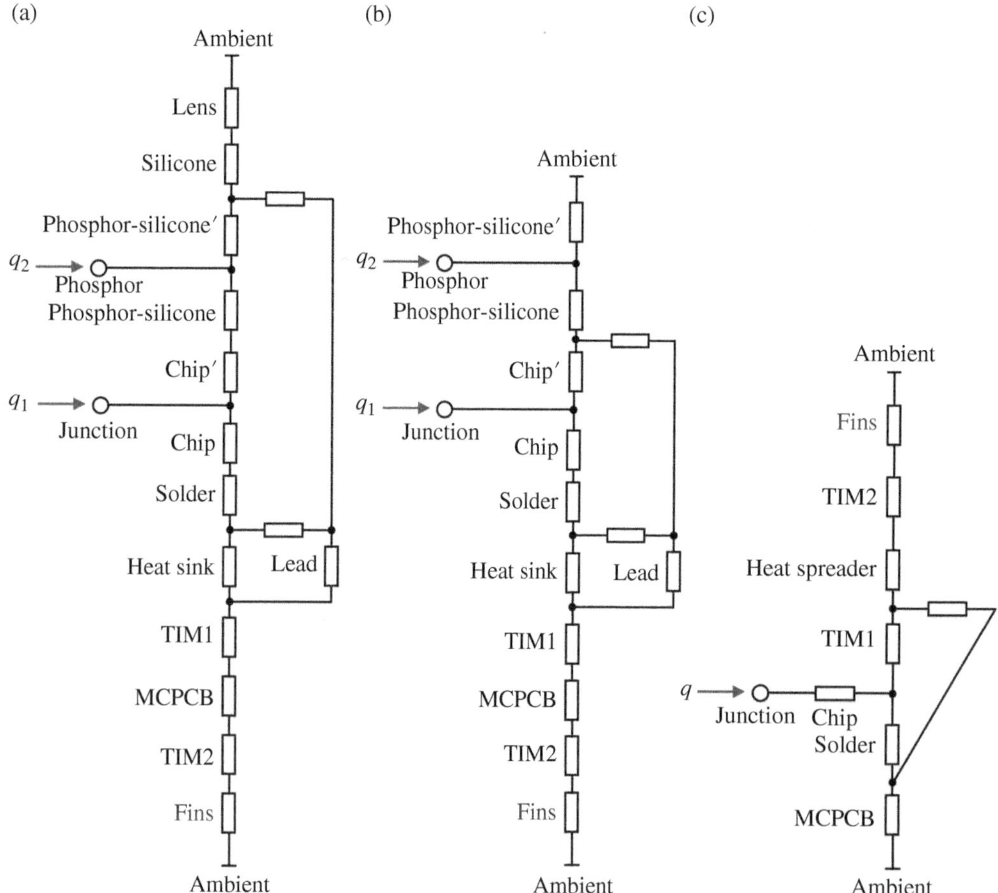

Figure 2.7 (a–c) Static thermal resistance networks according to the models shown in Figure 2.1.

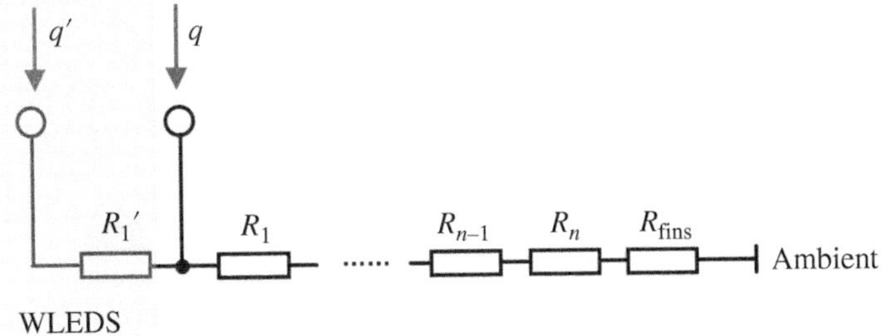

Figure 2.8 Simplified static thermal resistance network of the WLEDs and IC devices.

There are two heat inputs in the networks of the white light-emitting diodes (WLEDs) containing the phosphors-silicone luminous layer. About 75% of the electric input is converted to heat loss, and about 25% of the total heat loss comes from the light conversion in the phosphor-silicone. Unlike the WLEDs, the static thermal resistance network of the packaged IC device shown in Figure 2.7(c) has only one heat input from the chip.

Anyway, heat loss is mainly produced in the junction of the chips and spreads to the surfaces of the chips. The heat conducts to both the upper and lower packaging elements which have different thermal conductivities. The thermal resistances of the TIMs and the heat spreaders are about 0.1–0.2 $K \cdot W^{-1}$ and that of the solders vary a lot from each kind of internal packaging. The thermal resistances of the fins with a large interface for heat convection are about 0.5–1 $K \cdot W^{-1}$. The thermal resistances of the phosphor-silicone lays in the WLEDs are about 10–200 $K \cdot W^{-1}$. Based on the varied thermal conductivities of the packaging elements, most of the heat loss diffuses to the ambient from the low thermal resistance side.

In fact, the main heat transfer paths of the three networks are similar and located at the sides with the fins because the resistances of the phosphor-silicone, the lens, and the silicone in WLEDs are much greater than those of other packaging parts, as well as the metal core-printed circuit board (MCPCB) in the IC device. The simplified static thermal resistance network becomes a nearly linear structure as shown in Figure 2.8. It is worth mentioning that only one side of the chip in WLED is in contact with the heat sink because the other side of the chip coated with transparency materials that must enable light to pass through, and the IC chips can be cooled by fins in two directions.

It is worth mentioning that liquid cooling has become a main heat dissipation method. At present, most of the thermal package structures use the method of cooling the package shell, which is named as system-level liquid cooling as shown in Figure 2.9(a). The heat generated inside the chip often needs to be transported along the path of chip–solder–substrate–solder–shell–TIM-heat sink. The heat transfer path to the cooling fluid is long, resulting in large heat transfer resistance. Hence, package-level liquid cooling and chip-level liquid cooling have been put forward as shown in Figure 2.9(b) and (c). In a package-level liquid cooling system, coolant directly flows into the substrate inside the chip package, eliminating the bulk thermal resistance of the substrate and interfacial thermal resistance inside package structures. Insulated gate bipolar transistor (IGBT) is an important component in electric devices with high heat dissipation requirements. As shown in Figure 2.10, when compared with traditional system-level liquid cooling IGBT package, heat transfer path and thermal resistance are greatly simplified in package-level liquid cooling IGBT package. Moreover, in order to further reduce the thermal resistance, chip-level liquid cooling is proposed. Researchers impinged coolant on the surface of the chip [21] or brought coolant into the microchannels inside the chip and achieved a heat removal ability of 1700 $W \cdot cm^{-2}$ [22]. The package-level liquid cooling technology is a promising solution for high-power electronics, but the manufacturing, reliability, cost performance, lifetime, and compatibility still need further exploration and investigation.

2.2.5 Transient Thermal Conduction and Thermal Impedance

Considering that heat spreading is a dynamic process over time, the temperature quickly grows until the time $t = R \cdot C_p$ (R is the thermal resistance and C_p is the thermal capacitance), then gradually stabilizes at the temperature $T = P_{thermal} \cdot R$ ($P_{thermal}$ is the thermal power) in the single-resistance system. The temperature of this system follows the exponential temperature function:

$$T(t) = P_{\text{thermal}} \cdot R \cdot \left(1 - e^{-\frac{t}{\tau}}\right) \tag{2.13}$$

(a)

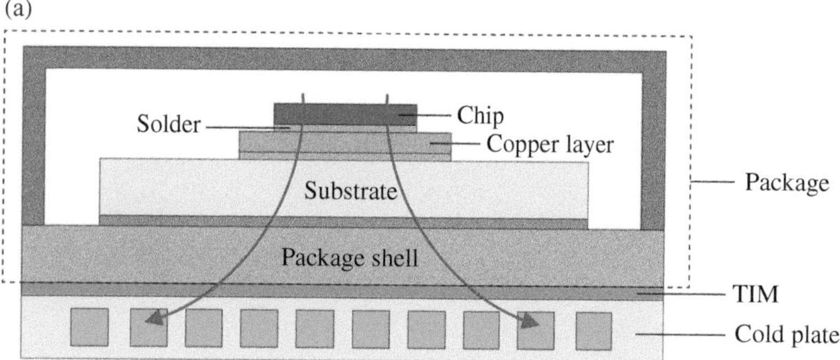

(b)

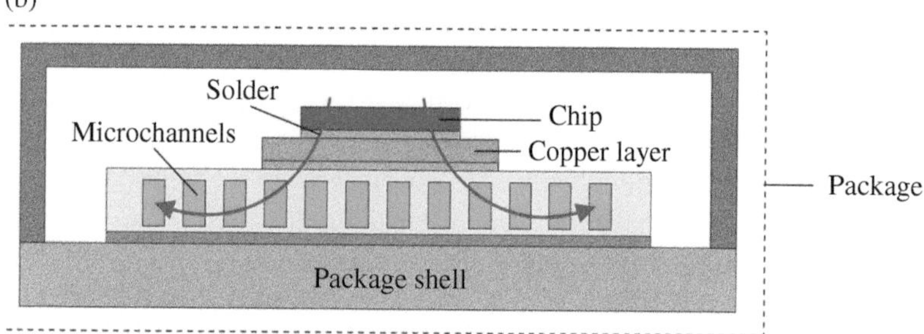

(c)

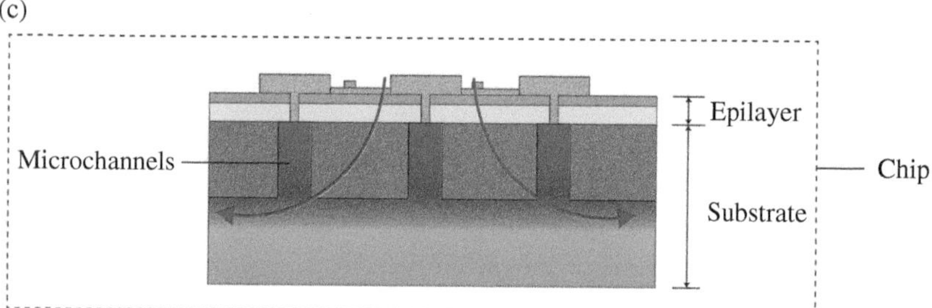

Figure 2.9 Different level liquid cooling. (a) System-level liquid cooling. (b) Package-level liquid cooling. (c) Chip-level liquid cooling.

The time constant τ:

$$\tau = R \cdot C_\mathrm{p} \tag{2.14}$$

Based on the static thermal resistance network earlier, a series of thermal capacitances are added to the thermal impedance network as shown in Figure 2.11, named the Cauer network. This model perfectly describes the time response of the thermal impedance by summing up all the exponential temperature functions:

$$T(t) = \sum_{i=1}^{n} P_{\mathrm{thermal}} \cdot R_i \cdot \left(1 - \mathrm{e}^{-\frac{t}{\tau_i}}\right) \tag{2.15}$$

When the power $P_{\mathrm{thermal}} = 1\ \mathrm{W}$, we get the thermal impedance function:

$$Z(t) = \sum_{i=1}^{n} R_i \cdot \left(1 - \mathrm{e}^{-\frac{t}{\tau_i}}\right) \tag{2.16}$$

The thermal impedance function can be generated from the time constants and the thermal resistances. The practical device described by the Cauer model contains several thermal resistances and thermal capacitances which can be calculated from the thermal impedance function $Z(t)$. Then the Cauer model can be presented in another way by the cumulative

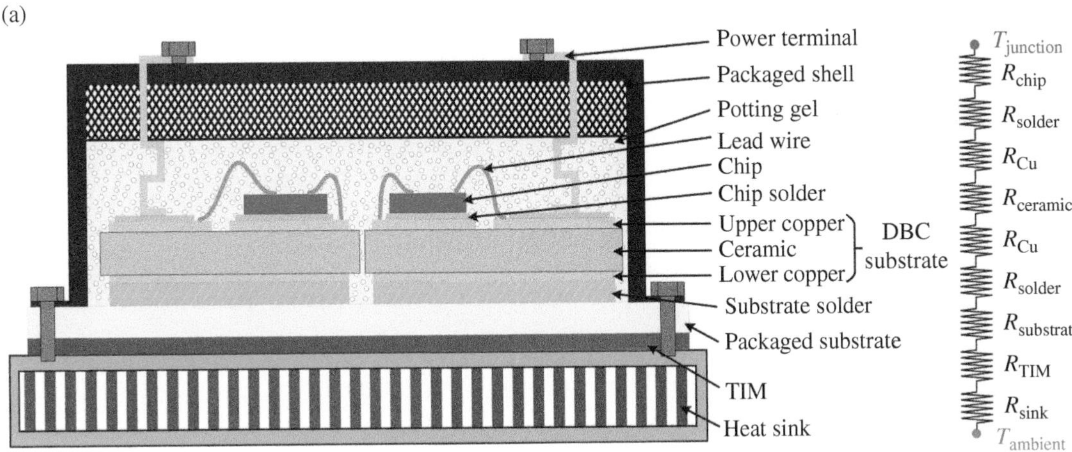

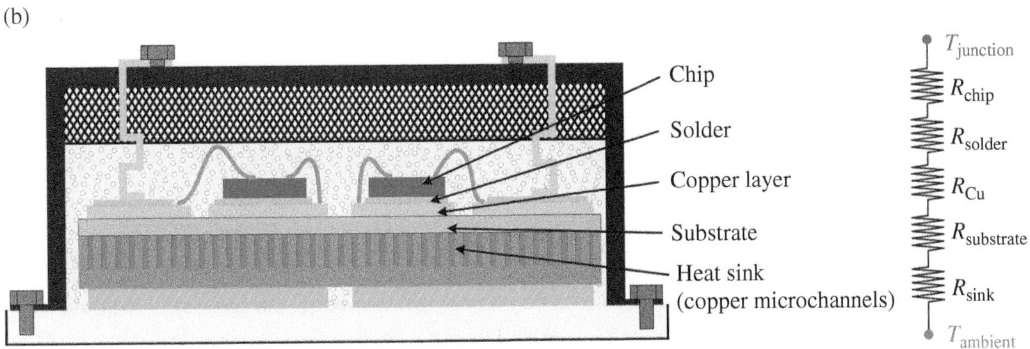

Figure 2.10 (a) Traditional system-level liquid IGBT package. (b) Package-level liquid cooling IGBT package.

thermal resistance ($R_\Sigma = \Sigma R_i$) and the cumulative thermal capacitance ($C_\Sigma = \Sigma C_i$). The cumulative thermal capacitance is a function of the cumulative thermal resistance, named the cumulative structure function (CSF):

$$\text{CSF} = C_\Sigma(R_\Sigma) \tag{2.17}$$

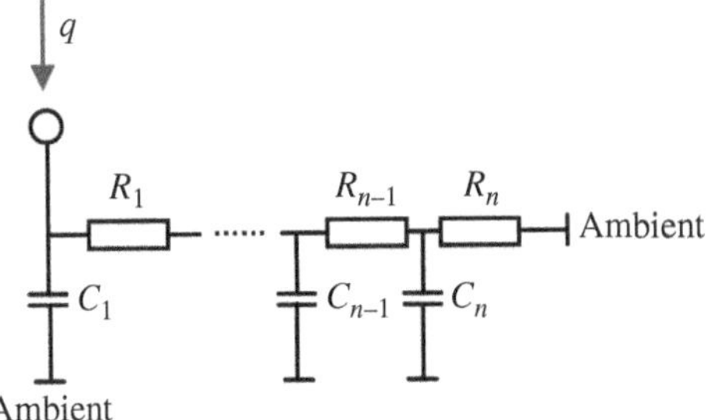

Figure 2.11 Simplified thermal resistance and capacity network of the WLEDs and IC devices.

The CSF starts from the junction and cumulates the partial thermal resistance and thermal capacitance along with the heat flow path. The heat conduction path ends in the ambient as a heat sink with infinite capacity, meaning that the CSF should end with a singularity. The distance of the singularity to the origin is equal to the junction-to-ambient thermal resistance. The CSF is used to analyze the physical structure. In low-gradient sections, a small amount of material having low capacitance causes a large change in thermal resistance. These regions have low TC or small cross-sectional area. Steep sections correspond to material regions of high TC or large cross-sectional area. Sudden breaks of the slope belong to material or geometry changes.

The CSF of a power LED with structural elements identified is shown in Figure 2.12. The steep regions are identified as the chip, sub-mount, and different parts of the heat slug. The flat sections represent different thermal interface layers (TIM_a, TIM_b, and TIM_c). The CSF realizes the direct investigation of the physical structures, reverse engineering, and failure analysis of the device instead of just viewing its thermal changes as time.

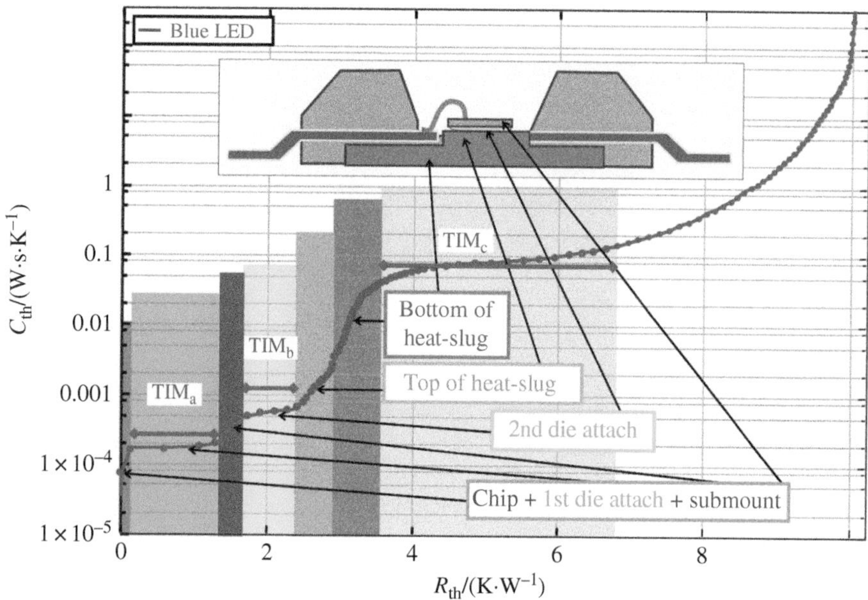

Figure 2.12 Cumulative structure function of a power LED with structural elements identified [23], reproduced with permission from Springer Nature.

2.3 High Thermal Conductivity Materials

A typical structure of a microprocessor is shown in Figure 2.1(c), which contains the substrate, solder, chip, TIM_1, heat spreader, TIM_2, and heat sink, from the bottom to top successively. When it is working, the heat is generated from the chip transfers toward the substrate and the heat sink simultaneously, as displayed in Figure 2.7(c). All the components mentioned in the microprocessor play important roles in the thermal management design. For example, the thermal contact resistances between different attached components are the bottleneck of the heat diffusion, but they are greatly reduced by the TIMs with high TC; heat spreader with high in-plane TC can uniformize the heat distribution from the chip with a much smaller surface than that of the heat sink and avoid the local heat area which leads to the thermal stress concentration and harms to the stability of the package; the coolants with high TC used in the liquid cooling system replace the air cooling for higher thermal diffusion demands. For decreasing the thermal resistance from the chip to the ambient in high-power electronic devices, a series of high thermal conductive materials for different components have been proposed and applied.

2.3.1 Structure and Materials of Chip

Chips are the main heat resources in high-power electronic devices, the aim of thermal management is to transfer the heat from chips to ambient and keep a safe working temperature for chips. A high thermal conductive chip is of great significance for decreasing the total thermal resistance from chip to ambient. A typical LED chip structure is shown in Figure 2.13, including substrate, buffer, P area, N area, and PN junction which make the shining layer [24]. When the LED is working by applying a voltage, the PN junction in the internal chip shines and generates heat that diffuses to the substrate. This part of the thermal resistance is directly related to the structure, material, and manufacturing process of the chip. LED chips, as the core component of LED electronic packaging, are divided into three categories in terms of structure: horizontal packaging, vertical packaging, and flip-chip. The three structure diagrams are shown in Figure 2.14. As the chip power grows, keeping the chip from high temperature is of great concern.

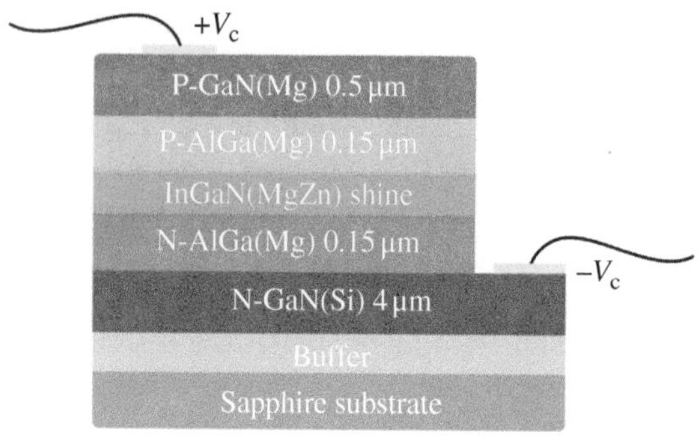

Figure 2.13 Typical LED chip structure.

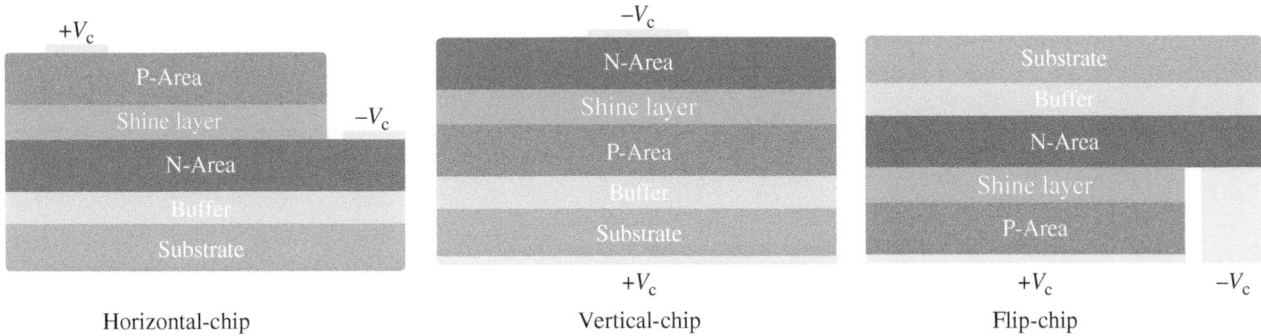

Figure 2.14 Three kinds of LED chip structures.

Specially, the structure and material of the chip play important roles in reducing its internal thermal resistance and improving the heat dissipation performance.

2.3.1.1 Structures of Chip

The horizontal chip has a traditional structure which is currently the most widely used chip type in LED packaging. Both the positive and negative electrodes are located on the top surface of the chip. The heat generated in the PN junction area of the chip is mainly downward transferred to the chip substrate and then transferred to the packaged substrate by TIM. The commonly used sapphire chip–substrate in the horizontal chip has a low TC of 25 $W \cdot m^{-1} \cdot K^{-1}$ and a large thermal resistance, which is not suitable for high-power LED.

In the vertical chip, the two electrodes of the chip are distributed on the upper and lower sides of the chip, respectively. The chip substrate of this structure must be made of conductive material, resulting in a complicated production process, high cost, and low yield. Based on the vertical conductive structure, the vertical chip allows the current to flow vertically through the entire LED chip, which improves the current distribution and the current injection ability of the chip compared to those of the horizontal chip with a lateral current flow. The vertical chip can be soldered on a silicon substrate with a metal layer to obtain a low thermal resistance and a good mechanical property, which is conducive to high-power devices.

The growth process of the flip-chip is the same as that of the horizontal chip, except that the chip–substrate needs to be thin to maintain a high light extraction rate, and the electrodes of the chip are placed on the bottom layer. The two electrodes on the bottom layer are directly welded on the packaged substrate with a wireless connection improving the stability of the chip and greatly reducing the thermal resistance of the heat from the PN junction to the packaged substrate. The emitting light no longer passes through the electrodes, so the electrodes can become thick to increase the current density of the chip for high-power devices.

2.3.1.2 Material of LED Chip

Different substrate materials in chips require different growth technologies, chip processing technologies, and packaging technologies. At present, three main substrate materials are commonly used: sapphire, silicon, and silicon carbide (SiC). In addition, some other materials can also be used as substrates, such as α-GaN, β-Ga_2O_3, ZnO, and $LiAlO_2$, which are usually selected according to the needs of the design. The choice of substrate material needs to consider many principles, including crystal structure matching, lattice matching, thermal expansion matching, thermal stability principle, large size principle, high heat dissipation performance, and cost performance. Table 2.1 lists several key properties of various widely used chip substrates.

2.3.1.3 Sapphire

Sapphire substrate is commonly used for GaN (semiconductor materials) growth: its melting point is 2042 °C; TC is 25.12 $W \cdot m^{-1} \cdot K^{-1}$; thermal shock coefficient is 790 $W \cdot m^{-1}$; thermal expansion coefficient is $5.8 \times 10^{-6}\,K^{-1}$. Besides, sapphire has many other advantages, such as chemical stability at high temperature (1000 °C), easy growth into large size, and low price. However, there is a large lattice mismatch (13.9%) and thermal expansion mismatch (30.3%) between sapphire and GaN, which results in a high dislocation density in the grown GaN epitaxial layer increasing the thermal resistance. Thermal stress will also be generated during the cooling process of the GaN epitaxial layer, ultimately reducing the

Table 2.1 Properties of various widely used chip substrates.

Properties CS	Sapphire (α-Al_2O_3) Trigonal	6H-SiC Wurtzite	α-GaN Wurtzite	Si Diamond	β-Ga_2O_3 Monoclinic	ZnO Wurtzite
LM/%	13.9	3.4	0	16.9	8.5	—
CLTE/(10^{-6} K^{-1})	$\alpha_c = 8.1$	$\alpha_c = 4.7$	$\alpha_c = 3.17$	2.6	$\alpha_c = \alpha_b = 4.2$	—
	$\alpha_a = 7.3$	$\alpha = 4.3$	$\alpha_a = 5.59$		$\alpha_a = 1.4$	
TMR/%	30.3	15.92129	0	53.49	24.86583	1.7
TC/($W \cdot m^{-1} \cdot K^{-1}$)	30	490	230	1300	13	54
ECB	×	√	√	√	√	√
B/eV	8.7	3	3.4	1.12	4.9	3.2

Test temperature of CLTE, TC, ECB, and BW is 300 K.
Abbreviations: CS, crystal structure; LM, lattice mismatch; CLTE, coefficient of linear thermal expansion; TMR, thermal mismatch rate; TC, thermal conductivity; ECB, electrical conductivity bandwidth; BW, bandwidth.
Adapted from Wang et al. [25] and Chen et al. [26].

product performance. The TC of sapphire is low compared to other substrate materials. In addition, sapphire is an electric insulator whose resistivity at room temperature is greater than 1011 Ω · cm, so it is impossible to fabricate devices with a vertical chip. However, even so, the sapphire substrate has shown its unique advantages in early use. The dislocation density of the GaN film grown is comparable to that of the SiC substrate [27]. Sapphire is grown by the melt method, which is more mature. Single crystals with lower cost, larger size, and high quality can be obtained. Hence, it is the earliest and most widely used substrate material in the LED industry.

2.3.1.4 Silicon

Due to those problems with sapphire substrates, many people directly grow GaN epitaxial layers on silicon substrates. Silicon material has become the most widely used and the most mature semiconductor material. Due to the high maturity of monocrystalline silicon material growth technology, it is easy to obtain low-cost, large-size (6–12 inches), high-quality backing, which can greatly reduce the cost of LED. Compared with sapphire, single crystal silicon has some advantages in performance: high TC, good electrical conductivity, suitable for vertical structure, and more suitable for high-power LED preparation. The thermal performance of the LED is also significantly improved because of the high TC of silicon, as well as the life of the device. However, the thermal expansion coefficient of Si is much less than that of GaN. Due to the thermal mismatch (57%) and large lattice mismatch (16.9%) between Si single crystal and GaN, the film will be subjected to huge thermal stress during epitaxial growth, resulting in a large number of defects and even tortoise-cracking in the epitaxial layer. Therefore, it is difficult to grow high-quality GaN films on a silicon substrate.

2.3.1.5 Silicon Carbide

In LED substrate materials, the market share of SiC is second only to sapphire. Of all the heteroepitaxy substrates described in this book (sapphire, SiC, silicon, and gallium oxide), SiC has the lowest lattice mismatch and thermal mismatch. Hence, SiC is the most suitable for the growth of high-quality GaN epitaxy layer. SiC has good electrical conductivity and can be used to make vertical structures. Compared with the horizontal structure prepared on sapphire substrate, the vertical structure prepared on SiC is simpler and can carry higher forward current. In addition, SiC has the best TC of the materials, which enables the device to have lower temperatures at high power. So, SiC is suitable for use under high current conditions. However, the wettability between SiC and GaN is poor. A buffer layer is usually used before GaN growth, which increases the difficulty of the process. In addition, it is difficult to grow high-quality, large-size, and single-crystal SiC. The lamellar structure of SiC is easy to cleavage, and the processing performance is poor. Hence, it is easy to introduce step-like defects on the substrate surface, which decrease the quality of the epitaxial layer. The price of SiC substrate of the same size is dozens of times that of sapphire substrate, limiting its large-scale application.

2.3.1.6 GaN

GaN is naturally the most suitable substrate for GaN epitaxial film growth. Homogeneous epitaxy growth fundamentally solves the lattice mismatch and thermal mismatch problems encountered by the use of heterogeneous substrate materials. The stress caused by the property difference between materials is minimized during the growth process. GaN substrate can grow GaN epitaxy layers with the highest quality. Due to its excellent TC, GaN is also suitable for applications in high-power devices. However, the small size and the high growth cost of single-crystal GaN impede its application. The research and development of single-crystal GaN preparation technology is an important development direction.

2.3.1.7 β-Ga_2O_3

As a new GaN substrate material, the lattice mismatch between β-Ga_2O_3 and GaN is only 8.5%. From the application of WLED, the light transmittance of β-Ga_2O_3 is comparable to that of sapphire. β-Ga_2O_3 is an N-type semiconductor with certain electrical conductivity. The electrical conductivity can be adjusted to match the properties of SiC. β-Ga_2O_3 has the light transmission properties of sapphire and the electrical conductivity of SiC. β-Ga_2O_3 can be grown in large sizes using the melt method. Hence, it can be considered as an ideal GaN substrate material to replace sapphire and SiC.

2.3.2 Solder

In LED, most of the heat is conducted to the external heat sink and heat dissipation system of the device through the path of "chip-welding layer-package substrate," so the performance of the welding layer is critical.

The welding layer not only plays a role in fixing the chip but also is the main heat transfer layer, which greatly impacts the thermal performance of LED. There are several kinds of solders used commonly, including thermal conductive glue, conductive silver paste, tin paste, and eutectic alloys. Thermal conductive glue is the most basic insulating bonding material, but with a low TC. The TC of conductive silver paste is about 15–20 $W \cdot m^{-1} \cdot K^{-1}$, but it contains toxic metals like lead. The TC of conductive tin paste is about 50 $W \cdot m^{-1} \cdot K^{-1}$. In order to meet the heat transfer requirements of high-power LED chips, eutectic alloy welding has been widely used. Eutectic alloy welding makes use of metal eutectic points to weld two metals together. The most commonly used LED solder alloy solder is Au/Sn. Kim et al. [28] have confirmed that Au/Sn eutectic welding can effectively improve the heat dissipation performance of the device compared to other solders such as silver paste. The TC of $Au_{20}Sn$ can reach 59.1 $W \cdot m^{-1} \cdot K^{-1}$and the thermal conductivities of Zn–Sn alloys can reach 100–106 $W \cdot m^{-1} \cdot K^{-1}$. Alloys from the Au/Sn, Au/Ge, Zn/Al, Zn/Sn, Bi/Ag, and Sn/Sb systems are potential replacements for Pb/Sn solders for high-temperature soldering applications [29]. Lead-free solder with high TC is the future development trend of welding technology.

2.3.3 Heat Spreader

As shown in Figure 2.12, the heat spreader is glued to the chip which is the heat source. The function of the heat spreader is to spread and dissipate heat from the chip to the whole surface to avoid the occurrence of hotspot. Therefore, a high in-plane TC is crucial for the heat spreader. Besides, metal films such as copper films, hexagonal boron nitride (h-BN) film, and graphene are the potential ideal materials for heat spreader [30]. In addition, vapor chamber is a kind of highly efficient heat transfer component using gas–liquid phase variable heat transfer, which has the advantages of high TC and low thermal diffusion resistance. It also exhibits good potential as a heat spreader.

2.3.3.1 Graphene

Precisely speaking, graphene is a single layer of graphite; its honeycomb-dimensional crystal structure, Brillouin zone, and dispersion spectrum are presented in Figure 2.15 [30]. As shown in Table 2.2, the in-plane TC of graphene is in the range of 2000–6000 $W \cdot m^{-1} \cdot K^{-1}$at room temperature (different results from different researchers). In addition, graphene possesses excellent mechanical properties, robust yet flexible [30]. It is notable that the "graphene" discussed here is not just graphene itself, but also graphene oxide (GO), which is exfoliated platelets derived from graphite oxide. And graphite oxide is the vital precursor for the production of chemically modified graphene. Tremendous work has been done to investigate the thermal properties of graphene and GO films. Table 2.2 lists the fabrication method, measurement, measured TC at room temperature (except particularly marked items) of graphene, and GO [31, 32]. Graphene is divided into single-layer graphene (SLG), few-layer graphene (FLG), and bilayer graphene (BG). In Table 2.2, "*n*" denotes the number of layers.

With respect to the synthesis, there are three main methods to prepare graphene and GO. The first method is chemical vapor deposition (CVD) growth for preparing epitaxial graphene. It is a process from bottom to up. Generally, at a very high

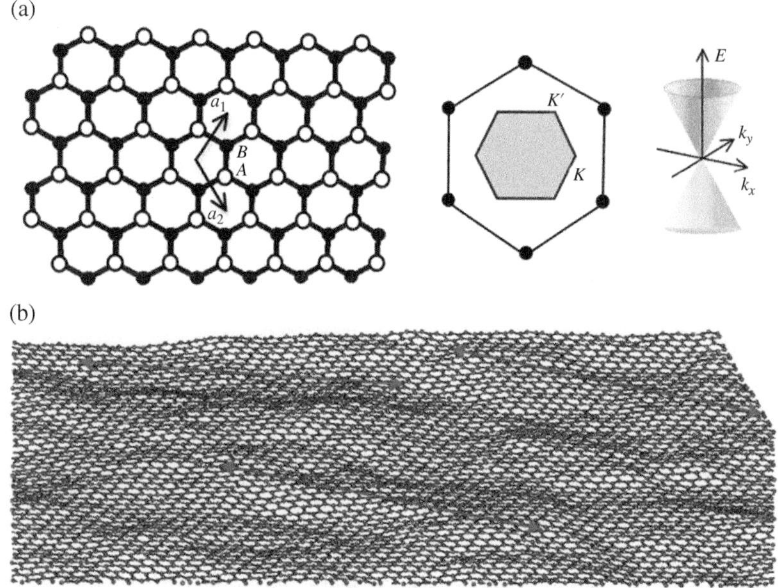

Figure 2.15 (a) Scheme of crystal structure, Brillouin zone, and dispersion spectrum of graphene; (b) "rippled graphene" from a Monte Carlo simulation. Zhu et al. [31]/with permission of John Wiley & Sons.

Table 2.2 Thermal conductivity of graphene and GO films.

Sample	Fabrication method	Measurement	TC/($W \cdot m^{-1} \cdot K^{-1}$)
SLG	Suspended; exfoliated	Raman optothermal	~3000–5000
SLG	Suspended; chemical vapor deposition (CVD)	Raman optothermal	2500
SLG	CVD; exfoliated	Electrical self-heating	310–530 ($T \sim 1000$ K)
FLG ($n = 2$–4)	Suspended; exfoliated	Raman optothermal	1300–2800
FLG ($n = 2..., 21$)	Encased within SiO_2	Heat-spreader method	50–970
FLG	Suspended and supported; polymeric residues on the surface	Electrical self-heating	150–1200
FLG ($n = 2$–8)	Suspended	Modified T-bridge	302–596
Bilayer graphene	Suspended; polymeric residues on the surface	Electrical self-heating	560–620
rGO	Blade-coating; evaporation	Laser flash	2600
rGO	Scrape coating; evaporation	Laser flash	1940
rGO	Vacuum filtration	Laser flash	1642
rGO	Evaporation	Laser flash	902
rGO	Roller coating	—	826
rGO	Filtration	—	373
rGO	Vacuum filtration	Angstrom method	220
rGO	Direct evaporation	Laser flash	61

Adapted from Zhu et al. [31] and Renteria et al. [32].

temperature (>1300 °C), precursor evaporation is deposited on a substrate (such as copper, nickel, and SiC) surface to grow epitaxial graphene. The second method is micromechanical exfoliation. This is an inverse process to the first one, which is from top to bottom. Apparently, it uses mechanical approach to exfoliate graphene from bulk graphite. High-quality graphene can be obtained by this method, but the size is limited. The third method is the exfoliation of graphite in solvents. In brief, the graphite powder is dispersed in various solvents and filtered in turn to obtain graphene. Its disadvantage is that the yield is relatively low.

2.3.3.2 h-BN

h-BN, which belongs to a hexagonal system, is a white block or powder and has a layered structure similar to the graphene lattice constant and similar characteristics, so is referred to as "white graphene" sometimes [33]. Its structure is shown in Figure 2.16. In each layer, B atoms and their adjacent N atoms are joined through strong covalent bonds. Adjacent layers are combined with weak van der Waals forces. The B—N bond length is 1.45 Å. The interlayer spacing is 0.333 nm. Different number of layers cause different TC of h-BN. At room temperature, h-BN bulk has been reported to possess a TC as high as $390\,\mathrm{W \cdot m^{-1} \cdot K^{-1}}$[34, 35]. Jo et al. [36] prepared and measured 5-layer and 11-layer h-BN, whose TCs are 250 and $360\,\mathrm{W \cdot m^{-1} \cdot K^{-1}}$, respectively. Due to the anisotropic structure, h-BN has distinct anisotropic TC between in-plane and cross-plane orientations. The TC along the *c*-axis of h-BN is as low as $2\,\mathrm{W \cdot m^{-1} \cdot K^{-1}}$.

A variety of methods have been utilized to prepare h-BN, including mechanical exfoliation, liquid exfoliation, CVD, and epitaxy [37]. Mechanical exfoliation of h-BN is just similar to that of graphene. It offers a feasible process to develop crystalline h-BN, but the h-BN layers are often randomly distributed with limited flake size and low yield. Liquid exfoliation is an efficient process to produce a large-scale practical production, without the use of any surfactants or organic solvents. h-BN nanosheets were directly exfoliated in water by cutting the pristine h-BN with the assistance of bath sonication [38]. The process of CVD and epitaxy method are analogous to that of graphene, with the advantage of controlling the layer number and large-scale growth of atomically thin h-BN nanosheets. It is notable that the substrates used in CVD were substantially classified into two types, substrates with a preferred orientation and amorphous substrates. The former can get higher crystallinity of monolayer h-BN, and the latter are more time-saving. By the way, there are some other novel synthetic approaches, such as ion-intercalation-assisted exfoliation [39] and physical vapor deposition [40], which can avoid the complex interrelation in growth parameters involved in the CVD process.

2.3.4 Package Substrate Materials

As shown in Figure 2.1(a) and (b), the package substrates in LEDs are required to quickly transfer the heat from the heat source to the heat sink with a large surface for thermal diffusion. The substrate needed in the high-power LEDs should have a high TC, a low thermal resistance, a matched coefficient of thermal expansion with the chip, a high insulation resistivity, a high package reliability, a good mechanical property, and an excellent machining property.

The substrate materials include Si, metal, metal alloy, ceramic, and composites. Their TCs and coefficients of thermal expansion (CTE) vary from each other as shown in Table 2.3. Among them, Si is the most early used substrate with a low cost and a high TC, but its low machining property leads to cracks easily. The main metal substrates (such as Al and Cu) need to be coated with electric insulative materials like resins. The poor TC of resins is against the heat dissipation requirement, and CTE of metal or metal alloy usually does not match with the chip materials. Ceramics substrate materials mainly contain Al_2O_3, BeO, AlN, Si_3N_4, and SiC. The Al_2O_3 substrates are the most widely applied ones due to their low costs and good mechanical properties. However, these ceramic substrates are difficult to be machined. The composite materials are made from several kinds of materials with different properties to attain a composite with a better comprehensive performance. For example, the Al/SiC substrate combines the high-thermal-conductivity metal Al and the low-thermal-expansion ceramic SiC, and, hence, possesses the comprehensive performance of a high TC, adjustable CTE (by the ratio

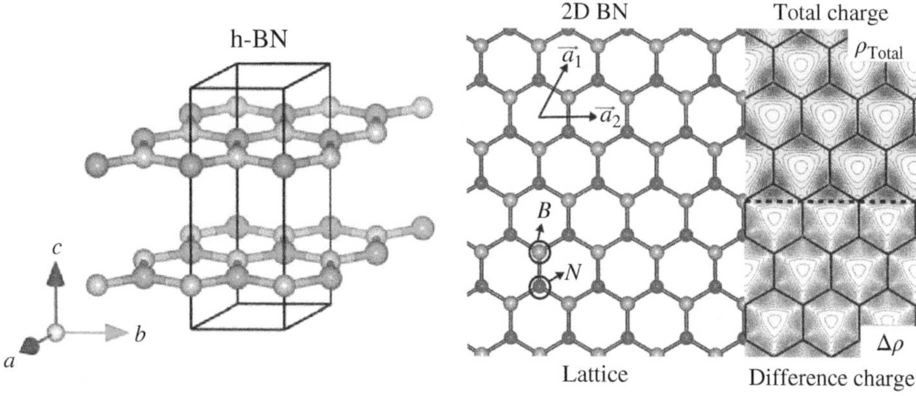

Figure 2.16 Structure of h-BN. Topsakal et al. [34]/with permission of the American Physical Society.

Table 2.3 Properties of the main materials in some substrates.

Substrates	Materials	Density/(g · cm^{-3})	CTE/(10^{-6} K^{-1})	TC/(W · m^{-1} · K^{-1})	DC (1 MHz)
Si	Si	2.33	2.49	120–150	11.8
Metal	Al	2.7	23.6	238	—
	Cu	8.96	17.8	400	—
	Cu–W	15.7–17.0	6.5–8.3	180–200	—
Ceramic	Al_2O_3	3.8	7.1	25	10.2
	BeO	2.85	6.3	285	6.7
	AlN	3.28	4.3	180–320	10
	Si_3N_4	3.20	2.8	200–320	9.4
	SiC	3.2	3.7	270	40
Composite	Al/SiC	3.0	6–9	250–280	—

Wang et al. [41]/with permission of Elsevier.

of SiC to Al), low density, good mechanical property, etc. To meet the requirements of the high-power substrate, many other substrates have been proposed and applied.

MCPCB developed from the printed circuit board (PCB) normally contains three layers: circuit layer, insulation layer, and metal layer. The high-TC metal layer can quickly transfer the heat to the heat sink. The TC of the MCPCBs varies based on the kinds of the core metal. The Golden Dragon series of LEDs promoted by OSRAM Lightbulb Company use the Al-core MCPCB with a TC of 1.3 W · m^{-1} · K^{-1}. The TC of the MCPCB (1–4 W · m^{-1} · K^{-1}) has increased a lot compared to that of the PCB (0.2–0.8 W · m^{-1} · K^{-1}). However, the mismatch of the CTE between each layer in MCPCB may decrease the reliability of the device.

Metal–ceramic substrates are prepared by coating the conventional ceramic with metal like copper and aluminum. Direct bonded copper substrates (DBCs) combine the ceramic substrate and the copper plate by high-temperature oxidation (over 1000 °C) and have a better TC and a considerable CTE with the Al_2O_3 or AlN ceramic substrates. The TCs of the DBC-Al_2O_3 and the DBC-AlN are 24 and 170 W · m^{-1} · K^{-1}, respectively. The fabrication theory of the direct aluminum-bonded substrates (DAB) is similar to that of the DBC. Besides, Al has a lower melting point (660 °C), lower price, and the excellent bonding strength and plasticity of Al can release the thermal stress resulted from the mismatch between the CTE of Al and ceramic. Direct plate copper substrates (DPCs) are prepared by coating the ceramic substrates with copper films by vacuum plating technology. The low manufacture temperature of the DPC not only avoids variation and destruction of materials but also decreases the manufacturing cost. The interface bond strength becomes the technical bottleneck of DPC substrates because the copper layer easily splits from the ceramic layer after thermal cycling. Xue [42] modified the chemical plating copper technology and prepared the DPC–AlN substrate with a TC of 147.29 W · m^{-1} · K^{-1}.

The flexible substrates are applied in portable flexible microelectronic applications. To increase the TC of the flexible substrate made from paper, boron nitride (BN) nanosheets with high TC and high electrical insulation were introduced. Zhu et al. [43] designed a flexible substrate based on the BN nanosheets and the nano-fibrillated cellulose. The prepared BN paper had a high TC along the surface (up to 145.7 W · m^{-1} · K^{-1} for 50 wt%), which was comparable to that of the Al alloy.

With the development of technology, thermal index densities of electronic devices have increased sharply, and these devices tend to miniaturization and high integration, which leads that thermal diffusion of electric devices being unprecedentedly challenged. The package substrates play an important role in quickly transferring the heat from the chip to the heat sink. The future substrates will still focus on the low cost, high TC, highly matched thermal expansion, and excellent mechanical properties (Table 2.4).

2.3.5 Thermal Conductive Polymer Composite for Encapsulation

The enclosure materials used in high-power electronic devices shall exhibit high dielectric constant (DC), high TC, low coefficient of thermal expansion, lightweight, and high mechanical properties. However, the best commercially available dielectric polymer using biaxially oriented polypropylene (PP) can only operate at temperatures below 105 °C [44].

Table 2.4 Thermal conductivities of different kinds of substrates.

Substrate types	Substrates	TC/($W \cdot m^{-1} \cdot K^{-1}$)
Metal–plastic	MCPCB-Al	1.3
	MCPCB-Al_2O_3	2
	MCPCB-Au	4
Metal–ceramic	DBC-Al_2O_3	24
	DBC-AlN	170
	DAB-Al_2O_3	32
	DPC-AlN	147.29
Flexible substrate	BN-paper	145.7

Therefore, thermally conductive polymer matrix composites with varied reinforcements, such as carbon fibers (CFs), woven, SiC particles, BN particles, titanium nitride particles, diamond particles, and whiskers, have been considered in electronic packaging.

These polymer matrixes are chosen because of the good wettability with reinforcements to couple their surfaces, the controllable or limited chemical reaction with reinforcements, the monolithization and configuration forming or shaping in regimes preventing thermo-degradation and mechanical fracture, and the continuity of a uniform matrix distribution over the whole interstitial space. However, the above polymers possess very low thermal conductivities and low glass transition temperatures, which both need to be improved a lot by the reinforcements for application in high-power electric devices. Most of the polymers can be synthesized and their properties can be adjusted to meet varied applications. Different kinds of reinforcements have been added and uniformly distributed in the polymer to prepare a composite with desired thermal, mechanical, and electrical properties. The properties of the polymer composite are strongly dependent on the reinforcements' properties, distributions, volume fractions, and alignments. For example, thermophysical properties of fiber-filled composites are normally anisotropic. Except for fiber-like and flake-like fillers, thermophysical properties of particle-filled polymers are normally isotropic in the case of uniform distribution.

Two basic strategies have been employed to fabricate dielectric thermal-conductive polymer composites. The first strategy is to add various ceramic nanofillers with high TC and DC into the polymer matrix. A high concentration of ceramic nanofillers not only guarantees the effective insertion of dipoles in highly dielectric polymer composites but also limits the thermal diffusion in the composites due to high interfacial thermal resistance. Li et al. [45] fabricated polymer composites with stable dielectric property over a broad temperature and frequency range by adding the BN nanosheets, whose TC is greatly improved (increased nine times). The second strategy is to include the incorporation of high thermally and electrically conductive nanofillers such as graphene, carbon nanotubes (CNTs), and metal into the polymer. Electrically conductive nanofillers contribute to efficient upgradation of dielectric properties because of the generated network of nanocapacitors by interfacial polarizations in the entire polymer matrix. The needed volume fractions of conductive nanofiller are low in high dielectric thermal-conductive polymer composite. In fact, if the volume fraction is higher than a certain value (percolation threshold), leakage of current will take place due to the contact of nanofillers. Another new strategy, called hybrid composite, has been researched by using small ratios of both kinds of fillers, the conductive and ceramic nanofillers, and incorporating them into the insulating polymer matrix. Ul Haq et al. [46] proposed the ternary polymer composite (polymethylmethacrylate [PMMA]/MXene/ZnO) by introducing the two-dimensional (2D) MXenes (transition metal ternary carbides or nitrides) nanosheets and ZnO nanoparticles into the PMMA and attained a substantial increment in DC and TC (14 times of increase).

2.3.6 Coolants

In conditions where air cooling is unable to satisfy the demand for heat dissipation, liquid cooling might be a suitable choice. Coolant is an essential part of liquid cooling. There are some general requirements for coolants and they may be slightly different according to the usage scenarios. The coolants for electronics must be non-flammable, nontoxic, inexpensive, and have excellent thermal properties including high TC, specific heat, and heat transfer coefficient as well as low viscosity. Besides, good chemical and thermal stability coolants must also be compatible with the materials of the vessel.

Table 2.5 Properties of several conventional coolants.

Coolants	BP/(°C)	FrP/(°C)	FlP/(°C)	TC/(W · m^{-1} · K^{-1})	SH/(kJ · kg^{-1} · K^{-1})	μ/(mPa · s)	ρ/(kg · m^{-3})
W	100	0	—	0.613	4.18	0.89	1000
EG	198	−11	125–138	0.26	2.84	19.83	1109
W/EG (50/50,v)	107	−37.8	—	0.37	3.285	3.8	1087
W/PG (50/50,v)	106	−35	—	0.36	3.40	6.4	1062
W/methanol (60/40,w)	79	−40	26	0.4	3.56	2.0	935
DEB	78	< −80	57	0.14	1.7	1	860
PAO	346	< −50	>175	0.137	2.15	9	770
Silicone	—	−111	42–55	0.11	1.6	1.4	850
FC-72	56	−90	—	0.054	1.09	0.65	1680
FC-77	97	< −100	—	0.06	1.17	1.13	1800
W/KFO (60/40,w)	—	−35	—	0.53	3.2	2.2	1250
Dynalene HC-30	112	−40	—	0.52	3.1	2.5	1275
Dynalene HC-50	118	−55	—	0.505	2.7	3.2	1340
Ga–In–Sn	—	−10	—	39	0.365	2.2	6363

Abbreviations and Greek letters in this table: BP, boiling point at 1 atm: FrP, freezing point: FlP, flash point; SH, specific heat; μ, viscosity; ρ, density. "50/50,v" denotes "50%/50%, volume ratio"; "60/40,w" denotes "60%/40%, weight ratio."
The thermal conductivity is the data at room temperature.
Murshed and De Castro [2]/with permission of Elsevier.

The widely used conventional coolants for electronics are substantially divided into two groups: dielectric and non-dielectric fluids. There are several types of usual dielectric coolants, including aromatics-based liquids (such as diethyl benzene, toluene, benzenes, and xylene), aliphatics polyalphaolefins-based liquids (such as aliphatic hydrocarbons), silicone-based liquids (such as silicone oils), and fluorocarbons-based fluids (such as FC-40, FC-72, FC-77, and FC-87). And, the common non-dielectric coolants include water (W), ethylene glycol (EG), a mixture of water and EG, propylene glycol (PG), water/methanol, water/ethanol, NaCl solution, potassium formate solution, and liquid metals. Table 2.5 lists several key properties of various widely used coolants [2].

Due to the recent advances in the heat flux density of electronic chips, conventional coolants cannot completely meet the requirements of heat dissipation. There is an urgent demand for new coolants with better thermal properties. Therefore, here comes nanofluids—a new class of heat transfer fluids, which is a mixture of the suspension of nanoparticles and conventional coolants [47, 48]. Extensive research work has been performed on the TC of nanofluids, which was found to be considerably higher than that of their base fluids. The following are some common nanoparticles in existing researches: Al, Al_2O_3, CNT, TiO_2, Cu, CuO, ZnO, SiC, Fe, Fe_2O_3, and CeO_2. Basically, TC of nanofluids will increase with the increase of nanoparticles concentration.

2.4 Thermal Interface Materials

2.4.1 Categories of Thermal Interface Materials

With the miniaturization, integration, and functionalization of electronics such as 3D chip stack architectures and LEDs, thermal dissipation has become more challenging and critical to ensure the performance, lifetime, and reliability of electronic devices. As shown in Figure 2.4, when two solids come in contact, the actual contacted area is only 1%–2% of the whole area, with other positions are filled by air between the interfaces. However, TC of air is as low as 0.027 W · m^{-1} · K^{-1} at 1 atm, 25 °C, which forms high thermal contact resistance and impedes heat transfer. Thus, TIMs are employed to fill the gap to substitute the air and enhance the thermal conduction. Some requirements must be satisfied by eligible TIMs, including flexibility, facile processability, electrical insulation, low cost, lightweight, and high TC. Polymer is chosen as an ideal material for TIMs on account of possessing most of the abovementioned requirements. However, most polymers have a low

TC in the range of 0.1–0.5 $W \cdot m^{-1} \cdot K^{-1}$ as shown in Table 2.6, which is not sufficient for the application as TIMs. In order to obtain TIMs with higher TC, polymer-based filler-reinforced composites are prepared by combining matrix and fillers. Besides polymers, PCMs and liquid metal are usually used as matrix for preparing TIMs. There are several kinds of TIMs, such as thermal grease, thermal pad, PCMs, thermal gel, thermal glue, solder, and liquid metal, as shown in Table 2.7.

Table 2.6 Thermal conductivity of commonly used polymer matrices.

Polymer matrices	TC/($W \cdot m^{-1} \cdot K^{-1}$)	Polymer matrices	TC/($W \cdot m^{-1} \cdot K^{-1}$)
High density polyethylene (HDPE)	0.33–0.53	Ultrahigh molecular weight polyethylene (UHMWPE)	0.41–0.51
Commercial thermotropic liquid crystalline polymers (LCP)	0.30–0.40	Polyoxymethylene (Homo) (POM)	0.30–0.37
Low density polyethylene (LDPE)	0.30–0.34	Poly(ethylene vinyl acetate) (EVA)	0.35
Polyphenylene sulfide (PPS)	0.30	Poly(butylene terephthalate) (PBT)	0.25–0.29
Polytetrafluoroethylene (PTFE)	0.27	Polyamide-6,6 (PA66)	0.24–0.33
Polyamide-6 (PA 6)	0.22–0.33	Polyetheretherketone (PEEK)	0.25
Polysulfone (PSU)	0.22	Polymethylmethacrylate (PMMA)	0.16–0.25
Polycarbonate (PC)	0.19–0.21	Urethane base TPE (TPU)	0.19
Poly(acrylonitrile-butadiene-styrene) copolymer (ABS)	0.15–0.20	Polyvinyl chloride (PVC)	0.13–0.29
Polyvinylidene difluoride (PVDF)	0.19	Styrene/polybutadiene copolymer (SB)	0.17–0.18
Styrene-acrylonitrile copolymer (SAN)	0.15–0.17	Poly(ethylene terephthalate) (PET)	0.15
Polystyrene (PS)	0.10–0.15	Polyvinylidene chloride (PVDC)	0.13
Polyisobutylene (PIB)	0.12–0.20	Polypropylene (PP)	0.11–0.17
Polyimide, thermoplastic (PI)	0.11	Epoxy (EP)	0.22
Polydimethylsiloxane (PDMS)	0.15 [49]	Silicone rubber (SR)	0.21
Polyvinyl alcohol (PVA)	0.22	Polyamide-imide (PAI)	0.21

Adapted from Chen et al. [50] and Zhang et al. [51].

Table 2.7 Commonly thermal interface materials.

TIMs	TC/($W \cdot m^{-1} \cdot K^{-1}$)	Fillers	Matrix	Features
Thermal grease	1–5	Al, Ag, Al_2O_3, AlN, SiC, etc.	Silicone oil	High thermal conductivity, thin thickness, easy processing, cost-effective, short lifespan, phase separation
Thermal pad	1.5–10	Al, Ag, BN, graphite, etc.	Polyurethane, silicone grease, etc.	Soft, good elasticity, corrosion resistance, short lifespan
Phase change materials	1.8–5	Al, Ni, carbon-based materials, etc.	Epoxy resin, polyolefin, paraffin	Good thermal stability, leakiness in molten state
Thermal gel	1–2.6	Al, Ag, Al_2O_3, AlN, SiC, etc.	Silicone oil (weakly crosslinked curing)	Good surface wettability, corrosion resistance, good insulation
Thermal glue	2–8	Ag, Cu, Al, Al_2O_3, graphite, etc.	Resin, rubber	Good adhesion, strong compression, deformation ability, low thermal conductivity
Solder	20–80	—	Tin, indium	High thermal conductivity, corrosion resistance, low pressure resistance, high thermal stress
Liquid metal	19–39	BN, Cu, diamond	Gallium, indium, thallium, mercury	High thermal conductivity, difficult to package, easy to leak, poor surface wettability

Table 2.8 Thermal conductivity of some fillers in TIMs.

Fillers	Category	TC/($W \cdot m^{-1} \cdot K^{-1}$)	Electrically conductive?
Carbon nanotube	Carbon-based	1000–4000	Yes
Carbon fiber	Carbon-based	10–1000	Yes
Graphene	Carbon-based	2000–6000	Yes
Graphite	Carbon-based	100–400	Yes
Aluminum	Metal	234	Yes
Copper	Metal	386–400	Yes
Silver	Metal	417–427	Yes
β-Si_3N_4	Ceramics	103–200	No
h-BN	Ceramics	185–400	No
AlN	Ceramics	100–319	No
Diamond	Ceramics	1000	No
β-SiC	Ceramics	120	—
α-Al_2O_3	Ceramics	30	No
BeO	Ceramics	230–330	No

Chen et al. [50]/with permission of Elsevier.

No matter what kind of matrix, thermally conductive fillers are indispensable to improve the TC of TIMs. Specifically, thermally conductive fillers can be divided into three categories as shown in Table 2.8 [50]: carbon-based fillers, metallic fillers, and ceramic fillers. Common carbon-based fillers include CNTs, CFs, graphite, and graphene. Common metallic fillers include aluminum, aluminum oxide, copper, copper oxide, silver, and nickel. Common ceramic fillers include BN, aluminum nitride (AlN), and beryllium oxide. According to the fillers, polymer-based TIMs can be divided into three types, carbon–polymer TIMs, metal–polymer TIMs, and ceramic–polymer TIMs.

2.4.1.1 Carbon–Polymer TIMs

Carbon-based materials, such as graphite, graphene, and CNTs, have extremely high TC. At room temperature, the reported values for TC of graphene are in the range of 2000–6000 $W \cdot m^{-1} \cdot K^{-1}$[52], TC of CNTs are in the range of 1000–4000 $W \cdot m^{-1} \cdot K^{-1}$. As typical one-dimensional (1D) materials, CFs and CNTs exhibit a high TC along the longitudinal direction. Importantly, the organic precursor and processing conditions used to determine the TC of CFs, varying from 10 to 1000 $W \cdot m^{-1} \cdot K^{-1}$. The two most important precursors for CFs are polyacrylonitrile (PAN) and mesophase pitch (MPP). On account of the highly crystalline graphitic structure and its high degree of orientation parallel to the fiber axis, MPP-based CFs exhibit a much higher TC (up to 1000 $W \cdot m^{-1} \cdot K^{-1}$) [50]. The high TC makes these carbon-based materials a suitable candidate for high-performance TIMs, by filling them in polymer matrix. The TC of composites can be significantly increased even with a very small load of carbon-based fillers. Another advantage is their light weight compared with metal or ceramic fillers. Table 2.9 gives the recent advances in TC of carbon–polymer composite TIMs and thermal conductivity enhancement (TCE) per wt% [53]. The TCE is measured by a term of TCE per wt%, which refers to the enhancement of TC by per weight content of carbon materials in composites.

2.4.1.2 Metal–Polymer TIMs

A variety of metal powders or particles are mixed into polymers to obtain high-performance TIMs, such as copper, nickel, aluminum, and silver [54]. In general, the TC of composites can be effectively increased with metal particles doping. However, due to the good electrical conductivity of metal particles, the electrical conductivity of the composites is dramatically enhanced and the dielectric breakdown voltage is reduced. Therefore, metal–polymer composites can only be utilized in environments where electrical insulation and dielectric breakdown strength are not essential [55].

Table 2.9 Thermal conductivity of carbon–polymer composite.

Composite	Carbon material content/(wt%)	TC/($W \cdot m^{-1} \cdot K^{-1}$)	TCE/(per wt%)
Py-PGMA-GNS/epoxy	3.8	1.91	225%
f-GFs/epoxy	10	1.53	66.5%
GnP-C750/epoxy	5	0.45	23.8%
DGEBA-f-GO/epoxy	4.64	0.72	52.3%
GS@Al_2O_3/PVDF	40	0.586	4.8%
Al2O3@GNP/epoxy	12	1.49	56.4%
GNP/PBT	20	1.98	61%
GNPs.PPS	37.8	4.414	49%
PI/SiCNWs-GSs	7	0.577	21%
GP/SR	0.72	0.3	69.4%
PA6/graphene-GO	10	2.14	56.9%
GNP/epoxy	25	2.67	49.4%
PVDF/FGS/ND	45	0.66	3.9%
ApPOSS-graphene/epoxy	0.5	0.348	115.8%
IL-G/PU	0.608	0.3012	55.9%
PA/TCA-rGO	5	5.1	357.8%
BE/graphene	2.5	0.542	73.7%
GNPs/silicone	16	~2.6	49.7%

Adapted from Balandin [53].

2.4.1.3 Ceramic–Polymer TIMs

Ceramic materials have both high TC and electrical insulation, which make them better than metals as fillers in TIMs. Similar to carbon materials, ceramic materials appear in multi-dimensional morphologies, including spherical, nanowire, nanotube, and nanosheet. The heat transfer of ceramic materials is predominantly through phonons due to lack of free electrons. Generally, except for BeO, most oxide fillers such as alumina (Al_2O_3) and silica (SiO_2) have a lower TC. Fortunately, some non-oxide fillers exhibit high TC, such as AlN, BN, silicon nitride (Si_3N_4), or SiC. This is attributed to their strong interatomic bonds and crystal structure, which can significantly reduce phonon scattering. Furthermore, BN and AlN are widely used as fillers for preparing thermally conductive and electrically insulating composites [50]. It is worth noting that BN has an extremely low TC in the through-plane direction, and AlN particles are easy to hydrolyze. Table 2.10 summarizes the TC, DC at 1 MHz, and electrical resistivity (ER) of ceramic fillers with different morphologies at room temperature [56].

Table 2.10 Thermal and electrical properties of ceramic fillers with different morphologies at room temperature.

Ceramics	Morphology	TC/($W \cdot m^{-1} \cdot K^{-1}$)	DC (at 1 MHz)	ER/($\Omega \cdot cm$)
AlN	Spherical	200–320	8.5–8.9	$>10^{14}$
Al_2O_3	Spherical	30–42	6.0–9.0	$>10^{14}$
SiC	Spherical	85–390	—	—
SiC	Nanowire	90	40	—
Si_3N_4	Spherical	86–155	8.3	$>10^{13}$
BN	Nanosheet	29–600	4.5	$>10^{13}$
BN	Nanotube	200–300	—	—

Hong et al. [56]/MDPI/CC BY 4.0.

2.4.2 Strategies for Enhancing TC of Thermal Interface Materials

When fillers are filled into the matrix, there are different factors influencing the TC of filler-reinforced composites, which include TC of matrix, TC of fillers, interfacial thermal resistance between fillers and fillers, interfacial thermal resistance between fillers and matrix, distribution of fillers in matrix, shape of fillers, and loading level of fillers. High loading of fillers is a direct way to improve the TC. However, with the increase of concentration, the mechanical property will decrease dramatically. Hence, there is a thermal–mechanical tradeoff relationship in composites. It is not advisable to blindly increase the concentration. From the view of thermal resistance, for a composite with fillers inside, the matrix is divided into many independent parts by fillers. There is a certain resistance for every part. Hence, the thermal resistance of composites can be regarded as a complicated serial–parallel network of fillers and matrix. It is well known that a big resistance will greatly increase the total resistance in a serial system and a low resistance can effectively decrease the total resistance in a parallel system. Hence, once fillers connect with each other and form thermal channels through composites, the TC of composites can be greatly increased. This is called the thermal percolation phenomenon. The minimum concentration of thermal percolation phenomena is known as the percolation threshold. As a result, many researches have focused on how to construct thermal networks or channels to realize TC in a low-filler loading. The main methods include filler hybridization, orientation, and network engineering. In addition, surface treatment can effectively decrease the interfacial thermal resistance between fillers and matrix, which is also conducive to the improvement of TC.

2.4.2.1 Surface Treatment

For filler dispersed in matrix, numerous filler/matrix interfaces are created. The interfaces impede the thermal conduction and the formation of thermal channels, forming interfacial thermal resistance. The incomplete contact and mismatch in the vibrational harmonics of energy carriers between different materials are the main reasons for forming interfacial thermal resistance. Suitable surface treatment of fillers is usually used to enhance the contact and coordinate the transfer of energy carriers, thus improving the TC of composites. Various surface modifiers have been used to modify filler surfaces including surfactants, coupling agents (organo-silanes and titanates), functional polymers, and inorganic coatings [50]. The chemical property of surface modifiers will influence the improvement of TC. In addition, for some situations with electrical insulation requirements, the insulative layer will be used to coat electrically conductive filler so as to inhibit electrical conductivity. Some nonmetal oxides are usually used as insulating layers like SiO_2.

2.4.2.2 Filler Hybridization

In general, the shape of thermal conductive fillers can be classified into four categories: zero-dimensional fillers (0D: Al_2O_3, Ag, and Cu), 1D fillers (CFs, CNTs, and SiC nanowires), 2D fillers (BN sheets and graphene [GP] sheets), and 3D fillers (GP sponge and foam). Specifically, 0D + 1D hybrid [57, 58], 0D + 2D hybrid [59], 1D + 2D hybrid [60–62], and another hybrid [21, 63, 64] have been deeply explored. The improvement of TC by the hybridization is mainly attributed to two reasons: ① the hybridization of different sizes can also effectively increase the filler concentration. ② The mixed filling of fillers with different shapes, sizes, dimensions, and morphologies can help the forming of thermal channels (Figure 2.17) [65].

2.4.2.3 Orientation and Network Engineering

For composites, high filler loading can help improve its TC but would decrease the mechanical property. Hence, achieving high TC with a low-filler loading has become a research hotspot. Based on the previous analysis, constructing thermal channels is an effective way. In addition, for some anisotropic fillers or special shapes, like CFs and BN sheets, orientation plays an important role in improving TC. Hence, many methods have been explored to help adjust the orientation of fillers or build thermal frameworks/channels. In general, these methods can be grouped into four categories: controlling fillers by electrical forces, controlling fillers by magnetic forces, controlling fillers by fluidic forces, and controlling fillers by other forces. For example, Uetani et al. [65] fabricated elastomeric TIMs with a high TC of $23.3\,W\cdot m^{-1}\cdot K^{-1}$ by electrostatic flocking. For electrical insulating fillers, special treatment needs to be carried out to attach charges on the surface of the filler, which ensures the response under an electric field. Similarly, magnetization is necessary for nonmagnetic fillers. Elements Fe, Go, and Ni are frequently used on account of their magnetism. Yuan et al. [67] attached Fe_3O_4 particles on the surface of h-BN to give a filler magnetic response. Then, magnetic fields are utilized to regulate the orientation of the h-BN platelets in the polymer matrix. As a result, the TC of the composite

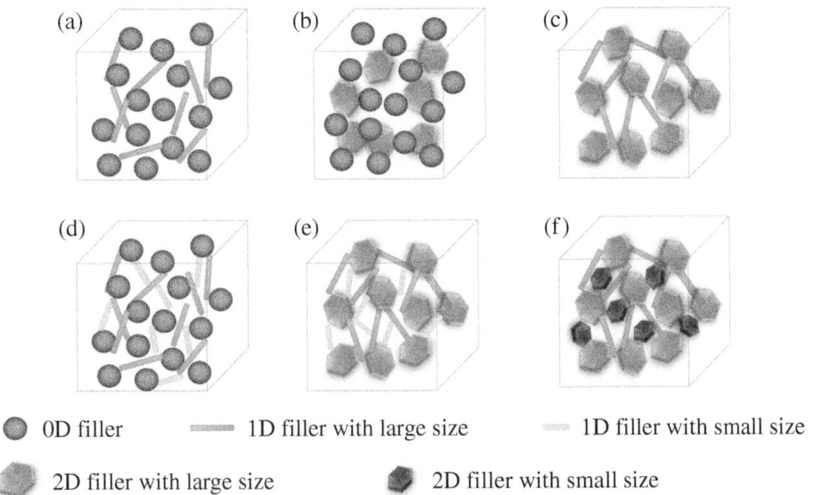

Figure 2.17 Schematic diagram of (a) 0D + 1D hydride fillers, (b) 0D + 2D hydride fillers, (c) 1D + 2D hydride fillers, (d) 0D + 1D + 1D hydride fillers, (e) 1D + 1D + 2D hydride fillers, and (f) 1D + 2D + 2D hydride fillers [66]. Reproduced with permission from Springer Nature.

with parallelly oriented h-BN was 0.357 $W \cdot m^{-1} \cdot K^{-1}$ at 9.14 vol%. Zhang et al. [68] make use of nickel to encase CF to obtain magnetic response. Benefitted by the strong response of nickel to magnetic fields, the oriented degree and fillers' concentration have greatly improved. In addition, the external magnetic field during the process ensures the connection of fillers. In terms of fluidic force, vacuum filtration [49, 69, 70], hot pressing [71, 72], and extrusion [73–75] have been deeply explored. The shear force and normal pressure of filler in fluid during the flow process drive filler orient along streamlines. Other methods (such as ice template method [76–80], foam template method [81], and bubble template method [82]) are also effective in enhancing the thermal property of composites, where thermal frameworks are built. In summary, orientation of fillers makes good use of fillers anisotropy. Construction of thermal framework provides continuous thermal channels inside composites. These microstructures promote the improvement of the thermal properties of composites.

2.4.3 Models for Thermal Conductivity of Thermal Interface Materials

Because of the diversity of materials, the combination of matrix and filler is endless. It is impossible to prepare and test all groups in the experimental method. Hence, to help estimate the properties of the candidate materials and select the optimal combination, theory and modeling play an important role in the process of screening potential compositions. Many researches have explored the theoretical model for TC of composites. The earliest study can be traced back to 1873, when Maxwell published *A Treatise on Electricity and Magnetism* [83]. Then, Maxwell Garnett proposed to use the effective DC to characterize the properties of composites in 1904 [84]. In 1935, Bruggeman used the mean-field theory to analyze the interaction between fillers with Bruggeman's equation proposed [85]. In 1940, Arnold Eucken proposed the Maxwell–Eucken equation [86]. Importantly, Kapitza discovered the interface thermal resistance of solid and liquid helium surfaces in 1941, which revealed the importance of interface thermal resistance [87]. Then, in 1962, Hamilton and Crosser analyzed the effect of fillers' geometry, shape, and aspect ratio on TC of composites [88]. In 1986, Agari and Uno considered the interfacial thermal resistance between fillers, forming an empirical equation in high filler loading [89]. In 1987, Hasselman and Johnson modified the Maxwell–Eucken equation by considering the interfacial thermal resistance [90]. In 1992, Every et al. modified Bruggeman's equation by including interface thermal resistance [91]. In 1997, Nan et al. derived a methodology for predicting the effective TC of arbitrary composites with interfacial thermal resistance in terms of an effective medium approach combined with the essential concept of Kapitza thermal contact resistance [92]. These models provide the basis for predicting the TC of composites. In addition, the finite-element method also can be used to calculate the TC of composites by building models of matrix and fillers. When the calculated scale is reduced to nanoscale or microscale, molecular dynamic simulations, the first principle calculation, and the Monte Carlo method will be more meaningful.

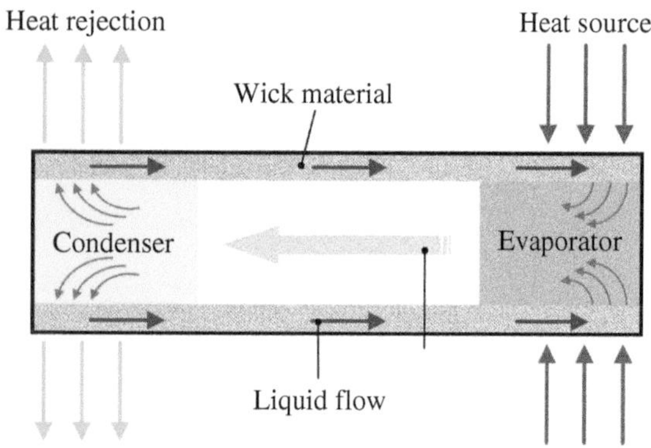

Figure 2.18 Schematic of the thermal conduction theory of a heat pop.

2.5 Heat Pipe and Vapor Chamber

2.5.1 Heat Pipe

Heat pipe (HP) is a two-phase flow heat transfer device widely used in thermal management. The TC of HPs is higher than the best solid conductor due to the accompanying latent heat during the closed two-phase cycle. Moreover, it has an excellent isothermal property, heat flux variability, reversible heat flow direction, constant temperature, thermal diode, and thermal switching property, good environmental adaptability, lightweight, low cost, etc. Because of the mentioned advantages, HP is considered as an effective thermal solution in many fields of engineering such as spacecraft thermal control, component cooling, temperature control and radiator design in satellites, dehumidification and air conditioning, heat exchangers, solar energy systems, and electronic cooling [93, 94].

HP uses latent heat of fluids to transfer energy from one place to another by simultaneous evaporation and condensation in a sealed container under a vacuum condition. As shown in Figure 2.18, it consists of an evaporator, where the working fluid absorbs heat, and a condenser, where the working fluid rejects heat with or without an adiabatic section in between them. As heat is added to the working fluid in the evaporator, it evaporates into vapor when it reaches its saturation temperature. It rises to the condenser with the assistance of buoyancy force and the vapor pressure difference between the two sections. The liquid condenses by giving out its enthalpy to the cooling water in the condenser section and returns to the evaporator for another cycle. Depending on the type, HP may have wick materials on its internal surface where the simultaneous evaporation and condensation take place. In such types of HP, the evaporator can be placed at the top, since the wick structure can return the condensate from the condenser by capillary effects and against gravity.

The recent researches on HPs include recent advances in working fluids (nanofluids, new refrigerants, etc.), wick structures (microgrooves, sintered, etc.), special types of HPs (variable conductance heat pipe [VCHP], pulsating HP, rotating HP, and electrokinetic force), and new applications (energy conservation and storage, reactors, spacecraft, renewable energy, food industries, cooling of electronic components, etc.).

2.5.2 Vapor Chamber

Unlike the cylindrical shape of common HPs, a vapor chamber is a special kind of HPs with a large and flat condensation structure which can dissipate heat from multiple heat spots, as shown in Figure 2.19. The vapor chamber spreads heat to the

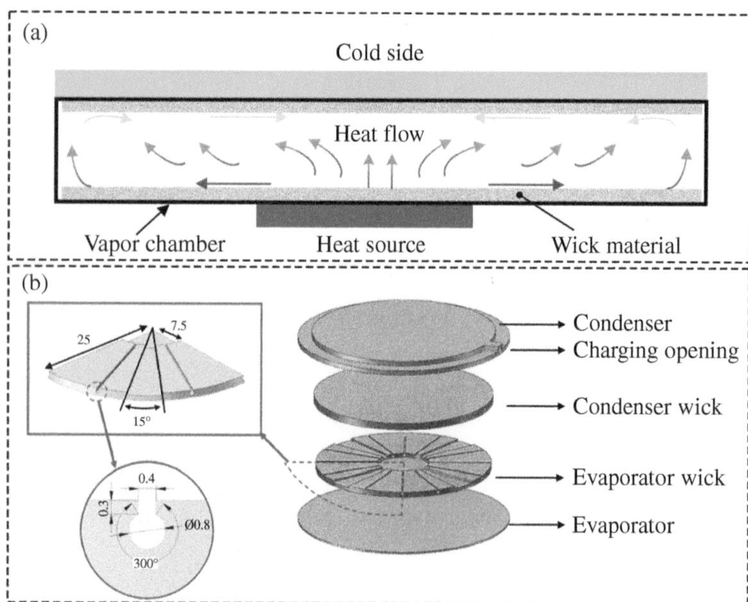

Figure 2.19 (a, b) Schematic of the structure of a vapor chamber. Chen et al. [94]/with permission of Elsevier.

cavity by the evaporation of the working fluid in the sealed space. At the condenser, the vapor of the working fluid is condensed into a liquid and then returned to the heat source by the action of capillary force and gravity. Besides, it features great uniform temperature distribution due to its short path between the evaporator and the condenser. Therefore, much attention has been paid to the vapor chambers for the thermal management of electronic components.

The wick structure is the core component of a vapor chamber. It provides capillary pressure for driving the two-phase circulation and serves as flow paths for the permeation and recirculation of working liquid. The start-up and thermal performance of vapor chambers thus rely significantly on the wick structures. Sintered powder and grooved wicks are two common wick structures in the vapor chambers [94, 95]. Large capillary pressure and excellent evaporation heat transfer performance can be provided by the sintered power wick. However, its small permeability and large flow resistance of the porous structures hinder its efficient recirculation of working liquid. On the other hand, the grooved wick features high permeability, whereas it provides relatively small capillary pressure.

In order to balance the capillary pressure and permeability, composite wicks have been developed in vapor chambers, such as sintered powder-mesh, groove-mesh, sintered powder-groove, sintered radial-multi-artery, and hybrid-evaporator and sintered rectangular-radial-grooves composite wicks.

2.6 Phase-Change Materials (PCMs)

2.6.1 Categories and Applications of PCMs

PCMs can save and absorb high heat energy within them by a phase-change mechanism. Three important categories are organic PCMs, inorganic PCMs, and eutectics, as shown in Figure 2.20. Inorganic PCMs are generally categorized into hydrated salts and metallic. The advantages of inorganic PCMs include considerable latent heat, wide temperature range, nontoxicity, thermal stability in high temperature, low cost, and good cyclic stability. Normally, these PCMs are denser and more energy efficient than organic PCMs. However, inorganic PCMs are usually corrosive and have a problem with phase separation. They are also hardly crystalline when they reach the freezing point. They are also hardly crystalline when they reach the freezing point. The organic PCMs can be divided to two groups: ① group of paraffin and ② group of non-paraffins (such as fatty acids, esters, glycolic acids, and alcohols). Paraffin is mainly composed of straight-chain alkanes, which can be expressed as C_nH_m, or $(CH–(CH_2)–CH_3)_n$. The fracture of (CH_3) chain can release a lot of phase change heat. The longer the chain is, the greater the latent heat and temperature of phase transition are. Usually, the right paraffin series are chosen as a PCM according to the application of the temperature range. However, considering the price, only a small part of paraffin can be used as a suitable PCM. Paraffin is safe, reliable, inexpensive, and non-corrosive. In addition, the volume of paraffin does not change at the time of melting and paraffin has a low vapor pressure. On account of these advantages, paraffin based thermal storage system s can carry out heat storage and cycle in a long time. However, paraffin also has some disadvantages: ① low TC and ② flammable. These shortcomings can be partially or completely eliminated by improving the properties of heat storage units and paraffin. Non-paraffin PCMs do not have similar thermal physical properties, resulting in many categories as PCMs. The disadvantages and advantages of varied chemistry categories of PCMs are listed in Table 2.11.

The materials of phase changes have found many applications due to the changing temperature. Materials that melt below 15 °C can be used to cool and ventilate room air are used. Materials that melt above 90 °C are used to reduce temperatures where temperatures can rise suddenly and prevent fire. The PCM, whose melting temperature is between these two values, is used to store solar energy. However, the limited TC of PCMs strongly constrains the conductive heat transfer process.

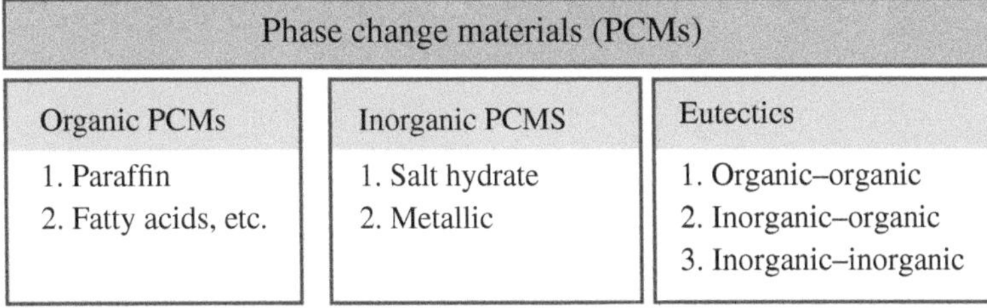

Figure 2.20 Different kinds of PCM.

Table 2.11 Various types of PCMs.

Classification	Benefits	Disadvantages
Inorganic PCMs	Considerable latent heat	Corrosive
	Wide temperature range	Supercoiling
	Nontoxicity	Phase separation
	Thermal stability	
	Good cyclic stability	
	Availability in low cost	
	Low-volume change	
	High thermal conductivity	
	High heat of fusion	
Organic PCMs	Safe and reliable	Flammability
	Inexpensive	Low thermal conductivity
	Non-corrosive	Relatively large volume change
	Good compatibility	
	Recyclable with other materials	
	No supercoiling	
	High heat of fusion	
	Availability in a large temperature range	
Eutectics	High volumetric thermal storage density	Lack of test data
	Sharp melting temperature	

2.6.2 Thermal Conductivity Enhancement of PCMs

Composite with porous materials is an effective method for TC enhancement of PCMs. Impregnation is a fast-growing technology. The porous material offers space for PCMs and the high TC of porous materials supports more effective heat transfer in composite. Expanded graphite, metal foam, and ceramic are the mostly adopted porous materials.

The enhancement of TC of PCMs by the addition of metal particles (especially nano-scale) has also been widely reported. Different from the obvious and established structure of compressed expanded graphite and metal foam, the distribution of particles is more like expanded graphite in composite. However, the effects are usually better in the case of nano-particles addition. The reinforcement effect by the dispersion of nano-particles in continuous PCMs should be attributed to the unique phenomenon at the microscopic scale, for example, reduce the internal resistance for heat transfer, which is also reported for the TCE of heat transfer fluid in literature.

Encapsulation of PCMs is also an effective method for heat transfer improvement in PCMs region. The mechanism may be explained as the reduction of heat transfer path as well as the increase of surface conducting heat transfer. Encapsulation of PCMs is to disperse PCMs in groups of small-sized particles, which are closed and surrounded by other materials or the derivatives of PCMs themselves after the procedure of treatment. So, the direct property of PCMs is not changed. The main research lies in the selection of raw material and the method of encapsulation. Some shapes of the encapsulated PCMs and methods of encapsulation are displayed in Figure 2.21 and Table 2.12, respectively.

2.7 Thermal Metamaterials

2.7.1 Concept of Thermal Metamaterials

Thermal metamaterials originated from the electromagnetic metamaterial have rapidly developed in the last decade, which have implemented some attractive thermal management functions, such as cloaking, focusing, reversal, camouflage, and active alteration of heat flux with basic complimentary [96, 97]. In a thermal metamaterial, TC paths can be elaborately

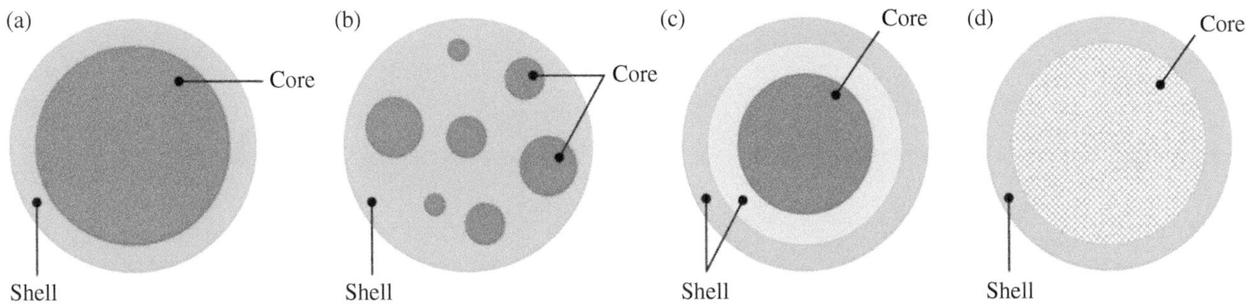

Figure 2.21 Sketch of encapsulated PCMs shapes. (a) Mononuclear, (b) polynuclear, (c) multi-wall, and (d) matrix.

Table 2.12 Methods of encapsulation of PCMs.

Physical methods	Chemical methods	Physic-chemical methods
Pan coating	Interfacial polymerization	Ionic gelation
Air-suspension coating	Suspension polymerization	Coacervation
Centrifugal extrusion	Emulsion polymerization	Sol–gel
Vibration nozzle		
Spray drying		
Solvent evaporation		

designed according to the unique thermal management function by orienting high TC and asymmetric particles in a preferred direction within a low TC matrix. Thus, thermal metamaterials are artificial anisotropic composites. There are rich studies on the experimental demonstrations of thermal metamaterials in both 2D [98–100] and 3D [101], as well as thermal–electrical multiphysics. Recently, detailed surface patterning of Cu traces by standard mass manufacturing methods has been applied in PCB for complex heat control functions.

2.7.2 Thermal Metamaterial Design

A coordinate transformation approach is often used to design the anisotropic TCs of the thermal metamaterial with the very heat control function. Typically, the steady-state diffusion equation within a predefined region in original Cartesian, (x, y, z), coordinates is modified by stretching or compressing into a designed transformed space in other coordinate systems or the original. Then, closed-form solutions of transformed material properties in the distorted space and coordinate system are obtained. For an anisotropic thermal conductive material, heat conduction is governed by Fourier's law and is written in a Cartesian coordinate system as

$$q_{\text{vol}} = -\frac{\partial}{\partial x_i}\left(\kappa_{ij}\frac{\partial T}{\partial x_j}\right) \tag{2.18}$$

where q_{vol} is the volumetric heat generation in the solid, T is the temperature, and κ_{ij} is the symmetric TC tensor which involves four conductivity coefficients in 2D, as follows:

$$\kappa_{ij} = \begin{bmatrix} \kappa_{11} & \kappa_{12} \\ \kappa_{21} & \kappa_{22} \end{bmatrix} \tag{2.19}$$

$$\kappa_{ij} = \kappa_{ji} \quad (i, j = 1, 2) \tag{2.20}$$

A composite theory [102, 103] is used to gain the TCs and orientation of the raw materials to manufacture the thermal metamaterials. For a planar thermal metamaterial involving a high TC inclusion like fiber and embedded in a low TC matrix like epoxy, the TC coefficients, κ_{ij}, are expressed as a function of the fiber orientation angle, θ, as

$$\kappa_{11} = \kappa_1 \cos^2\theta + \kappa_2 \sin^2\theta \tag{2.21}$$

$$\kappa_{22} = \kappa_1 \sin^2\theta + \kappa_2 \cos^2\theta \tag{2.22}$$

$$\kappa_{12} = \kappa_{21} = (\kappa_1 - \kappa_2) \sin\theta \cos\theta \tag{2.23}$$

where κ_1 and κ_2 are principle components of TC that are parallel and transverse to the fiber direction, respectively.

2.8 Chapter Summary

In this section, we introduce the thermal conduction and thermal resistances in electronic devices and present several solutions to enhance thermal transfer. First, we briefly introduce the concepts of thermal conduction, various thermal resistances, thermal resistance network, and thermal impedance. Thermal contact resistance easily leads to high-temperature rise, and thermal spreading resistance occurs when heat is transferred from a small source to a large plate. Then, we introduce the widely used high TC materials of different components (chips, solders, heat spreaders, substrates, encapsulation polymer composites, and coolants) in electronic packaging. In addition, we introduce the categories of TIMs, thermally conductive fillers, and polymers. The model for TC of TIMs and strategies for enhancing TC of TIMs are summed up and expounded. Besides, HP and vapor chamber with special structures and extra high thermal diffusion performances are introduced. Also, the application of PCMs in TCE is presented, as well as the thermal metamaterials with novel heat control abilities. The basic thermal conduction theory in high-power electronic devices and the corresponding thermal management materials and solutions have been explained.

References

1 iNEMI. (2014). Electronics manufacturing initiative technology roadmap.

2 Murshed, S.M.S. and De Castro, C.A.N. (2017). A critical review of traditional and emerging techniques and fluids for electronics cooling. *Renew. Sust. Energ. Rev.* 78: 821–833.

3 United States. Department of Defense (1982). *Reliability Prediction of Electronic Equipment*. Department of Defense.

4 Luo, X., Hu, R., Liu, S. et al. (2016). Heat and fluid flow in high-power LED packaging and applications. *Prog. Energy Combust. Sci.* 56: 1–32.

5 Yang, S. and Tao, W. (1998). *Heat Transfer*. Beijing: Higher Education Press.

6 Cui, Y., Li, M., and Hu, Y. (2020). Emerging interface materials for electronics thermal management: experiments, modeling, and new opportunities. *J. Mater. Chem. C* 8 (31): 10568–10586.

7 Prasher, R.S. (2001). Surface chemistry and characteristics based model for the thermal contact resistance of fluidic interstitial thermal interface materials. *J. Heat Transf. Trans. ASME* 123 (5): 969–975.

8 Ying, T.M. and Toh, K.C. (2000). A heat spreading resistance model for anisotropic thermal conductivity materials in electronic packaging. *The Seventh Intersociety Conference on Thermal and Thermomechanical Phenomena in Electronic Systems*, vol. 1, pp. 314–321.

9 Kennedy, D.P. (1960). Spreading resistance in cylindrical semiconductor devices. *J. Appl. Phys.* 31 (8): 1490–1497.

10 Kadambi, V. and Abuaf, N. (1985). An analysis of the thermal response of power chip packages. *IEEE Trans. Electron Dev.* 32 (6): 1024–1033.

11 Krane, M.J.M. (1991). Constriction resistance in rectangular bodies. *J. Electron. Packag.* 113 (4): 392–396.

12 Ellison, G.N. (2003). Maximum thermal spreading resistance for rectangular sources and plates with nonunity aspect ratios. *IEEE Trans. Compon. Packag. Technol.* 26 (2): 439–454.

13 Muzychka, Y.S., Stevanovic, M., and Yovanovich, M.M. (2001). Thermal spreading resistances in compound annular sectors. *J. Thermophys. Heat Transf.* 15 (3): 354–359.

14 Muzychka, Y.S., Yovanovich, M.M., and Culham, J.R. (2006). Influence of geometry and edge cooling on thermal spreading resistance. *J. Thermophys. Heat Transf.* 20 (2): 247–255.

15 Muzychka, Y.S., Bagnall, K.R., and Wang, E.N. (2013). Thermal spreading resistance and heat source temperature in compound orthotropic systems with interfacial resistance. *IEEE Trans. Compon. Packag. Manuf. Technol.* 3 (11): 1826–1841.

16 Bagnall, K.R., Muzychka, Y.S., and Wang, E.N. (2013). Application of the Kirchhoff transform to thermal spreading problems with convection boundary conditions. *IEEE Trans. Compon. Packag. Manuf. Technol.* 4 (3): 408–420.
17 Yang, K.-S., Chung, C.-H., Tu, C.-W. et al. (2014). Thermal spreading resistance characteristics of a high power light emitting diode module. *Appl. Therm. Eng.* 70 (1): 361–368.
18 Yang, C.T., Liu, W.C., and Liu, C.Y. (2012). Measurement of thermal resistance of first-level Cu substrate used in high-power multi-chips LED package. *Microelectron. Reliab.* 52 (5): 855–860.
19 Cheng, T., Luo, X., Huang, S. et al. (2010). Thermal analysis and optimization of multiple LED packaging based on a general analytical solution. *Int. J. Therm. Sci.* 49 (1): 196–201.
20 Ha, M. and Graham, S. (2012). Development of a thermal resistance model for chip-on-board packaging of high power LED arrays. *Microelectron. Reliab.* 52 (5): 836–844.
21 Wei, T.W., Oprins, H., Cherman, V. et al. (2020). Experimental and numerical investigation of direct liquid jet impinging cooling using 3D printed manifolds on lidded and lidless packages for 2.5D integrated systems. *Appl. Therm. Eng.* 164: 114535.
22 van Erp, R., Soleimanzadeh, R., Nela, L. et al. (2020). Co-designing electronics with microfluidics for more sustainable cooling. *Nature* 585 (7824): 211–216.
23 Farkas, G. and Poppe, A. (2014). Thermal testing of LEDs. In: *Thermal Management for LED Applications* (ed. C.J.M. Lasance and A. Poppe), 73–165. New York, NY: Springer New York.
24 Cheng, T. (2009). Research on heat dissipation in high power white led lighting devices. PhD thesis. Huazhong University of Science and Technology, Wuhan.
25 Wang, R., Chen, Z., and Hu, G. (2003). Feature comparison and research status of several LED substrate materials. *Sci. Technol. Eng.* 6 (2): 1671–1815.
26 Chen, W., Tang, H., Luo, P. et al. (2014). Research progress of GaN-based light-emitting diode substrate materials. *Acta Phys. Sin.* 63 (06): 288–300.
27 Ambacher, O. (1998). Growth and applications of group III-nitrides. *J. Phys. D Appl. Phys.* 31 (20): 2653.
28 Kim, J.S., Choi, W.S., Kim, D. et al. (2007). Fluxless silicon-to-alumina bonding using electroplated Au–Sn–Au structure at eutectic composition. *Mater. Sci. Eng. A* 458 (1–2): 101–107.
29 Zeng, G., McDonald, S., and Nogita, K. (2012). Development of high-temperature solders. *Microelectron. Reliab.* 52 (7): 1306–1322.
30 Song, H., Liu, J., Liu, B. et al. (2018). Two-dimensional materials for thermal management applications. *Joule* 2 (3): 442–463.
31 Zhu, Y., Murali, S., Cai, W. et al. (2010). Graphene and graphene oxide: synthesis, properties, and applications. *Adv. Mater.* 22 (35): 3906–3924.
32 Renteria, J.D., Nika, D.L., and Balandin, A.A. (2014). Graphene thermal properties: applications in thermal management and energy storage. *Appl. Sci.* 4 (4): 525–547.
33 Gong, F., Li, H., Wang, W. et al. (2018). Recent advances in graphene-based free-standing films for thermal management: synthesis, properties, and applications. *Coatings* 8 (2): 63.
34 Topsakal, M., Aktürk, E., and Ciraci, S. (2009). First-principles study of two- and one-dimensional honeycomb structures of boron nitride. *Phys. Rev. B* 79 (11): 115442.
35 Wang, J., Ma, F., and Sun, M. (2017). Graphene, hexagonal boron nitride, and their heterostructures: properties and applications. *RSC Adv.* 7 (27): 16801–16822.
36 Jo, I., Pettes, M.T., Kim, J. et al. (2013). Thermal conductivity and phonon transport in suspended few-layer hexagonal boron nitride. *Nano Lett.* 13 (2): 550–554.
37 Zhang, K., Feng, Y., Wang, F. et al. (2017). Two dimensional hexagonal boron nitride (2D-hBN): synthesis, properties and applications. *J. Mater. Chem. C* 5 (46): 11992–12022.
38 Lin, Y., Williams, T.V., Xu, T.-B. et al. (2011). Aqueous dispersions of few-layered and monolayered hexagonal boron nitride nanosheets from sonication-assisted hydrolysis: critical role of water. *J. Phys. Chem. C* 115 (6): 2679–2685.
39 Zeng, Z., Sun, T., Zhu, J. et al. (2012). An effective method for the fabrication of few-layer-thick inorganic nanosheets. *Angew. Chem. Int. Ed.* 51 (36): 9052–9056.
40 Sutter, P., Lahiri, J., Zahl, P. et al. (2013). Scalable synthesis of uniform few-layer hexagonal boron nitride dielectric films. *Nano Lett.* 13 (1): 276–281.
41 Wang, W., Wang, S., Zhang, D. et al. (2016). Research progress of high-power LED packaging substrates. *Mater. Guide* 30 (17): 44–50.
42 Xue, S. (2014). Preparation and performance of the metalized ceramic substrate for high-power light emitting dioxides heat dissipation. PhD thesis. Chongqing University, Chongqing.

43 Zhu, H., Li, Y., Fang, Z. et al. (2014). Highly thermally conductive papers with percolative layered boron nitride nanosheets. *ACS Nano* 8 (4): 3606–3613.

44 Rabuffi, M. and Picci, G. (2002). Status quo and future prospects for metallized polypropylene energy storage capacitors. *IEEE Trans. Plasma Sci.* 30 (5): 1939–1942.

45 Li, Q., Chen, L., Gadinski, M.R. et al. (2015). Flexible high-temperature dielectric materials from polymer nanocomposites. *Nature* 523 (7562): 576–579.

46 Ul Haq, Y., Murtaza, I., Mazhar, S. et al. (2020). Investigation of improved dielectric and thermal properties of ternary nanocomposite PMMA/MXene/ZnO fabricated by in-situ bulk polymerization. *J. Appl. Polym. Sci.* 137 (40): 49197.

47 Murshed, S.M.S. and De Castro, C.A.N. (2014). *Nanofluids: Synthesis, Properties, and Applications*. Nova Science Publishers, Incorporated.

48 Murshed, S.M.S. and de Castro, C.A.N. (2012). Nanofluids as advanced coolants. In: *Green Solvents I: Properties and Applications in Chemistry* (ed. A. Mohammad), 397–415. Springer.

49 Zhang, X., Xie, B., Zhou, S. et al. (2022). Radially oriented functional thermal materials prepared by flow field-driven self-assembly strategy. *Nano Energy* 104: 107986.

50 Chen, H.Y., Ginzburg, V.V., Yang, J. et al. (2016). Thermal conductivity of polymer-based composites: Fundamentals and applications. *Prog. Polym. Sci.* 59: 41–85.

51 Zhang, H., Zhang, X., Fang, Z. et al. (2020). Recent advances in preparation, mechanisms, and applications of thermally conductive polymer composites: a review. *J. Compos. Sci.* 4 (4): 180.

52 Nika, D.L., Pokatilov, E.P., Askerov, A.S. et al. (2009). Phonon thermal conduction in graphene: role of Umklapp and edge roughness scattering. *Phys. Rev. B* 79 (15): 155413.

53 Balandin, A.A. (2011). Thermal properties of graphene and nanostructured carbon materials. *Nat. Mater.* 10 (8): 569–581.

54 Li, A., Zhang, C., and Zhang, Y.-F. (2017). Thermal conductivity of graphene-polymer composites: mechanisms, properties, and applications. *Polymers* 9 (9): 437.

55 Mamunya, Y.P., Davydenko, V.V., Pissis, P. et al. (2002). Electrical and thermal conductivity of polymers filled with metal powders. *Eur. Polym. J.* 38 (9): 1887–1897.

56 Hong, H., Kim, J.U., and Kim, T.-i. (2017). Effective assembly of nano-ceramic materials for high and anisotropic thermal conductivity in a polymer composite. *Polymers* 9 (9): 413.

57 Cao, J.P., Zhao, J., Zhao, X.D. et al. (2013). High thermal conductivity and high electrical resistivity of poly(vinylidene fluoride)/polystyrene blends by controlling the localization of hybrid fillers. *Compos. Sci. Technol.* 89: 142–148.

58 Teng, C.-C., Ma, C.-C.M., Chiou, K.-C. et al. (2012). Synergetic effect of thermal conductive properties of epoxy composites containing functionalized multi-walled carbon nanotubes and aluminum nitride. *Compos. Part B* 43 (2): 265–271.

59 Liu, C.Q., Chen, C., Wang, H.M. et al. (2019). Synergistic effect of irregular shaped particles and graphene on the thermal conductivity of epoxy composites. *Polym. Compos.* 40: E1294–E1300.

60 An, D., Cheng, S.S., Xi, S. et al. (2020). Flexible thermal interfacial materials with covalent bond connections for improving high thermal conductivity. *Chem. Eng. J.* 383: 123151.

61 Raza, M.A., Westwood, A.V.K., Stirling, C. et al. (2015). Effect of boron nitride addition on properties of vapour grown carbon nanofiber/rubbery epoxy composites for thermal interface applications. *Compos. Sci. Technol.* 120: 9–16.

62 Feng, Y.Z., Li, X.W., Zhao, X.Y. et al. (2018). Synergetic improvement in thermal conductivity and flame retardancy of epoxy/silver nanowires composites by incorporating "branch-like" flame-retardant functionalized graphene. *ACS Appl. Mater. Interfaces* 10 (25): 21628–21641.

63 Yan, R., Su, F., Zhang, L. et al. (2019). Highly enhanced thermal conductivity of epoxy composites by constructing dense thermal conductive network with combination of alumina and carbon nanotubes. *Compos. Part A Appl. Sci. Manuf.* 125: https://doi.org/10.1016/j.compositesa.2019.105496.

64 Zeng, X.L., Sun, J.J., Yao, Y.M. et al. (2017). A combination of boron nitride nanotubes and cellulose nanofibers for the preparation of a nanocomposite with high thermal conductivity. *ACS Nano* 11 (5): 5167–5178.

65 Uetani, K., Ata, S., Tomonoh, S. et al. (2014). Elastomeric thermal interface materials with high through-plane thermal conductivity from carbon fiber fillers vertically aligned by electrostatic flocking. *Adv. Mater.* 26 (33): 5857–5862.

66 Ma, H., Gao, B., Wang, M. et al. (2020). Strategies for enhancing thermal conductivity of polymer-based thermal interface materials: a review. *J. Mater. Sci.* 56 (2): 1064–1086.

67 Yuan, C., Duan, B., Li, L. et al. (2015). Thermal conductivity of polymer-based composites with magnetic aligned hexagonal boron nitride platelets. *ACS Appl. Mater. Interfaces* 7 (23): 13000–13006.

68 Zhang, X.F., Zhou, S.L., Xie, B. et al. (2021). Thermal interface materials with sufficiently vertically aligned and interconnected nickel-coated carbon fibers under high filling loads made via preset-magnetic-field method. *Compos. Sci. Technol.* 213: 108922.

69 He, X.W., Gao, W.L., Xie, L.J. et al. (2016). Wafer-scale monodomain films of spontaneously aligned single-walled carbon nanotubes. *Nat. Nanotechnol.* 11 (7): 633–638.

70 Zhang, J., Wang, X., Yu, C. et al. (2017). A facile method to prepare flexible boron nitride/poly(vinyl alcohol) composites with enhanced thermal conductivity. *Compos. Sci. Technol.* 149: 41–47.

71 Jung, H., Yu, S., Bae, N.S. et al. (2015). High through-plane thermal conduction of graphene nanoflake filled polymer composites melt-processed in an L-shape kinked tube. *ACS Appl. Mater. Interfaces* 7 (28): 15256–15262.

72 Hu, M., Feng, J., and Ng, K.M. (2015). Thermally conductive PP/AlN composites with a 3-D segregated structure. *Compos. Sci. Technol.* 110: 26–34.

73 Niu, H., Guo, H., Kang, L. et al. (2022). Vertical alignment of anisotropic fillers assisted by expansion flow in polymer composites. *Nanomicro Lett.* 14 (1): 153.

74 Compton, B.G. and Lewis, J.A. (2014). 3D-printing of lightweight cellular composites. *Adv. Mater.* 26 (34): 5930–5935.

75 Zhang, C., Deng, K., Li, X. et al. (2023). Thermally conductive 3D-printed carbon-nanotube-filled polymer nanocomposites for scalable thermal management. *ACS Appl. Nano Mater.* 6 (14): 13400–13408.

76 Bo, Z., Zhu, H.R., Ying, C.Y. et al. (2019). Tree-inspired radially aligned, bimodal graphene frameworks for highly efficient and isotropic thermal transport. *Nanoscale* 11 (44): 21249–21258.

77 Liu, D., Lei, C., Wu, K. et al. (2020). A multidirectionally thermoconductive phase change material enables high and durable electricity via real-environment solar-thermal-electric conversion. *ACS Nano* 14 (11): 15738–15747.

78 Zeng, X.L., Yao, Y.M., Gong, Z.Y. et al. (2015). Ice-templated assembly strategy to construct 3D boron nitride nanosheet networks in polymer composites for thermal conductivity improvement. *Small* 11 (46): 6205–6213.

79 Wang, C.H., Chen, X., Wang, B. et al. (2018). Freeze-casting produces a graphene oxide aerogel with a radial and centrosymmetric structure. *ACS Nano* 12 (6): 5816–5825.

80 Xu, W.Z., Xing, Y., Liu, J. et al. (2019). Efficient water transport and solar steam generation via radially, hierarchically structured aerogels. *ACS Nano* 13 (7): 7930–7938.

81 Shen, X., Wang, Z.Y., Wu, Y. et al. (2018). A three-dimensional multilayer graphene web for polymer nanocomposites with exceptional transport properties and fracture resistance. *Mater. Horiz.* 5 (2): 275–284.

82 Xie, B., Wang, Y.J., Liu, H.C. et al. (2022). Targeting cooling for quantum dots by 57.3°C with air-bubbles-assembled three-dimensional hexagonal boron nitride heat dissipation networks. *Chem. Eng. J.* 427: 130958.

83 Maxwell, J.C. (1873). *A Treatise on Electricity and Magnetism*. Oxford: Oxford University Press.

84 Garnett, J.C.M. XII (1904). Colours in metal glasses and in metallic films. *Philos. Trans. R. Soc. Lond. A* 203 (359–371): 385–420.

85 Bruggeman, D.A.G. (1935). Berechnung verschiedener physikalischer Konstanten von heterogenen Substanzen. I. Dielektrizitätskonstanten und Leitfähigkeiten der Mischkörper aus isotropen Substanzen. *Ann. Phys.* 416 (7): 636–664.

86 Levy, F.L. (1981). A modified Maxwell-Eucken equation for calculating the thermal conductivity of two-component solutions or mixtures. *Int. J. Refrig.* 4 (4): 223–225.

87 Kapitza, P.L. (1941). Heat transfer and superfluidity of helium II. *Phys. Rev.* 60 (4): 354.

88 Hamilton, R.L. and Crosser, O.K. (1962). Thermal conductivity of heterogeneous two-component systems. *Ind. Eng. Chem. Fundam.* 1 (3): 187–191.

89 Agari, Y. and Uno, T. (1986). Estimation on thermal conductivities of filled polymers. *J. Appl. Polym. Sci.* 32 (7): 5705–5712.

90 Hasselman, D.P.H. and Johnson, L.F. (1987). Effective thermal conductivity of composites with interfacial thermal barrier resistance. *J. Compos. Mater.* 21 (6): 508–515.

91 Every, A.G., Tzou, Y., Hasselman, D.P.H. et al. (1992). The effect of particle size on the thermal conductivity of ZnS/diamond composites. *Acta Metall. Mater.* 40 (1): 123–129.

92 Nan, C.-W., Birringer, R., Clarke, D.R. et al. (1997). Effective thermal conductivity of particulate composites with interfacial thermal resistance. *J. Appl. Phys.* 81 (10): 6692–6699.

93 Maydanik, Y.F., Vershinin, S.V., Korukov, M.A. et al. (2005). Miniature loop heat pipes-a promising means for cooling electronics. *IEEE Trans. Compon. Packag. Technol.* 28 (2): 290–296.

94 Chen, L., Deng, D., Huang, Q. et al. (2020). Development and thermal performance of a vapor chamber with multi-artery reentrant microchannels for high-power LED. *Appl. Therm. Eng.* 166: 114686.

95 Wiriyasart, S. and Naphon, P. (2020). Thermal management system with different configuration liquid vapor chambers for high power electronic devices. *Case Stud. Therm. Eng.* 18: 100590.

96 Narayana, S. and Sato, Y. (2012). Heat flux manipulation with engineered thermal materials. *Phys. Rev. Lett.* 108 (21): 214303.

97 Vemuri, K.P., Canbazoglu, F.M., and Bandaru, P.R. (2014). Guiding conductive heat flux through thermal metamaterials. *Appl. Phys. Lett.* 105 (19): 193904.

98 Dede, E.M., Nomura, T., Schmalenberg, P. et al. (2013). Heat flux cloaking, focusing, and reversal in ultra-thin composites considering conduction-convection effects. *Appl. Phys. Lett.* 103 (6): 63501.

99 Han, T., Bai, X., Liu, D. et al. (2015). Manipulating steady heat conduction by sensu-shaped thermal metamaterials. *Sci. Rep.* 5 (1): 10242.

100 Chen, F. and Yuan, L.D. (2015). Experimental realization of extreme heat flux concentration with easy-to-make thermal metamaterials. *Sci. Rep.* 5 (1): 11552.

101 Xu, H., Shi, X., Gao, F. et al. (2014). Ultrathin three-dimensional thermal cloak. *Phys. Rev. Lett.* 112 (5): 054301.

102 Hiroshi, H. and Minoru, T. (1986). Equivalent inclusion method for steady state heat conduction in composites. *Int. J. Eng. Sci.* 24 (7): 1159–1172.

103 Hasselman, D.P.H., Bhatt, H., Donaldson, K.Y. et al. (1992). Effect of fiber orientation and sample geometry on the effective thermal conductivity of a uniaxial carbon fiber-reinforced glass matrix composite. *J. Compos. Mater.* 26 (15): 2278–2288.

3

Thermal Convection and Solutions

Thermal convection is a powerful and efficient heat transfer mechanism that is already massively applied in the thermal management of electronics. Most of the thermal management applications use the air cooling methods owing to their convenience and low cost. Although air cooling is still the mainstream thermal management method, the low thermal conductivity, low specific heat, and low density of air limit its application in the next-generation electronics with much larger power and heat flux. Therefore, researchers and engineers are focusing on single-phase and multiple-phase liquid cooling methods, which are more efficient, more compact, and have better cooling performance. However, in spite of the fact that the liquid cooling method is more popular to cool electronic components, air cooling is still the powerful and the ultimate solution to dissipate the heat to the environment in the current state.

3.1 Basic Knowledge of Convection Heat Transfer

3.1.1 Basic Concepts of Convection Heat Transfer

The heat transfer process occurring when the fluid flows through the surface of the solid with a different temperature is termed as convection heat transfer. It is an incorporation of macroscopic heat convection and microscopic heat conduction. When the fluid flows through the surface of the solid, the fluid touching the surface is still and only heat conduction happens, the fluid runs away from the surface and heat convection works.

In order to systematically analyze the complex process of convection heat transfer, several group methods are commonly used. For example, it is classified as natural convection heat transfer and forced convection heat transfer depending on different causes of the fluxion of fluid, inner convection heat transfer and outer convection heat transfer depending on the contact situation of the fluid and the solid surface, laminar convection heat transfer and turbulent convection heat transfer depending on the state of the current, and single-phase convection heat transfer and multi-phase convection heat transfer depending on whether there is a phase change or various phases exist during the heat transfer process.

Based on Newton's law of cooling equation (3.1), the density of convection heat transfer can be worked out as:

$$q_c = h\nabla T \tag{3.1}$$

In Equation (3.1), q_c, $\mathrm{W \cdot m^2}$, is the convection heat transfer density, h, $\mathrm{W \cdot m^{-2} \cdot K^{-1}}$, is the convection heat transfer coefficient, and ∇T, K or C°, is the temperature difference between the fluid and the solid, which is always positive to show the direction of the heat flow in common with that of the temperature decline. Convection heat transfer coefficient presents the convection heat transfer ability, which equals to the value of the heat transfer amount per unit time per unit area of heat transfer per unit temperature difference. As mentioned earlier, the heat transfer way of the fluid contacting the solid is heat conduction, $q_w = -\kappa \frac{\partial T}{\partial y}\Big|_{y=0}$, so that, according to the energy conservation law, $q_c = q_w$, Equation (3.2) comes into existence

Thermal Management for Opto-electronics Packaging and Applications, First Edition. Xiaobing Luo, Run Hu, and Bin Xie.

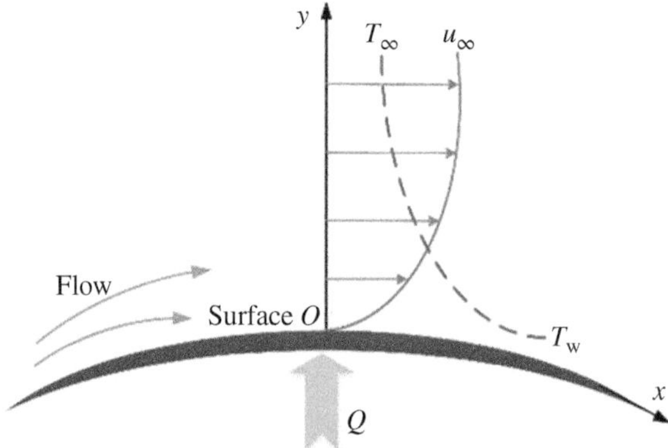

Figure 3.1 Schematic of the velocity and temperature distribution of the fluid flow in the vertical direction of the solid surface.

and shows how to acquire the h by the differential calculation of temperature yard of the fluid on the interface of fluid and solid.

$$h = -\frac{k}{\nabla T}\frac{\partial T}{\partial y}\bigg|_{y=0} \tag{3.2}$$

It should be noted that the situation of flow significantly influences the temperature distribution of the fluid during the heat transfer process, so the velocity profile is also on requirement. The governing equations of the temperature and velocity profile of fluid in convection heat transfer process include the continuity equation, momentum differential equation, and the energy differential equation.

And as for the common two-dimensional laminar convection heat transfer of incompressible fluid with constant physical properties, as shown in Figure 3.1, the governing equations (3.3) are the following:

$$\begin{cases} \dfrac{\partial u_x}{\partial x} + \dfrac{\partial u_y}{\partial y} = 0 \\ \rho\left(\dfrac{\partial u_x}{\partial t} + u_x\dfrac{\partial u_x}{\partial x} + u_y\dfrac{\partial u_x}{\partial y}\right) = F_x - \dfrac{\partial p}{\partial x} + \mu\left(\dfrac{\partial^2 u_x}{\partial x^2} + \dfrac{\partial^2 u_x}{\partial y^2}\right) \\ \rho\left(\dfrac{\partial u_y}{\partial t} + u_x\dfrac{\partial u_y}{\partial x} + u_y\dfrac{\partial u_y}{\partial y}\right) = F_y - \dfrac{\partial p}{\partial y} + \mu\left(\dfrac{\partial^2 u_y}{\partial x^2} + \dfrac{\partial^2 u_y}{\partial y^2}\right) \\ \rho C_p\left(\dfrac{\partial T}{\partial t} + u_x\dfrac{\partial T}{\partial x} + u_y\dfrac{\partial T}{\partial y}\right) = \kappa\left(\dfrac{\partial^2 T}{\partial x^2} + \dfrac{\partial^2 T}{\partial y^2}\right) \end{cases} \tag{3.3}$$

where u_x and u_y are velocities in x and y direction ($\mathrm{m \cdot s^{-1}}$), respectively. ρ is density ($\mathrm{kg \cdot m^{-3}}$). t is time (s). F_x and F_y are external forces in x and y direction (N). p is pressure (Pa). μ is dynamic viscosity ($\mathrm{N \cdot s \cdot m^{-2}}$). C_p is constant pressure heat capacity ($\mathrm{J{\cdot}kg^{-1}{\cdot}K^{-1}}$), T is temperature (K or °C) and κ is thermal conductivity ($\mathrm{W{\cdot}m^{-1}{\cdot}K^{-1}}$).

3.1.2 Basic Theories of Convection Heat Transfer

3.1.2.1 Similar Theory of Convection Heat Transfer

There are many variables in the convective heat transfer process, making the analysis of the heat transfer problem difficult. To reduce the number of variables, a similar theory classifies all the variables as physical quantities, geometrical quantities, and process quantities and combines them with non-dimensional parameters based on the eigenvalue of the physical process. The eigenvalue of the physical process is the characteristic parameter of the fluid profile, which contains the characteristic dimension, the characteristic velocity, and the qualitative temperature. Based on Equation (3.3), the non-dimensional differential equations (3.4) of convection heat transfer are the following:

$$\begin{cases} \dfrac{\partial u_{x\mathrm{dim}}}{\partial x_{\mathrm{dim}}} + \dfrac{\partial u_{y\mathrm{dim}}}{\partial y_{\mathrm{dim}}} = 0 \\ u_{x\mathrm{dim}}\dfrac{\partial u_{x\mathrm{dim}}}{\partial x_{\mathrm{dim}}} + u_{y\mathrm{dim}}\dfrac{\partial u_{x\mathrm{dim}}}{\partial y_{\mathrm{dim}}} = -\mathrm{Eu}\dfrac{\partial p_{\mathrm{dim}}}{\partial x_{\mathrm{dim}}} + \dfrac{1}{\mathrm{Re}}\left(\dfrac{\partial^2 u_{x\mathrm{dim}}}{\partial x_{\mathrm{dim}}^2} + \dfrac{\partial^2 u_{x\mathrm{dim}}}{\partial y_{\mathrm{dim}}^2}\right) \\ u_{x\mathrm{dim}}\dfrac{\partial u_{y\mathrm{dim}}}{\partial x_{\mathrm{dim}}} + u_{y\mathrm{dim}}\dfrac{\partial u_{y\mathrm{dim}}}{\partial y_{\mathrm{dim}}} = -\mathrm{Eu}\dfrac{\partial p_{\mathrm{dim}}}{\partial y_{\mathrm{dim}}} + \dfrac{1}{\mathrm{Re}}\left(\dfrac{\partial^2 u_{y\mathrm{dim}}}{\partial x_{\mathrm{dim}}^2} + \dfrac{\partial^2 u_{y\mathrm{dim}}}{\partial y_{\mathrm{dim}}^2}\right) \\ u_{x\mathrm{dim}}\dfrac{\partial T_{\mathrm{Edim}}}{\partial x_{\mathrm{dim}}} + u_{y\mathrm{dim}}\dfrac{\partial T_{\mathrm{Edim}}}{\partial y_{\mathrm{dim}}} = \dfrac{1}{\mathrm{Re}\cdot\mathrm{Pr}}\left(\dfrac{\partial^2 T_{\mathrm{Edim}}}{\partial x_{\mathrm{dim}}^2} + \dfrac{\partial^2 T_{\mathrm{Edim}}}{\partial y_{\mathrm{dim}}^2}\right) \end{cases} \tag{3.4}$$

where $u_{x\mathrm{dim}} = u_x/u_\infty$, $u_{y\mathrm{dim}} = u_y/u_\infty$, are dimensionless velocities in x and y direction. $x_{\mathrm{dim}} = x/l_{\mathrm{p}}$, $y_{\mathrm{dim}} = y/l_{\mathrm{p}}$, are dimensionless lengths in x and y direction. $p_{\mathrm{dim}} = p/(p_{\mathrm{in}} - p_{\mathrm{out}})$ is dimensionless pressure, $T_{\mathrm{Edim}} = (T - T_\infty)/(T_{\mathrm{w}} - T_\infty)$ is dimensionless excess temperature; and l_p is the length of the plate, u_∞ is the velocity of the flow, T_w and T_∞ are the temperature of the wall and the flow, respectively, and p_{in} and p_{out} are the pressure of the entrance and exit of the flow, respectively.

In Equation (3.4), several significant non-dimensional criteria come into being:

a) Euler number, $\mathrm{Eu} = \Delta p/(\rho u_\infty^2)$, shows the momentum loss rate during the flow process.
b) Reynolds number, $\mathrm{Re} = \rho u_\infty l_p/\mu$ is utilized to judge the stability of the flow state.
c) Prandtl number, $\mathrm{Pr} = \nu/\alpha$, reflects the comparison between the ability of momentum diffusion and heat diffusion.
d) Berkeley number, $\mathrm{Pe} = \mathrm{Re}{\cdot}\mathrm{Pr} = u_\infty L_p/a$ shows the comparison between the ability of heat convection and heat conduction.
e) Nusselt number, $\mathrm{Nu} = hl_p/\kappa$ shows the comparison between the heat change ability of the heat convection and the heat conduction.

With a similar theory, a series of convection heat transfer problems with similar geometry can be solved by deducing a certain functional relationship between several criteria.

3.1.2.2 Boundary Layer Theory of Convection Heat Transfer

The velocity of the fluid near the surface of the solid in the direction perpendicular to the surface will gradually change from the zero velocity at the surface to the flow velocity, as shown in Figure 3.2. This is because the fluid adjacent to the surface is motionless and the viscosity effect works in fluid. The velocity boundary layer is defined as the fluid layer whose velocity changes significantly in the vertical direction of the surface. And flow field with a small velocity change outside the boundary layer is regarded as potential flow region. Similarly, when there is a temperature difference between the fluid and the wall, if the viscosity and diffusion coefficient of the fluid are small, a thin layer with a significant temperature change can also be found in the vertical direction of the surface, which is called the temperature boundary layer.

According to the existence of the boundary layer, the corresponding boundary layer differential equation (3.5) can be obtained by simplifying the convective differential equation. $T_E = T - T_\infty$ is the excess temperature of the fluid. Under a given boundary value, the boundary layer problem can be solved by:

$$\begin{cases} \dfrac{\partial u_x}{\partial x} + \dfrac{\partial u_y}{\partial y} = 0 \\ \rho\left(u_x\dfrac{\partial u_x}{\partial x} + u_y\dfrac{\partial u_x}{\partial y}\right) = -\dfrac{\partial p}{\partial x} + \mu\dfrac{\partial^2 u_x}{\partial y^2} \\ \rho C_p\left(u_x\dfrac{\partial T_E}{\partial x} + u_y\dfrac{\partial T_E}{\partial y}\right) = \kappa\dfrac{\partial^2 T_E}{\partial y^2} \end{cases} \tag{3.5}$$

However, it is difficult to solve the boundary layer differential equations by analytical method. By direct integration of the boundary layer differential equation, the energy integral equation (3.6) and the momentum integral equation (3.7) of the plate boundary layer can be obtained, whose solution is the approximate solution of the boundary layer problem. p_{sw} is the shear stress of the partial wall at x.

$$\frac{d}{dx}\int_0^{\delta_{tb}} u_x(T_\infty - T)dy = \alpha\left(\frac{\partial T}{\partial y}\right)_{y=0} \tag{3.6}$$

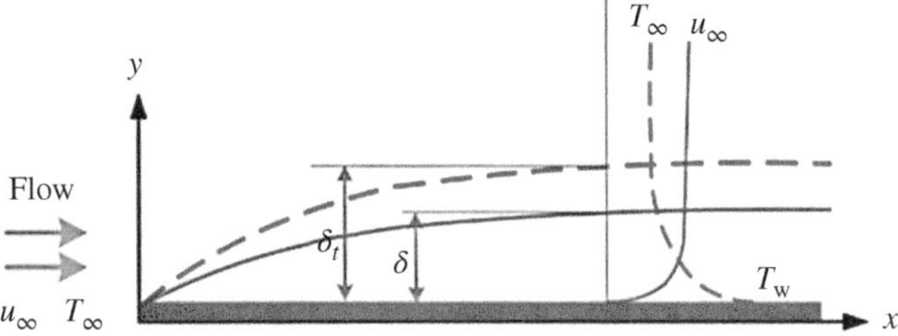

Figure 3.2 Schematic of the velocity and temperature boundary layer on the plate surface.

$$\rho \frac{d}{dx} \int_{0}^{\delta_b} u_x(u_\infty - u_x)dy = u_x \left(\frac{\partial u_x}{\partial y}\right)_{y=0} = p_{sw} \tag{3.7}$$

where δ_{tb} and δ_b are thicknesses of thermal boundary layer and boundary layers (m). α is thermal diffusivity ($m^2 \cdot s^{-1}$).

3.1.3 Basic Calculation of Convection Heat Transfer

3.1.3.1 Forced Convection Heat Transfer of a Fluid Over an Object

The average Nusselt number of fluid flow around the cylinder shown in Figure 3.3 can be obtained by using the criterion relational expression (3.8) of A.zhukaus-kas, where the values of the relevant constants k_{f1}, k_{f2} and k_{f3} are taken according to Table 3.1. The characteristic dimension is the outer diameter of the pipe d, and the characteristic velocity is the flow velocity. Except that, the qualitative temperature of Pr_w is the average temperature of the tube wall, and other qualitative temperature is the membrane temperature of the fluid. Note that the fluid flow around the cylinder discussed here refers to the orthogonal flush flow around 90° of the fluid and the cylinder axis.

$$\text{Nu} = k_{f1}\,\text{Re}^{k_{f2}}\,\text{Pr}^{k_{f3}} \left(\frac{\text{Pr}}{\text{Pr}_w}\right)^{0.25} \tag{3.8}$$

In engineering, heat transfer equipment usually utilizes a tube bundle which can be thought of as being arranged by a number of single pipes (or connected pipes) in a set manner. The common arrangement methods are fork row and forward row, as shown in Figure 3.4.

The calculation formula (3.9) of the average convection heat transfer coefficient of the fluid across the tube bundle (forward and fork) is as follows:

$$\text{Nu} = k_{f1}\,\text{Re}^{k_{f2}}\,\text{Pr}^{k_{f3}} \left(\frac{s_1}{s_2}\right)^{k_{f4}} \left(\frac{\text{Pr}}{\text{Pr}_w}\right)^{0.25} \tag{3.9}$$

The values of the coefficients k_{f1}, k_{f2}, k_{f3} and k_{f4} are determined by Reynolds number and pipe spacing, as given in Table 3.2. The flow velocity in the Reynolds number is the flow velocity (maximum flow velocity) of the narrowest flow region, and s_1 and s_2 are the horizontal pipe spacing in the vertical flow direction and the longitudinal pipe spacing along the flow, respectively. The correction factors k_z and k_β are determined by the number of pipe rows and the impact angle, as given in Tables 3.3 and 3.4.

In general, only considering bundle arrangement, convective heat transfer of fork row layout is stronger than that of in-line arrangement. However, in the design and calculation of heat exchange system, there are two main technical targets, namely the maximization of the heat transfer coefficient and the minimization of the flow resistance, which are often contradictory.

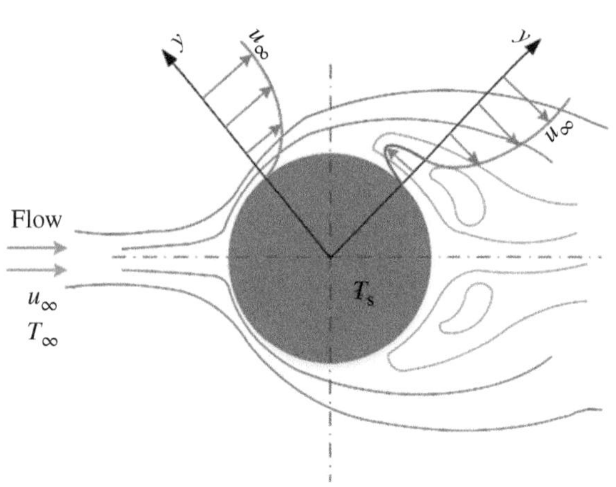

Figure 3.3 Velocity distribution of the flow around the cylinder.

3.1.3.2 Forced Convection Heat Transfer in the Duct

There are three types of pipe flow: laminar flow, turbulent flow, and transitional flow. Figure 3.5 shows the velocity distribution diagram of laminar flow and turbulent flow in a pipe. When the fluid starts to enter the pipe, the laminar boundary layer flow forms and develops near the pipe wall. Following the flow, the thickness of the boundary layer increases until the meeting of the boundary layer flow around the wall in the axis of the pipe happens, then the pipe flow has become stable flow in a sense. The pipe region from the pipe inlet to the boundary layer convergence is called the import flow region, and, after the import region, it is called the fully developed flow region.

The criterion for judging the flow form in the pipe is Re, where $\text{Re} = u_m D/\mu_k$ and u_m is the average velocity of the pipe section. When Re in the pipe is not more than 2300, it is the laminar flow. When Re is larger than 10^4, it is a turbulent flow. When Re is between 2300 and 10^4, it is the transition flow with unstable flow and heat transfer, which needs to be avoided in engineering.

Table 3.1 Value table of constants k_{f1}, k_{f2} and k_{f3} in formula (3.8).

Range of application		k_{f1}	k_{f2}	k_{f3}	Simplified formula of air or smoke
Pr = 0.60–350	Re = 5–10^3	0.5	0.5	0.38	$Nu = 0.44Re^{0.5}$
	Re = 10^3–2×10^5	0.26	0.6	0.38	$Nu = 0.22Re^{0.6}$
	Re = 2×10^5–2×10^6	0.023	0.8	0.37	$Nu = 0.02Re^{0.8}$

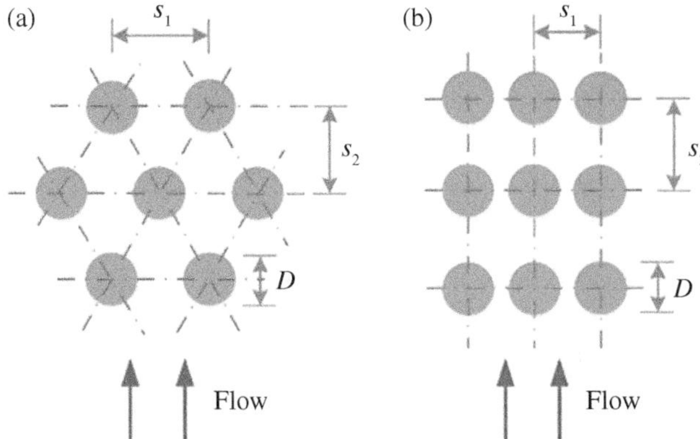

Figure 3.4 Common arrangements of pipes: (a) the fork row and (b) the forward row.

Table 3.2 The value table of the constants k_{f1}, k_{f2}, k_{f3} and k_{f4} in formula (3.9).

			The coefficients k_{f1}, k_{f2}, k_{f3} and k_{f4}				Simplified formula of air and smoke
	Range of application		k_{f1}	k_{f2}	k_{f3}	k_{f4}	Pr = 0.7
Forward	Re = 10^5–2×10^5		0.27	0.63	0.36	0	$Nu = 0.24Re^{0.63}$
	Re > 2×10^5		0.021	0.84	0.36	0	$Nu = 0.018Re^{0.84}$
Fork	Re = 10^3–2×10^5	$s_1/s_2 \le 2$	0.35	0.60	0.36	0.2	$Nu = 0.31Re^{0.60}(s_1/s_2)^{0.2}$
		$s_1/s_2 > 2$	0.40	0.60	0.36	0	$Nu = 0.35Re^{0.60}$
	Re > 2×10^5		0.022	0.84	0.36	0	$Nu = 0.019Re^{0.84}$

Table 3.3 The value table of the correction factor k_z.

Row	1	2	3	4	5	6	7	8	9	10
Forward	0.64	0.80	0.87	0.90	0.92	0.94	0.96	0.98	0.99	1.00
Fork	0.68	0.75	0.83	0.89	0.92	0.95	0.97	0.98	0.99	1.00

Table 3.4 The value table of the correction factor k_β.

β	80°–90°	70°	60°	45°	30°	15°
Forward	1.00	0.97	0.94	0.83	0.70	0.41
Fork	1.00	0.97	0.94	0.78	0.53	0.41

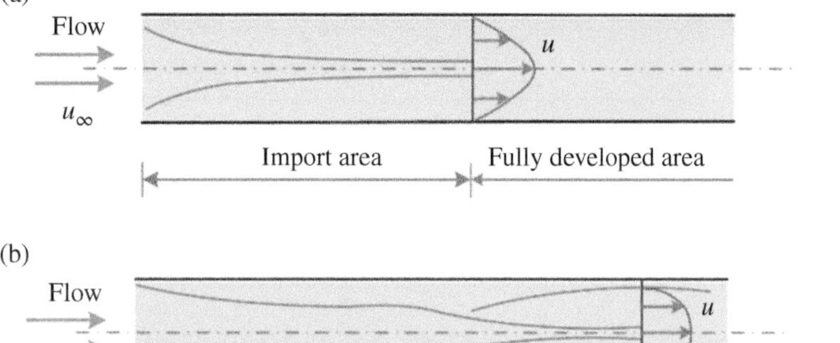

Figure 3.5 Schematic of the velocity distribution diagram of (a) laminar flow and (b) turbulent flow in pipe.

For laminar flow in the tube, the average Nusselt number of convective heat transfer of the fully developed region is 4.36 for the constant heat flow and 3.66 for the constant wall temperature. Sieder Tate proposed an average Nusselt number formula (3.10) for both the import region and fully development region, with Re < 2200, Pr > 0.6, and $\mathrm{Re} \cdot \mathrm{Pr}\frac{D}{l} > 10$. The characteristic size is the tube diameter D, and the characteristic velocity is the average velocity u_m. The qualitative temperature is the arithmetic mean temperature of the fluid, that is, $T_f = (T'_f + T''_f)/2$, and the qualitative temperatures of dynamic viscosity μ_f and μ_w are the average fluid temperature T_f, T'_f and T''_f are the average fluid temperature at inlet and outlet, and the average wall temperature T_w, respectively. And l is the entire length of the straight pipe.

$$\mathrm{Nu} = 1.86\left(\mathrm{Re} \cdot \mathrm{Pr}\frac{D}{l}\right)^{1/3}\left(\frac{\mu_f}{\mu_w}\right)^{0.14} \tag{3.10}$$

where μ_f and μ_w are dynamic viscosity of average fluid temperature and average wall temperature ($\mathrm{N \cdot s \cdot m^{-2}}$).

In the case of turbulence in the pipe, the average Nusselt number can be calculated by the Dittus–Boeleter formula (3.11), with $10^4 < \mathrm{Re} < 12 \times 10^4$, $0.7 < \mathrm{Pr} < 120$, and $l/D > 60$. The value of k_{f2} is related to the heating and cooling of the fluid: if the fluid is heated, $k_{f2} = 0.4$; if the fluid is cooled, $k_{f2} = 0.3$. In addition, the formula has certain limitations on the temperature difference ΔT between the fluid temperature and the wall temperature, which should not be more than 50 °C for gas, less than 20–30 °C for water, and less than 10 °C for oil.

$$\mathrm{Nu} = 0.023\,\mathrm{Re}^{0.8}\mathrm{Pr}^{k_{f2}} \tag{3.11}$$

Non-circular section pipes such as rectangular duct are also used in engineering. The cross-section of a rectangular duct is characterized by its aspect ratio $k_{asp} = l_y/l_z$, as shown in Figure 3.6 with the flow direction along the x-axis (perpendicular to the plane of paper).

For the boundary condition of constant peripheral wall temperature, the formula (3.12) for fully developed laminar flow is approximated by the following:

$$\mathrm{Nu} = 8.235\left(1 - 2.0421\alpha + 3.0853\alpha^2 - 2.4765\alpha^3 + 1.0578\alpha^4 - 0.1861\alpha^5\right) \tag{3.12}$$

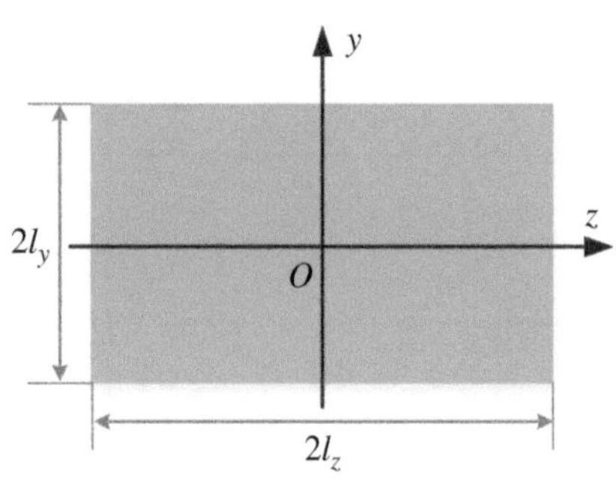

Figure 3.6 Schematic of a rectangular duct.

3.1.3.3 Natural Convection Heat Transfer of Vertical Plate

The uneven distribution of flow temperature results in the uneven distribution of flow density which produces the flow, namely the natural convection. There is a heat exchange between the natural convection flow and the wall due to the existing temperature difference, namely natural convection heat transfer. The driving force of natural convection is the buoyancy lift caused by the density difference of flow. Moreover, the judgment of the situation of natural convection, laminar or turbulent, is not Re, the ratio of the inertia force and viscous force, but another dimensionless number, the Grashof number Gr (3.13), reflecting the ratio of the buoyancy lift and viscous force.

$$\mathrm{Gr} = \frac{g k_{\mathrm{ve}} \Delta T l^3}{\mu_{\mathrm{K}}^2} \tag{3.13}$$

where g is the gravity $(\mathrm{m \cdot s^{-2}})$, k_{ve} is volume expansion coefficient, μ_{K} is kinematic viscosity $(\mathrm{m^2 \cdot s^{-1}})$.

The rule of Nu(Gr, Pr) is used to calculate the average heat transfer coefficient of the natural convection heat transfer. For the natural convection heat transfer of vertical plate in large space, as shown in Figure 3.7, Churchill and Chu put forward the following formula (3.14) for the case of constant wall temperature.

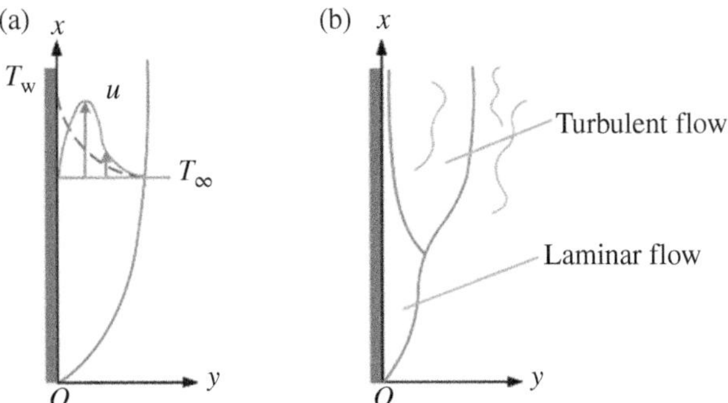

Figure 3.7 Schematic of natural convection heat transfer of vertical plate. (a) The velocity and temperature distribution; (b) laminar and turbulent flow state.

$$\mathrm{Nu} = \left\{ 0.825 + \frac{0.378(\mathrm{Gr \cdot Pr})^{1/6}}{\left[1 + (0.492/\mathrm{Pr})^{9/16}\right]^{8/27}} \right\}^2 \tag{3.14}$$

In the case of constant heat flux with unknown wall temperature, Grashof number Gr cannot be obtained. It needs to introduce a modified Grashof number Gr* (3.15) and the corresponding modified rule of $\mathrm{Nu} = k_{\mathrm{f5}}(\mathrm{Gr^* \cdot Pr})^{k_{\mathrm{f2}}}$

$$\mathrm{Gr^*} = \mathrm{Gr \cdot Nu} = \frac{g k_{\mathrm{ve}} q_{\mathrm{w}} l^4}{\kappa \mu_{\mathrm{K}}^2} \tag{3.15}$$

For vertical plate laminar flow with $\mathrm{Gr^* \cdot Pr} = 10^5 - 10^{11}$: $\mathrm{Nu} = 0.60(\mathrm{Gr^* \cdot Pr})^{1/5}$ (3.16)

For vertical plate turbulence flow with $\mathrm{Gr^* \cdot Pr} = 2 \times 10^{13} - 10^{16}$: $\mathrm{Nu} = 0.17(\mathrm{Gr^* \cdot Pr})^{1/4}$ (3.17)

3.1.3.4 Pool Boiling Convection Heat Transfer

Boiling heat transfer refers to the heat transfer from the liquid state to the vapor state, which must be accompanied by the heat exchange of latent heat of phase change in the process of heat transfer. Generally, latent heat of phase change carries much more heat than that of obvious heat, so its heat transfer intensity is much higher than that of convective heat transfer of single-phase fluid. In Figure 3.8, the boiling curve of a large container boiling experiment can help us further understand the process of boiling and its basic characteristics, which displays the relationship between the heat flux density q and the superheat of the heater surface.

The boiling curve is divided into four regions as shown in Figure 3.8:

a) Natural convection boiling region, *AB* segment. The average temperature of the fluid in the container is lower than the saturation temperature, producing a small number of bubbles and being in an excessively cold boiling state.

Figure 3.8 The boiling curve of large container.

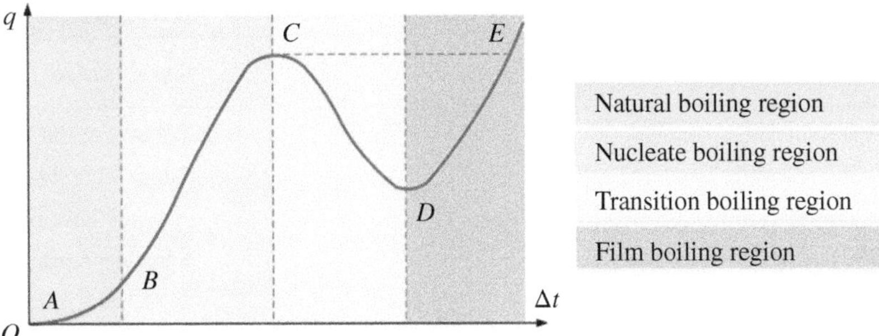

b) Nucleate boiling region, *BC* segment. The bubbles depart and regenerate at high frequency, which makes the liquid scour the wall continuously and maintain a certain degree of superheat. Meanwhile, the intense disturbance of bubbles makes the heat transfer intensity reach the maximum, and heat flow density will reach the maximum point *C*, which is called critical heat flux q_c.
c) Transition boiling region, *CD* segment. The bubble forms a gas film, hindering heat transfer from the hot surface to the fluid, and the heat transfer intensity drops sharply as well as the heat flux reaches the minimum point *D*. The boiling process is in an unstable transition zone where nuclear and membrane boiling coexist.
d) Film boiling region, *DE* segment. A stable gas film is formed on the whole wall and the heat transfer coefficient is reduced to the minimum. The surface temperature rises abruptly, so the radiation heat transfer effect appears and the heat flux density rises. At point E, the heat flux is again the same as at point *C*, but the superheat is probably a hundred times greater than that at point *C*.

In engineering, it is usually desirable to control the boiling point before point *C*, i.e. before the critical heat flux q_c. According to the convection analogy model, Rohsenow obtained the relationship (3.18) between the nuclear boiling heat flux q and the superheat of heated surface by using experimental data. The calculation formula of the critical heat flux (3.19) is also revealed. Generally speaking, it ought to be nuclear boiling when the superheat is not very high and the heat flux is less than the critical heat flux.

$$\frac{C_p(T_w - T_s)}{LH_{liq}} = k_{wl}\left\{\frac{q}{\mu_l LH_{liq}}\left[\frac{\zeta_s}{g(\rho_l - \rho_v)}\right]^{1/2}\right\}^{1/3} Pr_l^{k_{f2}} \tag{3.18}$$

$$q_c = \frac{\pi}{24} LH_{liq}\rho_v \left[\frac{\sigma_s g(\rho_l - \rho_v)}{\rho_v^2}\right]^{1/4}\left(1 + \frac{\rho_v}{\rho_l}\right)^{1/2} \tag{3.19}$$

where, T_w is the temperature of the wall (K or °C), T_s is the saturation temperature (K or °C), LH_{liq} is the latent heat of liquefaction under T_s ($J \cdot kg^{-2}$), μ_l is the dynamic viscosity of the saturation liquid ($N \cdot s \cdot m^{-2}$), ζ_s is the surface tension of the liquid ($N \cdot m^{-1}$), ρ_l and ρ_v are the density of the liquid and vapor, respectively, the qualification temperature of the vapor is the average temperature $T_m = (T_w + T_s)/2$, Pr_l is the Prandtl number under the saturation temperature, and k_{f2} equals 1.7 for most liquid but 1 for water. k_{wl} is a coefficient determined by the material of surface and liquid, as referred to in Table 3.5.

For the film boiling outside the horizontal pipe, the calculation formula (3.20) of the stable heat transfer coefficient is as follows:

$$h = 0.62\left[\frac{gLH_{liq}\rho_v(\rho_l - \rho_v)\kappa_v}{\mu_v D(T_w - T_s)}\right]^{1/4} \tag{3.20}$$

where ρ_l and LH_{liq} are determined by the saturation temperature, other physical properties take the average temperature T_m as the qualitative temperature (K or °C), and D is the outside diameter of the pipe (m), κ_v and μ_v are the thermal conductivity ($W \cdot m^{-1} \cdot K^{-1}$) and dynamic viscosity ($N \cdot s \cdot m^{-2}$) of the vapor.

Concerning the high temperature of the heated surface in film boiling causing considerable thermal radiation, the total heat transfer coefficient (3.21) can be described as:

$$h^{4/3} = h_c^{4/3} + h_r^{4/3} \tag{3.21}$$

Table 3.5 Value table of C_{wl}.

Liquid–surface	C_{wl}	Liquid–surface	C_{wl}
Water–nickel	0.006	Water–ground stainless steel	0.0080
Water–cuprum	0.013	Water–chemical erosion stainless steel	0.0133
Water–platinum	0.013	Alcohol–chromium	0.027
Water–cuprum	0.0128	Benzene–chromium	0.010
35%KOH–cuprum	0.0054	50%KOH–cuprum	0.0027

3.2 Air Cooling

With the continuous development of opto-/electronic devices toward high power and miniaturization size, the problem of heat dissipation becomes a major obstacle to their development. The stability and reliability of the system will be seriously affected, and the high-density integration technology cannot be applied, if the heat dissipation problem cannot be effectively solved, which becomes the critical bottleneck hindering the development of opto-/electronic devices. Therefore, more effective heat dissipation technology becomes an indispensable part of the development of opto-/electronic devices.

Cooling solutions for electronic packaging and applications use either passive or active cooling. Passive cooling occurs when heat is transferred without any artificially imposed force and extra energy consumption, such as natural convection. However, not all applications of opto-/electronic devices can be passively cooled. Higher power opto-/electronic devices usually require higher levels of cooling than passive cooling can provide. In such cases, active cooling methods are preferred due to the great enhancement in heat dissipation ability, which can meet the high-density heat dissipation requirements. Active cooling needs an input power or imposed force, such as forced air cooling, liquid cooling (microchannel cooling or micro-jet cooling), semiconductor refrigeration, ultrasonic heat dissipation, and superconducting cooling [1].

Air cooling is the most common type of heat dissipation due to its simple structure, high reliability, and low cost. Generally, the forced air cooling effect is much better than natural air cooling, and reliability is also high while the complexity is much lower than water cooling and oil cooling. So, it is the main cooling method for power opto-/electronic devices ranging from hundreds of watts to hundreds of kilowatts. Forced air cooling is often necessary for applications in which the volume/surface area is small or the ambient temperature is high, and thus natural convection is inadequate to dissipate the heat in time [2–4]. Forced convection is driven by externally powered components, such as a fan, which can generate a high airflow rate and achieve one or two orders of magnitude higher heat transfer coefficient compared to that in natural convection. A heat sink is usually coupled in the active cooling systems to provide a highly efficient, highly reliable cooling solution for opto-/electronic devices. Therefore, forced air cooling can endure a lower temperature difference over the ambient according to Newton's cooling law. Forced air cooling systems can help provide the needed heat dissipation levels for high-power opto-/electronic devices, but not as quietly, dependably, or reliably as passive air cooling systems. In addition, they are more expensive than passive cooling systems due to their complexity and power consumption. Therefore, a balance between performance and reliability needs to be considered when choosing passive or active cooling.

Generally, the heat generated in opto-/electronic packaging and applications is typically transferred to a heat sink by heat conduction, and then to the ambient by natural, mixed, or forced convection. Low efficiency of heat removal of the heat sink possibly causes damage to the electronic components as the temperature rises [5]. On the contrary, the smallest size of the opto-/electronic devices such as light-emitting diodes (LEDs) increases the overall flow resistance of the system and eventually suppresses the fluid flow between the fins of the heat sink. This significantly influences the fan's performance and affects its heat removal capacity. Therefore, the heat sink must be designed properly to promote heat transfer and to avoid overheating of the electronic element. The common popular coolant of electronic systems is air due to the simplicity, high reliability, and low cost of the required equipment [6]. Under normal circumstances, heat sinks with large heat dissipation area and fans with large air volume can reduce the thermal resistance of the heat sink to the ambient and improve the heat dissipation capacity. However, limited by the device volume, weight, and noise specifications, the maximum heat dissipation capacity of the air cooling systems highly depends on the heat sink design and optimization.

The studies on heat sink, whether applied in passive cooling (natural convection) or in active cooling (forced convection and forced air cooling), mainly focus on the parameter optimization, new materials or structures, coupling with some specific opto-/electronic devices, etc.

3.2.1 Heat Sink Design and Optimization

Heat sink is a kind of heat exchangers used for cooling the opto-/electronic devices with advantages of the simplicity of fabrication, low cost, and high reliability. To increase the heat dissipation area, the extended surfaces from the heat sinks are designed as either flat-plate fins or pin fins shapes. In the last decades, intensive attention has been spent on miniaturizing the opto-/electronic devices, such as high-power lasers, LEDs, and other opto-/electronic devices [7]. Meanwhile, it poses high requirement for thermal design and optimization of heat sink, including size minimization, weight cut, and high heat dissipation capacity.

Although the heat dissipation capacity of the fin heat sink is usually limited, it is the most widely used cooling solution. In addition to passive cooling, the fin heat sink is also used in active cooling, e.g. fins are also used to cool the heated working liquid in microchannels and microjets. Placing the packaging onto a metallic board and then attaching it to a heat sink is recognized as a common method. Heat generated in devices is thus dissipated to the ambient from the attached heat sink via either natural convection or forced convection. So far, studies of the fin heat sink can be classified into three categories.

a) Fin design and optimization. To design a proper heat sink that meets the thermal requirements, researchers have optimized the fin parameters, including spacing, height, configurations, and orientation, through theoretical analysis, numerical simulation, and experiments. Numerous heat sink design methods were proposed [8–13] and many fin optimization methods were developed, such as the least energy method [14], the least entropy generation method [15–20], and the minimum entransy dissipation-based thermal resistance principle [21–27].

   Costa and Lopes [28] presented a numerical study on optimizing the fins number, thickness, length (radial dimension), and height of the heat sink (Figure 3.9). They applied the heat sink to a LEDs lamp operating under natural convection conditions, and a minimum temperature of the heat source of 65 °C was achieved by increasing the fins number and height of fins and decreasing the fin thickness. Luo et al. [29] developed a semi-empirical algorithm to optimize plate fin heat sink for LED street lamp application, of which the basic optimization flowchart is shown in Figure 3.10. The fin height, fin thickness, and fin spacing of horizontal plate fin heat sinks were designed and optimized with the aim of maximal heat dissipation and the least material. The method is based on empirical equations and is relatively simple for engineers.
b) Fin heat sink design in specific packaging and applications. Considering the working characteristics of opto-/electronic applications, such as chip distribution, orientation, heat flux, and system space, some specific thermal designs were developed by simulations or experiments. Shyu et al. [30] experimentally investigated a 270 W LED backlight panel with a plate fin heat sink. Rather than the fin parameters, they analyzed the effect of shroud clearance and obstructions at the entrance or exit on the overall performance of the heat sink. Luo et al. [31] designed a heat dissipation structure for a 16 W LED bulb based on thermal behavior of a 4 W LED bulb. The simulation results show that thermal performance of the 16 W LED bulb is as good as expected in general environments.
c) Special fin heat sink design with new material or structure. Some researchers improved fin efficiency by changing the working principles of the conventional fin heat sink. Yang et al. [32] developed a thermosyphon heat sink that utilized the phase change of the working liquid inside the heat sink. Their design takes advantage of the heat pipe and conventional heat sink, and thus good performance was achieved in LED street lamps. Huang et al. [33] developed a plastic fin sink which uses the enhancement of surface emissivity and geometry surface area to compensate for the deterioration of heat dissipation ability caused by the plastic materials in a 7 W LED lamp. Figure 3.11 shows a fin heat sink comparison with a conventional aluminum fin heat sink. Although a 3 °C increase in maximal temperature was observed, the material weight and cost will be sharply decreased. The plastic heat sink with enough surface area and better surface finish can remove the heat effectively and can thus replace aluminum heat sink due to enhanced heat convection and radiation.

Jang et al. [34] focused on a radial heat sink for LED lighting applications, in which both convection and radiation are considered. They also investigated the effect of the orientation on the natural convection and radiation for a cylindrical heat

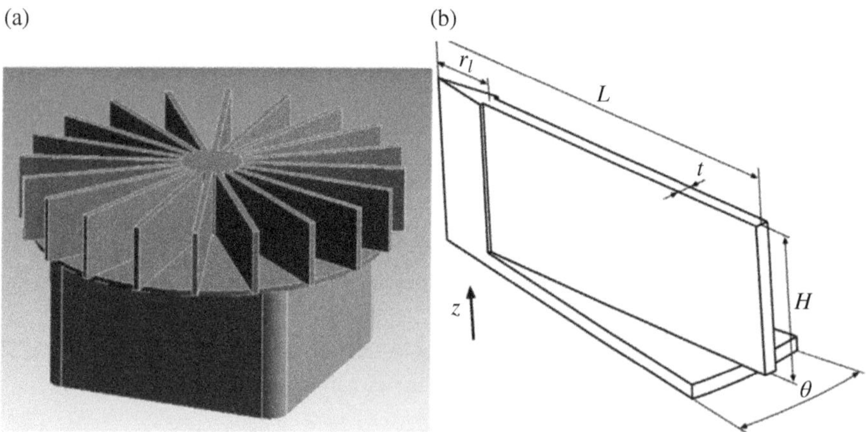

Figure 3.9 Schematic of the heat sink. (a) 3D view of the ensemble; (b) fins parameters. Costa and Lopes [28]/with permission of Elsevier.

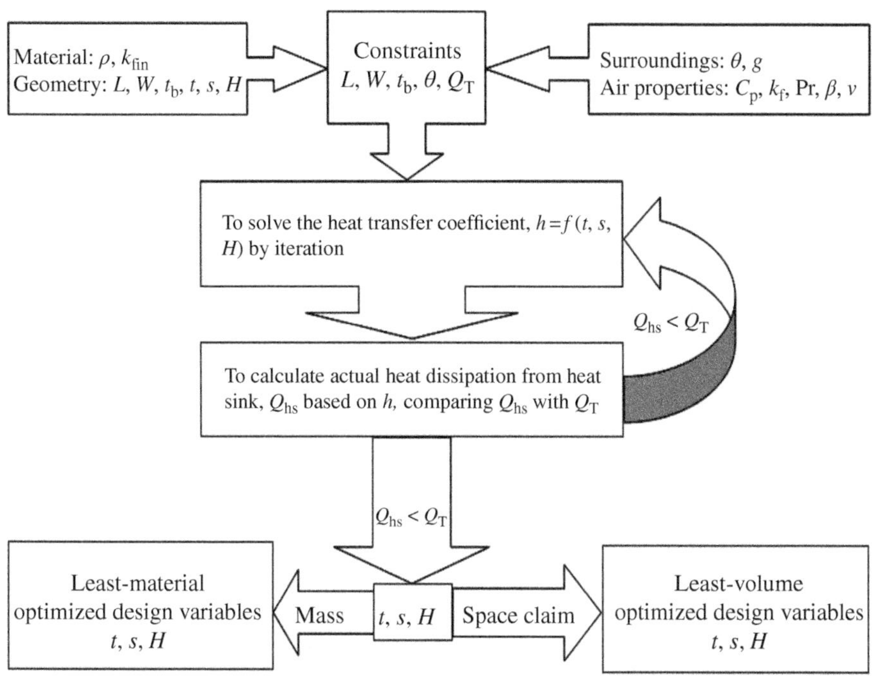

Figure 3.10 Plate-fin optimization flowchart. Luo et al. [29]/with permission of IEEE.

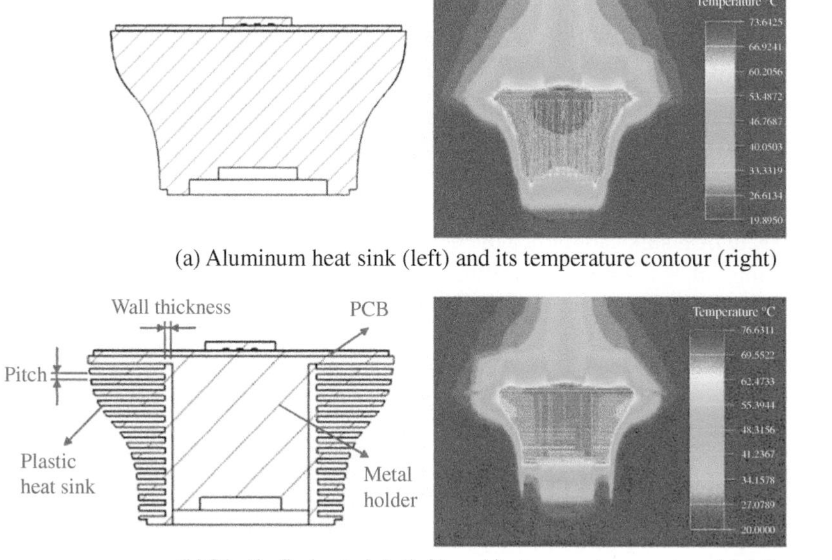

(a) Aluminum heat sink (left) and its temperature contour (right)

(b) Plastic fin heat sink (left) and its temperature contour (right)

Figure 3.11 Comparisons between (a) aluminum heat sink and (b) plastic fin heat sink for LED cooling. Huang et al. [33]/with permission of IEEE.

sink used to cool an LED light bulb. As a result, the stagnation points and flow separation appeared when the inclination angle increased. Thus, the drag coefficient increased steeply with increasing the orientation angle by increasing either the number of fins or the fin length, and, eventually, the orientation effect was intensified. Furthermore, there is no significant effect of the fin height on the orientation effect. They derived a correlation to predict the Nu number around an inclined cylindrical heat sink [35].

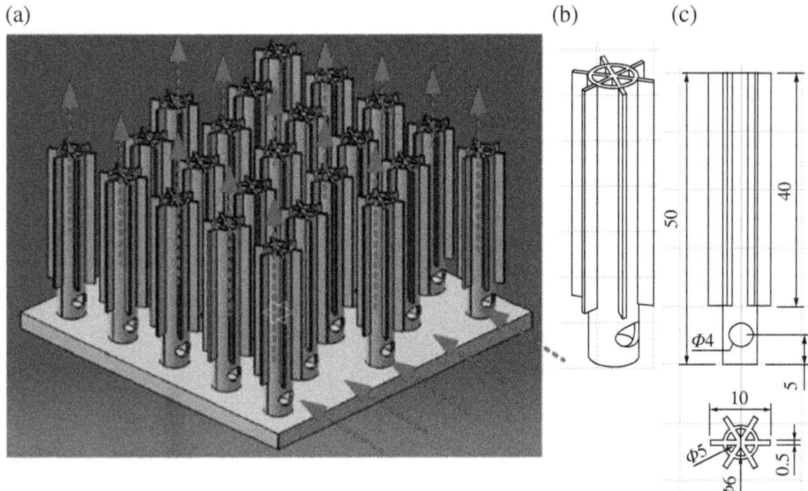

Figure 3.12 Smart heat sinks (a) a 3D view, (b) structure of hybrid pin fin, and (c) a 2D view.

Kim et al. [36] reported hybrid pin fins with internal channels to cool high-power LEDs under natural convection or forced convection. As shown in Figure 3.12, they drilled some holes near the bottom of the heat sink to inhale the air. The heated air flows upward and the chimney effect was utilized to enhance heat dissipation efficiency greatly. Although it is clear that the heat transfer coefficient will increase greatly due to the chimney effect, it may also raise the manufacturing costs.

Wang et al. [37] proposed novel cannelure fin structures with oblique dimples, straight grooves, full oblique grooves, half oblique grooves with dimple/cavity, and full oblique groove with dimple/cavity as well as flat plate fins. The results indicated that the fins with full oblique groove with dimple/cavity showed approximately 25% increase in heat transfer performance and accomplished with a friction reduction of about 20%.

Qu et al. [38] explored a hybrid heat sink with parallel plate fins sintered on the top of the substrate with three copper metal foam–paraffin composites and pure paraffin saturated in the hollow basement of the substrate as shown in Figure 3.13. Compared to the pure paraffin, the copper metal foam–paraffin composite has better thermal performance since the thermal conduction augmentation by the metal foam overcomes its suppression of natural convection of melted paraffin. The copper metal foam–paraffin composite heat sink can reduce the device temperature and the time required to reach the melting point of the paraffin by reducing either foam porosity or foam pore density.

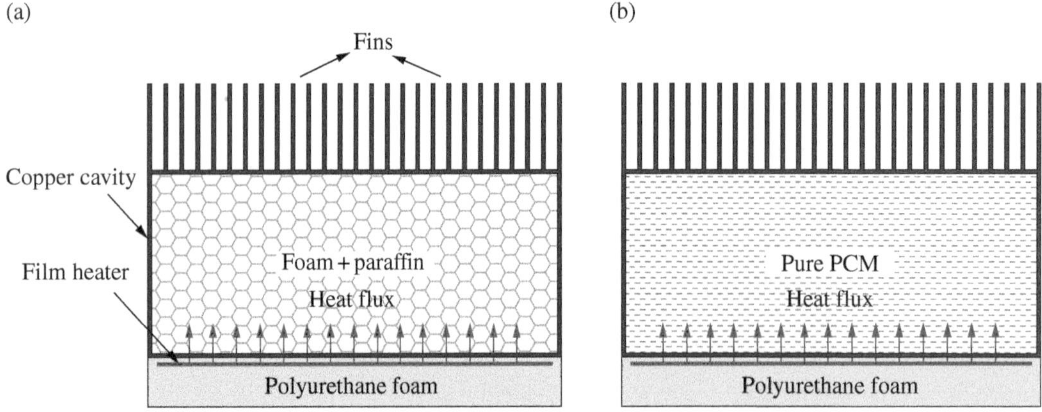

Figure 3.13 Experimental setup of the three dissipation modes: (a) metal foam–paraffin composite; (b) pure paraffin. Qu et al. [38]/with permission of Elsevier.

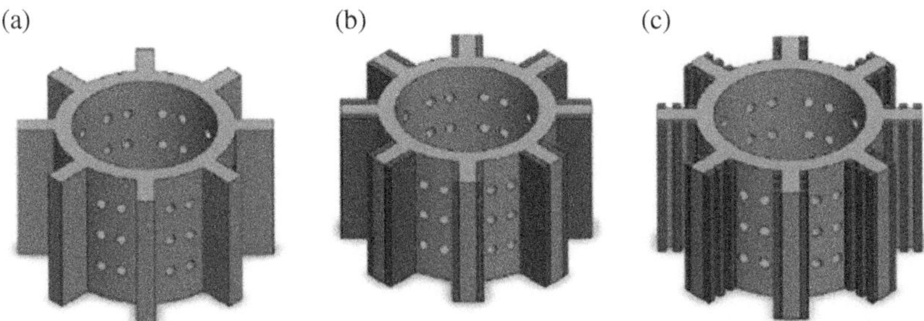

Figure 3.14 Plate shape heat sink. (a) Model A: perforated HS without porous media, (b) Model B: perforated sintered-bronze-beads layers porous heat sink, and (c) Model C: perforated sintered-bronze-beads layers porous heat sink.

Akhilesh et al. [39] optimized the dimensions of the composite heat sink. When the dimensions of the composite heat sink were less than a critical dimension, all of the phase change material (PCM) completely melted before the composite heat sink reached the set point temperature. They proposed a correlation for chosen material properties within 10% average deviation.

Jeng and Tzeng [40] tested sintered copper beads smoothly with the radial plate fins of the copper heat sink by thin layers at high temperature to form an LED cooling device. They examined perforated and not perforated heat sinks (Model A) as shown in Figure 3.14 (only the perforated types). Two modules were investigated; the plate-shape sintered-metal bead (Model B) and strip-shape sintered-metal-heat sink (Model C) were compared with the standard case of the pure copper-finned heat sink (Model A). It was demonstrated that the thermal resistances of Models B and C were 29% and 16% higher than that of Model A at corresponding conditions without a fan (pure natural convection heat transfer). It observed that the natural convection heat transfer could not be strengthened by using the sintered-metal-bead layers. The R_{th} was reduced by about 58%, 78%, and 50% when a heat sink with the fan was used for the above three models, respectively. Under mixed convection heat transfer conditions (heat sink with fan), the thermal resistance of Model B was 31% lower than that of Model A.

The techniques, which are used for improving the thermal performance of heat sinks, include the optimization of the thermal design of the heat sinks by examining the shapes and orientations of fins, the inlet/outlet of the heat sinks, the material of the heat sinks, and composite construction.

3.2.2 Piezoelectric Fan Cooling

These piezoelectric materials are widely used as actuator and sensor components because they have both electrical and mechanical characteristics. In recent years, researchers have tried to explore the potential applications of piezoelectric fans cooling due to the advantages of their lightweight, small volume, low energy consumption, low noise, long life expectancy, and adaptability in small spaces [41–47].

Piezoelectric fans are micro-vibrating machines used as airflow generators to help dissipate heat. As shown in Figure 3.15, they consist of a lightweight cantilever beam (typically Mylar) bonded with a piezoelectric patch (typically PZT) near their base ends. When the piezoelectric patch is subjected to an electric field, randomly oriented ions go into alignment and induce a slight deformation of the piezoelectric patch. As the generated displacement is small, piezoelectric patches are

Figure 3.15 Operation of a piezoelectric fan.

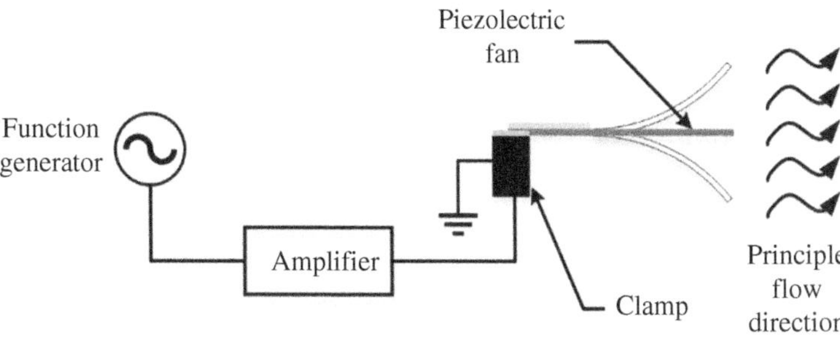

commonly compounded with lightweight cantilever beams to further amplify the displacement. As a result, an input signal to the piezoelectric material finally causes oscillatory motion at the free end of the beam, and the signal could induce the surrounding airflow.

Acikalin et al. [48] reported that piezoelectric fans are a feasible technology for cooling electronic devices because of their small size, low noise, and low energy consumption. Piezoelectric fans are also easily machined and easily installed in a cooling system. Although piezoelectric fans cannot replace rotary fans that are currently used for heat transfer, they can provide even better cooling performance in high-temperature areas or in areas where rotary fans are inefficient. Furthermore, they experimentally optimized the design parameters of a single piezoelectric fan, including fan amplitude, working distance between the fan and heat source, fan length, and fan frequency offset from resonance. They found that applying the piezoelectric fan results in a temperature drop at the LED by more than 36.4 °C.

Wait et al. [49] simulated and measured the flows of piezoelectric fans with three different blade lengths to compare the effects of different resonance frequencies on flow field and energy consumption. Their simulation and experimental results showed that although a high resonance frequency improves heat exchange by providing the complex flow field needed for sufficient mixing of the cooling fluid, it also increases energy consumption. Therefore, a piezoelectric fan for electronic devices requires an appropriate resonance frequency.

Sufian et al. [43] investigated the heat dissipation ability of piezoelectric fans on LED array by numerical and experimental methods. As shown in Figure 3.16, two piezoelectric fans were vertically oriented to the LED package, i.e. edge-to-edge arrangement and face-to-face arrangement. The results show that the single fan enhances the average heat transfer coefficient by ~1.8 times for the LED package, and the dual fans enhance the coefficient by ~2.3 times for edge-to-edge configuration and ~2.4 times for face-to-face configuration. Ma and Li [45] applied a dual-sided multiple fans system with a piezoelectric actuator in the thermal management of 30 W LEDs. Several parameters were discussed, including vibrating amplitude, power input, and arrangements. Ma et al. [44, 50] fabricated a micro multiple piezoelectric magnetic fan (m-MPMF) to provide an innovative cooling solution to the thermal management of LED lighting. The results showed that the m-MPMF system could keep the temperature of a 9 W LED substrate below 55 °C, and a T-shape m-MPMF could further reduce the temperature to 45.2 °C with only 0.3 W power consumption.

The implementation of piezoelectric fan technology will lead to a significant reduction in the power demand by the thermal management systems of power electronics. Further, the most recent articles and publications have concluded that the critical point has been reached, where rapid innovation in cooling systems must occur to keep up with the advancements in the power density of electronics. Piezoelectric fans provide an elegant solution for the immediate future of air cooling innovation, and further optimization of piezoelectric fans will widen the range of applications in which they are viable.

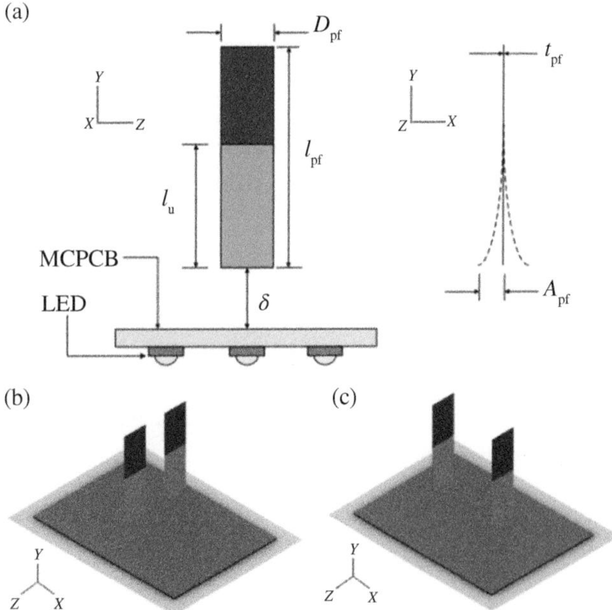

Figure 3.16 (a) Schematic of the experimental setup for piezoelectric fans oriented vertically to the bottom of the LED package with different arrangements: (b) edge-to-edge and (c) face-to-face.

3.3 Liquid Cooling

3.3.1 Microchannel Liquid Cooling

Figure 3.17 shows the schematic diagram of the microchannel liquid cooling. The fluid (coolant) flow through the multi-micro-channels is manufactured in the heat sink and carries the heat away from the heat source. The high heat transfer performance exhibited by the microchannel liquid cooling heat sink is contributed by expanding the overall heat transfer area to volume ratio. Therefore, the microchannel liquid cooling heat sink has been widely applied to the thermal management of electronics due to its small and compact geometry size, low coolant flow requirements, and large heat transfer coefficient. In 1981, Tuckerman and Pease [51] first proposed the microchannel liquid cooling as an advanced cooling solution for high heat flux electronics and they achieved a very high heat flux of 790 W/cm^2 from an integral water-cooled microchannel with a temperature rise of 71 °C. The heat transfer coefficient of the microchannel liquid cooling can be described as

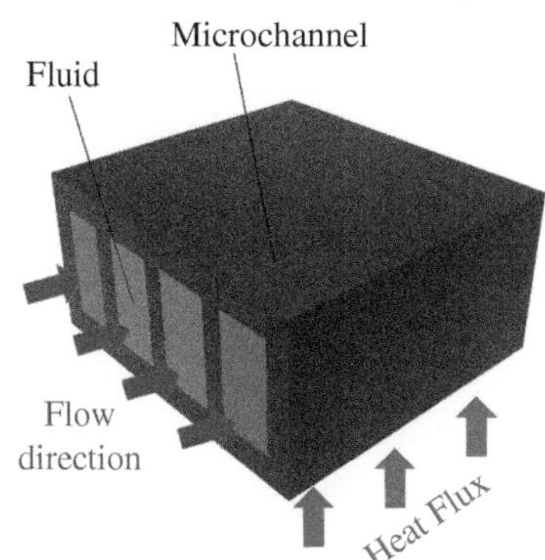

Figure 3.17 Schematic diagram of the microchannel liquid cooling.

$$h_{\text{microchannel}} = \kappa_{\text{f}} \cdot \frac{\text{Nu}_{\infty}}{D_{\text{h}}} \tag{3.22}$$

where $h_{\text{microchannel}}$ is the forced convection heat transfer coefficient, κ_{f} is the fluid thermal conductivity, and Nu is the Nusselt number. D_{h} is the so-called hydraulic diameter, which is the characteristic width of the channel, given as

$$D_{\text{h}} = \frac{2W_{\text{ch}}H_{\text{f}}}{W_{\text{ch}} + H_{\text{f}}} \tag{3.23}$$

The term microchannel is used when the hydraulic diameter of the designed heat sink is about ten to several hundred of micrometers. In the past decades, numerous studies about the microchannel liquid cooling have been done by the researchers. The microchannel liquid cooling technique is widely used in the thermal management of electronic devices.

In order to improve the overall thermal performance of the microchannel liquid cooling, various microchannel geometries were used, such as rectangular, trapezoidal, circular, diamond, triangular, and occasionally hexagonal shapes [52] as shown in Figure 3.18. The purpose of designing various microchannel geometries is to increase the effective area available for heat transfer, reduce the thermal boundary layer thickness, and reduce the flow resistance.

In the past decades, most researches on microchannel liquid cooling focused on maintaining the maximum junction temperature under a critical value, regardless of the temperature uniformity [53–56]· However, due to the different thermal expansion coefficients, a large temperature gradient increases the thermal stress in the chip and even causes interfacial delamination, resulting in performance degradation and low reliability [57]. On the other hand, single-phase microchannel liquid cooling usually has uniform heat dissipation capability, which means that when hotspots are kept under critical temperature, the other places (backgrounds) along the straight channels may be subcooled, resulting as a waste of energy [58]. For example, it is investigated that roughly 33% of the total electricity consumed is allocated to thermal management systems of electronic devices [59]. Temperature gradient is caused not only by non-uniform heat source but also by temperature rise of the flowing fluid in microchannel, which is the inherent disadvantage of single-phase liquid cooling.

Figure 3.18 Schematic diagram of different microchannel geometries.

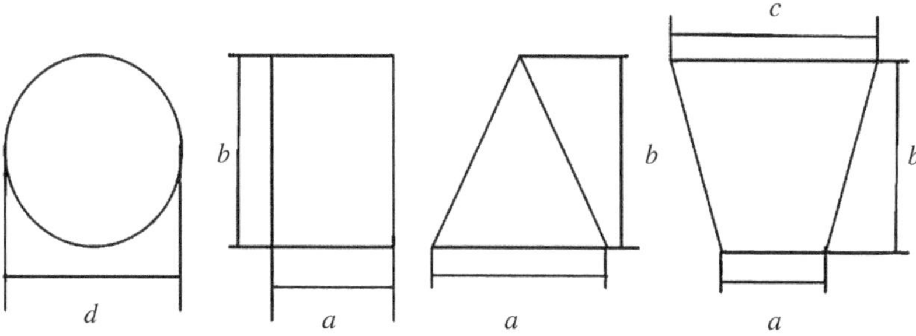

Recently, extensive researches have been done to solve the temperature uniformity problem including two-phase flow boiling and single-phase microchannel liquid cooling with various local channel densities. Although two-phase flow boiling presents much larger heat dissipation capability, the instability and critical heat flux problem are hard to handle and are still far from the industrial application [60–62]. Luo and Mao designed a tree-like microchannel to deal with temperature uniformity of multiple heat sources [57], and the temperature difference can be controlled within 1.3 °C. Lorenzini et al. used microgaps with variable pin fin clustering in the microchannel and had a great cooling effect on the hotspot with high heat flux [63]. Sharma et al. used the manifold microchannel with alternately fine and coarse channels to cool the electronics with multiple heat sources [64]. Although locally finer channels can provide larger local heat transfer capability owing to the larger heat transfer area, the flow rate may decrease due to the larger flow resistance and maldistribution effect, on the contrary, resulting in higher surface temperature. So the structure of the microchannel, such as how wide the local channel is, has a significant influence on the surface temperature that cannot be simply described by experiences. Therefore, an accurate model and optimal method are urgently required for the microchannel structural design.

3.3.2 Impingement Jet Liquid Cooling

Jet impingement liquid cooling has been studied extensively due to its high heat transfer rates and its widespread application. Extremely thin hydrodynamic and thermal boundary layers are formed when the jet impinges on the object's surface due to the jet deceleration and the resulting increase in pressure. As a result, extremely high heat transfer coefficients are obtained, especially within the stagnation zone. As shown in Figure 3.19, the flow field of the impinging jet can be divided into three separate regions with unique characteristics: free jet region, impingement region, and wall jet region. The flow in the free jet region is hardly affected by the object's surface so that the flow velocity is mainly in the axial direction. The free jet region consists of two sub-regions, the potential core and the shear layer. The flow velocity of fluid in the potential core is equal to the jet exit velocity and the fluid in the shear layer has lower velocity due to the entrainment of the surrounding fluid. The shear layer progressively grows and gradually replaces the potential core, and eventually reaches the jet centerline. The fluid touches and is affected by the object's surface, forming the impingement region. In this region, the flow is decelerated in the axial direction and accelerated in the radial direction. The radial acceleration of the fluid makes the thickness of the hydrodynamic and thermal boundary layer extremely thin and uniform, resulting in high heat transfer coefficient.

Jet impingement liquid cooling can be classified into three categories as shown in Figure 3.20: free-surface jet, submerged jet, and submerged confined jet. The free-surface jet represents the liquid flows into an air environment. The submerged jet represents the liquid flows within the same fluid in the same state. The submerged confined jet represents the submerged jet with confined walls. As electronic devices usually require compact sizes, the submerged confined jet impingement is widely applied in the thermal management of electronic devices due to its compact configuration. The performance of jet impingement depends on a numerous parameter, such as nozzle diameter, jet velocity, nozzle spacing, impact angle, nozzle length, nozzle-to-target spacing, and fluid properties. Compared to other liquid cooling technologies, higher heat flux (more than 1000 W/cm^2) can be achieved by jet impingement cooling technology [65–67].

Single-phase confined jet impingement cooling has been massively studied during the past decades. Jörg et al. presented an approach of direct single jet impingement liquid cooling of a typical metal-oxide-semiconductor field-effect transistor (MOSFET) power module [68]. Heat transfer coefficients up to 12,000 $W \cdot m^{-2} \cdot K^{-1}$ were achieved by using only 10.8 cm^2 assembly space for the cooling device. The single jet impingement can only be used to cool a small heated surface because of the temperature non-uniformity. When it comes to the high-power electronics with relatively large heated surface, jet array impingement must be applied. However, in the jet array impingement, the jet interference between adjacent jets prior to impingement on the surface and interaction due to the collision of surface flows make the heat transfer coefficient aggressively weakened [69]. In order to eliminate the interaction of fluids between neighboring nozzles, Huber and Viskanta proposed a distributed array jet cooling technique. This allows the spent fluid flows through the distributed extraction returns to exit

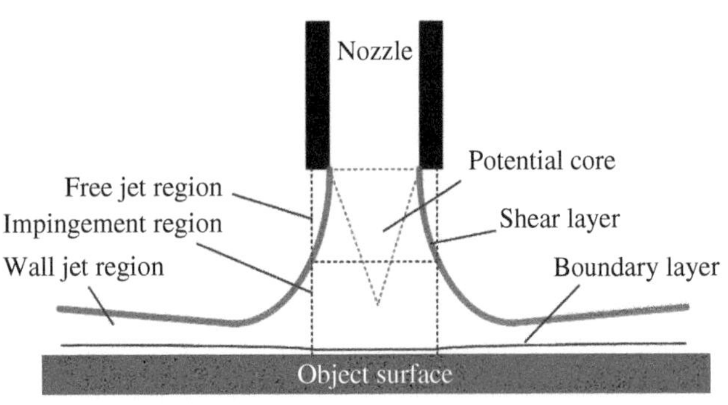

Figure 3.19 Schematic diagram of impingement jet liquid cooling.

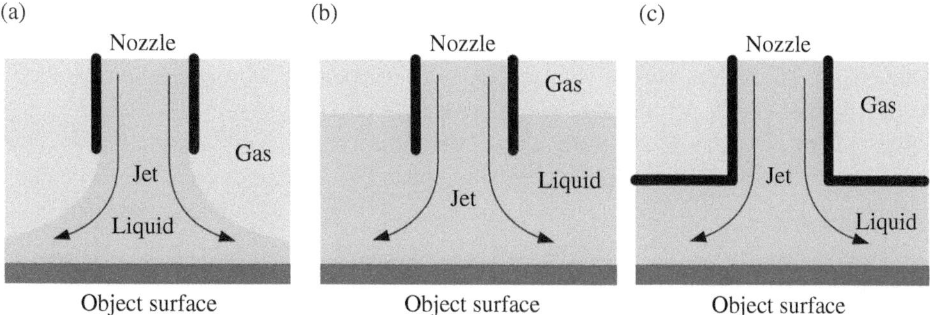

Figure 3.20 Jet impingement liquid cooling: (a) free-surface jet, (b) submerged jet, and (c) submerged confined jet.

the impingement domain without flowing past the surrounding jets [70]. After this, plenty of researches were done to study the jet array impingement with distributed extraction returns [71–75]. Bandhauer et al. developed a jet impingement direct liquid cooling solution for high-performance integrated circuits (ICs), and the average heat transfer coefficient of 13,100 $W \cdot m^{-2} \cdot K^{-1}$ was achieved [75].

3.3.3 Flow Boiling

In the past three decades, electronic components have been developed toward miniaturization and high integration to reach higher computing speed, resulting in increasing power density and local heat flux density. With the rapid heat accumulation, it is difficult for conventional air cooling technologies to meet the thermal control requirements, which leads to negative effect on lifetime and performance of devices [76]. Therefore, heat dissipation has become the bottleneck of development for high heat flux devices. By the early 1980s, this situation caused a switch from forced air convection cooling depending on fan-based heat sink to single-phase liquid cooling technologies relying on liquid coolants for heat exchange. Nevertheless, heat dissipation requirement in supercomputer chips reached 100 $W \cdot cm^{-2}$ by the mid-1980s, which exceeded the ability of single-phase liquid cooling technologies [77]. Afterward, two-phase cooling technologies have attracted the interest of researchers due to their low-level energy recovery and high heat transfer coefficient.

Flow boiling phenomenon is one of the most important processes involved in the field of heat transfer with various industrial applications such as high heat fluxes cooling in mechanics, aerospace, electronics, and biochemistry. As an efficient cooling technology, it has various advantages for cooling electronic chips and elements: ① flow boiling is capable of reaching fine heat transfer performance even at low liquid flow rates. ② The temperature of the surface can be kept almost uniform, which will enhance the durability of the electronic device. ③ With nucleate boiling being considered, the performance of the cooling system is much better, and the cost is reduced [78]. With the advantages listed earlier, flow boiling has been widely used in many industrial applications.

The experiment [79] showed that the air cooling technique costs 44% of the rated energy while the water cooling method just consumes 16.7% to achieve a 15% improvement in the overall energy efficiency (Figure 3.21). What's more, two-phase

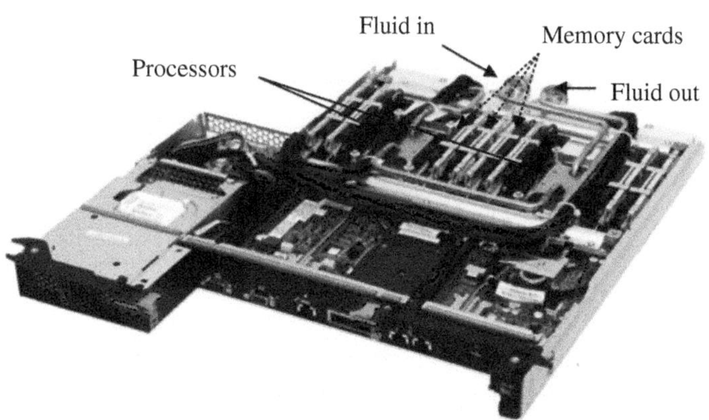

Figure 3.21 Liquid cooled IBM iDataplex 360 Server. Berkeley and Mahdavi [79]/Energy Technologies Area.

cooling technique is used to cool insulated gate bipolar transistors (IGBTs), which are widely used in vehicles, power supplies, and motors. Saums [80] compared three cooling techniques (air cooling, single-phase water cooling, and two-phase pumped loop cooling) used for cooling IGBT. The result showed two-phase cooling had a better performance than the others in two aspects: ① higher heat dissipation rates and ② compact size and small weight of cooling system.

It is well known that the flux of heat and mass, pressure, and the quality of vapor have an effect on flow boiling. And the performance of flow boiling is associated with those factors, especially the fluid mass flux and vapor quality. Flow boiling consists of five regions as shown in Figure 3.22.

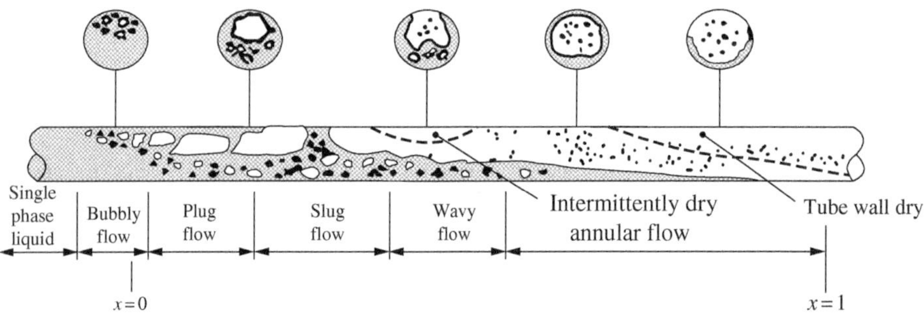

Figure 3.22 Illustration of flow boiling regimes. Collier and Thome [81]/with permission of Oxford University Press.

In order to get better performance, some changes on the surfaces, including intrinsic surfaces, porous coated surfaces, and porous foams, have been made [82]. Guo et al. [83] found that the boiling heat transfer can be reinforced with a big surface area and a high jet impingement velocity on a micro-pin-finned silicon chip. Compared to the plain surface, heat transfer was greatly improved by up to 90% for the surface with porous coatings. The porous coatings increased heat transfer coefficients attributed to the increased nucleation sites and bubble departure frequency which are the dominant factors affecting the performance of flow boiling [84–87]. Lastly, porous foams are used for flow boiling to enhance heat transfer. The mixing of fluid flow is reinforced. The heat transfer rate is increased until the mass flux is up to $72\,\text{kg}\cdot\text{m}^{-2}$. What is more, it is demonstrated that the higher the porosity is and the larger the size of pore is, the higher the heat transfer rate reached is [88]. Lu and Zhao [89] found that the high pore density and the small pore size greatly enhance the heat transfer, which is caused by the increase in surface area and strong flow mixing.

Flow boiling is a very efficient cooling method that enhances the heat transfer mainly by increasing the density of the nucleation site. Besides, the coefficient of heat transfer is also affected by surface wettability [90, 91].

3.3.4 Spray Cooling

The actual heat dissipation requirements of electronic components have created an urgent need for innovative cooling technologies to solve the cooling problem of high-power electronics [92]. In recent decades, spray cooling has been widely applied to many industrial applications, as shown in Figure 3.23, such as supercomputers, spacecraft, large radars, and laser transmitters [93]. This method uses the spray nozzle to atomize high-pressure liquid into fine droplets. With thin films forming and fine droplets impacting on the heated surface, the heated surface can be cooled through droplets impact, liquid film convection, evaporation, and boiling. Therefore, spray cooling has numerous advantages, such as high heat dissipation capacity with less flow rate demand and low superheat. Besides, there are no temperature overshoot and no contact thermal resistance with the heated surface [94–96].

However, the research on spray cooling is also limited by various factors. For one thing, the mechanism of spray cooling is complex on account of the combined effect of diversified heat transfer modes. Thus, there is no convincing and reliable conclusion about spray cooling mechanism yet. For another, the high possibility of clogging in the spray nozzle will cause a sharp rise in surface temperature and even damage the device. Besides, even slight differences between spray nozzles during the process of fabrication will show great differences in the effect of spray cooling [96–98]. Therefore, the research on how these factors influence the final effect of spray cooling is essential.

The influence factors in spray cooling can be divided into the following four categories: fluid characteristics (fluid properties and additives); surface characteristics (surface roughness, surface coating, and surface geometry); spray characteristics

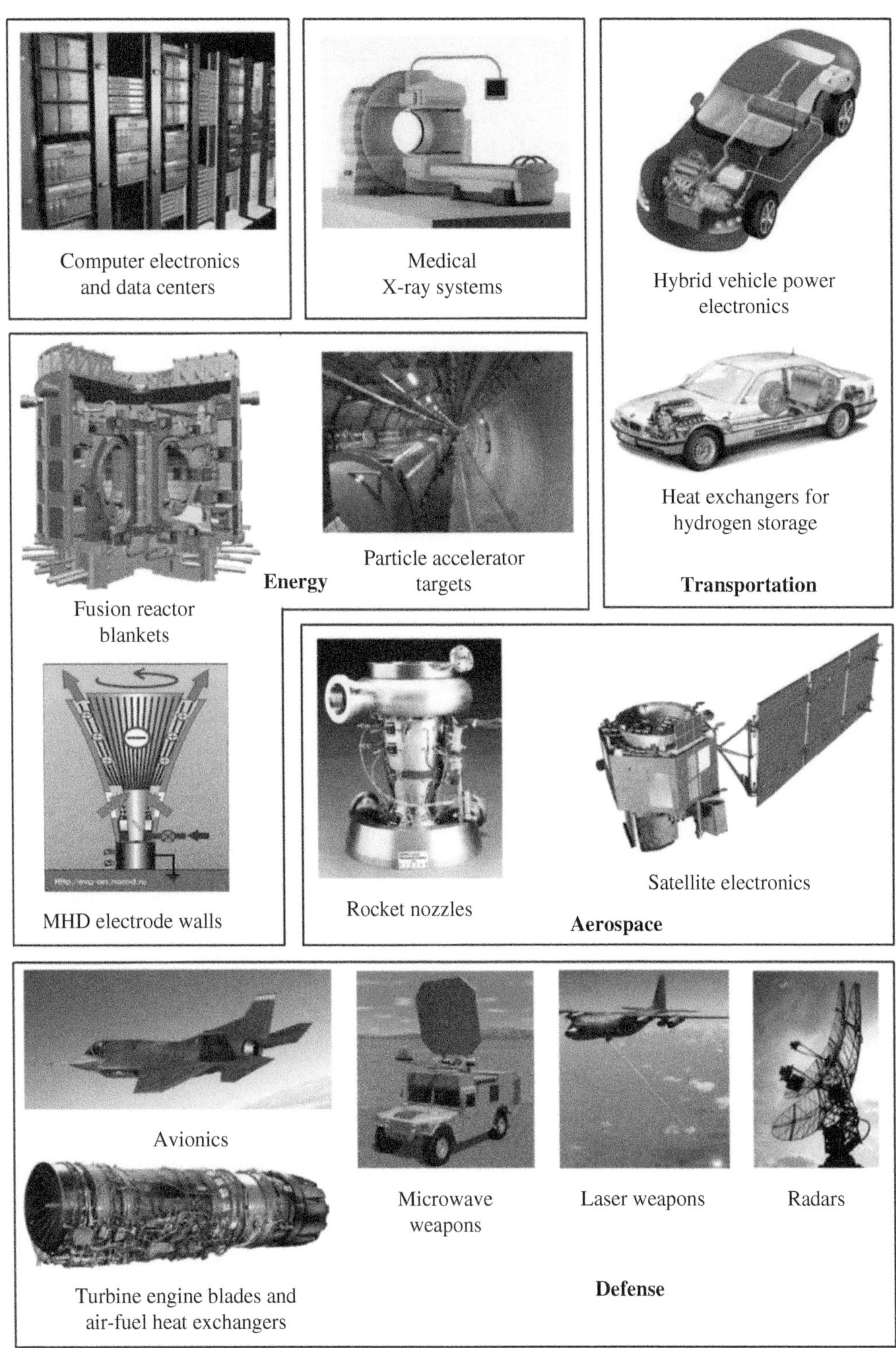

Figure 3.23 Applications demanding high heat flux cooling systems [93].

(droplet velocity, droplet Sauter mean diameter, droplet flux, spray flow rate, spray angle, etc.); and the external environment characteristics (non-condensable gas, microgravity, etc.).

Fluid characteristics refer to the viscosity, surface tension, thermal and electrical conductivity, etc. For different applications, pure water, alcohol, R134a, liquid ammonia, and other kinds of fluids are commonly used in spray cooling [99–101].

However, the cooling performance of pure coolants is still sometimes unsatisfactory. Thus, adding additives is a common method to improve the cooling capacity. The additives used in spray cooling usually are divided into three categories: surfactant, soluble additive, and nanoparticle. Compared with soluble additive, surfactant mainly exists on the surface or in the interface of the liquid. According to experimental results, alcohol surfactant can decrease surface tension and then result in smaller droplet diameter, while soluble salt additive will increase surface tension and cause intense bubble boiling. Both two kinds of additives will extremely improve heat transfer efficiency [102–104]. However, the mechanisms are not clear. Different from surfactant and soluble additive, nanoparticles are insoluble particles with the size of 1–100 nm, finally forming a stable suspension in the fluid. Compared with normal fluid, nanofluid has many advantages, such as higher thermal conductivity and less corrosion. However, nanoparticles usually deposit on the solid surface, thus increasing the possibility of clogging in spray nozzles and the thermal contact resistance. Anyway, the study in nanofluid spray cooling is still in the preliminary stage, and more researches will be carried out in the future [105, 106].

Surface characteristics such as roughness, coating, and geometry will influence heat transfer properties through changing heat transfer process, area, and time. Pais et al. [107] conducted heat transfer experiments on various roughness surfaces obtained by sanding with a spectrum of grit sizes between 0.3 and 22 μm. They found that the smoothest surface with 0.3 μm roughness, where the heat transfer is dominated by film conduction or evaporation, provided the best cooling performance. However, Sehmbey et al. [108, 109] investigated the effects of surface roughness with values ranging from 0.05 to 0.15 μm on cooling performance for liquid nitrogen pressure nozzles. Unlike Pais et al., they noted that increasing roughness greatly enhanced heat transfer performance. Besides, surfaces with microstructures have been applied to improve heat transfer properties on account of providing more nucleation sites. As to surface coating, although it will increase thermal contact resistance, the cooling property is generally enhanced due to the increase in nucleation sites, evaporation area, and residence time of liquid film. Various surface coatings such as diamond and metal particles have been studied to enhance heat transfer capability at present [110–112]. Moreover, different surface geometries including straight fins, pyramids, dimples, radial fins, porous tunnels, and combined with one another will exhibit different heat transfer capabilities. According to experimental results, surfaces with thin straight fins and porous tunnels showed the best heat transfer performance due to more nucleation sites, larger contact area, and longer contact time with liquid [113–119].

Spray characteristics such as droplet velocity, droplet Sauter mean diameter, droplets flux, spray flow rate, and spray angle are largely determined by the spray nozzle. In general, atomizer nozzles and pressure nozzles are normally used in spray cooling [120, 121]. Pressure nozzles, which only rely on the momentum of the liquid to achieve the droplet breakup, are better than atomizer nozzles, which use a secondary gas stream to obtain fine droplets. Nozzle size is also an important factor that determines droplets Sauter mean diameter, while Weber number and Reynolds number depend on outlet conditions and fluid properties. Furthermore, spray height and nozzle orientation largely affect cooling performance. Generally, tiny and uniform droplets with high velocity are conducive to heat transfer performance. The nonuniform spray characteristics will lead to the nonuniform surface temperature, which can cause the local high temperature and damage electronic chips.

External environment characteristics mainly include non-condensable gas and microgravity. According to experimental studies, non-condensable gas will increase the condensing thermal resistance on the heated surface, leading to a negative influence on cooling performance [122]. Therefore, a closed system is better than an open system. In addition, gravity is an essential factor that affects heat transfer performance. Experimental studies show that heat transfer performance will deteriorate under microgravity due to the difficulty in liquid film convection and bubble separation.

Although there are several heat transfer modes in spray cooling, resulting in the complexity and difficulty during researches, it has become the most representative cooling technology in high heat flux condition.

3.3.5 Nanofluid

Fluids play a dominant role in the liquid cooling system. There are plenty of kinds of cooling fluids widely applied in various cooling applications of electronic devices. As investigated and discussed in the previous works [123], the liquid coolants for liquid cooling of electronics must be nontoxic, non-flammable, inexpensive, and have the properties of high thermal conductivity, high specific heat, and low viscosity.

Among all the coolants, water is the most popular and effective coolant owing to its largest thermal conductivity and specific heat. Therefore, water is considered as the best coolant for the thermal management of electronic devices. However, the two major drawbacks that limit the application of water as coolant in the liquid cooling system are the electrical conductivity and the corrosiveness. Therefore, dielectric and non-corrosive fluids are often used in the liquid cooling system.

Unfortunately, the thermal conductivity of the dielectric coolant is much lower than water. A number of work has been carried out on adding nanoparticles with high thermal conductivity to the coolant, compositing the so-called nanofluids.

Because thermal conductivity is one of the most important properties of the liquid cooling coolant, most of the researches have been focused on this key property of nanofluids. Massive researches have been done on the thermal conductivity of nanofluids that was found to be obviously higher than that of the base fluids [124–128]. Numerous nanoparticles with high thermal conductivity have been investigated to increase the thermal conductivity of base fluids, such as Cu, Al_2O_3, carbon nano-tubes (CNT), CuO, TiO_2, SiO_2, graphene, and MgO.

3.4 Chapter Summary

In this chapter, the thermal convection was introduced in detail, including concepts, theories, and calculations of convection heat transfer. Typical convection heat transfer modes were presented, including natural convection, forced convection, and pool boiling convection heat transfer. Present and future developments of convection cooling technologies were introduced, such as air cooling based on fan and heat sink, liquid cooling based on microchannel, jet impingement, flow boiling, spray boiling, and nanofluid.

References

1 Luo, X.B., Hu, R., Liu, S. et al. (2016). Heat and fluid flow in high-power LED packaging and applications. *Prog. Energy Combust. Sci.* 56: 1–32.

2 Morgan, V.T. (1975). The overall convective heat transfer from smooth circular cylinders. In: *Advances in Heat Transfer* (ed. T.F. Irvine and J.P. Hartnett), 199–264. Elsevier.

3 Moores, K.A. and Joshi, Y.K. (2003). Effect of tip clearance on the thermal and hydrodynamic performance of a shrouded pin fin array. *J. Heat Transf. Trans. ASME* 125 (6): 999–1006.

4 Short, B.E., Raad, P.E., and Price, D.C. (2002). Performance of pin fin cast aluminum coldwalls, part 1: friction factor correlations. *J. Thermophys. Heat Transf.* 16 (3): 389–396.

5 Li, H.-Y., Chiang, M.-H., Lee, C.-I. et al. (2010). Thermal performance of plate-fin vapor chamber heat sinks. *Int. Commun. Heat Mass Transf.* 37 (7): 731–738.

6 Li, H.-Y. and Chao, S.-M. (2009). Measurement of performance of plate-fin heat sinks with cross flow cooling. *Int. J. Heat Mass Transf.* 52 (13): 2949–2955.

7 Ahmed, H.E., Salman, B.H., Kherbeet, A.S. et al. (2018). Optimization of thermal design of heat sinks: a review. *Int. J. Heat Mass Transf.* 111 (2018): 129–153.

8 Bar-Cohen, A., Iyengar, M., and Kraus, A.D. (2003). Design of optimum plate-fin natural convective heat sinks. *J. Electron. Packag.* 125 (2): 208–216.

9 Narasimhan, S. and Majdalani, J. (2002). Characterization of compact heat sink models in natural convection. *IEEE Trans. Compon. Packag. Technol.* 25 (1): 78–86.

10 Morrison, A.T. (1992). Optimization of heat sink fin geometries for heat sinks in natural convection. *Proceedings of the [1992 Proceedings] Intersociety Conference on Thermal Phenomena in Electronic Systems*, F (5–8 February 1992).

11 Yovanovich, M. and Jafarpur, K. (1993). *Models of Laminar Natural Convection from Vertical and Horizontal Isothermal Cuboids for All Prandtl Numbers and All Rayleigh Numbers Below 10^{11}, vol. 111, 264. ASME-Publications-Ltd.*

12 Bar-Cohen, A. and Iyengar, M. (2002). Design and optimization of air-cooled heat sinks for sustainable development. *IEEE Trans. Compon. Packag. Technol.* 25 (4): 584–591.

13 Culham, J.R., Khan, W.A., Yovanovich, M.M. et al. (2007). The influence of material properties and spreading resistance in the thermal design of plate fin heat sinks. *J. Electron. Packag.* 129 (1): 76–81.

14 Bar-Cohen, A., Bahadur, R., and Iyengar, M. (2006). Least-energy optimization of air-cooled heat sinks for sustainability-theory, geometry and material selection. *Energy* 31 (5): 579–619.

15 Culham, J.R. and Muzychka, Y.S. (2001). Optimization of plate fin heat sinks using entropy generation minimization. *IEEE Trans. Compon. Packag. Technol.* 24 (2): 159–165.

16 Khan, W.A., Culham, J.R., and Yovanovich, M.M. (2005). Optimization of pin-fin heat sinks using entropy generation minimization. *IEEE Trans. Compon. Packag. Technol.* 28 (2): 247–254.

17 Bejan, A. (1982). *Entropy Generation Through Heat and Fluid Flow*. New York: Wiley.

18 Bejan, A., Tsatsaronis, G., and Moran, M. (1996). *Thermal Design and Optimization*. New York: Wiley.

19 Bejan, A. (1996). Entropy generation minimization: the new thermodynamics of finite-size devices and finite-time processes. *J. Appl. Phys.* 79 (3): 1191–1218.

20 Bejan, A. (1997). *Advanced Engineering Thermodynamics*, 2nde. New York, Chichester: Wiley ISBN 0-471-1-4880-6 £65.00. *Eur. J. Eng. Educ.* 1998; 23(2): 274.

21 Guo, Z., Cheng, X., and Xia, Z. (2003). Least dissipation principle of heat transport potential capacity and its application in heat conduction optimization. *Chin. Sci. Bull.* 48 (4): 406–410.

22 Guo, Z.-Y., Zhu, H.-Y., and Liang, X.-G. (2007). Entransy—A physical quantity describing heat transfer ability. *Int. J. Heat Mass Transf.* 50 (13): 2545–2556.

23 Liu Xiong-Bin, G.Z.-Y. (2009). A novel method for heat exchanger analysis. *Acta Phys. Sin.* 58 (7): 4766.

24 Liu, X., Meng, J., and Guo, Z. (2009). Entropy generation extremum and entransy dissipation extremum for heat exchanger optimization. *Chin. Sci. Bull.* 54 (6): 943–947.

25 Guo, Z.Y., Liu, X.B., Tao, W.Q. et al. (2010). Effectiveness–thermal resistance method for heat exchanger design and analysis. *Int. J. Heat Mass Transf.* 53 (13): 2877–2884.

26 Li, Q. and Chen, Q. (2012). Application of entransy theory in the heat transfer optimization of flat-plate solar collectors. *Chin. Sci. Bull.* 57 (2): 299–306.

27 Chen, L. (2012). Progress in entransy theory and its applications. *Chin. Sci. Bull.* 57 (34): 4404–4426.

28 Costa, V.A.F. and Lopes, A.M.G. (2014). Improved radial heat sink for led lamp cooling. *Appl. Therm. Eng.* 70 (1): 131–138.

29 Luo, X., Xiong, W., Cheng, T. et al. (2009). Design and optimization of horizontally-located plate fin heat sink for high power LED street lamps. *Proceedings of the 2009 59th Electronic Components and Technology Conference*, F (26–29 May 2009).

30 Shyu, J.-C., Hsu, K.-W., Yang, K.-S. et al. (2011). Thermal characterization of shrouded plate fin array on an LED backlight panel. *Appl. Therm. Eng.* 31 (14): 2909–2915.

31 Luo, X, Mao, Z, and Liu, S. (2010). Thermal design of a 16W LED bulb based on thermal analysis of a 4W LED bulb. *Proceedings of the 2010 Proceedings 60th Electronic Components and Technology Conference (ECTC)*, F (1–4 June 2010).

32 Yang, K., Chen, Y., Lin, B. et al. (2011). Thermal design based on thermosyphon heat sink of high-power LED street light. *Proceedings of the Proceedings of 2011 International Conference on Electronics and Optoelectronics*, F (29–31 July 2011).

33 Huang, L., Chen, E., and Lee, D. (2012). Thermal analysis of plastic heat sink for high power LED lamp. *Proceedings of the 2012 7th International Microsystems, Packaging, Assembly and Circuits Technology Conference (IMPACT)*, F (24–26 October 2012).

34 Jang, D., Yu, S.-H., and Lee, K.-S. (2012). Multidisciplinary optimization of a pin-fin radial heat sink for LED lighting applications. *Int. J. Heat Mass Transf.* 55 (4): 515–521.

35 Jang, D., Park, S.-J., Yook, S.-J. et al. (2014). The orientation effect for cylindrical heat sinks with application to LED light bulbs. *Int. J. Heat Mass Transf.* 71: 496–502.

36 Kim, H., Kim, K.J., and Lee, Y. (2012). Thermal performance of smart heat sinks for cooling high power LED modules. *Proceedings of the 13th InterSociety Conference on Thermal and Thermomechanical Phenomena in Electronic Systems*, F (30 May–1 June 2012).

37 Wang, C.-C., Yang, K.-S., Liu, Y.-P. et al. (2011). Effect of cannelure fin configuration on compact aircooling heat sink. *Appl. Therm. Eng.* 31 (10): 1640–1647.

38 Qu, Z.G., Li, W.Q., Wang, J.L. et al. (2012). Passive thermal management using metal foam saturated with phase change material in a heat sink. *Int. Commun. Heat Mass Transf.* 39 (10): 1546–1549.

39 Akhilesh, R., Narasimhan, A., and Balaji, C. (2005). Method to improve geometry for heat transfer enhancement in PCM composite heat sinks. *Int. J. Heat Mass Transf.* 48 (13): 2759–2770.

40 Jeng, T.-M. and Tzeng, S.-C. (2014). Heat transfer measurement of the cylindrical heat sink with sintered-metal-bead-layers fins and a built-in motor fan. *Int. Commun. Heat Mass Transf.* 59: 136–142.

41 Li, H.Y., Chao, S.M., Chen, J.W. et al. (2013). Thermal performance of plate-fin heat sinks with piezoelectric cooling fan. *Int. J. Heat Mass Transf.* 57 (2): 722–732.

42 Jang, D. and Lee, K.S. (2014). Flow characteristics of dual piezoelectric cooling jets for cooling applications in ultra-slim electronics. *Int. J. Heat Mass Transf.* 79: 201–211.

43 Sufian, S.F., Fairuz, Z.M., Zubair, M. et al. (2014). Thermal analysis of dual piezoelectric fans for cooling multi-LED packages. *Microelectron. Reliab.* 54 (8): 1534–1543.

44 Ma, H.K., Tan, L.K., and Li, Y.T. (2014). Investigation of a multiple piezoelectric–magnetic fan system embedded in a heat sink. *Int. Commun. Heat Mass Transf.* 59: 166–173.

45 Ma, H.K. and Li, Y.T. (2015). Thermal performance of a dual-sided multiple fans system with a piezoelectric actuator on LEDs. *Int. Commun. Heat Mass Transf.* 66: 40–46.

46 Li, H.Y. and Wu, Y.X. (2016). Heat transfer characteristics of pin-fin heat sinks cooled by dual piezoelectric fans. *Int. J. Therm. Sci.* 110: 26–35.

47 Maaspuro, M. (2016). Piezoelectric oscillating cantilever fan for thermal management of electronics and LEDs — A review. *Microelectron. Reliab.* 63: 342–353.

48 Acikalin, T., Garimella, S.V., Petroski, J. et al. (2004). Optimal design of miniature piezoelectric fans for cooling light emitting diodes. *Proceedings of the Ninth Intersociety Conference on Thermal and Thermomechanical Phenomena in Electronic Systems (IEEE Cat No04CH37543)*, F (1–4 June 2004).

49 Wait, S.M., Basak, S., Garimella, S.V. et al. (2007). Piezoelectric fans using higher flexural modes for electronics cooling applications. *IEEE Trans. Compon. Packag. Technol.* 30 (1): 119–128.

50 Ma, H.K., Su, H.C., Liu, C.L. et al. (2012). Investigation of a piezoelectric fan embedded in a heat sink. *Int. Commun. Heat Mass Transf.* 39 (5): 603–609.

51 Tuckerman, D.B. and Pease, R.F.W. (1981). High-performance heat sinking for VLSI. *IEEE Electron Dev. Lett.* 2 (5): 126–129.

52 Adham, A.M., Mohd-Ghazali, N., and Ahmad, R. (2013). Thermal and hydrodynamic analysis of microchannel heat sinks: a review. *Renew. Sust. Energ. Rev.* 21: 614–622.

53 Chein, R. and Chuang, J. (2007). Experimental microchannel heat sink performance studies using nanofluids. *Int. J. Therm. Sci.* 46 (1): 57–66.

54 Koo, J.M., Im, S., Jiang, L. et al. (2005). Integrated microchannel cooling for three-dimensional electronic circuit architectures. *J. Heat Transf.* 127 (1): 49–58.

55 Zhang, H.Y., Pinjala, D., Wong, T.N. et al. (2005). Single-phase liquid cooled microchannel heat sink for electronic packages. *Appl. Therm. Eng.* 25 (10): 1472–1487.

56 Bello-Ochende, T., Liebenberg, L., and Meyer, J.P. (2007). Constructal cooling channels for micro-channel heat sinks. *Int. J. Heat Mass Transf.* 50 (21–22): 4141–4150.

57 Luo, X. and Mao, Z. (2012). Thermal modeling and design for microchannel cold plate with high temperature uniformity subjected to multiple heat sources. *Int. Commun. Heat Mass Transf.* 39 (6): 781–785.

58 Kheirabadi, A.C. and Groulx, D. (2016). Cooling of server electronics: a design review of existing technology. *Appl. Therm. Eng.* 105: 622–638.

59 Garimella, S.V., Yeh, L.T., and Persoons, T. (2012). Thermal management challenges in telecommunication systems and data centers. *IEEE Trans. Compon. Packag. Manuf. Technol.* 2 (8): 1307–1316.

60 Thome, J.R. (2006). State-of-the-art overview of boiling and two-phase flows in microchannels. *Heat Transf. Eng.* 27 (9): 4–19.

61 Zhang, T.J., Wen, J.T., Julius, A. et al. (2011). Stability analysis and maldistribution control of two-phase flow in parallel evaporating channels. *Int. J. Heat Mass Transf.* 54 (25–26): 5298–5305.

62 Cheng, L. and Xia, G. (2017). Fundamental issues, mechanisms and models of flow boiling heat transfer in microscale channels. *Int. J. Heat Mass Transf.* 108: 97–127.

63 Lorenzini, D., Green, C., Sarvey, T.E. et al. (2016). Embedded single phase microfluidic thermal management for non-uniform heating and hotspots using microgaps with variable pin fin clustering. *Int. J. Heat Mass Transf.* 103: 1359–1370.

64 Sharma, C.S., Tiwari, M.K., Zimmermann, S. et al. (2015). Energy efficient hotspot-targeted embedded liquid cooling of electronics. *Appl. Energy* 138: 414–422.

65 Laloya, E., Lucia, O., Sarnago, H. et al. (2015). Heat management in power converters: from state of the art to future ultrahigh efficiency systems. *IEEE Trans. Power Electron.* 31 (11): 7896–7908.

66 Ebadian, M.A. and Lin, C.X. (2011). A review of high-heat-flux heat removal technologies. *J. Heat Transf.* 133 (11): 110801.

67 Gould, K., Cai, S.Q., Neft, C. et al. (2014). Liquid jet impingement cooling of a silicon carbide power conversion module for vehicle applications. *IEEE Trans. Power Electron.* 30 (6): 2975–2984.

68 Jörg, J., Taraborrelli, S., Sarriegui, G. et al. (2017). Direct single impinging jet cooling of a MOSFET power electronic module. *IEEE Trans. Power Electron.* 33 (5): 4224–4237.

69 Li, W., Zhang, X., Cheng, S. et al. (2012). Thermal optimization for a HSPMG used for distributed generation systems. *IEEE Trans. Ind. Electron.* 60 (2): 474–482.

70 Huber, A.M. and Viskanta, R. (1994). Effect of jet-jet spacing on convective heat transfer to confined, impinging arrays of axisymmetric air jets. *Int. J. Heat Mass Transf.* 37 (18): 2859–2869.

71 Natarajan, G. and Bezama, R.J. (2007). Microjet cooler with distributed returns. *Heat Transf. Eng.* 28 (8–9): 779–787.

72 Rattner, A.S. (2017). General characterization of jet impingement array heat sinks with interspersed fluid extraction ports for uniform high-flux cooling. *J. Heat Transf.* 139 (8): 082201.

73 Han, Y., Lau, B.L., Zhang, H. et al. (2014). Package-level Si-based micro-jet impingement cooling solution with multiple drainage micro-trenches. *2014 IEEE 16th Electronics Packaging Technology Conference (EPTC)*, 330–334. IEEE.

74 Husain, A., Al-Azri, N.A., Al-Rawahi, N.Z.H. et al. (2015). Comparative performance analysis of microjet impingement cooling models with different spent-flow schemes. *J. Thermophys. Heat Transf.* 30 (2): 466–472.

75 Bandhauer, T.M., Hobby, D.R., Jacobsen, C. et al. (2018). Thermal performance of micro-jet impingement device with parallel flow, jet-adjacent fluid removal. *ASME 2018 16th International Conference on Nanochannels, Microchannels, and Minichannels*. American Society of Mechanical Engineers Digital Collection.

76 Mudawar, I. (2001). Assessment of high-heat-flux thermal management schemes. *IEEE Trans. Compon. Packag. Technol.* 24 (2): 122–141.

77 Anderson, T.M. and Mudawar, I. (1989). Microelectronic cooling by enhanced pool boiling of a dielectric fluorocarbon liquid. *J. Heat Transf. Trans. ASME* 111: 752–759.

78 Karayiannis, T.G. and Mahmoud, M.M. (2016). Flow boiling in microchannels: Fundamentals and applications. *Appl. Therm. Eng.* 115: 1372–1397.

79 Berkeley, L. and Mahdavi, R. (2014). *Liquid Cooling V.* Air Cooling Evaluation in the Maui High Performance Computing Centre, Federal Energy Management Program, FEMP.

80 Saums, D.L. (2010). Vaporizable dielectric fluid cooling of IGBT power semiconductors. *Integrated Power Electronics Systems (CIPS), 2010 6th International Conference on*. IEEE.

81 Collier, J.G. and Thome, J.R. (1994). *Convective Boiling and Condensation*. Clarendon Press.

82 Leong, K.C., Ho, J.Y., and Wong, K.K. (2016). A critical review of pool and flow boiling heat transfer of dielectric fluids on enhanced surfaces. *Appl. Therm. Eng.* 112: 999–1019.

83 Guo, D., Wei, J.J., and Zhang, Y.H. (2011). Enhanced flow boiling heat transfer with jet impingement on micro-pin-finned surfaces. *Appl. Therm. Eng.* 31 (11–12): 2042–2051.

84 Rainey, K.N., Li, G., and You, S.M. (2001). Flow boiling heat transfer from plain and microporous coated surfaces in subcooled FC72. *J. Heat Transf.* 123 (5): 918–925.

85 Ammerman, C.N. and You, W.M. (2001). Enhancing small-channel convective boiling performance using a microporous surface coating. *J. Heat Transf.* 123 (5): 10.

86 Sun, Y., Zhang, L., Xu, H. et al. (2011). Flow boiling enhancement of FC-72 from microporous surfaces in minichannels. *Exp. Thermal Fluid Sci.* 35 (7): 1418–1426.

87 Bai, P., Tang, T., and Tang, B. (2013). Enhanced flow boiling in parallel microchannels with metallic porous coating. *Appl. Therm. Eng.* 58 (s 1–2): 291–297.

88 Kim, D.W., Bar-Cohen, A., and Han, B. (2008). Forced convection and flow boiling of a dielectric liquid in a foam-filled channel. *Thermal and Thermomechanical Phenomena in Electronic Systems, 2008. ITHERM 2008. 11th Intersociety Conference on*. IEEE.

89 Lu, W. and Zhao, C.Y. (2009). Numerical modelling of flow boiling heat transfer in horizontal metal-foam tubes. *Adv. Eng. Mater.* 11 (10): 832–836.

90 Wu, W., Bostanci, H., Chow, L.C. et al. (2010). Nucleate boiling heat transfer enhancement for water and FC-72 on titanium oxide and silicon oxide surfaces. *Int. J. Heat Mass Transf.* 53 (9–10): 1773–1777.

91 An, S., Kim, D.Y., Lee, J.G. et al. (2016). Supersonically sprayed reduced graphene oxide film to enhance critical heat flux in pool boiling. *Int. J. Heat Mass Transf.* 98: 124–130.

92 Kandlikar, S.G. and Bapat, A.V. (2007). Evaluation of jet impingement, spray and microchannel chip cooling options for high heat flux removal. *Heat Transf. Eng.* 28 (11): 911–923.

93 Mudawar, I. (2013). Recent advances in high-flux, two-phase thermal management. *J. Thermal Sci. Eng. Appl.* 5 (2): 021012.

94 Kim, J. (2007). Spray cooling heat transfer: the state of the art. *Int. J. Heat Fluid Flow* 28 (4): 753–767.

95 Cader, T., Westra, L.J., and Eden, R.C. (2004). Spray cooling thermal management for increased device reliability. *IEEE Trans. Device Mater. Reliab.* 4: 605–613.

96 Silk, E.A., Golliher, E.L., and Selvam, R.P. (2008). Spray cooling heat transfer: technology overview and assessment of future challenges for micro-gravity application. *Energy Convers. Manag.* 49 (3): 453–468.

97 Shedd, T.A. (2007). Next generation spray cooling: high heat flux management in compact spaces. *Heat Transf. Eng.* 28 (2): 87–92.

98 Hall, D.D. and Mudawar, I. (1995). Experimental and numerical study of quenching complex-shaped metallic alloys with multiple, overlapping sprays. *Int. J. Heat Mass Transf.* 38 (7): 1201–1216.

99 Mohapatra, S.S., Andhare, S., Chakraborty, S. et al. (2012). Experimental study and optimization of air atomized spray with surfactant added water to produce high cooling rate. *J. Enhanc. Heat Transf.* 19 (5): 397–408.

100 Li, Q., Tie, P., and Xuan, Y. (2015). Investigation on heat transfer characteristics of R134a spray cooling. *Exp. Thermal Fluid Sci.* 60: 182–187.

101 Bostanci, H., Rini, D.P., Kizito, J.P. et al. (2012). High heat flux spray cooling with ammonia: investigation of enhanced surfaces for CHF. *Int. J. Heat Mass Transf.* 55 (13–14): 3849–3856.

102 Mohapatra, S.S., Jha, J.M., Srinath, K. et al. (2014). Enhancement of cooling rate for a hot steel plate using air-atomized spray with surfactant-added water. *Exp. Heat Transf.* 27 (1): 72–90.

103 Qiao, Y.M. and Chandra, S. (1998). Spray cooling enhancement by addition of a surfactant. *J. Heat Transf.* 120 (1): 92–98.

104 Ravikumar, S.V., Jha, J.M., Sarkar, I. et al. (2013). Achievement of ultrafast cooling rate in a hot steel plate by air-atomized spray with different surfactant additives. *Exp. Thermal Fluid Sci.* 50: 79–89.

105 Ravikumar, S.V., Haldar, K., Jha, J.M. et al. (2015). Heat transfer enhancement using air-atomized spray cooling with water–Al_2O_3 nanofluid. *Int. J. Therm. Sci.* 96: 85–93.

106 Duursma, G., Sefiane, K., and Kennedy, A. (2009). Experimental studies of nanofluid droplets in spray cooling. *Heat Transf. Eng.* 30 (13): 1108–1120.

107 Pais, M.R., Chow, L.C., and Mahefkey, E.T. (1992). Surface roughness and its effects on the heat transfer mechanism in spray cooling. *J. Heat Transf.* 114 (1): 211.

108 Sehmbey, M.S., Chow, L.C., Hahn, O.J. et al. (1995). Effect of spray characteristics on spray cooling with liquid nitrogen. *J. Thermophys. Heat Transf.* 9 (4): 757–765.

109 Sehmbey, M.S., Chow, L.C., Hahn, O.J. et al. (1995). Spray cooling of power electronics at cryogenic temperatures. *J. Thermophys. Heat Transf.* 9 (1): 123–128.

110 Sehmbey, M.S., Pais, M.R., and Chow, L.C. (1992). A study of diamond laminated surfaces in evaporative spray cooling. *Thin Solid Films* 212 (1): 25–29.

111 Thiagarajan, S.J., Narumanchi, S., and Yang, R. (2014). Effect of flow rate and subcooling on spray heat transfer on microporous copper surfaces. *Int. J. Heat Mass Transf.* 69: 493–505.

112 Srikar, R., Gambaryan-Roisman, T., Steffes, C. et al. (2009). Nanofiber coating of surfaces for intensification of drop or spray impact cooling. *Int. J. Heat Mass Transf.* 52 (25–26): 5814–5826.

113 Chan, Y., Charbel, F., Ray, S.S., and Yarin, A.L. (2010). Hydrodynamics of drop impact and spray cooling through nanofiber mats.

114 Silk, E.A., Kim, J., and Kiger, K. (2004). Investigation of enhanced surface spray cooling. *ASME International Mechanical Engineering Congress & Exposition.*

115 Silk, E.A., Kim, J., and Kiger, K. (2005). Spray cooling trajectory angle impact upon heat flux using a straight finned enhanced surface. *ASME Summer Heat Transfer Conference Collocated with the ASME Pacific Rim Technical Conference & Exhibition on Integration & Packaging of MEMS.* American Society of Mechanical Engineers.

116 Silk, E.A., Kim, J., and Kiger, K. (2005). Impact of cubic pin finned surface structure geometry upon spray cooling heat transfer. *Proceedings of the ASME 2005 Pacific Rim Technical Conference and Exhibition on Integration and Packaging of MEMS, NEMS, and Electronic Systems Collocated with the ASME 2005 Heat Transfer Summer Conference*, 1–9. American Society of Mechanical Engineers.

117 Silk, E.A., Kim, J., and Kiger, K. (2006). Spray cooling of enhanced surfaces: impact of structured surface geometry and spray axis inclination. *Int. J. Heat Mass Transf.* 49 (25–26): 4910–4920.

118 Silk, E.A., Kim, J., and Kiger, K. (2006). Enhanced surface spray cooling with embedded and compound extended surface structures. *Thermal and Thermomechanical Phenomena in Electronics Systems, 2006. ITHERM '06. The Tenth Intersociety Conference on.* IEEE.

119 Silk, E.A. (2008). *Investigation of Pore Size Effect On Spray Cooling Heat Transfer with Porous Tunnels.* American Institute of Physics.

120 Visaria, M. and Mudawar, I. (2009). *Application of Two-Phase Spray Cooling for Thermal Management of Electronic Devices*, 784–793. IEEE.

121 Cheng, W.L., Han, F.Y., Liu, Q.N. et al. (2011). Spray characteristics and spray cooling heat transfer in the non-boiling regime. *Energy* 36 (5): 3399–3405.

122 Jiang, S. and Dhir, V.K. (2004). Spray cooling in a closed system with different fractions of non-condensibles in the environment. *Int. J. Heat Mass Transf.* 47 (25): 5391–5406.

123 Mohapatra, S.C. and Loikits, D. (2005). Advances in liquid coolant technologies for electronics cooling. *Semiconductor Thermal Measurement and Management IEEE Twenty First Annual IEEE Symposium, 2005*, 354–360. IEEE.

124 Murshed, S.M.S., Leong, K.C., and Yang, C. (2008). Thermophysical and electrokinetic properties of nanofluids–a critical review. *Appl. Therm. Eng.* 28 (17–18): 2109–2125.

125 Yu, W., France, D.M., Routbort, J.L. et al. (2008). Review and comparison of nanofluid thermal conductivity and heat transfer enhancements. *Heat Transf. Eng.* 29 (5): 432–460.

126 Murshed, S.M.S. and De Castro, C.A.N. (2014). Superior thermal features of carbon nanotubes-based nanofluids–A review. *Renew. Sust. Energy Rev.* 37: 155–167.

127 Hassan, M., Sadri, R., Ahmadi, G. et al. (2013). Numerical study of entropy generation in a flowing nanofluid used in micro- and minichannels. *Entropy* 15 (1): 144–155.

128 Murshed, S.M.S., Leong, K.C., and Yang, C. (2008). Investigations of thermal conductivity and viscosity of nanofluids. *Int. J. Therm. Sci.* 47 (5): 560–568.

4

Thermal Radiation and Solutions

All objects in nature emit thermal radiation outwardly at anytime and anywhere in a broadband, non-selective, incoherent, diffusive, and reciprocal manner [1, 2]. Thanks to the fast development of thermal metamaterials and meta-surfaces in recent years, thermal radiation has been demonstrated to be engineered with comprehensive control of spectral, directional, and dynamic characteristics, enabling higher-efficiency radiative heat transfer than the thermal radiation of natural objects [3]. Among them, the regulation of emissivity spectrum can realize passive regulation of object temperature and has shown excellent effect in the fields of thermal camouflage (TC) [4, 5], radiation cooling (RC) [6, 7], near-field radiation control [8], etc.

In this chapter, we begin with the theoretical concept of thermal radiation. Then, the methods of regulating the thermal radiation spectrum are introduced, including manual design and machine-learning optimization. In addition, this section also introduces an emissivity spectral regulation algorithm framework, which can efficiently design structures of radiative cooling. Moreover, due to some aesthetic considerations, colored radiative cooling has been discussed. Finally, the application prospect of near-field thermal radiation in thermal management is introduced.

4.1 Concept of Thermal Radiation

Thermal radiation can be defined as the heat energy transmitted outward by an object in the form of electromagnetic waves, which are generated by the vibrational and rotational movements of atoms [9]. The intensity of the thermal radiation is determined by the temperature of the object and its surface properties [10]. Any objects whose temperature is above absolute zeros are able to emit thermal radiation which is in proportion to their temperature.

Unlike thermal conduction and convection, in the process of energy transmission via thermal radiation, the material medium is not necessarily in demand. Therefore, the transmission loss of thermal radiation varies greatly under different transmitting medium conditions. In the case of a vacuum, since there are no particles of matter in it, thermal radiation can pass through the medium without attenuation. However, if there are matter particles in the medium (e.g. gas, liquid, or solid) as shown in Figure 4.1, the thermal radiation will be reflected or absorbed to different degrees during the transmission depending on the type and density of the particles in the medium, which results in the loss of thermal radiation energy. Scholars defined reflectivity γ as the fraction reflected, absorptivity Abs as the fraction absorbed, and transmissivity Tran as the fraction transmitted. Thus

$$\gamma + \text{Abs} + \text{Tran} = 1 \tag{4.1}$$

Theoretically, the spectrum of thermal radiation can cover the whole wavelength band. For low-temperature objects, the thermal radiation energy is mainly concentrated in the infrared band, and, based on Wien's displacement law, as the temperature of the object increases, the main radiative energy band will shift to a short-wavelength band. In the process of thermal radiation transmission, an object that can absorb thermal radiation in all directions without reflection is called a blackbody. In 1859, German physicist Kirchhoff found that the ratio of the emissive power of a body to the emissive power of a blackbody at the same temperature is equal to the absorptivity of the body. This ratio is defined as emissivity ε, which links the radiation of any object to the blackbody. Hence, it is critical to figure out the characteristics of blackbody radiation.

Thermal Management for Opto-electronics Packaging and Applications, First Edition. Xiaobing Luo, Run Hu, and Bin Xie.

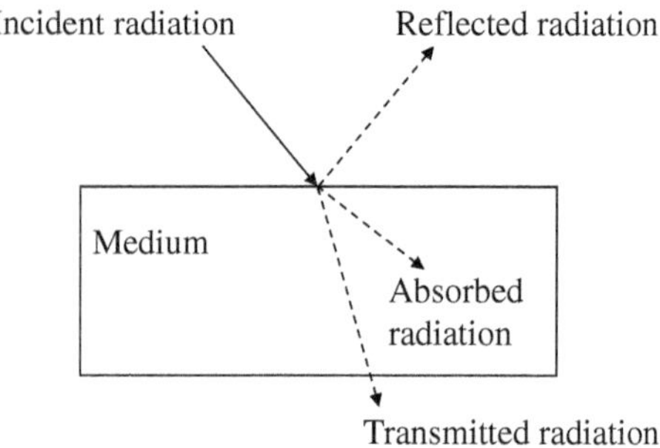

Figure 4.1 The transmitting process of thermal radiation through a medium.

British physicists Rayleigh and Kins derived the Rayleigh–Kins formula from the energy equipartition theorem to describe the blackbody radiation spectrum, which fits well with the experimental results in the long waveband but suffers enormous deviations in the short waveband. In contrast, Wien's formula, derived by the German physicist Wien, applies only to the short waveband part of the blackbody radiation spectrum. Such incongruity was not resolved until 1900, when Planck's formula for blackbody radiation was put forward from the perspective of energy quantization, which perfectly solved the problem of the blackbody radiation spectrum. Since then, blackbody radiation theory has become the basic theory in the field of thermal radiation. Planck's formula for the energy E_λ radiated per unit volume by a cavity of a blackbody in the wavelength interval λ to $\lambda + \Delta\lambda$ ($\Delta\lambda$ denotes an increment of wavelength) can be written as:

$$E_\lambda = \frac{8\pi k_P c}{\lambda^5} \times \frac{1}{\exp(k_P c / k_B T\lambda) - 1} \tag{4.2}$$

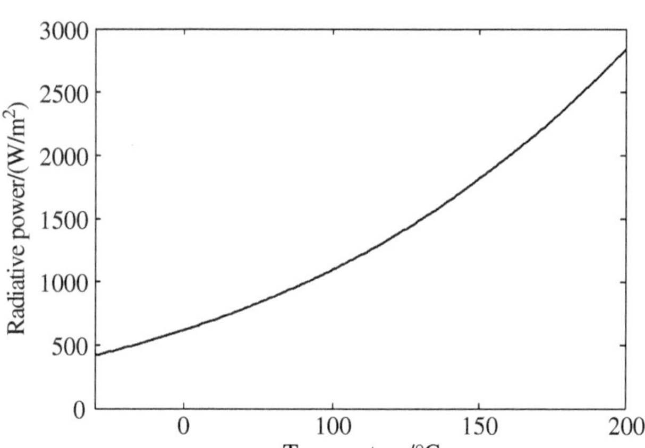

Figure 4.2 The relationship between radiative power and the temperature of a blackbody.

where k_P denotes Planck's constant, c denotes the speed of light, k_B denotes the Boltzmann constant, and T denotes the absolute temperature. In the late nineteenth century, Stefan and Boltzmann successively discovered and theorized the relationship between thermal radiation energy and temperature of the black body, which is expressed as the well-known Stefan–Boltzmann law as

$$q_{\text{rad}} = \sigma A_r T^4 \tag{4.3}$$

where q_{rad} denotes the rate of thermal radiation energy, A_r is the radiative surface area, and σ is the Stefan–Boltzmann constant. As shown in Figure 4.2, owing to the biquadrate relationship, the radiative power increases rapidly with the rise in temperature, which makes radiative heat dissipation become a potential way of thermal management for objects with high temperature.

In reality, few objects can achieve perfect blackbody radiation. As mentioned above, the radiation of real objects can be accurately described by emissivity ε and blackbody radiation. Based on this, the radiative power of a real object can be given by

$$q_{\text{rad}} = \varepsilon\sigma A_r T^4 \tag{4.4}$$

4.2 Atmospheric Transparent Window

In recent years, with the development of photonic engineering technology, radiative cooling has shown its potential for applications in highly efficient passive cooling, which can release heat energy from a hot object to the largest cold object, the space. However, the transmission medium, sky atmosphere, weakens thermal radiation over most of the wavelength bands owing to its low transmissivity. In order to increase the radiative heat dissipation from the devices into deep space, it is necessary to enhance the emission of devices in atmospheric high transmissivity wavelength bands. As shown in Figure 4.3, in the wavelength bands ranging from 8 to 13 μm, called the atmospheric transparent window (AW), the atmosphere is highly transparent for thermal radiation [11]. At normal temperatures, the wavelength bands of strongest thermal radiation by terrestrial objects coincide with the AW. Therefore, the primary goal of radiative cooling is to improve the emissivity of radiative coolers in AW.

4.3 Spectra-Regulation Thermal Radiation

The fine regulation of thermal radiation is more suitable for practical applications, in which the spectral regulation of thermal radiation can fine-design the energy distribution of thermal radiation at different wavelengths, so as to better regulate the energy of thermal radiation. The spectral regulation of thermal radiation is to control the thermal radiation intensity of an object at different wavelengths by changing its spectral emissivity through structural design or surface modification.

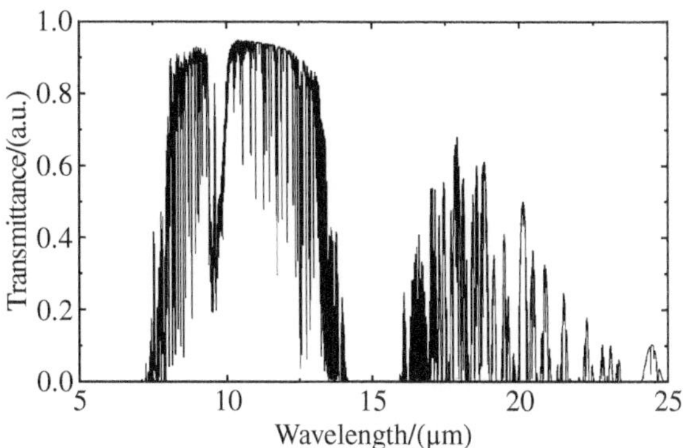

Figure 4.3 Atmospheric transparent window.

Emissivity spectral regulation depends on the modulation of electromagnetic waves by subwavelength structure [1, 12]. According to the variation of subwavelength topological structure dimensions, it can be divided into a one-dimensional structure [13–25] and high-dimensional structure [4, 26–32]. The one-dimensional structure is mainly multi-layer, including metal Fabry–Perot cavity [13–15], Bragg mirror [16, 17], Tam thermal radiator [18–21], metal–media–metal [22, 33], and photonic crystal [34, 35], which can realize the spectral regulation of thermal radiation. High-dimensional structures include two-dimensional structures represented by grating [26, 27], surface symmetric structure array [28, 29], deformed multilayer [30], and three-dimensional structures represented by surface asymmetric structure array [31] and composite structure [4, 32], which can realize the regulation of the spectral characteristics, spatial angle, and polarization state [1, 12].

The commonly used structural design routes mainly include manual design and structural improvement [18–21], and machine-learning optimization design [36–49]. The former is mainly based on optical resonances to carefully select materials and design structures for the spectral target to be achieved, and the design results are highly interpretable. However, this design method has a certain threshold of design and is subject to the existing theory and experience, which makes it difficult to explore the new structure style. The latter is an emerging design method in recent years, and the design routes mainly include: ① input a certain amount of data "structure-thermal radiation property (or corresponding evaluation index)" into the optimization algorithm as training data; ② the implicit relationship of "thermal radiation property (or corresponding evaluation index) → corresponding structure" is fitted with a machine learning algorithm; ③ the design structure is obtained by substituting the spectral properties of the target thermal radiation. The core of this optimization design method lies in the design of the optimization model and the objective function, which together determine whether the fitting effect of the implicit relationship between the thermal radiation spectral property and the structure is accurate. In addition, the objective function of the optimization design should also quantify the objective of the thermal radiation spectrum regulation as comprehensively as possible. Since this design method does not use the existing optical knowledge, it can get rid of the limitations of the existing theory and experience and find new structural patterns. The newly discovered structures often have far more thermal radiation properties than artificially designed structures and have stronger thermal radiation spectrum regulation.

However, designers still have to conduct extensive searches in existing work to determine suitable materials and initial structural parameters for their design goals before machine-learning optimization. Consequently, researchers, following their prior knowledge of materials or structures for different applications, either fix materials to design structures [41, 42, 50] or fix structures to optimize material arrangement [36, 51] to reduce the optimization space and improve the design efficiency. Hence, one open question is: Can we offer a general framework for spectral regulation of emissivity for different applications without a prior knowledge of materials and structures?

4.3.1 Deep Q-Learning Network for Emissivity Spectral Regulation

Deep learning has attracted increasing attention in various domains, such as natural language processing, computer vision, image processing, speech recognition, and material structure optimization [43]. Through establishing the artificial neural networks and data-driven method, deep learning obtains the mapping relationship between data pairs, that is, from emissivity spectra to design parameters of the emitters. However, challenges such as the one-to-many mapping problem, analysis from complex spectra to design parameters, along with the dataset acquisition, collectively render most neural network models inefficient for addressing the emitter design within an enormous optimization space that concurrently encompasses material selection and structural optimization simultaneously. Fortunately, deep reinforcement learning (DRL), which combines deep learning and reinforcement learning, promises to address the above challenges. It does not directly parse the mapping relationship between data pairs from the precollected dataset, but constantly interacts with the current environment to make decisions to update the state of the environment, and uses historical experience as the dataset to learn and optimize the deployment of decisions, so as to maximize the accumulated reward value [44]. Consequently, it has been proven to be capable of solving large-scale and complicated tasks, such as Go and Chess [45]. Based on a deep Q-learning network (DQN), we propose a general design framework for the design and optimization of spectral regulation of emissivity without a prior knowledge of materials and structures, as shown in Figure 4.4.

The whole optimization process can be described as an interactive process with the environment. The state of the environment, which consists of the material ID number and the thicknesses of each layer, represents the materials and structural parameters of the current multilayer. Here, we set up the multilayer emitters as a five-layer structure composed of alternating two materials. Naturally, the setting of design parameters is flexible and can be adjusted according to design objectives, including the kinds of materials, layer count, and other structural parameters. It is worth mentioning that, while increasing the number of layers and materials may meet more rigorous emissivity spectrum requirements, it also significantly expands the optimization space by several orders of magnitude, requiring greater computing power and longer design time. Consequently, according to the structural configuration set earlier, the state can be represented by a 1×7 vector containing material and structure information. The two materials are selected from the self-built material library, as shown in Table 4.1, which contains eight commonly used materials for emissivity spectral regulation. These candidate materials cover most optical properties. Regarding the substrate material, it needs to be selected according to specific design goals. Each layer thickness is varied within the range of 20–1000 nm with a uniform step size of 20 nm, which results in a total of 50 possible steps for each layer. Considering the eight available materials, this structural configuration leads to $8 \times 7 \times 50^5 = 1.75 \times 10^{10}$ potential candidate structures. The demand for simultaneous material selection and structure optimization, together with the sheer volume of optimization space, renders manual design impractical and presents significant challenges to conventional machine learning methods.

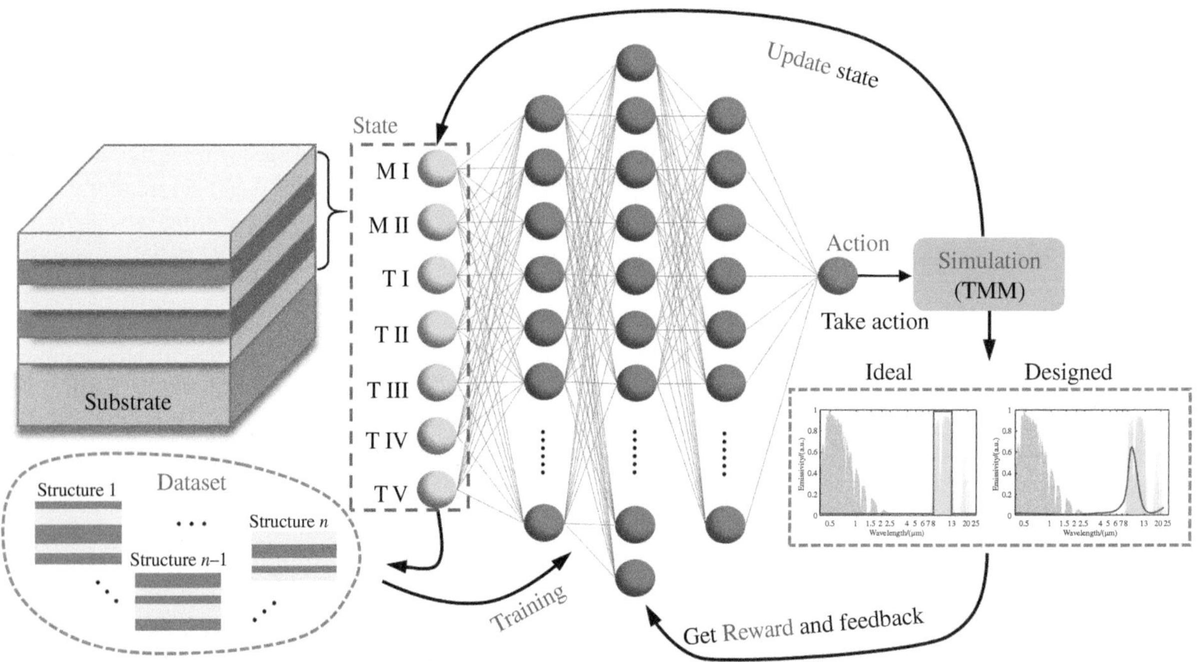

Figure 4.4 Schematic for the multilayer structure and DQN framework. The state consists of two materials and five thicknesses of the multilayer, then the state parameters are fed into the DQN to generate an action. Then take the action to update the state. Transfer matrix method (TMM) is adopted to simulate the new state, and reward is obtained to feed back to neural network (agent). The new state is fed into the DQN for the next iteration. Each pair of state, action, and reward is recorded as dataset to train the neural network, so that it can take the action that increases accumulated reward and, finally, get the corresponding state with the maximum reward.

After the physical information of the multilayer structure is encoded into digital information, it is inputted into an artificial neural network. The network, called "agent" in DQN, consists of an input layer, three fully connected layers, and an output layer. The number of neurons in the three fully connected layers is 24, 48, and 24, respectively. These layers perform computations on the input data, extracting relevant features, and learning patterns from the encoded structural information. The output layer of the agent is referred to as the "action" layer. It generates a single value and each value corresponds to a policy that can be applied to update the current state (structure). More details about the actions and their corresponding policies can be found in Table 4.2, which provides a mapping between the output values of the action layer and the structural modifications they represent. Then the transfer matrix method (TMM) is adopted to simulate the radiation characteristics of the new state (new structure) and obtain its emissivity according to Kirchhoff's law. To evaluate the performance of the new state, a reward is obtained from the emissivity spectra. The reward serves as feedback for the agent and plays a crucial role in determining the convergence direction of the DQN model. The specific definition of the reward will depend on the desired application or emissivity target.

Table 4.1 Material library of multilayer structures for emissivity spectral regulation.

Material ID	Material
1	Ge
2	ZnSe
3	Si
4	SiO_2
5	TiO_2
6	ZnS
7	Si_3N_4
8	MgF_2

In the DQN, a Q-function $Q(s, a)$ is defined to represent the expected cumulative reward for taking the action a on state s and following the optimal policy thereafter. The agent is trained to approximate the Q-function to make the best choice of action to achieve a higher reward by utilizing the replay buffer, which stores historical experiences (state, action, reward, and next state) during the interaction with the environment. To enhance the stability of the training process, the dual network structure is utilized, where the main network (agent) is used to collect experiences, and the target network, a copy of agent, is used to calculate the target Q-value based on Bellman equation as follows [46]:

$$y_t = r_t + \mathrm{DF} \cdot Q(s_{t+1}, a^*; \mathrm{wei}_{\mathrm{tar}}) \tag{4.5}$$

where r_t is the reward, DF is the discount factor, and $a^* = \arg\max_a Q(s_{t+1}, a; \mathrm{wei}_{\mathrm{main}})$ represents the action selected by the main network that maximizes the Q-value. $\mathrm{wei}_{\mathrm{tar}}$ and $\mathrm{wei}_{\mathrm{main}}$ are the weights of the target network and the main network, respectively. The update of the network parameters is achieved by the back-propagation algorithm to minimize the loss function, which is the mean squared error between the predicted Q-value and the target Q-value, as follows:

$$\mathrm{loss} = \mathrm{MSE}[y_t - Q(s, a; \mathrm{wei}_{\mathrm{main}})] \tag{4.6}$$

Table 4.2 Definitions of actions used in DQN.

Action no.	Action definition
0	Decrease the material ID of Material I by 1 (min 1)
1	Increase the material ID of Material I by 1 (max 8)
2	Decrease the material ID of Material II by 1 (min 1)
3	Increase the material ID of Material II by 1 (max 8)
4	Decrease the thickness I by 20 (min 20)
5	Increase the thickness I by 20 (max 1000)
6	Decrease the thickness II by 20 (min 20)
7	Increase the thickness II by 20 (max 1000)
8	Decrease the thickness III by 20 (min 20)
9	Increase the thickness III by 20 (max 1000)
10	Decrease the thickness IV by 20 (min 20)
11	Increase the thickness IV by 20 (max 1000)
12	Decrease the thickness V by 20 (min 20)
13	Increase the thickness V by 20 (max 1000)

In addition, the epsilon greedy exploration (EGE) algorithm is employed to balance exploration and exploitation. Initially, DQN tends to generate action randomly, but gradually, as epsilon decreases, it relies on the Q-function for decision-making. Finally, it is crucial to design an appropriate initialization method for the state to make DQN capable of multilayer optimization with high efficiency. Here we randomly initialize two materials of the state from the material library, with the thickness of each layer randomly generated with the range described above. Additionally, we introduce an iteration threshold, which serves to evaluate whether the iteration should continue. When the reward of a state exceeds the iteration threshold, the state with the highest historical reward is chosen as the initial structure for the next iteration. For each iteration, DQN continues to accept the state, take the action, simulate the emissivity spectra, feedback, and then accept the next state. Once the reward of a new state falls below the iteration threshold, the structure will be reinitialized for the next iteration. It is important to note that the "train from buffer" mechanism results in the number of simulations or the number of calculated structures is not equal to the number of iterations. In simple terms, the design and optimization process of DQN can be likened to playing a game. The game will continue until the mission fails, at which point, it needs to be initialized and restarted. An ingenious initialization method can help achieve higher scores efficiently. This framework demonstrates high accuracy and efficiency as well as flexibility and scalability in design parameters and applications. Two multilayer emitters for applications, including TC and radiative cooling, are designed and optimized by the framework, which are then experimentally fabricated and measured, matching with the designed emissivity spectra. The selection of materials and the design of the structure are independently completed by DQN within the extensive optimization space. Here, the design of radiative cooling radiators is taken as an example to briefly introduce the advantages of this algorithm framework in emissivity spectral regulation.

4.3.2 Design and Optimization of Radiative Cooling Radiators Based on DQN

According to the spectral emissivity profiles, the radiative coolers are divided into broadband radiative cooling radiators and selective radiative cooling radiators (SR). The former has emissivity like a blackbody within the entire emission band of the atmosphere except the main solar spectral band from 300 nm to 2.5 μm. The latter comprises unity emission only within the wavelengths of 8–13 μm, selectively spanning over the AW. For applications where a device itself produces a lot of heat or gains a lot of heat from other sources, the device temperature can reach high above the ambient temperature, where radiative cooling below the ambient temperature may not be possible. In such a case, a broadband radiator can be more useful than a selective radiator to decrease the device temperature rather than cooling it below the ambient temperature [47, 48]. Conversely, for devices that require sub-ambient cooling, the ability of the SR is vastly superior to that of a broadband radiator. The reason behind this is that a broadband radiator operating considerably below the ambient temperature may emit less radiative power than it receives from the incoming atmospheric radiation at the ambient temperature inhibiting it to cool further without any net cooling power. Therefore, considering different application requirements, it is necessary to regulate the emission spectrum more reasonably.

For designing an SR as an example, the objective is to maximize the emissivity within the AW, while minimizing it in the solar band so as to achieve maximum net energy power outflow. Considering the SR has an area A_r and a temperature T, facing the sky in the normal direction toward the zenith. In addition, the complex environment of the emitter is simplified, only considering the standard sun and breeze environment. With such a setup, its cooling power P_{cooling} can be described as:

$$P_{\text{cooling}}(T) = P_{\text{rad}}(T) - P_{\text{atm}}(T_{\text{amb}}) - P_{\text{solar}}(\theta) - P_{\text{cond+conv}}(T) \tag{4.7}$$

where P_{rad} is the power emitted from the emitter. P_{atm} is the input power from the atmosphere absorbed by the emitter. P_{solar} is the incident solar power, and the power of the non-radiative heat transfer due to the conductive and convective is described by the $P_{\text{cond+conv}}$. T and T_{amb} are the temperature of the emitter and the ambient air, respectively. θ is the angle of solar radiation. P_{rad} is given by:

$$P_{\text{rad}}(T) = A_r \int d\Omega \int_0^{\infty} I_{\text{BB}}(T,\lambda)\varepsilon(\lambda,\theta)\cos\theta d\lambda \tag{4.8}$$

where ε (λ, θ) is the emissivity of the emitter at wavelength λ and angle θ. $\int d\Omega = \int_0^{\pi/2} d\theta \sin\theta \int_0^{2\pi} d\varphi$ is the solid angle integration over a hemisphere and I_{BB} (T, λ) is the spectral radiance of a blackbody at temperature T and wavelength λ, which is given by Planck's law:

$$I_{BB}(\lambda, T) = \frac{2k_P c^2}{\lambda^5} \frac{1}{\exp(k_P c/\lambda k_B T) - 1} \tag{4.9}$$

where k_P is Planck's constant, k_B is the Boltzmann constant, and c is the speed of light. The input power from the atmosphere radiation in Equation (4.7) is given by:

$$P_{atm}(T_{amb}) = A_r \int d\Omega \int_0^{\infty} I_{BB}(T_{amb}, \lambda) \mathrm{Abs}_{emi}(\lambda, \theta) \varepsilon_{atm}(\lambda, \theta) \cos\theta d\lambda \tag{4.10}$$

where Abs_{emi} (λ, θ) is the absorptivity of the emitter at wavelength λ and angle θ. $\varepsilon_{atm}(\lambda, \theta)$ is the emissivity of the atmosphere, which can be calculated as: $\varepsilon_{atm}(\lambda, \theta) = 1 - \mathrm{Tran}_{atm}(\lambda)^{1/\cos\theta}$, here $\mathrm{Tran}_{atm}(\lambda)$ is the transmittance of the atmosphere in the zenith direction. The input solar power is given by:

$$P_{solar}(\theta) = A_r \cdot G \int_0^{\infty} \varepsilon(\lambda, \theta) I_{AM1.5}(\lambda) / \int_0^{\infty} I_{AM1.5}(\lambda) d\lambda \tag{4.11}$$

where $I_{AM1.5}(\lambda)$ is the standard AM 1.5 spectrum of solar radiation, and G is the total solar irradiance at $1\,\mathrm{kW/m^2}$. The $P_{cond+conv}$ is given by:

$$P_{cond+conv} = A_r \cdot h_{non\text{-}rad}(T_{amb} - T) \tag{4.12}$$

where $h_{non\text{-}rad}$ is a non-radiative heat transfer coefficient that combines the effective conductive and convective heat exchange. By integrating the above equations separately, we can obtain the cooling power $P_{cooling}$ of the emitter at different temperatures T. When the emitter reaches a thermal equilibrium state, the $P_{cooling}$ is zero, and the corresponding steady temperature T_{ste} can be obtained.

In the following calculation, the conjugate heat transfer coefficient in $P_{cond+conv}$ is set as $h_{non\text{-}rad} = 5\,\mathrm{W \cdot m^{-2} \cdot K^{-1}}$, and the ambient temperature is kept at $T_{amb} = 25\,°\mathrm{C}$ to simulate a breeze situation. Obviously, the greater the cooling power, the better the performance of the designed SR. However, it seems not intuitive to use cooling power as reward, and it is difficult to set a suitable iteration threshold. Therefore, the reward is set as the difference between the steady-state temperature (T_{ste}) of the SR and the ambient temperature, namely the temperature drop below the T_{amb}. If the $P_{cooling}$ is positive at the initial temperature T_0 ($T_0 = T_{amb}$), the SR starts to cool down. As the temperature of the cooler decreases, the cooling power $P_{cooling}$ also reduces until $P_{cooling}(T_{ste}) = 0$. At that time, the SR reaches an equilibrium state and the T_{ste} can be obtained from Equation (4.7). Some studies have shown that the temperature difference ($\Delta T = T_{amb} - T_{ste}$) can reach 8 °C or even higher [7, 49], so the iteration threshold is set as 5 °C. The reward less than 5 will be mandatorily modified to −5. The optimization is implemented five times with 1000 iterations each.

The design and optimization results of the SR are presented in Figure 4.5(a). SiO_2 and TiO_2 are finally chosen as the materials for the optimal SR. The layer thickness of the optimal SR also exhibits irregular and aperiodic. The emissivity spectra of the designed and fabricated structures are shown in Figure 4.5(b). It can be seen that the designed SR exhibits near zero emissivity in the solar spectrum band, allowing it to reflect the incident solar radiation energy. In contrast, a high emissivity is obtained within the AW, enabling it to radiate heat efficiently to outer space. Due to the differences between the thickness of the fabricated sample and the designed values, their emissivity spectra are not completely consistent. The reward R of the optimal RC emitter is 16.99, which means it can maintain 16.99 °C below the ambient temperature at thermal equilibrium in theory. The cooling power at the initial temperature is 132.40 $\mathrm{W \cdot m^{-2}}$. The equilibrium temperature difference and cooling power both exhibit excellent performance of the designed SR.

The optimization process is quantitatively shown in Figure 4.6(a). In the early stage of optimization, the reward increases sharply, which means that DQN can quickly identify suitable materials for the SR and perform optimization under this material combination until the optimization process tends to be smooth (as shown in Figure 4.7). The material combination

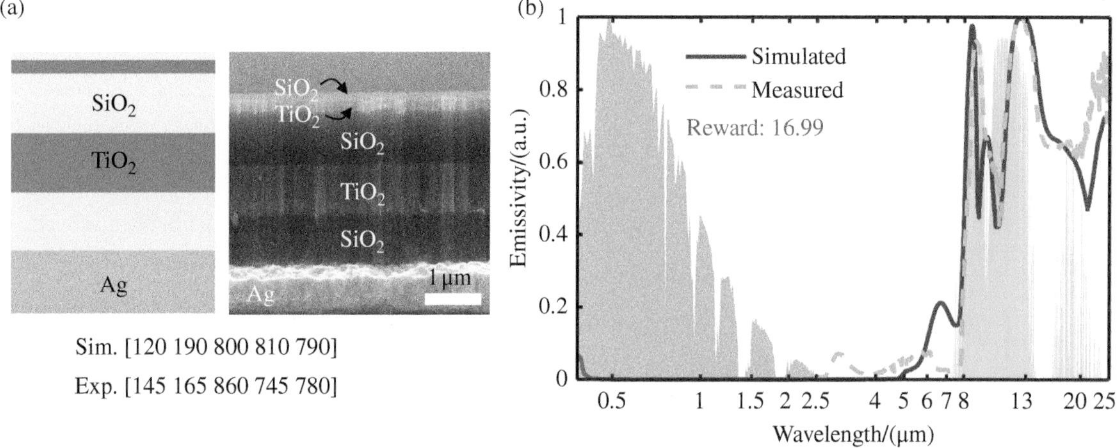

Figure 4.5 Results of SR designed by DQN. (a) Schematic and the scanning electron microscope (SEM) image of the optimal SR structure: TiO_2 and SiO_2 are chosen as the materials and the layer thicknesses of simulation and experiment are presented; (b) emissivity spectrum of the SR.

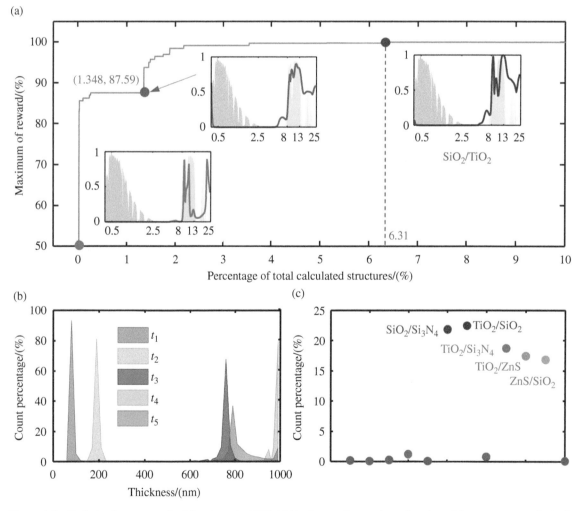

Figure 4.6 Optimization process for RC emitters by DQN. (a) Maximum of reward as a function of the percentage of calculated structures for RC; (b) parametric distribution curves of each layer thickness; (c) distribution of material combinations except SiO_2/TiO_2.

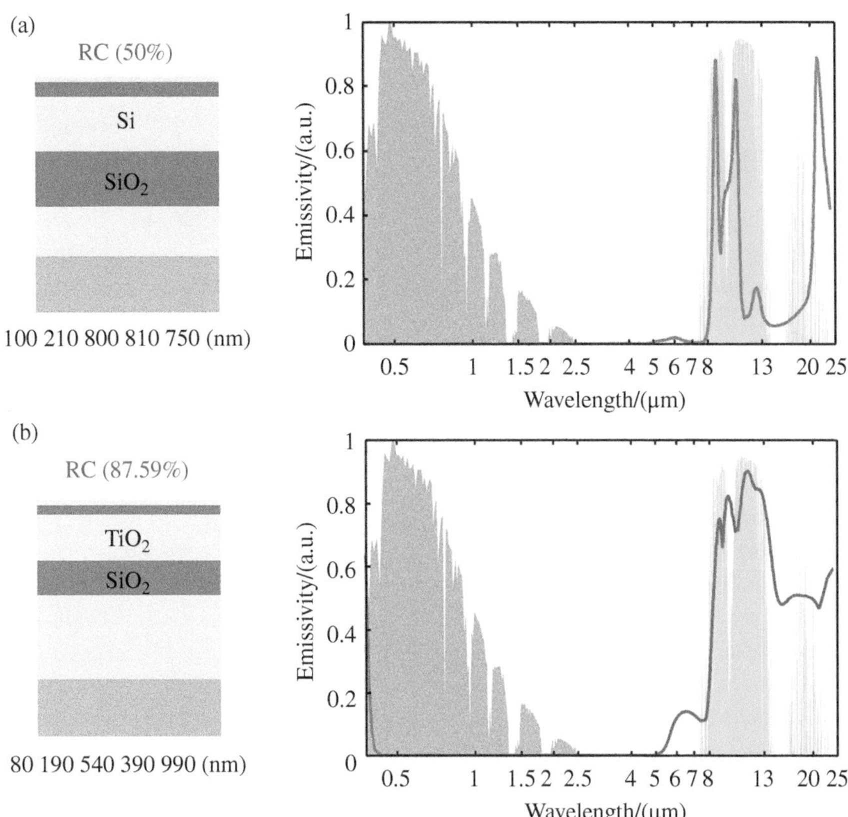

Figure 4.7 The structures obtained during the optimization process for RC. (a) The structure with the 50% of the maximum reward, which is composed of Si and SiO. (b) the structure with the 87.59% of the maximum reward, consisting of three layers of TiO_2 and two layers of SiO_2, which is different from the optimal structure (Figure 4.5) of three layers of SiO_2 and two layers of TiO_2. It is intuitive to see that the material and the structure are simultaneously optimized by DQN to achieve the optimal structural design.

of the structure yielding 50% of the maximum reward is Si/SiO_2, which indicates that DQN replaces Si with TiO_2 to achieve better cooling performance, as shown in Figure 4.7(a). The structure with the 87.59% of the maximum reward, as shown in Figure 4.7(b), consisting of three layers of TiO_2 and two layers of SiO_2, which is different from the optimal structure (Figure 4.5) of three layers of SiO_2 and two layers of TiO_2. It is intuitive to see that the material and the structure are simultaneously optimized by DQN to achieve the optimal structural design. During the smooth optimization period, the thickness of each layer is continuously optimized to further enhance the radiative cooling performance. When calculating less than 2% of the candidate structures, the RC emitter could reach a temperature drop of 14.94 °C below the ambient temperature in a steady state. After 1000 iterations, only 6.31% of structures need to be calculated to find the structure of the RC emitter with the maximum reward. To further exhibit the details of the optimization, the parametric distribution curves of each layer thickness are shown in Figure 4.6(b). In addition, except for the material combination of the optimal RC emitter, other material combinations are shown in Figure 4.6(c). It can be seen that SiO_2 and Si_3N_4 also exhibit potential as the materials of RC emitter, in addition to TiO_2 and SiO_2. The occurrence of less frequent material combinations can be explained by the random initialization of the DQN and the random selection of the EGE algorithm used in DQN.

4.3.3 Colored Radiative Cooling

In Section 4.3.2, in order to maximize cooling performance, the structures designed are mostly white to minimize the absorption of solar energy. However, a new problem arises that the radiative cooler always appears in white color so as to reflect the incident solar light as much as possible, but white colors are not always welcome in some scenarios for aesthetic reasons. Therefore, the colored radiative cooler (CRC) is more in demand with more practical and broader applications.

To achieve CRC, two strategies were proposed [52]. The first one is the dye rendering strategy, which directly applies dye materials onto the surface of the traditional cooler. For example, Chen et al. combined a layer of commercial paints with a porous poly(vinylidene fluoride-*co*-hexafluoropropene) (P(VdF-HFP)) or titanium dioxide (TiO_2)/polymer composite membrane to demonstrate paintable bilayer coatings [53]. Although this strategy can be applied in flexible materials like coating and fabric, the superlayer dye will absorb not only the radiation energy caused by color display in the visible band but also part of radiation energy in the infrared band, which will greatly weaken the cooling performance. The second strategy is the structural color strategy, which enables color decoration of the cooler through structure design [54–56]. In principle, an object's color can be calculated by the incident radiation spectrum, the reflectance of the object in the visible band, and the color-matching function of the human eye. These three factors represent the process of light projected onto the surface of an object, light reflected into human's eyes by the surface, and the optical signal converted to an electrical signal by human's cones. While the radiation spectrum of sunlight and the color-matching function of the human eye are fixed and untunable, only through altering the visible reflectance of object can one manipulate the color display. Based on this principle, the CRC is realized by designing specific visible reflectance which is based on multi-band emissivity regulation. Compared to the dye rendering strategy, a structural color strategy with a complicated and subtle structure can be designed elaborately to avoid extra energy absorbed in the infrared waveband, thus enabling better cooling performance. However, it suffers from complicated structure design, difficulty in and cost of manufacture as well as poor ductility. Lee et al. designed a multi-layer CRC which is the first one based on photonic structure to achieve a good color for CRC with yellow, cyan, and magenta colors. When convective heat transfer coefficient $h_{\text{non-rad}} = 6\ \text{W}\cdot\text{m}^{-2}\cdot\text{K}^{-1}$, the cooler could get a 3.9 °C lower temperature than the ambient temperature on average [57]. Sheng et al. presented a CRC with no angular dependence by adopting optical Tamm resonance, which got a better cooling effect of 5–6 °C below ambient when $h_{\text{non-rad}} = 6\ \text{W}\cdot\text{m}^{-2}\cdot\text{K}^{-1}$ [58]. Li et al. predicted the theoretic limit temperature of radiative cooling and heating corresponding to several kinds of colors and carried out a pink CRC and heater. However, it still lacked practical structures for other colors, and the theoretical potentials of cooling power for CRC remain unsolved [59].

In this section, we take a metal–dielectric–metal (MDM) [33, 60] colored radiative cooler structure as a representative to explore how the structural parameters influence the cooling performance and color display, respectively. Different roles of the parameters in the color display module are identified through compressive parameter exploration. Moreover, the reflectance spectra of the CRC are categorized into peak and valley types, whose characteristics are analyzed thoroughly. Angle dependency of color exhibited by the proposed structure is also studied. Finally, the correlations between radiative cooling power and CIE-LCH color space parameters (lightness, chroma, and hue) are also calculated to offer an intuitional relation between color display and radiative cooling performance.

4.3.3.1 Color Display Characterization

The calculation method of radiative power is referred to in Section 4.3.2, where the color display characterization is briefly introduced here. Generally, the color of an object depends on three factors: the spectral power distribution of the incident illuminant, the reflectivity of the object in the visible spectrum, and color-matching functions. The first one relies on the environment, such as the sun and the specific light source. Here, we adopt the standard D65 illumination to represent the typical outdoor daytime sunlight environment. The second one can be adjusted by the materials and structures, which are characterized by the TMM here for the CRC. The last one is the numerical description of the chromatic response of the human eye, which is experimentally characterized by International Committee on Illumination-XYZ color space (CIE-XYZ) in 1931. However, the description of color differences of CIE-XYZ color space is slightly different from human eyes. For a better description of the color perceived by human eyes, the CIE-LAB and CIE-LCH color spaces are put forward by the nonlinear transformation of the CIE-XYZ color space. CIE-LAB color space has three parameters: L^*, a^*, and b^*. L^* is used to represent the lightness, while a^* and b^* are used to represent chromaticity. These parameters can be transformed from the three tristimulus values X, Y, Z in the CIE-XYZ color space numerically as

$$L^* = 116f(Y/Y_0) - 16 \tag{4.13}$$

$$a^* = 500[f(X/X_0) - f(Y/Y_0)] \tag{4.14}$$

$$b^* = 200[f(Y/Y_0) - f(Z/Z_0)] \tag{4.15}$$

where X_0, Y_0, and Z_0 are the tristimulus values corresponding to the color white, and

$$f(s) = \begin{cases} t^{\frac{1}{3}}, t > \left(\dfrac{24}{116}\right)^3 \\ \left(\dfrac{841}{108}\right)t + \dfrac{16}{116}, t \leq \left(\dfrac{24}{116}\right)^3 \end{cases} \tag{4.16}$$

CIE-LCH color space is just essentially another equivalent form of CIE-LAB color space. It uses lightness, chroma, and hue to describe colors, which can be expressed by a combination of the L^*, a^*, and b^*, respectively. We chose the CIE-LCH color space because it fits better with human visual experience, along with the advantages of wider color space and unique color coding.

4.3.3.2 Influence of Structural Parameters on Colored Radiative Cooler

To simplify the analysis process, we adopt the 1D MDM photonic crystal structures to design the CRC with limited parameters like layer materials, thickness, and configurations. As shown in Figure 4.8(a), the CRC can be divided into two parts according to different functionalities, namely the radiative cooling module and color display module. Both the structure and materials are chosen deliberately after a comprehensive review of the existing literature [61–66]. Here, a typical radiative cooling module consists of the SiO_2/TiO_2 photonic crystal with varying layers of thickness in Figure 4.8(a), while the Ag/SiO_2 layer pair denotes the color display module, whose thicknesses are set to be 50 and 125 nm. Without the color display module, the rest structure can be regarded as a conventional white-color radiative cooler (WRC for short hereinafter). Both the CRC and WRC structures are mounted on a silver substrate for reflection in case. The material properties of TiO_2 are referred from Sarkar's work [62] and Palik's handbook [65], while SiO_2 are referred from Palik's handbook [66]. The optical constants of bulk Ag are referred from Yang's work [63], while optical constants of the thin film Ag come from Kim's work [64]. The emissivity and reflectivity spectra are calculated by TMM [67, 68], which is a common lightweight method to solve Maxwell's equations. The emissivity spectra of the WRC and CRC at 0.38–25 μm are shown in Figure 4.8(b). It is seen that the emissivity spectra overlap mostly and the only difference occurs in the visible spectrum. WRC exhibits near zero emissivity in the solar incident band to reflect the input solar radiation energy as much as possible to achieve higher cooling power. Such high reflectance in the visible band generates a white color display as a result. In contrast, the CRC exhibits a higher emissivity and a lower reflectivity at ~0.51 μm, thus producing a light-magenta color. Consequently, the absorbed solar radiation energy P_{solar} increases from 9.08 to 35.15 $W{\cdot}m^{-2}$, leading to a 25.8% reduction of the net RC power from 97.24 to 72.14 $W{\cdot}m^{-2}$. Basically, the increase of the absorbed solar radiation will weaken the RC power, implying the cooling power and the color display are competing. At thermal equilibrium, the WRC maintains at a low temperature below T_{amb} by 12.76 °C, while the CRC maintains at a relatively higher temperature below T_{amb} by 9.44 °C. It should be noted that we use a lower steady-state temperature T_{ste} of the CRC to represent a higher cooling power for convenience in this section. A detailed explanation is shown in Sections 4.3.2 and 4.3.3.1. From this preliminary comparison, we note that the main difference between the CRC and WRC lies in the diverse emissivity/reflectivity in the visible band for color display, and the high emissivity (low reflectivity) in the visible band will cause the decrease in cooling power. Therefore, the competing

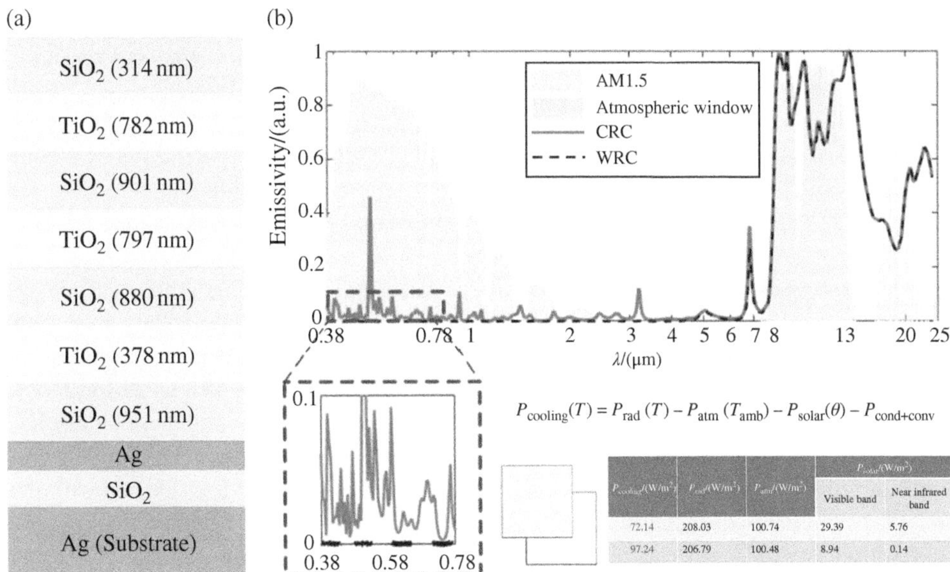

$P_{cooling}$/(W/m^2)	P_{rad}/(W/m^2)	P_{atm}/(W/m^2)	P_{solar}/(W/m^2)	
			Visible band	Near infrared band
72.14	208.03	100.74	29.39	5.76
97.24	206.79	100.48	8.94	0.14

Figure 4.8 (a) Schematic of a typical CRC and (b) emissivity spectra comparison of WRC and CRC and the corresponding detailed power comparation.

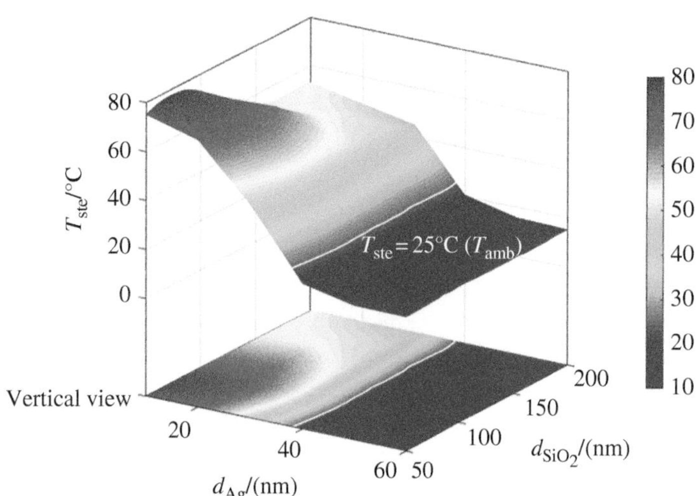

Figure 4.9 Steady-state temperature T_{ste} versus structural parameters.

role between the cooling power and the color display is coupled by the emissivity spectrum. The remaining challenge is how to balance the cooling power and color display by adjusting the structural parameters in the CRC design, and the first step is to reveal how parameters influence cooling and color.

Figure 4.9 shows the variation of CRC steady-state temperature T_{ste} with a fixed radiative cooling module and by varying d_{Ag} and d_{SiO_2} in the color display module. The white solid line denotes the ambient temperature $T_{amb} = 25\,°C$. The thickness of Ag layer d_{Ag} ranges from 10 to 60 nm with a step of 10 nm and the thickness of SiO_2 layer d_{SiO_2} ranges from 50 to 200 nm with a step of 5 nm. It is seen in Figure 4.9 that when the thickness of the Ag layer is smaller than 40 nm, the CRC absorbs, rather than radiates, more heat from the ambient, balancing at a rather high steady-state temperature. When the thickness of the Ag layer is larger than 40 nm, the CRC tends to show the net cooling effect with a relatively low temperature. Compared to Ag layer thickness, SiO_2 layer thickness is a secondary factor to the cooling effect, which has a limited influence on the cooling power. Therefore, it is advantageous to tune the thickness of the Ag layer for better cooling performance.

Although the SiO_2 layer thickness is not sensitive to the cooling power, it is the major factor for the color display. The reflectance maps, corresponding to the two CRC structures with the same $d_{SiO_2} = 125$ nm, and different $d_{Ag} = 10$ and 60 nm, are shown in Figure 4.10(a) and (b) respectively. Each horizontal line represents a reflectance spectrum of different d_{SiO_2}, while color denotes the value of reflectivity. It can be seen that the mean reflectance is low when $d_{Ag} = 10$ nm but near unity when $d_{Ag} = 60$ nm. This indicates that the thicker the Ag layer is, the better the cooling effect will be, which agrees well with the aforementioned discussion. When altering SiO_2 layer thickness d_{SiO_2}, the reflectance can be adjusted according to Figure 4.10(c) and (d) and causes diverse colors. To analyze concretely, Figure 4.10(c) and (d) show the corresponding reflectance spectra of two CRC structures as typical examples with the same $d_{SiO_2} = 125$ nm, and different $d_{Ag} = 10$ and 60 nm, respectively, which matches the black dashed line in Figure 4.10(a) and (b). It is seen that when $d_{Ag} = 10$ nm, the total reflectance is high only between 380 and 510 nm. Therefore, the structure will display a superposed color by mixing the light with a wavelength between 380 and 510 nm. When $d_{Ag} = 60$ nm, the reflectance spectra are high almost at the whole wavelength band except for $\lambda = 510$ nm, which corresponds to the complementary light-magenta color according to the additive color-mixing principle.

From the analysis earlier, the influence of Ag/SiO_2 layer thicknesses should be discussed in two cases: a thin Ag layer and a thick one. In the first case when $d_{Ag} < 40$ nm, the reflectance in the visible band is relatively small except for some reflectivity peaks, which is called the peak-type reflectance. The peak-type reflectance has a low mean reflectance (high absorptance) which leads to more absorbed solar radiation heat and, hence, a rather poor overall radiative cooling performance. Altering SiO_2 layer thickness can shift the reflectivity peak and, thus, influence both the cooling power and the color display. For cooling power, as the main reflectivity peak deviates from the high energy waveband of the solar radiation spectrum, the input solar radiation P_{solar} will decrease and, hence, improve the cooling power. Such improvement is weakened with the increase of the Ag layer thickness, corresponding to the situation where the steady-state temperature is higher than T_{amb} in Figure 4.9. For color display, the main reflectivity peak determines the wavelengths of light reflected from the object and further determines the color of the display. In the second case, when $d_{Ag} \geq 40$ nm, the reflectance in the visible band is relatively large except for some reflectivity valleys, which is called the valley-type reflectance. The valley-type reflectance makes the structure reflect most of the solar radiation outward, leading to a better overall RC performance. Altering SiO_2

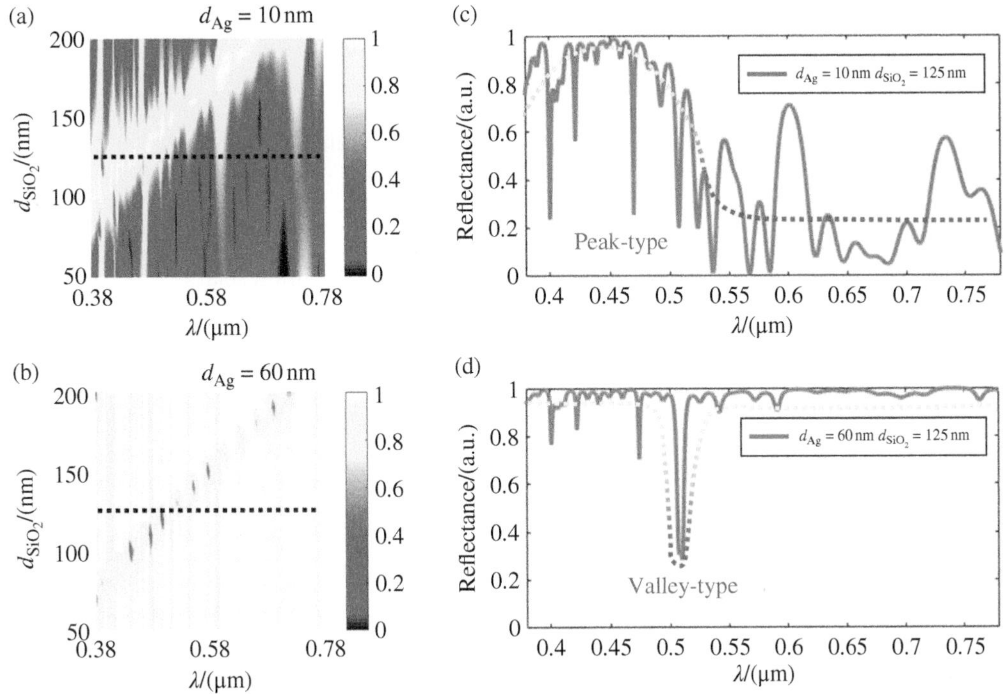

Figure 4.10 Reflectance corresponding to different SiO_2 thickness when (a) d_{Ag} = 10 nm and (b) 60 nm, respectively. (c) Peak-type reflectance when d_{Ag} = 10 nm and d_{SiO_2} = 125 nm; (d) valley-type reflectance when d_{Ag} = 60 nm and d_{SiO_2} = 125 nm.

layer thickness can also shift the reflectivity valley and change the color display, but the influence on radiative cooling power is neglectable, as shown in Figure 4.9. This is because no matter which wavelengths the reflectivity valley lies in, the amount of reflected solar radiation almost maintains the same and, hence, exhibits little influence on radiative cooling power. When comparing the peak-type and the valley-type reflectance in visible waveband, it is easy to notice that valley-type reflectance with a high mean reflectance is more likely to achieve higher cooling power but suffers from poor color display, while the peak-type reflectance exhibits more kinds of color but with limited cooling power.

To further demonstrate the competing role between radiative cooling and color display and find an intuitive way to gain a tradeoff between color display and cooling power, Figure 4.11 shows the perceived colors and the corresponding steady-state temperature T_s under different combinations of d_{SiO_2} and d_{Ag}. In general, large d_{Ag} accounts for higher cooling power while

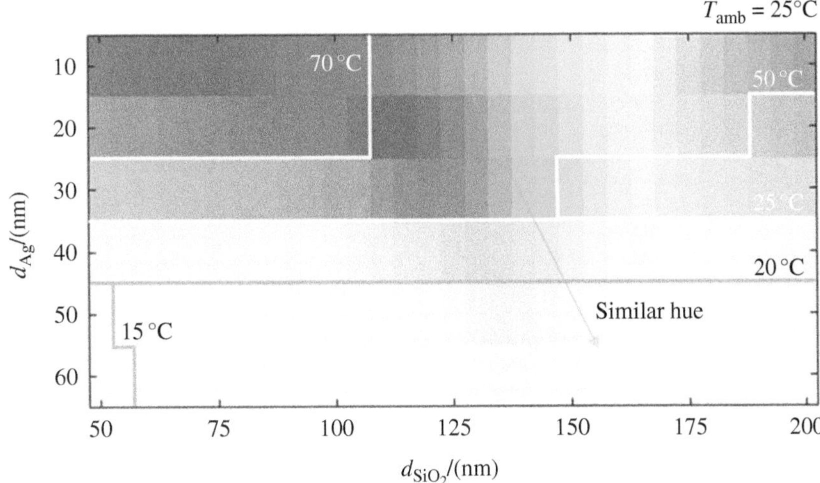

Figure 4.11 The color display and the steady-state temperature T_{ste} corresponding to different structural parameters.

d_{SiO_2} accounts for color display, which agrees well with previous analysis. Taking 25 °C as the environmental temperature, the structure with a lower T_{ste} shows limited colors that are rather light and white with poor aesthetics, especially in yellow, magenta, and cyan with the reflectivity valley around 450, 500, and 620 nm, respectively. On the contrary, more bright and colorful colors can be obtained with a higher T_{ste}, corresponding to reduced cooling power. Moreover, as the blue arrow in Figure 4.11 indicates, if a specific hue of color is desired, a matching of d_{SiO_2} and d_{Ag} will fulfill the need. Better color displays with brighter and more vivid colors can be easily achieved by reducing d_{Ag} at the expense of weakening radiative cooling power. Besides, good CRC can also be designed by well designing the color display module with two or even more reflectivity valleys while maintaining acceptable cooling power. For example, the MDM structure with reflectivity valleys at only 450 and 620 nm will display green after the superposition of yellow and cyan colors, but still reflect most solar radiation out and keep a good cooling performance. One possible candidate for the color display modules with multiple reflectivity valleys is 2D/3D photonic crystal. Special attention should be paid when combining the radiative cooling module and the color display module to prevent them from interfering with each other. Besides, there are still some problems such as structure stability to solve for the realization of richer color with a cooling effect. What is worth mentioning is the good angular independence of the colors. Figure 4.12 demonstrates the reflectance of two structures (d_{Ag} = 10 or 40 nm, d_{SiO_2} = 125 nm) in different incident angles θ. The corresponding colors under different θ are also shown in the left color bar. When d_{Ag} = 10 nm in Figure 4.12(a), the reflectivity peak still shows up in the visible band with a low overall reflectance, leading to the dark tone of color display. The reflectivity peak, which hardly blueshifts when θ increases until 40°, starts to shift out of the visible band and causes a lower overall reflectance as well as a darker shade color. A similar trend is also shown in the case of d_{Ag} = 40 nm in Figure 4.12(b), except that the reflectivity valley does not shift out of the visible band, hence, leading to the colors with a similar lighter tone under different θ. The consistency of color with little change in the range of 0°–40° can avoid information misleading in a specific situation such as warning color.

The correlation between color display and radiative cooling power has been analyzed. Here, the CIE-LCH color space is used to characterize color. As for cooling power, a lower steady-state temperature T_{ste} implies a better cooling performance and a higher cooling power; hence, we choose the temperature difference $T_{amb} - T_{ste}$ to represent cooling power for convenience. Figure 4.13(a) and (b) shows the dependence of the three LCH parameters and $T_{amb} - T_{ste}$ on the variation of d_{SiO_2}. All parameters are normalized to the same order of magnitude to find out which is the most influential. When d_{Ag} = 10 nm, both the lightness and $T_{amb} - T_{ste}$ grow gradually with the increase of d_{SiO_2}. While d_{Ag} = 40 nm, d_{SiO_2} plays a negligible role on these parameters. Moreover, the lightness has an extremely similar trend with $T_{amb} - T_{ste}$ in both cases. The Spearman rank-order correlation coefficients (k_{sro}) of *Lightness, Chroma, Hue* with $T_{amb} - T_{ste}$ are also calculated. The blue points in Figure 4.13(c) are approximately aligned linearly with k_{sro_L} = 0.92, which implies a strong linear correlation between lightness *Lightness* and $T_{amb} - T_{ste}$. The chroma is weakly negatively correlated with k_{sro_C} = −0.75. This can be seen from Figure 4.13(d) that the smaller chroma tends to have a better cooling performance. Figure 4.13(e) shows well-distributed points with k_{sro_H} = −0.30, which indicates that the hue has almost no correlation with cooling performance. Therefore,

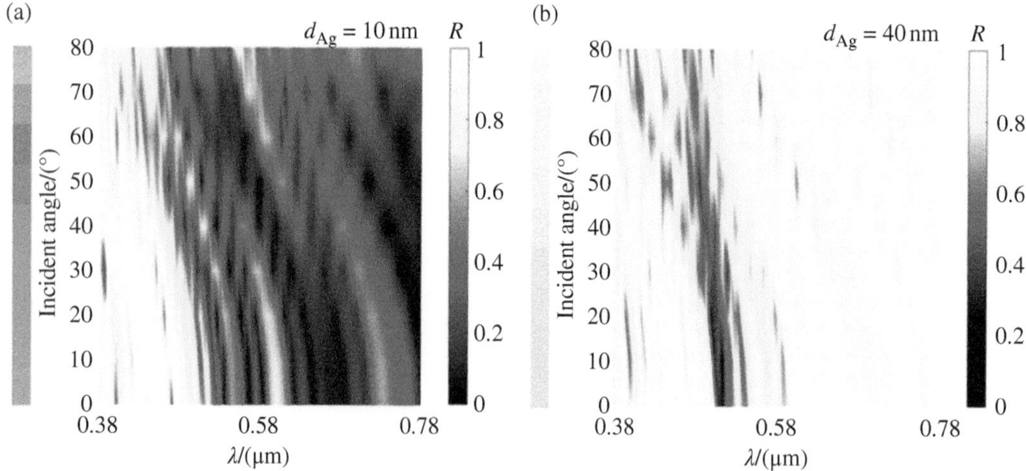

Figure 4.12 The angular dependency of reflectance and corresponding color. (a) d_{Ag} = 10 nm and d_{SiO_2} = 125 nm; (b) d_{Ag} = 40 nm and d_{SiO_2} = 125 nm.

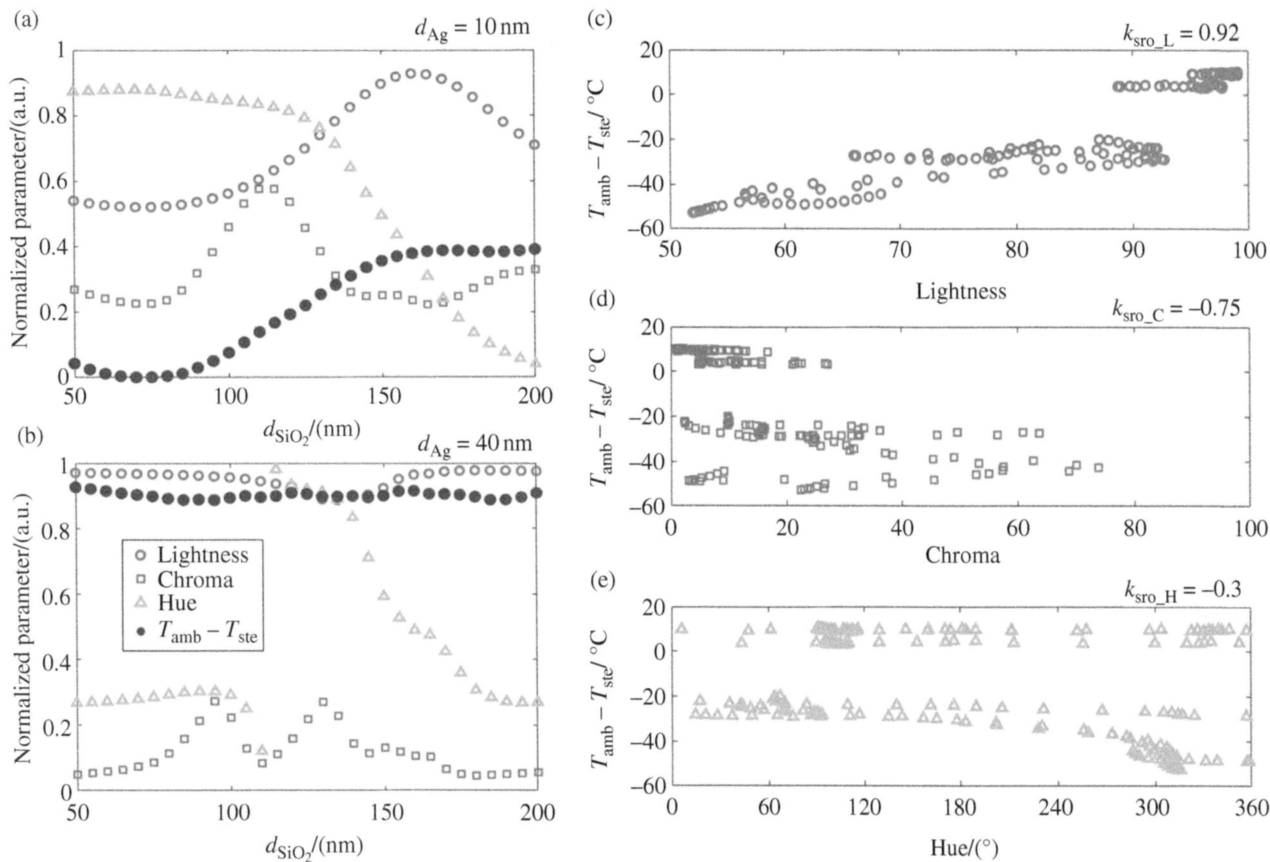

Figure 4.13 (a, b) The tendency of the lightness, chroma, hue, and $T_{amb} - T_{ste}$ changing with SiO_2 thickness when d_{Ag} = 10 and 40 nm, respectively. (c–e) The Spearman rank-order correlation coefficients (k_{sro}) of lightness, chroma, and hue with $T_{amb} - T_{ste}$, respectively.

color lightness plays a major role in the competitive relationship between color and cooling effect. Brighter and lighter colors stand for a better cooling effect, while structures with darker color even fail to maintain cooling effect, which reveals the basic principle of balancing the cooling power and aesthetic needs.

4.4 Near-Field Thermal Radiation in Thermal Management

The radiative cooling technologies mentioned above belong to the category of far-field radiation, which strictly obeys the Stefan–Boltzmann law. However, as the distance between the radiating bodies is reduced to a subwavelength range, the radiative heat can be dramatically enhanced, thus breaking the Stefan–Boltzmann limit. This radiation enhancement effect can efficiently improve the heat exchange between objects, which is of great significance to the thermal management of micro–nanoscale electronic devices. The near-field enhanced heat transfer mainly results from the evanescent coupling of thermally excited surface resonances; therefore, it is inversely proportional to the distance. The near-field distance limit depends on the materials of the objects and systematic geometry. For the application of thermal management, the radiating objects and heat sinks are usually made of different materials, thus reducing the resonant coupling, and the near-field distance limit is typically less than 100 nm [69, 70].

Owing to the strict requirements of parallelism and flatness of the andante in the near-field radiation, the minimum distance in the related experimental work was only 1 μm in the early years [71–73]. In recent years, with the development of nanotechnology and micro-scale processing, the limit distance in the experiment of near-field radiation is constantly shortened, which further expands the radiation enhancement effect. Song et al. [74] demonstrated a dramatic near-field radiation enhancement based on an experiment platform consisting of a heat-flow calorimeter with a resolution of 100 pW. The gap size between the spherical radiator and the planar receiver can be controlled precisely, ranging from 20 nm to 10 μm.

The experiment results show that as the gap size approaches the receiver film thickness, the near-field radiation can be significantly affected by the film and increases rapidly. Tiny distance is relatively easy to be achieve in the sphere-plane system. However, in this configuration, the effective heat transfer area of the objects is greatly reduced, resulting in low heat transfer efficiency. With regard to this, Lipson et al. [75] experimentally demonstrated near-field radiative heat transfer between parallel structures by using nanobeams monolithically integrated with electrostatic comb-drive actuators. As the gap size is reduced to the subwavelength range (<150 nm), the near-field heat transfer increases exponentially. At the minimum gap of 42 nm, an 82 times enhancement of heat transfer compared to the far-field limit is achieved. In order to further simplify the device and increase the heat transfer area, thus improving the application feasibility of near-field heat transfer, Francoeur and coworkers [76] fabricated a bonded device with surfaces in macroscale dimensions, which can independently support the gap spacing. The devices consist of two boron-doped silicon slabs separated by the micropillars. The ingenious pits design around the emitter and receiver enable the micropillars to be considerably longer than the gap size, thus effectively reducing the influence of parasitic thermal conduction without affecting the structural integrity. When the gap size reaches down to 110 nm, the near-field radiation can be enhanced nearly 28.5 times as much as the blackbody limit. These explorations of near-field radiation enhancement are crucial to optimizing the thermal management of nanoelectronic devices.

4.5 Chapter Summary

The development of photon engineering has promoted the flexible regulation of emissivity spectrum, which can achieve higher radiation heat transfer efficiency, and also makes it possible to utilize radiation for temperature regulation. In this chapter, the basic physical concepts and numerical formulas of thermal radiation are introduced. Then, it briefly introduces the means of regulating the thermal radiation spectrum, including manual design and machine-learning optimization. Among them, machine-learning optimization shows high design efficiency in the development of the current technology field, and a DQN framework is proposed to realize the design of radiative cooling. In addition, due to some aesthetic consideration, colored radiative cooling has been discussed and a CRC with Ag/SiO_2 color display module on top of a TiO_2/SiO_2 radiative cooler has been designed to comprehensively investigate the competing role of radiative cooling power and color display. Finally, the development status of near-field radiation enhancement technology is introduced. It is considered that the near-field radiation enhancement technology has a broad application prospect in the thermal management of nanoelectronic devices. In general, the development of these emerging fields further refines the concept of utilizing thermal radiation for thermal management and pushes radiation thermal management from the laboratory stage to practical applications.

References

1 Baranov, D.G., Xiao, Y.Z., Nechepurenko, I.A. et al. (2019). Nanophotonic engineering of far-field thermal emitters. *Nat. Mater.* 18 (9): 920–930.

2 Li, W. and Fan, S.H. (2018). Nanophotonic control of thermal radiation for energy applications invited. *Opt. Express* 26 (12): 15995–16021.

3 Wang, J., Dai, G.L., and Huang, J.P. (2020). Thermal metamaterial: fundamental, application, and outlook. *Iscience* 23 (10): 101637.

4 Zhu, H.Z., Li, Q., Tao, C.N. et al. (2021). Multispectral camouflage for infrared, visible, lasers and microwave with radiative cooling. *Nat. Commun.* 12 (1): 1805.

5 Xu, L.J., Wang, R.Z., and Huang, J.P. (2018). Camouflage thermotics: a cavity without disturbing heat signatures outside. *J. Appl. Phys.* 123 (24): 245111.

6 Cho, J.W., Lee, E.J., and Kim, S.K. (2022). Radiative cooling technologies: a platform for passive heat dissipation. *J. Korean Phys. Soc.* 81 (6): 481–489.

7 Hossain, M.M., Jia, B.H., and Gu, M. (2015). A metamaterial emitter for highly efficient radiative cooling. *Adv. Opt. Mater.* 3 (8): 1047–1051.

8 Biehs, S.A., Tschikin, M., and Ben-Abdallah, P. (2012). Hyperbolic metamaterials as an analog of a blackbody in the near field. *Phys. Rev. Lett.* 109 (10): 104301.

9 Sukhatme, S. (2005). *A Textbook on Heat Transfer*, 4the. Hyderabad: Universities Press (India) Private Limited.

10 Lienhard, J.H. (2019). *A Heat Transfer Textbook*, 5the. New York: Dover Publications.

11 Catalanotti, S., Cuomo, V., Piro, G. et al. (1975). The radiative cooling of selective surfaces. *Sol. Energy* 17: 83–89.

12 Liu, Y., Pan, D., Chen, W. et al. (2020). Radiative heat transfer in nanophotonics: from thermal radiation enhancement theory to radiative cooling applications. *Acta Phys. Sin.* 69 (3): 036501.

13 Dahan, N., Niv, A., Biener, G. et al. (2007). Enhanced coherency of thermal emission: beyond the limitation imposed by delocalized surface waves. *Phys. Rev. B* 76 (4): 045427.

14 Wang, L.P., Lee, B.J., Wang, X.J. et al. (2009). Spatial and temporal coherence of thermal radiation in asymmetric Fabry-Perot resonance cavities. *Int. J. Heat Mass Transf.* 52 (13–14): 3024–3031.

15 Celanovic, I., Perreault, D., and Kassakian, J. (2005). Resonant-cavity enhanced thermal emission. *Phys. Rev. B* 72 (7): 075127.

16 Shu, S.W., Zhan, Y.W., Lee, C. et al. (2016). Wide angle and narrow-band asymmetric absorption in visible and near-infrared regime through lossy Bragg stacks. *Sci. Rep.* 6: 27061.

17 Portnoi, M., Macdonald, T.J., Sol, C. et al. (2020). All-silicone-based distributed Bragg reflectors for efficient flexible luminescent solar concentrators. *Nano Energy* 70: 104507.

18 Wang, Z.Y., Clark, J.K., Ho, Y.L. et al. (2018). Narrowband thermal emission realized through the coupling of cavity and tamm plasmon resonances. *ACS Photon.* 5 (6): 2446–2452.

19 Zhu, H.Z., Luo, H., Li, Q. et al. (2018). Tunable narrowband mid-infrared thermal emitter with a bilayer cavity enhanced tamm plasmon. *Opt. Lett.* 43 (21): 5230–5233.

20 Yang, Z.Y., Ishii, S., Yokoyama, T. et al. (2016). Tamm plasmon selective thermal emitters. *Opt. Lett.* 41 (19): 4453–4456.

21 Kaliteevski, M., Iorsh, I., Brand, S. et al. (2007). Tamm plasmon-polaritons: possible electromagnetic states at the interface of a metal and a dielectric Bragg mirror. *Phys. Rev. B* 76 (16): 165415.

22 Zhou, H.C., Yang, G.A., Wang, K. et al. (2010). Multiple optical tamm states at a metal-dielectric mirror interface. *Opt. Lett.* 35 (24): 4112–4114.

23 Wang, L.P., Basu, S., and Zhang, Z.M. (2012). Direct measurement of thermal emission from a Fabry-Perot cavity resonator. *J. Heat Transf. Trans. ASME* 134 (7): 072701.

24 Yang, Z.Y., Ishii, S., Yokoyama, T. et al. (2017). Narrowband wavelength selective thermal emitters by confined tamm plasmon polaritons. *ACS Photon.* 4 (9): 2212–2219.

25 Wang, Z.Y., Clark, J.K., Ho, Y.L. et al. (2018). Narrowband thermal emission from tamm plasmons of a modified distributed Bragg reflector. *Appl. Phys. Lett.* 113 (16): 161104.

26 Wang, L.P. and Zhang, Z.M. (2009). Resonance transmission or absorption in deep gratings explained by magnetic polaritons. *Appl. Phys. Lett.* 95 (11): 111904.

27 Zhao, B. and Zhang, Z.M.M. (2014). Study of magnetic polaritons in deep gratings for thermal emission control. *J. Quant. Spectrosc. Radiat. Transf.* 135: 81–89.

28 Zhao, B., Wang, L.P., Shuai, Y. et al. (2013). Thermophotovoltaic emitters based on a two-dimensional grating/thin-film nanostructure. *Int. J. Heat Mass Transf.* 67: 637–645.

29 Liu, X.L., Tyler, T., Starr, T. et al. (2011). Taming the blackbody with infrared metamaterials as selective thermal emitters. *Phys. Rev. Lett.* 107 (4): 045901.

30 Cui, Y.X., Fung, K.H., Xu, J. et al. (2012). Ultrabroadband light absorption by a sawtooth anisotropic metamaterial slab. *Nano Lett.* 12 (3): 1443–1447.

31 Liu, B.A., Gong, W., Yu, B.W. et al. (2017). Perfect thermal emission by nanoscale transmission line resonators. *Nano Lett.* 17 (2): 666–672.

32 Kim, T., Bae, J.Y., Lee, N. et al. (2019). Hierarchical metamaterials for multispectral camouflage of infrared and microwaves. *Adv. Funct. Mater.* 29 (10): 1807319.

33 Li, Q., Li, Z.Z., Xiang, X.J. et al. (2019). Tunable perfect narrow-band absorber based on a metal-dielectric-metal structure. *Coatings* 9 (6): 393.

34 Joannopoulos, J.D., Villeneuve, P.R., and Fan, S.H. (1997). Photonic crystals: putting a new twist on light. *Nature* 386 (6621): 143–149.

35 Pralle, M.U., Moelders, N., McNeal, M.P. et al. (2002). Photonic crystal enhanced narrow-band infrared emitters. *Appl. Phys. Lett.* 81 (25): 4685–4687.

36 Hu, R., Song, J.L., Liu, Y.D. et al. (2020). Machine learning-optimized tamm emitter for high-performance thermophotovoltaic system with detailed balance analysis. *Nano Energy* 72: 104687.

37 Liu, Z.C., Zhu, D.Y., Rodrigues, S.P. et al. (2018). Generative model for the inverse design of metasurfaces. *Nano Lett.* 18 (10): 6570–6576.

38 Liu, D.J., Tan, Y.X., Khoram, E. et al. (2018). Training deep neural networks for the inverse design of nanophotonic structures. *ACS Photon.* 5 (4): 1365–1369.

39 So, S., Mun, J., and Rho, J. (2019). Simultaneous inverse design of materials and structures via deep learning: demonstration of dipole resonance engineering using core-shell nanoparticles. *ACS Appl. Mater. Interfaces* 11 (27): 24264–24268.

40 So, S., Lee, D., Badloe, T. et al. (2021). Inverse design of ultra-narrowband selective thermal emitters designed by artificial neural networks. *Opt. Mater. Express* 11 (7): 1863–1873.

41 He, M.Z., Nolen, J.R., Nordlander, J. et al. (2021). Deterministic inverse design of tamm plasmon thermal emitters with multi-resonant control. *Nat. Mater.* 20 (12): 1663–1669.

42 Sakurai, A., Yada, K., Simomura, T. et al. (2019). Ultranarrow-band wavelength-selective thermal emission with aperiodic multilayered metamaterials designed by Bayesian optimization. *ACS Cent. Sci.* 5 (2): 319–326.

43 Molesky, S., Lin, Z., Piggott, A.Y. et al. (2018). Inverse design in nanophotonics. *Nat. Photon.* 12 (11): 659–670.

44 Ma, W., Liu, Z.C., Kudyshev, Z.A. et al. (2021). Deep learning for the design of photonic structures. *Nat. Photon.* 15 (2): 77–90.

45 Mnih, V., Kavukcuoglu, K., Silver, D. et al. (2015). Human-level control through deep reinforcement learning. *Nature* 518 (7540): 529–533.

46 van Hasselt, H., Guez, A., Silver, D. et al. (2016). Deep reinforcement learning with double q-learning. 30th Association-for-the-Advancement-of-Artificial-Intelligence Conference on. *Artif. Intell.* 30 (1): 2094–2100.

47 Zhu, L.X., Raman, A., Wang, K.X. et al. (2014). Radiative cooling of solar cells. *Optica* 1 (1): 32–38.

48 Safi, T.S. and Munday, J.N. (2015). Improving photovoltaic performance through radiative cooling in both terrestrial and extraterrestrial environments. *Opt. Express* 23 (19): A1120–A1128.

49 Fan, S.H. and Li, W. (2022). Photonics and thermodynamics concepts in radiative cooling. *Nat. Photon.* 16 (3): 182–190.

50 Wang, Q.X., Huang, Z.Q., Li, J.Z. et al. (2023). Module-level polaritonic thermophotovoltaic emitters via hierarchical sequential learning. *Nano Lett.* 23 (4): 1144–1151.

51 Kim, S., Shang, W.J., Moon, S. et al. (2022). High-performance transparent radiative cooler designed by quantum computing. *ACS Energy Lett.* 7 (12): 4134–4141.

52 Srinivasarao, M. (1999). Nano-optics in the biological world: beetles, butterflies, birds, and moths. *Chem. Rev.* 99 (7): 1935–1961.

53 Chen, Y.J., Mandal, J., Li, W.X. et al. (2020). Colored and paintable bilayer coatings with high solar-infrared reflectance for efficient cooling. *Sci. Adv.* 6 (17): eaaz5413.

54 Zhu, L.X., Raman, A., and Fan, S.H. (2013). Color-preserving daytime radiative cooling. *Appl. Phys. Lett.* 103 (22): 223902.

55 Yalçin, R.A., Blandre, E., Joulain, K. et al. (2020). Colored radiative cooling coatings with nanoparticles. *ACS Photon.* 7 (5): 1312–1322.

56 Kim, H.H., Im, E., and Lee, S. (2020). Colloidal photonic assemblies for colorful radiative cooling. *Langmuir* 36 (23): 6589–6596.

57 Lee, G.J., Kim, Y.J., Kim, H.M. et al. (2018). Colored, daytime radiative coolers with thin-film resonators for aesthetic purposes. *Adv. Opt. Mater.* 6 (22): 1800707.

58 Sheng, C.X., An, Y.D., Du, J. et al. (2019). Colored radiative cooler under optical tamm resonance. *ACS Photon.* 6 (10): 2545–2552.

59 Li, W., Shi, Y., Chen, Z. et al. (2018). Photonic thermal management of coloured objects. *Nat. Commun.* 9: 4240.

60 Shin, H., Yanik, M.F., Fan, S.H. et al. (2004). Omnidirectional resonance in a metal-dielectric-metal geometry. *Appl. Phys. Lett.* 84 (22): 4421–4423.

61 Kecebas, M.A., Menguc, M.P., Kosar, A. et al. (2020). Spectrally selective filter design for passive radiative cooling. *J. Opt. Soc. Am. B Opt. Phys.* 37 (4): 1173–1182.

62 Sarkar, S., Gupta, V., Kumar, M. et al. (2019). Hybridized guided-mode resonances via colloidal plasmonic self-assembled grating. *ACS Appl. Mater. Interfaces* 11 (14): 13752–13760.

63 Yang, H.H.U., D'Archangel, J., Sundheimer, M.L. et al. (2015). Optical dielectric function of silver. *Phys. Rev. B* 91 (23): 235137.

64 Kim, J., Oh, H., Seo, M. et al. (2019). Generation of reflection colors from metal-insulator-metal cavity structure enabled by thickness-dependent refractive indices of metal thin film. *ACS Photon.* 6 (9): 2342–2349.

65 Ribarsky, M.W. (1997). Titanium dioxide (TiO_2) (rutile). In: *Handbook of Optical Constants of Solids* (ed. E.D. Palik), 795–804. Burlington: Academic Press.

66 Philipp, H.R. (1997). Silicon dioxide (SiO_2) (glass). In: *Handbook of Optical Constants of Solids* (ed. E.D. Palik), 749–763. Burlington: Academic Press.

67 Katsidis, C.C. and Siapkas, D.I. (2002). General transfer-matrix method for optical multilayer systems with coherent, partially coherent, and incoherent interference. *Appl. Opt.* 41 (19): 3978–3987.

68 Li, Z.Y. (2005). Principles of the plane-wave transfer-matrix method for photonic crystals. *Sci. Technol. Adv. Mater.* 6 (7): 837–841.

69 Iizuka, H. and Fan, S.H. (2012). Rectification of evanescent heat transfer between dielectric-coated and uncoated silicon carbide plates. *J. Appl. Phys.* 112 (2): 024304.

70 Laroche, M., Carminati, R., and Greffet, J.J. (2006). Near-field thermophotovoltaic energy conversion. *J. Appl. Phys.* 100 (6): 063704.

71 Hargreaves, C.M. (1969). Anomalous radiative transfer between closely-spaced bodies. *Phys. Lett. A* 30 (9): 491.

72 Kralik, T., Hanzelka, P., Zobac, M. et al. (2012). Strong near-field enhancement of radiative heat transfer between metallic surfaces. *Phys. Rev. Lett.* 109 (22): 224302.

73 Ottens, R.S., Quetschke, V., Wise, S. et al. (2011). Near-field radiative heat transfer between macroscopic planar surfaces. *Phys. Rev. Lett.* 107 (1): 014301.

74 Song, B., Ganjeh, Y., Sadat, S. et al. (2015). Enhancement of near-field radiative heat transfer using polar dielectric thin films. *Nat. Nanotechnol.* 10 (3): 253–258.

75 St-Gelais, R., Zhu, L.X., Fan, S.H. et al. (2016). Near-field radiative heat transfer between parallel structures in the deep subwavelength regime. *Nat. Nanotechnol.* 11 (6): 515–519.

76 DeSutter, J., Tang, L., and Francoeur, M. (2019). A near-field radiative heat transfer device. *Nat. Nanotechnol.* 14 (8): 751–755.

5

Opto-Thermal Coupled Modeling

This chapter describes the opto-thermal-coupled modeling in light-emitting diode (LED) packaging. First, we discuss the opto-thermal interactions in the LED chip, which is induced by the thermal droop. And then, we review the photo-electro-thermal (PET) theory considering the thermal droop for LED chip array. Second, we give discussions and modeling of the opto-thermal interaction in the light converting material, which is usually phosphor in typical white LED/LD (laser diode) application. Third, by considering the heat source in both chip and phosphor, we present a modified bidirectional thermal resistance model to estimate the junction and phosphor temperature for phosphor-converted light-emitting diode (pc-LED). Finally, we give a summary of the opto-thermal interactions in the LED packaging.

5.1 Opto-Thermal Modeling in Chips

5.1.1 Thermal Droop

Once free electrons and holes are injected into the LED active region, intensive recombination processes occur, which includes radiative recombination and non-radiative recombination. As discussed in Chapter 1, the physical mechanisms of the non-radiative recombination mainly includes the Shockley–Read–Hall (SRH) recombination and the Auger recombination [1]. Clearly, the normal functioning of LEDs relies on radiative recombination scheme as it results in photon emission (i.e. producing light), while the latter type only releases its energy as phonons to the crystal lattice (i.e. producing heat) [2].

Light emission intensity from an LED is a result of competition among different recombination rates for a given injection current. Recombination rates of different types are then directly related to the carrier concentration in the LED active region, which can be described by a simple but widely used model (commonly known as "ABC" model) [3, 4]. According to this model, if assuming both the electron and hole concentrations in the active region are n, then SRH recombination rate is $k_{\mathrm{SRH}}n$, radiative recombination rate is $k_{\mathrm{rad}}n^2$, and Auger recombination rate is $k_{\mathrm{auger}}n^3$ as they are essentially one-, two-, and three-carrier recombination processes, respectively, where k_{SRH} is the SRH recombination coefficient, k_{rad} is the radiative recombination coefficient, and k_{auger} is the Auger recombination coefficient. Since other non-radiative recombination mechanisms such as surface recombination are minimized, and leakage current escaping from the quantum well region is negligible for well-designed LED structures, the LED injection current density J can be expressed as [2]

$$J = e\delta_{\mathrm{LED}}\left(k_{\mathrm{SRH}}n + k_{\mathrm{rad}}n^2 + k_{\mathrm{auger}}n^3\right) \tag{5.1}$$

where δ_{LED} is the thickness of LED active region, and e is the electron charge. The carrier concentration can be expressed as

$$n^2 = \mathrm{BE_c BE_V}\exp\left(-\frac{E_g}{k_B T_j}\right)\exp\left(\frac{eV}{k_B T_j}\right) = n_i^2\exp\left(\frac{eV}{k_B T_j}\right) \tag{5.2}$$

where $\mathrm{BE_c}$ is the effective density of states at conduction band edge, $\mathrm{BE_V}$ is the effective density of states at valence band edge, k_B is the Boltzmann constant, and T_j is the junction temperature. It can be seen that the carrier concentration is negative exponential related to the junction temperature.

The external quantum efficiency (EQE) η_{EQ} is a key parameter to characterize LED performance, which is defined as a ratio between emitted number of photons and total injected number of carriers. It can be obtained by multiplying the internal quantum efficiency (IQE) η_{IQ}, light extraction efficiency (LEE) η_{ex}, and electrical injection efficiency η_{inj}:

$$\eta_{EQ} = \eta_{IQ} \times \eta_{ex} \times \eta_{inj} \tag{5.3}$$

η_{IQ} can be simply expressed by the carrier concentration in LED active region through the ABC model introduced earlier

$$\eta_{IQ} = \frac{k_{rad}n^2}{k_{SRH}n + k_{rad}n^2 + k_{auger}n^3} \tag{5.4}$$

η_{ex} is defined as the ratio of photons finally escaping the LED to all photons radiated from the active region, which mainly depends on the external packaging structure. η_{inj} mainly counts the leakage current tunneling through the LED active region without undergoing recombination, which is usually taken as small and thus omitted from the expression of η_{EQ} [2].

For a given chip structure, the EQE is mainly dependent on the injected current and temperature. Figure 5.1(a) shows the EQE of a commercial high-power LED as a function of driving current measured at several ambient temperatures [5]. The EQE peaks at a low forward current and then drops with increasing current. This phenomenon is commonly referred to as current droop at high current density. Several explanations have been proposed for the causes of current droop, including electron leakage due to polarization mismatch and poor hole injection caused by asymmetry of carrier-transport properties, delocalization of carriers, density-activated defect recombination, and Auger recombination [3, 5–9].

In addition, like most semiconductor electronics, GaN-based LEDs suffer from thermal degradation at a given current density, i.e. the IQE of the device reduces at elevated temperatures. This phenomenon is known as thermal droop [10–13]. Increasing temperature from 300 to 450 K results in the reduction of the EQE by about 30% of its peak value, indicating that the thermal droop can be more detrimental than the current droop [14]. The mechanism behind the thermal droop has been revealed in the reported literature [10, 14]. At lower current density, the SRH recombination plays a relatively large role in the total recombination and the reduce is related to the increased SRH recombination [10]. At higher current density, the increased carrier leakage rate with rising temperature is responsible for the thermal droop [14].

Figure 5.1(b) shows a typical normalized light output power as a function of temperature at several operating points [5]. It can be seen that at a given current, when the temperature increases, the optical power keeps decreasing as a result of reduced EQE. And this inevitably produces excessive heat to the device junction, leading to temperature rise. The overheating of LED further reduces its IQE and, thus, the output power by the thermal droop. In order to maintain the output power of LED, driving current density has to be increased and, hence, the IQE will drop further. This vicious cycle exactly

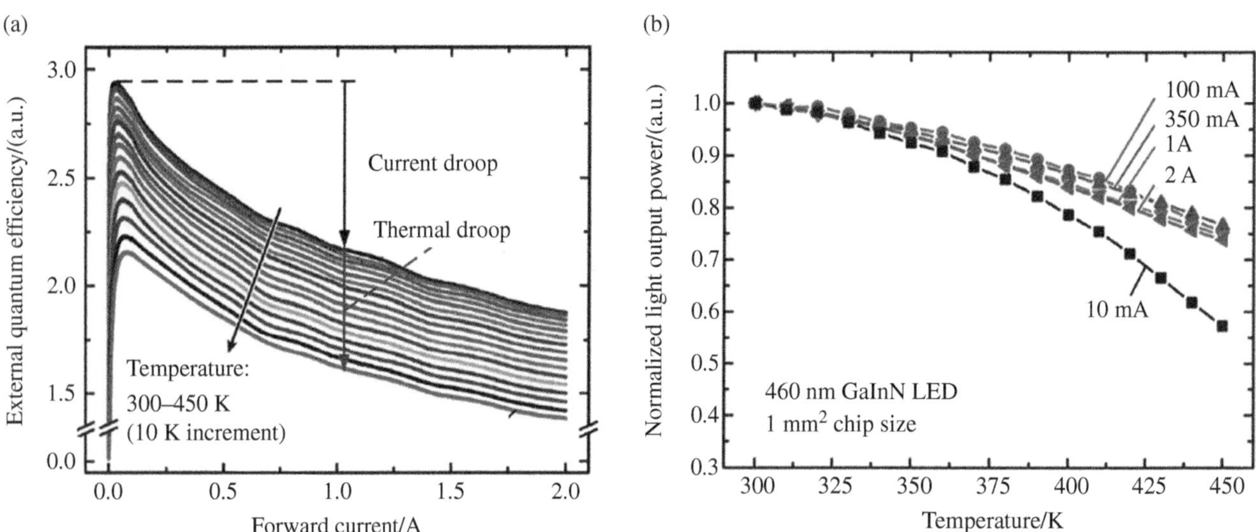

Figure 5.1 (a) The external quantum efficiency of a commercial high-power LED as a function of driving current measured at several ambient temperatures; (b) normalized light output power as a function of temperature at several operating currents. Meyaard et al. [5]/with permission of AIP Publishing.

illustrates the opto-thermal interaction in LED chip [2]. It is obvious that the interaction between the reduced IQE and increased junction temperature can lead to a final failure of the LED chip. In this case, good thermal management of LED is vital to eliminate the thermal droop and thus maintain high performance under high current.

5.1.2 Opto-Electro-Thermal Theory for LED

For LED chip, the photometric parameters such as luminous flux and luminous efficacy, electrical parameters such as electric power, current, and voltage of an LED, and thermal parameters such as junction and heat sink temperature and thermal resistance are concerned and closely linked together. In Refs. [15, 16], the relationship between the luminous output and thermal behavior has been reported. Reference [17] highlights the highly nonlinear thermal behavior of the junction-to-case thermal resistance of LED with electric power consumption of LED. The junction-to-case thermal resistance is affected by many factors, such as the mounting and cooling methods [18, 19], the size of the heat sink, and even the orientation of the heat sink [18]. Thus, analysis on the junction thermal resistance [18, 20, 21] and thermal management [22, 23] have been major LED research topics. By considering all these factors together, Hui et al. have proposed a general PET theory for LED system considering the thermal droop effect [24–26]. This theory consists of the following steps. First, simple procedures for optical and electrical power measurements are done to derive the temperature and current dependence of wall-plug efficiency, which connects the electrical and optical metrics. Next, a simple thermal resistance model is established to connect the optical and thermal metrics. By combining them, the electrical, optical, and thermal can be linked together. In the following, the PET theory will be illustrated.

We first focus on the radiometric metric for LED, i.e. the optical power P_{opt}. When the LED is working, driven by a constant input electrical power P_{ele}, the heat generation power P_{heat} within the LED can be calculated by:

$$P_{ph} = P_{in} - P_{out} \tag{5.5}$$

The wall-plug efficiency η_{wp} can be defined as the ratio of P_{opt} to P_{ele}:

$$\eta_{wp} = \frac{P_{opt}}{P_{ele}} \tag{5.6}$$

Similarly, the heat-dissipation coefficient k_{heat} can be defined as:

$$k_{heat} = \frac{P_{heat}}{P_{ele}} = 1 - \eta_{wp} \tag{5.7}$$

Based on the measured P_{opt} and P_{ele} at different junction temperatures T_j and current, as shown in Figure 5.2, the following relationship can be derived by assuming that η_{wp} is linear related to T_j and second order related to current:

$$\eta_{wp}(T_j, P_{ele}) = \frac{(a_1 T_j + a_2)(a_3 P_{ele}^2 + a_4 P_{ele} + a_5)}{a_6} \tag{5.8}$$

where the coefficients can all be obtained from Figure 5.2. It should be noted the current is replaced with P_{ele}, considering P_{ele} is approximately linearly proportional to the injection current at constant junction temperature.

Based on Equations (5.6) and (5.8), the optical power can be obtained as:

$$P_{opt} = \frac{(a_1 T_j + a_2)(a_3 P_{ele}^3 + a_4 P_{ele}^2 + a_5 P_{ele})}{a_6} \tag{5.9}$$

For an LED array with N number of LED chips mounted on the same heat sink, the simplified thermal resistance model under steady condition can be derived as Figure 5.3.

The heat sink temperature T_{hs} and junction temperature T_j can be, respectively, expressed as:

$$T_{hs} = T_a + R_{hs}(N \cdot P_{heat}) = T_a + NR_{hs}P_{ele}k_{heat} = T_a + NR_{hs}P_{ele}\left(1 - \eta_{wp}\right) \tag{5.10}$$

$$T_j = T_{hs} + R_{jc}P_{heat} = T_{hs} + \left(R_{jc} + NR_{hs}\right)P_{ele}k_{heat} = T_a + \left(R_{jc} + NR_{hs}\right)P_{ele}\left(1 - \eta_{wp}\right) \tag{5.11}$$

where R_{hs} is the heat sink to ambient thermal resistance, R_{jc} is the junction-to-case thermal resistance, and T_a is the ambient temperature.

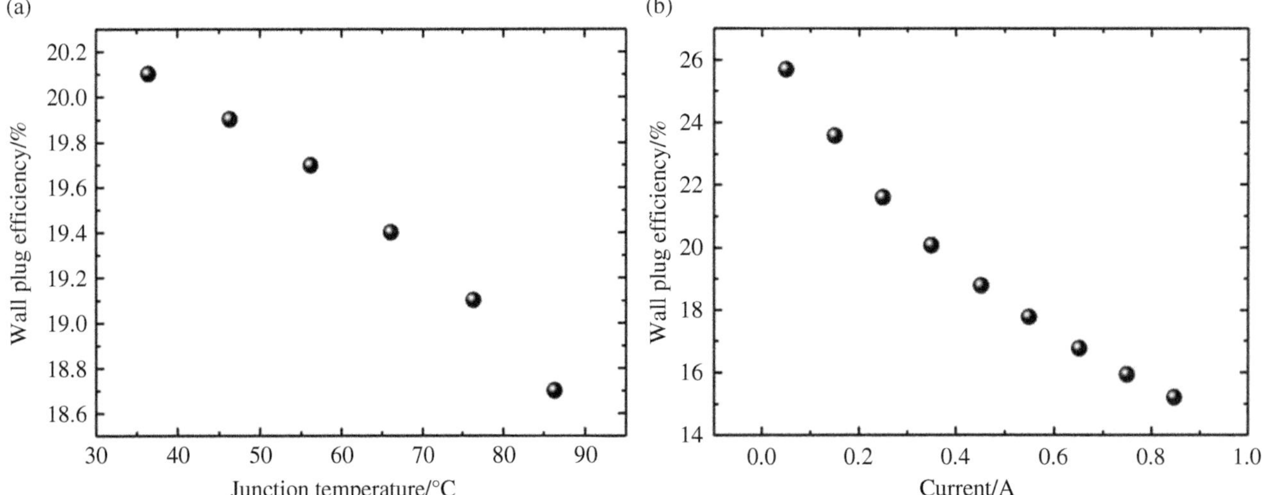

Figure 5.2 (a) Relation of wall-plug efficiency versus junction temperature under "constant electrical power" operation and (b) relation of wall-plug efficiency versus current under "constant junction temperature" operation. Chen et al. [24]/with permission of IEEE.

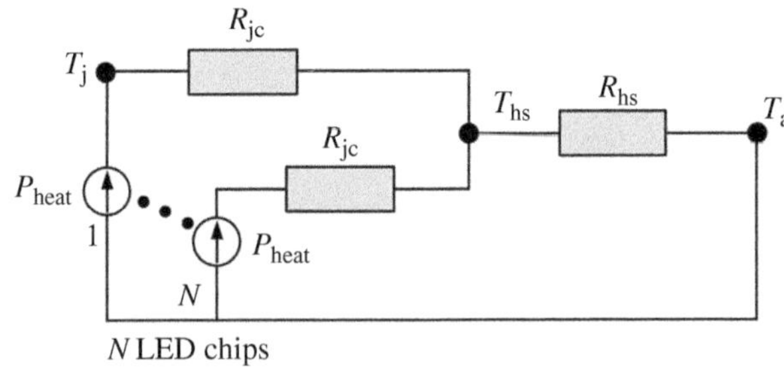

Figure 5.3 The simplified thermal resistance model of N LED chips mounted on the same heat sink. Chen et al. [24]/with permission of IEEE.

Then we focus on the photometric metrics, i.e. the luminous flux and luminous efficiency. The total luminous flux of an LED array consisting of N LED chips can be expressed as:

$$\Phi = N \times \eta_{lum} \times P_{ele} \tag{5.12}$$

where η_{lum} is the luminous efficiency and approximately decreasing linearly with junction temperature in the normal operating range (i.e. lower than 125 °C):

$$\eta_{lum} = \eta_0 \left[1 + k_{rs}\left(T_j - T_0\right)\right] \tag{5.13}$$

where η_0 is the rated efficiency at the rated temperature T_0 and k_{rs} is the relative slope. The related coefficients can be obtained by fitting the data curve provided in LED manufacturer data sheets.

Substituting Equations (5.8) and (5.11) into Equation (5.13) yields:

$$\eta_{lum} = \eta_0 \left\{ 1 + k_{rs}(T_a - T_0) + k_{rs}\left(R_{jc} + NR_{hs}\right)P_{ele}\left[1 - \frac{\left(a_1 T_j + a_2\right)\left(a_3 P_{ele}^2 + a_4 P_{ele} + a_5\right)}{a_6}\right]\right\} \tag{5.14}$$

The total luminous flux can be re-written as:

$$\Phi = N\eta_0 \left\{ \begin{array}{l} [1 + k_{rs}(T_a - T_0)]P_{ele} + \\ k_{rs}\left(R_{jc} + NR_{hs}\right)P_{ele}^2\left[1 - \dfrac{\left(a_1 T_j + a_2\right)\left(a_3 P_{ele}^2 + a_4 P_{ele} + a_5\right)}{a_6}\right] \end{array} \right\} \tag{5.15}$$

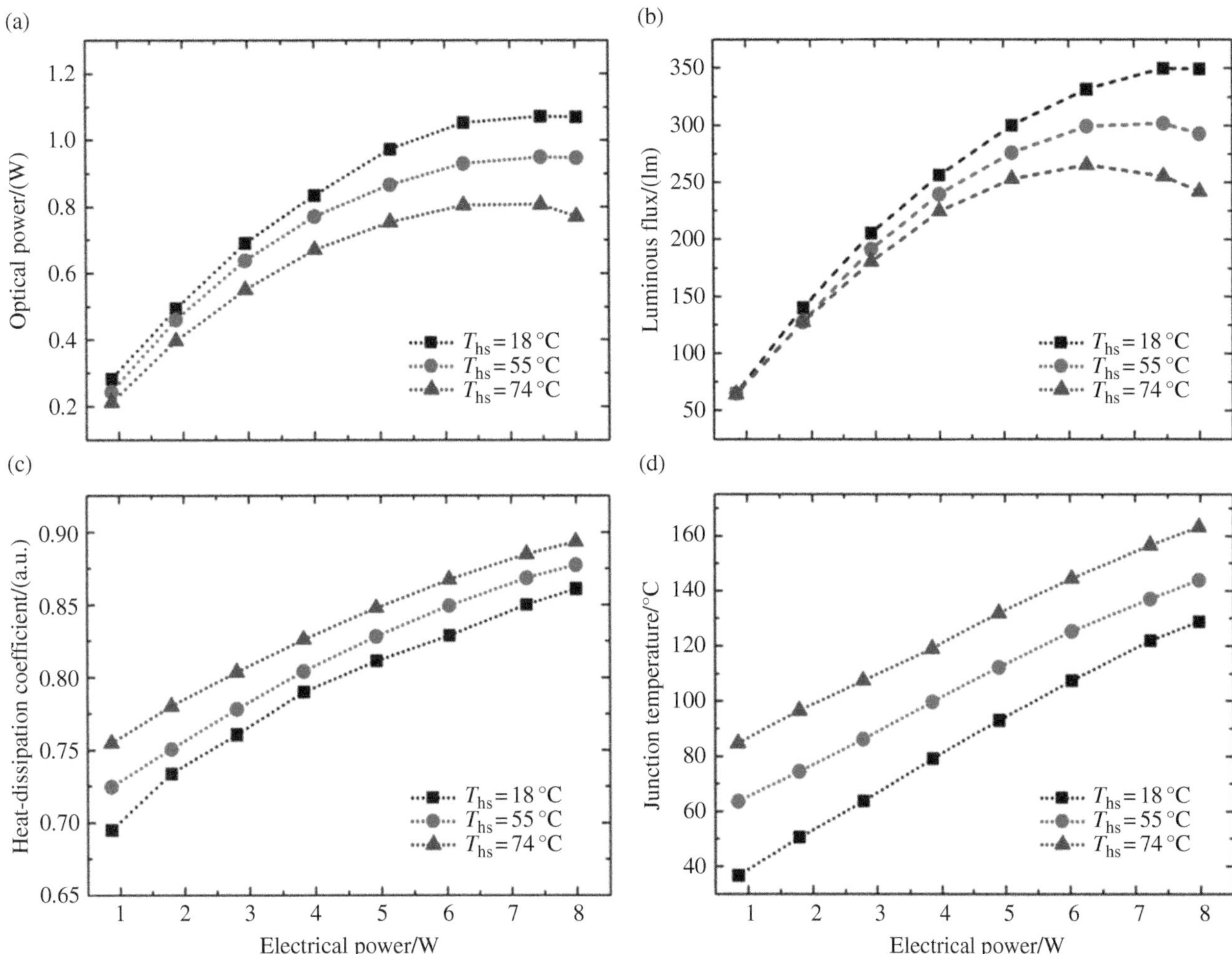

Figure 5.4 Calculated (a) optical power, (b) luminous flux, (c) heat-dissipation coefficient, and (d) junction temperature versus electrical power with different heat sink temperature. Chen et al. [24]/with permission of IEEE.

It can be seen that in Equations (5.9) and (5.15), the electrical metric (P_{ele}), optical metric (P_{opt} or Φ), and thermal metrics (T_j, R_{jc} and R_{hs}) are linked together.

Using the PET theory, Hui et al. have predicted the performances of a commercial Sharp 4.4 W LED, which is mounted on the heat sink with a thermal resistance of 7.8 K·W^{-1}. Figure 5.4 shows the calculated optical power, luminous flux, heat-dissipation coefficient, and junction temperature versus electrical power. It can be seen that the optical power and luminous flux rise almost linearly with electrical power when P_{ele} is small. As P_{ele} continues rising, the optical power and luminous flux reach a peak value and then drop. This can be explained by the constant increase of heat-dissipation coefficient and junction temperature with electrical power, resulting in the constant reduction of wall-plug efficiency and more heat generation. This thermal roll-over reflects the optical and thermal interaction and needs to be avoided through better thermal management.

5.2 Opto-Thermal Modeling in Phosphor

Light-converting material is a key part in white LED/LD applications. The most efficient LEDs or LDs emit light in the wavelength range of 440–460 nm. In this case, they are usually called blue LED/LDs. To cover the whole wavelength range of white light, a simple and efficient way is to combine the blue light with yellow light. Following this idea, researchers are using blue GaN/InGaN LED/LD to excite cerium-doped yttrium aluminum garnet (YAG:Ce) phosphor. Such a blue

LED/LD covered with a YAG:Ce phosphor layer is usually called as pc-LED or phosphor-converted laser diode (pc-LD). Besides LED/LD chip, phosphor is another vital part in white LED/LD packages. The particle diameter, geometry, concentration, and packaging form of the phosphor layer affect the overall optical and thermal performances. It should be pointed out that when the excitation optical energy density is very high (e.g. pc-LD), the local high phosphor temperature and even thermal quenching happen easily. In this case, there is an interaction between the optical and thermal performances, namely the opto-thermal interaction.

Therefore, in this part, we focus on the opto-thermal interaction in the phosphor. First, we present the phosphor heating phenomenon caused by the light-to-heat converting effect. Next, to analyze the finding, we present an optical model to calculate the heat generation power within the phosphor layer. In the following, based on the optical model, we further consider the optical–thermal-coupled effect, which is essentially the temperature-dependent phosphor quantum efficiency (QE), to analyze the phosphor thermal quenching phenomenon.

5.2.1 Phosphor Heating Phenomenon

For pc-LED, since only a part of the captured blue light is transferred into yellow light, there exists an optical energy loss during the color conversion process, and the lost optical energy is transformed into heat. Because of the low thermal conductivity of the encapsulant, local high temperature may appear in the phosphor layer, which would bring many disadvantages, such as peak shift of emitting spectrum, decrement of QE, and phosphor degradation [27, 28]. Some researchers paid attention to the phenomenon of phosphor self-heating. Hwang et al. [29] studied the effect of the relative position of the phosphor layer in the LED packaging on the lifetime of high power LED and found that the phosphor in the die contact case had a lower temperature than the remote phosphor case. Yan et al. [30] investigated thermal performance of phosphor-based white LEDs by simulation while considering the thermal and optical interaction. They found that the temperature of the phosphor is higher than the junction temperature of LED regardless of the phosphor placement. Hu et al. [31] studied the hotspot location shift in the high power phosphor-converted white LED packages by combining the Monte-Carlo optical simulation and finite-element simulation together. Both of them treated the phosphor layer, which is the mixture of phosphor particles and silicone, as a whole with uniform thermal conductivity obtained by the percolation theory [32], and gave the phosphor layer a volume thermal load. Arik et al. [33] studied the effects of localized heat generation caused by the color conversion in phosphor particles and different layers of high brightness LEDs. They found that there was a significant light output reduction because of the localized heating of the phosphor particles in the experiments and excessive temperatures appeared when a 3 mW heat was loaded on a 20 μm diameter spherical phosphor particle in the simulation.

Moreover, based on the comparison experiments, we validated that the silicone carbonizations observed in our experiments are caused by local high temperature, which results from the phosphor self-heating [34]. The LED modules used in our experiment are shown in Figure 5.5(a). Twelve LED chips are mounted on a substrate as an array of 2 × 6, in which six

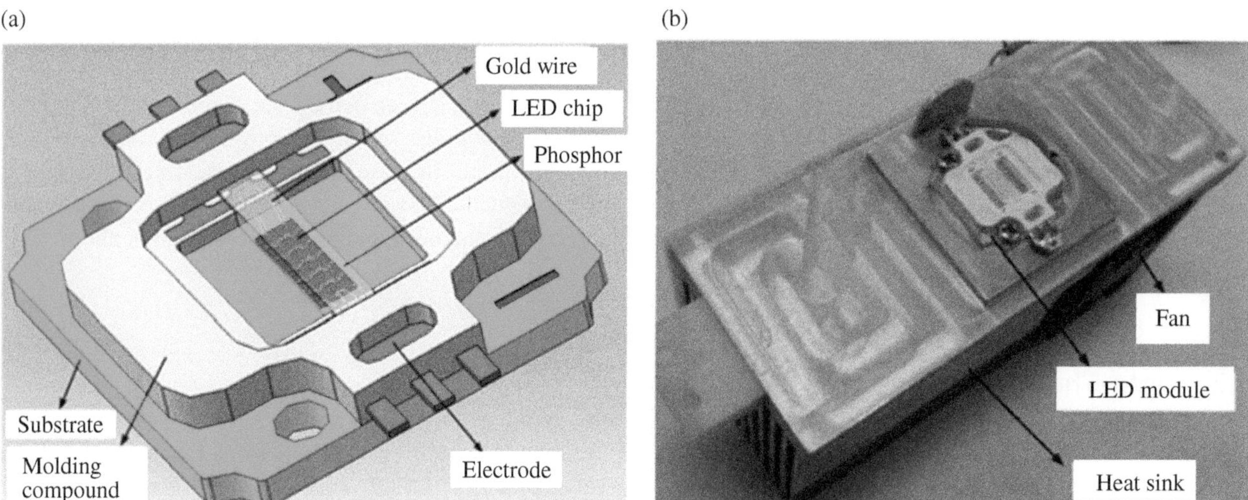

Figure 5.5 (a) Schematic of the multiple LEDs module. (b) Cooling system of the LEDs module.

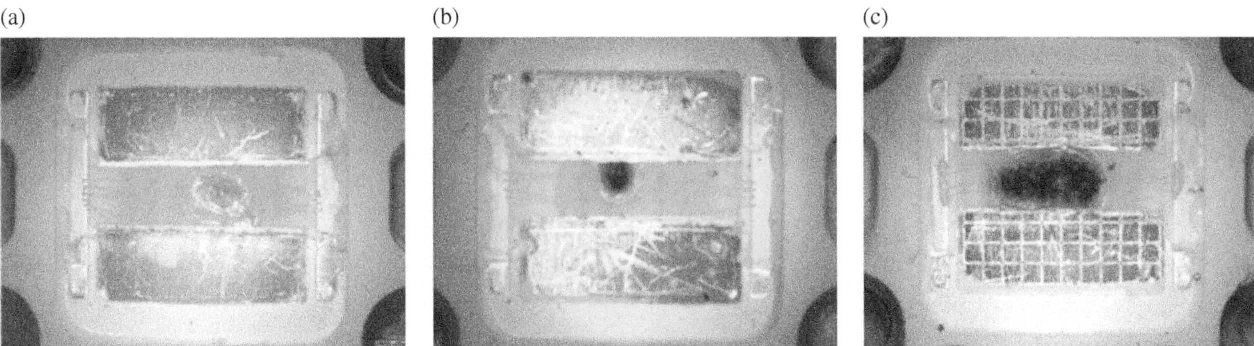

Figure 5.6 (a–c) Silicone carbonized modules. Luo et al. [34]/with permission of Elsevier.

LED chips are electronically connected in series and two rows of six chips are parallelly connected. The drive current of the module is 1.4 A and the total voltage is about 22 V. Therefore, the drive current for each LED chip is 700 mA, and the electric power for one chip is about 2.57 W. The phosphor layer, which is a mixture of silicone and $YAG:Ce^{3+}$ phosphor particles, is conformally coated on the chip. The height of the phosphor layer is 0.8 mm and the concentration of the phosphor layer is 0.087 g/g. The cooling system of the multiple LEDs module is indicated in Figure 5.5(b). The module is mounted on a heat sink, which is cooled by a fan. The ambient temperature is 19.0 °C, and the average temperature of the four corners of the top surface of the substrate is 29.2 °C when the module is light up. Based on tested temperature of the substrate, the dimensions, and the materials of the substrate and solder, it can be inferred that the maximum junction temperature of the LEDs is less than 80 °C based on the basic thermal resistance model [35–37]. However, it was found that a black spot appeared on the phosphor layer after the module lighted up for a while, and then the silicone around the black spot was carbonized. Figure 5.6 shows three phosphor carbonized modules, whose carbonization degrees are different because of different lighting time. For module (a) shown in Figure 5.6(a), the phosphor layer is just beginning to soften and make transformation, but a small black point appears in module (b). In module (c), most of the silicone is carbonized. In addition, it is noted that the carbonization point usually first appears in the middle part of the phosphor layer.

The phosphor used in our experiment is extremely thermally stable, it keeps stable even when it is sintered in the oven with a temperature of 1000 °C. As to the silicone, the thermal performance is totally different. Experiments were carried out to measure the carbonization temperature. Small cured pure silicone cuboid, as shown in Figure 5.7(a), was put in the oven with temperature ranging from 300 to 550 °C, and the temperature was maintained for a quarter of an hour with each increment of 50 °C. It was observed the transparent silicone cuboid turned to yellow and cracks appeared at the temperature of 450 °C, as indicated in Figure 5.7(b). The silicone cuboid was totally carbonized at the temperature of 500 °C, as shown in Figure 5.7(c). It was found the silicone cuboid started to carbonize at the temperature of 480 °C, so we can know the carbonization temperature of the present silicone is around 480 °C. To sum up the above experiments, the carbonization temperature of the silicone used is about 480 °C, the junction temperatures of the LEDs are inferred to be less than 80 °C. So the

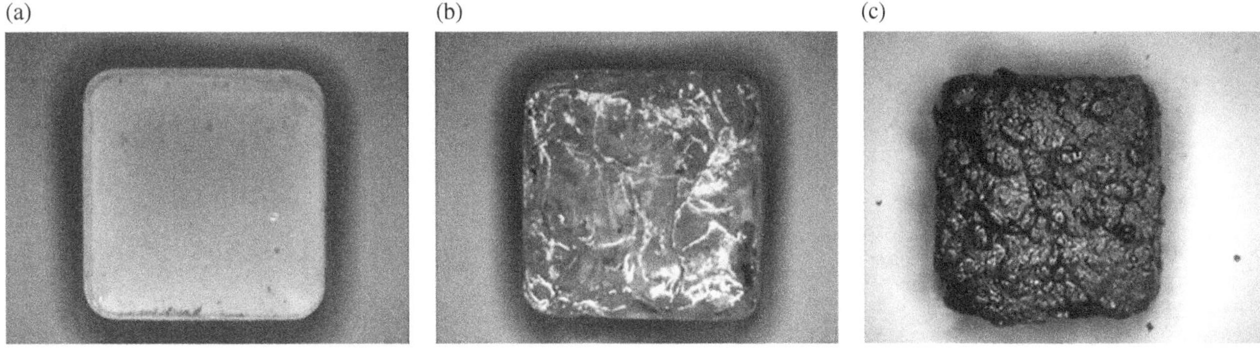

Figure 5.7 Images of cured pure silicone cuboid in different temperature: (a) ambient temperature, (b) 450 °C, and (c) 500 °C. Luo et al. [34]/with permission of Elsevier.

(a)

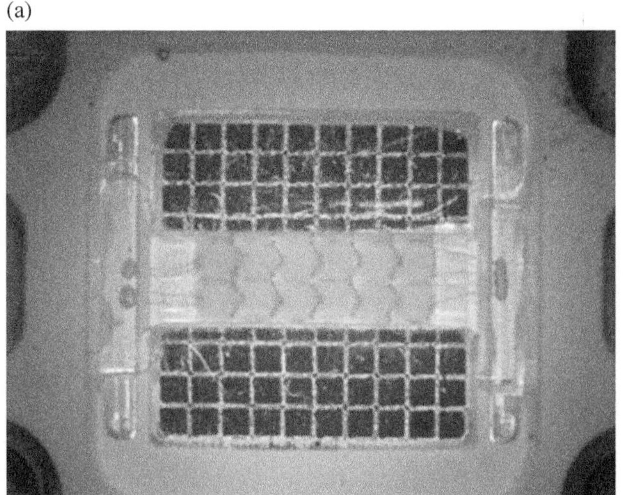

(b)

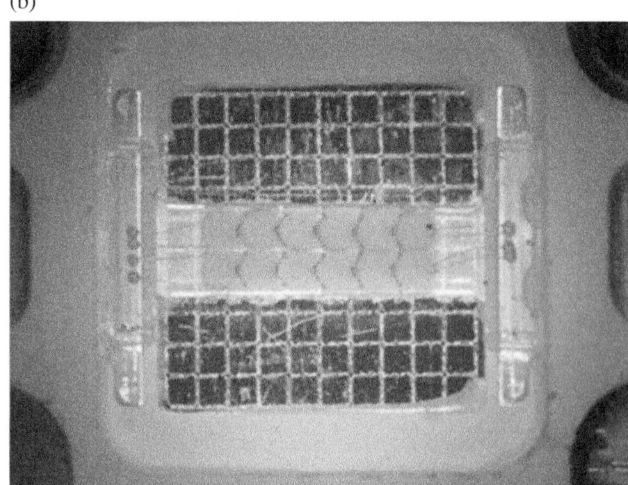

Figure 5.8 (a, b) Comparative modules at the same input conditions.

carbonization phenomenon could not be caused by the high temperature of the LED chip. The other heat source in the LED packaging is the phosphor particles, we may think that phosphor self-heating results in the silicone carbonization. To validate the supposition, comparative experiment was conducted. In the experiment, a pure silicone layer replaced the phosphor layer while all of the other parameters remained the same with the modules shown in Figure 5.6. The results are shown in Figure 5.8; it was found that no black spot appeared and no silicone was carbonized in the module with pure silicone layer, which proves that the phosphor carbonization is caused by the local high temperature due to the phosphor self-heating.

5.2.2 Phosphor Optical Model

Phosphor, as a wavelength converter, has been widely applied in the lighting industry, such as fluorescent lamps, pc-LEDs, and pc-LDs. Among them, pc-LEDs have almost dominated the lighting market with the advantages of energy saving, long lifetime, and high luminous efficiency [38]. It has been reported that the phosphor parameters and layer structures play an important role in determining the overall performances of pc-LEDs, e.g. luminous efficiency, color rendering index, and correlated color temperature (CCT) [39–41].

Light propagating within the phosphor layer in pc-LED exhibits simultaneous absorption, scattering, and fluorescence characteristics [42]. It is hard to observe the internal light propagation properties by experiments. Researchers have turned to developing theoretical and numerical methods to model the phosphor, including the Monte-Carlo ray-tracing simulations [39, 43–45], the diffusion-approximation (DA) method [46, 47], and a simplified one-dimensional model based on an extended Kubelka–Munk (KM) theory for fluorescence [48–54]. Monte-Carlo simulations can model the phosphor with high accuracy, but it often takes plenty of computational time and resources because usually a million rays need to be traced to get accurate results [55]. The DA method basically applies the first-order spherical harmonics approximation and it fails to model such as media with significant absorption characteristic [56]. The commonly used KM theory solves the forward and the backward scattering light intensity analytically by applying a two-flux approximation [48]. It has been a very fast and efficient method compared to the Monte-Carlo simulations. However, it also has some limitations: ① the scattering effect characterized by the scattering phase function is not taken into account, making it inapplicable when the scattering is strongly anisotropic; ② the Fresnel reflection at the interface caused by the refractive index mismatch is not considered; ③ only two fluxes travelling in the thickness direction are considered and the angular performance of the phosphor layer cannot be evaluated [57, 58]. Alternatively, an extended adding-doubling method for fluorescent materials has been proposed by Leyre et al. [59] to calculate the reflection and transmission characteristics of a combination of phosphor layers. In their method, the incident flux is divided into a number of channels and the lights propagating in the forward and backward directions are calculated for each channel. But the Fresnel reflection is also not taken into account. Hence, there is a need to develop a general and efficient model for strongly absorbing and strongly anisotropic scattering phosphor with accurate boundary conditions for actual applications in pc-LEDs.

In essence, these methods are all based on the radiative transfer equation (RTE). Light propagation in absorbing and scattering media has been extensively analyzed [60–65] and governed by the steady-state RTE. For fluorescing media, e.g. phosphor, the emission wavelength differs from the excitation wavelength. Fluorescence can be interpreted as inelastic volume scattering [66]. In theory, the elastic and inelastic scattering can be described by an extension of RTE integrated with fluorescence (FRTE), which has been used to model the biological tissues [66]. Therefore, it is feasible to apply the FRTE to model the phosphor.

It is known that the exact analytical solutions of the partial differential–integral equations of RTE only exist in very simple cases and many numerical solutions are usually applied, including the Monte-Carlo method [60], the finite-volume method [61], the finite-element method (FEM) [62], and the spectral element method [63, 64]. Among these methods, spectral element method possesses the advantages of both spectral and the FEMs and has been demonstrated to be a very efficient and accurate method to solve RTE [64]. Hence, it is a natural idea to extend this method to solve FRTE.

Hence, we presented a general optical model for a phosphor by solving FRTE using the spectral element method. The light absorption, anisotropic scattering, and fluorescence were considered. To model the actual phosphor layer in pc-LEDs, the Lambertian incident light intensity distribution for LED, the diffuse reflection at the substrate/reflector, and the Fresnel reflection at the interface between the phosphor and the surrounding air were taken into account. Through discretization in both the spatial domain and angular domain, the light intensity at an arbitrary location in any direction was calculated. Then, the LEE and angular CCT distributions were obtained, followed by experimental validations. The model presented was also compared with the previous KM theory.

For YAG:Ce phosphor, which absorbs the incident blue light and re-emits yellow light, the extended governing equations FRTEs can be expressed as follows [66]:

For blue light:

$$\boldsymbol{\Omega}\cdot\nabla I_{\mathrm{B}}(\boldsymbol{R},\boldsymbol{\Omega}) = -\left(k_{\mathrm{a}}^{\mathrm{B}} + k_{\mathrm{s}}^{\mathrm{B}}\right)I_{\mathrm{B}}(\boldsymbol{R},\boldsymbol{\Omega}) + \frac{k_{\mathrm{s}}^{\mathrm{B}}}{4\pi}\int_{4\pi} I_{\mathrm{B}}(\boldsymbol{R},\boldsymbol{\Omega})p(\boldsymbol{\Omega},\boldsymbol{\Omega}')\mathrm{d}\Omega' \tag{5.16}$$

For yellow light:

$$\begin{aligned}\boldsymbol{\Omega}\cdot\nabla I_{\mathrm{Y}}(\boldsymbol{R},\boldsymbol{\Omega}) = &-\left(k_{\mathrm{a}}^{\mathrm{Y}} + k_{\mathrm{s}}^{\mathrm{Y}}\right)I_{\mathrm{Y}}(\boldsymbol{R},\boldsymbol{\Omega}) + \frac{k_{\mathrm{s}}^{\mathrm{Y}}}{4\pi}\int_{4\pi} I_{\mathrm{Y}}(\boldsymbol{R},\boldsymbol{\Omega})p(\boldsymbol{\Omega},\boldsymbol{\Omega}')\mathrm{d}\Omega' \\ &+ \frac{\eta_{\mathrm{con}}}{4\pi}k_{\mathrm{a}}^{\mathrm{B}}\int_{4\pi} I_{\mathrm{B}}(\boldsymbol{R},\boldsymbol{\Omega})\mathrm{d}\Omega\end{aligned} \tag{5.17}$$

where $I(\boldsymbol{R}, \boldsymbol{\Omega})$ denotes the radiative intensity at spatial location $\boldsymbol{R}$ and direction $\boldsymbol{\Omega}$; k_a and k_s are the absorption and scattering coefficients, respectively; the scripts B and Y represent the blue and yellow lights, respectively. It should be mentioned that the spectral distribution is ignored, i.e. only the peak wavelengths for excitation and emission are considered. $p(\boldsymbol{\Omega}, \boldsymbol{\Omega}')$ is the scattering phase function, representing the probability of the energy scattering from direction $\boldsymbol{\Omega}'$ to differential solid angle $\mathrm{d}\Omega$ around direction $\boldsymbol{\Omega}$ [65]. For blue light in Equation (5.16), the expression of the left side denotes the gradient of the intensity in the specified direction $\boldsymbol{\Omega}$. The two terms on the right denote the intensity decrement due to absorption and outscattering and the intensity increment due to inscattering, respectively [65]. For yellow light in Equation (5.17), similar expressions can be interpreted by the same characteristics except for the contribution of the converted part from the absorbed blue light. This part is defined according to the steady-state rate equation [67] and characterized by the conversion efficiency η_{con} and the absorption coefficient for blue light. It should be noted that η_{con} accounts for the phosphor QE η_{qm} as well as the Stokes Shift loss, and it can be expressed as $\eta_{\mathrm{con}} = \eta_{\mathrm{qm}}\cdot\lambda_{\mathrm{Y}}/\lambda_{\mathrm{B}}$, where λ_{B} and λ_{Y} are wavelengths for excitation of blue light and emission of yellow light, respectively. It is assumed that the converted yellow light is emitted isotropically [66].

Figure 5.9 illustrates the schematic of a one-dimensional phosphor model and boundary conditions. It is assumed that the phosphor layer has an infinite length in the x- and y-directions and only the thickness (z-) direction is considered. Hence, the radiant intensity $I(\boldsymbol{R}, \boldsymbol{\Omega})$ for blue and yellow lights to be solved can be simplified as $I(z, \boldsymbol{\Omega})$ or $I(z, \theta, \varphi)$, in the units of W/sr, where $\theta (0 \le \theta \le \pi)$ and $\varphi (0 \le \varphi \le 2\pi)$ denote the polar and the azimuth angle of the solid angle $\boldsymbol{\Omega}$, respectively.

For the external LED light source (boundary 1), the typical Lambertian intensity distribution is applied, which can be written as:

$$I_{\mathrm{B,LED}}(0,\boldsymbol{\Omega}) = P_{\mathrm{in}}\cdot\cos\theta/\pi, \quad \boldsymbol{\Omega}\cdot\boldsymbol{n}_{\mathrm{w}} > 0 \tag{5.18}$$

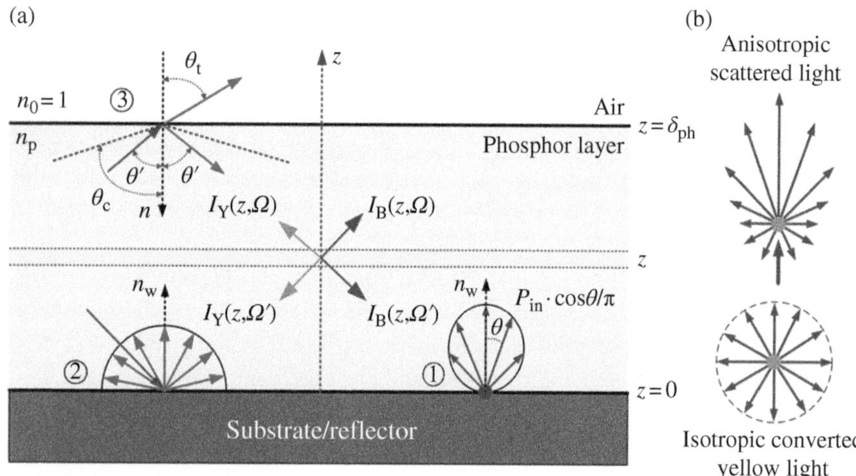

Figure 5.9 (a) The schematic of a one-dimensional phosphor model and boundary conditions. Conditions 1, 2, and 3 denote the incident LED Lambertian intensity, diffuse reflection at the substrate/reflector surface, and the Fresnel reflection at the phosphor–air interface, respectively. (b) The schematic of anisotropic scattered light and isotropic converted yellow light.

where P_{in} is the total output optical power from LED and θ denotes the polar angle of the solid angle $\boldsymbol{\Omega}$ with respect to the z-axis. Here, $\boldsymbol{n}_w$ is the unit inward normal vector at the substrate boundary ($z = 0$) and the incident light travels inward, which satisfies $\boldsymbol{\Omega}\cdot\boldsymbol{n}_w > 0$. It needs to be noted that the arbitrary incident intensity distribution with respect to θ and φ can also be easily applied.

When light hits the substrate or the reflector, part of light is absorbed and the other is reflected. For simplicity, the diffuse reflection for both blue and yellow lights at the substrate is applied (boundary 2) [49]:

$$I_{i,\mathrm{sub}}(0,\boldsymbol{\Omega}) = \frac{\gamma_i}{\pi}\int_{\boldsymbol{\Omega}'\cdot\boldsymbol{n}_w<0} I_{\mathrm{sub}}(0,\boldsymbol{\Omega}')\cdot|\boldsymbol{\Omega}'\cdot\boldsymbol{n}_w|\mathrm{d}\Omega',\quad \boldsymbol{\Omega}\cdot\boldsymbol{n}_w > 0,\quad i = B, Y \tag{5.19}$$

where γ_B and γ_Y are the diffuse reflectivity on the substrate surface for blue and yellow lights, respectively.

Considering the refractive index mismatch of the phosphor layer and the surrounding air, light will be specularly reflected back into the layer when attempting to travel out through the phosphor–air boundary [66, 68]. However, in previous models [49–54], the refractive indexes of the phosphor and the air are considered identical and equal to one. In this context, light will travel through the layer without deflection and the obtained angular intensity distribution may be misleading. In order to model the accurate condition, by assuming the phosphor–air interface to be a smooth and optically flat surface, the partly reflecting boundary for blue and yellow lights is applied at the phosphor–air interface (boundary 3) [68]:

$$I_{i,\text{ph-air}}(\delta_{ph},\boldsymbol{\Omega}) = \gamma(\boldsymbol{\Omega}'\cdot\boldsymbol{n})I_{i,\text{ph-air}}(\delta_{ph},\boldsymbol{\Omega}'),\quad \boldsymbol{\Omega}\cdot\boldsymbol{n} > 0,\quad i = B, Y \tag{5.20}$$

where δ_{ph} is the thickness of the phosphor plate, $\boldsymbol{n}$ is the unit inward normal vector at the phosphor–air interface, and $\boldsymbol{\Omega}$ is the specular reflection of $\boldsymbol{\Omega}'$, which points outward ($\boldsymbol{\Omega}'\cdot\boldsymbol{n} < 0$) and satisfies $\boldsymbol{\Omega}' = \boldsymbol{\Omega} - 2(\boldsymbol{\Omega}\cdot\boldsymbol{n})\boldsymbol{n}$. The reflectivity $\gamma(\boldsymbol{\Omega}'\cdot\boldsymbol{n})$ in Equation (5.20) is given by [68]:

$$\gamma(\cos\theta') = \begin{cases} \dfrac{1}{2}\left[\dfrac{\sin^2(\theta'-\theta_t)}{\sin^2(\theta'+\theta_t)} + \dfrac{\tan^2(\theta'-\theta_t)}{\tan^2(\theta'+\theta_t)}\right] & \text{for } \theta' \le \theta_c \\ 1 & \text{for } \theta' > \theta_c \end{cases} \tag{5.21}$$

The incident angle θ' and the refracted angle θ_t satisfy $\cos\theta' = \boldsymbol{\Omega}'\cdot\boldsymbol{n}$ and $n_p\cdot\sin\theta' = n_0\cdot\sin\theta_t$ (Snell's law), respectively. Here, n_p and n_0 are the refractive index of the phosphor layer and the air. The critical angle θ_c for total internal reflection ($R = 1$) is defined as $n_p\cdot\sin\theta_c = n_0$.

As a whole, the boundary conditions for both blue and yellow lights at input and output locations can be, respectively, given by:

$$I_B(0,\boldsymbol{\Omega}) = I_{B,\mathrm{LED}}(0,\theta) + I_{B,\mathrm{sub}}(0,\boldsymbol{\Omega}),\quad I_B(\delta_{ph},\boldsymbol{\Omega}) = I_{B,\mathrm{ph-air}}(\delta_{ph},\boldsymbol{\Omega}) \tag{5.22}$$

$$I_Y(0,\boldsymbol{\Omega}) = I_{Y,sub}(0,\boldsymbol{\Omega}), \quad I_Y(\delta_{ph},\boldsymbol{\Omega}) = I_{Y,ph-air}(\delta_{ph},\boldsymbol{\Omega}) \tag{5.23}$$

To solve the governing equations (5.16) and (5.17) together with the boundary conditions equations (5.18)–(5.23), the spectral element method based on the discrete-ordinates equation (DOE) is applied. The implementation of the spectral element method for solving RTE in an absorbing and scattering media has been extensively discussed in the previous literature [64]. A short summary of the spectral element method will be given and extended to solve FRTE.

Figure 5.10 shows the flowchart of solving FRTE using spectral element method. First, the input parameters appearing in the governing equations and boundary conditions need to be obtained. Then, the solution domain is discretized using spectral element approximation, and the solution nodes z_i and corresponding basis functions ξ_i can be generated using Chebyshev points and polynomials, respectively. Next, the total solid angle (4π sr) is divided uniformly in θ and φ directions and the corresponding weight ω for each angle is derived. This angle discretization method is termed the piecewise constant angular (PCA) approximation and the details can be found in [69]. In the following, the DOEs are built for one-dimensional FRTE described by Equations (5.16) and (5.17):

$$\mathrm{dir}^m \frac{dI_B^m}{dz} = -(k_a^B + k_s^B)I_B^m + \frac{k_s^B}{4\pi}\sum_{m'=1}^{M} I_B^{m'} p^{m',m}\omega^{m'} \tag{5.24}$$

$$\mathrm{dir}^m \frac{dI_Y^m}{dz} = -(k_a^Y + k_s^Y)I_Y^m + \frac{k_s^Y}{4\pi}\sum_{m'=1}^{M} I_Y^{m'} p^{m',m}\omega^{m'} + \frac{\eta_{con}}{4\pi}k_a^B\sum_{m=1}^{M} I_B^m\omega^m \tag{5.25}$$

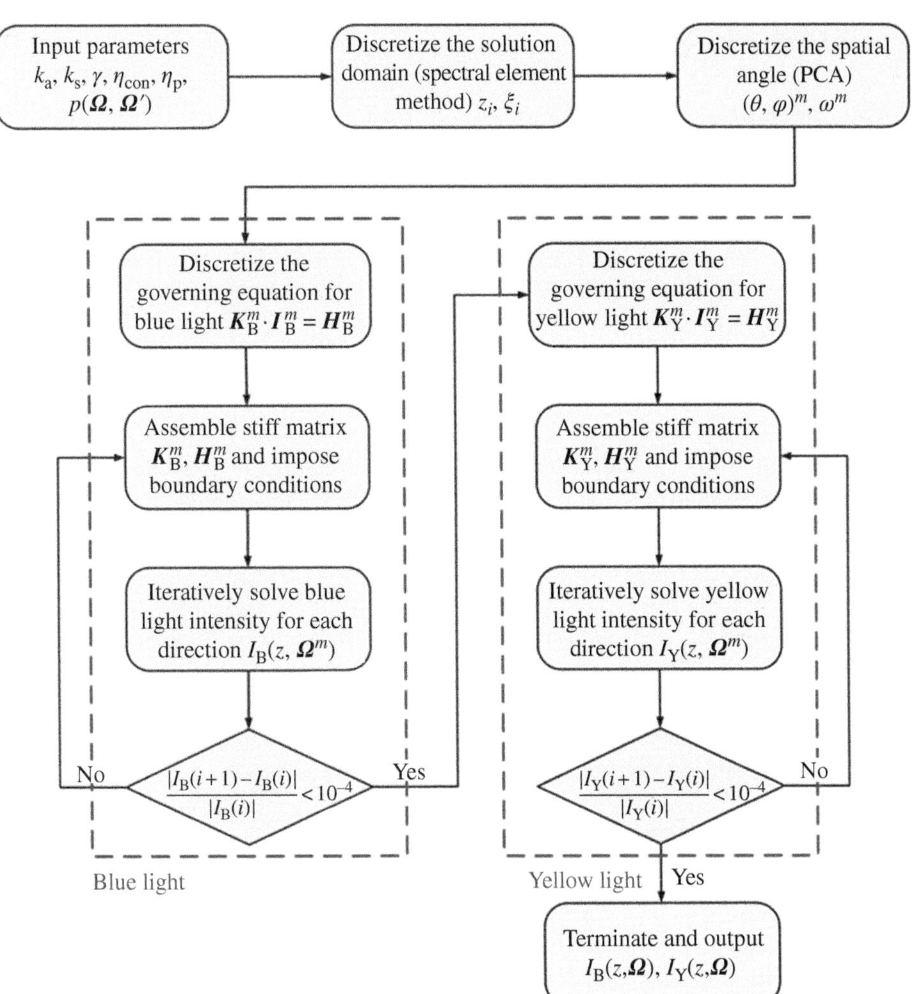

Figure 5.10 The flowchart of solving the fluorescent radiative transfer equations using the spectral element method.

where the superscript m denotes a specific angular direction $\boldsymbol{\Omega}^m$, M is the total angular number, dir^m is the direction cosine along the z direction, and ω^{m} is the angular weight corresponding to the incident direction $\boldsymbol{\Omega}^m$. For simplicity, the in-scattering terms, including the elastic and inelastic scattering, are treated implicitly as part of the source term. In this way, the system of partial differential–integral equations is transformed into partial differential equations [64]. By using global intensity approximation and integrating Equations (5.24) and (5.25) over the solution domain, the discretized system of linear equations is obtained as $\boldsymbol{K}^m \cdot \boldsymbol{I}^m = \boldsymbol{H}^m$. The global approximation intensity I_{g} can be expressed as:

$$I_{\mathrm{g}}(z) = \sum_{j=1}^{N_{\mathrm{sol}}} I_j \xi_j(z) \tag{5.26}$$

where I_j represents the radiative intensity at solution node j and N_{sol} is the total number of solution nodes. The detailed expression and assembling of the global stiffness matrix $\boldsymbol{K}^m$ and the vector $\boldsymbol{H}^m$ have been demonstrated in [64].

Then, begin to loop each angular direction for blue light and impose boundary conditions in Equation (5.22). The radiative intensity for blue light $I_{\mathrm{B}}(z, \boldsymbol{\Omega}^m)$ on each solution node z for each direction $\boldsymbol{\Omega}^m$ can be obtained by solving Equation (5.24). If the stop criterion (the maximum relative error of the radiative intensity below 10^{-4}) is satisfied, the iteration process terminates. Otherwise, go back to the loop. After solving the blue light intensity, the same steps can be repeated for yellow light $I_{\mathrm{Y}}(z, \boldsymbol{\Omega}^m)$ by solving Equation (5.25) with boundary conditions in Equation (5.23).

Finally, the optical performance of the phosphor can be extracted using the obtained radiative intensity of blue and yellow lights. The radiant flux for both blue and yellow lights can be derived as the integration of the total solid angle in each solution node [18]:

$$\Phi_i(z) = \int_{4\pi} I_i(z, \boldsymbol{\Omega}) \mathrm{d}\Omega = \sum_{m=1}^{M} I_i(z, \boldsymbol{\Omega}) \omega^m, \quad i = B, Y \tag{5.27}$$

When simulating a phosphor material, the optical performance metrics are always important. When the incident angle θ at the phosphor–air interface is below the critical angle θ_{c}, the light will transmit to the phosphor layer at the transmitted angle θ_{t} and the intensity is given by:

$$I_i(\delta_{\mathrm{ph}}, \theta_{\mathrm{t}}, \varphi) = [1 - \gamma(\cos\theta)] \cdot I_i(\delta_{\mathrm{ph}}, \theta, \varphi), \quad \theta < \theta_{\mathrm{c}}, \quad i = \mathrm{B}, \mathrm{Y} \tag{5.28}$$

In addition, the angular yellow to blue intensity ratio distribution can be applied to determine the angular CCT distribution [70]. The output optical power of blue and yellow lights can also be derived as:

$$P_{\mathrm{out},i} = \sum_{m=1}^{M'} I_i(\delta_{\mathrm{ph}}, \theta_{\mathrm{t}}^m) \varpi^m, \quad i = \mathrm{B}, \mathrm{Y} \tag{5.29}$$

where the total number of the transmitted angle is M' and the corresponding angular weight is ϖ obtained by the linear interpolation of the original weight ω. Similarly, the ratio of $P_{\mathrm{out,Y}}$ to $P_{\mathrm{out,B}}$ can also determine the average CCT. The LEE of the phosphor layer can be determined by the ratio of the total output power P_{out} to the input power P_{in}:

$$\mathrm{LEE} = P_{\mathrm{out}}/P_{\mathrm{in}} = (P_{\mathrm{out,B}} + P_{\mathrm{out,Y}})/P_{\mathrm{in}} \tag{5.30}$$

Moreover, based on the energy conservation law, the phosphor heating power P_{ph} can be further obtained by [71]:

$$P_{\mathrm{ph}} = P_{\mathrm{in}} - (P_{\mathrm{out,B}} + P_{\mathrm{out,Y}}) \tag{5.31}$$

Considering the light-to-heat conversion process, the heat generation density versus invasion depth $q(z)$ within the phosphor layer can be derived as the sum of the heat generated by the absorbed blue and yellow lights:

$$q(z) = k_{\mathrm{a}}^{\mathrm{B}} \cdot (1 - \eta_{\mathrm{con}}) \Phi_{\mathrm{B}}(z) + k_{\mathrm{a}}^{\mathrm{Y}} \Phi_{\mathrm{Y}}(z) \tag{5.32}$$

The obtained $q(z)$ can be coupled with the heat diffusion equation to further solve the temperature distribution of the phosphor layer.

To validate the one-dimensional model, a remote pc-LED module was fabricated. As shown in the inset in Figure 5.11, a commercial LED module bonded onto a substrate was assembled with a designed aluminum reflector with a reflectivity of 0.8 and then a planar phosphor plate was placed onto the top of the reflector. We fabricated two groups of phosphor samples with varying thicknesses of 0.6 and 1.0 mm, respectively. For each thickness, there are six different phosphor concentrations

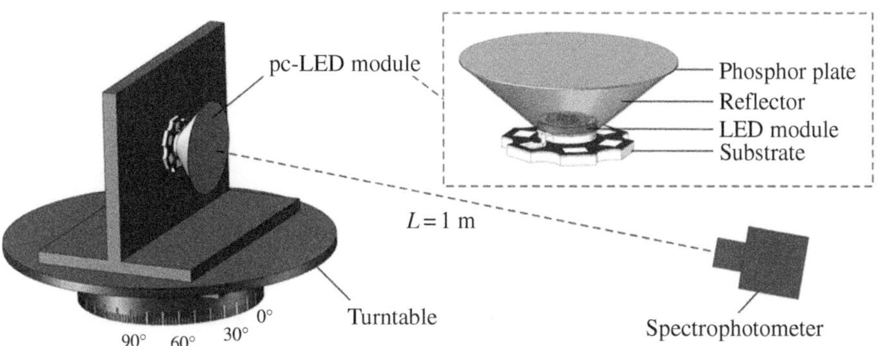

Figure 5.11 Schematic of the experimental setup for angular CCT distribution and the inset shows the remote pc-LED with planar phosphor plate.

n_c varying from 0.05 to 0.30 g·cm^{-3} with an interval of 0.05 g·cm^{-3}. The optical power and average CCT of the module without and with phosphor plate were measured by an integrating sphere (ATA-1000, Everfine Inc.). The corresponding output optical power under driving current of 350 mA was $P_{opt,1}$ and $P_{opt,2}$, respectively. The LEE could be derived as the ratio of the $P_{opt,2}$ to $P_{opt,1}$. It should be noted that each module was measured for 10 times at the same condition. The error was obtained by calculating the average value and standard deviation of these data. In addition, the angular CCT distribution was measured experimentally as depicted in Figure 5.11. The pc-LED module was fixed at a designed turntable, which can rotate from 0° to 360°. A spectrophotometer (XYC-I, Enci Co. Ltd.) was applied to measure the color coordinates and CCTs. Before test, the position of the spectrophotometer needs to be adjusted horizontally to the center of the LED. And then the spectrophotometer is fixed at the selected point, which was 1 m away from the module. During the test, the CCT distribution was measured by rotating the turntable from 0° to 90° with an increment of 5°.

To implement the model, it is necessary to obtain accurate estimates for the input parameters. The absorption and scattering coefficients of both blue and yellow lights are calculated by the Mie–Lorenz theory [72]. The phosphor QE η_{qm} is assumed to be 0.9 and the individual wavelengths of blue and yellow lights are measured to be 445 and 558 nm, respectively, corresponding to the conversion efficiency η_{con} of 0.718. The diffuse reflectivity γ at $z = 0$ for both blue and yellow lights is assumed to be equal to that of the used reflector. The refractive index n_p of the phosphor plate is set to be 1.53 [40]. In this work, we use the Henyey–Greenstein (HG) phase function [73], which has been widely applied for medical [74] and biological [75] applications as well as the phosphor [76]:

$$p(\cos\Theta) = \frac{1 - {k_{ap}}^2}{\left(1 + {k_{ap}}^2 - 2k_{ap}\cos\Theta\right)^{3/2}} \tag{5.33}$$

where Θ is the scattering angle which satisfies $\cos\Theta = \boldsymbol{\Omega}\cdot\boldsymbol{\Omega}'$. The anisotropy parameter k_{ap} is set to be 0.82 for YAG:Ce phosphor calculated by the Mie–Lorenz theory [72].

After the definition of parameters, the one-dimensional solution domain is discretized with N_{sol} Chebyshev–Gauss–Lobatto points. By using the PCA method, the total solid angle is discretized uniformly with N_θ and N_φ discrete angles in the polar and azimuth directions, respectively. The implementation of the model was conducted using the commercial software MATLAB using a 3.3-GHz Intel(R) Core(TM) i3-3220 processor. Figure 5.12 shows the effects of those three discrete numbers on computing time and the accuracy of the results. It should be noted that the accuracy was indicated by the change of the radiant flux at $z = 0$ and $z = \delta_{ph}$, i.e. $\Phi(0)$ and $\Phi(\delta_{ph})$, for blue and yellow lights. It can be easily understood that as the discrete number increases, the computing time keeps rising and the radiant flux tends to be stable. From the figure, we can see that the number of discrete angles poses a larger effect. As long as N_θ and N_φ are, respectively, greater than 20 and 30, the result can be regarded as accurate. Moreover, even when the total number of discrete numbers reaches 1200 ($N_\theta \times N_\varphi = 60 \times 20$), the computing time is only 45 seconds, which is more efficient than the Monto-Carlo ray-tracing simulations. In the following, we choose those three numbers as $N_{sol} = 40$, $N_\theta = 40$, and $N_\varphi = 40$. The calculated results will be presented and discussed.

Figure 5.13(a) and (b) shows the calculated angular intensity distribution at varying normalized invasion depths z/δ_{ph} of 0, 0.5, and 1.0 for blue and yellow lights, respectively. Considering the symmetry in the azimuth direction of the optical system, only the angular intensities in the polar direction are plotted. It should be noted that θ varies from 0° to 180° and the left part

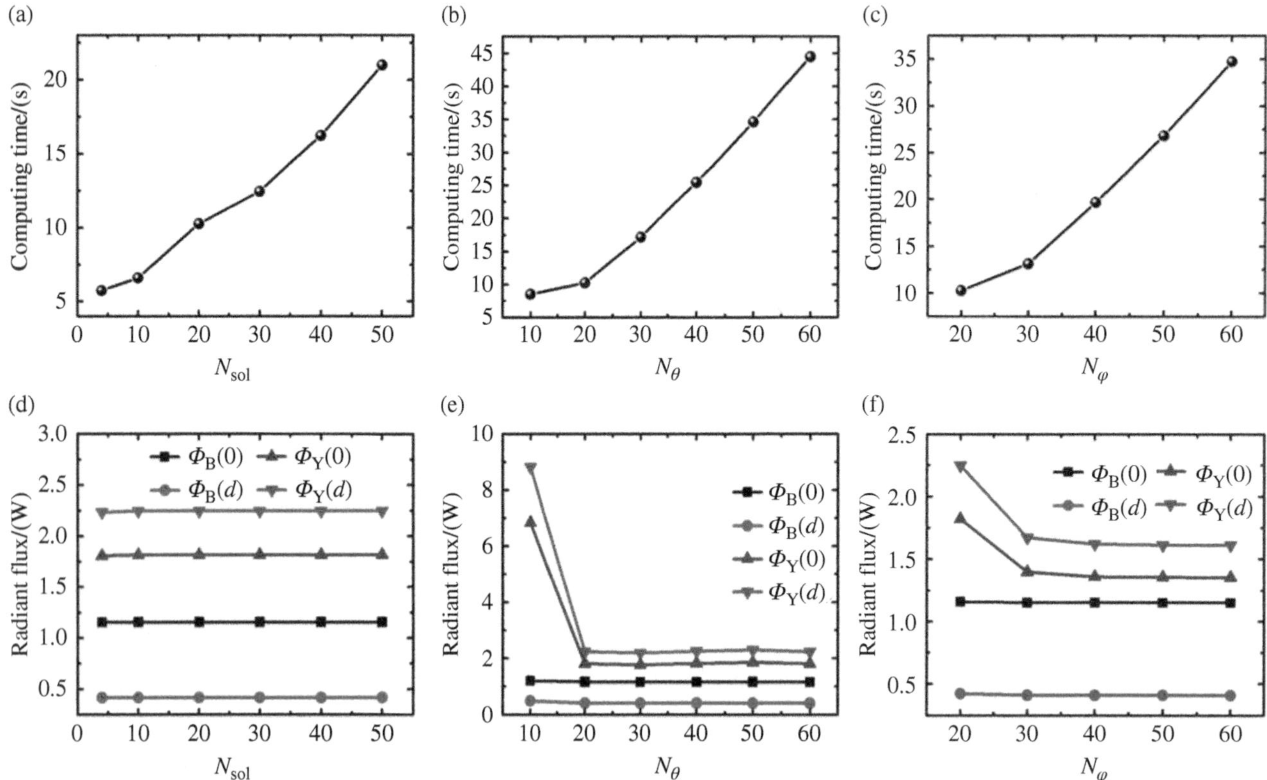

Figure 5.12 Upper: the effect of (a) N_{sol}, (b) N_θ, and (c) N_φ on the computing time. Bottom: the effect of (d) N_{sol}, (e) N_θ, and (f) N_φ on the radiant flux at $z = 0$ and $z = \delta_{ph}$ for both blue and yellow lights. The reference values for those three numbers are $N_{sol} = 20$, $N_\theta = 20$, and $N_\varphi = 20$, respectively.

of these two figures is just symmetric with the right part. The angle range from 0° to 360° is just labeled to match the usual cases.

For forward-scattering light ($\theta < 90°$), the yellow intensity distribution at the incident surface ($z/\delta_{ph} = 0$) is isotropic, representing the diffuse reflection boundary. However, for blue light, it is caused by the combination of Lambertian incidence and diffuse reflection. With the invasion depth increasing, the blue light intensity decreases in each direction in the similar shape. However, for the yellow light, intensity rises with depth at small angle range, but rises first then drops at high angle range (e.g. $\theta < 75°$). For backward-scattering light ($\theta > 90°$), the developing trends with depth are just the opposite with that of forward-scattering blue and yellow lights. It can be seen that there is a sudden intensity change for both lights. This behavior is related to the partially reflective boundary and can be explained by the plotted angular specular reflectivity at $z/\delta_{ph} = 1$ as shown in Figure 5.13(c). When the incident angle θ is below the critical angle ($\theta_c = 41°$ for $n_p = 1.53$), the reflectivity rises very fast as θ grows. Otherwise, the reflectivity equals to 1. In this case, the sudden change in backward-scattering intensity at $z/\delta_{ph} = 1$ can be interpreted by the product of the forward-scattering intensity and the angular reflectivity, with a corresponding critical angle of 139°. Figure 5.13(d) shows the output angular intensity of blue and yellow lights obtained by Equation (5.28). It can be seen that the blue and yellow lights have different intensity distributions. Moreover, the relative intensity distribution is used to determine the CCT distribution to compare with the experimental results in the following.

With the obtained angular intensities for blue and yellow lights in Figures 5.13(a) and (b), the radiant flux $\Phi(z)$ and the normalized phosphor heat generation density $q(z)/P_{in}$ can be further calculated using Equations (5.27) and (5.32), respectively. Figure 5.14 shows the calculated results. It can be seen that the radiant flux decreases and increases with invasion depth for blue and yellow lights, respectively. It is attributed to the constant absorption of blue light and conversion of yellow light as lights travel in the layer. As for the light-to-heat characteristic, we can see that $q(z)/P_{in}$ shows a monotonic decrease with invasion depth. Because the absorption of blue light accounts for the main part of the heat generation

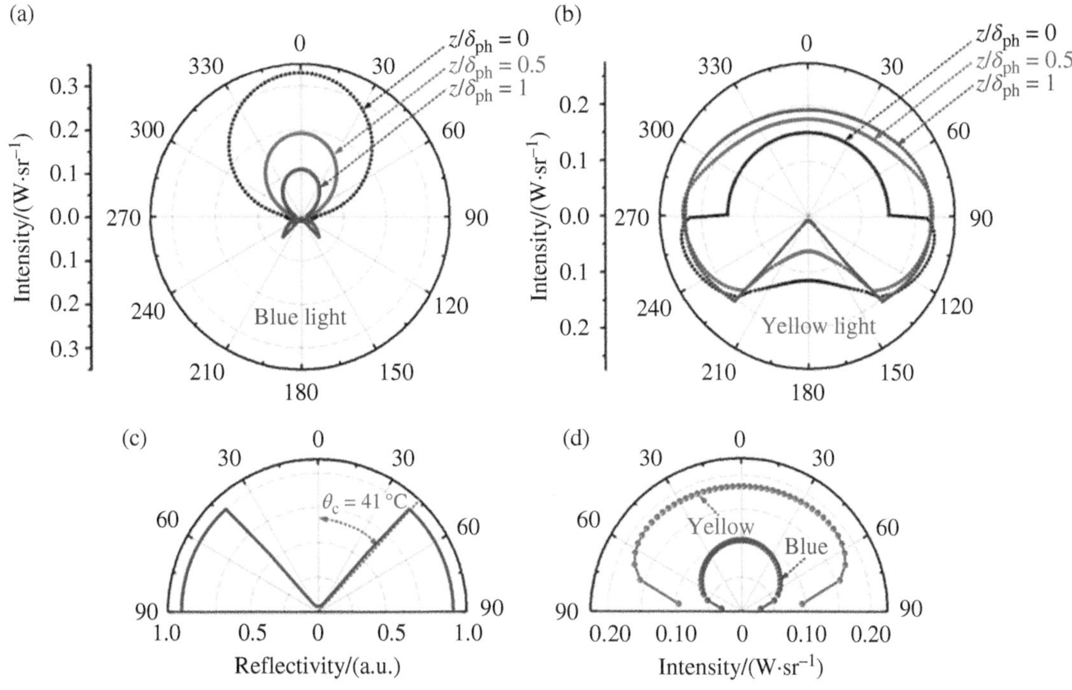

Figure 5.13 Angular intensity distribution under varying invasion depth for (a) blue and (b) yellow light, and (c) angular specular reflectivity distribution at the phosphor–air interface, (d) the output angular intensity of blue and yellow lights. The phosphor thickness and concentration used in this case are 0.6 mm and 0.15 g·cm^{-3}, respectively.

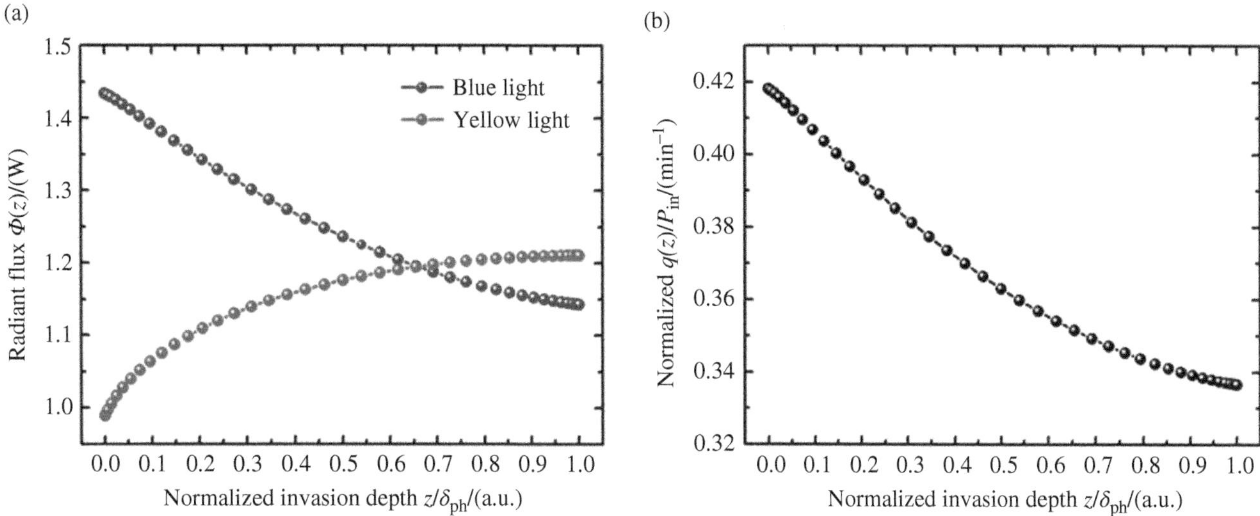

Figure 5.14 (a) The calculated radiant flux for blue and yellow lights and (b) the normalized phosphor heat generation density versus the normalized invasion depth.

(i.e. $k_a^B \gg k_a^Y$), $q(z)$ has a similar trend with $\Phi(z)$. This finding is consistent with our previous work [77]. It is worth noting that the temperature field of the phosphor can be obtained by inputting $q(z)$ into the heat diffusion equation. This coupled radiative and conductive heat transfer problems have been successfully resolved [78, 79].

Figure 5.15 shows the comparisons between the experimental results and the model predictions. It should be noted that the fabricated phosphor plate has the diameter of 30 mm and the thickness of 0.6 and 1.0 mm. In this context, the

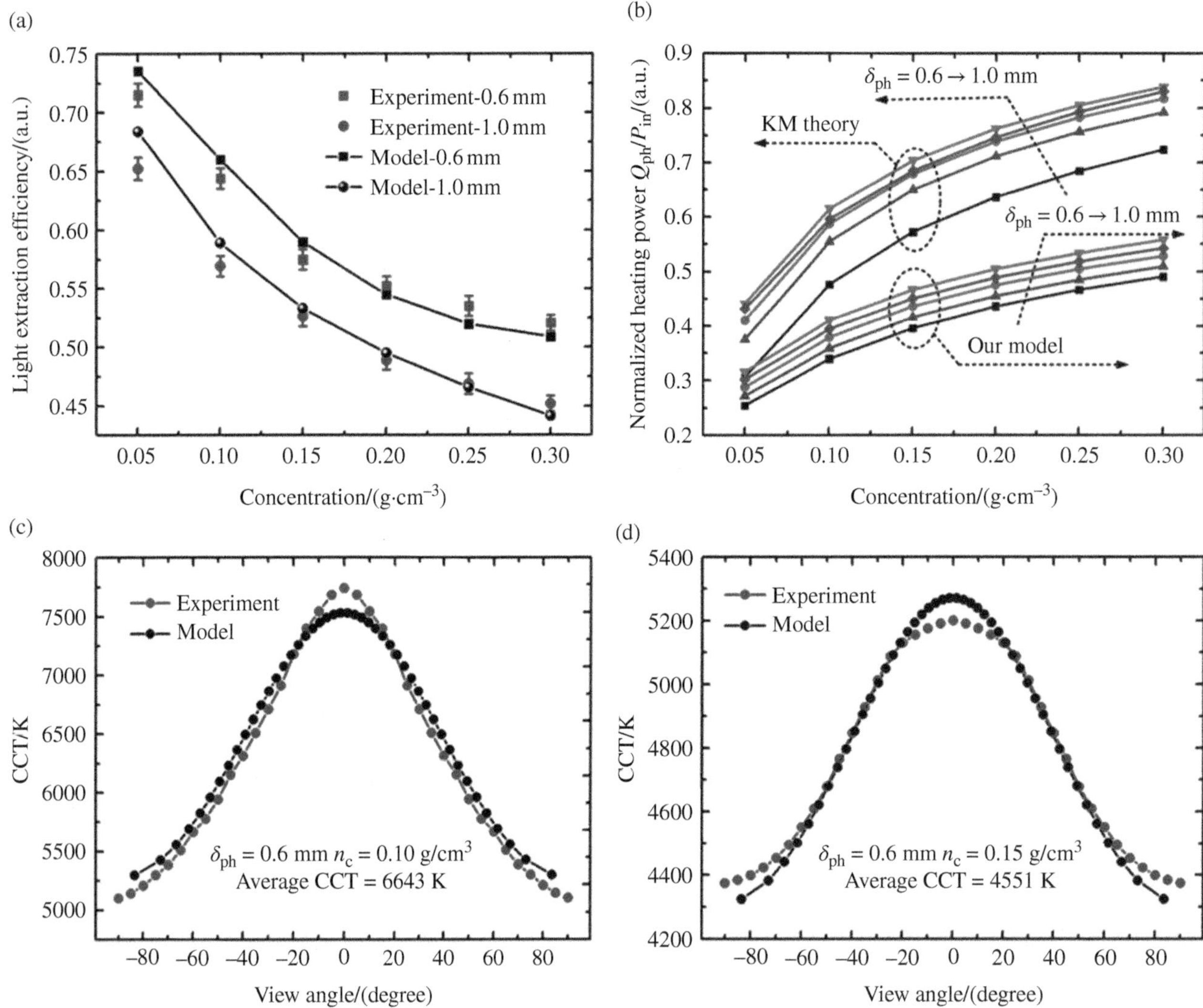

Figure 5.15 (a) Comparison of light extraction efficiency versus phosphor concentration between the experiment and the model for thicknesses of 0.60 and 1.0 mm; (b) comparison of the normalized phosphor heating power between KM and our model; comparison of angular CCT distribution between the experiment and the model for thicknesses of 0.6 mm and concentration of (c) 0.10 g·cm^{-3} and (d) 0.15 g·cm^{-3}.

one-dimensional assumption is feasible. Figure 5.15(a) plots the comparisons of LEE under phosphor plate thickness of 0.6 and 1.0 mm versus phosphor concentration varying from 0.05 to 0.30 g·cm^{-3} with an interval of 0.05 g·cm^{-3}. For both cases, LEE decreases with concentration due to more absorption of light and conversion to heat [54]. It can be seen that LEE also decreases with the thickness at each concentration. For comparison, the LEEs of experiment and model show good agreement in terms of developing trend and absolute value with the maximum deviation of 4.9%.

In addition, we also compare the normalized phosphor heating power P_{ph}/P_{in} calculated by our model and KM theory. As shown in Figure 5.15(b), the obtained P_{ph}/P_{in} by KM are obviously higher than that of our model. It is mainly because that the Lambertian incident light pattern of LED and blue light scattering effect inside the phosphor layer cannot be considered in the KM theory. In this case, the light energy in other directions may not contribute to the light extraction and be dissipated into heat. For further comparison, the angular performance is also compared between experiment and our model. Figure 5.15(c) and (d) plots the comparisons of the angular CCT distribution versus view angle under phosphor concentration of 0.10 and 0.15 g·cm^{-3}, with corresponding average CCTs of 6643 and 4551 K, respectively. As a whole, the calculated results agree well with the measured results for both cases, despite the deviation in the low view angle range (0°–15°) and high view angle range (70°–90°) with the maximum deviation of 3.7%. The main sources of the deviations may lie in that the input parameters, including the absorption, scattering coefficient, and the phase function, do not match perfect with the real case. Moreover, the smooth surface assumption may also cause deviation. As a whole, our model is accurate enough to model the light propagation within the phosphor.

5.2.3 Optical–Thermal Phosphor Model Considering Thermal Quenching

In the previous phosphor optical model, the phosphor QE is assumed to be a constant. In actual, the QE will start to decrease with rising phosphor temperature when reaching a critical temperature. This is also known as phosphor thermal quenching. For pc-LED, the optical power density is not so large due to the Lambertian pattern and the phosphor temperature is usually below the critical temperature. In this case, the constant QE assumption holds and thus the previous optical model is also feasible. However, when the phosphor is excited by a focused spot even with relatively low power, the local optical power density and phosphor temperature can be very high, leading to a reduction of QE. Laser excited remote phosphor (LERP) exactly belongs to this case.

The state-of-the-art LEDs are known to suffer from the "efficiency droop," i.e. a decrease of QE at high operating current density [80, 81]. In contrast, LDs can achieve higher efficiency at high current density, because the Auger recombination no longer grows after the threshold current [82, 83]. Moreover, LDs also exhibit other excellent characteristics, including directional beam pattern and small light-emitting area, enabling the capability of high-luminance and collimated lighting [83, 84]. Similar to white LEDs, pc-LDs gain more attention with their advantages of high efficiency, low cost, and compact size [85–87]. LERP has been commonly used in pc-LD packaging [88, 89].

In LERP, light emitted from the LD chip is usually focused onto a phosphor layer, and the luminance is usually much higher than that of conventional white LEDs [90]. Consequently, the phosphor temperature will be much higher than LEDs due to the extremely higher radiant power density from LDs. High phosphor temperature will result in the severe thermal quenching problem, which will decrease the efficiency, deteriorate the reliability, and shorten the lifetime of LERP [47]. Although thermal quenching has been regarded as a significant obstacle to the development of high-luminance pc-LDs, there are quite few efficient/accurate tools/methods for evaluation. Either the phosphor temperature or the heat flux generated by the phosphors is quite hard to measure in the experiments.

Monte-Carlo ray-tracing simulations together with FEM have been widely used to evaluate the optical and thermal performances of pc-LEDs [91–93]. In the most methods used for phosphor modeling in pc-LEDs, the optical and thermal effects are independent of each other and this may not lead to misunderstanding because phosphor temperature is relatively low and the thermal quenching effect is not severe. But for pc-LDs, the thermal quenching is too significant to be ignored. In general, the temperature dependence of phosphor QE was usually not considered, making it impossible to evaluate thermal quenching [45]. Actually, light scattering, absorption, conversion, and thermal quenching interact with one another, making it difficult and complicated for the numerical simulation. Moreover, the QE has complex dependencies on temperature and it is hard to establish the exact relationship between them. Recently, to tackle this problem, Correia et al. proposed a method to mesh the phosphor layer using tetrahedral element discretization and stored the optical and thermal flux. Despite its complexity in meshing, this method proved to be an effective way to characterize the overall performance of pc-LED/LDs [45]. Alternatively, Lenef et al. used a DA radiation transport model to calculate optical effects and then coupled with FEM to study the thermal effects of pc-LDs [46, 47]. In the previous part, we have established a phosphor optical model to analyze the phosphor heating effects in pc-LEDs [94]. We also built the thermal resistance model to predict the junction temperature of LEDs with high accuracy [71, 95]. Can we apply these models to evaluate the thermal quenching directly? The answer may be NO because ① the phosphor scattering model only considers the light-to-heat conversion part with a constant phosphor QE and ② the thermal resistance model only considers the heat-dissipation part. Actually, the essence of thermal quenching is the temperature dependence of phosphor QE. An intuitive but feasible way is to combine our previous two models together with considering the temperature-dependent phosphor QE simultaneously.

Therefore, we attempted to develop an optical–thermal coupling model to study phosphor quenching effects on optical/thermal performance of LERP. The interacted optical and thermal effects were coupled by introducing the temperature dependence of phosphor QE. In this way, the existing phosphor model could be extended to evaluate phosphor thermal quenching effects under extremely high radiant power density of LDs. Optical and thermal experiments were conducted to verify the model. Based on this model, we systematically studied the effects of various factors on critical radiant power against thermal quenching. Finally, practical guidelines were provided to enhance radiant limit for high-reliability LERPs.

Figure 5.16 illustrates the schematic of the optical–thermal model for LERP. A typical reflective LERP package consists of an LD chip, phosphor layer, mirror layer, bonding layer, and heat sink [46, 47]. The blue light and converted yellow light will

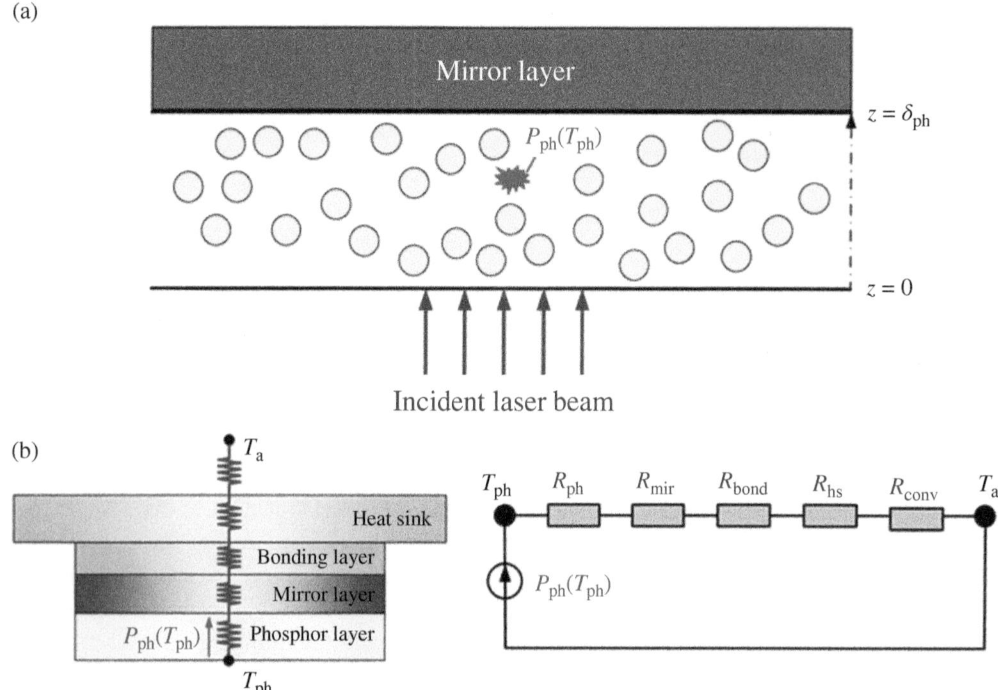

Figure 5.16 Schematic of the optical–thermal model for LERP comprising (a) phosphor optical model and (b) thermal resistance model.

be reflected on the mirror surface, and the output white light is in the opposite direction of incident light. Along with the light conversion and mixing process, there is also light-to-heat conversion known as phosphor heating [34]. Due to the relatively low thermal conductivity of phosphor–silicone mixture (~0.2 W·m^{-1}·K^{-1}), the heat generated within the phosphor layer may not be dissipated efficiently, resulting in local high phosphor temperature and thermal quenching problem. As shown in Figure 5.16, the present model consists of two sub-models, i.e. (a) phosphor optical model and (b) steady-state thermal resistance model, and they are connected through the interaction between phosphor heating power P_{ph} and phosphor temperature T_{ph}.

The first sub-model, i.e. phosphor optical model, has been established in the previous part. The corresponding boundary conditions are applied as specular reflection in the phosphor–mirror interface, Fresnel reflection in the incident phosphor surface, and collimated laser incidence. The radiance of blue and yellow lights along thickness in any angular direction can be obtained by solving RTE together with the boundary conditions. And then, the total heat generation power of the phosphor layer P_{ph} can be calculated using Equation (5.32).

After calculating P_{ph}, steady-state thermal resistance model is applied to calculate phosphor temperature T_{ph}. For reflective remote phosphor, the highest concentration of converted light was found to be very close to the incident surface, i.e. $z = 0$ [46]. Therefore, phosphor temperature node is assumed to be located at $z = 0$. It should be noted that the heat flow path from the incident surface to the ambient is ignored because the natural convective heat transfer is very weak with a relatively small heat transfer coefficient and area. In this case, the total thermal resistance between T_{ph} and the ambient temperature T_a is composed of a series thermal resistance, as shown in Figure 5.16(b), which can be expressed as:

$$R_{tot} = R_{ph} + R_{mir} + R_{bond} + R_{hs} + R_{conv} \tag{5.34}$$

where R_{ph}, R_{mir}, R_{bond}, and R_{hs} are conductive thermal resistance of phosphor layer, mirror layer, bonding layer, and heat sink, respectively, which can be calculated accordingly:

$$R_j = \frac{\delta_j}{\lambda_j A_j}, (j = \text{ph, mir, bond, and hs}) \tag{5.35}$$

and R_{conv} denotes the convective thermal resistance between heat sink and the ambient:

$$R_{conv} = \frac{1}{hA_{conv}} \tag{5.36}$$

However, the above thermal resistances are not enough without considering thermal spreading resistance. Figure 5.17 shows a circular phosphor plate with diameter of D_{ph} excited by a circular pump spot with diameter of D_{spot}. To achieve high-luminance lighting, D_{spot} is usually very small, resulting in a large ratio of D_{ph} to D_{spot}. In this case, the thermal spreading resistance from pump spot to the plate $R_{s,ph}$ plays a main role in the R_{tot} and needs to be included. $R_{s,ph}$ can be calculated based on the analytical solutions for an isotropic disk with circular heat source [95, 96] and the general form is:

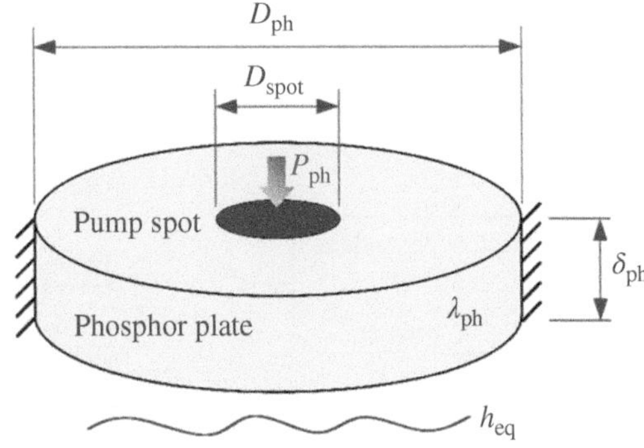

Figure 5.17 Schematic of thermal spreading resistance from spot to a circular phosphor plate.

$$R_{s,ph} = \frac{4}{\pi\varepsilon\lambda_{ph}D_{spot}}\sum_{n=1}^{\infty} A_n(n,\varepsilon)B_n(n,\xi)\frac{J_1(\delta_n\varepsilon)}{\delta_n\varepsilon} \tag{5.37}$$

where

$$A_n = -\frac{2\varepsilon J_1(\delta_n\varepsilon)}{\delta_n^2 J_0^2(\delta_n)}, \quad B_n = -\frac{\delta_n + B_i\tanh(\delta_n\xi)}{\delta_n\tanh(\delta_n\xi) + B_i} \tag{5.38}$$

More details can be found in Refs. [95, 96]. Similarly, the thermal spreading resistance from the bonding layer to the heat sink $R_{s,hs}$ can also be calculated using Equation (5.38). It should be noted that Equation (5.37) is also applicable for the rectangular plate after transforming the rectangular one into a circular one [95]. Finally, T_{ph} can be calculated by:

$$T_{ph} = T_a + P_{ph}\left(R_{s,ph} + R_{ph} + R_{mir} + R_{bond} + R_{s,hs} + R_{hs} + R_{conv}\right) \tag{5.39}$$

After defining the phosphor scattering and thermal resistance model, the optical–thermal coupling model is constructed by further introducing the temperature dependence of phosphor QE $\eta(T_{ph})$, which can be calculated as:

$$\eta(T_{ph}) = \frac{\tau_{nr}(T_{ph})}{\tau_{nr}(T_{ph}) + \tau_r} \tag{5.40}$$

The temperature dependence of $\tau_{nr}(T_{ph})$ is calculated by Equation (5.41) [97, 98]:

$$\tau_{nr}(T_{ph}) = \frac{1}{W_0}\exp\left(\frac{E_0}{k_B T_{ph}}\right) \tag{5.41}$$

For YAG:Ce, the fitted values for W_0 (4×10^{13} s^{-1}) and E_0 (6500 cm^{-1}) are all obtained according to the literature [97]. τ_r is assumed to be a constant due to its weaker temperature dependence than τ_{nr}. We can obtain a τ_r of 67 ns for YAG:Ce from the fitting of the reported experimental curve of lifetime [97, 99]. Figure 5.18 shows the flowchart of the optical–thermal model. When inputting the initial T_{ph} (e.g. $T_{ph} = T_a$), we can obtain η by Equation (5.40), P_{ph} by Equation (5.32), and, finally, an updated T_{ph} by Equation (5.39) in sequence. If the updated T_{ph} does not change significantly between consecutive steps (e.g. $T_{ph}(i+1) - T_{ph}(i) < 0.01$ K), the whole framework ends and output the ultimate $T_{ph}(i+1)$. Otherwise, the iterative calculations will be continued until thermal equilibrium is reached. Through this iteration, both accurate phosphor heating and temperature can be obtained. The whole calculations were done by coding using the commercial software MATLAB.

Optical and thermal measurements were performed to verify the calculated P_{out} and T_{ph}, respectively. Figure 5.19 shows the schematic of the (a) optical and (b) thermal test setups. A commercial LD (L450P1600MM, Thorlabs) was mounted onto a heat sink to lower the junction temperature and thus enable a stable output. The LD was driven by a current controller (LDC220C, Thorlabs). A pair of adjustable collimated and focused lenses was used to obtain a collimated laser spot with a diameter of about 1 mm. A circular phosphor plate (powder phosphor embedded in silicone matrix) was glued onto an aluminum mirror with a reflectivity of 95%. Finally, the mirror was bonded with a designed phosphor plate-finned heat sink by thermal grease.

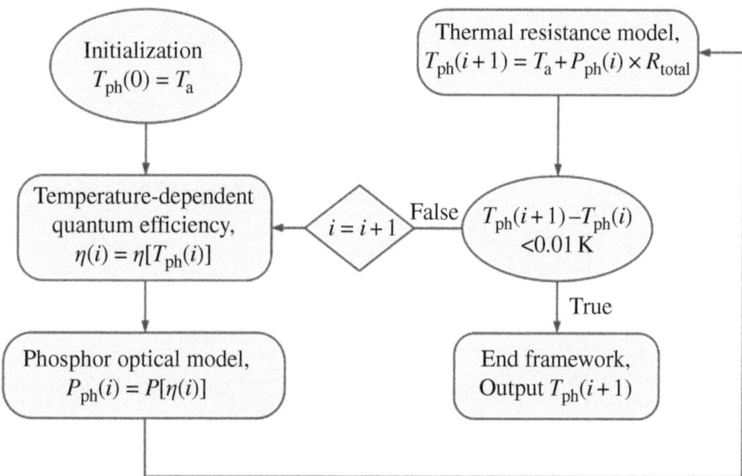

Figure 5.18 Flowchart of the optical–thermal model considering thermal quenching effects.

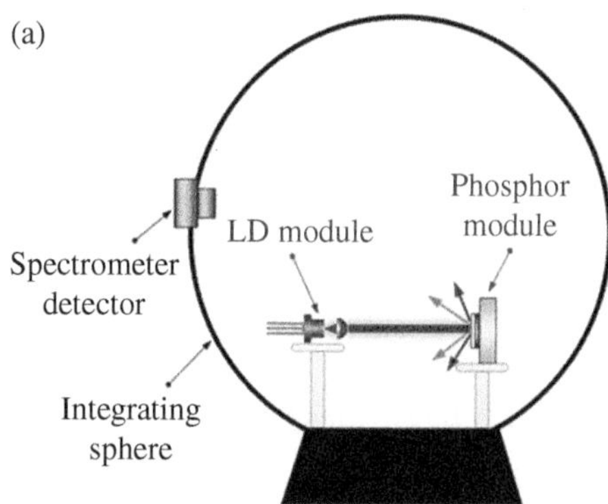

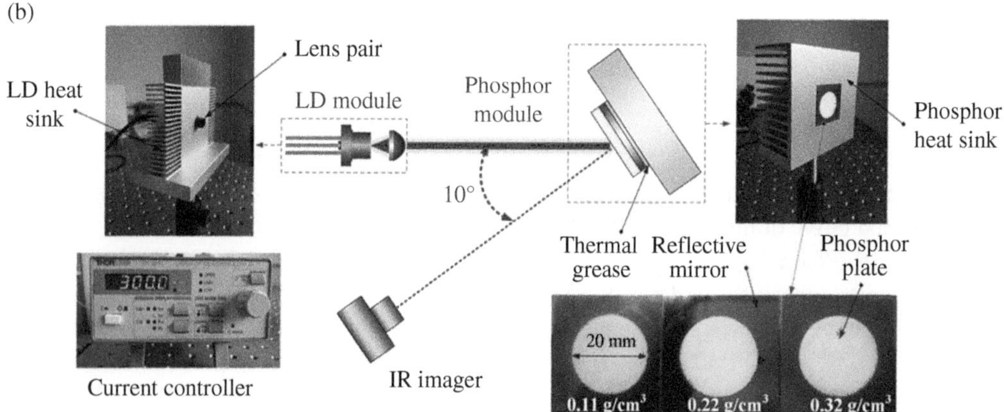

Figure 5.19 The schematic of the (a) optical and (b) thermal test setup. Ma et al. [54] / with permission of Elsevier.

For optical test, as shown in Figure 5.19(a), the whole setup was placed in an integrating sphere (ATA-1000, Everfine) and the output optical power P_{out} was measured. For thermal test, as shown in Figure 5.19(b), an infrared (IR) thermal imager (SC620, FLIR) was used to capture the surface temperature of the phosphor plate. Before the test, the emissivity of phosphor surface was calibrated. In the calibration process, a thermocouple was attached tightly onto the surface of phosphor layer by

Table 5.1 The thermal conductivities and geometric parameters of the components.

Components	Materials	Thermal conductivity ($W \cdot m^{-1} \cdot K^{-1}$)	Thickness (mm)	Dimensions (mm)
Phosphor plate	Phosphor/silicone	0.17	0.32	$D = 20$
Mirror	Al 7075-T6	130	0.45	$W \times L = 30 \times 30$
Thermal grease	—	0.67	0.1	$W \times L = 30 \times 30$
Heat sink	Al 6061-T6	167	—	$W \times L = 80 \times 80$

conductive adhesive tape. Then the phosphor layer was heated evenly by a heating plate. When it reached thermal steady state, the surface temperature was measured to be 54.1 °C. Then adjust the emissivity until the temperature of phosphor surface reached the same value. In this way, the calibrated emissivity of phosphor surface was 0.94. It should be noted that the laser beams were projected onto the plate in a small deflection angle (~10°) so that the IR could be placed normal to the phosphor target. To validate the change of P_{out} and T_{ph} with varying factors, several easy-to-implement variations were conducted. Driving current varying from 0 to 0.80 A with an interval of 0.05 A was used. Then three different phosphor plates were made with varying phosphor concentrations of 0.11, 0.22, and 0.32 $g \cdot cm^{-3}$, respectively. The thermal conductivities and geometric parameters of the components involved in the model are listed in Table 5.1.

For validation, P_{out} and T_{ph} need to be calculated using the presented model. Besides the parameters listed in Table 5.1, the natural convective coefficient h was calculated to be 0.52 $W \cdot m^{-2} \cdot K^{-1}$ based on the empirical equations for the designed rectangle fin arrays [100]. Figure 5.20 shows the measured IR images under 0.11 $g \cdot cm^{-3}$ and varying currents of 0.25 and 0.8 A, respectively. We can see that the maximum temperature is located at the center of the spot and regarded as the measured T_{ph}. When the current rises from 0.25 to 0.8 A, T_{ph} shows a great increase from 48.9 to 549.0 °C, resulting in thermal quenching and even silicone carbonization, which corresponds to the central blackening point as shown in Figure 5.21(b). A similar phenomenon was also observed in pc-LED array under total input electrical power of 31 W in our previous study [34].

In order to further understand the phenomenon and verify our model, the measured results versus different currents and concentrations are analyzed. Figure 5.22(a) shows the output power and voltage of LD versus varying currents from 0 to 0.80 A with an interval of 0.05 A. We can see that the LD threshold current is about 0.20 A and P_{in} rises approximately linearly with the current above threshold. Figure 5.22(b)–(d) demonstrates the comparisons between the measured and calculated T_{ph} and P_{out} versus P_{in} under varying phosphor concentrations of 0.11, 0.22, and 0.32 $g \cdot cm^{-3}$, respectively. Similar trends can be observed among them. We can see from Figure 5.22(b) that both the measured P_{out} and T_{ph} first increase with T_{in}, which can be easily understood by the increased input power and phosphor heat generation, respectively. However,

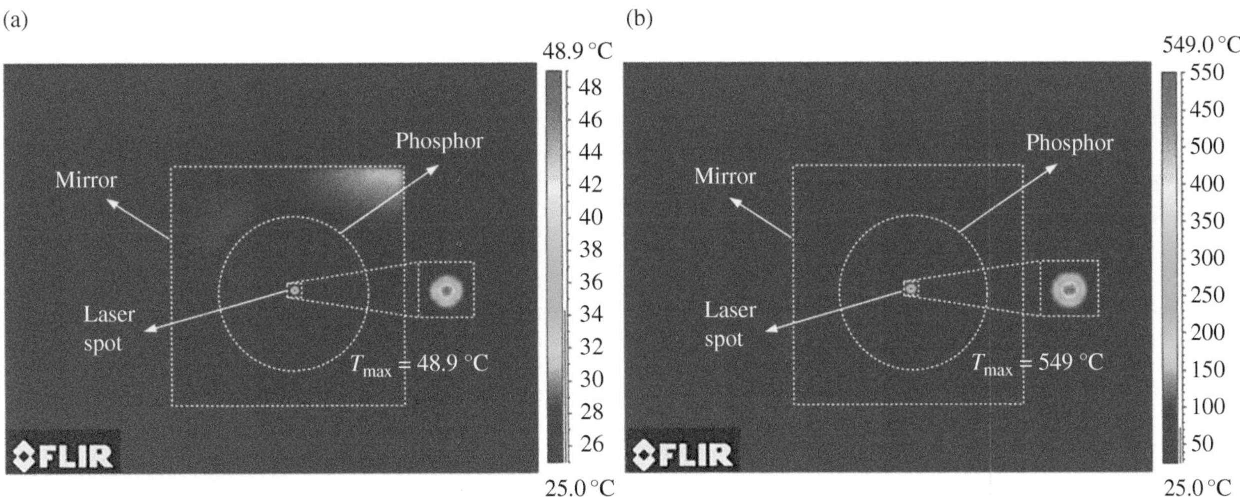

Figure 5.20 The measured IR images and the enlarged central spot (in the right rectangle) under 0.11 $g \cdot cm^{-3}$ and varying currents of (a) 0.25 A and (b) 0.8 A, respectively.

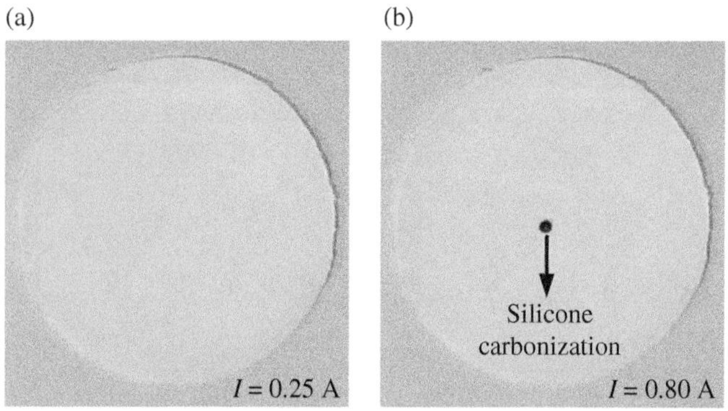

Figure 5.21 The photographs of phosphor plate under current of (a) 0.25 A and (b) 0.80 A.

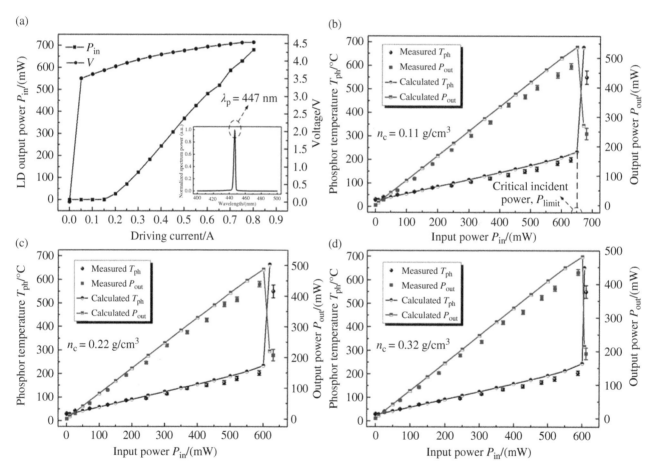

Figure 5.22 (a) The measured *P*–*I*–*V* curves of LD (the inset is the normalized spectrum power distribution of LD); and the comparison between the calculated and measured T_{ph} and P_{out} under varying phosphor concentrations of (b) 0.11 g·cm^{-3}, (c) 0.22 g·cm^{-3}, and (d) 0.32 g·cm^{-3}, respectively.

sudden changes happen when P_{in} exceeds a critical value. With P_{in} varying from 650 to 680 mW, T_{ph} shows a rapid rise from 198.0 to 549.0 °C, corresponding to a sudden drop of P_{out} from 473 to 243 mW. This observation can be explained by the temperature-dependent $\eta(T_{ph})$. Figure 5.23 plots the normalized QE versus T_{ph} obtained by Equation (5.40). When T_{ph} is approaching the onset quenching temperature T_c (i.e. $T_{0.95}$ corresponding to 95% of the peak QE shown in

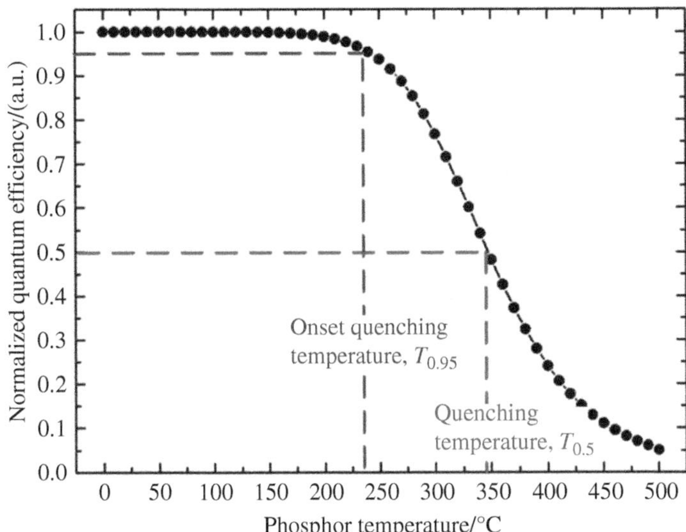

Figure 5.23 Normalized phosphor quantum efficiency versus phosphor temperature of YAG:Ce.

Figure 5.23) [100], the decrease of η leads to a rise of P_{ph} and T_{ph}, and conversely results in a decline in η [46]. This thermal runaway effect finally leads to a higher T_{ph} far exceeding quenching temperature $T_{0.5}$ corresponding to 50% of the peak QE [100], demonstrating the occurrence of the thermal quenching.

It can also be observed that under low P_{in}, the experimental results agree well with the calculations. However, the deviation can be large as P_{in} grows. We obtained that when P_{in} reaches 630 mW, the maximum difference between the measured and calculated P_{out} and T_{ph} is 48.7 mW and 21.5 °C, with corresponding errors of 10.1% and 9.7%, respectively. The deviations are within an acceptable range, especially in view of the same trends shared by the measurements and calculations. In addition, the measured P_{out} is always lower than the calculated one because part of output light is absorbed by the whole setup due to reflection loss, resulting in a decrease of P_{out}. It should also be mentioned that when thermal quenching occurs, the T_{ph} of test (549.0 °C) is obviously lower than the calculation (677.1 °C) because of the limitation of the IR imager. Moreover, in reality, the rapid increase of T_{ph} can occur quite quickly on millisecond time-scales, making it hard to probe the accurate phosphor temperature. In this case, the absolute value does not make much sense as long as thermal quenching occurs. Hence, this large deviation may not affect the feasibility of the model in evaluating thermal quenching effects.

In order to characterize and analyze the thermal quenching effects, we defined the input power corresponding to the turning point shown in Figure 5.22(b) as the critical incident power P_{limit}. It can be seen that only a few tens of milliwatts above P_{limit} can lead to thermal quenching. Obviously, for a given laser spot, the highest attainable output power and radiance are both limited by P_{limit}. Hence, it is vital to investigate the methods to enhance P_{limit}, and thus alleviate thermal quenching. For better understanding, P_{limit} can be approximately expressed as follows by transforming Equation (5.42):

$$P_{limit} = \frac{T_c - T_a}{(P_{ph}/P_{in})R_{tot}} \tag{5.42}$$

For a given type of phosphor, the critical temperature T_c can be regarded as a fixed value. Hence, P_{limit} is inversely proportional to the product of P_{ph}/P_{in} and R_{tot}. In the following, using the model, we conducted overall parameter analysis on P_{limit}. The involving parameters can be divided into optical, thermal, and optical–thermal factors. This classification is based on whether P_{ph}/P_{in} or R_{tot} is mainly affected.

Figure 5.24 illustrates effects of optical factors on P_{limit}, including phosphor concentration and mirror reflectivity γ_m ($\gamma_B = \gamma_Y$). One can observe that a rise of P_{limit} can be achieved by decreasing phosphor concentration or increasing γ_m. And the strong dependence is seen in the low concentration or high reflectivity region. When γ_m rises from 0.8 to 1.0, P_{limit} can be increased by 2.1 times. A similar observation was reported in Ref. [46]. The developing trend of P_{limit} can be understood, based on Equation (5.42), by the opposite variation of P_{ph}/P_{in}, with R_{tot} remaining a constant. Another key parameter to characterize phosphor heating is the QE. It is apparent that strengthening thermal stability, i.e. increasing onset

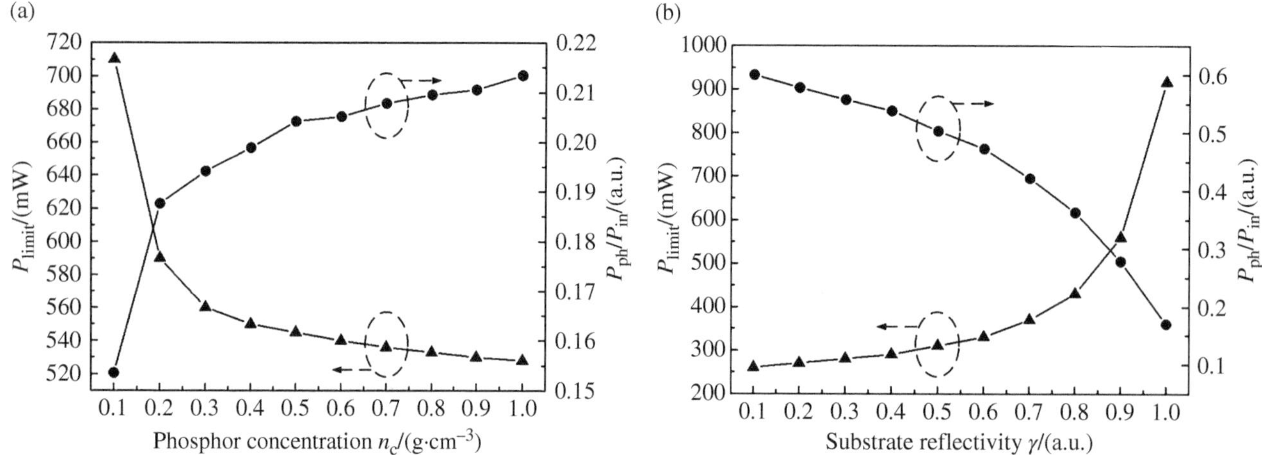

Figure 5.24 Effects of optical factors on P_{limit}: (a) phosphor concentration n_c and (b) mirror reflectivity γ_m.

quenching temperature T_c, can also dramatically enhance P_{limit}. Previous researches have reported that the single-crystal phosphor can withstand higher radiance compared with conventional powder phosphor, contributing to high luminance [85, 101].

Figure 5.25 shows effects of thermal factors on P_{limit}, including pump spot diameter D_{spot} and convective coefficient h. It should be noted that other thermal parameters, e.g. the thermal conductivity and thickness of mirror and bonding layer, are not discussed because of the corresponding relative small thermal resistance in the simulated case. It is seen that P_{limit} rises with both increasing D_{spot} and h, which can also be easily explained by the changing R_{tot}, with P_{ph}/P_{in} unchanged. Substantial enhancement of P_{limit} versus D_{spot} is also found, e.g. when D_{spot} increases from 0.5 to 3.0 mm, P_{limit} can be enhanced by 19 times. It is because that $R_{s,ph}$ dominates R_{tot} and small change in D_{spot} can result in large difference of $R_{s,ph}$. But increasing spot size may weaken the luminance [46], so there is a trade-off and this problem is worth further research. On the other hand, P_{limit} may not rise with h under high h, implying that there may exist limitation in increasing P_{limit} by enhancing convective heat transfer.

Figure 5.26 demonstrates the effect of optical–thermal factor on P_{limit}. As shown in Figure 5.26(b), with varying phosphor layer thicknesses d_{ph}, both P_{ph}/P_{in} and R_{tot} change. When d_{ph} increases, R_{tot} drops due to decreased $R_{s,ph}$ [95], but P_{ph}/P_{in} rises instead due to more absorbed light [34]. The combined effect leads to a first sudden and then slight increase of the product of them (shown in the inset figure), corresponding to the opposite trend of P_{limit} versus δ_{ph} as shown in

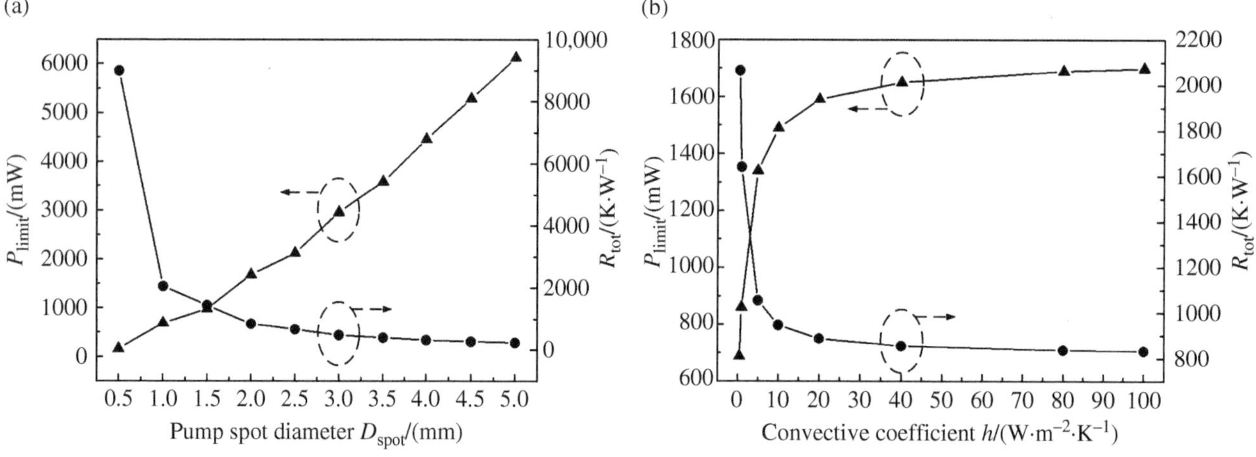

Figure 5.25 Effects of thermal factors on P_{limit}: (a) pump spot diameter D_{spot} and (b) convective coefficient h.

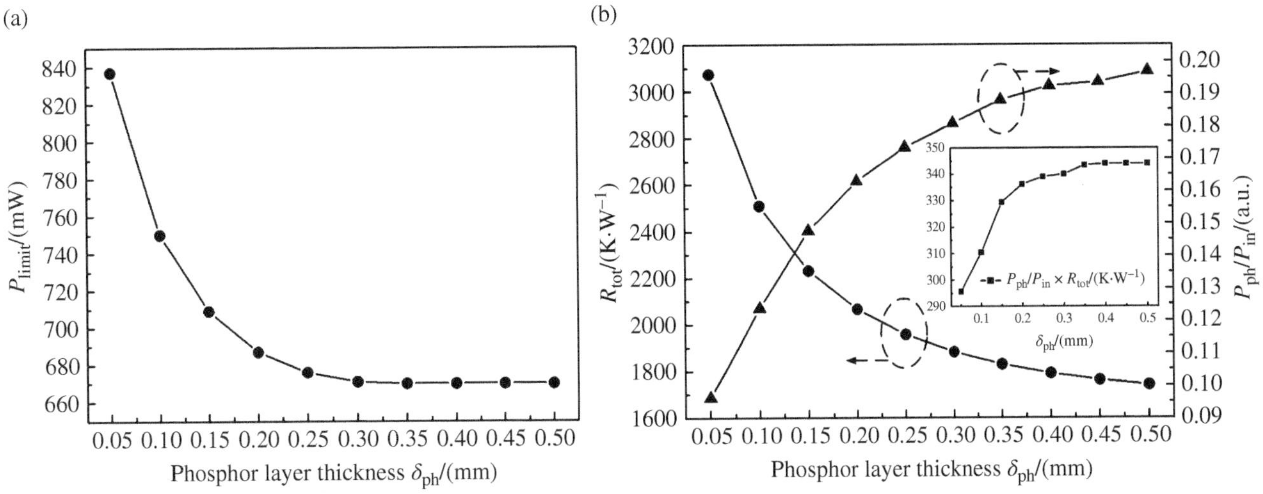

Figure 5.26 (a) The effect of optical–thermal factor phosphor layer thickness δ_{ph} and (b) the corresponding mechanism.

Figure 5.26(a). This implies that P_{ph}/P_{in} plays a dominant role, especially in low δ_{ph} region. So thinner phosphor layer is preferable in terms of small phosphor heating, but the reduced yellow to blue light ratio may be a concern.

In summary, simultaneous lower phosphor heating and total thermal resistance are preferred to enhance P_{limit}, and, finally, improve the optical–thermal performance of LERPs. This may be achieved by combining two or more methods presented earlier.

5.3 Opto-Thermal Modeling Applications in White LEDs

From the previous part, it can be seen that besides the junction temperature, the phosphor temperature is also an important factor to characterize the thermal performance of white pc-LEDs. However, the phosphor temperature is difficult to measure because the phosphor particles are dispersed in the silicone matrix and the phosphor diameter usually falls in the range of 13–15 μm. The measurement of phosphor temperature has remained a challenging problem for years. There was no experimental measurement of the phosphor temperature until Kim and Shin attempted to directly measure the phosphor temperature by a micro thermocouple [102]. At such circumstance, developing an alternative method to predict the phosphor temperature is meaningful and urgent.

Thermal resistance model is demonstrated to be an effective tool to predict junction temperature for LED packaging [95, 103–105]. In the conventional model [95, 103, 104], it only takes into account of the heat-dissipation path from the junction layer, to the heat slug, substrate, and the ambient. Chen et al. [106] proposed a bidirectional thermal resistance model considering the bidirectional heat flow on both sides of the LED packaging structure and improved the accuracy of junction temperature estimation. But they did not consider the heat generation of the phosphor layer in the model. Actually, for pc-LEDs, the heat produced in the phosphor layer cannot be ignored. To solve this problem, Juntunen et al. [107] developed an improved model and considered the heat generation of phosphor layer, but here they think both the heat from the phosphor and the chip transfers from the lead-frame to the ambient. Additionally, their model only focused on junction temperature estimation and did not calculate phosphor temperature. Recently, we proposed a modified bidirectional thermal resistance model considering both the heat generation of the phosphor layer and the heat flow through the phosphor layer to estimate the junction and phosphor temperature for pc-LEDs. Experimental measurements are conducted to validate the model.

The modified model is established based on the comparison of three LED packaging structures: (a) LED chip without coating, (b) with silicone coating, and (c) with phosphor coating (i.e. pc-LED), respectively, as shown in Figure 5.27. Figure 5.28 illustrates the schematic of the heat flow path and the corresponding thermal resistance model of these three packaging structures.

For LED chips without any coating, one-dimensional thermal resistance model is applied, as shown in Figure 5.28(a). The LED packaging structure is simplified by neglecting the internal structure and only considering a bare chip attached to a

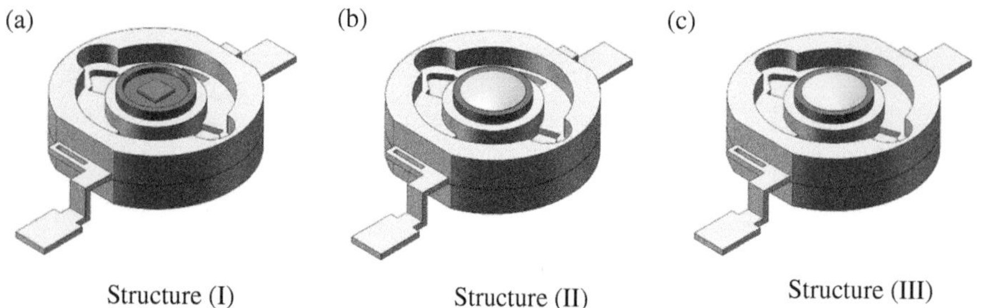

Figure 5.27 Schematic of three LED packaging structures: (a) structure I: LED chip without coating; (b) structure II: LED chip with silicone coating; and (c) structure III: LED chip with phosphor coating.

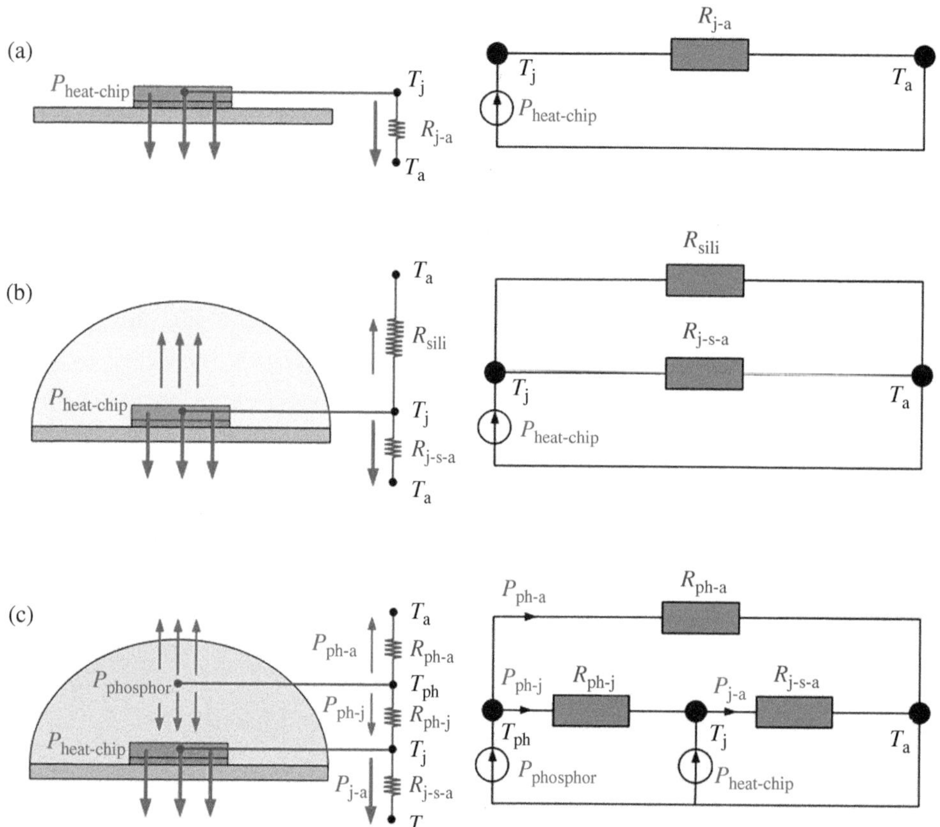

Figure 5.28 Schematic of the heat flow path (left) and the corresponding thermal resistance model (right) for three LED packaging structures: (a) LED chip without coating; (b) LED chip with silicone coating; (c) LED chip with phosphor coating.

substrate with die attach adhesive (DAA). Heat generated in the chip layer is transferred from the junction layer to the ambient through conduction and convection. In this case, the total junction-to-ambience thermal resistance $R_{1j\text{-}a}$ is introduced to define the ratio of the temperature difference between junction temperature T_j and ambient temperature T_a, to the total heat flux $P_{heat\text{-}chip}$, which can be expressed as:

$$R_{1j\text{-}a} = \frac{T_j - T_a}{P_{heat\text{-}chip}} \tag{5.43}$$

For LED chips with silicone coating, a bidirectional thermal resistance model is applied, as shown in Figure 5.28(b). There are two heat transfer paths from the junction to the ambient, namely, the lower path and the upper path. The lower path refers to the conventional pathway which is from junction through substrate to the ambient and the corresponding thermal resistance is $R_{\text{j-s-a}}$. And the upper path is from the junction layer through silicone coating layer to the ambient and the corresponding thermal resistance is R_{sili}. In this way, the thermal resistance of the adding silicone layer R_{sili} is connected with $R_{\text{j-s-a}}$ in parallel. It is noted that $R_{\text{j-s-a}}$ can be regarded as equal to $R_{1\text{,j-a}}$ for the same series of LEDs. Knowing the total junction-to-ambience thermal resistance $R_{1\text{,j-a}}$ and $R_{2\text{,j-a}}$, we can calculate R_{sili} as:

$$R_{\text{sili}} = \frac{1}{1/R_{2\text{,j-a}} - 1/R_{1\text{,j-a}}} \tag{5.44}$$

For LED chip with phosphor coating, a modified bidirectional thermal resistance model is proposed, as shown in Figure 5.28(c). Besides the chip heat generation $P_{\text{heat-chip}}$, there is another heat source, namely the phosphor heat generation P_{phosphor}. In order to express the added heat source, it is necessary to introduce the phosphor node T_{ph} which is defined as the highest temperature in the phosphor layer. We assume that all the heat dissipation of phosphor layer is generated at the phosphor node. Then P_{phosphor} is divided into two parts, namely, one is from the phosphor node to the ambient $P_{\text{ph-a}}$ through $R_{\text{ph-a}}$, and the other is from the phosphor node to the junction node $P_{\text{ph-j}}$ through $R_{\text{ph-j}}$. And the heat flux component $P_{\text{ph-j}}$ and $P_{\text{heat-chip}}$ gather into $P_{\text{j-a}}$, then continues conducting downward to the ambient node. Hence, there are three heat flow branches and the heat flux of each branch satisfies the following two equations:

$$P_{\text{ph-a}} + P_{\text{ph-j}} = P_{\text{phosphor}} \tag{5.45}$$

$$P_{\text{ph-j}} + P_{\text{heat-chip}} = P_{\text{j-a}} \tag{5.46}$$

In order to calculate T_{ph}, we should first determine several parameters, including T_{a}, $P_{\text{heat-chip}}$, P_{phosphor}, $R_{\text{j-s-a}}$, $R_{\text{ph-j}}$, and $R_{\text{ph-a}}$. The ambient temperature T_{a} is usually a constant which can be measured easily. $P_{\text{heat-chip}}$ and P_{phosphor} can be calculated by the output optical power comparison between packaging structures (II) and (III). As for the thermal resistance $R_{\text{j-s-a}}$, $R_{\text{ph-j}}$, and $R_{\text{ph-a}}$, indirect measurements can be applied to acquire these variables. Thermal transient tester (T3ster) is used for thermal characterization of those three packaging structures, of which the total junction-to-ambience thermal resistance is $R_{1\text{,j-a}}$, $R_{2\text{,j-a}}$, and $R_{3\text{,j-a}}$, respectively. We can assume that the thermal resistance from the junction through the substrate to the ambient $R_{\text{j-s-a}}$ of three packaging structures is all approximately equal to $R_{1\text{,j-a}}$, which can be expressed as:

$$R_{1\text{,j-s-a}} = R_{2\text{,j-s-a}} = R_{3\text{,j-s-a}} = R_{1\text{,j-a}} \tag{5.47}$$

For packaging structures (II) and (III), the thermal resistance of the added coating can be regarded approximately as equal, as long as two conditions are satisfied, namely, one is that both the coating share the same morphology and the other is that the phosphor volume fraction is not too high so that the thermal conductivity difference between the silicone and phosphor coating is negligible. According to Yuan's work [108, 109], thermal conductivity of the phosphor/silicone composite remains stable with a slight rise when phosphor volume fraction is below 40 vol%. In this case, the following relationships are obtained:

$$R_{\text{ph-j}} + R_{\text{ph-a}} = R_{\text{sili}} \tag{5.48}$$

The next step is to solve the two variables $R_{\text{ph-j}}$ and $R_{\text{ph-a}}$. Based on the proposed model, junction temperature T_{j} can be calculated as follows:

$$T_{\text{j}} = T_{\text{a}} + R_{1\text{,j-a}} \cdot P_{\text{j-a}} \tag{5.49}$$

In addition, the difference between T_{j} and T_{a} can be calculated by the product of $R_{3\text{,j-a}}$ and the total heat generation of the pc-LED, namely the sum of $P_{\text{heat-chip}}$ and P_{phosphor} [106]. Therefore, T_{j} can be acquired as follows:

$$T_{\text{j}} = T_{\text{a}} + R_{3\text{,j-a}} \cdot \left(P_{\text{heat-chip}} + P_{\text{phosphor}}\right) \tag{5.50}$$

Combining Equations (5.49) and (5.50), we can calculate $P_{\text{j-a}}$ as follows:

$$P_{\text{j-a}} = \frac{\left(P_{\text{heat-chip}} + P_{\text{phosphor}}\right) \cdot R_{3\text{,j-a}}}{R_{1\text{,j-a}}} \tag{5.51}$$

Substituting Equation (5.51) into Equation (5.46), we can acquire $P_{\text{ph-j}}$ as:

$$P_{\text{ph-j}} = P_{\text{j-a}} - P_{\text{heat-chip}} \tag{5.52}$$

Substituting Equation (5.52) into Equation (5.45), we can also acquire $P_{\text{ph-a}}$ as:

$$P_{\text{ph-a}} = P_{\text{phosphor}} - P_{\text{ph-j}} \tag{5.53}$$

According to the model, the phosphor temperature T_{ph} can be obtained by either Equation (5.54) or (5.55):

$$T_{\text{ph}} = T_{\text{a}} + P_{\text{ph-a}} \cdot R_{\text{ph-a}} \tag{5.54}$$

$$T_{\text{ph}} = T_{\text{j}} + P_{\text{ph-j}} \cdot R_{\text{ph-j}} \tag{5.55}$$

Combining Equations (5.48), (5.50), (5.54), and (5.55), $R_{\text{ph-a}}$ and $R_{\text{ph-j}}$ can be calculated as:

$$R_{\text{ph-a}} = \frac{R_{3,\text{j-a}} \cdot \left(P_{\text{phosphor}} + P_{\text{heat-chip}}\right) + R_{\text{sili}} \cdot P_{\text{ph-j}}}{P_{\text{phosphor}}} \tag{5.56}$$

$$R_{\text{ph-j}} = R_{\text{sili}} - R_{\text{ph-a}} \tag{5.57}$$

Till then, all the unknown variables are obtained and the junction and phosphor temperature for pc-LED can be calculated via Equations (5.49) and (5.54), respectively.

To validate the earlier model, three LED packaging structures were fabricated, as shown in Figure 5.27. Each packaging structure possesses five samples. A volume-controlled dip-transfer coating process was applied [110] to ensure that the geometry of the silicone and phosphor layer is consistent. Both the silicone and phosphor are directly coated onto the LED chip and the phosphor volume fraction is 8.5 vol%. To start with, the optical measurement was conducted by an integrating sphere to calculate $P_{\text{heat-chip}}$ and P_{phosphor}. The input electrical power P_{elec} and output optical power P_{opt} of packaging structures (II) and (III) under varying driving currents were directly measured. For those LED samples from the same batch, P_{elec} of both packaging structures under a same current is nearly equal. The output optical power of those two structures is donated as $P_{2,\text{opt}}$ and $P_{3,\text{opt}}$, respectively. Because the silicone and phosphor coating share the same geometry, the output optical power from LED chip and $P_{\text{heat-chip}}$ of both structures can be regarded as equal. Moreover, when the blue light penetrates the silicone and phosphor layer, optical power loss can be defined as the difference of $P_{2,\text{opt}}$ and $P_{3,\text{opt}}$. And it is assumed that all the optical loss is occurred in the phosphor layer and converted into heat P_{phosphor}, on the condition that the heat generation in substrate surface is negligible due to its high reflectance and low absorption [111]. So $P_{\text{heat-chip}}$ and P_{phosphor} can be calculated as follows:

$$P_{\text{heat-chip}} = P_{\text{elec}} - P_{2,\text{opt}} \tag{5.58}$$

$$P_{\text{phosphor}} = P_{2,\text{opt}} - P_{3,\text{opt}} \tag{5.59}$$

Second, T3ster was used to measure the total junction-to-ambience thermal resistance of three packaging structures. Figure 5.29 illustrates the experimental LED apparatus. The LED packages were attached to a designed heat sink by using the thermal grease, and the heat sink had a controlled temperature of 30 °C, which was close to the ambient temperature of 29.8 °C. The thermal resistance of LED models can be obtained by evaluating the distribution of resistance–capacitance (RC) networks [112]. Before measurement, voltage–temperature sensitive parameter calibration was conducted. A small current of 1 mA was applied to a temperature-controlled heat sink at different ambient temperatures from 25 to 75 °C with an increment of 10 °C. And the measured voltage–temperature coefficient was $-1.3\,\text{mV}\cdot\text{K}^{-1}$. The heating current for the packaging structure was from 0.05 to 0.65 A with an interval of 0.1 A and the heating/cooling time was 20 minutes to ensure that the thermal stabilization was reached. In this way, the junction temperature under different driving currents can be measured.

Figure 5.29 The schematic of the experimental LED apparatus.

To further verify the predicted phosphor temperature, an IR camera (FLIR SC620) was used as well. At first, the emissivity of the phosphor layer surface was calibrated and set to be 0.95. Surface temperature distribution under varying currents from 0.05 to 0.65 A with an interval of 0.1 A was obtained in the form of thermal image, in which the maximum temperature was regarded as the phosphor temperature. In this way, phosphor temperature can be measured.

Figure 5.30 shows the calculated $P_{heat\text{-}chip}$ and $P_{phosphor}$ for packaging structure (III) versus driving current based on Equations (5.58) and (5.59). It can be seen that $P_{heat\text{-}chip}$ is always higher than $P_{phosphor}$. Under the driving current of 0.65 A, $P_{heat\text{-}chip}$ is about 2.5 times more than $P_{phosphor}$, which proves that, compared with $P_{phosphor}$, most of the heat is generated in the chip layer for pc-LEDs. Figure 5.31 shows the cumulative structural function curves of three LED packaging structures under the driving current of 0.35 A. It is assumed that $R_{1,j\text{-}a}$, $R_{2,j\text{-}a}$, and $R_{3,j\text{-}a}$ are independent of the driving current. The measured results show that the average values of $R_{1,j\text{-}a}$, $R_{2,j\text{-}a}$, and $R_{3,j\text{-}a}$ are 12.37, 12.19, and 11.05 $K \cdot W^{-1}$, respectively.

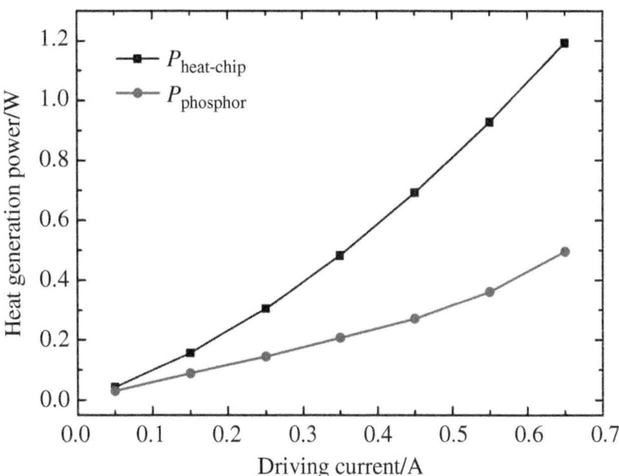

Figure 5.30 The calculated $P_{heat\text{-}chip}$ and $P_{phosphor}$ for packaging structure (III) versus driving current.

By means of the proposed model, the junction temperature and phosphor temperature can be calculated with the parameters obtained in Figures 5.30 and 5.31. Figure 5.32 illustrates the calculated and measured T_j and T_{ph} versus driving current. It can be seen that, for direct coted pc-LED, the calculated T_{ph} is always higher than the calculated T_j, which agrees with the findings of Yan's work [30]. And the rising rate of T_{ph} with current is obviously higher than that of T_j, which implies that phosphor temperature is more sensitive to current than junction temperature. It can be explained that $R_{ph\text{-}a}$ is 10 or more times than $R_{j\text{-}a}$, based on Equations (5.49) and (5.54); increase of $P_{ph\text{-}a}$ and $P_{j\text{-}a}$ caused by increased current will lead to a sharper increase of T_{ph} than T_j. Under the driving current of 0.65 A, the calculated T_{ph} can reach 134.1 °C, which is 85.5 °C higher than the calculated T_j. When the current keeps rising, T_{ph} will exceed thermal quenching temperature of phosphors, thus leads to the failure of LED. Hence, phosphor temperature should be paid more attention in LED design and manufacturing process. From the comparison between the calculated and measured T_j, it can be found that the maximum deviation between calculated and measured T_j was less than 1%, which means the calculation agrees pretty well with the measurement.

The measured T_{ph} obtained by the IR thermal imager (FLIR SC620) under different currents is shown in Figure 5.33. The red cursor is located at the center of the overview of the LED package and its temperature is denoted as measured T_{ph}. As for the comparison between the calculated and measured T_{ph}, it can be seen in Figure 5.32, when the driving current is below 0.25 A, the temperature difference between calculation and measurement is quite small. However, as the driving current is over 0.25 A, the difference increases with rising current. And the temperature difference can reach 12.5 °C with a

Figure 5.31 The cumulative structural function curves of three LED packaging structures.

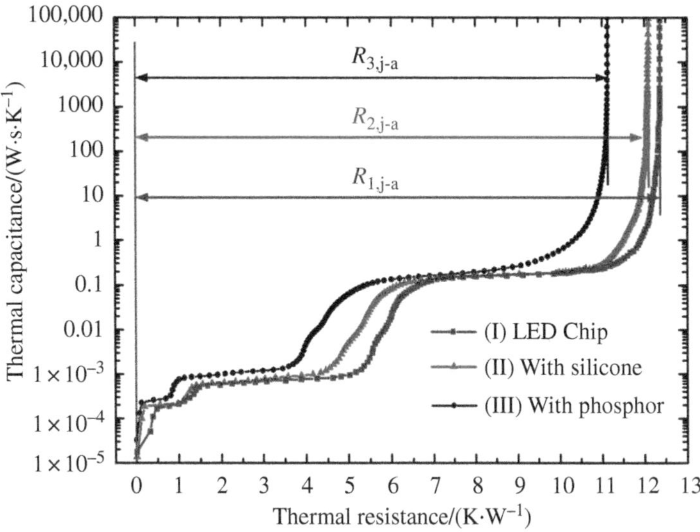

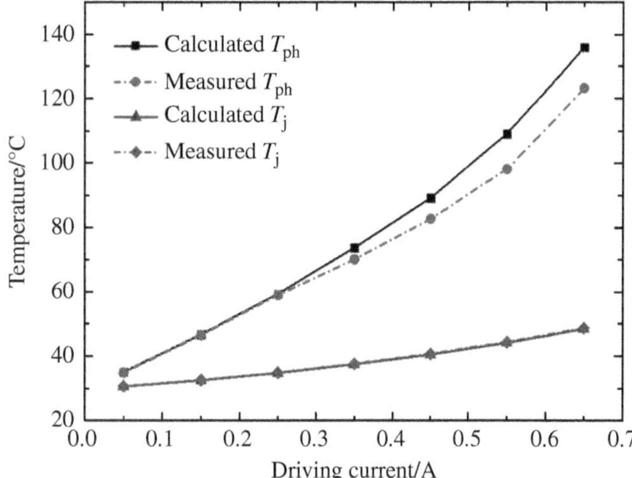

Figure 5.32 The calculated and measured T_j and T_{ph} versus driving current.

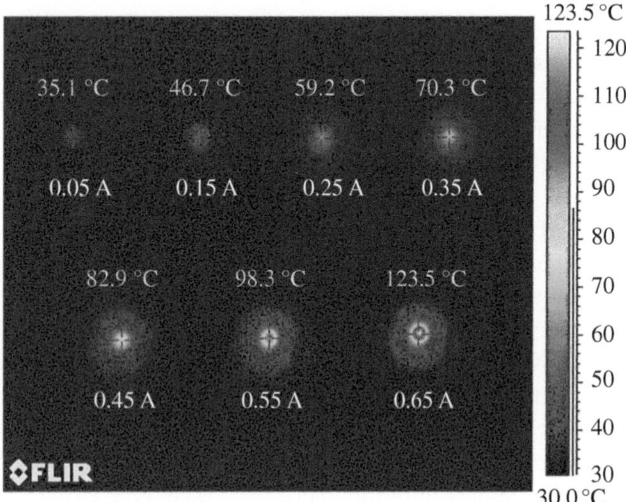

Figure 5.33 Measured temperature fields of packaging structure (III) under varying driving currents.

corresponding deviation of 9.2% when the current is 0.65 A. It can be explained that IR can only capture the surface temperature distribution of the phosphor layer, but the maximum phosphor temperature is usually located in the internal phosphor layer [30] due to the convective cooling effect on the outer surface. Therefore, the measured T_{ph} by IR is slightly lower than the calculated T_{ph}. We admit that it is better to calculate the surface temperature and to compare it with the measured phosphor temperature. But the surface temperature cannot be achieved by the present methods. It is very hard to calculate the field temperatures due to the thin phosphor layer and too many parameters coupling.

It should be noted that the temperature in this model is the maximum value; however, the temperatures at different locations in the phosphor layer are different, so this model cannot precisely reflect the effect of the location on the phosphor temperature. There are some factors affecting the accuracy of the model, e.g. the heat generation $P_{heat\text{-}chip}$ and $P_{phosphor}$, and the thermal resistances $R_{1,j\text{-}a}$, $R_{2,j\text{-}a}$ and $R_{3,j\text{-}a}$.

5.4 Chapter Summary

This chapter describes the opto-thermal interacted behaviors in white LED/LD packaging. The interactions are presented from two aspects, i.e. the chip and phosphor. For LED chip, the thermal droop induced by the negative temperature-dependent IQE is responsible for the opto-thermal interaction. It is indicated that good thermal management is essential to eliminate the thermal droop and thus maintain high performance of LED. Then, a general PET theory considering the thermal droop is given to accurately characterize the LED chip. For the phosphor, the phosphor thermal quenching and silicone carbonization phenomena induced by the phosphor heating are presented. The phosphor thermal quenching is a result of the interaction between increased phosphor temperature and reduced phosphor QE. To analyze the interaction, an opto-thermal model composing of a phosphor optical model based on fluorescent RTE and a steady-state thermal resistance model is presented by further including the negative temperature-dependent phosphor QE. It is suggested that selection of phosphor material with high thermal stability and thermal conductivity is important to relieve the thermal quenching. In addition, phosphor heat-dissipation design can also contribute a lot. Finally, by considering both the heat source in chip and phosphor, a modified bidirectional thermal resistance model is presented to estimate the junction and phosphor temperature simultaneously for pc-LED.

References

1 Schubert, E.F., Gessmann, T., and Kim, J.K. (2005). *Light Emitting Diodes*. Wiley Online Library.

2 Xue, J. (2017). *Towards the Efficiency Limit of Visible Light-Emitting Diodes*. Massachusetts Institute of Technology.

3 David, A. and Grundmann, M.J. (2010). Droop in InGaN light-emitting diodes: a differential carrier lifetime analysis. *Appl. Phys. Lett.* 96 (10): 103504.

4 Zhang, M., Bhattacharya, P., Singh, J., and Hinckley, J. (2009). Direct measurement of auger recombination in $In_{0.1}Ga_{0.9}N$/GaN quantum wells and its impact on the efficiency of $In_{0.1}Ga_{0.9}N$/GaN multiple quantum well light emitting diodes. *Appl. Phys. Lett.* 95 (20): 201108.

5 Meyaard, D.S., Shan, Q.F., Dai, Q. et al. (2011). On the temperature dependence of electron leakage from the active region of GaInN/GaN light-emitting diodes. *Appl. Phys. Lett.* 99 (4): 041112.

6 Verzellesi, G., Saguatti, D., Meneghini, M. et al. (2013). Efficiency droop in InGaN/GaN blue light-emitting diodes: physical mechanisms and remedies. *J. Appl. Phys.* 114 (7): 071101.

7 Kim, M.H., Schubert, M.F., Dai, Q. et al. (2007). Origin of efficiency droop in GaN-based light-emitting diodes. *Appl. Phys. Lett.* 91 (18): 183507.

8 Meyaard, D.S., Lin, G.B., Shan, Q.F. et al. (2011). Asymmetry of carrier transport leading to efficiency droop in GaInN based light-emitting diodes. *Appl. Phys. Lett.* 99 (25): 251115.

9 Iveland, J., Martinelli, L., Peretti, J. et al. (2013). Direct measurement of auger electrons emitted from a semiconductor light-emitting diode under electrical injection: identification of the dominant mechanism for efficiency droop. *Phys. Rev. Lett.* 110 (17): 177406.

10 Meyaard, D.S., Shan, Q.F., Cho, J. et al. (2012). Temperature dependent efficiency droop in GaInN light-emitting diodes with different current densities. *Appl. Phys. Lett.* 100 (8): 081106.

11 Hader, J., Moloney, J.V., and Koch, S.W. (2011). Temperature dependence of the internal efficiency droop in GaN-based diodes. *Appl. Phys. Lett.* 99 (18): 181127.

12 Piprek, J. (2010). Efficiency droop in nitride-based light-emitting diodes. *Phys. Status Solidi A* 207 (10): 2217–2225.

13 Pan, C.C., Gilbert, T., Pfaff, N. et al. (2012). Reduction in thermal droop using thick single-quantum-well structure in semipolar (2021) blue light-emitting diodes. *Appl. Phys. Express* 5 (10): 102103.

14 Cho, J., Schubert, E.F., and Kim, J.K. (2013). Efficiency droop in light-emitting diodes: challenges and countermeasures. *Laser Photon. Rev.* 7 (3): 408–421.

15 Garcia, J., Lamar, D.G., Costa, M.A. et al. (2008). An estimator of luminous flux for enhanced control of high brightness LEDs. *IEEE Power Electronics Specialists Conference*, Rhodes, Greece, 1852–1856.

16 Biber, C. (2008). LED light emission as a function of thermal conditions. *IEEE Semiconductor Thermal Measurement Management Symposium*, San Jose, CA, USA, 180–184.

17 Lalith, J., Gu, Y.M., and Nadarajah, N. (2006). Characterization of thermal resistance coefficient of high-power LEDs. *6th International Conference on Solid State Lighting*, San Diego, CA, USA, 63370–63377.

18 Yuan, L., Liu, S., Chen, M.X., and Luo, X.B. (2006). Thermal analysis of high power LED array packaging with microchannel cooler. *7th International Conference on Electronic Packaging Technology*, Shanghai, China, 1–5.

19 Petroski, J. (2004). Spacing of high-brightness LEDs on metal substrate PCB's for proper thermal performance. *9th Intersociety Conference on Thermal and Thermomechanical Phenomena Electronic Systems*, Las Vegas, NV, USA, 507–514.

20 Ma, Z.L., Zheng, X.R., Liu, W.J. et al. (2005). Fast thermal resistance measurement of high brightness LED. *6th International Conference on Electronic Packaging Technology*, Shenzhen, China, 614–616.

21 Siegal, B. (2006). Practical considerations in high power LED junction temperature measurements. *31st International Conference on Electronics Manufacturing and Technology*, Kuala Lumpur, Malaysia, 62–66.

22 Arik, M., Becker, C., Weaver, S., and Petroski, J. (2004). Thermal management of LEDs: package to system. *Proc. SPIE* 5187: 64–75.

23 Cheng, Q. (2007). Thermal management of high-power white LED package. *8th International Conference on Electronic Packaging Technology*, Shanghai, China, 1–5.

24 Chen, H.T., Tao, X.H., and Hui, S.Y.R. (2012). Estimation of optical power and heat-dissipation coefficient for the photo-electro-thermal theory for LED systems. *IEEE Trans. Power Electron.* 27 (4): 2176–2183.

25 Hui, S.Y.R., Chen, H.T., and Tao, X.H. (2012). An extended photoelectrothermal theory for LED systems: a tutorial from device characteristic to system design for general lighting. *IEEE Trans. Power Electron.* 27 (11): 4571–4583.

26 Hui, S.Y.R. and Qin, Y.X. (2009). A general photo-electro-thermal theory for light emitting diode (LED) Systems. *IEEE Trans. Power Electron.* 24 (8): 1967–1976.

27 Narendran, N., Gu, Y., Freyssinier, J.P. et al. (2004). Solid-state lighting: failure analysis of white LEDs. *J. Cryst. Growth* 268 (3–4): 449–456.

28 Fan, B., Wu, H., Zhao, Y. et al. (2007). Study of phosphor thermal-isolated packaging technologies for high-power white light-emitting diodes. *IEEE Photon. Technol. Lett.* 19 (15): 1121–1123.

29 Hwang, J., Kim, Y., Kim, J. et al. (2010). Study on the effect of the relative position of the phosphor layer in the LED package on the high power LED lifetime. *Phys. Status Solidi C* 7 (7–8): 2157–2161.

30 Yan, B., Tran, N.T., You, J., and Shi, F.G. (2011). Can junction temperature alone characterize thermal performance of white LED emitters. *IEEE Photon. Technol. Lett.* 23 (9): 555–557.

31 Hu, R., Luo, X.B., and Zheng, H. (2012). Hotspot location shift in the high power phosphor converted white light-emitting diode package. *Jpn. J. Appl. Phys.* 51: 09MK05.

32 Furgel, I.A., Molin, O.V., Borshch, V.E. et al. (1992). Thermal conductivity of polymer composites with a disperse filler. *J. Eng. Phys. Thermophys.* 62 (3): 335–340.

33 Arik, M., Weaver, S., Becker, C. et al. (2003). Effects of localized heat generation due to the color conversion in phosphor particles and layers of high brightness light emitting diodes. *International Electronic Packaging Technical Conference and Exhibition*, Maui, Hawaii, USA.

34 Luo, X.B., Fu, X., Chen, F., and Zheng, H. (2013). Phosphor self-heating in phosphor converted light emitting diode packaging. *Int. J. Heat Mass Transf.* 58: 276–281.

35 Luo, X.B., Cheng, T., Xiong, W. et al. (2007). Thermal analysis of an 80 W light emitting diode street lamp. *IET Optoelectron.* 1 (5): 191–196.

36 Luo, X.B., Xiong, W., Cheng, T., and Liu, S. (2009). Temperature estimation of high power LED street lamp by a multi-chip analytical solution. *IET Optoelectron.* 3 (5): 225–232.

37 Cheng, T., Luo, X.B., Huang, S.Y., and Liu, S. (2010). Thermal analysis and optimization of multiple LED packaging based on a general analytical solution. *Int. J. Thermal Sci.* 49 (1): 196–201.

38 Schubert, E.F., Kim, J.K., Luo, H., and Xi, J.Q. (2006). Solid state lighting–a benevolent technology. *Rep. Prog. Phys.* 69: 3069–3099.

39 Hung, C.H. and Tien, C.H. (2010). Phosphor-converted LED modeling by bidirectional photometric data. *Opt. Express* 18: A261–A271.

40 Luo, X.B., Hu, R., Liu, S., and Wang, K. (2016). Heat and fluid flow in high-power LED packaging and applications. *Prog. Energy Combust. Sci.* 56: 1–32.

41 Peng, Y., Li, R.X., Chen, H. et al. (2017). Facile preparation of patterned phosphor-in-glass with excellent luminous properties through screen-printing for high-power white light-emitting diodes. *J. Alloys Compd.* 693: 279–284.

42 Lenef, A., Kelso, J.F., and Piquette, A. (2014). Light extraction from luminescent light sources and application to monolithic ceramic phosphors. *Opt. Lett.* 39: 3058–3061.

43 Liu, Z.Y., Wang, K., Luo, X.B., and Liu, S. (2010). Precise optical modeling of blue light-emitting diodes by Monte Carlo ray-tracing. *Opt. Express* 18: 9398–9412.

44 Jeon, S.W., Noh, J.H., Kim, K.H. et al. (2014). Improvement of phosphor modeling based on the absorption of Stokes shifted light by a phosphor. *Opt. Express* 22: A1237–A1242.

45 Correia, A., Hanselaer, P., and Meuret, Y. (2016). An efficient optothermal simulation framework for optimization of high-luminance white light sources. *IEEE Photon. J.* 8: 1–15.
46 Lenef, A., Kelso, J., Tchoul, M. et al. (2014). Laser-activated remote phosphor conversion with ceramic phosphors. *Proc. SPIE* 9190: 919000C.
47 Lenef, A., Kelso, J., Zheng, Y., and Tchoul, M. (2013). Radiance limits of ceramic phosphors under high excitation fluxes. *Proc. SPIE* 8841: 884107.
48 Shakespeare, T. and Shakespeare, J. (2003). A fluorescent extension to the Kubelka–Munk model. *Color. Res. Appl.* 28: 4–14.
49 Kang, D.Y., Wu, E., and Wang, D.M. (2006). Modeling white light-emitting diodes with phosphor layers. *Appl. Phys. Lett.* 89: 231102.
50 Hu, R. and Luo, X.B. (2012). A model for calculating the bidirectional scattering properties of phosphor layer in white light-emitting diodes. *J. Lightwave Technol.* 30: 3376–3380.
51 Hu, R., Cao, B., Zou, Y. et al. (2013). Modeling the light extraction efficiency of bi-layer phosphor in white LEDs. *IEEE Photon. Technol. Lett.* 25: 1141–1144.
52 Hu, R., Wang, Y.M., Zou, Y. et al. (2013). Study on phosphor sedimentation effect in white LED packages by modeling multi-layer phosphors with the modified Kubelka-Munk theory. *J. Appl. Phys.* 113: 063108.
53 Hu, R., Zheng, H., Hu, J.Y., and Luo, X.B. (2013). Comprehensive study on the transmitted and reflected light through the phosphor layer in light-emitting diode packages. *J. Disp. Technol.* 9: 447–452.
54 Ma, Y.P., Lan, W., Xie, B. et al. (2018). An optical-thermal model for laser-excited remote phosphor with thermal quenching. *Int. J. Heat Mass Transf.* 116: 694–702.
55 Chen, J. and Intes, X. (2011). Comparison of Monte Carlo methods for fluorescence molecular tomography-computational efficiency. *Med. Phys.* 38: 5788–5798.
56 Tarvainen, T.J., Kolehmainen, V., Arridge, S.R., and Kaipio, J.P. (2011). Image reconstruction in diffuse optical tomography using the coupled radiative transport–diffusion model. *J. Quant. Spectrosc. Radiat. Transf.* 112: 2600–2608.
57 Vargas, W.E. and Niklasson, G.A. (1997). Applicability conditions of the Kubelka–Munk theory. *Appl. Opt.* 36: 5580–5586.
58 Cheong, W.F., Prahl, S.A., and Welch, A.J. (1990). A review on the optical properties of biological tissues. *IEEE J. Quantum Electron.* 26: 2166–2185.
59 Leyre, S., Durinck, G., Giel, B.V. et al. (2012). Extended adding-doubling method for fluorescent applications. *Opt. Express* 20: 17856–17872.
60 Howell, J.R. (1968). Application of Monte Carlo to heat transfer problems. *Adv. Heat Transf.* 5: 1–54.
61 Chai, J.C. (2003). One-dimensional transient radiation heat transfer modeling using a finite-volume method. *Numer. Heat Transf. B* 44: 187–208.
62 Liu, L.H. (2004). Finite element simulation of radiative heat transfer in absorbing and scattering media. *J. Thermophys. Heat Transf.* 18: 555–557.
63 Zhao, J.M. and Liu, L.H. (2007). Discontinuous spectral element method for solving radiative heat transfer in multidimensional semitransparent media. *J. Quant. Spectrosc. Radiat. Transf.* 107: 1–16.
64 Zhao, J.M. and Liu, L.H. (2006). Least-squares spectral element method for radiative heat transfer in semitransparent media. *Numer. Heat Transf. B* 50: 473–489.
65 Fiveland, W.A. (1988). Three-dimensional radiative heat-transfer solutions by the discrete-ordinates method. *J. Thermophys. Heat Transf.* 2: 309–316.
66 Klose, A.D. (2009). Radiative transfer of luminescence light in biological tissue. In: *Light Scattering Reviews*, vol. 4 (ed. A.A. Kokhanovsky), 293–345. Springer.
67 Sevick-Muraca, E.M. and Burch, C.L. (1994). Origin of phosphorescence signals reemitted from tissues. *Opt. Lett.* 19: 1928–1930.
68 Yudovsky, D. and Pilon, L. (2010). Modeling the local excitation fluence rate and fluorescence emission in absorbing and strongly scattering multilayered media. *Appl. Opt.* 49: 6072–6084.
69 Fiveland, W.A. and Jessee, J.P. (1995). Comparison of discrete ordinates formulations for radiative heat transfer in multidimensional geometries. *J. Thermophys. Heat Transf.* 9: 47–54.
70 McCamy, S. (1992). Correlated color temperature as an explicit function of chromaticity coordinates. *Color. Res. Appl.* 17: 142–144.0.
71 Ma, Y.P., Hu, R., Yu, X.J. et al. (2017). A modified bidirectional thermal resistance model for junction and phosphor temperature estimation in phosphor-converted light-emitting diodes. *Int. J. Heat Mass Transf.* 106: 1–6.

72 Liu, Z.Y., Liu, S., Wang, K., and Luo, X.B. (2010). Measurement and numerical studies of optical properties of YAG:Ce phosphor for white LED packaging. *Appl. Opt.* 49: 247–257.

73 Henyey, L.G. and Greenstein, J.L. (1941). Diffuse radiation in the galaxy. *Astrophys. J.* 93: 70–83.

74 Beek, J.F., Blokland, P., Posthumus, P. et al. (1997). In vitro double-integrating-sphere optical properties of tissues between 630 and 1064 nm. *Phys. Med. Biol.* 42: 2255–2261.

75 Saeys, W., Velazco-Roa, M.A., Thennadil, S.N. et al. (2008). Optical properties of apple ski and flesh in the wavelength range from 350 to 2200 nm. *Appl. Opt.* 47: 908–919.

76 Leyre, S., Durinck, G., Hofkens, J. et al. (2014). Experimental determination of the absorption and scattering properties of YAG:Ce phosphor, in light, energy and the environment. *OSA Technical Digest (online)*, Optical Society of America, paper DTu4C.4.

77 Luo, X.B. and Hu, R. (2014). Calculation of the phosphor heat generation in phosphor-converted light-emitting diodes. *Int. J. Heat Mass Transf.* 75: 213–217.

78 Zhao, J.M. and Liu, L.H. (2007). Spectral element approach for coupled radiative and conductive heat transfer in semitransparent medium. *J. Heat Transf.* 129: 1417–1424.

79 Sun, J., Yi, H.L., and Tan, H.P. (2016). Local RBF meshless scheme for coupled radiative and conductive heat transfer. *Numer. Heat Transf. A* 69: 1390–1404.

80 Ryu, H.Y., Kim, H.S., and Shim, J.I. (2009). Rate equation analysis of efficiency droop in InGaN light-emitting diodes. *Appl. Phys. Lett.* 95 (8): 081114.

81 Maur, M.A.D., Pecchia, A., Penazzi, G. et al. (2016). Efficiency drop in green InGaN/GaN light emitting diodes: the role of random alloy fluctuations. *Phys. Rev. Lett.* 116 (2): 027401.

82 Wierer, J.J., Tsao, J.Y., and Sizov, D.S. (2013). Comparison between blue lasers and light-emitting diodes for future solid-state lighting. *Laser Photon. Rev.* 7 (6): 963–993.

83 Wierer, J.J. and Tsao, J.Y. (2015). Advantages of III-nitride laser diodes in solid-state lighting. *Phys. Status Solidi A* 212 (5): 980–985.

84 Basu, C., Wollweber, M.M., and Roth, B. (2013). Lighting with laser diodes. *Adv. Opt. Technol.* 2 (4): 313–321.

85 Cantore, M., Pfaff, N., Farrell, R.M. et al. (2015). High luminous flux from single crystal phosphor-converted laser-based white lighting system. *Opt. Express* 24 (2): A215.

86 Xu, Y., Chen, L.H., Li, Y.Z. et al. (2008). Phosphor-conversion white light using InGaN ultraviolet laser diode. *Appl. Phys. Lett.* 92 (2): 021129.

87 George, A.F., Al-Waisawy, S., Wright, J.T. et al. (2016). Laser-driven phosphor-converted white light source for solid-state illumination. *Appl. Opt.* 55 (8): 1899–1905.

88 Denault, K.A., Cantore, M., Nakamura, S. et al. (2013). Efficient and stable laser-driven white lighting. *AIP Adv.* 3 (7): 072107.

89 Lee, D.H., Joo, J., and Lee, S. (2015). Modeling of reflection-type laser-driven white lighting considering phosphor particles and surface topography. *Opt. Express* 23 (15): 18872–18887.

90 Masui, S., Yamamoto, T., and Nagahama, S. (2015). A white light source excited by laser diodes. *Electron. Commun. Japan* 98 (5): 23–27.

91 Fulmek, P., Sommer, C., Hartmann, P. et al. (2013). On the thermal load of the color-conversion elements in phosphor-based white light-emitting diodes. *Adv. Opt. Mater.* 1 (10): 753–762.

92 Hu, R., Luo, X.B., and Zheng, H. (2012). Hotspot location shift in the high-power phosphor-converted white light-emitting diode packages. *Jpn. J. Appl. Phys.* 51: 09MK05.

93 Shih, B.J., Chiou, S.C., Hsieh, Y.H. et al. (2015). Study of temperature distributions in pc-WLEDs with different phosphor packages. *Opt. Express* 23 (26): 33861–33869.

94 Ma, Y.P., Wang, M., Sun, J. et al. (2018). Phosphor modeling based on fluorescent radiative transfer equation. *Opt. Express* 26 (13): 16442–16455.

95 Luo, X.B., Mao, Z.M., Yang, J., and Liu, S. (2012). Engineering method for predicting junction temperatures of high-power light-emitting diodes. *IET Optoelectron.* 6 (5): 230–236.

96 Muzychka, Y.S., Yovanovich, M.M., and Culham, J.R. (2003). Spreading thermal resistances in rectangular flux channels part I -geometric equivalences. *36th AIAA Thermophysics Conference*, Orlando, FL, USA, 4187.

97 Lyu, L.J. and Hamilton, D.S. (1991). Radiative and nonradiative relaxation measurements in Ce^{3+} doped crystals. *J. Lumin.* 48: 251–254.

98 Bachmann, V., Ronda, C., and Meijerink, A. (2009). Temperature quenching of yellow Ce^{3+} luminescence in YAG:Ce. *Chem. Mater.* 21 (10): 2077–2084.

99 Arjoca, S., Villora, E.G., Inomata, D. et al. (2015). Temperature dependence of Ce:YAG single-crystal phosphors for high-brightness white LEDs/LDs. *Mater. Res. Exp.* 2 (5): 055503.

100 Ueda, J., Dorenbos, P., Bos, A.J.J. et al. (2015). Insight into the thermal quenching mechanism for $Y_3Al_5O_{12}$:Ce^{3+} through thermoluminescence excitation spectroscopy. *J. Phys. Chem. C* 119 (44): 25003–25008.

101 Víllora, E.G., Arjoca, S., Inomata, D., and Shimamura, K. (2016). Single-crystal phosphors for high-brightness white LEDs/LDs. *Proc. SPIE* 9768: 976805. (*Light-Emitting Diodes - Materials, Devices, and Applications for Solid State Lighting XX*).

102 Kim, J.H. and Shin, M.W. (2015). Thermal behavior of remote phosphor in light-emitting diode packaging structures. *IEEE Electron Dev. Lett.* 36 (8): 832–834.

103 Ha, M. and Graham, S. (2012). Development of a thermal resistance model for chip-on-board packaging of high power LED arrays. *Microelectron. Reliab.* 52 (5): 836–844.

104 Fu, X., Hu, R., and Luo, X.B. (2014). An engineering method to estimate the junction temperatures of light-emitting diodes in multiple LED application. *J. Korean Phys. Soc.* 65 (2): 176–184.

105 Chen, H.T., Lu, Y.J., Gao, Y.L. et al. (2009). The performance of compact thermal models for LED package. *Thermochim. Acta* 488 (1-2): 33–38.

106 Chen, H.T., Tan, S.C., and Hui, S.Y.R. (2015). Analysis and modeling of high-power phosphor-coated white light-emitting diodes with a large surface area. *IEEE Trans. Power Electron.* 30 (6): 3334–3344.

107 Juntunen, E., Tapaninen, O., Sitomaniemi, A., and Heikkinen, V. (2013). Effect of phosphor encapsulant on the thermal resistance of a high-power COB LED packaging structure. *IEEE Trans. Compon. Packag. Manuf. Technol.* 3 (7): 1148–1154.

108 Yuan, C. and Luo, X.B. (2013). A unit cell approach to compute thermal conductivity of uncured silicone/phosphor composites. *Int. J. Heat Mass Transf.* 56 (1–2): 206–211.

109 Yuan, C., Xie, B., Huang, M.Y. et al. (2016). Thermal conductivity enhancement of platelets aligned composites with volume fraction from 10% to 20%. *Int. J. Heat Mass Transf.* 94: 20–28.

110 Yu, X.J., Xie, B., Shang, B.F. et al. (2016). A cylindrical tuber encapsulant geometry for enhancing optical performance of chip-on-board packaging light-emitting diodes. *IEEE Photon. J.* 8 (3): 1600709.

111 Xie, B., Hu, R., Yu, X.J. et al. (2016). Effect of packaging method on performance of light-emitting diodes with quantum dot phosphor. *IEEE Photon. Technol. Lett.* 28 (10): 1115–1118.

112 Farkas, G., Vader, Q., Poppe, A., and Bognar, G. (2005). Thermal investigation of high power optical devices by transient testing. *IEEE Trans. Compon. Packag. Technol.* 28 (1): 45–50.

6

Thermally Enhanced Thermal Interfacial Materials

Thermal interface materials (TIMs) are commonly used in electronics to reduce the contact resistance arising from the incomplete contact between two solid surfaces. After inserting TIM between the solid surfaces, a solid–TIM–solid joint forms at the surface. Thermal resistance at the joint (R_j) has two components: the contact resistance (R_c) at the TIM–solid interface arising from the incomplete wetting of the interface and the bulk resistance of TIM (R_{bulk}). Therefore, the researches on contact resistance (R_c) and the thermal conductivity (TC) of TIM (κ_{TIM}) are of great significance for enhancing heat transfer. In this chapter, we will first introduce the modeling of contact resistance and its validation by a thermal interface testing system. In terms of improving the TC of TIM, the microstructure design using magnetic field and the interfacial thermal transport manipulation will be introduced.

6.1 Modeling of TIM

TIMs are commonly used in the thermal management of microelectronics. Polymeric fluidic TIMs, such as thermal greases and phase change materials (PCMs), are often utilized to reduce the R_c between the die and the heat spreader, and between the heat spreader and the heat sink [1–3]. Although the polymeric TIMs with high TC lead to low R_{bulk}, thermal resistance at the TIM bond lines is often higher than expected. One of the most important reasons is that when filling the polymeric fluidic materials between the solid surfaces, an amount of air is normally entrapped inside the microcavities of the rough solid surface, resulting in an R_c arising at the liquid–solid interface [1, 4]. R_c plays an important role on the thin TIMs with high TC [3], while good physical-based analytical models for R_c at liquid–solid surface are still lacking in literature reviewed carefully in Ref. [3]. It is therefore necessary to develop an effective model to predict it.

Past literature abounds with the models of R_c in solid–solid contact conditions [5–8]. These studies have all concluded that R_c is a function of the surface roughness, asperity slope, mechanical properties of the solid bodies, apparent area of contact, and the compressive load between the contacting surfaces. However, these models are not suitable for the polymeric fluidic TIMs because the mechanical properties, such as Young's modulus or hardness, cannot be defined in the same way [4].

Two approaches have been used to predict R_c according to the limited literature. One way is to use a numerical model of heat transfer within the solid substrate and the liquid, and to account for temperature gradients within them [9, 10]. Bennett and Poulikakos [9] developed a model for estimating the thermal contact coefficient between the molten metal and substrate, and they found that the pockets of entrapped air cause indentations on the bottom surface of molten metal. Liu et al. [10] also applied the numerical method to conduct thermal contact analysis for a molten metal drop impacting on a substrate, their results suggest that the type of substrate material effects the thermal contact coefficient a lot. The second way is to follow the method at the solid–solid contact conditions [5–8] to build theoretical models at the liquid–solid interface by conducting topographical and mechanical analyses at the interface [4, 11, 12]. Prasher [4] developed a surface chemistry model for the fluidic TIMs–solid contacts based on the definition of R_c proposed by Madhusudana [13]. The advantage of such model is taking into consideration the wettability of the liquid on a rough surface and making mechanical analysis in detail. Thus, the model successfully illustrated the interface characteristics at the liquid–solid contact. The surface geometry assumption in this model is that the solid asperities are conical in shape with the same heights, and the spacing of the peaks and valleys is identical and equal to the roughness sampling interval [4]. However, roughness sampling interval is a profilometer parameter that is used to modify the roughness measurement conditions [14]. On the other hand, Hamasaiid et al.

Thermal Management for Opto-electronics Packaging and Applications, First Edition. Xiaobing Luo, Run Hu, and Bin Xie.

[11, 12] developed a predictive model for the casting–die interface based on the definition of R_c in the Cooper–Mikic–Yovanovich (CMY) model [6]. The advantages of this model are assuming the asperity heights follow a Gaussian distribution, which is more acceptable for exhibiting the surface geometry [6], and using mean asperity peak spacing (R_{sm}), measured by the profilometer, as the spacing of the peaks and valleys. Different from surface chemistry model, mechanical analysis in Hamasaiid et al.'s model is completed by evaluating the capillarity pressure of the molten alloy as a function of the applied pressure through experiments. Predictions of the model are found to agree with the experimental results.

According to the above review, Hamasaiid et al.'s model is found to have better topographical analysis, while wettability analysis in the surface chemistry model makes it characteristic-based and suitable for different types of liquid–solid interfaces. In this section, we will introduce an improved model based on Hamasaiid et al.'s model by adopting the wettability analysis at the liquid–solid interface. This improved model is compared with the experimental results of R_c at the TIMs–solid interface.

6.1.1 Model of Thermal Contact Resistance

6.1.1.1 Theoretical Background

Figure 6.1 schematically shows the formation of R_c at the fluidic TIM–solid interface. Heat flux passing from the solid to TIM or from TIM to solid needs to be constricted at the microcontact spots. At the macroscopic scale, it brings in a temperature difference at the interface resulting in a thermal resistance. Based on the CMY model [6], the thermal field at the interface can be split into several identical isothermal flux tubes as illustrated in Figure 6.1. The thermal resistance of a single microcontact spot can be determined from the following relationship [6]:

$$r_c = \frac{\left(1 - \frac{r_p}{r_t}\right)^{1.5}}{2\kappa_s r_p} \tag{6.1}$$

where $\kappa_s = 2\kappa_1\kappa_2/(\kappa_1 + \kappa_2)$ is the effective TC of the contacting bodies, r_p is the radius of the circular microcontact point, and r_t is the radius of the heat flux tube. Considering the microcontact density ρ_{md} of the interface, the overall thermal resistance is the sum of all the single microcontact resistances which are connected in parallel:

$$R_c = \frac{r_c}{\rho_{md}} = \frac{\left(1 - \frac{<r_p>}{<r_t>}\right)^{1.5}}{2\kappa_s\rho_{md} <r_p>} \tag{6.2}$$

where $<r_p>$ and $<r_t b_s>$ are the average radii of the microcontact points and the heat flux tubes, respectively. Equation (6.2) has been widely accepted in the field of R_c. Following the ways of developing the R_c model at the solid–solid contacts [5–8, 15–19], $<r_p>$, $<r_t>$, and ρ_{md} can be estimated by topographical and mechanical analyses at liquid–solid contacts. Here, asperities' resistance is neglected because it is small enough compared to R_c [12].

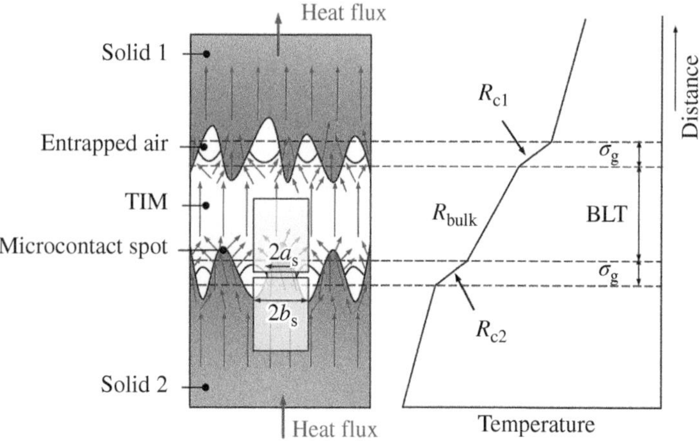

Figure 6.1 Formation of R_c and the resistance components in the solid–TIM–solid structure.

6.1.1.2 Topographical Analysis

To describe the solid surface profile, three key surface roughness parameters are considered: the arithmetic mean deviation of the profile R_a, root-mean-square deviation of the profile R_q, and R_{sm}. Hamasaiid et al. [12] have made three assumptions to simplify the solid surface profile:

① The solid surface profile consists of successive peaks and valleys with the bases on the mean plane. The spacing of the peaks and valleys is identical and equal to R_{sm}.

② The asperities are conical in shape with the same slope (k_{sl}).

③ The heights of asperities (H_a) follow a Gaussian distribution with a distribution function $\varphi(H_a)$ which is expressed as:

$$\varphi(H_a) = \frac{1}{\sqrt{2\pi}\sigma_g} \exp\left(\frac{-H_a}{2\sigma_g^2}\right) \quad H_a \in (-\infty, +\infty) \tag{6.3}$$

Based on the standard definitions, R_a and R_q can be expressed as:

$$R_a = \int_{-\infty}^{+\infty} \varphi(H_a)|H_a| \mathrm{d}H_a \tag{6.4}$$

$$R_q = \sqrt{\int_{-\infty}^{+\infty} \varphi(H_a){H_a}^2 \mathrm{d}H_a} = \sigma_g \tag{6.5}$$

Equation (6.5) shows that R_q is equal to the standard deviation of the Gaussian distribution σ_g.

For the convenience of mechanical analysis, the model proposed by Hamasaiid et al. [12] modifies the real solid surface profile to a new profile as shown in Figure 6.2. The modified surface profile has a common base for the asperities. Then all the valleys are brought to the same level ($H_a = 0$). The asperities are still conical in shape but with a new normalized slope (k_{sln}). To keep the initial surface roughness parameters (R_a, R_q, and R_{sm}) measured by profilometer constant, Hamasaiid et al. [12] propose a new height distribution of peaks $\varphi_B(y)$ for the modified profile which is expressed by the equation:

$$\varphi_B(H_a) = \frac{2}{\sqrt{2\pi}\sigma_g} \exp\left(\frac{-{H_a}^2}{2{\sigma_g}^2}\right) \quad H_a \in [0, +\infty) \tag{6.6}$$

After the transformation of the solid profile, the liquid–solid contact profile is shown in Figure 6.3. The authors [12] have successfully developed the expressions of $<r_p>$, $<r_t>$, and ρ_{md}, respectively, as follows:

$$< r_p > = \int_{H_a = Y}^{\infty} \frac{(H_a - Y)}{k_{sln}} \varphi_B(H_a) \mathrm{d}H_a = \frac{R_{sm}}{2}\left(\exp\left(-\frac{Y^2}{2{\sigma_g}^2}\right) - \sqrt{\frac{\pi}{2}}\frac{Y}{\sigma_g}\operatorname{erfc}\left(\frac{Y}{\sqrt{2}\sigma_g}\right)\right) \tag{6.7}$$

$$< r_t > = \frac{R_{sm}}{2} \tag{6.8}$$

$$\rho_{md} = \frac{8}{\varepsilon\pi^2}\left(\frac{1}{R_{sm}}\right)^2 \operatorname{erfc}\left(\frac{Y}{\sqrt{2}\sigma_g}\right) \tag{6.9}$$

where Y is the mean height of the entrapped air, ε is a factor determined statistically to be close to 1.5 [12], and k_{sln} is given by:

$$k_{sln} = 2\sqrt{\frac{2}{\pi}}\frac{\sigma_g}{R_{sm}} \tag{6.10}$$

By substituting Equations (6.7)–(6.9) into Equation (6.2), the thermal contact resistance is determined as follows:

$$R_c = \frac{\left(1 - \left(\exp\left(-\frac{Y^2}{2{\sigma_g}^2}\right) - \sqrt{\frac{\pi}{2}}\frac{Y}{\sigma_g}\operatorname{erfc}\left(\frac{Y}{\sqrt{2}\sigma_g}\right)\right)\right)^{1.5}}{\frac{8k_s}{1.5\pi^2 R_{sm}}\operatorname{erfc}\left(\frac{Y}{\sqrt{2}\sigma_g}\right)\left(\exp\left(-\frac{Y^2}{2{\sigma_g}^2}\right) - \sqrt{\frac{\pi}{2}}\frac{Y}{\sigma_g}\operatorname{erfc}\left(\frac{Y}{\sqrt{2}\sigma_g}\right)\right)} \tag{6.11}$$

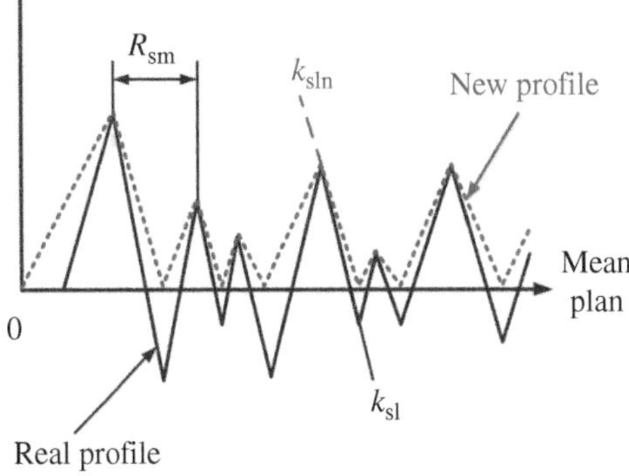

Figure 6.2 Real and new surface profiles.

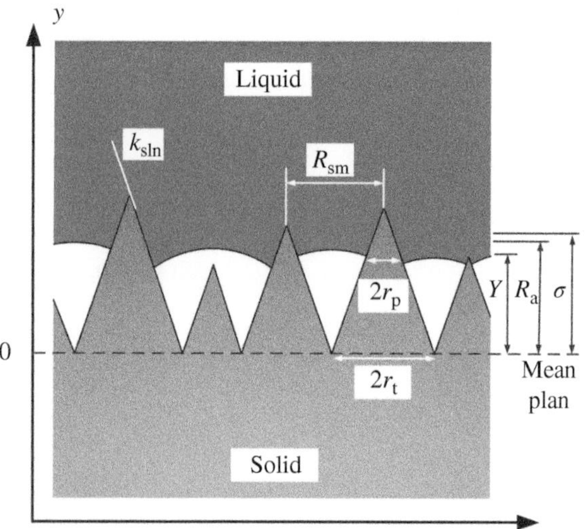

Figure 6.3 Modified profile of the liquid–solid interface.

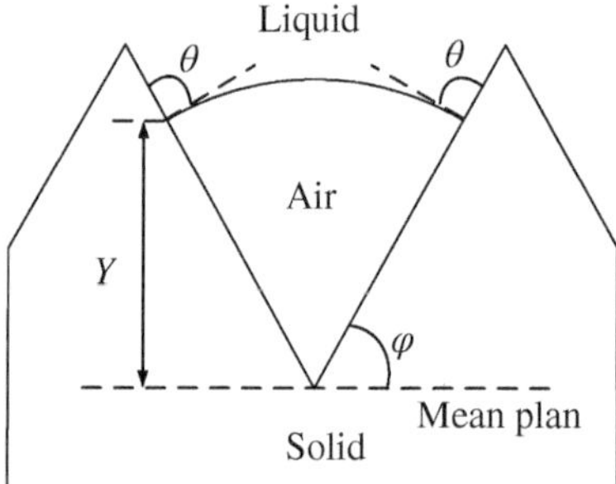

Figure 6.4 Spreading of a liquid material on the solid rough surface.

According to the above equations, R_c is a function of κ_s, surface roughness (r, R_{sm}), and Y. To determine Y, mechanical analysis should be done at the contacting interface.

6.1.1.3 Mechanical Analysis

Figure 6.4 shows the wetting of the liquid on the solid rough surface. When liquid material impinges on the solid surface, it spreads out and wets the microcavity inside which the air is then compressed. An equilibrium state is established at the liquid–solid interface among the applied pressure p_b, the capillarity pressure p_Γ due to the liquid surface energy, and the back pressure of air p_0. Assuming the entrapped air is an ideal gas and not allowed to escape, the ideal gas equation can express the equilibrium as follows:

$$\frac{(p_b \pm p_\Gamma)V_c}{T_c} = \frac{p_0 V_0}{T_0} \tag{6.12}$$

where V_0 and V_c are the volumes of the entrapped air before and after the liquid–solid contact, T_0 and T_c are the initial and contact temperature at the interface, and + or − means good or poor surface wetting, respectively. To solve Equation (6.12), p_Γ must be given first. In Hamasaiid et al.'s model [12], an empirical expression of p_Γ has been evaluated through experiments at the casting–die interface. In this section, the wettability analysis of surface chemistry model is adopted to solve this problem. In surface chemistry model [4], p_Γ is found to be determined by the surface tension of liquid, wetting conditions, and the topography of microcavity. For this problem, p_Γ is given by Ref. [4]:

$$p_\Gamma = \frac{2\Gamma_l \sin(\theta + \varphi)}{Y \cot(\varphi)} \tag{6.13}$$

where Γ_l is the surface energy of liquid, θ is the contact angle of liquid on solid surface, and φ is the angle between the cavity and mean plane which can be expressed as:

$$\varphi = \arctan(k_{sln}) \tag{6.14}$$

According to the above topographical analysis and the assumption that the entrapped air is conical in shape, the average values of V_0 and V_c can be expressed as [12]:

$$V_c = \frac{1}{3}\pi\frac{1}{k_{sln}^2}\int_{H_a=0}^{Y} \varphi_B(H_a){H_a}^3 dH_a \approx \frac{\pi Y^3}{3{k_{sln}}^2} \tag{6.15}$$

$$V_0 = \frac{1}{3}\pi\frac{1}{k_{sln}^2}\int_{H_a=0}^{\infty} \varphi_B(H_a){H_a}^3 dH_a = \frac{2\pi}{3}\sqrt{\frac{2}{\pi}\frac{{\sigma_g}^3}{k_{sln}^2}} \tag{6.16}$$

By substituting Equations (6.13), (6.15), and (6.16) into Equation (6.12), the equilibrium equation can be expressed as follows:

$$\left(p_b \pm \frac{2\Gamma_l \sin(\theta + \varphi)}{Y \cot(\varphi)}\right)\frac{1}{T_c} = \frac{p_0}{T_0}\frac{V_0}{V_c} = \frac{p_0}{T_0}\frac{2\sqrt{\frac{2}{\pi}\sigma_g^3}}{Y^3} \tag{6.17}$$

The contact angle of fluidic TIMs on substrate is less than 90°, which will be illustrated in detail in Section 6.1.2.3, this work deals with the good wetting condition. Thus, Equation (6.17) can be solved to determine the Y for the microcavities:

$$Y \approx \frac{1}{\Gamma + \dfrac{2\Gamma_1 \sin(\theta + \varphi)}{3\Gamma p_0 Y_0^3 \chi \cot(\varphi)}} \tag{6.18}$$

Where

$$\Gamma = \left(\left(\frac{p_b^2}{4p_0^2 Y_0^6 \chi^2} - \frac{8\Gamma_1^3 \sin^3(\theta + \varphi)}{27 p_0^3 Y_0^9 \chi^3 \cot^3(\varphi)} \right)^{1/2} + \frac{p_b}{2p_0 Y_0^3 \chi} \right)^{1/3} \tag{6.19}$$

$$\chi = \frac{T_c}{T_0} \tag{6.20}$$

Equations (6.18)–(6.20) show that Y depends on applied pressure (p_b), wetting ability (Γ_1, θ), topography of the solid surface (σ_g, k_{sln}), and temperature (T_c, T_0). These parameters together with R_{sm} and κ_s determine the R_c according to Equation (6.11). In the following, we will introduce the validation of the improved model via experiments and discuss the influences of σ_g, R_{sm}, κ_s, and p_b on R_c in detail.

6.1.2 Experiment for the Measurement of R_c

6.1.2.1 Experimental Principles

Experimental measurements are conducted on three layers structure as shown in Figure 6.1. The total resistance R_{tot} at the TIM bond lines can be expressed as:

$$R_{tot} = \frac{\Delta T}{P_h} = R_{bulk} + R_{c1} + R_{c2} \tag{6.21}$$

where ΔT is the temperature difference of the bond lines, P_h is the heat transfer rate through the bond lines, R_{bulk} is equal to the ratio of bond line thickness (BLT) to TC of TIM (κ_{TIM}), and R_{c1} is equal to R_{c2} when the roughness of upper and lower substrates is assumed to be identical. Thus, R_{tot} can be given by:

$$R_{tot} = \frac{1}{\kappa_{TIM}} \text{BLT} + 2R_c \tag{6.22}$$

From Equation (6.22), Figure 6.5 shows the plots of R_{tot} versus BLT for multiple materials. It can be seen that collecting R_{tot} data over a range of BLT will create a linear relation with a slope $1/\kappa_{TIM}$ and a vertical axis intercept $2R_c$. Thus, based on this principle, both κ_{TIM} and R_c can be measured.

6.1.2.2 Thermal and BLT Measurement

The R_c and κ_{TIM} were measured using a TIM testing system which is based on the ASTM D 5470 standard [20]. Figure 6.6 illustrates the sketch of testing system consisting of five main components: heat flux meters, heating and cooling units, load fixture, insulation materials, and the camera to measure the BLT.

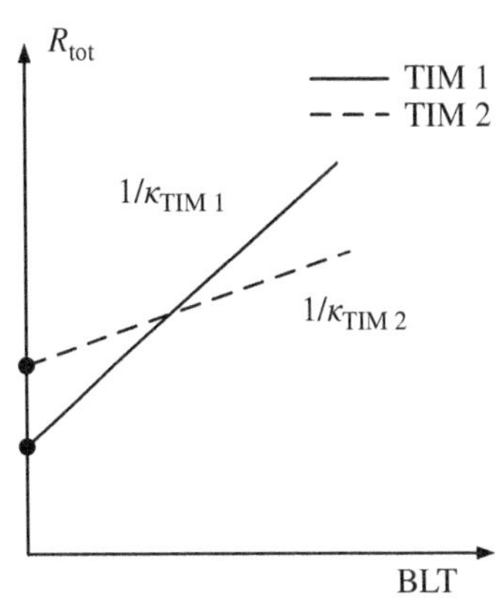

Figure 6.5 Plots of R_{tot} versus BLT for multiple materials.

Heat flow through the specimen is measured by the heat flux meters with a 30 mm × 30 mm cross-sectional area. It is accomplished by measuring the linear temperature gradient dT/dx in the heat flux meters and using Fourier's law of heat conduction $q = \kappa(dT/dx)$. The centerline temperature is measured at 10 mm intervals along each of the flux meters using five 1.5 mm diameter platinum resistance temperature detectors (RTDs).

Power is applied by a heater block embedded with four wire-wound cartridge heaters capable of 120 W. A water-cooled heat sink is designed to remove the heat rejected from the upper heat flux meter. The design uses a micro-pump with a maximum flow rate of 2.3×10^{-5} m$^3 \cdot$ s^{-1} to circulate water. Consequently, heat flow in flux meter is assured to be one dimensional. The contact pressure on the specimen is controlled by the lead screws. A load cell with a resolution of 3.3 kPa continuously monitors the applied load.

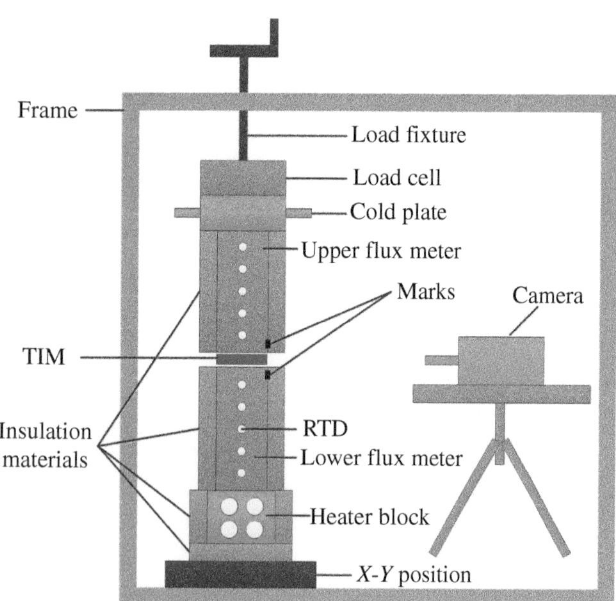

Figure 6.6 Sketch of the testing system.

Insulation tapes with a TC of $0.034\,\mathrm{W \cdot m^{-1} \cdot K^{-1}}$ and a thickness of 6 mm are placed around four sides of the heat flux meters to ensure the heat flow through the lower and upper meters varies by less than 5%. A 10-mm-thick firebond insulation is placed around four sides of heat block to eliminate heat leakage to the environment. Heat loss from the bottom is minimized by attaching it to a 10-mm thick FR-4 epoxy material.

A camera with a microscope lens is implemented to measure BLT, as well as to detect the parallelism of the interface before the start of each test. Two 1 mm × 1 mm square marks are attached to the upper and lower flux meters located approximately 2 mm from the flux meter edges. These marks are used as targets and the camera measures the distance between the centers of the targets. These marks are scanned in an approximately 5 mm field of view. The camera provides 1024 × 1024 bit resolution and each pixel is approximately 5 μm in length, so the resolution of BLT measurement is limited to 5 μm. There are two steps to measure BLT. The first step is to apply a 1 MPa load between the heat flux meters without TIM and measure the targets distance L_0. The second step is to separate the flux meters, fill the TIM between the flux meters, and measure the targets distance L_1 at the desired pressure. Then BLT is calculated by subtracting the L_0 from L_1.

For the measurement of R_c and κ_{TIM}, BLT of TIMs needs to be controlled so that R_{tot} can be measured for a range of BLT as mentioned earlier. Here, stainless steel shims are mixed into the TIMs to control the thickness as shown in Figure 6.7. The shims are 1.5 mm in diameter, and the thermal conduction through them is negligible which has been verified by Prasher et al. [21]. Although the pressure from lead screws is applied on the sample, the shims take most of the pressure.

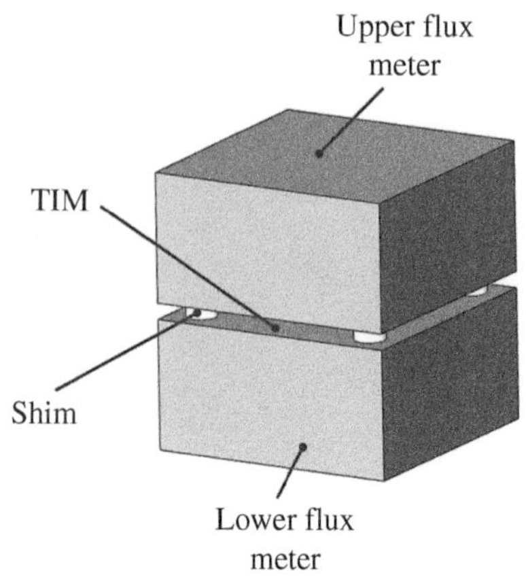

Figure 6.7 Use of shims to control BLT.

6.1.2.3 Sample Preparation

Aluminium-6061 T4 blocks were chosen as the solid substrates, as well as the heat flux meters. The TC is $154\,\mathrm{W \cdot m^{-1} \cdot K^{-1}}$. Four couples of blocks were machined with different surface roughness. For each couple, surface roughness parameters are measured by a Mitutoyo SJ-401 surface profilometer for 12 times, and the average value is set as the roughness of the couple. Shin Etsu KF96H silicone oil and Dow Corning TC5121 thermally conductive grease are prepared for the TIMs. Because this experiment can measure the TC, the average values of experimental results are set to be their TCs. Eight experimental combinations of substrates roughness and TIMs are shown in Table 6.1.

The surface energy of the silicone oil was provided by the supplier and that of the grease and aluminum substrate is measured with standard two liquid methods

Table 6.1 Experimental combinations for the measurement of R_c.

	Materials	Silicone oil ($\kappa_{TIM\,1}$)	Grease ($\kappa_{TIM\,2}$)
Aluminium	$\sigma_g = 0.23\ \mu m$ $R_{sm} = 99.2\ \mu m$	1	5
	$\sigma_g = 0.34\ \mu m$ $R_{sm} = 90.8\ \mu m$	2	6
$154\ W \cdot m^{-1} \cdot K^{-1}$	$\sigma_g = 0.72\ \mu m$ $R_{sm} = 132.1\ \mu m$	3	7
	$\sigma_g = 1.13\ \mu m$ $R_{sm} = 166.4\ \mu m$	4	8

Table 6.2 Surface energy of TIMs and aluminium block, and the contact angle of TIMs on the block.

	Silicone oil	Thermal grease	Aluminium block
$\Gamma_p\ (mN \cdot m^{-1})$	—	24.2	41.6
$\Gamma_d\ (mN \cdot m^{-1})$	—	14.0	4.2
$\Gamma\ (mN \cdot m^{-1})$	21.3	38.2	45.8
θ (deg)	22.3	45.0	—

using water and diiodomethane (CH_2I_2) based on the ASTM D 7490 standard [22]. This method can measure the polar and dispersion components of the material surface energy, Γ_p and Γ_d, respectively. The total surface energy of the material is the sum of Γ_p and Γ_d. The contact angle of silicone oil was directly measured on the smoothest aluminum substrate by the sessile drop method [23]. However, because thermal grease has extremely high viscosity, the method cannot be used for grease. The contact angle of grease on the aluminum substrate can be calculated by using the following equation:

$$(1 + \cos\theta)\Gamma_l = 2\sqrt{\Gamma_{lp}\Gamma_{sp} + \Gamma_{ld}\Gamma_{sd}} \tag{6.23}$$

where the subscripts l and s refer to the liquid and the substrate, respectively. Because Γ_p and Γ_d of the grease and substrate have been measured, the contact angle can be calculated. Table 6.2 shows the surface energy of silicone oil, thermal grease, and aluminum block, and the contact angle of silicone oil and thermal grease on the aluminum block.

6.1.2.4 Error Analysis

According to Equation (6.22), the error in R_c is given as Equation (6.24)

$$\frac{\Delta R_c}{R_c} = \pm\sqrt{\left(\frac{\Delta R}{(R_{tot} - BLT/\kappa_{TIM})/2}\right)^2 + \frac{1}{\kappa_{TIM}^2}\left(\frac{\Delta BLT}{(R_{tot} - BLT/\kappa_{TIM})/2}\right)^2} \tag{6.24}$$

Following the method in Ref. [24], the errors in R for the silicone oil and grease are measured with 3.8×10^{-6} and $1.3 \times 10^{-6}\ m^2 \cdot K \cdot W^{-1}$, respectively. The error in BLT is 5 μm which is equal to the pixel size of the camera.

6.1.3 Validation and Discussion

6.1.3.1 Comparison of Experimental Data with the Model

For the all experimental combinations, experiments were conducted for four different BLTs to find out R_c and κ_{TIM}. Figures 6.8(a) and 6.9(a) shows the plots of thermal resistance R versus BLT for the silicone oil and thermal grease, respectively. It is shown that R is linearly dependent on BLT, thus R_c and κ_{TIM} can be computed by the linear least square method. The solved linear regression equations and correlation coefficients are given in Table 6.3. According to Equation (6.22), TC of silicone oil can be obtained by taking the average of the 1/slope of the groups 1–4 linear equations, which is equal to $0.21\ W \cdot m^{-1} \cdot K^{-1}$. In the same way, TC of grease can be computed, which is equal to $2.44\ W \cdot m^{-1} \cdot K^{-1}$. Meanwhile, R_c for each group can be obtained by taking half of the intercept of the linear regression equations. Figure 6.8(b) shows the experimental results of R_c for the silicone oil and thermal grease, respectively, as well as the error bars. According to Equation (6.24), the biases of silicone oil and grease from the experiments are 2.41×10^{-5} and $2.43 \times 10^{-6}\ m^2 \cdot K \cdot W^{-1}$, respectively.

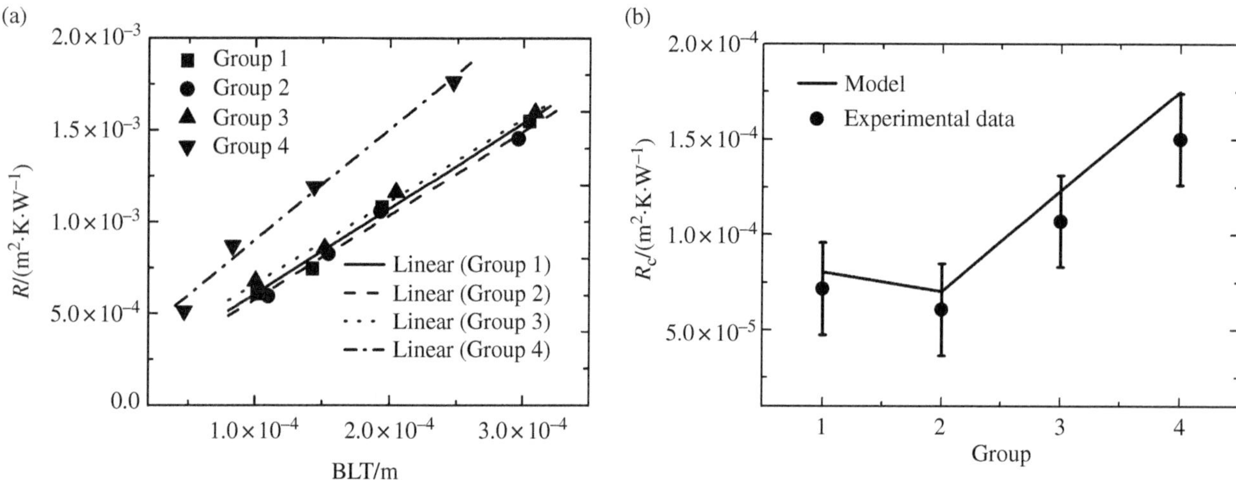

Figure 6.8 (a) Plots of thermal resistance R versus BLT; (b) comparison of the model results with experimental data for silicone oil.

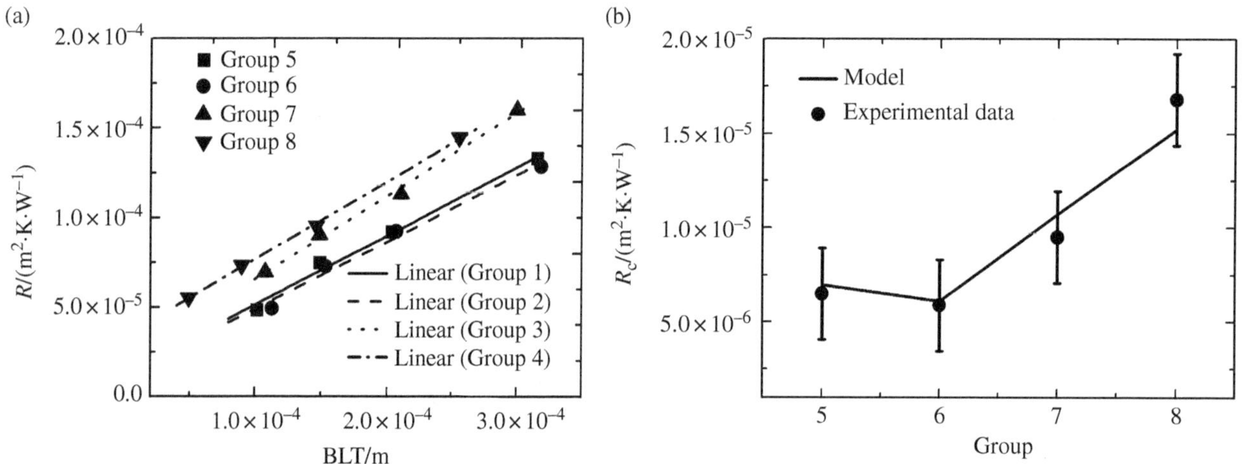

Figure 6.9 (a) Plots of thermal resistance R versus BLT; (b) comparison of the model results with experimental data for thermal grease.

Table 6.3 Linear regression equations and correlation coefficients of all groups.

Group	Linear regression equations	Correlation coefficients r^2
1	$R = 4.64 \times \text{BLT} + 1.431 \times 10^{-4}$	0.9970
2	$R = 4.57 \times \text{BLT} + 1.217 \times 10^{-4}$	0.9899
3	$R = 4.49 \times \text{BLT} + 2.140 \times 10^{-4}$	0.9949
4	$R = 6.02 \times \text{BLT} + 3.003 \times 10^{-4}$	0.9872
5	$R = 0.383 \times \text{BLT} + 1.299 \times 10^{-5}$	0.9911
6	$R = 0.373 \times \text{BLT} + 1.175 \times 10^{-5}$	0.9856
7	$R = 0.465 \times \text{BLT} + 1.898 \times 10^{-5}$	0.9952
8	$R = 0.431 \times \text{BLT} + 3.369 \times 10^{-5}$	0.9991

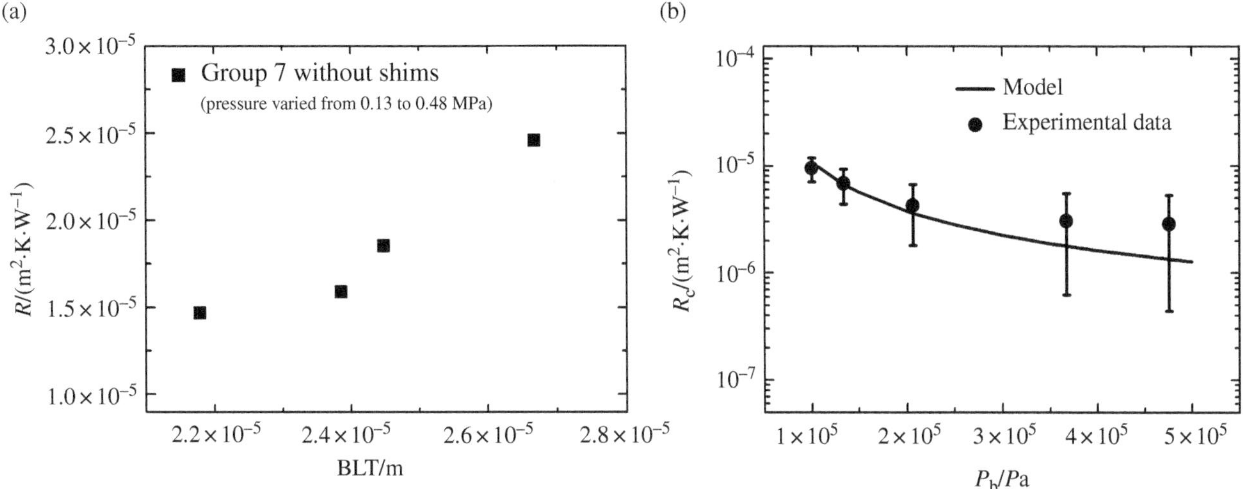

Figure 6.10 (a) Scatters of thermal resistance R versus BLT for group 7 with different applied pressures; (b) comparison of the model results with experimental data.

As mentioned earlier, predicting R_c needs to obtain the parameters as follows: topography parameters (σ_g, R_{sm}), wettability parameters (Γ_l, θ), mechanical parameter (p_b), and thermal parameters (κ_s, T_0, T_c). In the experiments of groups 1–8, p_b is 0.1 MPa, T_0 is 288 K, and T_c is approximately 323 K by taking the average temperature of the TIMs in the testing. Other parameters can be obtained in Tables 6.1 and 6.2. Figures 6.8(b) and 6.9(b) show the model results of R_c for all groups. Compared with the experimental results, it can be found that the improved model matches well with experimental data. The deviation between the experimental and model results is less than 14.3%.

In order to study the influence of p_b on R_c, different pressures varied from 0.13 to 0.48 MPa were set on group 7 without the shims controlling the BLT. Figure 6.10(a) shows the scatters of measured R versus measured BLTs. Based on Equation (6.22), the values of R_c with various pressures can be computed. Figure 6.10(b) shows the comparison between the experimental and model results. At lower pressure, the model matches well with experimental results. But the model predictions tend to deviate at higher pressure owing to the higher relative error in thin BLT measurement.

6.1.3.2 Influence of the Parameters on the Model Results

In Ref. [12], the authors have highlighted the effect of surface roughness on R_c in liquid–solid interface. According to their predictive model, σ does not have any influence on R_c, while R_{sm} has more effect on R_c. Similar results can be found in the improved model. Using the parameters of group 7, Figure 6.11 presents the curves of R_c versus σ_g and R_{sm} when one of them keeps constant. It is shown that R_c changes hardly with σ_g, but increases with R_{sm} obviously. In order to explain the results, Figure 6.12 shows the plots of $<r_p>/<r_t>$ and $<r_p> \times \rho_{md}$ as a function of σ_g and those as a function of R_{sm}. When R_{sm} keeps constant, σ_g nearly has no significant influence on $<r_p>/<r_t>$ or $<r_p> \times \rho_{md}$. Thus, R_c does not change with σ according to Equation (6.2). Contrary to σ_g, the larger the value of R_{sm}, the smaller the $<r_p> \times \rho_{md}$. And the $<r_p>/<r_t>$ still remains constant so that R_c turns larger according to Equation (6.2). Comparing the R_c between groups 1 and 2 in Figure 6.8(b), or between groups 5 and 6 in Figure 6.9(b), the experimental results seem to be accorded with the claim that R_{sm} has more important impact on R_c. However, the considerable experimental bias and the limited experimental samples fail to make it credible enough to verify the point. Thus, it is desired to investigate this work in the future to validate the results of the improved R_c model.

Comparing the predictive results of R_c between the groups having the same roughness parameters, such as groups 1 and 5 and groups 2 and 6, it can be found that R_c is approximately inversely proportional to TC of TIMs. TC of aluminum is much larger than κ_{TIM} and this makes κ_s approximately equal to $2\kappa_{TIM}$. Then Equation (6.2) shows R_c and varies in an inverse proportion to κ_{TIM}.

As illustrated in Figure 6.10(b), R_c decreases with increasing p_b. To better understand the effect of p_b on R_c, Y is plotted as a function of p_b and it is shown in Figure 6.13. It shows that Y decreases with increasing p_b. Thus, higher p_b makes the liquid and solid contact better and results in a lower R_c at the interface.

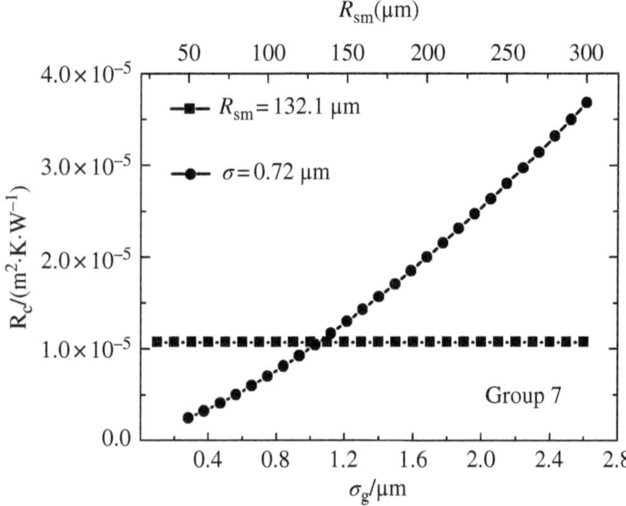

Figure 6.11 Variation of R_c with σ_g and R_{sm}, respectively, when one of them is kept constant.

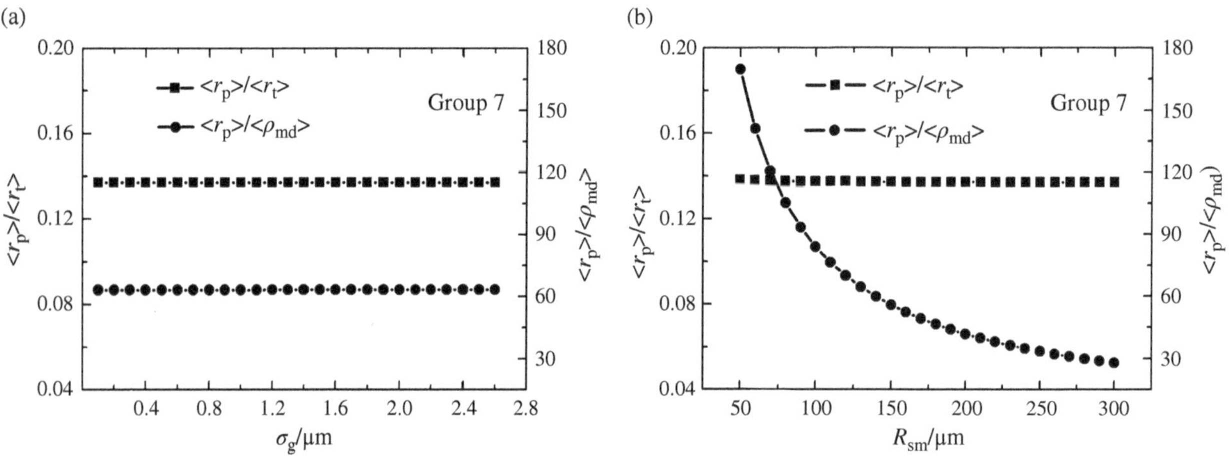

Figure 6.12 Variation of <r_t>/<r_p> and <r_t> × ρ_{md} with (a) σ_g when R_{sm} is kept constant; (b) R_{sm} when σ_g is kept constant.

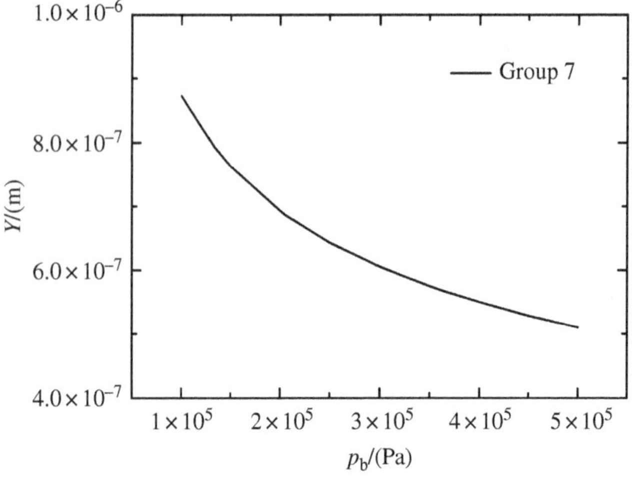

Figure 6.13 Variation of Y as a function p_b.

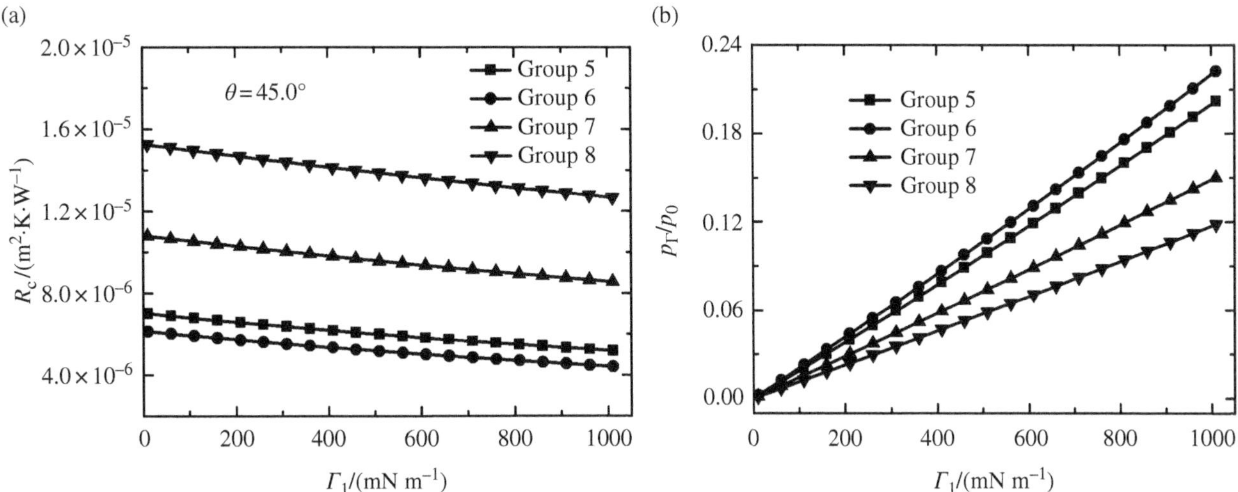

Figure 6.14 (a) Variation of R_c and (b) p_Γ/p_0 with Γ_1 when θ is kept constant.

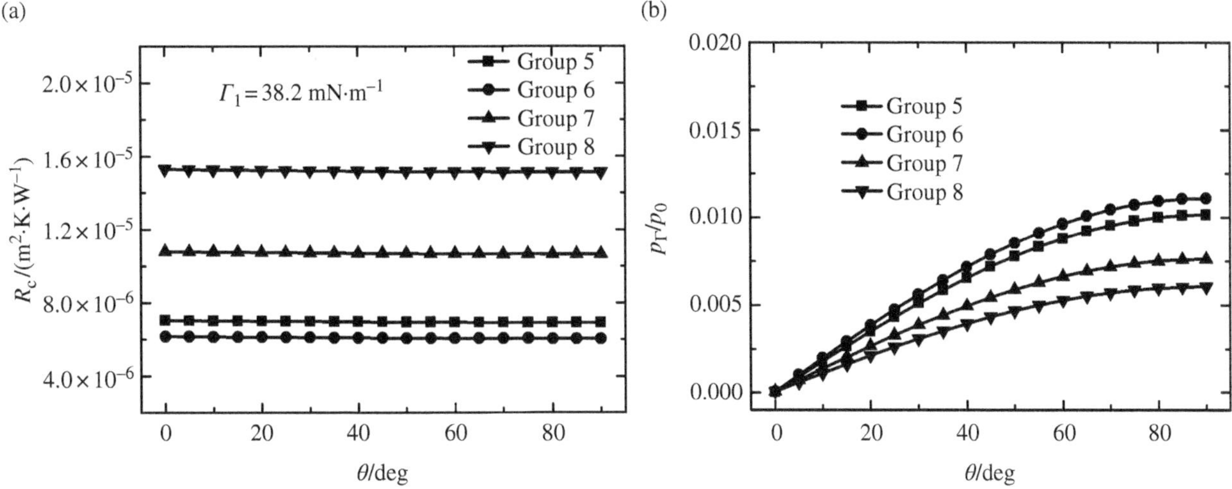

Figure 6.15 (a) Variation of R_c and (b) p_Γ/p_0 with θ when Γ_1 is kept constant.

Finally, the effect of wettability parameters (Γ_1, θ) on R_c is investigated. Figures 6.14(a) and 6.15(a), respectively, present the curves of R_c versus Γ_1 and θ for the groups 5–8. From Figure 6.14(a), it is seen that R_c tendentially decreases with Γ_1 increasing when θ is equal to 45.0°. Figure 6.14(b) plots p_Γ/p_0 as a function of Γ_1. The figure shows that p_Γ increases with Γ_1 increasing. Then Y decreases according to Equation (6.18) and, finally, R_c decreases. From the figure, it can also be found that the larger the R_{sm} is, the lower the p_Γ is. On the other hand, Figure 6.15(a) shows a very slow decrease of R_c with θ increasing when Γ_1 is relatively small and equal to 38.2 $\text{mN} \cdot \text{m}^{-1}$. In order to explain the results, p_Γ/p_0 is plotted as a function of θ as shown in Figure 6.15(b). It shows that p_Γ increase with θ increasing. However, the value of p_Γ is much smaller than p_0 which means p_Γ makes less contribution to Y. So, for the materials with small Γ_1, R_c hardly changes with θ.

6.2 Thermal Conductivity Tunability of TIM

Hexagonal boron nitride (hBN) platelets are widely used as the reinforcing fillers for enhancing the TC of polymer-based composites. Since hBN platelets have high aspect ratio and show a highly anisotropic thermal property, the TC of the hBNs-filled composites should be strongly associated with the platelets' orientation. However, the orientation effect has been explored less frequently due to the technical difficulties in precontrol of the platelets' orientation in the polymer matrix. In Section 6.2.1, we will introduce the use of magnetic fields to assemble the platelets into various microstructures and to study the TCs of the designed composites. The experimental results showed that TCs are dramatically different among these composites. For instance, the TCs of the composites with platelets oriented parallel and perpendicular to the heat flux direction are, respectively, 44.5% higher and 37.9% lower than that of unaligned composites at the volume fraction of 9.14%. The results were also analyzed by a theoretical model. The model suggests that the orientation of the hBN platelets is the main reason for the variance in the TC.

In order to enhance the materials heat transport capability, the hBN platelets are expected to be assembled into a well-ordered structure. Such structure has been achieved in practice by the magnetic alignment approach. However, this approach is limited to the composites loaded with low-volume fraction of platelets (<10%). In Section 6.2.2, we will introduce the use of combined mechanical and magnetic stimuli to fabricate the well-aligned composites at the volume fraction of 10% to 20%. The platelets in the resulting composites exhibit a high degree of alignment. For instance, in the 10 vol% composite, the angle of 95.3% of platelets is greater than 45°, only ~5% of platelets fall into the horizontal direction. TC of the composites was investigated experimentally. It exhibited a strong correlation with the platelet alignment. The measured TC of 10 vol% aligned composite is 74% higher than that of unaligned composite. TC was also analyzed by a theoretical model. Thermal boundary resistance (R_b), arising at the platelets–matrix interface, was extracted by fitting the measured TC to model prediction. R_b is found to decrease with the increase of alignment degree. The results suggest that assembling the platelets into a well-ordered structure can greatly enhance the heat transport capability due to the formation of conductive networks and the reduction of R_b.

Local heat source in electronic device is likely to produce hotspots which can degrade the reliability and performance of the device. Various materials have been attempted to enhance the heat dissipation of local heat source. Many theoretical studies have demonstrated that the heterogeneous composite materials with fillers concentrated at the preferential paths of heat flux are effective in cooling the local heat source. However, this unique control of microstructure and property of polymer-based composites has been less achieved in practice due to the technical difficulties in controlling the fillers positions. In Section 6.2.3, we will introduce a locally reinforced heterogeneous composite with conductive particles concentrated at the preferential path of heat flux to cool the local heat source. The local reinforcement was achieved by using magnetically responsive particles as reinforcing elements and a specific magnetic field to organize the elements into the predefined structure. To evaluate the thermal performance of the proposed material, the comparative thermal tests were performed. The results show that compared to the homogeneous composites, the present composites with local reinforcement can significantly enhance the heat dissipation of local heat source. When the heat flux is 5840 $W \cdot m^{-2}$, the locally reinforced composites with a fillers volume fraction of 5% reduced the average and maximum temperature of heater 7.7 and 8.7 °C, respectively.

Carbon fibers (CFs) are widely used as the reinforcing fillers to enhance the TCs of polymer-based composites because of their extremely high TC in the axial direction. However, conventional methods of CFs cannot take full advantage of their high TCs because heat conductive channels in composite are not efficiently built due to the greatly higher viscosity of composite caused by CFs under high filler concentration. To solve this problem, in Section 6.2.4, we introduce a magnetic field-based and viscosity-independent method to fabricate the nickel-coated carbon fibers (NICFs)-filled polydimethylsiloxane (PDMS) composites with the highspeed through-plane heat conductive channels under high filler concentration. The NICFs-composite shows 69 times enhanced through-plane TC (10.50 $W \cdot m^{-1} \cdot K^{-1}$) compared to that of pure PDMS (0.15 $W \cdot m^{-1} \cdot K^{-1}$) and a low coefficient of thermal expansion (CTE) of 55.14 $ppm \cdot °C^{-1}$ at 51.54 wt%. Compared to commercial TIMs, the NICFs-composites exhibit better thermal performance, demonstrating potential application prospects in electronic device cooling area.

Inside filler-reinforced polymer composites, thermally conductive fillers are usually uniformly distributed or unidirectionally oriented in polymer matrix to improve thermal performances. However, the ever-shrinking and spatially distributed heat sources in three-dimensional (3D), high-density packaged electronic devices have created the localized "hotspot" problem, which raises a new challenge and stricter requirement for the composite thermal materials. Inspired by the amazing radial microstructures in ginkgo leaf, in Section 6.2.5, we will introduce a flow field-driven self-assembly strategy to fabricate functional thermal materials with radially oriented CFs. To quantitatively evaluate the orientation, an orientation algorithm based on microscale image identification was developed, and an evaluation criterion was introduced. The underlying orientation mechanisms of fillers under the driving of flow field were revealed by visual simulation of vacuum filtration. Thanks to the well-oriented fillers architecture, the composites demonstrated an ultrahigh in-plane TC of 35.5 $W \cdot m^{-1} \cdot K^{-1}$ with a TC anisotropy of 19.8, which enables rapid and efficient heat dissipation pathways toward localized hotspots.

In addition, this flow field-driven self-assembly strategy provides a promising self-design ability that is expected to solve the heat dissipation of arbitrary-shaped heat sources and shed light on other application scenarios like efficient solar–thermal–electric conversion.

6.2.1 Thermal Conductivity Enhancement of BN-Composites Using Magnetic Field

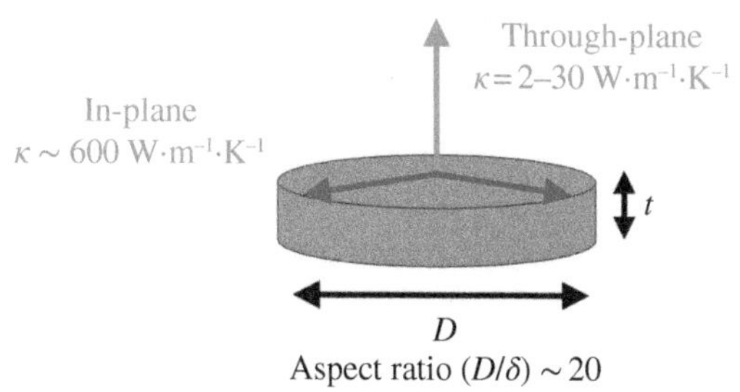

Figure 6.16 Representation of hBN platelets geometry and anisotropic properties.

Thermally conducting but electrically insulating polymer-based composites have been widely used in electronics for die attachments, encapsulations, and dielectrics [25–28]. These materials are predominantly manufactured by introducing the highly thermally conducting particles, such as ceramics, metals, or metal oxides into the polymer matrix [1, 3, 29–31]. Nowadays, hBN has been receiving significant attention as the new form of reinforcing filler due to its excellent thermal conductive and electrically insulating properties. As schematically shown in Figure 6.16, hBN is a platelet-shaped particle with a high aspect ratio (D/δ) and shows a highly anisotropic thermal property: the in-plane TC is about $600\ \mathrm{W \cdot m^{-1} \cdot K^{-1}}$[32], whereas the through-plane TC is only $2\text{–}30\ \mathrm{W \cdot m^{-1} \cdot K^{-1}}$[32, 33]. Owing to its significant anisotropies in shape and TC, TC of the hBNs-filled composites should be strongly associated with hBNs orientation. However, the orientation effect has less been studied and the thermal properties of hBNs-filled composites have often been assumed to be isotropic [34–36].

The orientation effect is less explored due to the technical difficulties in precontrol of the orientation of platelets in the polymer matrix. Several synthetic approaches have been proposed to obtain the well-ordered microstructures in platelet-reinforced composites, including electrical fields [37], tape-casting [38], and freeze-casting followed by sintering and polymer infiltration [39]. These approaches usually yield composites that can exhibit highly anisotropic properties but normally require multiple processing steps. The squeezing process is a simple but effective method to control the platelets orientation [35]. The platelets are preferentially aligned parallel to the flow direction of the polymer matrix [35]. However, this method is only for the in-plane alignment and limited to the thin films.

Recently, an attractive strategy was proposed to control filler orientation or position in a matrix [40]. The approach relies on coating the reinforcing particles with superparamagnetic nanoparticles to make them magnetically responsive [40, 41]. These coated particles exhibit an ultrahigh magnetic response (UHMR) [40] that enables remote control over their orientation under low external magnetic fields in low-viscosity suspending fluids. Such fluids can be then consolidated to fix the magnetically imposed orientation. With this strategy, various microstructures have been achieved in the polymer-based composites [32, 40].

In this section, the TC of hBN-filled composites was investigated as a function of the orientation of the hBN. In the beginning, the composites with through-plane- and in-plane-aligned hBN platelets were both fabricated with the magnetic alignment method. For the purpose of comparison, the composite with randomly oriented platelets was also prepared without the magnetic control. Then the morphology of those composites was characterized by scanning electron microscopy (SEM), and the degree of orientation (ORI) was estimated by X-ray diffraction (XRD) measurements. After that, through-plane TC (κ_T) was measured for those composites. A theoretical model was finally used to fit and analyze the measured results.

6.2.1.1 Fabrication of the Composites

Preparation of Magnetically Responsive hBN (mhBN) Platelets hBN platelets are coated with superparamagnetic iron oxide nanoparticles to make them responsive to magnetic fields via a previously reported procedure [40]. hBN platelets, with an averaged diameter D of 5 μm, were kindly provided by Momentive. About 4 g of hBN platelets were first suspended in 200 ml of deionized water at pH = 7. Under continuous stirring, 400 μl of EMG-605 ferrofluid (Ferrotec, USA) diluted with 5 ml of deionized water was added dropwise. The EMG-605 ferrofluid is an aqueous suspension containing iron oxide nanoparticles coated with a cationic surfactant. At pH = 7, the hBN platelets have a negative surface charge. Thus, the iron oxide nanoparticles were attached to platelets surface through the electrostatic interaction between the positively charged nanoparticles and negatively charged platelets. The suspension was incubated for 1 hour to coat the platelets with all the iron oxide nanoparticles. After that, the coated platelets were washed three times with deionized water, by repeatedly changing the supernatant solution after the platelets have settled onto the bottom of a glass vial. Subsequently, the magnetized platelets were dried for 12 hours at 90 °C in vacuum.

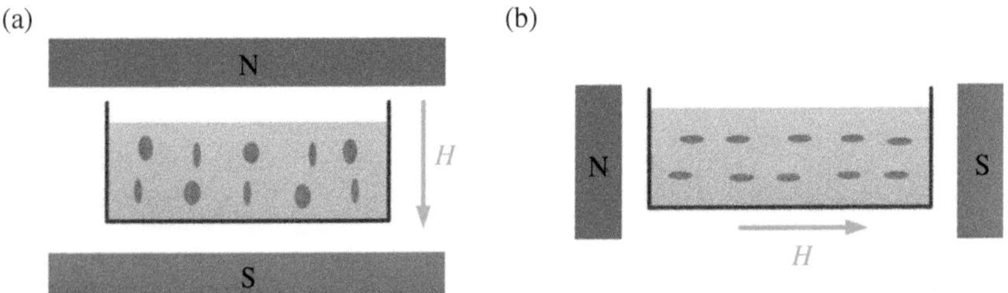

Figure 6.17 Schematic for (a) through-plane and (b) in-plane alignments of mhBN platelets at the linear and uniform magnetic fields.

Preparation of Composites with Controlled Orientation of Reinforcements Composites with controlled orientation of reinforcements were prepared using the polymer matrix and magnetically responsive hBN (mhBN) platelets. Here, silicone gel (OE-6550, Dow Corning) was selected as the matrix due to its high thermal and chemical stabilities, and low viscosity which is benefit for the filler alignment. mhBN platelets were first suspended in silicone gel and stirred for 30 minutes to fully disperse. Then, the curing catalyst was added with stirring. The bubbles introduced during the stirring process were removed by applying alternating cycles of vacuum. After that, the polymer suspension was poured onto a 30 mm × 30 mm × 2 mm Teflon mold. To obtain the composites with vertical or horizontal aligned platelets, the mold was placed into the magnetic field with corresponding direction. The linear, uniform magnetic field was applied using two parallel arranged custom 200 mm × 100 mm rectangular solenoids. Figure 6.17 schematically shows the alignment of mhBN platelets with the magnetic field control. Samples were heated with the magnetic field at 60 °C for 6 hours to lower down the viscosity of silicone gel for efficient filler alignment. An annealing step at 150 °C for 6 hours was conducted to ensure the fully curing of the composites. For the comparative purposes, the polymer suspension was also cured without magnetic field to obtain the unaligned composites.

6.2.1.2 Characterization and Analysis

Field-emission scanning electron microscopy (FSEM, Quanta 450 FEG) was used to characterize the morphology of mhBN platelets at an accelerating voltage of 20 kV. The platelets were first sputtered with a thin layer of graphite before visualization in the instrument. To confirm the presence of iron oxide particles on mhBN platelets, XRD analysis of the platelets was carried out with X'Pert PRO using Cu Kα radiation (40 kV and 40 mA). To confirm the attachment of iron oxide nanoparticles to the surface of hBN platelet, SEM was carried out using the FEI Nova Nano SEM 450.

The morphology of mhBN–silicone composite surfaces was observed using focused-ion beam scanning electron microscopy (FIB-SEM, Quanta 3D FEG) at an accelerating voltage of 30 kV. Before visualization, a focused-ion beam was used to remove selectively the outer layer of silicone gel to expose the inner mhBN platelets. Then the platelets can be easily visualized from SEM images. XRD analysis of mhBN–silicone composites was also conducted to evaluate the orientation of mhBN platelets in composites. Before the tests, the composites were cut to expose the inner surfaces.

TC is calculated by $\kappa = \alpha \times C_p \times \rho$, where α, C_p, and ρ are thermal diffusivity, heat capacity, and density of sample, respectively. α of the composite was measured with a laser flash method using an LFA 457 (Netzsch) at 25 °C. The geometry of testing sample is a 10 mm × 10 mm × 2 mm cuboid which was cut from the prepared cured composites. Before the test, a thin graphite film is applied on the composite surfaces to increase the energy absorption and the emittance of the surfaces. During the test, heat propagates from the bottom to the top surface of the material. C_p was tested by a differential scanning calorimetry (DSC) (Diamond DSC, PerkinElmer Instruments). ρ was calculated from composite weight and dimensions of the composite.

Characterization of mhBN Platelets Figure 6.18(a) shows the FSEM images of mhBN platelets. From this figure, large aspect ratio (D/δ) is found for the platelets. Since the thickness (δ) of platelets is estimated to be 250 nm [32], D/δ is approximately equal to 20. The left side of Figure 6.18(b) shows a digital photograph of mhBN platelets dispersion in acetone. The right side of Figure 6.18(b) presents a facile separation of mhBN platelets from the acetone by an external magnetic field. So, this figure illustrates that mhBN platelets can respond to the external magnetic field. The presence of iron oxide particles is verified by the XRD pattern of mhBN platelets (Figure 6.18(c)), in which the diffraction peaks [(220), (311), (511), (440)] from iron oxide are observed. Figure 6.18(d) shows the nanostructure of mhBN platelet. It reveals that the size of iron oxide nanoparticles is about 20 nm and further confirms the attachment of nanoparticles to the platelet surface.

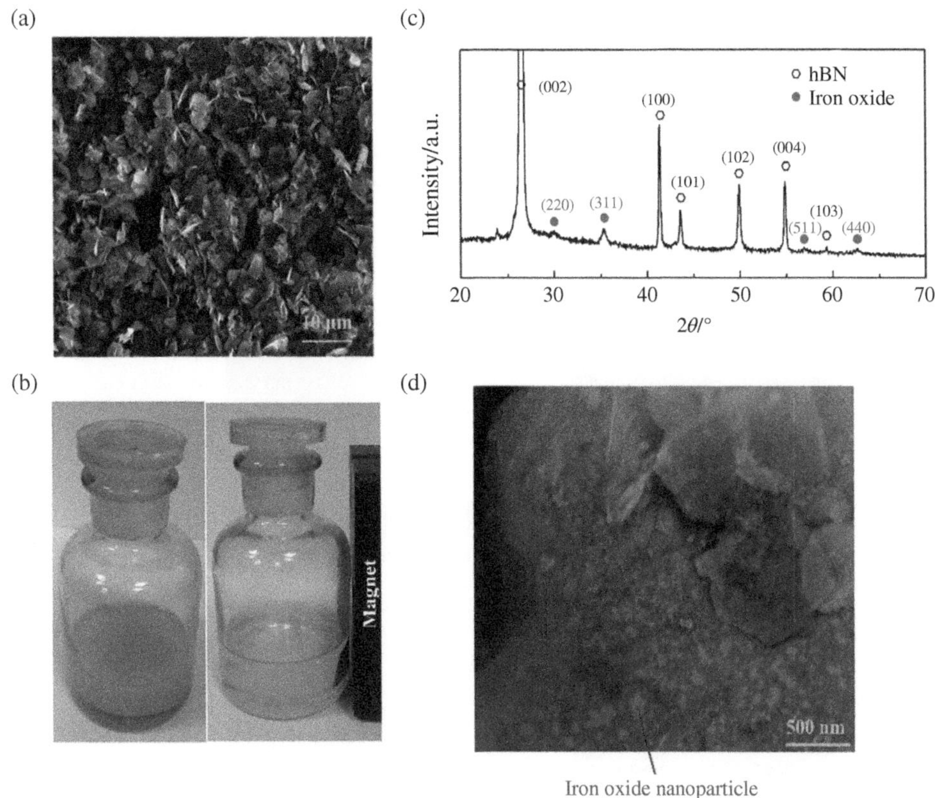

Figure 6.18 mhBN platelets characterization. (a) FSEM image of mhBN platelets. (b) Images of mhBN platelets dispersion in acetone and the separation of platelets from the acetone by an external magnetic field. (c) XRD pattern of mhBN platelets; the black hexagons refer to hBN peaks, and the red circles refer to iron oxide nanoparticles peaks. (d) SEM image shows the nanostructure of mhBN platelet. (a, b, d) Yuan et al. [42]/with permission of American Chemical Society.

Characterization of the Composites with Controlled Orientation of mhBN Platelets Through-plane-aligned mhBN-silicone (TmhBN-silicone) and in-plane-aligned mhBN-silicone (ImhBN-silicone) were prepared by applying a through-plane and an in-plane magnetic field, respectively. For the comparative purposes, randomly oriented mhBN-silicone (RmhBN-silicone) was fabricated without the magnetic control. Figure 6.19(a)–(c) schematically show the microstructure of the TmhBN-silicone, ImhBN-silicone, and RmhBN-silicone composites, respectively, and Table 6.4 summarizes the representative morphologies of mhBN platelets at cross-section and top view of those composites. FIB-SEM images of 9.14 vol% mhBN-silicone composites are shown in Figure 6.19(d)–(i). Figure 6.19(d) and (g) give the cross-sectional and top-view FIB-SEMs of TmhBN-silicone, respectively. Figure 6.19(d) illustrates that mhBN platelets align along the orientation of magnetic field. Two representative morphologies of platelets exist in the cross-sectional view: 1D vertical rods and 2D plates, which are indicated by green and blue arrows respectively. In the top view (Figure 6.19(g)), most platelets are projected as 1D rods with random orientation. Schematic pictures of cross-section and top-view are also given in the insets. It is found that the morphology observed in FIB-SEM is very close to the schematic pictures. This demonstrates the effective magnetic alignment of mhBN in silicone. As given in Table 6.4, the ideal morphology for the ImhBN-silicone is that the platelets should be projected as the 1D horizontal rods in the cross-sectional view and 2D plates in the top-view. The cross-sectional and top-view FIB-SEMs of the ImhBN-silicone are given in Figure 6.19(e) and (h), respectively. It is shown that except for some unexpected platelets indicated by red arrows, the observed morphology accords with the ideal morphology. In contrast, in the RmhBN-silicone, the mhBN platelets orientated randomly and both through-plane and in-plane alignment characterizations can be found in the FIB-SEMs. For example, 1D vertical and horizontal rods which are separately observed in cross-section of TmhBN-silicone and ImhBN-silicone exist simultaneously in the cross-section of RmhBN-silicone (Figure 6.19(f)). Moreover, in the top-view FIB-SEM (Figure 6.19(i)), both 2D plates and 1D rods can be found abundantly.

To further determine the orientation of mhBNs in silicone matrix, XRD analysis was carried out. Figure 6.20(a) gives the XRD patterns. It is shown that the peak intensities for the three composites are dramatically different. As reported

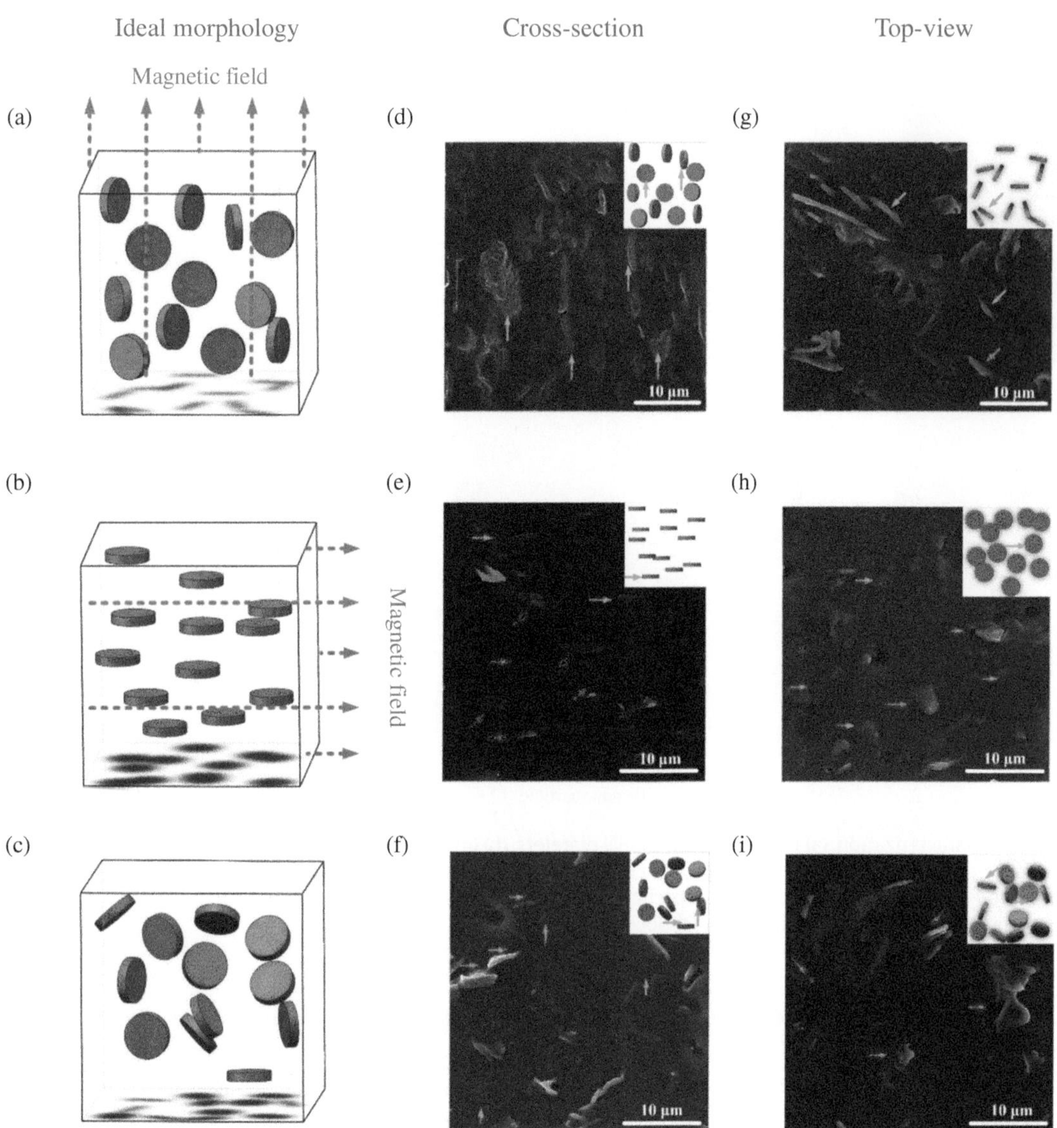

Figure 6.19 Alignment of mhBN platelets in composites with a volume faction of 9.14%. (a) Schematic and (d and g) cross-sectional and top-view FIB-SEMs of composites with through-plane-oriented platelets. (b) Schematic and (e and h) cross-sectional and top-view FIB-SEMs of composites with in-plane-oriented platelets. (c) Schematic and (f and i) cross-sectional and top-view FIB-SEMs of composites with randomly oriented platelets. 1D rods and 2D plates are indicated by green and blue arrows, respectively, and the unexpected morphology is indicated by red arrows. (d–i) Yuan et al. [42]/with permission of American Chemical Society.

Table 6.4 Representative morphologies of mhBN platelets in TmhBN-silicone, RmhBN-silicone, and ImhBN-silicone composites.

	TmhBN-silicone	ImhBN-silicone	RmhBN-silicone
Cross-sectional	1D vertical rods and 2D plates	1D horizontal rods	Randomly oriented plates or rods
Top-view	Randomly oriented 1D rods	2D plates	

previously [43] and schematically illustrated in Figure 6.20(b), the horizontally and vertically oriented hBNs are responsible for the (002) and (100) peaks, respectively. Thus, the difference in peak intensities suggests the various degrees of orientation (ORI) of mhBNs. ORI is estimated by comparing the relative intensity (I) of the (100) peak to the sum of the relative intensities of the (002) and (100) peaks.

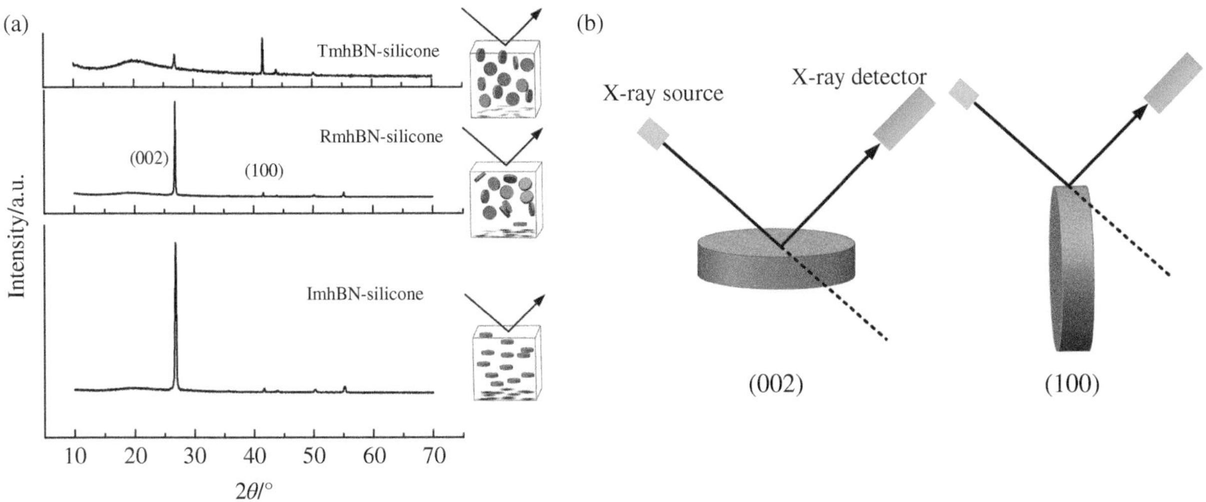

Figure 6.20 XRD analysis of mhBN-silicone composites. (a) XRD patterns of TmhBN-silicone, RmhBN-silicone, and ImhBN-silicone composites. (b) Illustration of hBN orientation effect on XRD pattern.

$$\text{ORI} = \frac{I_{100}}{I_{002} + I_{100}} \times 100\% \tag{6.25}$$

Table 6.5 summarizes the values of ORI for the three composites. It is found that ORI of TmhBN-silicone is more than two times larger than that of RmhBN-silicone, which suggests the larger amount of through-plane-oriented mhBNs in TmhBN-silicone. Then, ORI is compared between the RmhBN-silicone and ImhBN-silicone. The result is that ORI of ImhBN-silicone is approximately five times lower than that of RmhBN-silicone. The lower value of ORI demonstrates that ImhBN-silicone has a larger amount of ImhBNs and lower amount of TmhBNs.

Table 6.5 Degree of mhBNs orientation of TmhBN-silicone, RmhBN-silicone, and ImhBN-silicone composites, respectively.

Composites	ORI (%)
TmhBN-silicone	65.6
RmhBN-silicone	27.2
ImhBN-silicone	5.9

6.2.1.3 Thermal Properties of Composites

Table 6.6 gives the measured data of ρ, C_p, and through-plane thermal diffusivity (α_T) for the three composites at the volume fraction (*f*) of 5%, 7.5%, and 9.14%, respectively. From the data, κ_T can be calculated. Figure 6.21 shows the measured κ_T of TmhBN-silicone, RmhBN-silicone, and ImhBN-silicone, respectively. The results demonstrate the same tendency for all composites that κ_T increases with the increase of *f*. This tendency can be explained by the greater contribution of thermal transport through mhBNs at higher *f*. However, κ_T is dramatically different among the three composites with the same *f*. Figure 6.21 gives the thermal enhancements of TmhBN-silicone and ImhBN-silicone referring to RmhBN-silicone, respectively. TmhBN-silicone exhibits 35.4%, 43.6%, and 44.5% higher κ_T for the *f* at 5%, 7.5%, and 9.14%, respectively. Higher κ_T of

Table 6.6 Density, heat capacity and through-plane thermal diffusivity of mhBN-silicone composites at volume fraction ranging from 5% to 9.14%.

Volume fraction *f* (%)	Density ρ (g · cm^{-3})	Heat capacity C_p (J · g^{-1} · K^{-1})	Through-plane thermal diffusivity α_T (mm^2 · s^{-1})		
			TmhBN	RmhBN	ImhBN
5	1.20	1.36	0.237	0.175	0.123
7.5	1.22	1.33	0.311	0.215	0.135
9.14	1.24	1.31	0.357	0.247	0.154

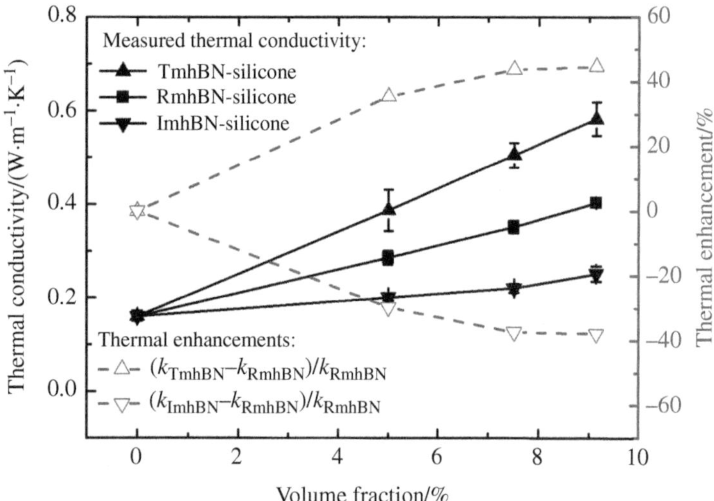

Figure 6.21 Thermal conductivities of composites corresponding thermal enhancements.

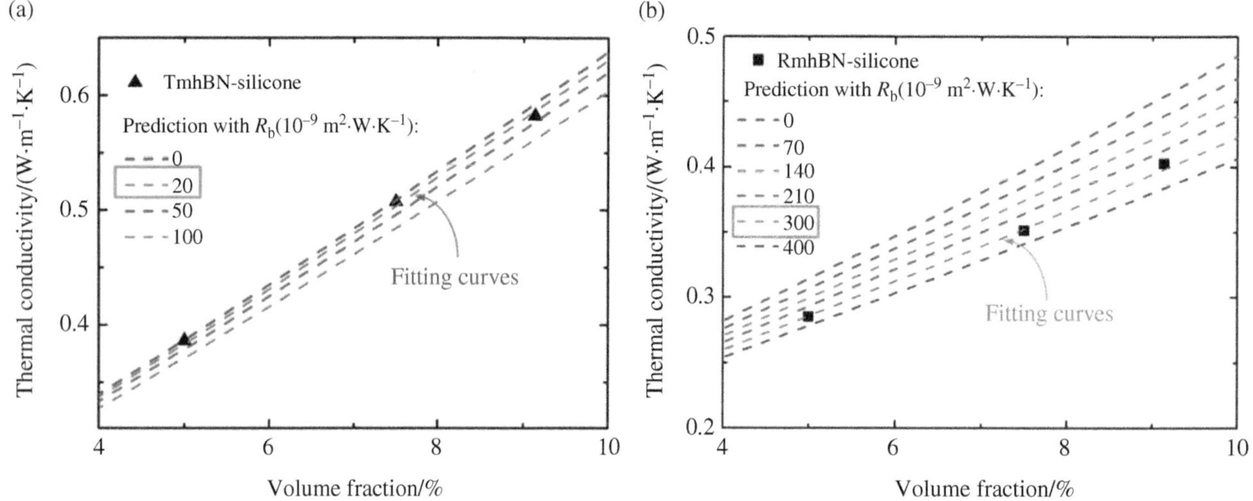

Figure 6.22 Data fitting to extract the thermal boundary resistance (R_b) of (a) TmhBN-silicone and (b) RmhBN-silicone composites. Black dots: measured through-plane thermal conductivity (κ_T) of mhBN-silicone composites; colored dash lines: predicted κ_T from effective medium approximation (EMA) with different values of R_b.

TmhBN-silicone is owing to the formation of efficient thermal pathways made by the mhBN platelets oriented parallel to the heat flux. In sharp contrast, κ_T of the composites containing the ImhBNs at those volume fractions are 29.7%, 37.4%, and 37.9% lower compared to the RmhBN-silicone. The results illustrate that κ_T will be suppressed when the mhBNs orient perpendicular to the heat flux.

A theoretical model is applied to analyze the experimental data. The modified effective medium approximation (EMA) [44] is very useful in predicting κ_T of composites containing the platelet-shaped fillers. This model takes into account filler geometry and orientation, volume fraction, and thermal boundary resistance (R_b) at the filler–matrix interface. It is well-known [1, 3, 31, 44] that R_b arises from the combination of a poor mechanical or chemical adherence at the filler–matrix interface and a thermal expansion mismatch. Here, the added iron oxide nanoparticles on hBN surface are another factor creating R_b at the interface [32]. Figure 6.22 gives the predicted κ_T from EMA with the consideration of R_b, it can be observed that R_b has the adverse impact on κ_T. By fitting the measured κ_T to EMA prediction, the actual R_b of the composites can be extracted. Figure 6.22(a) and (b) give the fitting results for the TmhBN-silicone and RmhBN-silicone, respectively. R_b for the TmhBN-silicone is found to be $20 \times 10^{-9}\ m^2 \cdot W \cdot K^{-1}$, which is comparable with the reported value for the polymer composites with through-plane-oriented mhBNs [32]. For the RmhBN-silicone, R_b is fitted to be $300 \times 10^{-9}\ m^2 \cdot W \cdot K^{-1}$, which

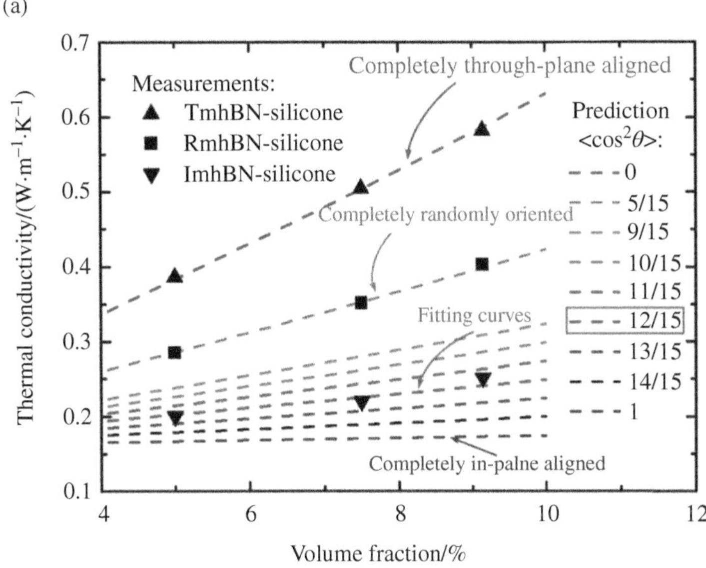

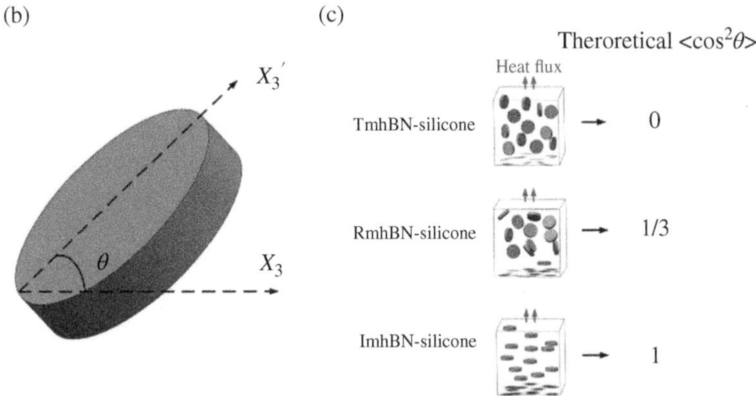

Figure 6.23 (a) Comparison between the measured and predicted through-plane thermal conductivity (κ_T) of the composites and data fitting to extract $<\cos^2\theta>$ of ImhBN-silicone composites. Black dots: measured κ_T of TmhBN-silicone, RmhBN-silicone, and ImhBN-silicone composites; colored dash lines: predicted κ_T from EMA with the $<\cos^2\theta>$ ranging from 0 to 1. (b) Schematic of a platelet with the angle θ. (c) Theoretical values of $<\cos^2\theta>$ for the composites containing completely through-plane, in-plane, and randomly oriented platelets, respectively.

is much larger than that of TmhBN-silicone. Such anisotropic R_b has also been observed in the hBN-epoxy composites [32]. In addition, it can be found that with the large R_b, thermal transport in RmhBN-silicone composites is suppressed: κ_T of the composites at the volume fraction of 5%, 7.5%, and 9.14% are reduced by 10.1%, 11.6%, and 11.4%, respectively.

Figure 6.23(a) gives the comparison between the measured and predicted κ_T of the composites. For the TmhBN-silicone and RmhBN-silicone, the predictions surely match well with the experiments owing to the fitting. However, it is interesting to note that for the ImhBN-silicone, the predicted κ_T is lower than that observed in the experiments. This deviation could be attribute to the inappropriate assumption of mhBNs orientation in the ImhBN-silicone composites. The EMA prediction assumes that the platelets are completely in-plane oriented in the matrix. However, according to the FIB-SEM observations, some mhBN platelets are not aligned to the horizontal direction. These unexpected platelets increase the thermal transport in the composites and, hence, increase κ_T.

The theoretical model is used to further investigate the effect of orientation on κ_T. In this model, the orientational characteristic of platelets in the matrix is presented by the parameter, $<\cos^2\theta>$, which is given by [44]:

$$<\cos^2\theta> = \frac{\int \rho(\theta)\cos^2\theta \sin\theta d\theta}{\int \rho(\theta)\sin\theta d\theta} \tag{6.26}$$

where $\rho(\theta)$ is a distribution function. As schematically shown in Figure 6.23(b), θ is the angle between the material axis X_3 and the local platelet symmetric axis X_3'. As reported previously [44] and schematically illustrated in Figure 6.23(c), $<\cos^2\theta>$ is theoretically equal to 0, 1/3, and 1 for the composites containing the completely through-plane, randomly, and in-plane oriented platelets. Figure 6.23(a) gives the predicted κ_T from EMA with the $<\cos^2\theta>$ ranging from 0 to 1. It tells clearly that κ_T decreases with the increase of $<\cos^2\theta>$. By fitting the measured κ_T to EMA prediction, the actual $<\cos^2\theta>$ of ImhBN-silicone composites can be extracted. The fitting result is shown in Figure 6.23(a). $<\cos^2\theta>$ is found to be 12/15.

According to the above analysis, TC of the composites is strongly associated with the magnetic alignment of mhBN platelets. Aligning the platelets parallel to the direction of heat flux can greatly enhance the thermal transport in the composites due to the formation of conductive networks and the reduced R_b. Whereas, when the orientation of platelets is perpendicular to the heat flux direction, thermal transport will be highly suppressed.

6.2.2 Thermal Conductivity Enhancement of BN-Composites Using Combined Mechanical and Magnetic Stimuli

As schematically shown in Figure 6.24(a), hBN is a platelet-shaped, high aspect ratio (D/δ) particle and possesses high in-plane thermal conductivity. In most thermal management applications, the hBN-filled composite is expected to provide efficient heat transfer rate (P_h) from the heat source to the heat sink in the direction normal to the composite layer. So, the ideal orientation of hBN platelets in polymeric matrix is parallel to the direction of P_h (Figure 6.24(b)I). Such well-ordered structures can make the composite take full advantage of platelets high in-plane TC. On the contrary, the isotropic (Figure 6.24(b)II) orientation is much less favorable. Therefore, control of the platelets orientation can enable the enhancement of composite heat transport capability.

The well-ordered structures have been achieved in practice by several approaches, including tape-cast [45], spin-cast [46], shear alignment [47], electrical alignment [48], and magnetic alignment [32, 42, 49]. Among these approaches, magnetic alignment has been recently highlighted due to the remote control of fillers' orientation and possibility of aligning fillers at arbitrary directions [32, 40]. This approach relies on coating the hBN platelets with magnetic nanoparticles, such as Ni, Fe, Co [49], and Fe_3O_4 [32, 42]. These modified hBN (mohBN) platelets exhibit a high magnetic response which enables the alignment of them in low-viscosity suspending fluids under a linear, uniform magnetic field [32, 42]. Then the magnetically imposed alignment can be fixed by consolidating the suspending fluids.

However, this approach is limited to the composites loaded with low-volume fraction of mhBN platelets. When f increases above a percolation threshold, the steric interactions between the platelets will dominate, which hinders the platelets alignment. For instance, applying such approach, Lin et al. [32] prepared the aligned composites at f from 3% to 28%. When f was below 13%, the platelets could be assembled into highly ordered structures in composites. While, as f increases above 13%, a significant amount of unaligned mhBN platelets were found in composites. In addition, in our previous [42] and Boussaad's [49] works, the examples of well-aligned composites obtained by this approach had a maximum f of 9.14% and 10%, respectively.

To overcome the strong steric hindrances between platelets and thus achieve a high degree of alignment, one possible method is to provide additional energy to platelets during magnetic alignment. Mechanical vibration is such a candidate as the additional energy, which is often applied for the arrangement of disordered particles [50–52]. Furthermore, the

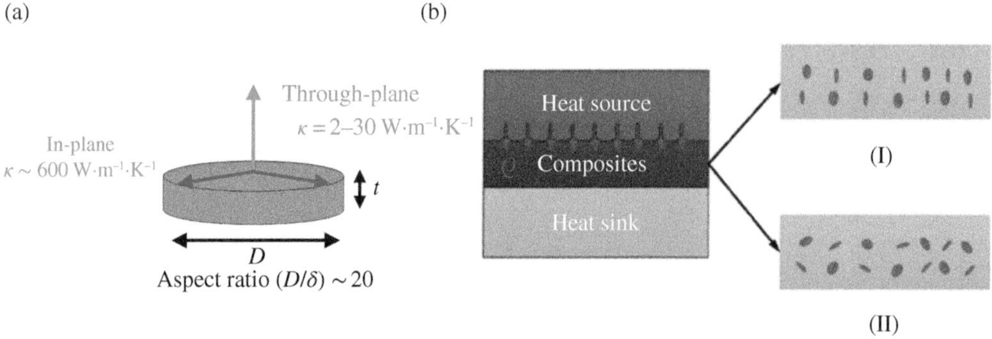

Figure 6.24 (a) Representation of hBN platelet geometry and thermal property. (b) Schematic showing the utilization of hBN-filled composite for the heat dissipation from heat source to heat sink. Orientations of hBN platelets in the composite layer: (I) through-plane and (II) isotropic.

former investigation [53] has shown the possibility in assembling the high concentrations of alumina platelets into ordered architectures using magnetic fields and mechanical vibration. The results show that the mechanical properties of the alumina-reinforced composites can be tuned deliberately. Therefore, it is possible to tailor the thermal property of the platelets-reinforced composites with such a strategy.

In this section, we will introduce the investigation of the TC of polymer-based composites containing high concentrations of aligned hBN platelets. The outline of this section is as follows. In the beginning, the fabrication of aligned composites using combined mechanical and magnetic stimuli is introduced. For the purpose of comparison, the composites were also prepared with the former magnetic approach [32, 42]. Then, the cross-section of those composites was imaged by SEM and the degree of platelets alignment was quantitatively examined by measuring the platelets angles θ with the SEM images. After that, XRD analysis was carried out to further evaluate the degree of alignment. Finally, thermal conductivities of those composites were compared and analyzed in light of platelets alignment degree.

6.2.2.1 Fabrication of the Composites

hBN platelets (AC6041), with an averaged diameter (D) of 5 μm, were purchased from Momentive. The thickness (δ) of platelets is estimated to be 250 nm [32], leading to an average D/δ of 20. Epoxy resin, supplied by Huntsman, was selected as the polymer matrix. It is composed of a low viscosity bisphenol-A-based liquid resin (Araldite LY1564) and an amine-based hardener (Aradur 3487). The mix ratio is 100:34 by weight. The initial viscosity (η) of the mixture is about 0.27 $\text{Pa} \cdot \text{s}^{-1}$ at 25 °C. The aqueous-based EMG-605 ferrofluid used to magnetize the hBN platelets was kindly supplied by Ferrotec. It contained 3.9 vol% iron oxide nanoparticles coated with a cationic surfactant. The average diameter of nanoparticles is 20 nm [42].

Preparation of Magnetically Responsive hBN (mhBN) Platelets hBN platelets were coated with superparamagnetic iron oxide nanoparticles to make them magnetized via a previously reported method [32, 40, 42]. Specifically, 4 g of hBN platelets were dispersed in 200 ml of deionized water at pH = 7, while 400 μl of ferrofluid diluted in 5 ml of deionized water was added dropwise. In the deionized water, the hBN platelets have a negative surface charge, causing the positively charged magnetic nanoparticles to electrostatically adsorb on the hBN surface. The adsorption was considered complete when the supernatant was clear. After that, the coated platelets were rinsed three times with deionized water. Finally, the magnetized platelets were dried in an oven at 90 °C for 12 hours. The SEM image (Figure 6.18(d)) provided by the previous work [42] has shown the nanostructure of the obtained platelet, which confirmed the successful attachment of iron oxide nanoparticles to the platelet surface.

Preparation of Composites with Aligned mhBN Platelets Composites were prepared using the epoxy resin and mhBN platelets. mhBN platelets were first suspended in the mixture of Araldite LY1564 resin and Aradur 3487 hardener. The resulting suspension was mechanically stirred for 10 min to fully disperse. Next, the suspension was sent into a vacuum chamber to remove the bubbles introduced during the mixing step. After that, the polymer suspension was cast into a 10 mm × 10 mm × 2 mm Teflon mold.

Figure 6.25(a) schematically shows the way of aligning mhBN platelets in matrix. The Teflon mold containing the suspension was fixed with a vibrating table and placed under a 40 mm × 40 mm × 20 mm rare earth magnet which was connected to a motor. The table could provide the continuous mechanical vibration to the sample with a power of 550 W and an amplitude of 0.5 mm. The magnet, rotated with a fixed frequency (w), produced a rotating magnetic field of about 8000 Gs in the plane normal to the suspension. According to Ref. [54], in order to achieve the vertical alignment of platelets, w should be larger than a critical frequency (w_c). In this study, w_c was calculated to be about 0.3 Hz, and then w was set to be 5 Hz. The mechanical vibration lasted for 30 minutes. Then, the sample was kept in the rotating magnetic field for 6 hours to form the ordered structures. Afterwards, the sample was placed in an oven at 60 °C for 4 hours. During this period, the sample was still exposed to a linear, uniform, vertical magnetic field to keep the obtained structures. The vertical magnetic field was produced by two parallel arranged 200 mm × 100 mm custom rectangular magnets. Finally, an annealing step at 120 °C for 2 hours was conducted to ensure the full curing of the sample.

For comparison, the composites were also prepared with the former magnetic approach [32, 42]. Figure 6.25(b) depicts the process of fabrication. First, the Teflon mold containing the suspension was positioned in the uniform, vertical magnetic field for 4 hours at 25 °C. Afterwards, the sample was heated with the magnetic field at 60 °C for 4 hours and 120 °C for 2 hours.

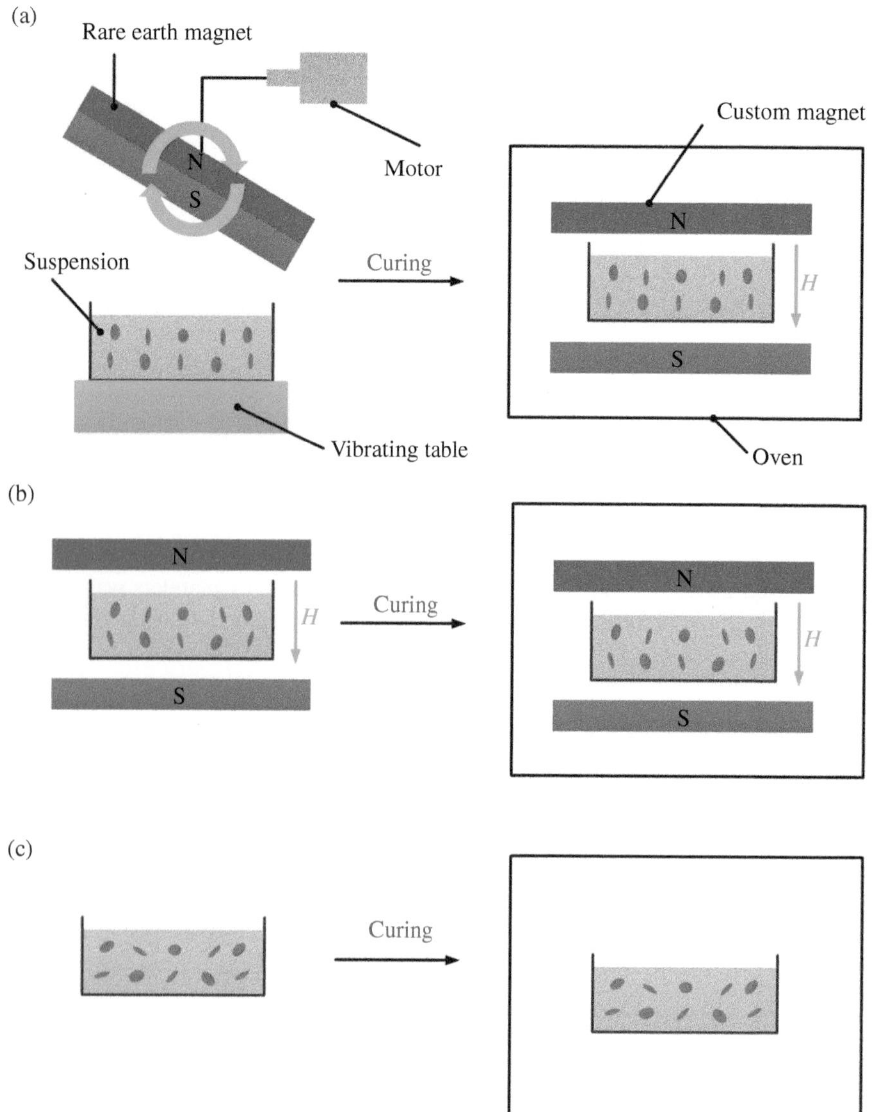

Figure 6.25 Description for the preparation of mhBN-resin composites with (a) combined mechanical and magnetic stimuli, (b) magnetic stimuli, and (c) no stimuli.

The composites were also fabricated without any stimuli to obtain the unaligned composites. Figure 6.25(c) schematically shows the process of fabrication. In the beginning, the Teflon mold containing the suspension was incubated for 4 hours at 25 °C. Then, the sample was placed in the oven at 60 °C for 4 hours and 120 °C for 2 hours.

6.2.2.2 Characterization and Analysis

The morphology of mhBN platelets was observed by imaging the polished surfaces of samples using a SEM (Quanta 200) at an accelerating voltage of 20 kV. To get the polished surfaces, samples embedded in the epoxy resin was ground with silicon carbide foils (grits: 600, 1200, 1500, and 2000) and polished with diamond suspension (grain size: 7, 3, and 1.5 μm). Before visualization, the polished surfaces were sputtered with a thin layer of gold for better imaging. The angle distributions of mhBN platelets were analyzed with ImageJ.

XRD analysis of mhBN-resin composites was also conducted to evaluate the orientation of mhBN platelets in composites. This analysis was carried out with D/MAX-RB using Cu Kα radiation (40 kV and 40 mA). Before the tests, the samples were ground with silicon carbide foils (600, 1500, and 2000) to expose the inner surfaces.

Here, the resin-based composites, prepared with combined mechanical and magnetic stimuli, uniform magnetic stimuli, and no stimuli, are named as MM-mhBN-resin, M-mhBN-resin, and R-mhBN-resin, respectively. Figure 6.26(a) illustrates the ideal microstructure of the aligned composites. In the 3D view, all platelets orient to the vertical orientation. In the cross-sectional view, the platelets are projected as two representative morphologies: 1D vertical rods and 2D plates, which are indicated by green and blue arrows, respectively. Cross-sectional SEM images of 10 vol% mhBN-resin composites are shown in Figure 6.27(a)–(c). Figure 6.27(a) shows that, in MM-mhBN-resin, a large number of platelets are projected as the two representative morphologies. Few platelets orient to the horizontal direction. In contrast, in the cross-sectional view of

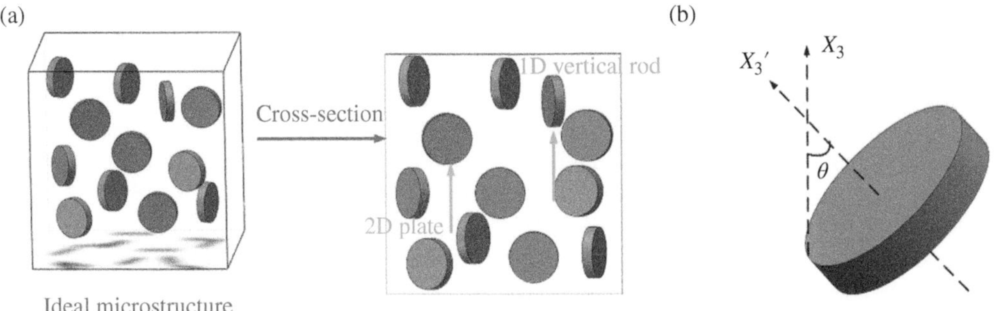

Figure 6.26 (a) Ideal microstructure of composites contacting the well-aligned platelets and the representative morphologies of platelets at the cross-sectional view. (b) Schematic of a platelet with the angle θ.

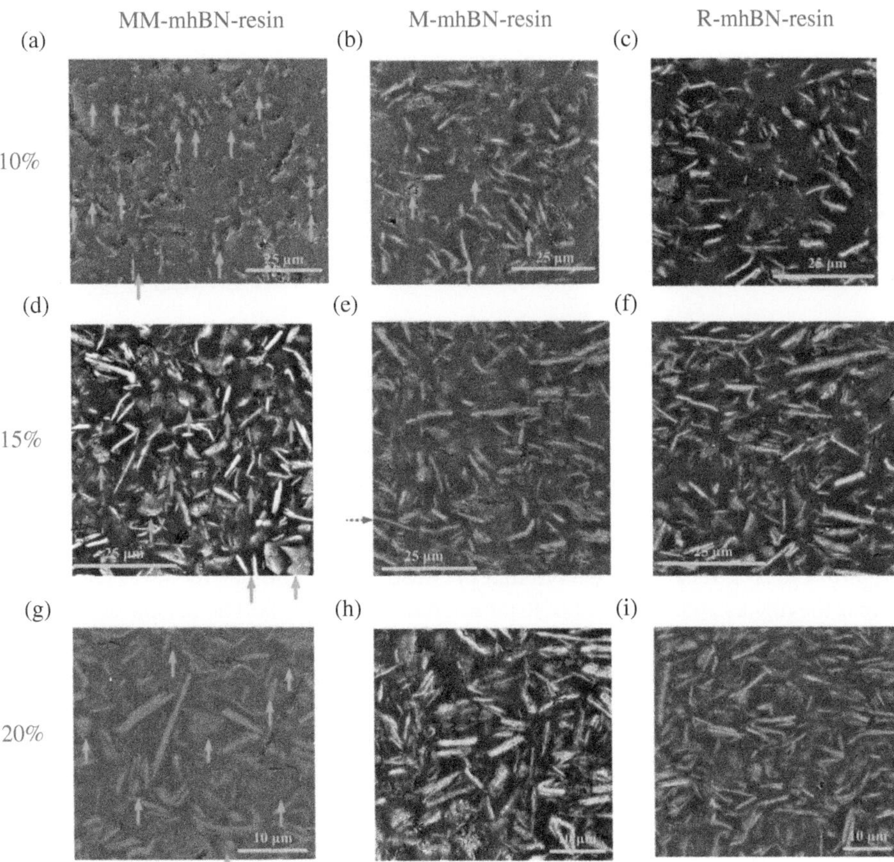

Figure 6.27 Cross-sectional SEM images of MM-mhBN-resin composites at f of (a) 10%, (d) 15%, and (g) 20%; M-mhBN-resin composites at f of (b) 10%, (e) 15%, and (h) 20%; and R-mhBN-resin composites at f of (c) 10%, (f) 15%, and (i) 20%. 1D vertical rods and 2D plates are indicated by green and blue arrows, respectively, and the unexpected morphology is indicated by red arrows.

M-mhBN-resin (Figure 6.27(b)), although the majority of platelets are aligned along the vertical direction, a number of platelets, indicated by red arrows, orient to the horizontal direction. Figure 6.27(c) gives the cross-sectional view of R-mhBN-resin. Both vertical and horizontal alignment characterizations can be easily found in the SEM image.

Figure 6.27(d) and (g) show the cross-sectional SEM images of 15 and 20 vol% MM-mhBN-resin composites. Quantities of platelets are projected as the representative morphologies. However, with the increase of platelets loading, some unexpected platelets, which orient to the horizontal direction, are observed in the MM-mhBN-resin. Parts (e, h) and (f, i) of Figure 6.27 give the cross-sectional views of M-mhBN-resin and R-mhBN-resin composites, respectively. In M-mhBN-resin, the horizontally oriented platelets also increase with the increase of loading. In R-mhBN-resin, both vertically and horizontally oriented mhBN platelets can be found abundantly. So, we assume that mhBN platelets orient randomly in R-mhBN-resin composites.

The degree of platelets alignment is examined by measuring the platelets angles (θ) with the SEM images. As schematically shown in Figure 6.26(b), θ is the angle between the composite axis X_3 and the local particle symmetric axis X_3'. When the platelet aligns along the vertical or horizontal direction, θ is equal to 90° or 0°, respectively. The results are represented as a distribution histogram in Figure 6.28 and the parameter, P, is defined as population fraction of the platelets being above 45°. Table 6.7 summarizes the values of P for the MM-mhBN-resin and M-mhBN-resin composites. It is found that in the 10 vol% MM-mhBN-resin, θ of 95.3% of platelets is greater than 45°, only ~5% of platelets have fallen into the horizontal direction. While, in the M-mhBN-resin with the same f, p decreases to 74.7%. This means that the ratio of in-plane-oriented platelets increases. In addition, for the composites at f of 15% and 20%, the MM-mhBN-resin also exhibits higher value of p.

XRD analysis is carried out to further evaluate the alignment of mhBNs in resin matrix. Figure 6.29 gives the XRD patterns for the 10 vol% MM-mhBN-resin, M-mhBN-resin, and R-mhBN-resin composites. The XRD patterns are compared to show their relative alignment degrees. It is observed that the peak intensities are dramatically different for the three composites. As reported previously [32, 42], the vertically and horizontally oriented hBNs are responsive for the (100) and (002) peaks, respectively. Thus, we use the ratio of (100) peak intensity to (002) peak intensity (I_{100}/I_{002}) to estimate the degree of vertical alignment of platelets. Table 6.8 provides the values of I_{100}/I_{002} for the 10 vol% mhBN-resin composites. It is found that I_{100}/I_{002} of MM-mhBN-resin is more than 6 times and 28 times larger than that of M-mhBN-resin and R-mhBN-resin, respectively. Table 6.8 also provides the I_{100}/I_{002} values for the composites with higher loading. I_{100}/I_{002} of MM-mhBN-resin is still higher than other two composites with the same f. The higher value suggests the larger amount of vertically aligned mhBNs in MM-mhBN-resin.

6.2.2.3 Thermal Properties of the Composites

TC was calculated according to the formula:

$$k = \alpha C_p \rho \tag{6.27}$$

where C_p, ρ, and α are heat capacity, density, and thermal diffusivity of sample, respectively. C_p was measured by a DSC (Diamond DSC, PerkinElmer Instruments). ρ was calculated according to the sample weight and dimensions. α was measured by a laser flash apparatus (LFA457, Netzsch) at 25 °C. Before the test, the sample surfaces were coated with a thin graphite film to increase the energy absorption and the emittance of the surfaces. In the test, heat propagates from the bottom to the top surface of the sample.

It has been illustrated that the degree of platelets alignment in matrix is greatly different among the composites fabricated with different methods. Figure 6.30(a) provides the measured through-plane TC (κ_T) of MM-mhBN-resin, M-mhBN-resin, and R-mhBN-resin composites at f from 10% to 20%. It can be observed that κ_T is also different among the three composites with the same f. κ_T of MM-mhBN-resin is the highest, whereas that of R-mhBN-resin is the lowest. The results demonstrate that there is a positive correlation between the degree of alignment and TC. To future study the correlation, Figure 6.30(b) gives the thermal enhancements of MM-mhBN-resin and M-mhBN-resin referring to R-mhBN-resin, respectively. For the M-mhBN-resin, thermal enhancement is 36% at 10 vol%, but drops rapidly to 14% and 6% at 15 and 20 vol%, respectively. The drop of thermal enhancement at high mhBN loadings is due to the decreased degree of mhBN alignment. Thermal enhancement of the MM-mhBN-resin also decreases with the increase of f due to the same reason. However, the MM-mhBN-resin composites achieve greater thermal enhancement. It is found that the thermal enhancement reaches to 74% at 10 vol% and maintains at 20% at higher f. The great thermal enhancement is owing to the formation of conductive networks.

6.2.2.4 Theoretical Analysis of Thermal Conductivity

The modified EMA [44] is applied to analyze the experimental data. It is famous for predicting TC of composites containing the platelet-shaped fillers [3, 32, 42]. It illustrates that TC is determined by the intrinsic thermal properties of fillers and matrix, fillers loading, geometries, orientations, and the thermal boundary resistance (R_b) at the filler–matrix interface.

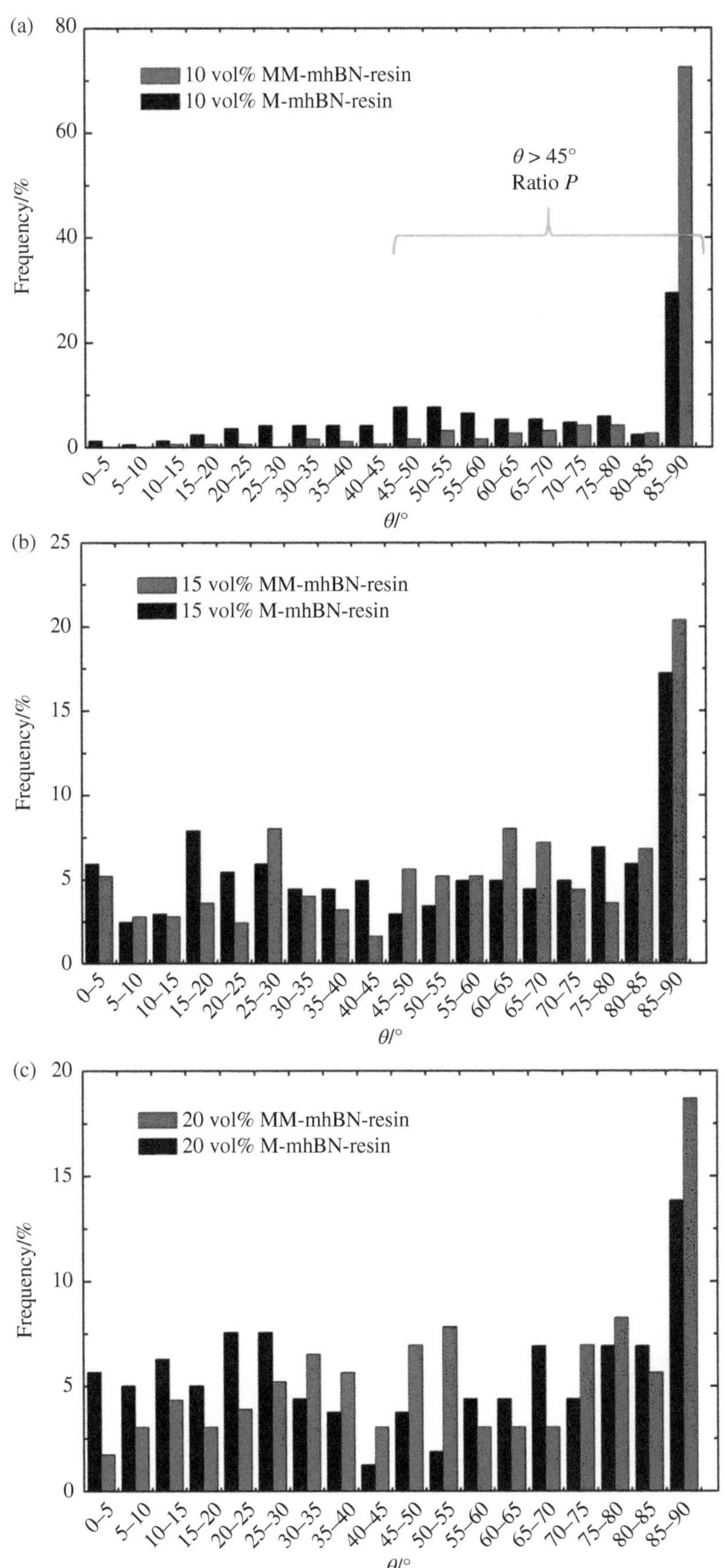

Figure 6.28 Angle distribution of mhBNs platelets in MM-mhBN-resin and M-mhBN-resin composites at *f* of (a) 10%, (b) 15%, and (c) 20%.

Table 6.7 Values of P for the MM-mhBN-resin and M-mhBN-resin composites.

Volume fraction f (%)	P (%)	
	MM-mhBN-resin	M-mhBN-resin
10	95.3	74.7
15	66.4	55.7
20	63.5	53.6

In this model, the orientational characteristic of fillers in matrix is presented by the parameter, $<\cos^2\theta>$, which is expressed as [44]:

$$<\cos^2\theta> = \frac{\int \rho(\theta)\cos^2\theta \sin\theta d\theta}{\int \rho(\theta)\sin\theta d\theta} \tag{6.28}$$

where θ is the angle which has been schematically shown in Figure 6.26(b) and $\rho(\theta)$ is the distribution function of θ. With the aid of distribution histograms (Figure 6.28), $<\cos^2\theta>$ of the MM-mhBN-resin and M-mhBN-resin composites can be calculated using the discretization method. For the R-mhBN-resin, the platelets are assumed to be completely randomly oriented in matrix. Thus, $<\cos^2\theta> = 1/3$ [44]. Table 6.9 summarizes the values of $<\cos^2\theta>$ for all mhBN-resin composites.

In addition to orientation, R_b is another key parameter governing the thermal transport in composites. R_b arises from the poor mechanical or chemical adherence at the interface and the thermal expansion mismatch [1, 3, 31, 44]. Here, the additional iron oxide nanoparticles on the hBN surface also result in R_b at the interface. In general, the actual R_b can be extracted by fitting the measured TC to model prediction [32, 42, 55]. For R-mhBN-resin composites, Figure 6.30(c) shows the predicted κ_T under different values of R_b and the measured κ_T of 10, 15, and 20 vol% composites. R_b is found to be $360 \times 10^{-9}\ \mathrm{m^2 \cdot W \cdot K^{-1}}$. For the MM-mhBN-resin and M-mhBN-resin composites, $<\cos^2\theta>$ is different for each composite. Therefore, R_b of each composite is extracted according to its own $<\cos^2\theta>$. Figure 6.30(d) shows the predicted κ_T of 10 vol% MM-mhBN-resin under different values of R_b. R_b is fitted to be $11 \times 10^{-9}\ \mathrm{m^2 \cdot W \cdot K^{-1}}$. The way of extracting R_b for other MM-mhBN-resin and M-mhBN-resin composites are the same as that for 10 vol% MM-mhBN-resin composite. All the

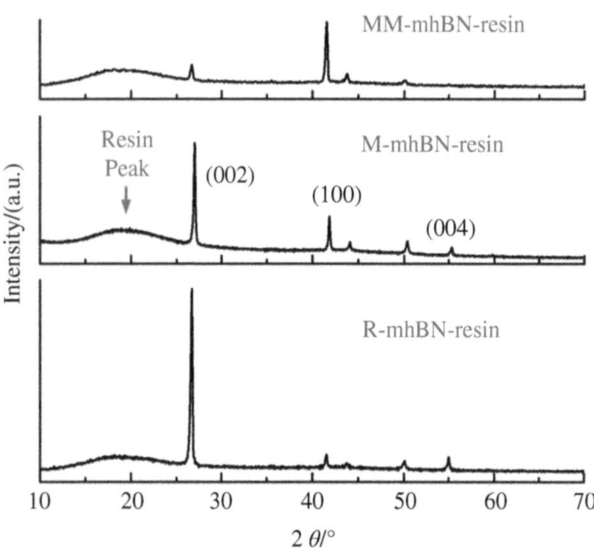

Figure 6.29 XRD analysis of 10 vol% MM-mhBN-resin, M-mhBN-resin, and R-mhBN-resin composites.

Table 6.8 Values of I_{100}/I_{002} for the MM-mhBN-resin, M-mhBN-resin, and R-mhBN-resin composites.

Volume fraction f (%)	I_{100}/I_{002}		
	MM-mhBN	M-mhBN	R-mhBN
10	2.68	0.39	0.10
15	0.15	0.08	0.05
20	0.12	0.07	0.05

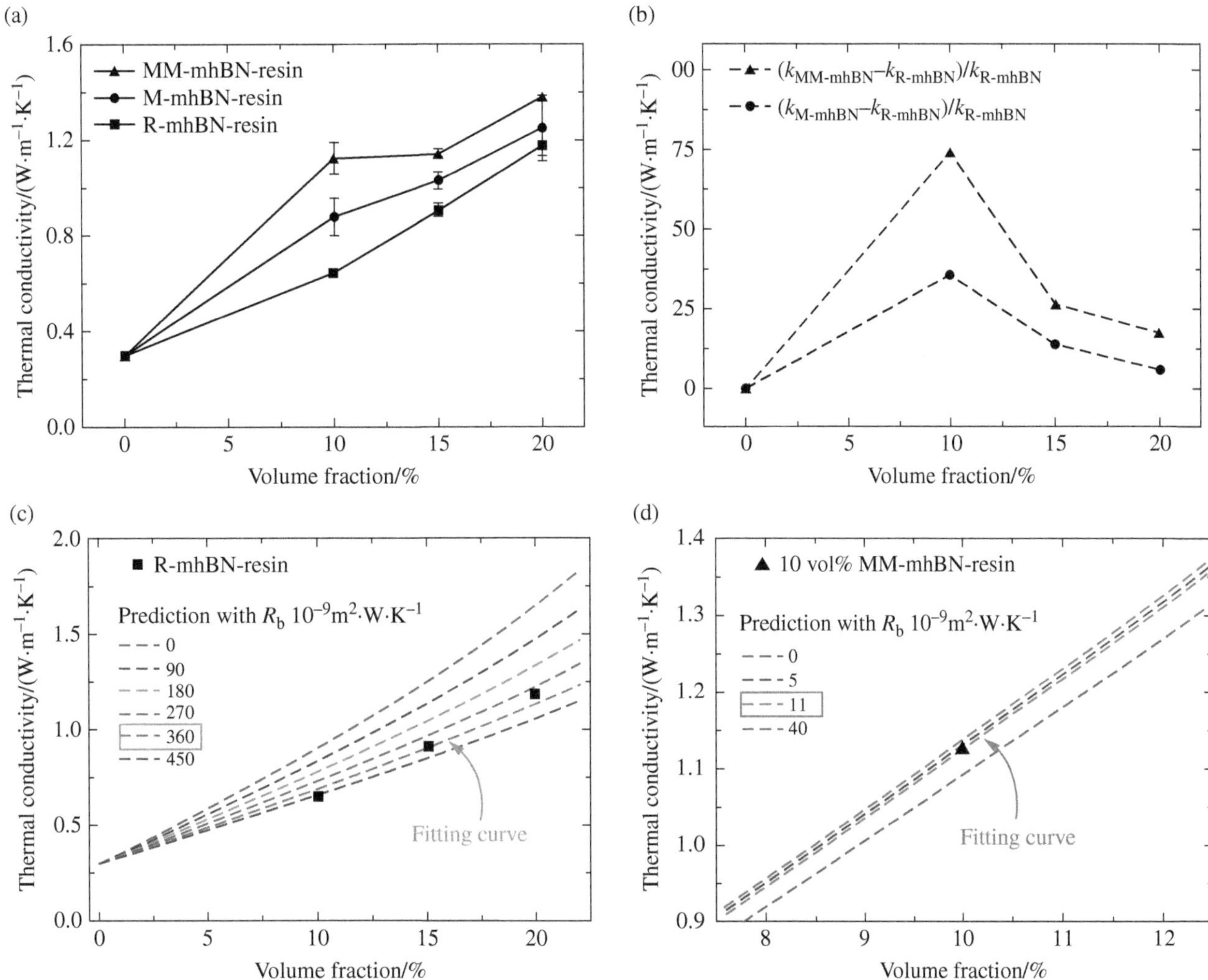

Figure 6.30 (a) Thermal conductivities of mhBN-resin composites. (b) Thermal enhancements of MM-mhBN-resin and M-mhBN-resin referring to R-mhBN-resin, respectively. Data fitting to extract R_b of (c) R-mhBN-resin and (d) 10 vol% MM-mhBN-resin composites. Black dots: measured κ_T of composites; colored dash lines: predicted κ_T from effective medium approximation (EMA) with different values of R_b.

Table 6.9 Values of $<\cos^2\theta>$ and R_b for the MM-mhBN-resin, M-mhBN-resin, and R-mhBN-resin composites.

	$<\cos^2\theta>$			$R_b \times 10^{-9}$ ($m^2 \cdot W \cdot K^{-1}$)		
Volume fraction *f* (%)	**MM-mhBN**	**M-mhBN**	**R-mhBN**	**MM-mhBN**	**M-mhBN**	**R-mhBN**
10	0.0530	0.2137	0.3333	11	145	360
15	0.2316	0.2625	0.3333	186	275	360
20	0.2529	0.2724	0.3333	232	317	360

fitting results are summarized in Table 6.9. It is interesting to note that R_b decreases with the reduction of $<\cos^2\theta>$, which means that the higher degree of platelets alignment leads to the lower value of R_b. The same results have also been observed in other hBN–polymer interface [32, 42] and the interfaces with graphite or carbon nanotubes (CNTs) [56, 57], which have a similar atomic structure to hBN. In addition, theoretical calculation can also explain this point. Prasher [58] theoretically calculated R_b between an anisotropic material and an isotropic substrate. Figure 6.31 schematically shows two contact situations. For the anisotropic material, such as hBN, graphite, and CNTs, effective velocity of phonons at the in-plane direction

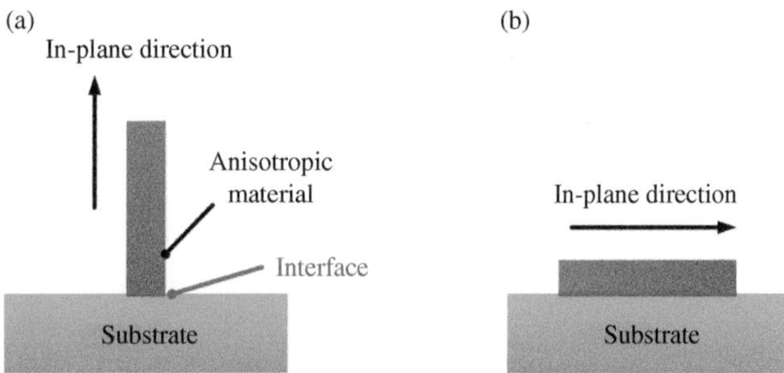

Figure 6.31 Schematic showing the two contact situations: (a) vertical contact and (b) horizontal contact.

is larger than that at the through-plane direction [58]. Due to the larger phonon velocity at the in-plane direction, R_b for the vertical contact is smaller than that for the horizontal contact [58]. Accordingly, in the mhBN-resin composite, the higher degree of platelets alignment makes R_b reduced.

According to the earlier analysis, the TC of the composites is strongly associated with the degree of platelets alignment. Assembling the platelets into a well-ordered structure can greatly enhance the heat transport capability of composites due to the formation of conductive networks and the reduction of R_b.

6.2.3 Magnetic-Tuning TIMs for Local Heat Dissipation

Over the past few decades, the revolution in electronics has resulted in the packaging of multiple functional units on the same chip [59, 60]. These units create a non-uniform distribution of heat source throughout the chip. The local heat source with high power is likely to produce hotspots. These hotspots can lead to excessive stresses on the chip, which degrade the reliability and performance of electronic devices [59–62]. Thermal management of local heat source is a big challenge. Various materials [63–65] have been attempted to enhance the heat dissipation. Thermally conductive polymer-based composites, fabricated by incorporation of highly conductive fillers, are potential for the thermal management materials [1, 3, 66, 67]. However, applying the conventional composites that possess uniform thermal properties for the local heat source cooling is still insufficient and not cost-effective [68].

Many theoretical studies [69–73] have demonstrated that the heterogeneous composite materials with fillers concentrated at the preferential paths of heat flux are effective in cooling the local heat source. For instance, some topology designs [69–71] have been applied to the composite material which is loaded with local heat source, to minimize the heat source temperature. These designs are accomplished by locating the fillers in matrix with optimal distribution using the optimization algorithm. The designed results show that concentrating the fillers at the position with large temperature gradient is most efficient to decrease the heat source temperature. In addition, the percolation theory [72, 73] can further illustrate the benefit of fillers local concentration. Percolation happens when the high conductivity particles form at least one continuous chain in composites from heat source to sink. At that time, there is a sudden increase in thermal performance [73]. Figure 6.32(a) schematically shows a homogenous composite with low fillers loading assembled with local heat sources. Percolation hardly takes place in this composite since all particles are well separated. While, as seen in Figure 6.32(b), concentrating the fillers makes it more probable to form the continuous chains resulting in a local reinforcement at the region with local heat source.

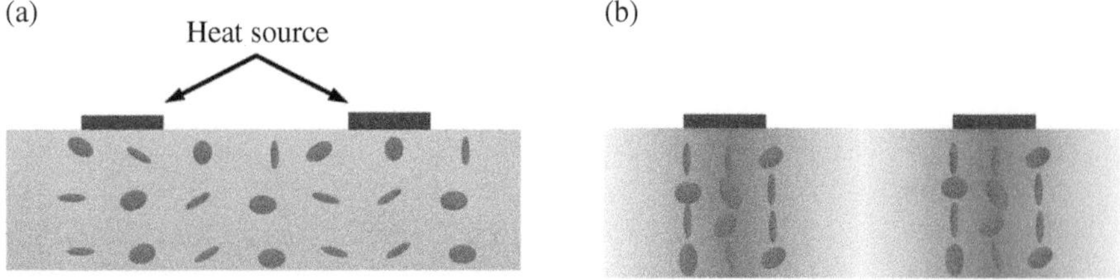

Figure 6.32 Schematic of (a) homogenous composite and (b) heterogeneous composite with local reinforcement.

The aforementioned theoretical studies well illustrate the benefit of fillers local concentration. However, this unique control of microstructure and property for polymer-based composites has less been achieved in practice due to the technical difficulties in controlling the fillers positions. Squeezing [74–76] is a process which can make fillers entrapped and compacted in matrix. The heterogeneity of composites occurs when the rate of matrix flow through the fillers is greater than that of composites deformation [75, 76]. However, this method is limited to the thin films and demands highly accurate control of the squeeze rate [75, 76]. Electric field [77] is another approach to achieve the unique microstructure, in which the localization of fillers is induced by electric field concentration. But this approach requires ultrahigh direct-current (DC) electric field (1 kV) [77] and multiple processing steps. Recently, an attractive strategy is proposed to control the distribution of fillers in the matrix [40, 41]. The approach relies on coating the non-magnetic reinforcing particles with superparamagnetic nanoparticles. These coated particles exhibit an UHMR [40] which enables remote control over their distribution under low external magnetic fields in low-viscosity suspending fluids. Such fluids can then be consolidated to fix the magnetically imposed distribution and thus produce the composites with deliberately tuned properties. With this strategy, the former investigations [40, 78] have shown the possibility in fabricating the polymeric substrate with locally tuned mechanical properties. A careful review of literature indicates that this strategy has not been applied for tailoring the composites thermal properties. In addition, although the reports [76, 77] have fabricated the composite with deliberately controlled microstructure, experiments have not been conducted to evaluate its thermal performance.

In this section, we will introduce the polymer-based composites applied for heat dissipation from local heat source reinforced by concentrating the thermally conducting fillers at the preferential paths of heat flux. The locally reinforced composite is fabricated by using the magnetically responsive thermally conducting particles as reinforcing elements and a specific magnetic field to control the elements distribution. Then, a thermal testing system is built to evaluate the composites' thermal performance. Finally, a simulation on the constructed composite is conducted to thoroughly investigate its thermal property.

6.2.3.1 Fabrication of the Composites

Polymer-based composite is composed of polymer matrix and reinforcing particles. Here, the silicone gel (OE-6550, Dow Corning) is employed, with a viscosity of 4 Pa · s and a relatively low TC ($0.16\,\mathrm{W \cdot m^{-1} \cdot K^{-1}}$) [31], as the soft matrix. hBN platelets (AC-6041, Momentive), with an average diameter of 5 μm and a much higher in-plane TC ($600\,\mathrm{W \cdot m^{-1} \cdot K^{-1}}$) [32, 42], were chosen as the reinforcing elements. To gain magnetic control of the reinforcements, the platelets were coated with 2 wt% superparamagnetic iron oxide nanoparticles via a previously reported procedure [32, 42]. hBN platelets (4 g) were first stirred in deionized water (200 ml) at pH = 7. Then, EMG-605 ferrofluid (Ferrotec, USA) (400 μl) diluted with deionized water (5 ml) was added dropwise to the suspension under vigorous stirring. The pH of suspension was held at 7 to keep a negative charge on the surface of hBN platelets. The EMG-605 ferrofluid is an aqueous suspension containing iron oxide nanoparticles coated with a cationic surfactant, which enables electrostatic adsorption of the positively charged magnetic nanoparticles on the platelets surface. The suspension was incubated for 1 hour to allow the bonding between the platelets and iron oxide nanoparticles. After that, the coated platelets were filtered and dried for 12 hours at 90 °C in vacuum.

Reinforced composites were produced by adding the magnetically responsive platelets to the fluidic silicone gel followed by mold casting and curing through polymerization. Magnetically responsive platelets in the powder form were first added to the silicone gel and stirred for 30 minutes to fully disperse. Bubbles introduced during the stirring process were removed by applying alternating cycles of vacuum. The resulting suspension was then poured into a 26 mm × 26 mm × 2 mm (length × width × height) Teflon mold. During the mold casting, a magnetic field should be applied on the cast sample to concentrate the platelets to the specific region of interest. Figure 6.33(a) gives the diagram which schematically shows the approach of locally concentrating the platelets. Two 45 mm × 3 mm × 3 mm rare earth magnets are placed symmetrically below and above the sample. The distance between the magnet and sample was 6 mm. These magnets can create a magnetic field gradient on the sample: the strength of magnetic field between the magnets is the largest and decreases gradually toward the sides of the sample. The largest strength of magnetic field is 26 mT measured by a Gaussmeter. According to Ref. [78], the magnetically responsive platelets can be attracted to the place with a maximum magnetic field. After 1-hour incubation, the magnetically responsive platelets were highly concentrated between the magnets. Samples with a field were heated at 60 °C for pre-cure. An annealing step at 150 °C for 6 hours was conducted to ensure the good adhesion between the hBN platelets and silicone matrix.

Figure 6.33(b) shows a fabricated locally reinforced composite with a filler volume fraction of 5%. A 6.8 mm-wide linear stripe with a high local concentration of platelets is observed at the composite. For comparative purpose, Figure 6.33(c)

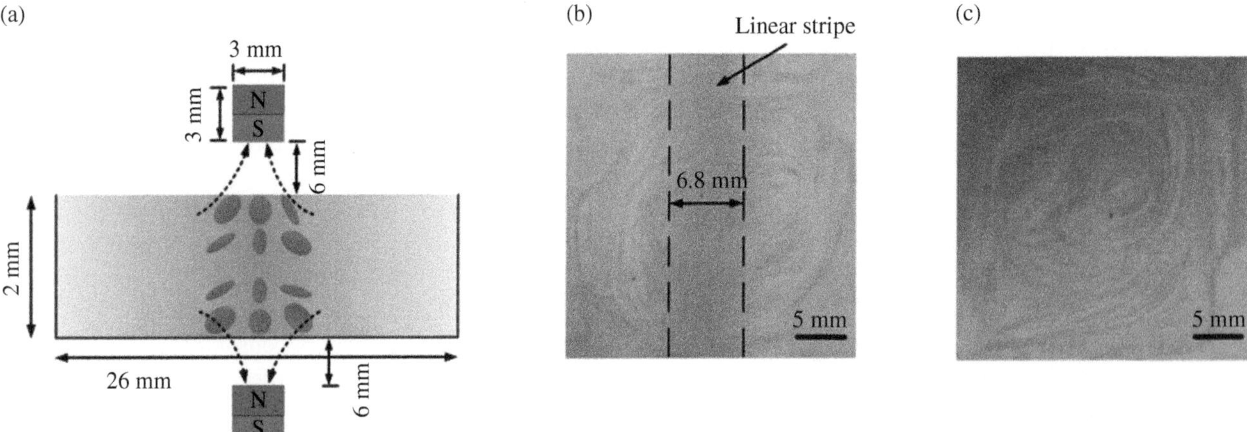

Figure 6.33 (a) Schematic diagram showing the approach of locally concentrating the platelets; photographs of prepared (b) locally structured and (c) homogeneous composites.

presents the homogeneous composites with the same volume fraction prepared without magnetic field. According to the figure, there is no sign of platelets concentration.

6.2.3.2 Evaluation for Thermal Performance of the Composites

When a specific region of composite is applied to a local heat source, the accumulation of platelets in that region is expected to improve the heat dissipation and, hence, decrease the temperature of heat source. To demonstrate such potential, comparative thermal tests were performed on the homogeneous and locally structured samples, which were applied to the same local heat source. In the tests, the temperature response of heat source to the two samples was compared.

Figure 6.34 schematically illustrates the testing system and the experimental processes. The system consists of six components: heater, samples, heat sink, supporting plates, rare earth magnets, and infrared radiation (IR) thermometer. To create a local heat source, a small polyimide film heater with dimensions of 26 mm × 8 mm × 0.2 mm was attached on the samples with dimensions of 26 mm × 26 mm × 2 mm and energized with certain power. Since it was difficult to mount the heater on the solid-state composites, the uncured composites were employed as the testing samples. Good contact between the heater and samples was achieved by the adhesive force from fluidic samples. Another advantage of using fluidic samples was that the comparative tests could be conducted at the same base material, ensuring the same testing conditions. As seen in Figure 6.34, a block (aluminium-6061 T4) with a square groove on the surface was applied to hold the samples. And the lower part of block acted as the heat sink. The overall dimension of the block is 29 mm × 29 mm × 7 mm, and the dimension of groove is 26 mm × 26 mm × 2 mm. Two plates made of aluminum were used to support the block. The rare earth magnets were placed below and above the sample when necessary. Temperature field was measured by the IR

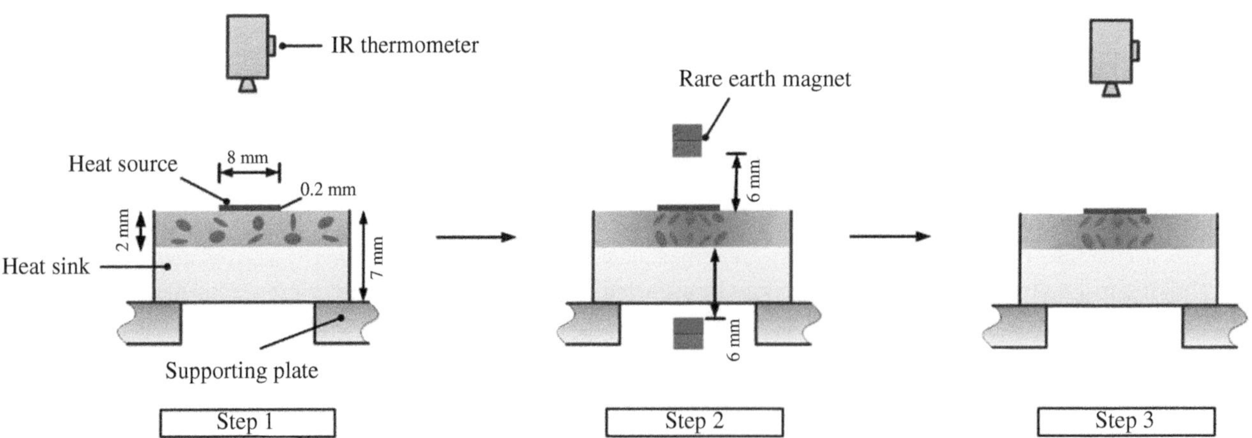

Figure 6.34 Descriptions of the testing system and experimental processes.

thermometer (SC620, FLIR) fixed on top of the heat source with a height of 30 cm. The temperature resolution of this instrument is ±0.1 °C and the emissivity of heat source (polyimide) was calibrated and equal to 0.95. All the experiments were conducted at the environmental temperature of 15 °C.

The experiment processes can be generalized as three steps:

① A constant heat flux is applied on the fluidic homogeneous composites. It takes about 10 minutes to reach a steady state. After that, measure the steady-state temperature field of heater.

② Place the rare earth magnets above and below the sample. After about 10 minutes, the local concentration of platelets is formed.

③ Take away the magnets and measure the steady-state temperature field of heater with the same input power.

Generally, the silicone will cure when heated. It is possible that thermal properties change after silicone curing and some additional heat is released to sample due to the exothermic polymerization. However, in this experiment, the silicone is hard to cure since its temperature is relative low (<60 °C) and the heating time is less than 30 minutes. Normally, it will take 48 hours for silicone to cure completely at 60 °C. Therefore, the probable effect resulting from silicone curing can be neglected. Platelets settlement is a potential effect on platelet distribution and the resulting thermal performance of sample. The density of hBN platelet and silicone is 2270 and 1140 $kg \cdot m^{-3}$, respectively. The platelets will settle in silicone matrix in the case of no magnetic force (at step 1). Referring to the researches [79, 80], the time for the platelets settling to bottom based on density difference, viscosity of silicone, and dimensions of platelet are estimated. The result is about 17 hours. Thus, the probable effect resulting from settlement can also be neglected since step 1 spends only 10 minutes. The small settlement velocity is due to the small density difference and platelet dimensions.

6.2.3.3 Thermal Properties of the Composites

Figure 6.35(a) and (b) give the steady-state temperature fields of the testing system before and after applying the magnetic field on the 5 vol% composite, respectively. In this case, the heat flux given by the heater is 3200 $W \cdot m^{-2}$. Regions where the contours are white and red indicate the high temperature, while the blue contours represent the low value. The rectangular area indicated in the figures is regarded as the approximate region of heat source. Average temperature of heat source (T_{ave}) is then calculated. T_{ave} for the homogeneous and locally reinforced samples are 36.8 and 34.0 °C, respectively. So the reinforced sample reduces T_{ave} with 2.8 °C. The lower T_{ave} indicates that the local reinforcement brings in a substantial promotion for the heat conduction from heat source to sink. To further characterize the impact of local reinforcement, the variations of temperature on the heater at the horizontal direction are investigated. Hotspots in the temperature contours have been indicated by the blue points. Line A lies along the heater which is attached on the homogeneous sample, and Line B is along the heater with locally reinforced sample. The variations of temperature along these two lines are plotted and shown in Figure 6.35(g). It can be found that the temperature varies along the two lines with a similar trend. The temperature at the left of the lines (edge of heater) is very close and low and increases gradually until reaching a maximum temperature (T_{max}) and then decreases toward the right edge of heater. Although the temperature variations of the two lines show a similar trend, T_{max} and temperature gradients across the heater are much different. T_{max} of heat source attached on the reinforced sample is 38.3 °C. There is 3.8 °C reduction compared to that (42.1 °C) of the homogeneous case. Furthermore, there is a smoother change of temperature on the heater assembled with the reinforced composites. Therefore, the locally reinforced composite shows greater ability for the heat dissipation from local heat source.

Figure 6.35(c)–(f) shows the temperature contours of those experiments. Similar results can be found. At the heat flux of 4560 $W \cdot m^{-2}$, the reduction of T_{ave} and T_{max} (ΔT_{ave} and ΔT_{max}) is 5.9 and 6.3 °C, respectively. Furthermore, ΔT_{ave} and ΔT_{max} increase to 7.7 and 8.7 °C when the heat flux is set to be 5840 $W \cdot m^{-2}$. Figure 6.35(g) also compares the temperature variations along the lines through the hotspots. Lower temperature gradients are also found at the heater assembled with the locally reinforced composites.

6.2.3.4 Finite-Element Analysis of Composites Loaded with Local Heat Source

Figure 6.36(a) and (b) shows the representative heat source-composite layer-heat sink models. The composite layer is divided into three bodies and the width of body 1 (wid_1) is equal to that of linear stripe. There are two cases to be simulated according to the comparative experiments. The first case is that particles homogeneously disperse in matrix at the composite layer (Figure 6.36(a)). This is corresponding to the homogeneous case of experiments. For this case, bodies 1, 2, and 3 have the same particle volume fraction (5%). The second is heterogeneous case (Figure 6.36(b)). Body 1 is assumed to homogeneously contain all the platelets. Correspondingly, bodies 2 and 3 are regarded as the pure matrix without any reinforcing elements. Since the width of composite is wid_c (26 mm), body 1 has the volume fraction of 5% × wid_c/wid_1 (19%). κ of those

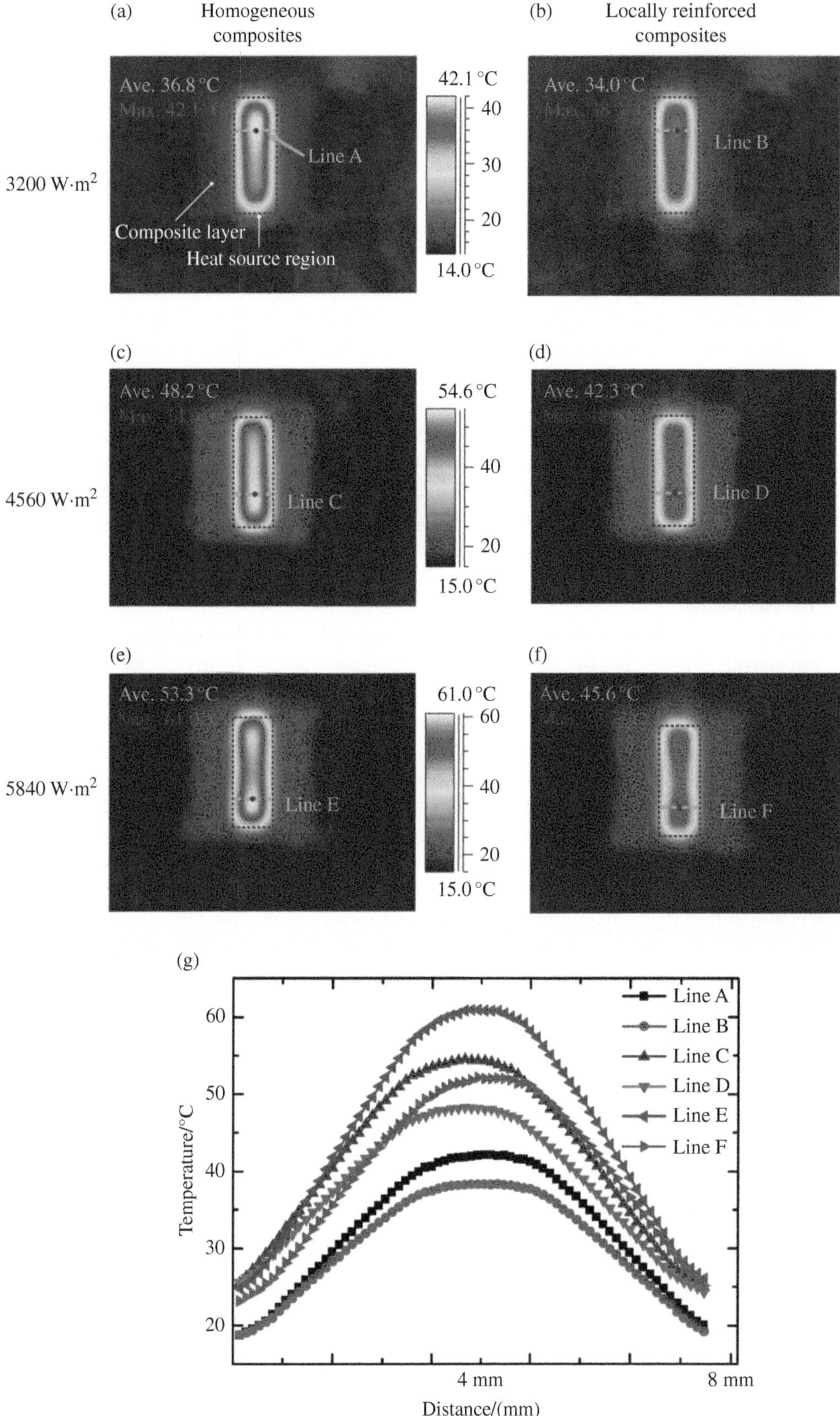

Figure 6.35 Steady-state temperature fields of testing system with homogenous composites at the heat flux density of (a) 3200 W · m^{-2}, (c) 4560 W · m^{-2}, and (e) 5840 W · m^{-2}; steady-state temperature fields of testing system with the locally reinforced composites at the heat flux density of (b) 3200 W · m^{-2}, (d) 4560 W · m^{-2}, and (f) 5840 W · m^{-2}. (g) Temperature variations along the lines through the hotspots.

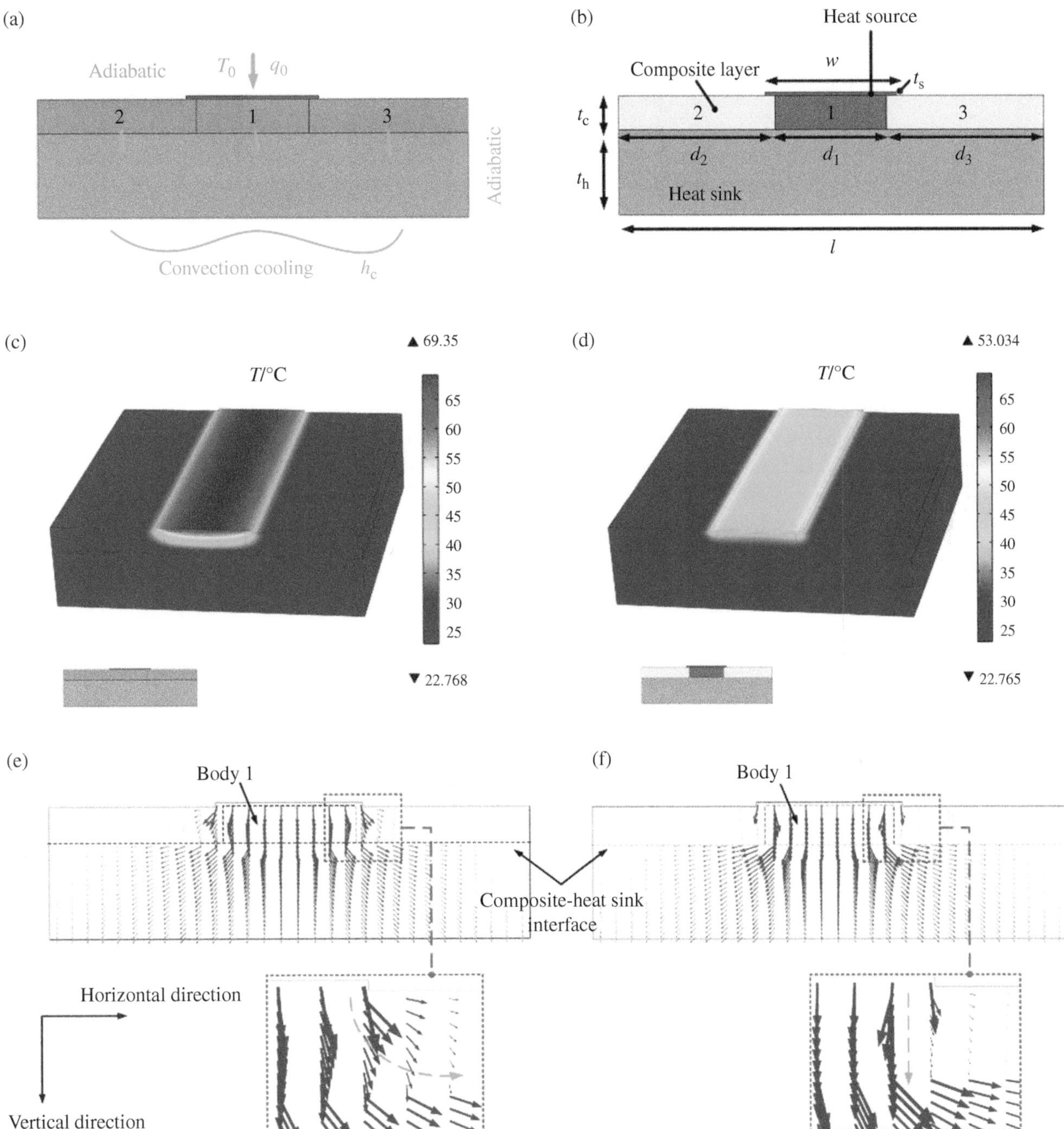

Figure 6.36 Representative heat source-composite layer-heat sink models for simulation: (a) homogeneous case and (b) heterogeneous case; (c) simulated steady-state temperature field of homogeneous case, T_{ave} and T_{max} are equal to 65.5 and 69.4 °C, respectively; (d) simulated steady-state temperature field of heterogeneous case, T_{ave} and T_{max} are equal to 51.0 and 53.0 °C, respectively. (e and f) Simulated cut plane vectors of heat flux inside the whole material domain of the two cases, respectively.

bodies are the key parameters for the simulation. Here, a modified EMA [44] to predict κ of the bodies according to their volume fraction is used. With EMA, κ of homogeneous composites is estimated to be $0.29\ \mathrm{W \cdot m^{-1} \cdot K^{-1}}$; for the heterogeneous case, κ of body 1 is calculated to be $0.70\ \mathrm{W \cdot m^{-1} \cdot K^{-1}}$, while κ of bodies 2 and 3 are equal to that of silicone gel ($0.16\ \mathrm{W \cdot m^{-1} \cdot K^{-1}}$). In this simulation, the heat flux density of heater (q_0) is fixed to be $5840\ \mathrm{W \cdot m^{-2}}$.

Figure 6.36(c) and (d) shows the simulated temperature fields of the two cases, respectively. It is found that the heat source attached to the locally reinforced composites exhibits lower T_{ave} and T_{max}. The results accord with the experimental results.

With the simulated temperature fields, ΔT_{ave} and ΔT_{max} are calculated to be 14.5 and 16.4 °C, respectively. It is interesting to note that the values of ΔT_{ave} and ΔT_{max} estimated by simulation are larger than those derived from experimental results. There could be two reasons lead to the overprediction of simulation. The first one is due to the assumption that at heterogeneous case, all particles are concentrated at the space of body 1. However, as shown in Figure 6.33(b), the magnetic field gradient cannot attract all particles to body 1. This assumption leads to the overprediction of filler volume fraction of body 1 and, hence, overprediction of TC. The overprediction of TC leads to overprediction of ΔT_{ave} and ΔT_{max} in the end. Second, the boundary conditions of simulation are not the same with the experiments. For example, heat transfer from the heater to ambient is neglected in simulation resulting in a larger q_0 compared to the experiments. And the larger q_0 will lead to the increase of ΔT_{ave} and ΔT_{max}.

Furthermore, Figure 6.36(e) and (f) present the simulated vectors of heat flux inside the whole material domain. Through the vectors, the direction of heat propagation in the composite layer can be observed. The vectors reveal that heat generated from heat source inclines to pass through the body 1 to heat sink and goes to ambient ultimately. However, the vectors near the edges of body 1 for the two cases show some difference. Figure 6.36(e) and (f) also give the partial enlarged drawings of the vectors near the edges of body 1. It is observed that in the homogeneous composites, the heat flux at the regions tends to propagate in the horizontal direction. Whereas, in the locally reinforced composites, the orientation of heat flux at the regions is still normal to the plane of the composite layer. The results illustrate that the locally reinforced composites can make the most of heat pass through the shortest path to reach the heat sink.

With the simulation results, a qualitative analysis to explain why the heat source attached on the locally reinforced composites exhibits lower temperature is conducted. Heat flux density (q_1, q_2, and q_3) at the three bodies are calculated from the simulation results, respectively. Since heat transfer from the heater and composites to ambient are neglected in the model, heat flux density of the source q_0 is the total of q_1, q_2, and q_3. q_1/q_0 is calculated as 71.7% and 86.5% for homogeneous and heterogeneous cases, respectively. It is found that q_1 takes the most proportion of q_0 in the two cases. Thus, a 1D heat conduction of body 1 is assumed. According to Fourier's law,

$$q_1 = \kappa_1 \Delta T / \delta_1 \tag{6.29}$$

where κ_1 is the TC of body 1; δ_1 is the thickness of composite layer; neglecting the heat conduction at the thickness of heat source, ΔT is defined as temperature difference between the heat source (T_0) and the upper surface of heat sink (T_1) and expressed as:

$$\Delta T = T_0 - T_1 \tag{6.30}$$

Thus, T_0 can be expressed as:

$$T_0 = q_1 \delta_1 / \kappa_1 + T_1 \tag{6.31}$$

Here, δ_1 is a constant. According to the simulation results, T_1 is kept constant for the two cases, too. Now, T_0 is decided by the value of q_1/κ_1. For the heterogeneous and homogeneous cases, $(q_1/q_0)/\kappa_1$ is equal to 1.24 and 2.47, respectively. Therefore, the lower value of q_1/κ_1 can explain the lower T_0 of heat source attached on the locally structured composites.

6.2.4 Thermal Conductivity Enhancement of CFs-Composites Using Preset Magnetic Field

As the rapid development of microelectronics technology, the smaller size and the higher power density of electronic device have resulted in much higher working temperature, threating the stability, reliability, and lifetime of electronic components. Therefore, the high temperatures of electronic components have become a critical factor limiting the performances of electronic devices. To further improve the performances of electronic devices, heat dissipation has become an important issue for devices. TIMs between heat source and heat sink are important to guarantee stable and sufficient heat dissipation from heat source to heat sink. Thermal pad usually used as TIMs is a kind of composite that consists of the polymer matrix (silicon, epoxy, or rubber) and fillers. However, polymer matrix generally has a low TC (usually less than 0.4 $W \cdot m^{-1} \cdot K^{-1}$) because of the random arrangement of their macromolecular chains [81, 82].

Normally, there are two ways to improve the TC (κ) of thermal pad. The first one is to blend the polymer with high thermal conductive fillers such as metal, metal oxides (e.g. ZnO and Al_2O_3) [83–87], ceramics (e.g. BN, AlN) [85, 88–96], and carbon-based fillers. Recently, a number of studies have reported the enhancement of thermal conductivities of TIMs using carbon-based fillers, including graphene or graphene oxide [97–109], graphite [110–113], CNTs [114–119], and CFs [81, 120–122]. In addition to high TC, carbon-based fillers have advantages such as low thermal expansion, high mechanical strength, flexibility, and low weight [123]. The second one is to regulate the fillers including regulating the orientation of fillers and constructing a 3D thermally conducive network. Several synthetic approaches have been applied to obtain the well-ordered fillers in polymer matrix, including electrical fields (electrostatic flocking and electrospinning), tape-casting, and magnetic field. Uetani et al. [120] fabricated elastomeric TIMs with a high TC of 23.3 $W \cdot m^{-1} \cdot K^{-1}$ by electrostatic flocking; however, electrostatic force was so weak that special-sized CFs had to be chosen to ensure fillers to stand up under electrostatic flocking. Ma et al. [81] fabricated the 3D-CFs/epoxy composites as TIMs with an enhanced TC of 2.84 $W \cdot m^{-1} \cdot K^{-1}$ at 13.0 vol% by vertical freezing the solution of CFs to assemble the through-plane skeleton. Unfortunately, this method normally requires multiple processing steps, and the addition of hydroxyethyl cellulose (HEC) as binders has a bad impact on the thermal conductivities of composites. Tian et al. [124] fabricated three dimensionally interconnected hierarchical porous boron nitride (BN)/epoxy composites by the direct foaming method for thermal interface applications. The BN/epoxy composite exhibited a high through-plane TC of 3.48 $W \cdot m^{-1} \cdot K^{-1}$ at 24.4 wt%. However, the direct foaming method is complex, costly, and time-consuming. Yuan et al. [42] utilized magnetic fields to regulate the orientation of the hBN platelets in the polymer matrix. As a result, the TC of the composite with parallelly oriented hBN was 0.357 $W \cdot m^{-1} \cdot K^{-1}$ at 9.14 vol%. Although this method can improve the TC of composite effectively, the high thermal conductivities of the fillers have not been made full use. There are two main reasons why the thermal conductivities of the composites could not be improved greatly. In the case of low concentration, the fillers filled in a polymer matrix are insufficient to form the internal heat conductive channels. In the case of high concentration, the viscosity of the composite will greatly increase after fillers incorporation. However, the high viscosity of the composite results in a greater resistance for the orientation of fillers. At the same time, the fillers of high concentration will interfere with one another, leading to the restrained deflection and poor orientation.

In this study, we reported a preset-magnetic-field method to prepare the composites with sufficiently aligned fillers of high concentration. The thermal conductivities of nickel-coated carbon fibers-filled composites (NICFs-composites) were investigated. Different from traditional methods, we first arranged fillers in air instead of a polymer matrix by magnetic fields and then filled the intervals between fillers with a polymer matrix. As the viscosity of the polymer matrix no longer restrained the alignment performance of fillers, we could greatly increase the filler concentration without causing deflection difficulties. For the purpose of comparison, two traditional technological processes were used to prepare the composites with different filler concentrations, including mixing followed by arrangement method and directly mixing without arrangement method. The surface states of fillers were characterized on X-ray photoelectron spectroscopy (XPS). Different arrangement performances were observed in different samples by SEM. The thermal conductivities of through-plane (κ_{th}) were measured for these composites. At last, we compared the heat dissipation abilities of the fabricated NICFs-composites and two types of commercial TIMs in a simulated practical application.

6.2.4.1 Fabrication of the Composites

PDMS obtained from Dow Corning Co., Ltd. was used as polymer matrix. Two different NICFs were used. Polyacrylonitrile (PAN)-based CFs (t700 sc-12000, with the diameter of 10 μm and the length of 100 μm) coated with 80 wt% nickel (PAN-NICFs) were purchased from Cangzhou Zhongli New Material Technology Co., Ltd. Pitch-based CFs (NGF, HC-600-15M, with the diameter of 10 μm and the length of 200 μm) coated with 60 wt% nickel (Pitch-NICFs) were purchased from Shanghai Kajite Chemical Technology Co. Ltd., China. The thermal conductivities of PAN-based CFs and Pitch-based CFs are approximately 15 $W \cdot m^{-1} \cdot K^{-1}$ and greater than 600 $W \cdot m^{-1} \cdot K^{-1}$, respectively. Rectangularly sintered NdFeB magnets (250 mT) were used to make magnetic fields.

Preparations of the NICFs-Composites with Different Technological Processes *Arrangement followed by mixing method* (*AM-composite*): as shown in Figure 6.37(a), the AM-composites through the arrangement followed by mixing method were prepared by arranging fillers first and then mixing with a polymer matrix. First, NICFs were added to a Teflon mold. Second, the mold was placed into a vertical magnetic field which ensures NICFs are completely deflecting to the vertical direction to build the vertically heat conductive channels. Third, the mold with magnets were rotated 90° and PDMS was filled into the gap between channels from an entrance in the side edge of mold. Fourth, the bubbles in the mixture were removed by applying alternating cycles of vacuum. At last, a thermal treatment step at 70 °C for 3 hours was conducted to ensure the full curing of the composites. AM-composites were prepared in a high concentration range from 37 to 62.2 wt%.

Mixing followed by arrangement method (MA-composite) and directly mixing without arrangement method (DM-composite): Figure 6.37(b) and (c) schematically show two traditional methods to fabricate the NICFs-composites: the mixing followed

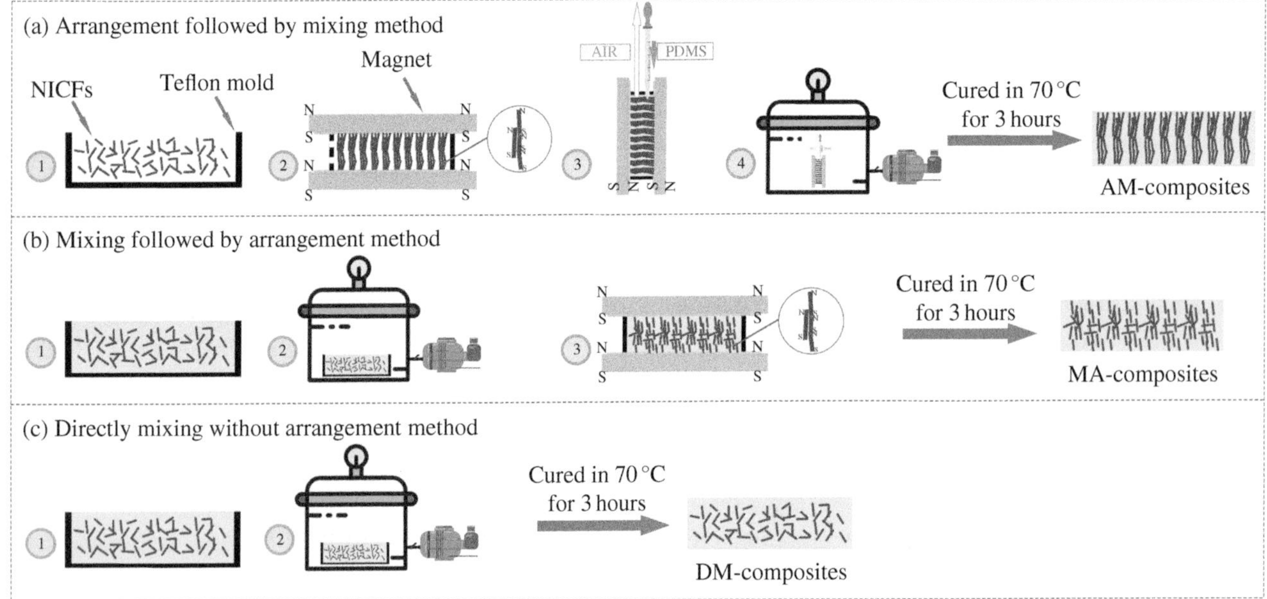

Figure 6.37 Scheme illustrations for the fabrications of (a) AM-composites, (b) MA-composites, and (c) DM-composites.

by arrangement method to prepare the MA-composites and the directly mixing without arrangement method to prepare the DM-composites. Both kinds of composites were prepared at the concentration range from 3 to 59 wt%. For MA-composites, NICFs were first added to PDMS and stirred for 30 minutes to fully disperse. Subsequently, the bubbles introduced during the stirring process were removed by applying alternating cycles of vacuum. Then, the mold was placed into a vertical magnetic field for 10 minutes to ensure the deflection of NICFs. At last, a thermal treatment step at 70 °C for 3 hours was conducted to ensure the full curing of the composites. For DM-composites, after the fully dispersing of NICFs in PDMS and the completely discharging of bubbles, the mixture was thermal cured at 70 °C for 3 hours.

6.2.4.2 Characterization and Analysis

The chemical composition of NICFs fillers was measured by XPS (AXIS-ULTRA DLD-600W, Kratos) to prove the existence of nickel. FSEM (Nova NanoSEM 450) was used to characterize the morphologies of NICFs at an accelerating voltage of 10 kV to observe the nickel on the surface of CFs. The FSEM images of the cross profiles of different samples (AM-, MA-, and DM-composites) with the filler concentration of 53 and 59 wt% were shown to intuitively compare their orientations.

Characteristics of NICFs Figure 6.38 shows the morphologies of Pitch-NICFs and PAN-NICFs with average lengths of 100 and 200 μm, respectively. (a) and (c) in Figure 6.38 are SEM images (under 400 times) of Pitch-NICFs and PAN-NICFs, respectively. Nickel coating on the surface of CFs can be observed from Figure 6.38(b) and (d). Compared with the surfaces of Pitch-NICFs in Figure 6.38(b), thicker nickel can be observed on the surfaces of PAN-NICFs in Figure 6.38(d), which means different magnetic responses of the NICFs to the same magnetic field. The surface chemical elements were identified by XPS. As shown in Figure 6.38(e) and (f), the binding energies were referenced to 285 eV as determined by the locations of the maximum peaks on the C1s spectra of the hydrocarbon associated with adventitious contamination. The Ni 2p peak was observed at 856 eV, proving the existence of nickel element. In addition to Ni, C, and O elements, there were also a small amount of pollution elements such as Mn, Fe, and Co which came from the solution during plating of nickel.

Characteristics of NICFs-Composites To compare the orientation of fillers in different composites made by different methods, their cross-sectional SEM images are shown in Figure 6.39(a) and (d) for the PAN-AM-composite, Figure 6.39(b) and (e) for the PAN-MA-composite, and Figure 6.39(c) and (f) for the PAN-DM-composite at the filler concentrations of 53 and 59 wt%, respectively. PAN-NICFs were used to prepare the above NICFs-composites. At the same filler concentration, PAN-AM-composite shows the most orderly orientation of NICFs in the vertical direction than that of the PAN-MA-composite and the PAN-DM-composite. At the filler concentration of 53 wt%, the arrangement of NICFs in PAN-MA-composite in

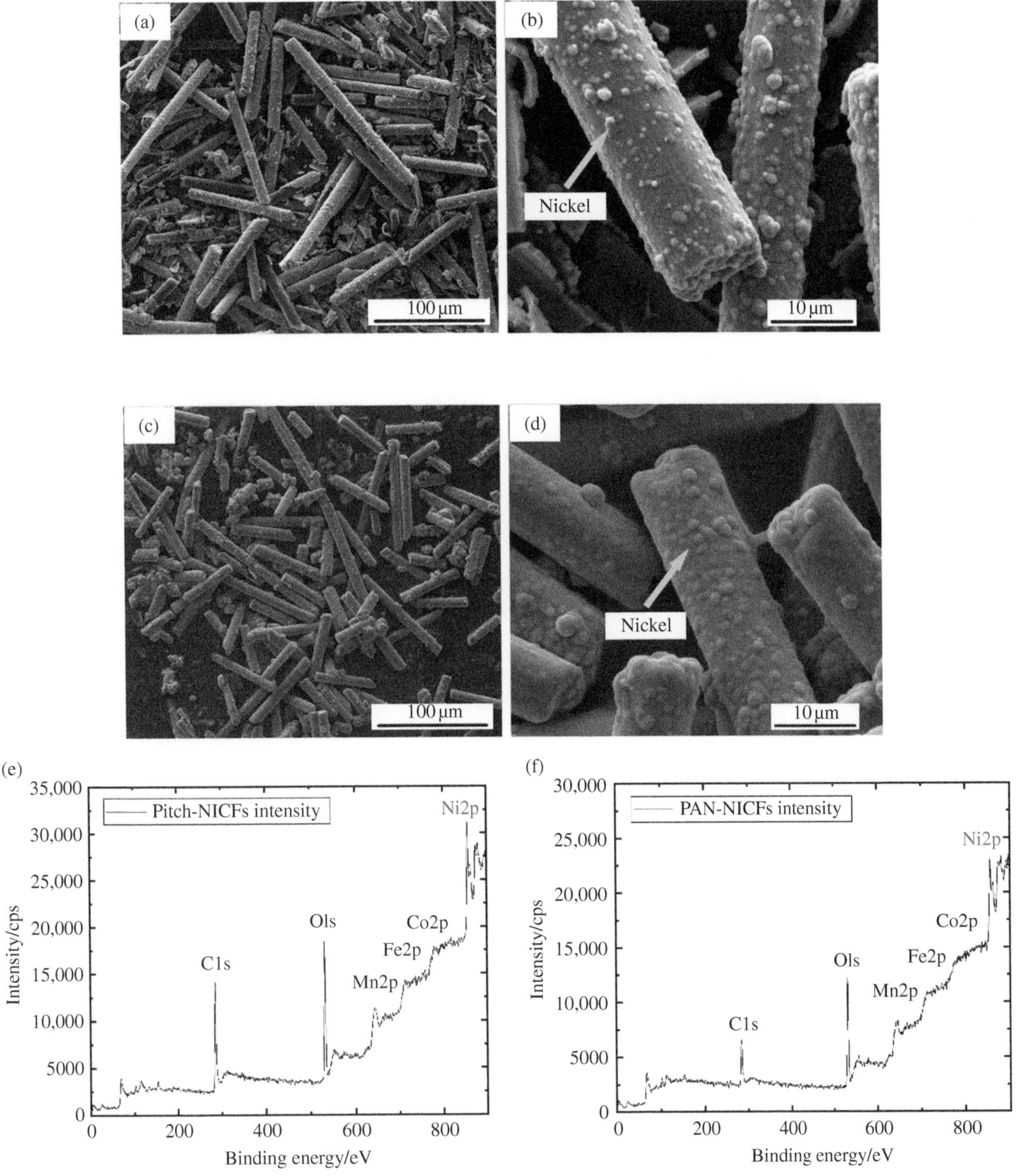

Figure 6.38 SEM images of Pitch-NICFs under (a) 400 times and (b) 3000 times. SEM images of PAN-NICFs under (c) 400 times and (d) 3000 times. (a–d) Zhang et al. [125]/with permission of Elsevier. XPS broad spectra of (e) Pitch-NICFs and (f) PAN-NICFs.

Figure 6.39(b) was obvious better than the disorderly and unsystematic NICFs in PAN-DM-composites in Figure 6.39(c) because of the deflection by the magnetic field.

However, at a higher filler concentration of 59 wt%, the orientations of NICFs in PAN-MA-composite became a disordered as that of the PAN-DM-composite. It is worth noticing that the PAN-AM-composite maintained good vertical orientation of NICFs at the high filler concentration of 59 wt% and even the higher of 62.2 wt%, as shown in Figure 6.39(d) and (g),

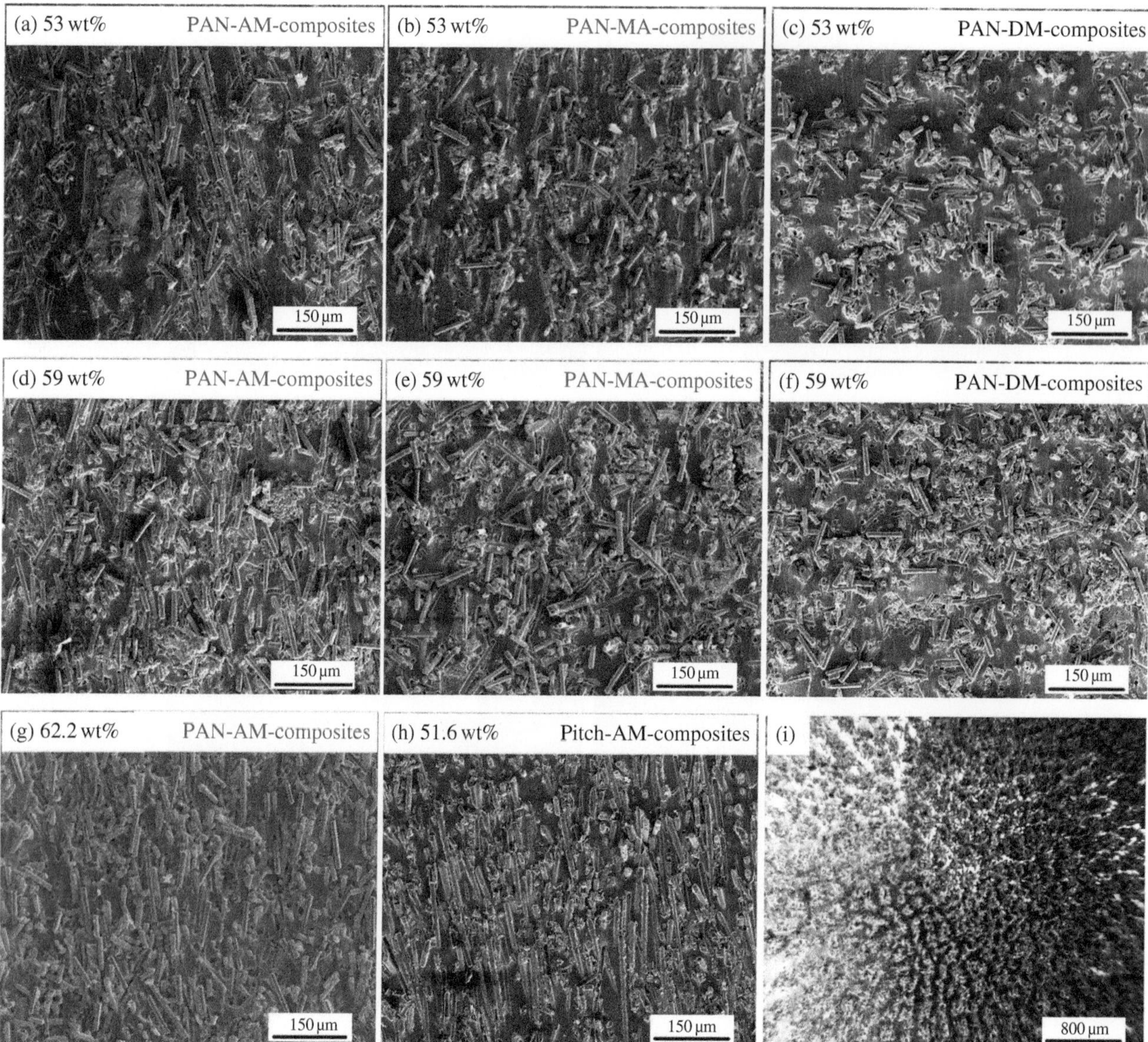

Figure 6.39 Cross-sectional SEM images of (a) PAN-AM-composites, (b) PAN-MA-composite, and (c) PAN-DM-composite made with PAN-NICFs under 400 times in 53 wt%, respectively. Cross-sectional SEM images of (d) PAN-AM-composites, (e) PAN-MA-composite, and (f) PAN-DM-composite made with PAN-NICFs under 400 times in 59 wt%. Cross-sectional SEM images of (g) PAN-AM-composites made with PAN-NICFs under 400 times in 62.2 wt% and (h) Pitch-AM-composites made with Pitch-NICFs under 400 times in 51.6 wt%. (i) Top-view optical image of PAN-NICFs arranged in the air regulated by magnets. (a–i) Zhang et al. [125]/with permission of Elsevier.

respectively. Figure 6.39(i) shows the optical image of the top view of vertically aligned PAN-NICFs in air by a vertical magnetic field. Only the ends of NICFs are observed in the top-view image. In Figure 6.39(i), the magnetized NICFs by magnets were endowed with the NS poles at two ends. The magnetized NICFs with high length-diameter ratios tended to arrange vertically in a vertical magnetic field. In addition, magnetized fillers got together because the NS poles between the magnetized fillers would attract and then connect with one another. The vertically aligned magnetized fillers built vertical heat conductive channels by the connection between the magnetic ends of the fillers. The arrangement of NICFs in AM-composites was carried out in air and mainly determined by the magnetic force. It is practical to increase the concentration of fillers to improve the thermal properties of composites. However, the deflection and connection of NICFs in MA-composites happened in viscous PDMS and was restrained by the viscous force of mixture. In addition, due to the increase of the fillers load in PDMS, the interaction between fillers was enhanced, which worsened the fillers orientation in the magnetic field [126].

To compare the effects of different types of CF, Pitch-AM-composite with Pitch-NICFs of concentration of 51.6 wt% was prepared, as shown in Figure 6.39(h). Pitch-AM-composite exhibited better directional alignment than others due to their different contents of nickel and length.

6.2.4.3 Thermal and Mechanical Properties of Composites

Thermal conductivities are calculated by $\kappa = \alpha \times \rho \times C_p$, where α, ρ, and C_p are thermal diffusivities, heat capacities, and densities of the samples, respectively. α of the composites were measured by a laser flash method using an LFA 467 (Netzsch) at 25 °C. The geometry of the tested sample is a cylinder with a diameter of 12.7 mm and a thickness of 1.5 mm, which was cut from the prepared cured composite. Densities of composites were measured by a densimeter to compare the theoretical densities. C_p of the composites were measured by DSC (Diamond DSC, PerkinElmer Instruments). The CTE of the composites in the through-plane direction was measured by the thermomechanical analysis (TMA) (model no. TMA402 F1 Hyperion, Netzsch) within the temperature range from 35 to 250 °C.

The variations of the surface temperature of different composites were recorded by infrared thermograph (FLIR SC620). The variations of the surface temperature of ceramic heater with different TIMs were measured by thermocouples and recorded by a data collector.

We investigated the thermal conductivities of the PAN-AM-composites in comparison with that of PAN-MA-composites and PAN-DM-composites. To calculate the thermal conductivities of composites, the densities (ρ) and specific heats C_p of composites were measured and shown in Figure 6.40(a) and (b), respectively. The measured values were in good agreement with the theoretical values, demonstrating all bubbles in composites were removed. Thermal diffusivities of composites were also measured. The thermal conductivities of all composites are shown in Figure 6.40(c).

The black line with squares in Figure 6.40(c) shows that the thermal conductivities of PAN-DM-composites increased with the mass fraction of NICFs, but it is not increased greatly due to the disorder of NICFs in composites. By applying magnetic fields in the preparation process, PAN-MA-composites (red line with circles) exhibited higher thermal conductivities than those of PAN-DM-composites. However, this advantage was only shown at low concentrations. At filler concentrations higher than about 50 wt%, the thermal conductivities of PAN-MA-composites decreased with concentrations because of worse orientation. Specifically, the TC enhancements of PAN-MA-composites to PAN-DM-composites, shown in Figure 6.40(d), indicated that the biggest enhancement is located at a middle concentration about 37 wt%. NICFs were not orderly aligned in PAN-MA-composites with a high filler concentration because the higher viscosity of the mixture with a higher filler concentration severely restrained the deflection of filler in the mixture by the magnetic field. More seriously, more NICFs got together and blocked one another, resulting in more difficult deflection.

However, the fillers in the PAN-AM-composites were orientated in air before mixing with PDMS, which means the viscosity of PDMS hardly limited the deflection of filler. The blue line with rhombuses in Figure 6.40(c) shows the thermal conductivities of PAN-AM-composites increase with concentration even at a high concentration. The PAN-AM-composites were superior to both the PAN-MA-composites and PAN-DM-composites in the thermal conduction at a high concentration range. For instance, at the concentration of 62.2 wt%, the TC of PAN-AM-composite reached 1.95 $W \cdot m^{-1} \cdot K^{-1}$, 12 times higher than that of the pure PDMS (0.15 $W \cdot m^{-1} \cdot K^{-1}$). Pitch-based CFs (more commonly used than PAN-based CFs) have a TC greater than 600 $W \cdot m^{-1} \cdot K^{-1}$. At the concentration of 51.54 wt% (21.5 vol%), the TC of Pitch-AM-composite reached 10.50 $W \cdot m^{-1} \cdot K^{-1}$. Table 6.10 lists previously reported thermal conductivities of polymer composites with other fillers, including boron nitride nanosheets (BNNs), graphene, and CNT. The TC of Pitch-based CFs in the axial direction is greater than 600 $W \cdot m^{-1} \cdot K^{-1}$, much larger than that in the radial direction. Based on the high axial TC of Pitch-based CFs, their high ORI in through-plane direction is conducive to improve the thermal conductivities of composites in through-plane direction. The data shows the Pitch-AM-composite exhibits the highest TC due to the following points: ① fillers exhibited better orientation because they were oriented in air, meaning deflection was not restrained by the viscous force of PDMS. ② The presence of magnetic field throughout all the preparation process ensured NICFs are always in touch with each other, which reduces the interface between NICFs and PDMS, forming a heat conductive channel. ③ The concentration can be improved greatly because deflection was not restrained.

In order to exhibit different thermal conductive performances of above composites, four different samples were prepared with the same thickness of 2.1 mm including ① PAN-AM-composite of 69 wt%, ② PAN-MA-composite of 53 wt%, ③ PAN-DM-composite of 53 wt%, and ④ pure PDMS. As shown in Figure 6.41(a), four different composites were placed on a heating plate with a constant temperature of 100 °C simultaneously. The infrared thermograph was used to record the changes of surface temperature with time. Figure 6.41(b) shows the corresponding infrared thermal images of above four composites.

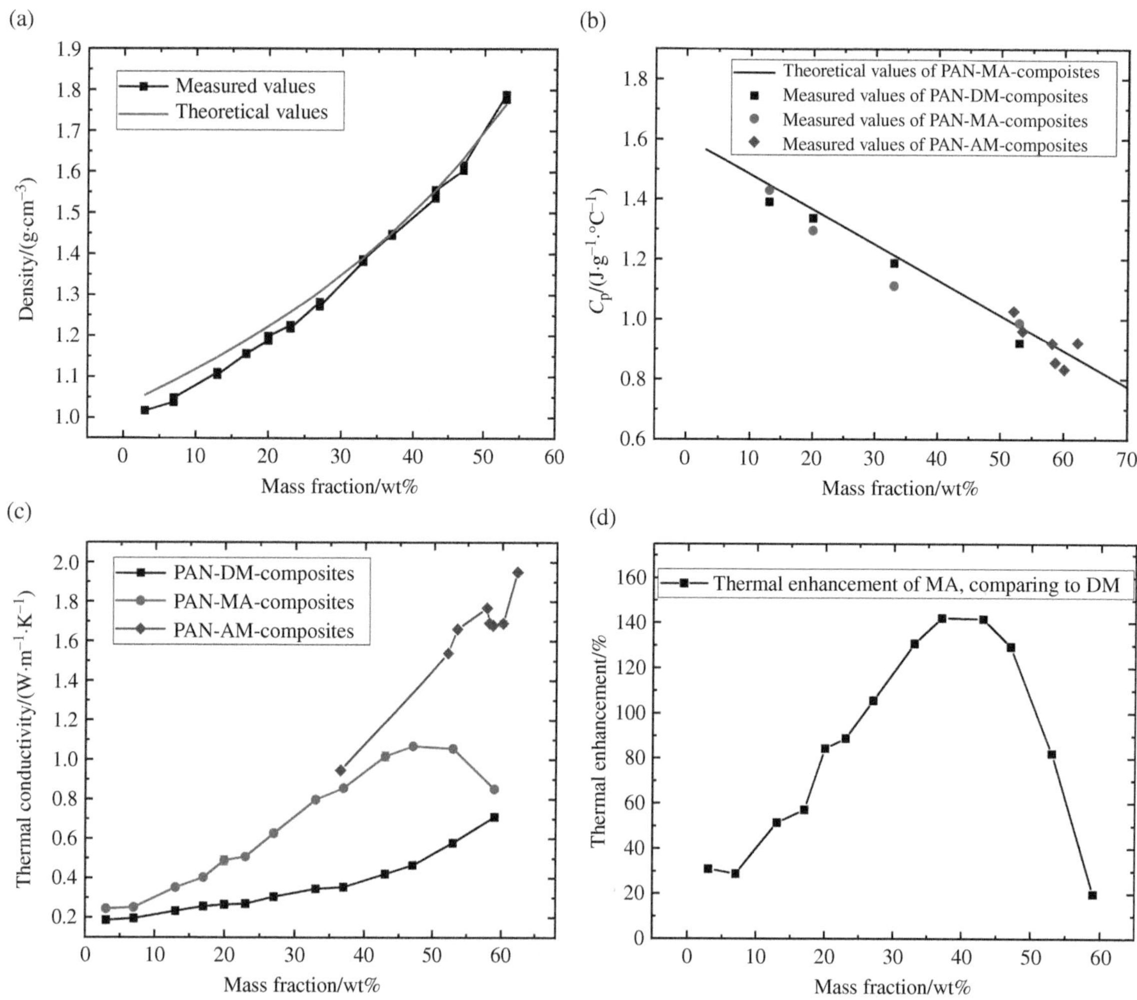

Figure 6.40 (a) The densities of PAN-MA-composites and PAN-DM-composites. (b) The specific heat capacities of PAN-DM-composites, PAN-MA-composites, and PAN-AM-composites, respectively. (c) The thermal conductivities of PAN-DM-composites, PAN-MA-composites, and PAN-AM-composites. (d) The thermal enhancement of MA-composites in thermal conductivities comparing to DM-composites.

Table 6.10 Comparison of through-plane thermal conductivity of Pitch-AM-composites with other thermally conductive polymer composites.

Fillers	Matrix	Loading	Thermal conductivity $(W \cdot m^{-1} \cdot K^{-1})$	References and year
BNNs	Silicone gel	9.14 vol%	0.58	2015 [42]
BNNs	Epoxy	50 wt%	3.59	2016 [127]
BNNS	PDMS	28.7 wt%	1.94	2017 [82]
BNNs	Epoxy	24.4 wt%	5.19	2018 [124]
Aligned CNT	Epoxy	16.7 vol%	4.87	2011 [128]
Graphene	Epoxy	25 vol%	10	2015 [109]
Graphite platelets	Poly(vinyl pyrrolidone)	60 wt%	8.9	2018 [123]
Aligned CNT	Epoxy	5.4 wt%	9.62	2020 [103]
Pitch-based CFs	Epoxy	13 vol%	2.84	2020 [81]
Pitch-NICFs	PDMS	51.54 wt%	10.50	This work

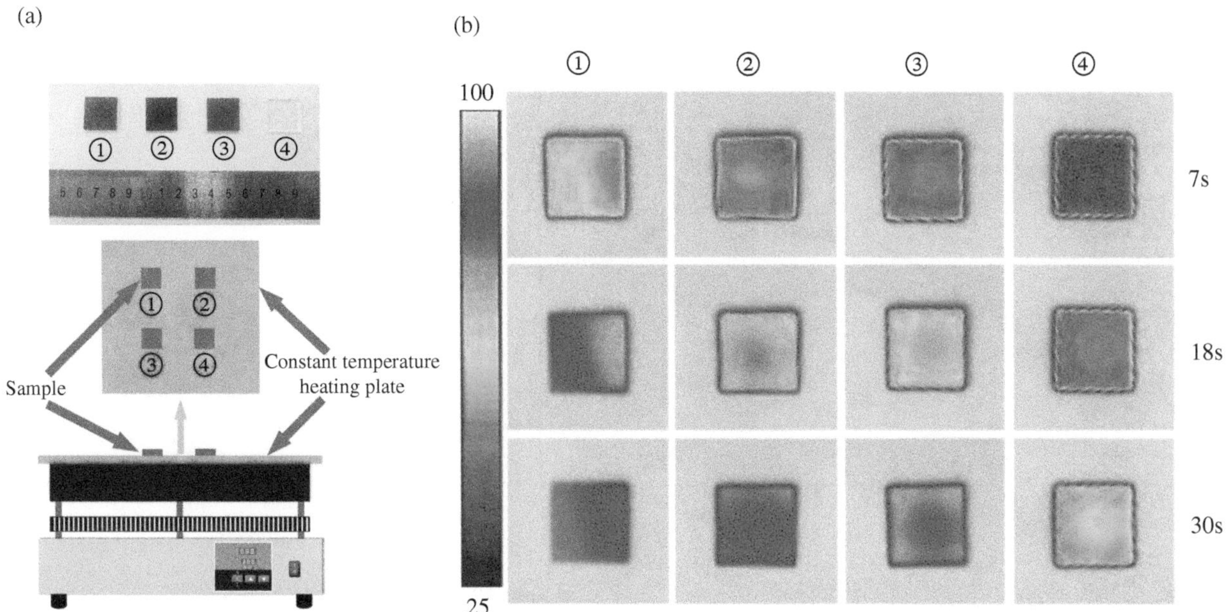

Figure 6.41 (a) Experimental equipment diagram including constant temperature heating plate. Zhang et al. [125]/with permission of Elsevier. (b) The changes of surface temperature and the corresponding infrared thermal image of the composites.

Sample ①, the PAN-AM-composite of 69 wt%, exhibited the highest temperature on the surface than other samples because of its highest TC.

In addition, a low CTE is also important for TIMs to keep a better dimensional thermal stability with a lower thermal expansion stress which avoids the damage to device. The length of expansion of PAN-AM-composites in through-plane direction as a function of temperature is shown in Figure 6.42(a) with the temperature ranging from 35 to 250 °C. Figure 6.42(b) reveals that there is a steady decline in the CTEs of PAN-AM-composites with the increase of PAN-NICFs content. However, the CTEs of PAN-MA-composites show different tendencies. The main reason is connected with the ORI. CF shows a negative CTE in the axial direction [129, 130]. High ORI in through-plane direction is conducive to decrease the CTEs of composites in through-plane direction. As the fillers loadings increase, the CTEs of PAN-MA-composites decrease first and then increase with the inflection point at a concentration of 37 wt%. The phenomenon confirms that fillers were not orderly aligned in PAN-MA-composites with a high filler concentration. The reason is that the higher viscosity of the mixture with a higher filler concentration severely restrained the deflection of filler in the mixture by the magnetic field. More seriously, more NICFs got together and blocked each other, resulting in more difficult deflection in high filler concentration. However, the steady decline in the CTE of PAN-AM-composites confirms that PAN-AM-composites have great orientation even in high concentration. The CTEs of PAN-AM-composites were much lower than that of pure PDMS. For example, at the concentration of 58.2 wt%, the CTE of PAN-AM-composite was 41.99 $\text{ppm} \cdot {}^{\circ}\text{C}^{-1}$, but the CTE of pure PDMS was 261.92 $\text{ppm} \cdot {}^{\circ}\text{C}^{-1}$. Pitch-AM-composite also exhibited a low CTE of 55.14 $\text{ppm} \cdot {}^{\circ}\text{C}^{-1}$ at 51.54 wt%. AM-composites have the lower CTE mainly attributed to the well-aligned structure and low thermal expansion of NICFs.

A Pitch-AM-composite and two commercial TIMs were applied in the thermal management of high-temperature electronic devices like a computer's central processing unit (CPU). Figure 6.43(a) shows the optical photograph of experimental facility. Figure 6.43(c) shows the specific experimental set-up. A ceramic heating sheet (24 V, 25 W) was used to simulate the computer's CPU as a heat source. A DC power supply provided the voltage of 24 V to the ceramic heating sheet. Three TIMs were used to connect the heat source to fins, respectively. A customized polyphenylene sulfide (PPS) board was placed on the heat source, fixed to the fins with bolts and nuts. A torque of 0.3 $\text{N} \cdot \text{m}$ was supplied on the bolts and nuts to ensure different TIMs sustaining the same pressure. A data collector collected the temperature data of the surface of the ceramic heating sheet by thermal couples.

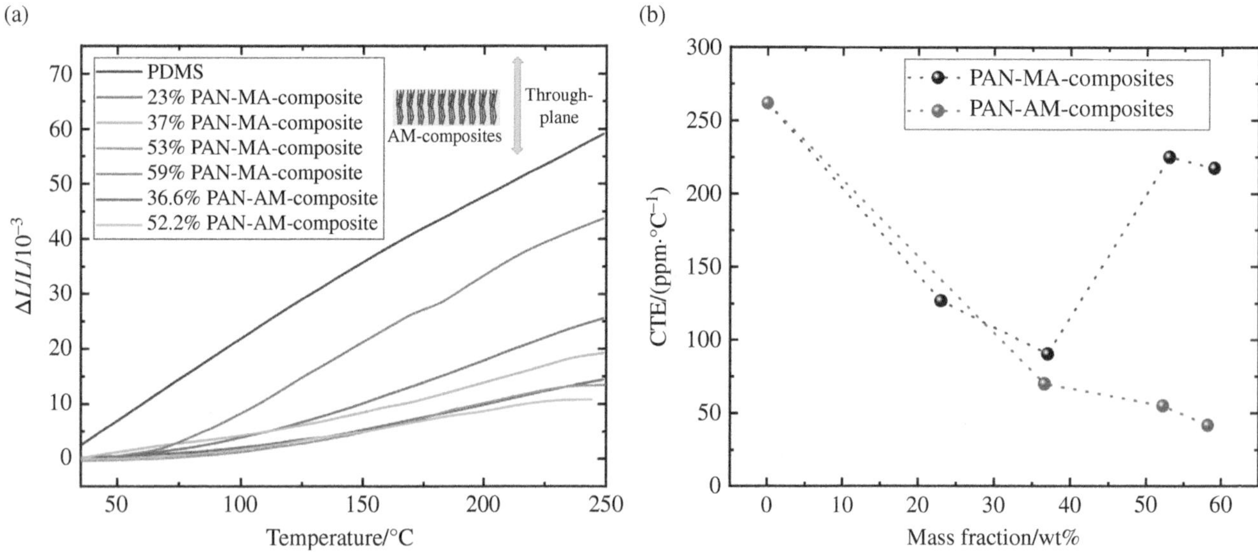

Figure 6.42 (a) The length expansion of PAN-AM-composites as a function of temperature. (b) The CTE of AM-composites.

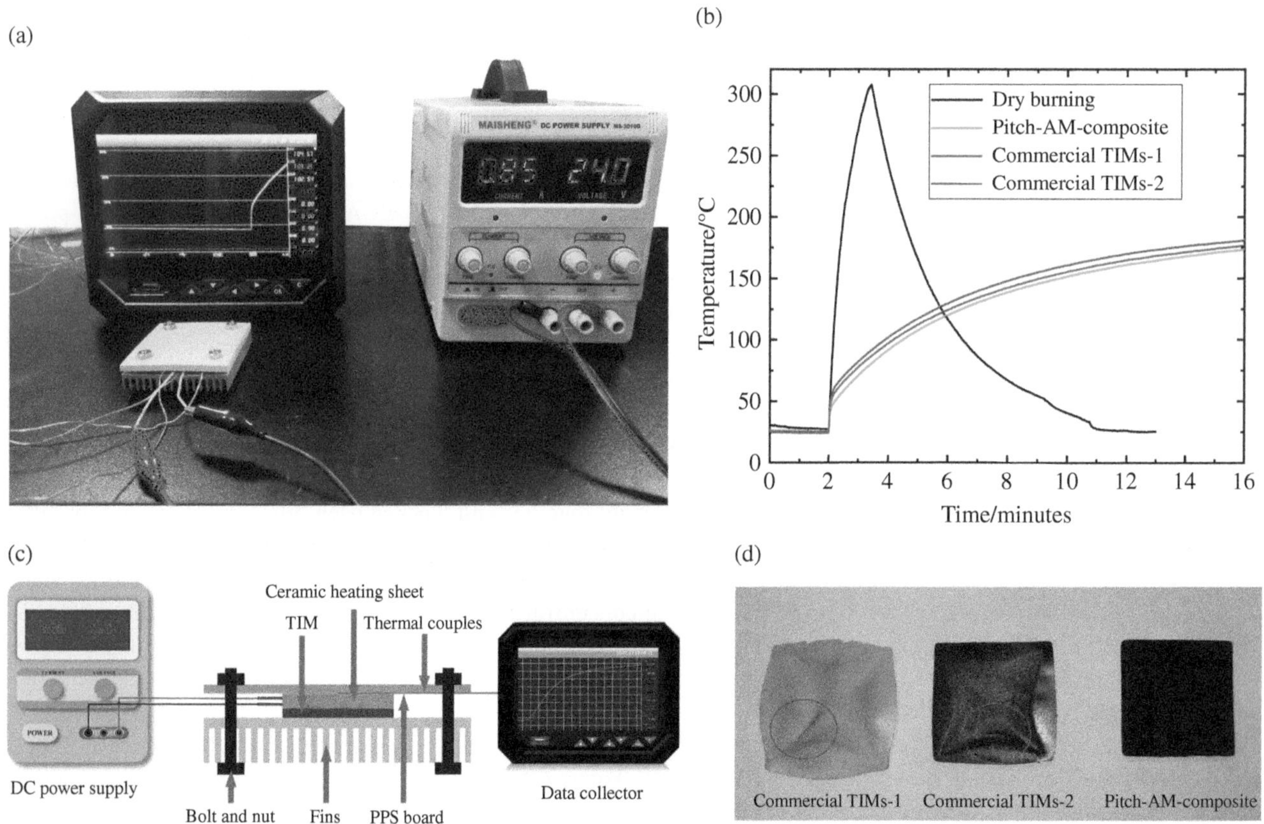

Figure 6.43 (a) The optical photograph of an experimental facility. (b) Graph of temperature changes over time with different TIMs. (c) The specific experimental set-up for the simulated experiment of applying TIMs to a computer's CPU. (d) Picture of TIMs after the test. (a, c, d) Zhang et al. [125]/with permission of Elsevier.

As shown in Figure 6.43(b), the temperature of the dry-burning ceramic heating sheet rapidly raises over 300 °C. After using TIMs, the surface temperature of the ceramic heating sheet decreases to about 180 °C. Specifically, the Pitch-AM-composite in the green line exhibits the lower surface temperature of ceramic heating sheet than those of the commercial TIMs-1 and TIMs-2 in the red and blue lines, respectively, indicating its superior thermal performance as TIMs. The three TIMs after the test are shown in Figure 6.43(d). Cracks occurred in both the commercial TIMs-1 and TIMs-2 due to their low mechanical properties and CTEs. However, the Pitch-AM-composite maintained its original shape with no deformation, proving its excellent mechanical property and thermal stability. In short, the Pitch-AM-composite TIMs with superior thermal conductive ability and thermal–mechanical stability have huge application prospects in the thermal management of high-temperature electric devices.

6.2.5 Self-Assembly Design of TIMs for Hotspot Problem

The artificial composites have been widely applied in people's lives. Polymers are usually used as the matrix of artificial composites because of their promising heat transport ability, superior flexibility, and excellent durability [28, 131–134], which has attracted widespread attentions from thermal [135, 136], mechanical [137], biochemistry [138], food scientific [139], and other fields. However, in thermal management field, the application of polymers is usually limited by the low thermal conductivities (TCs) of polymers when polymers are used as TIMs. Currently, a common strategy to improve the thermal performance of polymer composites is to directly blend kinds of special fillers with polymers, including metal or metal oxides (e.g. Al, Al_2O_3, and CuO) [140–142], ceramics (e.g. BN and AlN) [47, 143–146], and carbon-based fillers (e.g. graphene and CFs) [147]. Importantly, besides the attributes of fillers, the macroscopic performance and function of polymer composites depend on fillers' microstructure inside the polymer, especially for anisotropic fillers, such as CFs and hBN [148]. Thus, a growing number of researches have focused on the microstructural regulation of fillers inside polymers by electrical force [120], magnetic force [42], and shear force [149, 150], aiming to manipulate fillers and fabricate vertically and horizontally oriented structures or 3D networks [151–153]. These polymer/fillers microstructures bring a remarkable improvement in thermal performance in a specific direction and thus broaden their practical applications, typically in heat dissipation systems [154–156]. However, the development of next-generation 5G cellular devices and military communications bring new challenges for composite thermal materials [62]. Specifically, the integral chips are becoming miniaturized and exhibit higher power density, up to 300 $W \cdot cm^{-2}$, even next-generation devices could exceed 1 $kW \cdot cm^{-2}$ [28]. Besides the power density, the spatial distribution of heat at chip level is also of concern and can have thermal management implications. Localized functional areas in modern chips can incur "hotspot" – regions where local temperatures are significantly higher than the average temperature of chips [28]. The localized accumulation of heat always leads to fast aging or even failure of the core chip because the overall reliability of the chip is determined by the hottest region on the chip rather than the average temperature [28, 157]. Therefore, the heat caused by hotspots needs to be rapidly diffused by targeted and regional-level heat channels and then dissipated to the environment by the heat sink. The improvement of thermal performance along a specific direction cannot solve this problem well. Hence, it is urgently needed to elaborately manipulate the fillers to form special microstructures to provide embedded multidirectional phonon pathways in composites for adapting to the thermal flow field of the hotspot, which remains a challenge.

Fortunately, there are a lot of fantastic microstructures among creatures in nature to meet different requirements, which provide significant inspirations for the designing of artificial thermal materials. For example, the balsa shows a regular circinate structure of the trunk, whose water-delivery xylem vessels grow vertically to ensure sufficient and rapid moisture vertical transportation, as shown in Figure 6.44(a) and (b). In addition, the ginkgo leaf grows the radially aligned microchannels to ensure efficient moisture transportation from the root to the edge of leaves, as shown in Figure 6.44(c) and (d). Similarly, the radial structure of sea urchin shell provides excellent mechanical strength, as shown in Figure 6.44(e). Ingenious natural prototypes undoubtedly provide novel ideas for constructing microstructures in composites [158, 159]. For example, nacre-bionic nanocomposite membranes were fabricated by vacuum filtration method, showing superior thermal conductivities and mechanical performances on account of the orderly stacked arrangement [160]. Inspired by these biological structures, we came up with the idea that composites with a radial structure exhibit great potential as TIMs in dealing with the hotspot problem. To realize the rapid and accurate heat transfer for hotspots, we intended to mimic leaves to prepare composites with radially oriented CFs (one-dimensional material) to utilize their excellent TC in the axial direction. In previous studies, a similar structure has been accomplished and applied in different fields including solar energy collection [161] and water purification and transportation [162] by the multidirectional ice-template assembly method [163, 164]. For solar energy collection, radially oriented and overlapped BN forms continuous heat transfer channels, which contributes to

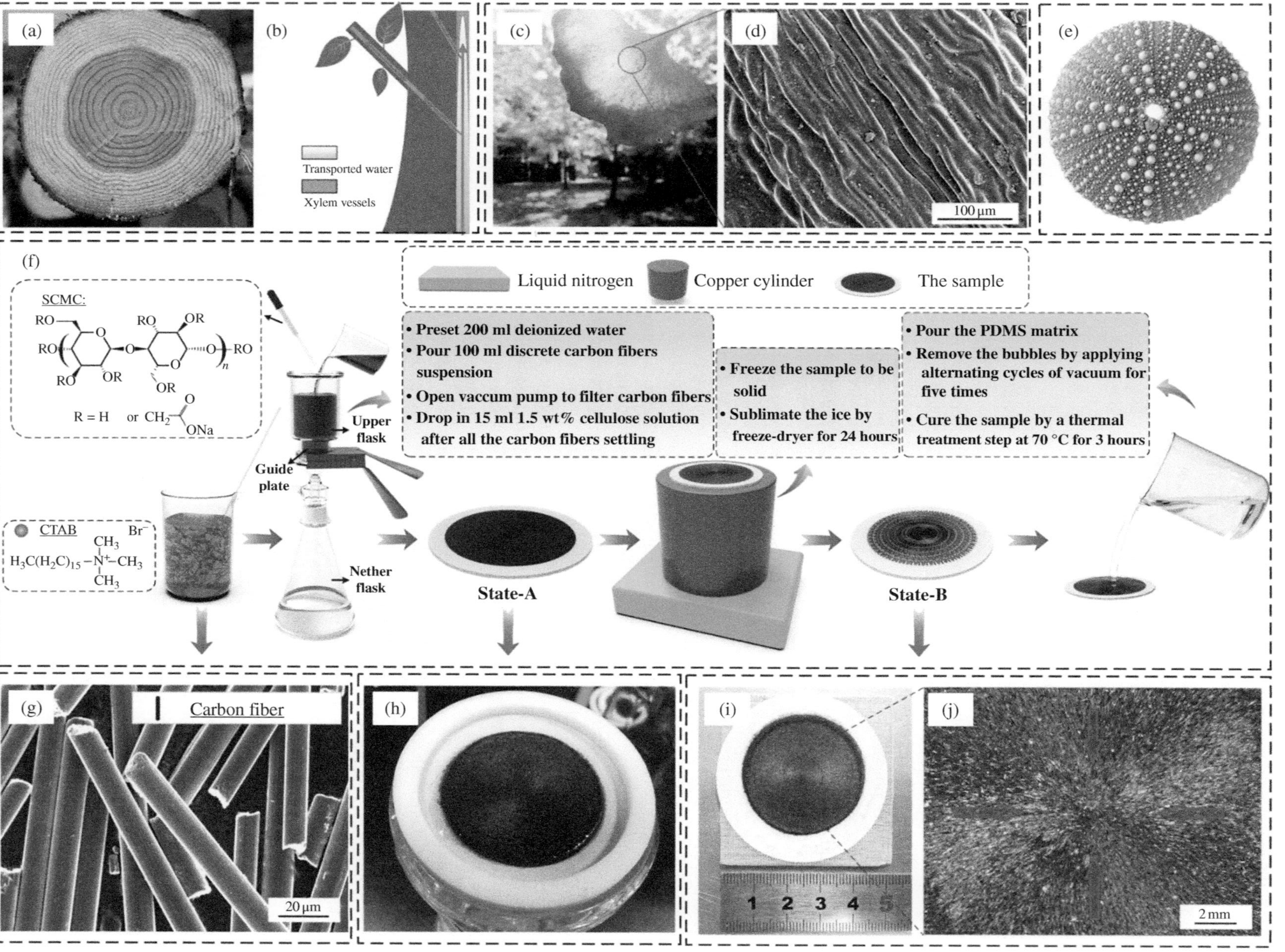

Figure 6.44 (a) Optical image of balsa. (b) Diagram illustration of water transportation within trees. (c) Optical image of a ginkgo leaf. (d) Superficial SEM image of a ginkgo leaf. (e) Optical photo of the shell. (f) Schematic diagram for the preparation of composites with radially oriented CFs. (g) SEM image of CFs. Optical photo of state-A (h) after vacuum filtration and state-B (i) after freeze-drying. (j) Optical image of the center of state-B.

the rapid conduction of localized light sources. However, the ice-template assembly method is unsuitable for the preparation of radial structure for 1D materials, because it cannot control the axial orientation of 1D materials such as CFs. To obtain a better orientation of CFs, a more effective method for 1D materials is needed.

Here, we proposed a flow field-driven self-assembly strategy, which combines the vacuum filtration method [160, 165–167] and fluid control method [168], to prepare thermally conductive composites with radial and centrosymmetric structures. Different from the traditional vacuum filtration setup, a guide plate is designed and placed on the upper flask, to control fluid flow above the filtrating membrane and form the special flow field, which drove the CFs to arrange along the streamlines by the action of shear forces. Combining the vacuum filtration, CFs are blocked and deposited on the filtrating membrane with CFs keeping the regular and special arrangement when the liquid flow through the filtrating membrane, which forms a self-assembly process. Specifically, we prepared centrosymmetric composites with radial aligned (Ra) by designing the guide plate, which are beneficial to the heat diffusion of hotspots. To quantitively evaluate the ORI, we proposed an orientation algorithm based on microscale image identification and an evaluation criterion with no need of XRD or wide-angle X-ray scattering (WAXS) [151]. In addition, the flow field of vacuum filtration was simulated and analyzed, demonstrating the flow under the guide plate and interpreting the separation phenomenon discovered in experiments. In the Ra-composites, fast, straight-through, and efficient phonon pathways are built by the radial arrangement and connection of CFs, which maximally utilizes CFs' anisotropic TC. The Ra-composites exhibit excellent arrangement and ultrahigh in-plane thermal conductivities (in-plane TC) with the highest value reaching 35.5 $W \cdot m^{-1} \cdot K^{-1}$ (236 times that of pure PDMS [0.15 $W \cdot m^{-1} \cdot K^{-1}$]), which, therefore, enables a powerful heat diffusion and uniform temperature distribution when applied in the hotspot. Meanwhile, the highest anisotropy of Ra-composites reaches 19.8, which indicates CFs' anisotropy is fully utilized. Derived from this fast and radially uniform heat-transfer characteristic of radial structure, the hotspot temperature with Ra-composites was 33.4 °C lower than commercial TIMs in the experiment of thermal diffusion (TD) for hotspots. Moreover, the heat dissipation system with Ra-composites exhibited a better cooling effect than that with commercial TIMs. Furthermore, this flow field-driven self-assembly strategy demonstrates a promising self-design ability for the arrangement of arbitrary two-dimensional shapes and a promising application in hotspots. Composites with more fantastic arranged structures are expected to be applied in arbitrary-shaped heat sources and other fields like solar–thermal–electric conversion.

6.2.5.1 Fabrication of the Composites

As shown in Figure 6.44(g), CFs with the typical 1D structure and high axial thermally conductive characteristic (~600 $W \cdot m^{-1} \cdot K^{-1}$) are selected as the basic unit. The flow field-driven self-assembly strategy is intended to engineer the radial architecture, which is schematically illustrated in Figure 6.44(f). Given that CFs are poorly dispersed and easy to agglomerate in water, herein, hexadecyl trimethyl ammonium bromide (CTAB, cationic surface-active agent) is first added into CFs suspension, to promote the dispersion of CFs. After all the CFs are settled in the process of vacuum filtration, sodium carboxymethyl cellulose (SCMC) solution is added into the suspension to fix the CFs. After vacuum filtration finishes, CFs are deposited on the filtrating membrane with SCMC and a small amount of water (state-A), as shown in Figure 6.44(h). After freeze-drying by liquid nitrogen in a freeze-dryer, the remaining water is removed with SCMC and CFs remaining (state-B, CFs-scaffold), as shown in Figure 6.44(i) and (j). The detailed description of composites preparation can be seen in the experiment section. The difference between flow field-driven self-assembly strategy and traditional vacuum filtration method is illustrated in detail in Figure 6.45(a) and (b). In the flow field-driven self-assembly strategy, a guide plate is designed and placed 3 mm above the filtrating membrane to control the flow field, as shown in Figure 6.45(a). To form a radial flow field, the guide plate is made into a 2 mm disc with an aperture in the center. The fluid above the guide plate is compelled to flow through the aperture under the negative pressure of the nether flask and then spreads out from the center to the edge, which forms the diffusion flow. In this process, CFs are driven to orientate along the streamlines by the shear force of fluid. Simultaneously, the water is pumped out through the filtrating membrane because of the negative pressure in the nether flask, enforcing CFs to be deposited on the filtrating membrane, where CFs maintain the special arrangement. Importantly, the filtrating membrane pore and guide plate aperture can be adjusted to control the flow field above the filtrating membrane and thus determine the arrangement of CFs. A certain amount of whirlpool forming above the diffusion flow does not influence the arrangement of the filtrating membrane because the diffusion flow exists until all the CFs are deposited. Ultimately, CFs form a radially oriented structure, resembling the flow field. However, in traditional vacuum filtration method, the flow field above filtrating membrane is random and changeful, resulting in random and chaotical deposition of CFs. Herein, Ra/0.45/5-composite was prepared using a filtrating membrane with 0.45 μm pore and a guide plate with 5 mm aperture by flow field-driven self-assembly strategy. For comparison, we prepared the Ro/0.45/blank-composite with random orientation (Ro) using 0.45 μm pore filtrating membrane without the guide plate by traditional vacuum filtration method.

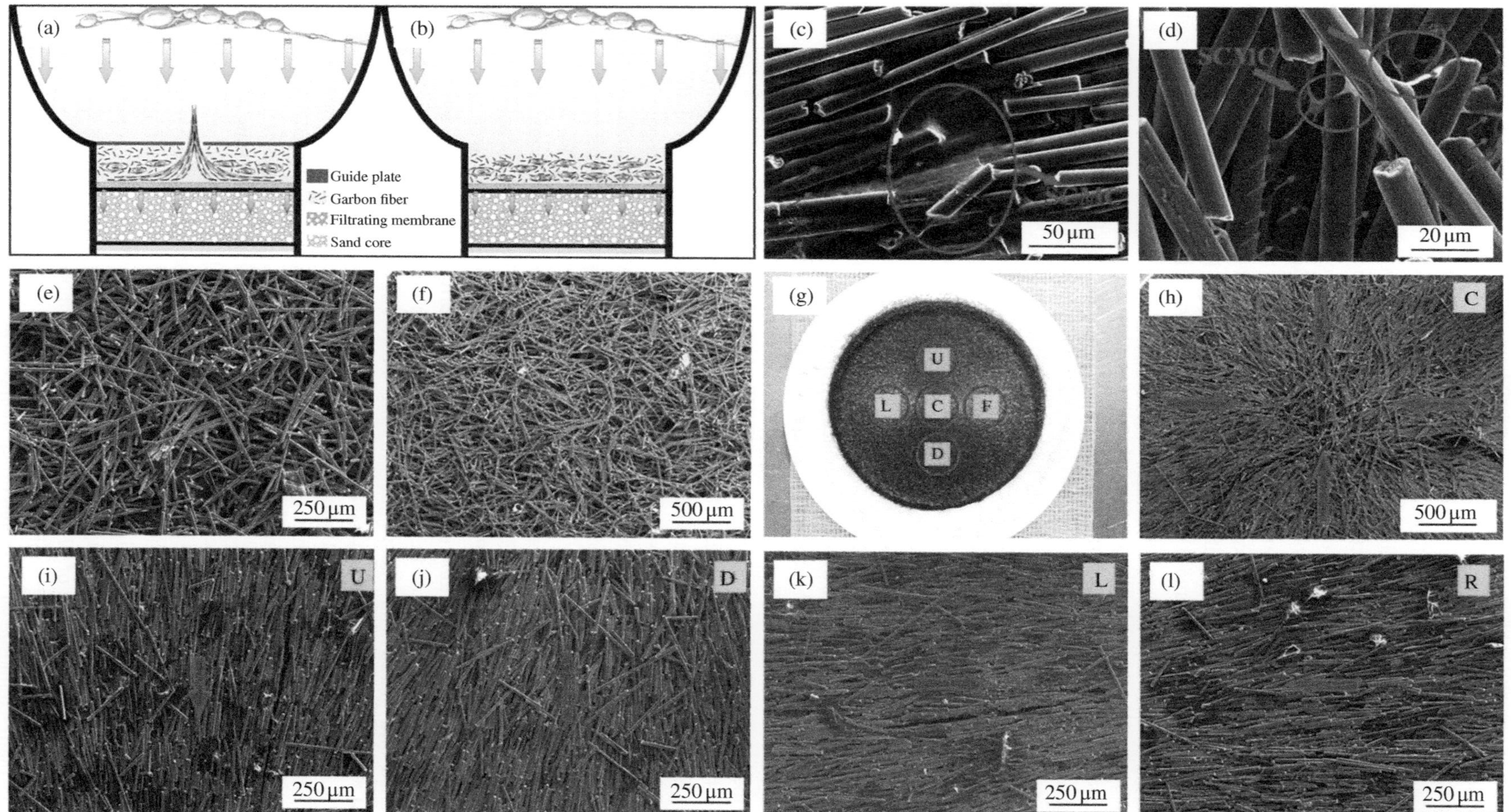

Figure 6.45 Schematic diagram of flow field-driven self-assembly strategy (a) and traditional vacuum filtration method (b). (c and d) SEM images of SCMC binding CFs. (e and f) SEM images of Ro/0.45/blank-composite in state-B. (g) Optical image of Ra/0.45/5-composite in state-B of different areas. SEM images of the corresponding region of C (h), U (i), D (j), L (k), and R (l) in (g), representing the center, up, down, left, and right, respectively.

Figure 6.45(g)–(l) presents the detailed structural information of Ra/0.45/5/composite in state B. Specifically, Figure 6.45 (h)–(l) is the partially enlarged SEM images of the corresponding position related to Figure 6.45(g), where C, U, D, L, and R represent the center, up, down, left, and right, respectively. Obviously, CFs in Ra/0.45/5/composite form the radial arrangement, in which most of the CFs are oriented toward the center. For comparison, the SEM images of Ro/0.45/blank-composite (control group) are shown in Figure 6.45(e) and (f), where the guide plate is removed with other parameters unchanged. It can be clearly seen from Figure 6.45(e) and (f) that CFs intermesh chaotically without any order. The reason for the difference between Ra/0.45/5/composite and Ro/0.45/blank-composite is that the guide plate constrains the flow of fluid, forming the diffusion flow under the guide plate. The detailed flow process has been interpreted in the former section. In addition, Figure 6.45(c) and (d) present the detailed SEM information of SCMC in state-B of Ra/0.45/5-composite. The white SCMC sticks and wraps around the CFs to fix them spatially, protecting the deposited CFs skeleton from damage in the process of matrix filling.

6.2.5.2 Characterization, Analysis, and Optimization

Evaluation of Composites' Orientation It is worth noting that the diameter of aperture will affect the flow velocity of liquid through the aperture, and the filtrating membrane with different pores will determine the overall speed of filtration. Hence, the diameter of aperture and the filtrating membrane pore will influence the flow field above the filtrating membrane, which will influence the arrangement of CFs. To obtain the optimal parameters, we explored the influence of different parameters including the diameter of aperture (5, 7, and 9 mm) and the filtrating membrane pore (0.22, 0.45, and 0.80 μm). Figure 6.46(a)–(f) presents the upper surface SEM image of composites with six different parameters. It is clear that different degrees of orientation can be observed. To quantitively evaluate the ORI, an orientation algorithm based on microscale image identification was developed, and an evaluation criterion was proposed. In this evaluation criterion, the degree of orientation in the vertical direction (V-ORI) is supposed to evaluate how close the orientation of CFs is to the vertical direction. The process and schematic diagram of calculating V-ORI are shown in Figure 6.46(i)–(m) in detail. First, the initial SEM image (Figure 6.46(i)) was handled by edge extraction, smoothing, and binarization to obtain the binary image (Figure 6.46(j)), which was actually a pixel matrix (containing white pixel and black pixel). Figure 6.46(k) clearly presents that the white pixels are derived from the edge of CFs because their brightness is different from other locations. Figure 6.46(l) presents the three different conditions of CFs. It is interesting that there are more continuous white pixels in the vertical direction when CFs are closer to the vertical direction, which is similar to the horizontal direction. According to this rule, the algorithm for calculating ORI is proposed, as shown in Figure 6.46(m). For each white pixel (base chunk), the same operation is executed. Specifically, in the vertical and horizontal directions, the pixel blocks on the top and right of the base chunk are traversed and checked whether it is a white pixel step by step upward and rightward until a black pixel is encountered, respectively. The numbers of continuous white pixels are defined as n and m in the vertical and horizontal directions, respectively. Particularly, the sustainability of base chunk in the vertical and horizontal directions is defined as SV and SH, respectively, which are calculated with weight function ($F(n)$, defined to expand the discrimination between SV and SH). Hence, SV and SH are calculated as follows:

$$\mathrm{SV} = \sum_{s=0}^{n} F(s) \text{ and } \mathrm{SH} = \sum_{s=0}^{m} F(s) \tag{6.32}$$

For the binary image, the sustainability of the image in the vertical and horizontal direction is defined as SSV and SSH, respectively. All the SV and SH are added up to SSV and SSH and calculated as:

$$\mathrm{SSV} = \sum_{\text{each white pixel block}} \mathrm{SV} \text{ and } \mathrm{SSH} = \sum_{\text{each white pixel block}} \mathrm{SH} \tag{6.33}$$

It is obviously true that there are more continuous white pixels in the vertical direction resulting in the larger SSV and smaller SSH when the CFs are closer to the vertical direction. Hence, V-ORI is defined as follows:

$$\text{V-ORI} = \left(1 - \frac{\mathrm{SSH}}{\mathrm{SSV}}\right) \times 100\% = \left(1 - \frac{\sum_{\text{each white pixel block}} \sum_{s=0}^{m} F(s)}{\sum_{\text{each white pixel block}} \sum_{s=0}^{n} F(s)}\right) \times 100\% \tag{6.34}$$

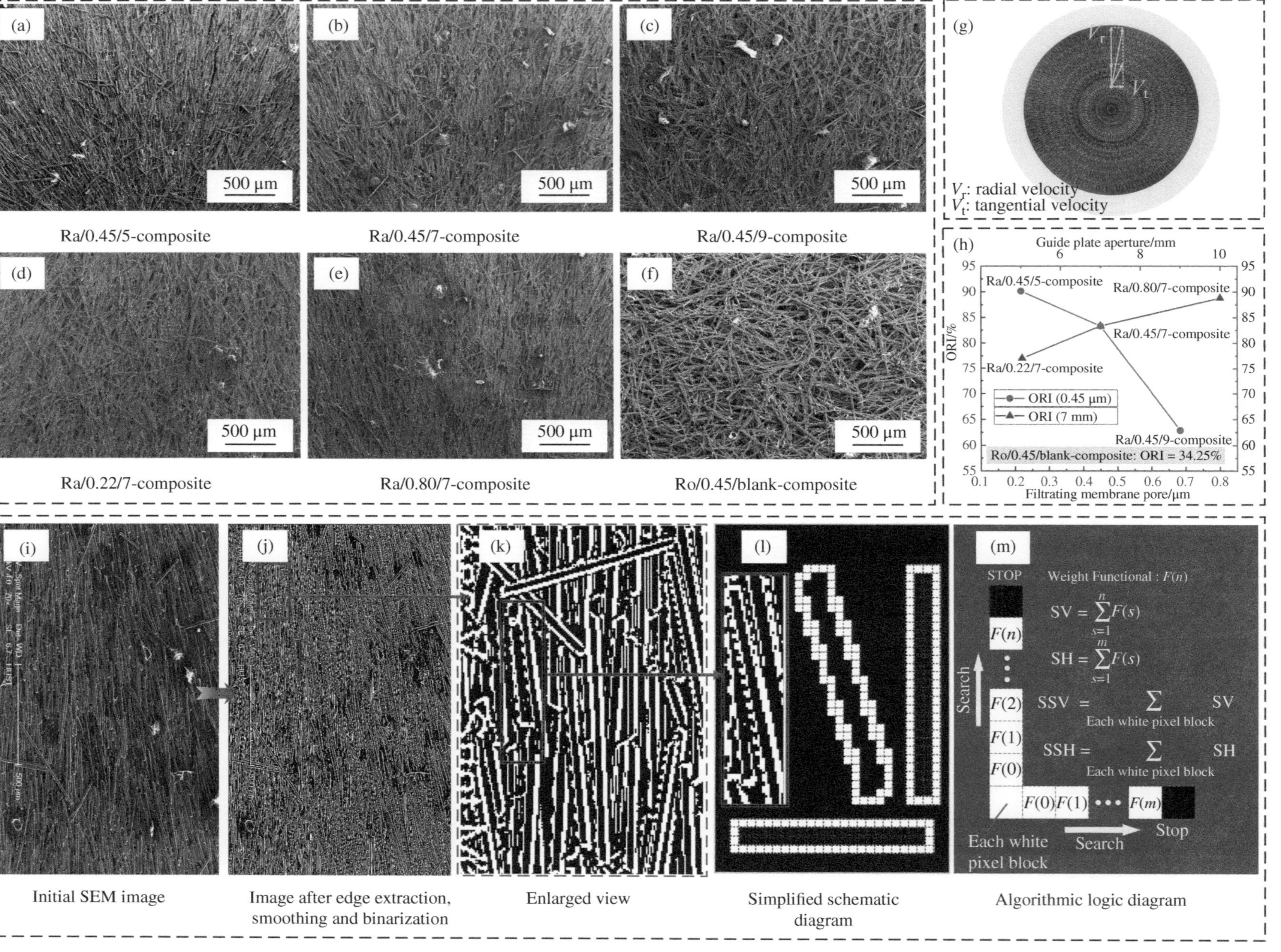

Figure 6.46 SEM images of six composites with different parameters, Ra/0.45/5-composite (a), Ra/0.45/7-composite (b), Ra/0.45/9-composite (c), Ra/0.22/7-composite (d), Ra/0.80/7-composite (e), and Ro/0.45/blank-composite (f). (g) Schematic diagram of radial and tangential velocity. (h) The calculated result of composites' ORI. (i–m) The schematic diagram of calculating V-ORI based on microscale image identification.

To evaluate the real orientation, the V-ORI(i) (V-ORI of the $(i)_{th}$ rotation) is calculated by 180 times, where the initial SEM image is rotated by one angle for each calculation. The maximum of V-ORI(i) is regarded as the final ORI. As a result, the ORI and the holistically oriented direction of all CFs can be recognized. The ORI of all composites is calculated and shown in Figure 6.46(h). It is worth noticing that the ORI declines with the increase of aperture diameter (the red line with circles). The main reason is that the in-plane-diffused velocity (u_i) declines with the increase of aperture diameter. The higher radial velocity makes the resultant velocity tend toward the radial direction when the stochastic disturbance is considered, as shown in Figure 6.46(g). Similarly, the ORI increases with the increase of filtrating membrane pore (blue line with triangles), because the bigger pore brings less resistance, resulting in higher in-plane-diffused velocity. As a result, the diameter of aperture and filtrating membrane pore influence the ORI by impacting the in-plane-diffused velocity. In addition, Ro/0.45/blank-composite shows an ORI of 34.25%. It is hard to accomplish a completely random arrangement (ORI equals 0%) because the random flow also exists kind of orientation in a small area. However, with respect to the whole area of random composite, no obvious regulation can be discovered in Ro/0.45/blank-composite, as shown in Figure 6.46(f). According to abundant measurements, when the ORI is lower than 40%, the composite can be regarded as a randomly oriented composite.

Simulation of the Self-Assembly Process Interestingly, it is clearly observed that different from Ra/0.80/7-composite, an obvious boundary occurred in Ra/0.45/5-composite, which distinguishes two regions with different ORI (separation phenomenon), as shown in Figure 6.47(a). In the area near the boundary, region A shows a better arrangement with an ORI of 74.9%, and region B shows an ORI of 45.2%, which can almost be considered as no orientation. Hence, besides the ORI, the size of region A needs to be considered to obtain the optimal parameters. To interpret the separation phenomenon, the finite-element simulation (FES) of vacuum filtration was conducted. As shown in Figure 6.47(b), 3D models were established with four different apertures (9, 7, 5, and 3 mm) to explore the change of flow field with different aperture's diameters. In the process of self-assembly, laminar flow will drive CFs to form a well-organized structure on account of the regular streamlines. In reverse, the turbulent flow will cause chaos because of the disordered flow. Figure 6.47(c)–(f) presents the flow fields of different apertures at 0.1 mm above the filtrating membrane, respectively, which could be regarded as the arrangement of CFs because of the tiny distance from the filtrating membrane. With the decrease of aperture diameter, the velocity of flow gradually increases, and the ordered area gradually shrinks with the edges starting to get disordered. In detail, as shown in Figure 6.47(c), the streamlines start from the center point and end at the edge, which means that the fluid enters the aperture and leaves through the filtrating membrane. Different from Figure 6.47(c), the flow field in Figure 6.47(f) is divided into two parts by a boundary, which is in accordance with the experimental results. In the center region, the diffusion flow is still maintained with the streamlines starting from the center point. However, the streamlines of diffusion flow end at the boundary. In the edge of flow domain, the streamlines do not disappear directly, but reflux when crashing to the edge. To analyze the flow in detail, the flow fields of more cross-sections are shown in Figure 6.47(g)–(l). Figure 6.47(g) and (j) presents the longitudinal section views and their enlarged views of 3 and 9 mm, respectively. Obviously, the flow is divided into upper and lower parts, as shown in Figure 6.47(j). The lower part forms an ordered flow covering the whole filtrating membrane and the upper part forms vortexes, which will not influence the arrangement of CFs. However, as shown in Figure 6.47(g), the stable and ordered flow only covers the region in the center. In the region near the edge, the fluid flow gets disordered and chaotic forming different flow fields. Figure 6.47(h) and (k) presents the A–A cross-sectional views (at 1 mm above the filtrating membrane). In Figure 6.47(h), the vortexes are formed because the fluid cannot flow down but reflux when the flow crash to the edge. In contrast, most fluid flows regularly with a small number of disordered flows in Figure 6.47(k). These disordered flows do not originate from the edge but from the vortexes of the upper part shown in Figure 6.47(j), which would not influence the flow of lower part. As a result, the reason for this phenomenon is that bigger in-plane velocity makes the time from the center to the edge decrease. However, the overall velocity of vacuum filtration stays the same for different apertures because the filtrating membrane pore does not change. Hence, the fluid cannot flow down fast enough through the filtrating membrane for a smaller aperture, which results in the fluid reflux but not flowing down when the flow crash to the edge. Hence, the backflow and the coming flow crash into one another, forming the complex flow with a distinct boundary. In addition, Figure 6.47(i) and (l) presents the streamlines of flow domain under the guide plate above the filtrating membrane, which are similar to the Ra/0.45/5-composite and Ra/0.80/7-composite, respectively, validating the accuracy of the simulation. Hence, during the self-assembly process, the diameter of aperture and the filtrating membrane pore will influence the size of the laminar flow area and determine whether there is a stable and ordered flow, by influencing the in-plane diffused velocity. Although the higher in-plane-diffused velocity will give a better arrangement, the higher in-plane-diffused velocity will reduce the size of the laminar flow area. Overall, 0.80 μm filtrating membrane pore and 7 mm aperture are chosen as the optimal parameters.

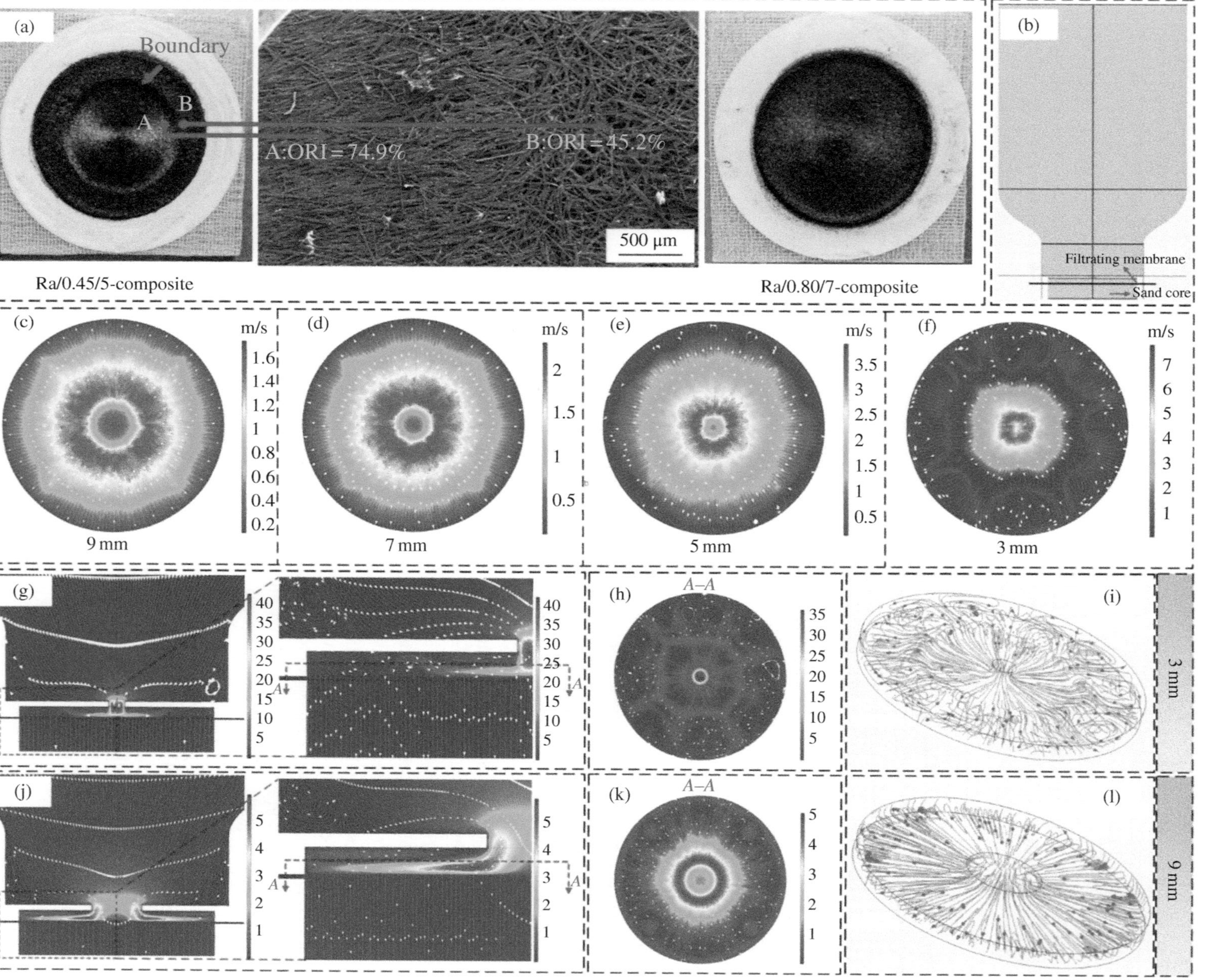

Figure 6.47 (a) Separation phenomenon with an obvious boundary. (b) Geometric model of simulation. (c–f) The flow field of different apertures at 0.1 mm above the filtrating membrane. The longitudinal section views and their enlarged views of 3 mm (g) and 9 mm (j). (h and k) The *A–A* cross-sectional views of corresponding position. The streamlines of flow domain above filtrating membrane under the guide plate of 3 mm (i) and 9 mm (l), respectively.

6.2.5.3 Thermal and Mechanical Properties of Composites

To analyze the thermal and mechanical properties of the prepared composites, a series of measurements, including laser flash analysis (LFA), DSC, thermal gravimetric analysis (TG), thermal mechanical analysis (TMA), dynamic mechanical analysis (DMA), and tensile test, were carried out to explore the influence of parameters, respectively. First, the in-plane TD coefficients of different composites obtained from LFA measurement are shown in Figure 6.48(a). The in-plane TD of Ro/0.45/blank-composite is 6.879 $mm^2 \cdot s^{-1}$, which is significantly higher than the pure PDMS (0.09 $mm^2 \cdot s^{-1}$), because the CFs tend to lie in plane and the axial TD of CFs highly exceeds the radial TD of CFs. However, in-plane random arrangement restricts the improvement of TD. With the addition of the guide plate, the TD highly increases with the highest TD reaching 23.1 $mm^2 \cdot s^{-1}$ because of the radial structure. The guide plate aperture and filtrating membrane pore will affect the in-plane-diffused velocity of flow, the ORI of CFs, the mass fraction of CFs, and the integrity of composites. As a result, it is hard to summarize a general rule in the change of TD. Then, the thermogravimetric curves of some composites are shown in Figure 6.48(b) from 25 to 1000 °C to calculate mass fractions of components. According to the weight loss ratio in the temperature range of 25–620 °C and 25–1000 °C, the CFs' mass fractions of Ra/0.80/5-composite, Ra/0.80/7-composite, Ra/0.80/9-composite, and Ro/0.45/blank-composite are 55.1%, 50.3%, 50.2%, and 34.0%, respectively. Ro/0.45/blank-composite shows a much lower mass fraction because random arrangement results in more holes in the scaffold, which partly accounts for the lower TD. Compared to that, in Ra/0.80/7-composite, more CFs can be deposited on the filtrating membrane because of the regular and ordered arrangement, which brings a higher mass fraction than the other three composites. Then the heat capacities and densities of some composites were measured to calculate the TC, as shown in Figure 6.48(c). Particularly, the in-plane TC of Ra/0.80/7-composite reaches 35.5 $W \cdot m^{-1} \cdot K^{-1}$, showing 3.5 times that of Ro/0.45/blank-composite and 236 times that of pure PDMS. The highest anisotropy reaches 19.8, showing 2.8 times that of Ro/0.45/blank-composite. These improvements are attributed to the higher mass fraction of CFs and the excellent orientation of Ra/0.80/7-composite. Note that the thermal conductivity enhancement efficiencies (TCEE) of Ra/0.80/7-composite and Ro/0.45/blank-composite describe the efficiency of CFs per unit volume in contributing to the thermal enhancement, which is defined as:

$$\mathrm{TCEE} = \frac{\kappa - \kappa_{\mathrm{m}}}{100 f \kappa_{\mathrm{m}}} \tag{6.35}$$

where κ and κ_m are the TC of the composite and pure PDMS, respectively. f represents the volume fraction of CFs in the composite. The in-plane TCEE of Ra/0.80/7-composite (6.5) is highly larger than that of Ro/0.45/blank-composite (3.2), validating the better heat transfer characteristic of CFs' pathways in Ra/0.80/7-composite. In addition, Figure 6.48(d) presents the TD's changing curves of composites with the increase of temperature. Compared to Ro/0.45/blank-composite, the greater decrease in the in-plane TD of Ra/0.80/7-composite demonstrated that the anisotropy of CFs is taken full advantage of by the radial structure. It is worth mentioning that, in the composites made by vacuum filtration, CFs tend to lie along the in-plane direction, although they are disordered, which makes the through-plane TC of composites show little difference.

Moreover, the mechanical properties were measured and compared between composites with radial and random structures. Figure 6.48(e) illustrates the measuring result of DMA, showing the universal decrease of composites compared with pure PDMS because of the incorporation of CFs. Interestingly, Ra/0.80/7-composite shows less sacrifice in storage modulus than other composites because of the more regular orientation. Figure 6.48(f) illustrates the CTE of some composites between 50 and 250 °C, which reflects the thermal stability. It is unusual that the CTE of composites is numerically close, showing a little enhancement compared to the PDMS. It is likely explained that CFs are handled to be hydrophilic in the process of vacuum filtration, resulting in structural defects inside the composites when the matrix is filled in. All in all, via the above analysis of experimental results, Ra/0.80/7-composite shows better thermal and mechanical performances than others because of its better orientation.

Inspired by biological structures, the radial structure was proposed to solve the hotspot problem. To verify the effect of the radial structure, the FES (a more direct method to visualize the process of heat propagation) and TD experiments were carried out. As shown in Figure 6.49(a)–(c), temperature fields of composites with different CFs' structures were simulated, including random arrangement (a), radially oriented arrangement without connection (b), and radially oriented arrangement with sectional connection (c). A 1.5 mm diameter heat source was set in the center. Figure 6.49(d) and (e) shows the curves of maximum temperature and maximum temperature difference of three kinds of composites over time. Figure 6.49(c) shows the lowest temperature in these composites, indicating their excellent heat transfer capacity in hotspots. To be specific, when the CFs connect with one another partially, continuous heat conduction channels are established, resulting in smaller contact thermal resistance (removing the thermal resistance between CFs and matrix), which maximally takes advantage of CFs' high axial TC. Moreover, Figure 6.49(c) shows the lowest temperature difference, indicating that the radial structure is much more beneficial to uniform heat conduction. The heat flow in Figure 6.49(c) was

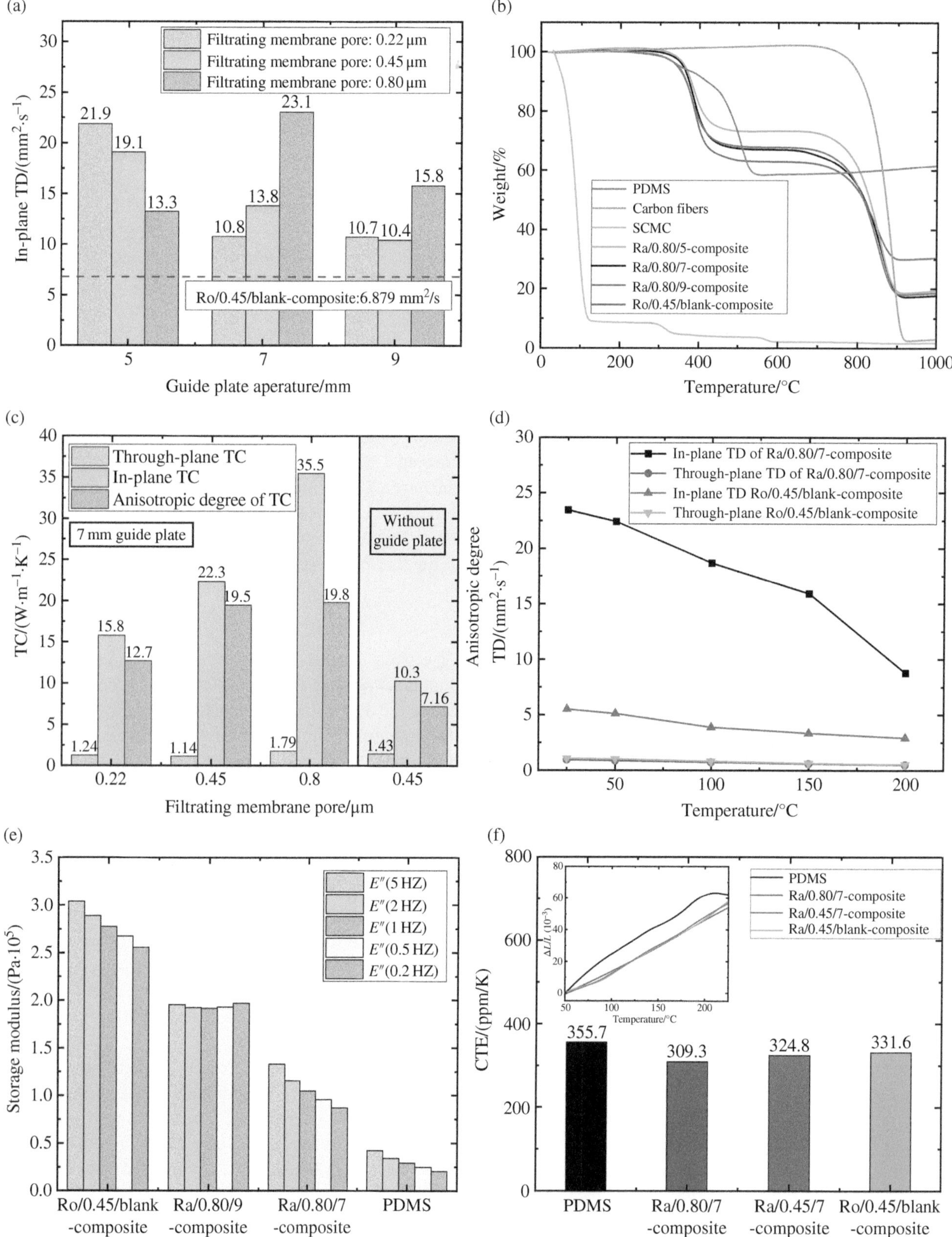

Figure 6.48 (a) TD of different parameters. (b) The thermogravimetric curves of some composites. (c) TC of some composites. (d) The TD's changing curves of composites as the increase of temperature. (e) The storage modulus of some composites in different frequencies. (f) The CTE of PDMS, Ra/0.80/7-composite, Ra/0.45/7-composite, and Ro/0.45/blank-composite between 50 and 250 °C.

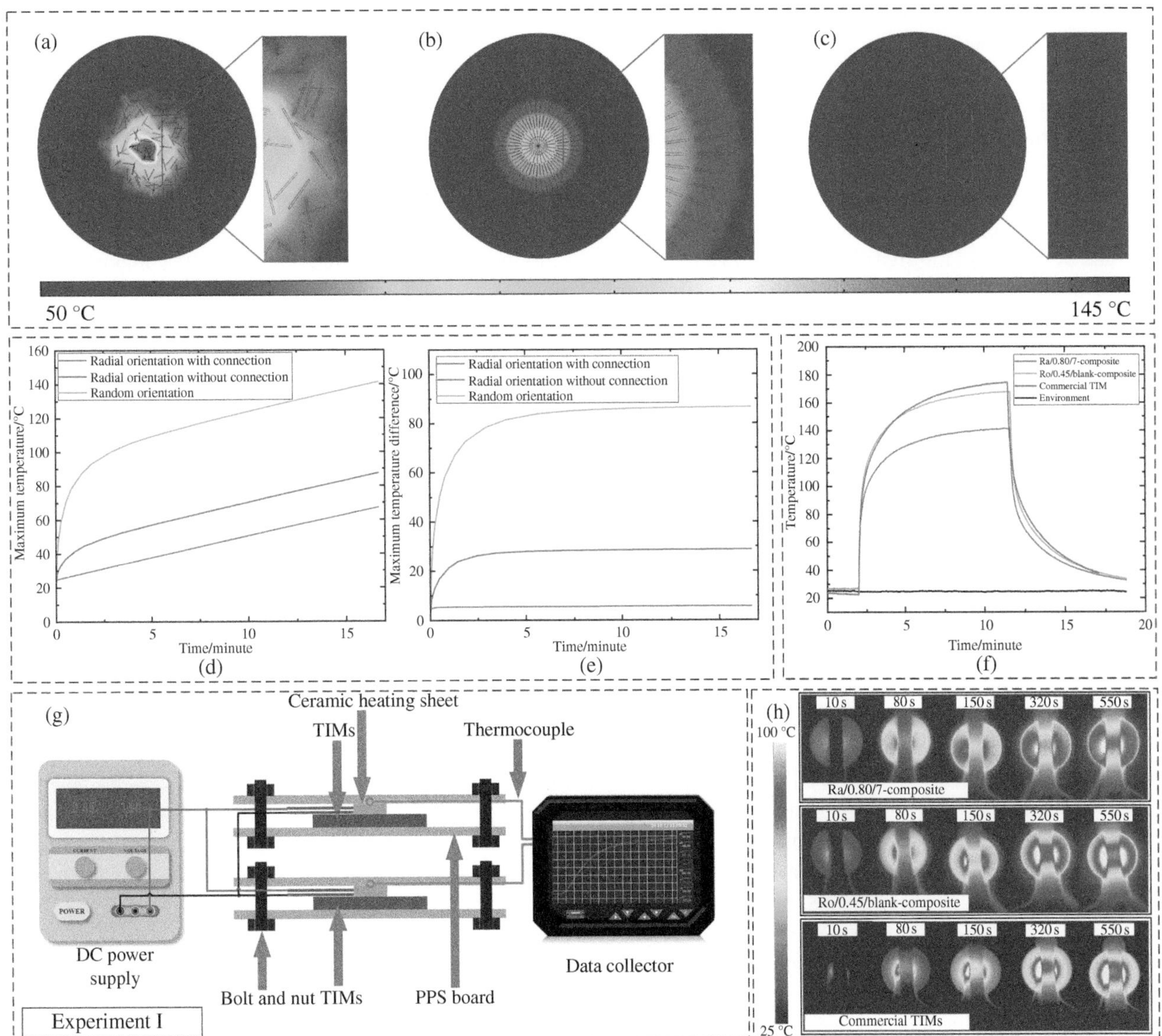

Figure 6.49 Temperature field of composites with different CFs' structures including random arrangement (a), radially oriented arrangement without connection (b), and radially oriented arrangement with sectional connection (c). The curve of maximum temperature (d) and maximum temperature difference (e) of three kinds of composites over time. (f) The core temperature curve of the ceramic heating sheet over time. (g) The experimental setup for comparing the actual heat transfer effect. (h) The infrared thermograms of composites' surface over time.

uniformly and quickly transferred from the center hotspot to the edge. However, the heat flow was restricted in a localized region in Figure 6.49(a). As a result, the composite embedded with the radial oriented and connected structure of CFs demonstrated a fast and uniform heat-conduction capability, which is much superior to the random structure. To compare the actual heat transfer effect, experiment I was designed and prepared to compare Ra/0.80/7-composite, Ro/0.45/blank-composite, and commercial TIMs, whose experimental setup is shown in Figure 6.49(g). 5 mm diameter ceramic heating sheets were used to simulate the hotspots and connected to a power supply in parallel. The surface temperatures of composites were recorded by a thermal infrared camera, and the core temperature of ceramic heating sheet was detected by thermocouples. Figure 6.49(f) presents the core temperature curves of ceramic heating sheets over time. The ceramic heating sheets with Ra/0.80/7-composite show the lowest core temperature, which was 33.4 °C lower than that with commercial TIMs finally. In addition, Figure 6.49(h) shows the infrared thermograms of composites' surfaces over time. It is worth noting that Ra/0.80/7-composite shows the most uniform temperature field; however, Ro/0.45/blank-composite and commercial TIMs show a bigger temperature difference between the center and the edge. As a result, this experiment shows consistent results to the simulation, indicating that Ra/0.80/7-composite (radial structure) has an excellent heat diffusion capacity for hotspots because of the establishment of heat pathways.

To verify the actual effect in heat dissipation system, experiment II was designed and conducted. Figure 6.50(a) shows the typical heat dissipation device for chips with hotspots. In this experiment, a 9 -mm diameter ceramic heating sheet was selected to simulate the chip. To compare the heat dissipation effect in actual devices, four different materials were used in system as TIMs to connect the chip and heat sink, including PDMS, Ra/0.80/7-composite (the photo of Ra/0.80/7-composite is shown in Figure 6.50(c)), Ro/0.45/blank-composite, and commercial TIMs. Thermocouples were used to detect the chip's core temperature, as shown in Figure 6.50(e). The infrared thermograms of composites' surfaces were recorded by a thermal infrared camera, as shown in Figure 6.50(b). Obviously, the system with Ra/0.80/7-composite shows the lowest chip temperature and the best temperature uniformity due to the radially oriented structure of CFs, which is conducive to guaranteeing the performance and lifetime of chips in practical application. As a result, Ra/0.80/7-composite shows terrific application potentiality in the heat dissipation of hotspots because of the radial structure inside.

In addition, the self-assembly strategy which combines vacuum filtration method and fluid control method can not only be used in the heat dissipation of local hotspots but also is adaptive to arbitrary-shaped heat sources. It is interesting that, as shown in Figure 6.50(f), the special flow fields of "H," "U," "S," and "T" were designed by machining the holes of corresponding glyphs. In theory, the arrangement of arbitrary 2D shapes can be designed for heat sources of special shapes. Besides the hotspot problem, as shown in Figure 6.50(d), the radial structure is also potential to be applied to solar–thermal–electric conversion because this special structure is contributory to absorbing concentrated solar energy and transferring heat [161]. When PCM is incorporated in composites, the heat can spread rapidly and be absorbed in quantity. All in all, the flow field-driven self-assembly strategy proposed in this study is potential to be applied in lots of application prospects when the structure is designed properly.

6.2.5.4 Experiment Section

CFs with a length of 200 μm (K223HM) and axial thermal conductivities of 600 $W \cdot m^{-1} \cdot K^{-1}$ were purchased from Mitsubishi Chemical. CTAB was purchased from Aladdin and used as a cationic surface-active agent. The SCMC was purchased from Aladdin and used as a binder to bind CFs. PDMS was purchased from Dow Corning Co., Ltd. and used as the polymer matrix. All chemicals were directly used without further purification. Mixed cellulose esters (MCE, made of nitrocellulose and cellulose acetate) filtrating membranes of different pores were purchased from Shanghai Xinya purification device manufacturer, which have a good mechanical strength, toughness, and temperature stability (maximum operating temperature of 130 °C). Guide plates were obtained by CNC machining.

Synthesis of Radially and Randomly Oriented-CFs Scaffolds First, to promote the hydrophilia and dispersibility of CFs in solution, 0.25 g CFs and 0.02 g CTAB were added into 100 ml deionized water and stirred in magnetic stirring apparatus for 10 minutes, forming mixture A. Second, 200 ml of deionized water was added into the upper flask as preset water to slow down the sedimentation of CFs and create a stable initial flow field under the guide plate. Third, mixture A was added into the preset water of the upper flask, forming mixture B. After a while, until the flow of mixture B weakens (to relieve the influence of the existing flow on the flow field under the guide plate), the vacuum pump was started to force deionized water flow through the filtrating membrane. Meanwhile, CFs were forced to be deposited on the filtrating membrane, obeying the feature of flow field. Fourth, after all CFs had been deposited, 15 ml of SCMC solution (1.5 wt%) was dropped into the upper flask with vacuum filtration sustaining 15 minutes, which ensured SCMC enwind the CFs. Fifth, after all deionized water was pumped out, the composite of state-A was frozen upon a copper cylinder with liquid nitrogen in the bottom. After that, the composite of state-A was transferred to a freeze-dryer and sustained 24 hours of vacuum-assisted freeze-drying (−60 °C, <10 Pa), obtaining the CFs scaffold (state-B). Specifically, the radially oriented-CFs scaffold was obtained when the guide plate was assembled into the upper flask. The randomly oriented-CFs scaffold was obtained when the guide plate was removed.

Fabrication of CFS/PDMS Composites First, PDMS was prepared by mixing component A and component B by 10:1. Then alternating cycles of vacuum were applied in the mixed PDMS to remove bubbles. Second, PDMS was slowly poured into the scaffold, and continuous degassing was carried out every 5 minutes. This process was carried out repeatedly five times to ensure that PDMS was fully impregnated into the scaffold. In this process, freeze-dried SCMC played a role in fixing CFs and preventing the scaffold from disintegrating due to air bubbles. Third, when air bubbles were no longer produced, the redundant PDMS upon the scaffold was discreetly scraped off and squeezed out. After that, a thermal treatment step

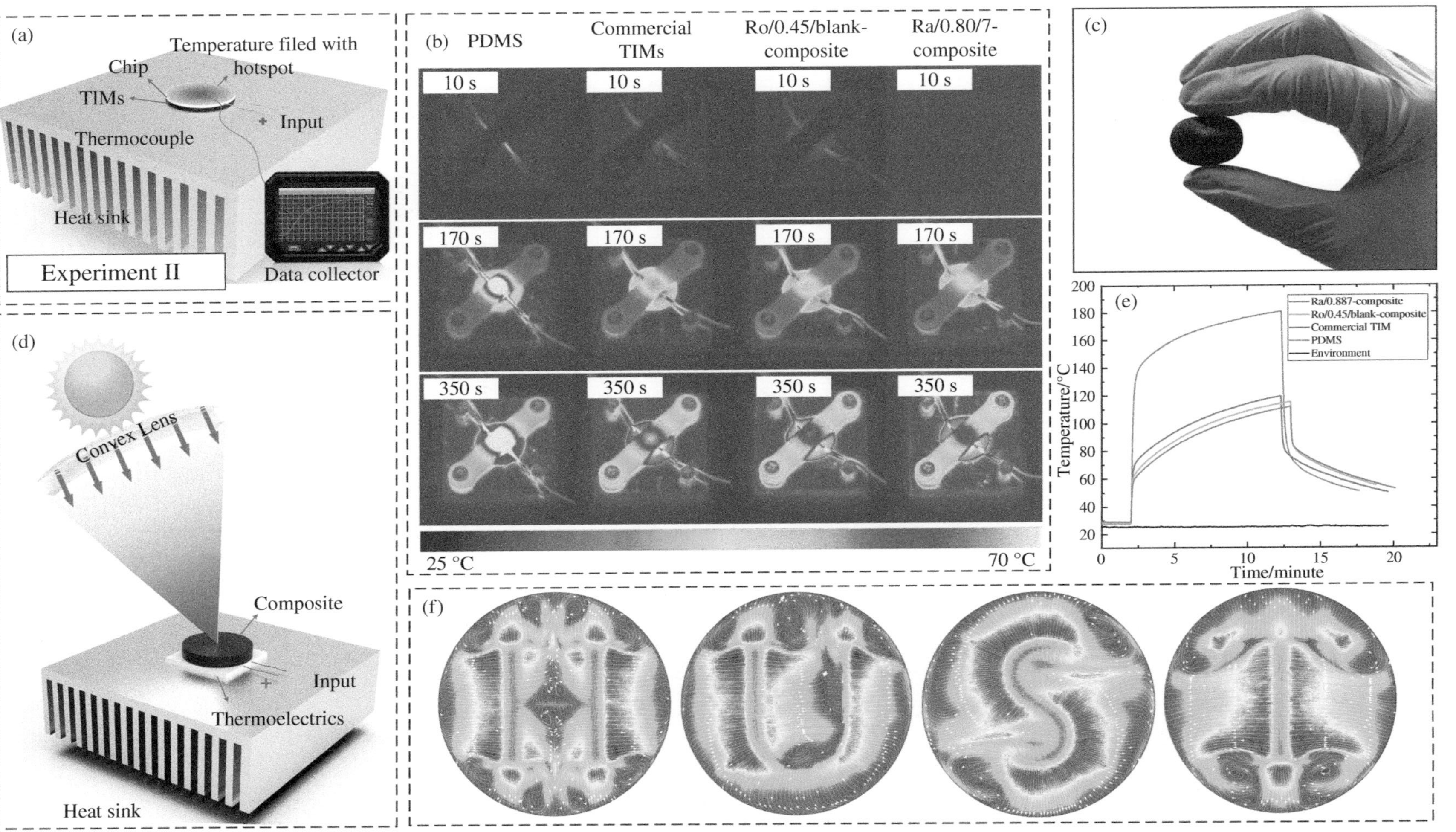

Figure 6.50 (a) The experimental setup for comparing actual heat dissipation effect. (b) The infrared thermograms of composites' surface over time. (c) The photo of Ra/0.80/7-composite. (d) Schematic of the device for solar–thermal–electric conversion. (e) The core temperature curves of ceramic heating sheets over time. (f) The special flow fields of "H," "U," "S," and "T."

at 70 °C for 3 hours was conducted to ensure the full curing of the composites. Composites with radially oriented CFs and randomly oriented CFs were prepared, respectively.

Characterization The microscopic morphology was tested by a FSEM (Sirion 200) to characterize the morphologies of scaffolds and cross-section of composites at an accelerating voltage of 10 kV. The SEM images of scaffolds were used to calculate the ORI by applying an orientation algorithm based on microscale image identification. TG (TGA8000) was carried out to verify the mass fraction of each component, which can be calculated according to the equation: $\Delta M = W_{PDMS} \times \Delta M_{PDMS} + W_{SCMC} \times \Delta M_{SCMC} + W_{CFs} \times \Delta M_{CFs}$, where ΔM, ΔM_{PDMS}, ΔM_{SCMC}, and ΔM_{CFs} correspond to the weight loss ratio of each component in a temperature range, respectively. W_{PDMS}, W_{SCMC}, and W_{CFs} correspond to the mass fraction of each component. In-plane TC and through-plane TC were calculated according to the equation $\kappa = \alpha \times \rho \times C_p$, where κ, α, ρ, and C_p correspond to TC, TD, density, and specific heat capacity, respectively. The in-plane and through-plane TD of composites were measured by LFA 467 (Netzsch). The specific heat capacities of composites were measured by DSC (Diamond DSC, PerkinElmer Instruments). In addition, the CTE of the composites in the through-plane direction was measured by static TMA (Q400EM) with temperature ranging from 35 to 250 °C. The storage moduli of composites were measured by DMA (Diamond DMA) in a compressed mode. The tensile tests of composites were measured with a loading rate of 1 $mm \cdot min^{-1}$ by Instron 5943.

6.3 Interfacial Thermal Transport Manipulation of TIM

Assembling hard particles, such as ceramics, metals, or metal oxides, into soft materials often results in large thermal resistance at the hard/soft material interface, which favors the degradation of thermal properties. Examples of particulate-filled composites whose thermal properties are impaired by interfacial resistance are numerous and range from polymer-based thermal interfaces and encapsulant materials in electronic packaging [1, 3, 28, 42, 66, 95, 169, 170], to nanofluids in thermal storage and sensor applications [171, 172], to nanoparticle-assisted therapeutics in medicine [173, 174]. In some applications, this issue can be circumvented by organizing the particles into heterogeneous microstructures in which the particles aggregate or connect with one another to allow for rapid heat flow over the highly thermally conductive media [175, 176]. However, for nanofluids, the use of large aggregates in fluids leads to a dramatic increase in viscosity, resulting in the impairment of fluidic properties [176].

Directly tuning the interfacial thermal transport properties is a promising strategy. Phonon spectra match of the materials constituting the interface is generally recognized as a critical factor that influences interfacial thermal transport [177]. Due to their distinct compositions and bond natures, a large mismatch is usually observed in the phonon spectra of hard and soft materials, resulting in a low interfacial thermal conductance, G_{int}. A former study [178] demonstrated that bridging the acoustic mismatch between Au and an alkane-based polymer, polyethylene, with an alkanethiol self-assembled monolayer (SAM) can greatly enhance G_{int}. Polyethylene and the alkanethiol SAM have very similar chemical compositions, which enables a favorable acoustic match. However, for other commonly used soft materials with complex structures, like epoxy resin, it is quite difficult to obtain similar structured SAMs to match their phonon spectra.

Theoretical studies [179–181] also suggest that G_{int} can be tailored by an order of magnitude by controlling the interfacial bond energy. Such an idea has been put into practice in inorganic solid/solid [182, 183] and solid/water interfaces [184, 185] by chemically introducing a strongly bonding SAM. However, this unique control of interfacial bonds and thermal transport properties has less often been achieved in practice at hard/soft interfaces. In addition, although the nexus between G_{int} and interfacial bonds has been clarified through solid/water interfaces [184, 185], the formation and characters of the bonds are yet to be explored or understood.

In this section, a general strategy aiming for interfaces which are incompatible with the previous strategy will be introduced. Copper (Cu) and epoxy resin are chosen as representative hard and soft materials, respectively. The measurement results show that G_{int} can be enhanced by as much as 11-folds by the proposed strategy. To study the correlation between interfacial bond energy and G_{int}, another interface system where the SAM is chosen to be of the alkanethiol type adopted in Ref. [178] is created. Through a series of characterization techniques, the formation and characters of the bonds in the two interface systems are explored. Finally, a theoretical prediction is made to show the great importance of the proposed strategy in manipulating the thermal transport properties of nanocomposites.

6.3.1 Synthesis of Interface Systems

As illustrated in Figure 6.51(a) and (b), Cu/epoxy interfaces were created by dispersing spherical Cu microspheres into an epoxy resin solution and consolidating the suspending fluids. To achieve interfaces with SAM connection, the Cu microspheres were first modified with SAMs (see Figure 6.51(a)) using a chemical solution deposition method. Two Cu/SAM/epoxy systems are built by varying the SAM's end-group functionalities (schematically shown in Figure 6.51(c)). The first system employs 11-amino-1-undecanethiol hydrochloride (SAM-NH_2), and the other employs an alkanethiol type SAM (dodecanethiol, SAM-CH_3), which is adopted in Ref. [178]. Both SAMs possess the thiol functionality (—SH) at the o-end permitting strong bonding of the molecules to the Cu surface. Different a-end-group chemistries (—CH_3, —NH_2) ensure large variations in bond strength at the SAM/epoxy interface. The —NH_2 group is expected to provide strong covalent bonds, and weak van der Waals bonds will be provided by the —CH_3 group.

6.3.2 Measurement of Interfacial Thermal Conductance

The Cu/SAM/epoxy systems are assumed to be homogenously embedded in the epoxy-based composites. Here, the G_{int} of the systems was evaluated by fitting the measured TC data of the composites to the results predicted by an analytical model. A great number of models [3, 44, 186–189] have shown that the TC of two-component homogeneous composites is determined by the intrinsic properties of the fillers and the matrix, the geometry and loading of the fillers, and G_{int}. For spherical particles, one of the most prominent models is the modified Bruggeman asymmetric model (MBAM) which is expressed as [189]:

$$(1-f)^3 = \left(\frac{\kappa_{\text{m}}}{\kappa_{\text{eff}}}\right)^{(1+2\beta)/(1-\beta)} \left(\frac{\kappa_{\text{eff}} - \kappa_{\text{p}}(1-\beta)}{\kappa_{\text{m}} - \kappa_{\text{p}}(1-\beta)}\right)^{3/(1-\beta)} \tag{6.36}$$

where κ_{eff} is the effective TC of the composite; κ_{m} and κ_{p} are the matrix and filler thermal conductivities, respectively; f is the volume fraction of the filler; r_β is called the Kapitza radius defined as:

$$r_\beta = \frac{\kappa_{\text{m}}}{G_{\text{int}} d/2} \tag{6.37}$$

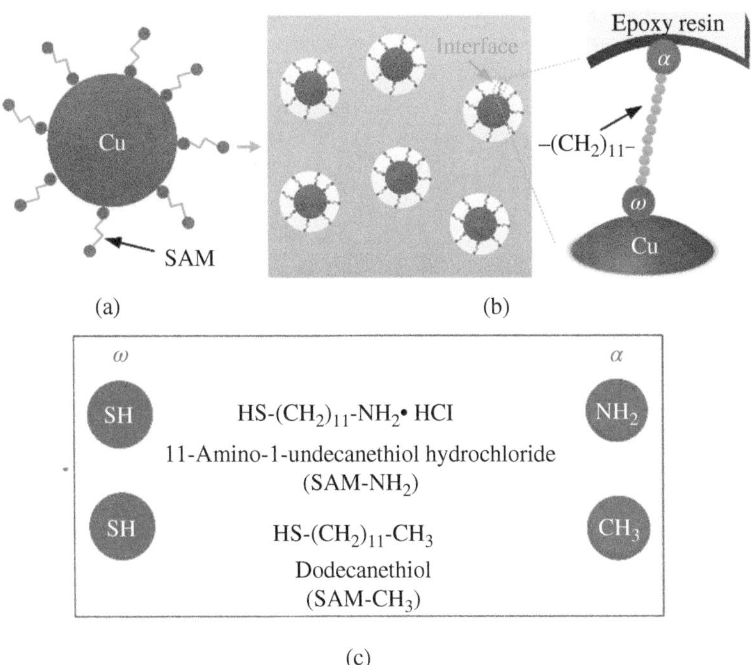

Figure 6.51 Depiction of (a) Cu surface modified with SAM and (b) Cu/SAM/epoxy systems; (c) list of SAM chemistries studied and abbreviations used in the text.

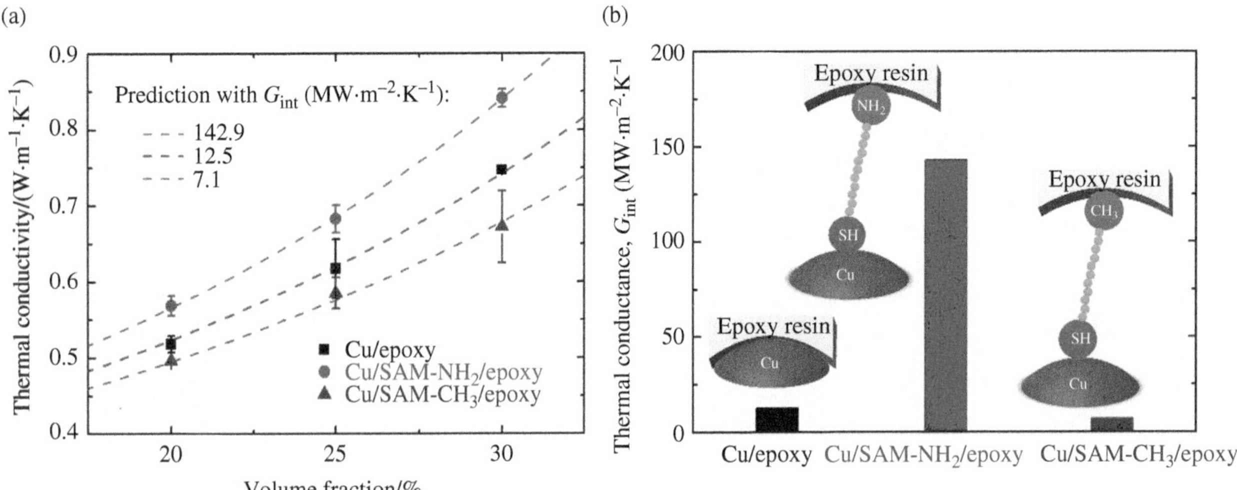

Figure 6.52 (a) Data fitting to extract G_{int} for Cu/epoxy, Cu/SAM-NH$_2$/epoxy, and Cu/SAM-CH$_3$/epoxy systems. Dots: measured thermal conductivity of composites; colored dash lines: predicted thermal conductivity from modified Bruggeman asymmetric model (MBAM). Adapted from Moore and Shi [28]. (b) Extracted results of G_{int}.

where d is the filler diameter. When the dispersed phase is much more conducting than the matrix ($\kappa_p/\kappa_m \gg 1$), Equation (6.36) can be simplified and written as:

$$\kappa_{eff} = \frac{\kappa_m}{(1-f)^{3(1-r_\beta)/(1+2r_\beta)}} \qquad (6.38)$$

The Cu/epoxy composite is very suitable as a simplified case considering that epoxy has very low conductivity and Cu has very high conductivity.

Therefore, according to Equations (6.37) and (6.38), G_{int} can be extracted once κ_{eff} is measured and other variables are obtained. Figure 6.52(a) gives the fitting results for the Cu/epoxy, Cu/SAM-NH$_2$/epoxy, and Cu/SAM-CH$_3$/epoxy systems and Figure 6.52(b) provides the evaluated values. The G_{int} of the Cu/epoxy interface is extracted to be 12.5 MW · m^{-2} · K^{-1}. It is of the same order as the G_{int} of other hard/soft interfaces dominated by weak van der Waals interaction, such as CNT/organic solvent (~12 MW · m^{-2} · K^{-1}) [190], hBN/epoxy (13.2 MW · m^{-2} · K^{-1}) [32], gold/hexadecane (28 MW · m^{-2} · K^{-1}) and gold/paraffin wax (25 MW · m^{-2} · K^{-1}) [178] interfaces. Functionalizing the Cu/epoxy interface with SAM-NH$_2$ obviously enhances the G_{int}. The G_{int} is improved from 12.5 to 142.9 MW · m^{-2} · K^{-1}, exhibiting more than 11-fold increase. It is interesting to note that in comparison with Cu/epoxy, the interface modified with SAM-CH$_3$ shows an approximately 2-fold lower G_{int}, 7.1 MW · m^{-2} · K^{-1}.

6.3.3 Characterization of Interfacial Bonds

It is generally believed that the enhancement of G_{int} results from the formation of covalent bonds at both the Cu/SAM-NH$_2$ and SAM-NH$_2$/epoxy interfaces. Independent experimental measurements to characterize the bonds and elucidate their formation are conducted. In addition, for the Cu/SAM-CH$_3$/epoxy system, although covalent bonds are believed to occur between thiols and Cu, the SAM-CH$_3$/epoxy interfacial interaction should be still of the van der Waals type according to the low G_{int}. Therefore, the bond type at such an interface is experimentally verified.

The presence of chemically bonded thiols on Cu is verified by FESM and XPS. Figure 6.53(a) and (d) and (b), (c), and (f) provide the surface topography of Cu microspheres before and after modification with SAM-NH$_2$ and SAM-CH$_3$. The images with higher magnification show that the SAMs modify the Cu with nanomesh structures (Figure 6.53(e)) or nanoparticles (Figure 6.53(f)). These nanostructures bundle together covering the whole microspheres.

Figure 6.53(g) gives the XPS spectra for the pure microspheres and microspheres treated with SAMs. On comparing with the control spectrum, significantly higher S and N content are measured from the SAM-NH$_2$-treated sample and higher S content is measured from the SAM-CH$_3$-treated sample. Due to spin–orbit splitting, S 2p peaks are separated into S 2p$_{3/2}$ and S 2p$_{1/2}$ components. As discussed in literature [191], for the free alkanethiols SAM, the S 2p$_{3/2}$ component should

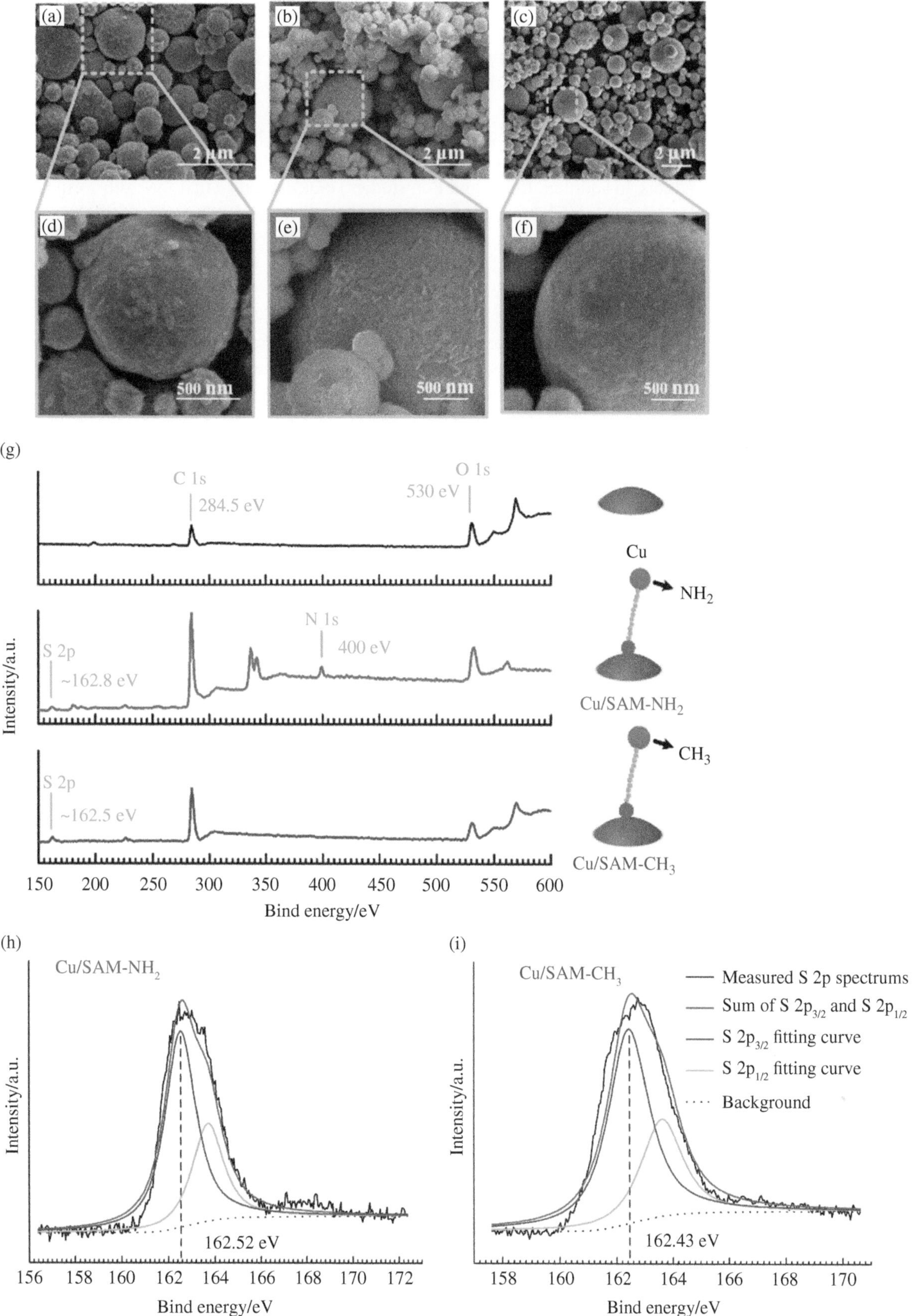

Figure 6.53 FESM images of (a and d) pure Cu, (b and e) SAM-NH_2-modified Cu and (c and f) SAM-CH_3-modified Cu. (g) XPS spectra of pure Cu and Cu treated with SAMs. High-resolution S 2p spectra of (h) SAM-NH_2 and (i) SAM-CH_3-modified Cu, and the corresponding fitting curves of S $2p_{3/2}$ and S $2p_{1/2}$.

appear at 163.5–163.8 eV, whereas, for the alkanethiols chemisorbed on Cu, S $2p_{3/2}$ peak will move downward to ~162 eV. Figure 6.53(h) and (i) provide the measured high-resolution S 2p spectra of SAM-NH_2- and SAM-CH_3-modified Cu, and the corresponding fitting curves of S $2p_{3/2}$ and S $2p_{1/2}$. S $2p_{3/2}$ components are observed at 162.52 and 162.43 eV, which compare very well with the typical S $2p_{3/2}$ bind energy of alkanethiols chemisorbed on Cu. This XPS analysis suggests that chemically bonded thiols are present on Cu.

Next, the mechanism of S—Cu formation is analyzed. Since Cu modification experiments were conducted on ambient environment, Cu got oxidized before reacting with SAM. As discussed in literature [192], oxidized surfaces of Cu are still active to the chemisorption of alkanethiols. In addition, Keller and Ron studies [193, 194] prove that the oxide formed on a Cu surface can be reduced to metallic Cu by —SH, and their reactions can result in the chemisorbed copper thiolate layers. The most likely oxide on Cu surface is Cu_2O [192, 194, 195], so the reaction can be described by the equation [194, 196]:

$$2RSH + Cu_2O \rightarrow 2RSCu(surface) + H_2O \tag{6.39}$$

where R represents $NH_2(CH_2)_{11}$ and $(CH_2)_{12}$ for SAM-NH_2 and SAM-CH_3, respectively.

Compared to Cu—S bonds, it is more difficult to characterize the bonds at SAM/epoxy interfaces, since they are embedded in the composites. The tensile strength tests [197–199] are adopted to conduct a qualitative analysis for the interfacial bonds. As schematically shown in Figure 6.54(a), tensile strength tests are applied on the composite samples. The measured results

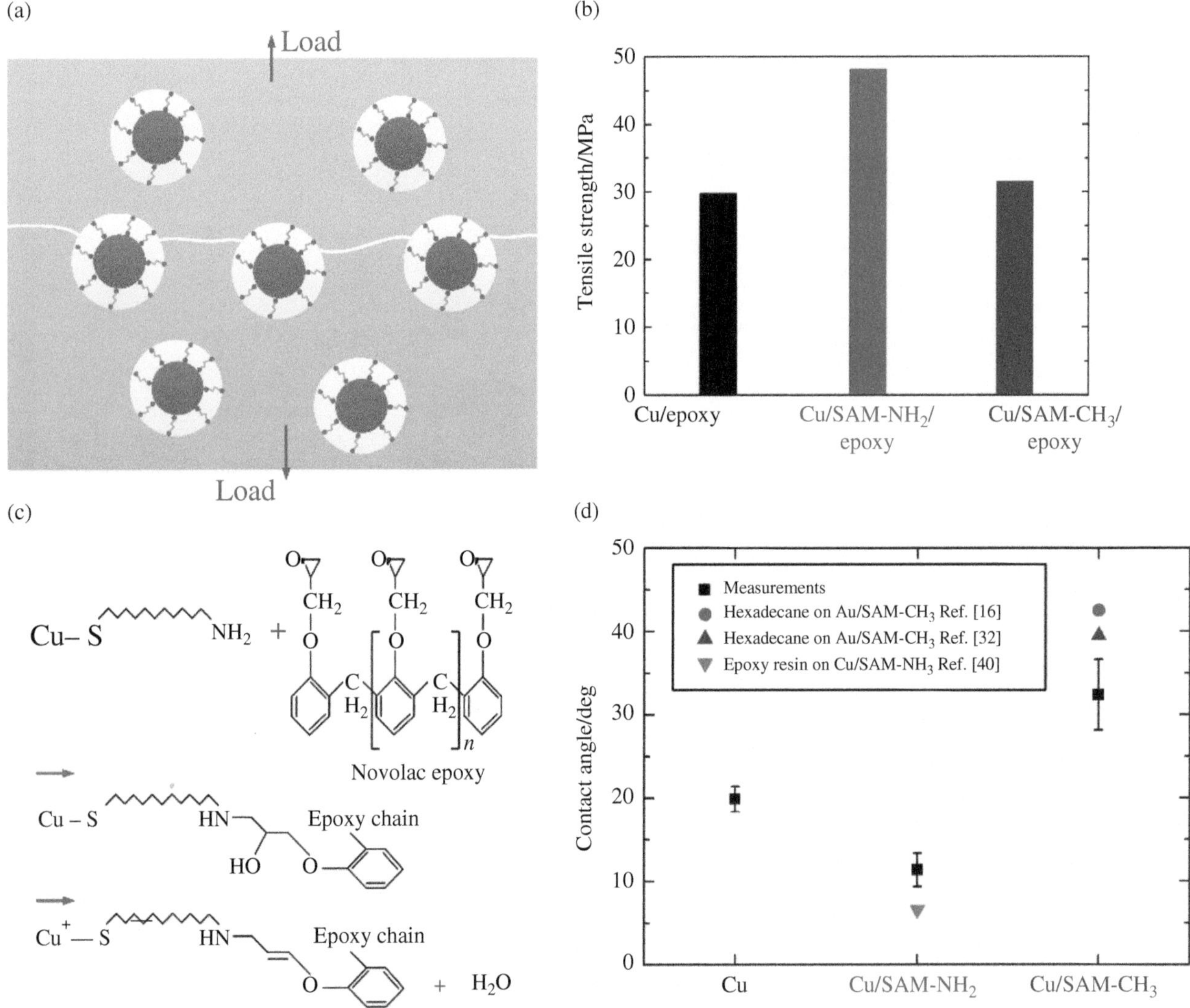

Figure 6.54 (a) Schematic of the tensile test. (b) Measured tensile strength for Cu/epoxy, Cu/SAM-NH_2/epoxy, and Cu/SAM-CH_3/epoxy composites. (c) Nucleophilic reaction between SAM-NH_2-treated Cu and a novolac epoxy. (d) Contact angle of polymeric materials (epoxy and hexadecane) on the pure and SAM-modified surfaces.

for Cu/epoxy, Cu/SAM-NH_2/epoxy, and Cu/SAM-CH_3/epoxy composites are 29.79, 48.17, and 31.53 MPa, respectively (Figure 6.54(b)). Compared to Cu/epoxy composite, Cu/SAM-NH_2/epoxy exhibits 61.7% larger tensile strength. The enhanced tensile property is attributed to the interaction between SAM-NH_2 and epoxy compound during composite polymerization. While interacting with epoxy, the alkane chains extend and expose the —NH_2 groups to the epoxy molecules. Figure 6.54(c) illustrates the nucleophilic reaction between SAM-NH_2-treated Cu and a novolac epoxy, which was proposed and proved with the experiments in Ref. [196]. The reaction shows that the —NH_2 group in SAM-NH_2 opens the oxirane in novolac epoxy, forming the secondary amine with covalent bonds. In contrast, there is no obvious enhancement in tensile strength for the composite modified with SAM-CH_3, although Cu is covalently bonded with S molecule. The poor tensile property is a consequence of the non-polar nature of the SAM-CH_3 molecule [196]. The non-polar tail (—CH_3) hinders formation of hydrogen bonding with those ingredients inside the epoxy, resulting in a weak van der Waals interaction between them.

For some solid–liquid interfaces [184, 185], the wettability of liquid on solid material is used to characterize the interfacial bonding energy. A thermodynamic parameter, W_{SL}, expresses the relationship between the bonding energy and surface wettability (or contact angle) [184, 185]:

$$W_{SL} = \zeta_{AW}(1 + \cos\theta) \tag{6.40}$$

where ζ_{AW} is the air–liquid surface tension and θ is the contact angle.

According to Equation (6.40), the low θ means a strong interfacial bonding. Thus, we characterized the interfacial bonding energy by measuring the θ of fluidic epoxy resin on surfaces before and after modification with SAM-NH_2 and SAM-CH_3. As shown in Figure 6.54(d), the surface consisting of pure Cu microspheres has a θ of 19.9°, whereas the SAM-NH_2-treated surface exhibits a much lower θ (~11°). θ of SAM-NH_2-treated surface was also measured in Ref. [200] and exhibits a much smaller value (6.6°). Conversely, the wettability of SAM-CH_3-treated surfaces is poorer than that of other two surfaces. The θ was measured to be 32.4°. Figure 6.54(d) also presents θ of another polymeric material, hexadecane [178, 192], on SAM-CH_3-treated surfaces. The large values demonstrate that the wettability is poor, too. The wettability tests consolidate the conclusions that interfacial bonding is stronger between SAM-NH_2/epoxy interface and lower between SAM-CH_3/epoxy interface.

The simulation results obtained in Ref. [196] support the experimental observations. By the molecular dynamics (MD) simulations, the interfacial bonding energy is calculated for Cu/epoxy and Cu/epoxy modified with cystamine dihydrochloride and hexadecanethiol, respectively. The calculated results are 0.535, 0.838, and 0.055 $J \cdot m^{-2}$, respectively. The structures and chemistry properties of cystamine dihydrochloride and hexadecanethiol are similar to those of SAM-NH_2 and SAM-CH_3, respectively. All of them have the thiol functionality (—SH), the long alkane chain in the middle, and —CH_3 or —NH_2 on α-end. The simulation results prove the existence of covalent bonds between Cu and epoxy when Cu is modified with amine SAM, instead of alkanethiol SAM.

6.3.4 Importance of Covalent Bonds

For solid–water interface [184, 185], the researchers find that G_{int} increases directly with interfacial bonding energy. The same tendency is found for the Cu/epoxy interface. However, G_{int} at some weakly bonded interface (Au/alkanethiol SAM/polyethylene) is not decreased but enhanced significantly [178]. The reason has been explained in the Introduction. The acoustic match between the two components makes thermal transport efficient. As for the SAM-CH_3/epoxy interface, their phonon spectra cannot match, since there is an obvious distinction between the two components' structures. Therefore, when the structure of soft material is complex, the enhancement of G_{int} should be induced by covalent bonds rather than by acoustic match.

6.3.5 Manipulation of the Thermal Properties of Nanocomposites

Two types of Cu/epoxy composites are investigated. One is assumed to be modified with SAM-NH_2 (G_{int} = 142.9 $MW \cdot m^{-2} \cdot K^{-1}$) and another is assumed to be without any surface modification (G_{int} = 12.5 $MW \cdot m^{-2} \cdot K^{-1}$). Figure 6.55 gives the predicted composites thermal conductivities under different Cu microspheres diameters. The results show that for the 30 vol% composite without modification, TC decreases from 0.74 to 0.39 $W \cdot m^{-2} \cdot K^{-1}$ by 47.3% and to 0.20 $W \cdot m^{-2} \cdot K^{-1}$ by 73.0%, when 1 μm of diameter decreases by 10 and 100 times, respectively. Particularly, TC of 10 nm Cu-filled composite will not increase with volume fraction due to the low G_{int}. While, after the surface modification with

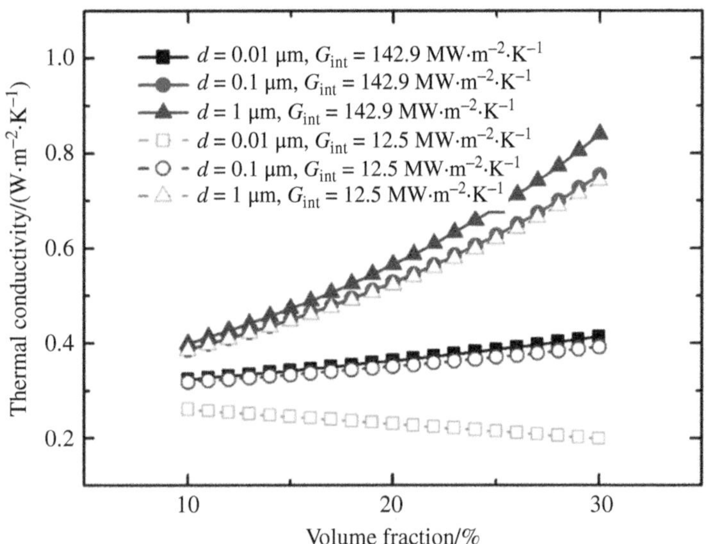

Figure 6.55 Predicted thermal conductivities of Cu/epoxy and Cu/SAM-NH_2/composites versus volume fraction at different Cu diameters.

SAM-NH_2, the extent to which the TC decreases with diameter is weakened. The percentage of reduction decreases down to 10.7% and 51.2%, respectively. Thus, with the dimension decreasing to nanosize, filling the soft matrix with high loading particles will not contribute to the TC any more. However, functionalizing the interfaces inside the composites with strongly bonding SAMs will relieve the adverse impact of interfacial thermal resistance on TC.

6.4 Chapter Summary

TIMs are commonly used in electronics to reduce the contact resistance arising from the incomplete contact between two solid surfaces. However, a solid–TIM–solid joint will form at the surface after inserting TIM between the solid surfaces, which induces large thermal resistance. The thermal resistance at the joint (R_j) mainly has two components: the contact resistance (R_c) at the TIM–solid interface arising from the incomplete wetting of the interface and the bulk resistance of TIM (R_{bulk}). Therefore, the researches of contact resistance (R_c) and the TC of TIM (κ_{TIM}) are of great significance for enhancing heat transfer. In this chapter, the modeling of contact resistance and its validation, the microstructure design using magnetic field, and the interfacial thermal transport manipulation are introduced.

In Section 6.1, an improved model for predicting thermal contact resistance R_c at the liquid–solid interface based on Hamasaiid et al.'s model is introduced. Through conducting the wettability analysis on the liquid–solid contact, an explicit expression is derived to predict the height of entrapped air between the liquid and the solid. Experimental measurements have been conducted on TIMs–aluminum interface. The results show the model matches well with the experimental data.

In Section 6.2, the microstructure design of TIM using magnetic field is introduced. In Section 6.2.1, hBNs were coated with magnetic iron oxide nanoparticles to be magnetically responsive. Then, various microstructures were achieved in the hBNs-filled composites by applying external magnetic fields. TC was measured for the resulting composites. The modified EMA was applied to analyze the experimental data. Both experimental and modelling results illustrate that TC of the composites is strongly associated with the magnetic alignment of mhBN platelets. By fitting the measured TC to EMA prediction, thermal boundary resistance (R_b) at the hBN–silicone interface was extracted. It is found that R_b also changes with hBNs orientation. In Section 6.2.2, the use of combined mechanical and magnetic stimuli to fabricate the composites containing well-aligned mhBN platelets is introduced. The platelets in the resulting composites exhibit a high degree of alignment. TC of the resulting composites was also measured. The results show that TC can be greatly enhanced by the platelets alignment. The modified EMA was applied to analyze the experimental data. By fitting the measured TC to the EMA prediction, the thermal boundary resistance (R_b) at the mhBN–resin interface was extracted. It is found that the higher degree of platelets alignment results in a lower value of R_b. In Section 6.2.3, the fabrication of the locally reinforced composites applied for efficient heat dissipation of local heat source is introduced. The local reinforcement was achieved by using the mhBN platelets as reinforcing elements and a magnetic field gradient to concentrate the elements below the heat source. To evaluate the thermal performance of the locally reinforced composites, the comparative thermal tests on the homogeneous and locally reinforced samples were performed. The results show that the locally reinforced composites can greatly reduce the average and maximum temperatures of heat source. Finally, a simulation study on the composites applied with local heat source was conducted. The simulation results show that the temperature of heat source is primarily dependent on the heat conduction below the heat source. Thus, concentrating the platelets at that place can enhance the heat conduction and, hence, decrease the temperature of heat source.

In Section 6.3, an effective strategy to tune thermal transport across hard/soft materials interfaces is introduced. The proposed strategy relies on using a strongly bonding SAM to covalently connect hard and soft materials. The thermal measurements show that for Cu/epoxy interface, G_{int} can be enhanced by as much as 11-folds. Moreover, this study experimentally illustrates that when the structure of soft material is complex, interfacial thermal transport should be tuned by covalent bonds rather than by phonon spectra match. The proposed strategy shows great potential in material design for heat transfer-critical applications like electronic packaging, thermal storage, sensors, and medicine.

References

1 Prasher, R.S., Shipley, J., Prstic, S. et al. (2003). Thermal resistance of particle laden polymeric thermal interface materials. *J. Heat Transf. Trans. ASME* 125 (6): 1170–1177.

2 Prasher, R.S. (2005). Rheology based modeling and design of particle laden polymeric thermal interface materials. *IEEE Trans. Compon. Packag. Technol.* 28 (2): 230–237.

3 Prasher, R. (2006). Thermal interface materials: historical perspective, status, and future directions. *Proc. IEEE* 94 (8): 1571–1586.

4 Prasher, R.S. (2001). Surface chemistry and characteristics based model for the thermal contact resistance of fluidic interstitial thermal interface materials. *J. Heat Transf.* 123 (5): 969–975.

5 Tien, C.L. (1968). A correlation for thermal contact conductance of nominally flat surfaces in vacuum. *Proceedings of the 7th Thermal Conductivity Conference*, 755–759.

6 Cooper, M.G., Mikic, B.B., and Yovanovich, M.M. (1969). Thermal contact conductance. *Int. J. Heat Mass Transf.* 12 (3): 279–300.

7 Thomas, T.R. and Probert, S.D. (1970). Thermal contact resistance: the directional effect and other problems. *Int. J. Heat Mass Transf.* 13 (5): 789–807.

8 Mikić, B.B. (1974). Thermal contact conductance; theoretical considerations. *Int. J. Heat Mass Transf.* 17 (2): 205–214.

9 Bennett, T. and Poulikakos, D. (1994). Heat transfer aspects of splat-quench solidification: modelling and experiment. *J. Mater. Sci.* 29: 2025–2039.

10 Liu, W., Wang, G.X., and Matthys, E.F. (1995). Thermal analysis and measurements for a molten metal drop impacting on a substrate: cooling, solidification and heat transfer coefficient. *Int. J. Heat Mass Transf.* 38 (8): 1387–1395.

11 Hamasaiid, A., Dour, G., Loulou, T. et al. (2010). A predictive model for the evolution of the thermal conductance at the casting–die interfaces in high pressure die casting. *Int. J. Therm. Sci.* 49 (2): 365–372.

12 Hamasaiid, A., Dargusch, M.S., Loulou, T. et al. (2011). A predictive model for the thermal contact resistance at liquid–solid interfaces: analytical developments and validation. *Int. J. Therm. Sci.* 50 (8): 1445–1459.

13 Madhusudana, C.V. (1996). *Thermal Contact Conductance*. Springer.

14 *SJ-401 Surface Roughness Tester User's Manual*. Japan: Mitutoyo Corporation.

15 Sridhar, M.R. and Yovanovich, M.M. (1994). Review of elastic and plastic contact conductance models-comparison with experiment. *J. Thermophys. Heat Transf.* 8 (4): 633–640.

16 Sridhar, M.R. and Yovanovich, M.M. (1996). Elastoplastic contact conductance model for isotropic conforming rough surfaces and comparison with experiments. *J. Heat Transf.* 118: 3–9.

17 Bahrami, M., Yovanovich, M.M., and Culham, J.R. (2004). Thermal joint resistances of nonconforming rough surfaces with gas-filled gaps. *J. Thermophys. Heat Transf.* 18 (3): 326–332.

18 Bahrami, M., Culham, J.R., Yovanovich, M.M. et al. (2004). Thermal contact resistance of nonconforming rough surfaces, part 1: contact mechanics model. *J. Thermophys. Heat Transf.* 18 (2): 209–217.

19 Bahrami, M., Culham, J.R., Yovanovich, M.M. et al. (2004). Thermal contact resistance of nonconforming rough surfaces, part 2: thermal model. *J. Thermophys. Heat Transf.* 18 (2): 218–227.

20 Standard, A. (2006). *Standard Test Method for Thermal Transmission Properties of Thermally Conductive Electrical Insulation Materials*. ASTM International.

21 Prasher, R.S., Koning, P., Shipley, J. et al. (2003). Dependence of thermal conductivity and mechanical rigidity of particle-laden polymeric thermal interface material on particle volume fraction. *J. Electron. Packag.* 125 (3): 386–391.

22 ASTM (2013). *D7490--13 Standard Test Method for Measurement of the Surface Tension of Solid Coatings, Substrates and Pigments Using Contact Angle Measurements*. Pennsylvania: ATSM International.

23 Wu, S. (2017). *Polymer Interface and Adhesion*. Routledge.

24 Prasher, R.S., Simmons, C., and Solbrekken, G. (2000). *Thermal Contact Resistance of Phase Change and Grease Type Polymeric Materials*, 461–466. American Society of Mechanical Engineers.

25 Lu, D. and Wong, C.P. (2009). *Materials for Advanced Packaging*. New York, NY: Springer.

26 Wong, C.P., Moon, K.-S., and Li, Y. (2010). *Nano-Bio-Electronic, Photonic and MEMS Packaging*. New York, NY: Springer.

27 Liu, S. and Luo, X. (2011). *LED Packaging for Lighting Applications: Design, Manufacturing, and Testing*. Wiley.

28 Moore, A.L. and Shi, L. (2014). Emerging challenges and materials for thermal management of electronics. *Mater. Today* 17 (4): 163–174.

29 Otiaba, K.C., Ekere, N.N., Bhatti, R.S. et al. (2011). Thermal interface materials for automotive electronic control unit: trends, technology and R&D challenges. *Microelectron. Reliab.* 51 (12): 2031–2043.

30 Mallik, S., Ekere, N., Best, C. et al. (2011). Investigation of thermal management materials for automotive electronic control units. *Appl. Therm. Eng.* 31 (2–3): 355–362.

31 Yuan, C. and Luo, X. (2013). A unit cell approach to compute thermal conductivity of uncured silicone/phosphor composites. *Int. J. Heat Mass Transf.* 56 (1–2): 206–211.

32 Lin, Z., Liu, Y., Raghavan, S. et al. (2013). Magnetic alignment of hexagonal boron nitride platelets in polymer matrix: toward high performance anisotropic polymer composites for electronic encapsulation. *ACS Appl. Mater. Interfaces* 5 (15): 7633–7640.

33 Takahashi, F., Ito, K., Morikawa, J. et al. (2004). Characterization of heat conduction in a polymer film. *Japan. J. Appl. Phys. 1* 43 (10): 7200–7204.

34 Li, T.L. and Hsu, S.L.C. (2010). Enhanced thermal conductivity of polyimide films via a hybrid of micro- and nano-sized boron nitride. *J. Phys. Chem. B* 114 (20): 6825–6829.

35 Li, T.L. and Hsu, S.L.C. (2011). Preparation and properties of thermally conductive photosensitive polyimide/boron nitride nanocomposites. *J. Appl. Polym. Sci.* 121 (2): 916–922.

36 Sato, K., Horibe, H., Shirai, T. et al. (2010). Thermally conductive composite films of hexagonal boron nitride and polyimide with affinity-enhanced interfaces. *J. Mater. Chem.* 20 (14): 2749–2752.

37 Lin, T.H., Huang, W.H., Jun, I.K. et al. (2010). Bioinspired assembly of surface-roughened nanoplatelets. *J. Colloid Interface Sci.* 344 (2): 272–278.

38 Libanori, R., Münch, F.H.L., Montenegro, D.M. et al. (2012). Hierarchical reinforcement of polyurethane-based composites with inorganic micro- and nanoplatelets. *Compos. Sci. Technol.* 72 (3): 435–445.

39 Munch, E., Launey, M.E., Alsem, D.H. et al. (2008). Tough, bio-inspired hybrid materials. *Science* 322 (5907): 1516–1520.

40 Erb, R.M., Libanori, R., Rothfuchs, N. et al. (2012). Composites reinforced in three dimensions by using low magnetic fields. *Science* 335 (6065): 199–204.

41 Erb, R.M., Son, H.S., Samanta, B. et al. (2009). Magnetic assembly of colloidal superstructures with multipole symmetry. *Nature* 457 (7232): 999–1002.

42 Yuan, C., Duan, B., Li, L. et al. (2015). Thermal conductivity of polymer-based composites with magnetic aligned hexagonal boron nitride platelets. *ACS Appl. Mater. Interfaces* 7 (23): 13000–13006.

43 Zhi, C., Bando, Y., Tan, C. et al. (2005). Effective precursor for high yield synthesis of pure BN nanotubes. *Solid State Commun.* 135 (1–2): 67–70.

44 Nan, C.W., Birringer, R., Clarke, D.R. et al. (1997). Effective thermal conductivity of particulate composites with interfacial thermal resistance. *J. Appl. Phys.* 81 (10): 6692–6699.

45 Xie, B.-H., Huang, X., and Zhang, G.-J. (2013). High thermal conductive polyvinyl alcohol composites with hexagonal boron nitride microplatelets as fillers. *Compos. Sci. Technol.* 85: 98–103.

46 Tanimoto, M., Yamagata, T., Miyata, K. et al. (2013). Anisotropic thermal diffusivity of hexagonal boron nitride-filled polyimide films: effects of filler particle size, aggregation, orientation, and polymer chain rigidity. *ACS Appl. Mater. Interfaces* 5 (10): 4374–4382.

47 Song, W.L., Wang, P., Cao, L. et al. (2012). Polymer/boron nitride nanocomposite materials for superior thermal transport performance. *Angew. Chem. Int. Ed.* 51 (26): 6498–6501.

48 Han, Y.W., Lv, S.M., Hao, C.X. et al. (2012). Thermal conductivity enhancement of BN/silicone composites cured under electric field: stacking of shape, thermal conductivity, and particle packing structure anisotropies. *Thermochim. Acta* 529: 68–73.

49 Boussaad, S. (2012). Hexagonal boron nitride compositions characterized by interstitial ferromagnetic layers, process for preparing, and composites thereof with organic polymers. Google patents.

50 Olafsen, J.S. and Urbach, J.S. (1998). Clustering, order, and collapse in a driven granular monolayer. *Phys. Rev. Lett.* 81 (20): 4369–4372.

51 Olafsen, J.S. and Urbach, J.S. (2005). Two-dimensional melting far from equilibrium in a granular monolayer. *Phys. Rev. Lett.* 95 (9): 098002.

52 Yu, A.B., An, X.Z., Zou, R.P. et al. (2006). Self-assembly of particles for densest packing by mechanical vibration. *Phys. Rev. Lett.* 97 (26): 265501.

53 Libanori, R., Erb, R.M., and Studart, A.R. (2013). Mechanics of platelet-reinforced composites assembled using mechanical and magnetic stimuli. *ACS Appl. Mater. Interfaces* 5 (21): 10794–10805.

54 Erb, R.M., Segmehl, J., Charilaou, M. et al. (2012). Non-linear alignment dynamics in suspensions of platelets under rotating magnetic fields. *Soft Matter* 8 (29): 7604–7609.

55 Yan, H.Y., Tang, Y.X., Long, W. et al. (2014). Enhanced thermal conductivity in polymer composites with aligned graphene nanosheets. *J. Mater. Sci.* 49 (15): 5256–5264.

56 Hirotani, J., Ikuta, T., Nishiyama, T. et al. (2011). Thermal boundary resistance between the end of an individual carbon nanotube and a Au surface. *Nanotechnology* 22 (31): 315702.

57 Schmidt, A.J., Collins, K.C., Minnich, A.J. et al. (2010). Thermal conductance and phonon transmissivity of metal-graphite interfaces. *J. Appl. Phys.* 107 (10): 104907.

58 Prasher, R. (2008). Thermal boundary resistance and thermal conductivity of multiwalled carbon nanotubes. *Phys. Rev. B* 77 (7): 075424.

59 Nakayama, W. (2013). Study on heat conduction in a simulated multicore processor chip—Part I: analytical modeling. *J. Electron. Packag.* 135 (2): 021002.

60 Nakayama, W. (2013). Study on heat conduction in a simulated multicore processor chip—Part II: case studies. *J. Electron. Packag.* 135 (2): 021003.

61 Kim, Y.J., Joshi, Y.K., Fedorov, A.G. et al. (2010). Thermal characterization of interlayer microfluidic cooling of three-dimensional integrated circuits with nonuniform heat flux. *J. Heat Transf. Trans. ASME* 132 (4): 041009.

62 Bar-Cohen, A. and Wang, P. (2012). Thermal management of on-chip hot spot. *J. Heat Transf. Trans. ASME* 134 (5): 051017.

63 Yi, P., Awang, R.A., Rowe, W.S.T. et al. (2014). PDMS nanocomposites for heat transfer enhancement in microfluidic platforms. *Lab Chip* 14 (17): 3419–3426.

64 Yan, Z., Liu, G., Khan, J.M. et al. (2012). Graphene quilts for thermal management of high-power GaN transistors. *Nat. Commun.* 3 (1): 827.

65 Kong, Q.Q., Liu, Z., Gao, J.G. et al. (2014). Hierarchical graphene-carbon fiber composite paper as a flexible lateral heat spreader. *Adv. Funct. Mater.* 24 (27): 4222–4228.

66 Sun, X., Sun, H., Li, H. et al. (2013). Developing polymer composite materials: carbon nanotubes or graphene? *Adv. Mater.* 25 (37): 5153–5176.

67 Zhou, Y., Yao, Y., Chen, C.-Y. et al. (2014). The use of polyimide-modified aluminum nitride fillers in AlN@ PI/epoxy composites with enhanced thermal conductivity for electronic encapsulation. *Sci. Rep.* 4 (1): 4779.

68 Chhasatia, V., Zhou, F., Sun, Y. et al. (2008). Design optimization of custom engineered silver-nanoparticle thermal interface materials. *2008 11th IEEE Intersociety Conference on Thermal and Thermomechanical Phenomena in Electronic Systems*, vols 1–3, 419–427.

69 Li, Q., Steven, G.P., Querin, O.M. et al. (1999). Shape and topology design for heat conduction by evolutionary structural optimization. *Int. J. Heat Mass Transf.* 42 (17): 3361–3371.

70 Xia, Z.Z., Cheng, X.G., Li, Z.X. et al. (2004). Bionic optimization of heat transport paths for heat conduction problems. *J. Enhanced Heat Transf.* 11 (2): 119–131.

71 Bruns, T.E. (2007). Topology optimization of convection-dominated, steady-state heat transfer problems. *Int. J. Heat Mass Transf.* 50 (15–16): 2859–2873.

72 Stauffer, D. and Aharony, A. (1992). *Introduction to Percolation Theory*. Washington, DC: Taylor & Francis.

73 Devpura, A., Phelan, P.E., and Prasher, R.S. (2000). Percolation theory applied to the analysis of thermal interface materials in flip-chip technology. *ITHERM 2000: Seventh Intersociety Conference on Thermal and Thermomechanical Phenomena in Electronic Systems, Proceedings*, vol. I, 21–28.

74 Delhaye, N., Poitou, A., and Chaouche, M. (2000). Squeeze flow of highly concentrated suspensions of spheres. *J. Non-Newtonian Fluid Mech.* 94 (1): 67–74.

75 Chaari, F., Racineux, G., Poitou, A. et al. (2003). Rheological behavior of sewage sludge and strain-induced dewatering. *Rheol. Acta* 42 (3): 273–279.

76 Rae, D.F., Borgesen, P., and Cotts, E.J. (2011). The effect of filler-network heterogeneity on thermal resistance of polymeric thermal bondlines. *JOM* 63 (10): 78–84.

77 Fujihara, T., Cho, H.B., Kanno, M. et al. (2014). Three-dimensional structural control and analysis of hexagonal boron nitride nanosheets assembly in nanocomposite films induced by electric field concentration. *Jpn. J. Appl. Phys.* 53 (2): 02BD12.

78 Erb, R.M., Cherenack, K.H., Stahel, R.E. et al. (2012). Locally reinforced polymer-based composites for elastic electronics. *ACS Appl. Mater. Interfaces* 4 (6): 2860–2864.

79 Choi, C., Yatsuzuka, K., and Asano, K. (2001). Dynamic motion of a conductive particle in viscous fluid under DC electric field. *IEEE Trans. Ind. Appl.* 37 (3): 785–791.

80 Wang, Y., Zheng, H., Hu, R. et al. (2014). Modeling on phosphor sedimentation phenomenon during curing process of high power LED packaging. *J. Solid State Light.* 1: 1–9.

81 Ma, J.K., Shang, T.Y., Ren, L.L. et al. (2020). Through-plane assembly of carbon fibers into 3D skeleton achieving enhanced thermal conductivity of a thermal interface material. *Chem. Eng. J.* 380: 122550.

82 Chen, J., Huang, X.Y., Sun, B. et al. (2017). Vertically aligned and interconnected boron nitride nanosheets for advanced flexible nanocomposite thermal interface materials. *ACS Appl. Mater. Interfaces* 9 (36): 30909–30917.

83 Lee, S.-K., Tuan, W.-H., Wu, Y.-Y. et al. (2013). Microstructure–thermal properties of Cu/Al_2O_3 bilayer prepared by direct bonding. *J. Eur. Ceram. Soc.* 33 (2): 277–285.

84 Mao, D., Chen, J., Ren, L. et al. (2019). Spherical core-shell Al@Al_2O_3 filled epoxy resin composites as high-performance thermal interface materials. *Compos. A: Appl. Sci. Manuf.* 123: 260–269.

85 Kim, Y.-K., Chung, J.-Y., Lee, J.-G. et al. (2017). Synergistic effect of spherical Al_2O_3 particles and BN nanoplates on the thermal transport properties of polymer composites. *Compos. A: Appl. Sci. Manuf.* 98: 184–191.

86 Sim, L.C., Ramanan, S.R., Ismail, H. et al. (2005). Thermal characterization of Al_2O_3 and ZnO reinforced silicone rubber as thermal pads for heat dissipation purposes. *Thermochim. Acta* 430 (1–2): 155–165.

87 Yang, F., Zhao, X., and Xiao, P. (2010). Thermal conductivities of YSZ/Al_2O_3 composites. *J. Eur. Ceram. Soc.* 30 (15): 3111–3116.

88 Zhang, D.-L., Zha, J.-W., Li, C.-Q. et al. (2017). High thermal conductivity and excellent electrical insulation performance in double-percolated three-phase polymer nanocomposites. *Compos. Sci. Technol.* 144: 36–42.

89 Yu, J., Huang, X., Wu, C. et al. (2012). Interfacial modification of boron nitride nanoplatelets for epoxy composites with improved thermal properties. *Polymer* 53 (2): 471–480.

90 Zhi, C., Bando, Y., Tang, C. et al. (2009). Large-scale fabrication of boron nitride nanosheets and their utilization in polymeric composites with improved thermal and mechanical properties. *Adv. Mater.* 21 (28): 2889–2893.

91 Wang, F., Zeng, X., Yao, Y. et al. (2016). Silver nanoparticle-deposited boron nitride nanosheets as fillers for polymeric composites with high thermal conductivity. *Sci. Rep.* 6: 19394.

92 Huang, X., Iizuka, T., Jiang, P. et al. (2012). Role of interface on the thermal conductivity of highly filled dielectric epoxy/AlN composites. *J. Phys. Chem. C* 116 (25): 13629–13639.

93 Xiao, C., Guo, Y.J., Tang, Y.L. et al. (2020). Epoxy composite with significantly improved thermal conductivity by constructing a vertically aligned three-dimensional network of silicon carbide nanowires/boron nitride nanosheets. *Compos. B Eng.* 187: 107855.

94 Kim, K. and Kim, J. (2016). Magnetic aligned AlN/epoxy composite for thermal conductivity enhancement at low filler content. *Compos. B Eng.* 93: 67–74.

95 Yuan, C., Xie, B., Huang, M. et al. (2016). Thermal conductivity enhancement of platelets aligned composites with volume fraction from 10% to 20%. *Int. J. Heat Mass Transf.* 94: 20–28.

96 Mercado, E., Yuan, C., Zhou, Y. et al. (2020). Isotopically enhanced thermal conductivity in few-layer hexagonal boron nitride: implications for thermal management. *ACS Appl. Nano Mater.* 3 (12): 12148–12156.

97 Renteria, J., Nika, D., and Balandin, A. (2014). Graphene thermal properties: applications in thermal management and energy storage. *Appl. Sci.* 4 (4): 525–547.

98 Shahil, K.M. and Balandin, A.A. (2012). Graphene-multilayer graphene nanocomposites as highly efficient thermal interface materials. *Nano Lett.* 12 (2): 861–867.

99 Park, W., Guo, Y., Li, X. et al. (2015). High-performance thermal interface material based on few-layer graphene composite. *J. Phys. Chem. C* 119 (47): 26753–26759.

100 Nika, D.L. and Balandin, A.A. (2017). Phonons and thermal transport in graphene and graphene-based materials. *Rep. Prog. Phys.* 80 (3): 036502.

101 Teng, C.-C., Ma, C.-C.M., Lu, C.-H. et al. (2011). Thermal conductivity and structure of non-covalent functionalized graphene/epoxy composites. *Carbon* 49 (15): 5107–5116.

102 Balandin, A.A. (2011). Thermal properties of graphene and nanostructured carbon materials. *Nat. Mater.* 10 (8): 569–581.

103 Hu, Y., Chiang, S.W., Chu, X.D. et al. (2020). Vertically aligned carbon nanotubes grown on reduced graphene oxide as high-performance thermal interface materials. *J. Mater. Sci.* 55 (22): 9414–9424.

104 Wang, Z., Cao, Y., Pan, D. et al. (2020). Vertically aligned and interconnected graphite and graphene oxide networks leading to enhanced thermal conductivity of polymer composites. *Polymers* 12 (5): 1121.

105 Fu, Y.X., He, Z.X., Mo, D.C. et al. (2014). Thermal conductivity enhancement of epoxy adhesive using graphene sheets as additives. *Int. J. Therm. Sci.* 86: 276–283.

106 Shahil, K.M.F. and Balandin, A.A. (2012). Thermal properties of graphene and multilayer graphene: applications in thermal interface materials. *Solid State Commun.* 152 (15): 1331–1340.

107 Lian, G., Tuan, C.C., Li, L.Y. et al. (2016). Vertically aligned and interconnected graphene networks for high thermal conductivity of epoxy composites with ultralow loading. *Chem. Mater.* 28 (17): 6096–6104.

108 Dai, W., Ma, T.F., Yan, Q.W. et al. (2019). Metal-level thermally conductive yet soft graphene thermal interface materials. *ACS Nano* 13 (10): 11561–11571.

109 Jung, H., Yu, S., Bae, N.S. et al. (2015). High through-plane thermal conduction of graphene nanoflake filled polymer composites melt-processed in an L-shape kinked tube. *ACS Appl. Mater. Interfaces* 7 (28): 15256–15262.

110 Veca, L.M., Meziani, M.J., Wang, W. et al. (2009). Carbon nanosheets for polymeric nanocomposites with high thermal conductivity. *Adv. Mater.* 21 (20): 2088–2092.

111 Ganguli, S., Roy, A.K., and Anderson, D.P. (2008). Improved thermal conductivity for chemically functionalized exfoliated graphite/epoxy composites. *Carbon* 46 (5): 806–817.

112 Kalaitzidou, K., Fukushima, H., and Drzal, L.T. (2007). Multifunctional polypropylene composites produced by incorporation of exfoliated graphite nanoplatelets. *Carbon* 45 (7): 1446–1452.

113 Debelak, B. and Lafdi, K. (2007). Use of exfoliated graphite filler to enhance polymer physical properties. *Carbon* 45 (9): 1727–1734.

114 Biercuk, M.J., Llaguno, M.C., Radosavljevic, M. et al. (2002). Carbon nanotube composites for thermal management. *Appl. Phys. Lett.* 80 (15): 2767–2769.

115 Tong, T., Zhao, Y., Delzeit, L. et al. (2007). Dense, vertically aligned multiwalled carbon nanotube arrays as thermal interface materials. *IEEE Trans. Compon. Packag. Technol.* 30 (1): 92–100.

116 Xu, J. and Fisher, T.S. (2006). Enhancement of thermal interface materials with carbon nanotube arrays. *Int. J. Heat Mass Transf.* 49 (9–10): 1658–1666.

117 Ji, T.X., Feng, Y.Y., Qin, M.M. et al. (2016). Thermal conducting properties of aligned carbon nanotubes and their polymer composites. *Compos. A: Appl. Sci. Manuf.* 91: 351–369.

118 Marconnet, A.M., Panzer, M.A., and Goodson, K.E. (2013). Thermal conduction phenomena in carbon nanotubes and related nanostructured materials. *Rev. Mod. Phys.* 85 (3): 1295–1326.

119 Qiu, L., Wang, X.T., Su, G.P. et al. (2016). Remarkably enhanced thermal transport based on a flexible horizontally-aligned carbon nanotube array film. *Sci. Rep.* 6: 21014.

120 Uetani, K., Ata, S., Tomonoh, S. et al. (2014). Elastomeric thermal interface materials with high through-plane thermal conductivity from carbon fiber fillers vertically aligned by electrostatic flocking. *Adv. Mater.* 26 (33): 5857–5862.

121 Yu, Z.F., Wei, S., and Guo, J.D. (2019). Fabrication of aligned carbon-fiber/polymer TIMs using electrostatic flocking method. *J. Mater. Sci. Mater. Electron.* 30 (11): 10233–10243.

122 Ji, T.X., Feng, Y.Y., Qin, M.M. et al. (2018). Thermal conductive and flexible silastic composite based on a hierarchical framework of aligned carbon fibers-carbon nanotubes. *Carbon* 131: 149–159.

123 Chung, S.H., Kim, H., and Jeong, S.W. (2018). Improved thermal conductivity of carbon-based thermal interface materials by high-magnetic-field alignment. *Carbon* 140: 24–29.

124 Tian, Z.L., Sun, J.J., Wang, S.G. et al. (2018). A thermal interface material based on foam-templated three-dimensional hierarchical porous boron nitride. *J. Mater. Chem. A* 6 (36): 17540–17547.

125 Zhang, X., Zhou, S., Xie, B. et al. (2021). Thermal interface materials with sufficiently vertically aligned and interconnected nickel-coated carbon fibers under high filling loads made via preset-magnetic-field method. *Compos. Sci. Technol.* 213: 108922.

126 Ren, L.Q., Zhou, X.L., Xue, J.Z. et al. (2019). Thermal metamaterials with site-specific thermal properties fabricated by 3D magnetic printing. *Adv. Mater. Technol.* 4 (7): 1900296.

127 Xu, S., Liu, H., Li, Q. et al. (2016). Influence of magnetic alignment and layered structure of BN&Fe/EP on thermal conducting performance. *J. Mater. Chem. C* 4 (4): 872–878.

128 Marconnett, A.M., Yamamoto, N., Panzer, M.A. et al. (2011). Thermal conduction in aligned carbon nanotube-polymer nanocomposites with high packing density. *ACS Nano* 5 (6): 4818–4825.

129 Cho, D., Choi, Y., Park, J.K. et al. (2004). Thermal conductivity and thermal expansion behavior of pseudo-unidirectional and 2-directional quasi-carbon fiber/phenolic composites. *Fibers Polym.* 5 (1): 31–38.

130 Lalet, G., Kurita, H., Heintz, J.M. et al. (2014). Thermal expansion coefficient and thermal fatigue of discontinuous carbon fiber-reinforced copper and aluminum matrix composites without interfacial chemical bond. *J. Mater. Sci.* 49 (1): 397–402.

131 Liu, B.C., Li, Y.B., Fei, T. et al. (2020). Highly thermally conductive polystyrene/polypropylene/boron nitride composites with 3D segregated structure prepared by solution-mixing and hot-pressing method. *Chem. Eng. J.* 385: 123829.

132 Zhang, X., Wu, K., Liu, Y.H. et al. (2019). Preparation of highly thermally conductive but electrically insulating composites by constructing a segregated double network in polymer composites. *Compos. Sci. Technol.* 175: 135–142.

133 Ning, N.Y., Fu, S.R., Zhang, W. et al. (2012). Realizing the enhancement of interfacial interaction in semicrystalline polymer/filler composites via interfacial crystallization. *Prog. Polym. Sci.* 37 (10): 1425–1455.

134 Suh, D., Moon, C.M., Kim, D. et al. (2016). Ultrahigh thermal conductivity of interface materials by silver-functionalized carbon nanotube phonon conduits. *Adv. Mater.* 28 (33): 7220–7227.

135 Stankovich, S., Dikin, D.A., Dommett, G.H.B. et al. (2006). Graphene-based composite materials. *Nature* 442 (7100): 282–286.

136 Chen, H.Y., Ginzburg, V.V., Yang, J. et al. (2016). Thermal conductivity of polymer-based composites: fundamentals and applications. *Prog. Polym. Sci.* 59: 41–85.

137 Coleman, J.N., Khan, U., Blau, W.J. et al. (2006). Small but strong: a review of the mechanical properties of carbon nanotube-polymer composites. *Carbon* 44 (9): 1624–1652.

138 Garlotta, D. (2001). A literature review of poly(lactic acid). *J. Polym. Environ.* 9 (2): 63–84.

139 Tao, Y.B., Wang, H.L., Li, Z.L. et al. (2017). Development and application of wood flour-filled polylactic acid composite filament for 3D printing. *Materials* 10 (4): 339.

140 Anis, A., Elnour, A.Y., Alam, M.A. et al. (2020). Aluminum-filled amorphous-PET, a composite showing simultaneous increase in modulus and impact resistance. *Polymers* 12 (9): 2038.

141 Huang, C.H., Wang, S.B., and Liu, L.G. (2007). Experimental study on the mechanical properties of metal oxides filled PA1010 composites. *Prog. Fract. Strength Mater. Struct., 1-4* 353–358: 1346–1349.

142 Minakshi, P., Mohan, H., Manjeet et al. (2020). Organic polymer and metal nano-particle based composites for improvement of the analytical performance of electrochemical biosensor. *Curr. Top. Med. Chem.* 20 (11): 1029–1041.

143 Al-Jawoosh, S., Ireland, A., and Su, B. (2018). Fabrication and characterisation of a novel biomimetic anisotropic ceramic/polymer-infiltrated composite material. *Dent. Mater.* 34 (7): 994–1002.

144 Wu, Y., Xue, Y., Qin, S. et al. (2017). BN nanosheet/polymer films with highly anisotropic thermal conductivity for thermal management applications. *ACS Appl. Mater. Interfaces* 9 (49): 43163–43170.

145 Wang, J., Wu, Y., Xue, Y. et al. (2018). Super-compatible functional boron nitride nanosheets/polymer films with excellent mechanical properties and ultra-high thermal conductivity for thermal management. *J. Mater. Chem. C* 6 (6): 1363–1369.

146 Wang, J., Li, Q., Liu, D. et al. (2018). High temperature thermally conductive nanocomposite textile by “green” electrospinning. *Nanoscale* 10 (35): 16868–16872.

147 Kasar, A., Xiong, G.P., and Menezes, P.L. (2018). Graphene-reinforced metal and polymer matrix composites. *JOM* 70 (6): 829–836.

148 Lei, C.X., Zhang, Y.Z., Liu, D.Y. et al. (2021). Highly thermo-conductive yet electrically insulating material with perpendicularly engineered assembly of boron nitride nanosheets. *Compos. Sci. Technol.* 214: 108995.

149 Hansson, J., Nilsson, T.M.J., Ye, L. et al. (2017). Novel nanostructured thermal interface materials: a review. *Int. Mater. Rev.* 63 (1): 22–45.

150 Ma, H., Gao, B., Wang, M. et al. (2020). Strategies for enhancing thermal conductivity of polymer-based thermal interface materials: a review. *J. Mater. Sci.* 56 (2): 1064–1086.

151 Lei, C., Xie, Z., Wu, K. et al. (2021). Controlled vertically aligned structures in polymer composites: natural inspiration, structural processing, and functional application. *Adv. Mater.* 33 (49): 2103495.

152 Zeng, X.L., Yao, Y.M., Gong, Z.Y. et al. (2015). Ice-templated assembly strategy to construct 3D boron nitride nanosheet networks in polymer composites for thermal conductivity improvement. *Small* 11 (46): 6205–6213.

153 Wang, J., Liu, D., Li, Q. et al. (2019). Lightweight, superelastic yet thermoconductive boron nitride nanocomposite aerogel for thermal energy regulation. *ACS Nano* 13 (7): 7860–7870.

154 Ge, X., Zhang, J.Y., Zhang, G.Q. et al. (2020). Low melting-point alloy-boron nitride nanosheet composites for thermal management. *ACS Appl. Nano Mater.* 3 (4): 3494–3502.

155 Tutika, R., Kmiec, S., Haque, A.B.M.T. et al. (2019). Liquid metal-elastomer soft composites with independently controllable and highly tunable droplet size and volume loading. *ACS Appl. Mater. Interfaces* 11 (19): 17873–17883.

156 Liu, P., Luo, Y.Y., Liu, J.M. et al. (2021). Laminar metal foam: a soft and highly thermally conductive thermal interface material with a reliable joint for semiconductor packaging. *ACS Appl. Mater. Interfaces* 13 (13): 15791–15801.

157 Han, J.K., Du, G.L., Gao, W.W. et al. (2019). An anisotropically high thermal conductive boron nitride/epoxy composite based on nacre-mimetic 3D network. *Adv. Funct. Mater.* 29 (13): 1900412.

158 Wang, Y.J., Xia, S., Li, H. et al. (2019). Unprecedentedly tough, folding-endurance, and multifunctional graphene-based artificial nacre with predesigned 3D nanofiber network as matrix. *Adv. Funct. Mater.* 29 (38): 1903876.

159 Wegst, U.G.K., Bai, H., Saiz, E. et al. (2015). Bioinspired structural materials. *Nat. Mater.* 14 (1): 23–36.

160 Wang, J., Liu, D., Li, Q. et al. (2021). Nacre-bionic nanocomposite membrane for efficient in-plane dissipation heat harvest under high temperature. *J. Mater.* 7 (2): 219–225.

161 Liu, D., Lei, C., Wu, K. et al. (2020). A Multidirectionally thermoconductive phase change material enables high and durable electricity via real-environment solar-thermal-electric conversion. *ACS Nano* 14 (11): 15738–15747.

162 Xu, W.Z., Xing, Y., Liu, J. et al. (2019). Efficient water transport and solar steam generation via radially, hierarchically structured aerogels. *ACS Nano* 13 (7): 7930–7938.

163 Wang, C.H., Chen, X., Wang, B. et al. (2018). Freeze-casting produces a graphene oxide aerogel with a radial and centrosymmetric structure. *ACS Nano* 12 (6): 5816–5825.

164 Bo, Z., Zhu, H.R., Ying, C.Y. et al. (2019). Tree-inspired radially aligned, bimodal graphene frameworks for highly efficient and isotropic thermal transport. *Nanoscale* 11 (44): 21249–21258.

165 Xie, K., Liu, Y.H., Tian, Y.X. et al. (2021). Improving the flexibility of graphene nanosheets films by using aramid nanofiber framework. *Compos. A: Appl. Sci. Manuf.* 142: 106265.

166 Zeng, X.L., Sun, J.J., Yao, Y.M. et al. (2017). A combination of boron nitride nanotubes and cellulose nanofibers for the preparation of a nanocomposite with high thermal conductivity. *ACS Nano* 11 (5): 5167–5178.

167 He, X.W., Gao, W.L., Xie, L.J. et al. (2016). Wafer-scale monodomain films of spontaneously aligned single-walled carbon nanotubes. *Nat. Nanotechnol.* 11 (7): 633–639.

168 Krishnan, V., Kasuya, Y., Ji, Q. et al. (2015). Vortex-aligned fullerene nanowhiskers as a scaffold for orienting cell growth. *ACS Appl. Mater. Interfaces* 7 (28): 15667–15673.

169 Luo, X., Hu, R., Liu, S. et al. (2016). Heat and fluid flow in high-power LED packaging and applications. *Prog. Energy Combust. Sci.* 56: 1–32.

170 Yuan, C., Li, L., Duan, B. et al. (2016). Locally reinforced polymer-based composites for efficient heat dissipation of local heat source. *Int. J. Therm. Sci.* 102: 202–209.

171 Zheng, R., Gao, J., Wang, J. et al. (2011). Reversible temperature regulation of electrical and thermal conductivity using liquid–solid phase transitions. *Nat. Commun.* 2 (1): 289.

172 Wen, D., Lin, G., Vafaei, S. et al. (2009). Review of nanofluids for heat transfer applications. *Particuology* 7 (2): 141–150.

173 Barreto, J.A., O'Malley, W., Kubeil, M. et al. (2011). Nanomaterials: applications in cancer imaging and therapy. *Adv. Mater.* 23 (12): H18–H40.

174 Huang, X., El-Sayed, I.H., Qian, W. et al. (2006). Cancer cell imaging and photothermal therapy in the near-infrared region by using gold nanorods. *J. Am. Chem. Soc.* 128 (6): 2115–2120.

175 Prasher, R., Evans, W., Meakin, P. et al. (2006). Effect of aggregation on thermal conduction in colloidal nanofluids. *Appl. Phys. Lett.* 89 (14): 143119.

176 Evans, W., Prasher, R., Fish, J. et al. (2008). Effect of aggregation and interfacial thermal resistance on thermal conductivity of nanocomposites and colloidal nanofluids. *Int. J. Heat Mass Transf.* 51 (5–6): 1431–1438.

177 Swartz, E.T. and Pohl, R.O. (1989). Thermal boundary resistance. *Rev. Mod. Phys.* 61 (3): 605.

178 Sun, F., Zhang, T., Jobbins, M.M. et al. (2014). Molecular bridge enables anomalous enhancement in thermal transport across hard-soft material interfaces. *Adv. Mater.* 26 (35): 6093–6099.

179 Hu, L., Zhang, L., Hu, M. et al. (2010). Phonon interference at self-assembled monolayer interfaces: molecular dynamics simulations. *Phys. Rev. B* 81 (23): 235427.

180 Hu, M., Keblinski, P., and Schelling, P.K. (2009). Kapitza conductance of silicon–amorphous polyethylene interfaces by molecular dynamics simulations. *Phys. Rev. B* 79 (10): 104305.

181 Prasher, R. (2009). Acoustic mismatch model for thermal contact resistance of van der Waals contacts. *Appl. Phys. Lett.* 94 (4): 041905.

182 Losego, M.D., Grady, M.E., Sottos, N.R. et al. (2012). Effects of chemical bonding on heat transport across interfaces. *Nat. Mater.* 11 (6): 502–506.

183 O'Brien, P.J., Shenogin, S., Liu, J. et al. (2013). Bonding-induced thermal conductance enhancement at inorganic heterointerfaces using nanomolecular monolayers. *Nat. Mater.* 12 (2): 118–122.

184 Shenogina, N., Godawat, R., Keblinski, P. et al. (2009). How wetting and adhesion affect thermal conductance of a range of hydrophobic to hydrophilic aqueous interfaces. *Phys. Rev. Lett.* 102 (15): 156101.

185 Harikrishna, H., Ducker, W.A., and Huxtable, S.T. (2013). The influence of interface bonding on thermal transport through solid–liquid interfaces. *Appl. Phys. Lett.* 102 (25): 251606.

186 Phelan, P.E., Prasher, R.S., and Devpura, A. (2001). Size effects on the thermal conductivity of polymers laden with highly conductive filler particles. *Microscale Thermophys. Eng.* 5 (3): 177–189.

187 Li, L., Zheng, H., Yuan, C. et al. (2016). Study on effective thermal conductivity of silicone/phosphor composite and its size effect by Lattice Boltzmann method. *Heat Mass Transf.* 52: 2813–2821.

188 Ganapathy, D., Singh, K., Phelan, P.E. et al. (2005). An effective unit cell approach to compute the thermal conductivity of composites with cylindrical particles. *J. Heat Transf. Trans. ASME* 127 (6): 553–559.

189 Every, A.G., Tzou, Y., Hasselman, D.P.H. et al. (1992). The effect of particle size on the thermal conductivity of ZnS/diamond composites. *Acta Metall. Mater.* 40 (1): 123–129.

190 Huxtable, S.T., Cahill, D.G., Shenogin, S. et al. (2003). Interfacial heat flow in carbon nanotube suspensions. *Nat. Mater.* 2 (11): 731–734.

191 Ang, T.P., Wee, T.S.A., and Chin, W.S. (2004). Three-dimensional self-assembled monolayer (3D SAM) of n-alkanethiols on copper nanoclusters. *J. Phys. Chem. B* 108 (30): 11001–11010.

192 Parikh, A.N. and Nuzzo, R.G. (1991). Comparison of the structures and wetting properties of self-assembled monolayers of n-alkanethiols on the coinage metal surfaces, Cu, Ag, Au. *J. Am. Chem. Soc.* 113: 7152–7167.

193 Keller, H., Simak, P., Schrepp, W. et al. (1994). Surface chemistry of thiols on copper: an efficient way of producing multilayers. *Thin Solid Films* 244 (1–2): 799–805.

194 Ron, H., Cohen, H., Matlis, S. et al. (1998). Self-assembled monolayers on oxidized metals. 4. Superior n-alkanethiol monolayers on copper. *J. Phys. Chem. B* 102 (49): 9861–9869.

195 Gong, Y.S., Lee, C., and Yang, C.K. (1995). Atomic force microscopy and Raman spectroscopy studies on the oxidation of Cu thin films. *J. Appl. Phys.* 77 (10): 5422–5425.

196 Wong, C.K.Y., Fan, H., and Yuen, M.M.F. (2008). Interfacial adhesion study for SAM induced covalent bonded copper-EMC interface by molecular dynamics simulation. *IEEE Trans. Compon. Packag. Technol.* 31 (2): 297–308.

197 Xu, H., Zhang, X., Liu, D. et al. (2016). Cyclomatrix-type polyphosphazene coating: improving interfacial property of carbon fiber/epoxy composites and preserving fiber tensile strength. *Compos. Part B* 93: 244–251.

198 Zang, J., Wan, Y.J., Zhao, L. et al. (2015). Fracture behaviors of TRGO-filled epoxy nanocomposites with different dispersion/interface levels. *Macromol. Mater. Eng.* 300 (7): 737–749.

199 Liu, K., Zhang, X., Takagi, H. et al. (2014). Effect of chemical treatments on transverse thermal conductivity of unidirectional abaca fiber/epoxy composite. *Compos. A: Appl. Sci. Manuf.* 66: 227–236.

200 Wong, C.K.Y. and Yuen, M.M.F. (2009). Hydrophobic self assembly molecular layer for reliable Cu-epoxy interface. *Proceedings of the 59th Electronic Components and Technology Conference*, 1816–1823, San Diego, CA: IEEE.

7

Packaging-Inside Thermal Management for Quantum Dots-Converted LEDs

Quantum dots (QDs) nanocrystals, as a promising material for absorbing and converting light energy, have attracted numerous scientific and industrial attention, due to their unique optical properties such as narrow and size-tunable emission spectra, high quantum efficiency, and broad absorption spectra [1–3]. Over the past two decades, QDs have been utilized in many applications, such as light-emitting diodes (LEDs), solar cells, and biosensors [4–6]. Especially, QDs are considered promising alternatives for next-generation high-quality lighting and display devices [7, 8]. It has been theoretically and experimentally proved that white light-emitting diodes (WLEDs) composed of blue LED chip, yellow phosphor, and red QDs can achieve both high luminous efficiency (LE) and excellent color quality [9–17]. Figure 7.1(a) shows the schematic fabrication process of typical QDs-WLEDs package. First, the QDs solution is mixed with phosphor gel, and the bubbles in the mixed gel are removed by applying alternating cycles of vacuum. Then, the mixed gel is coated onto the surface of the LED chip, followed by a thermal curing process in vacuum oven. Figure 7.1(b) shows the schematic working principle of QDs-WLEDs. First, the LED chip emits blue light when electrically driven. Then, the blue light is incident into the mixed phosphor/QDs gel. In this process, part of the blue light is absorbed by phosphor and QDs particles to be converted to yellow and red light. Finally, unconverted blue light mixes with the converted yellow and red light and produces white light. At the same time, there exists significant heat generation during the light-emitting process. In LED chip, heat is generated from the energy loss derived from the electroluminescence (EL) process. In a QD, when an electron transits between the discrete energy levels, it emits phonon as radiative process and generates heat as nonradiative process [18]. When QDs' temperature exceeds critical value, significant opto/electronic degradation of QDs has been reported frequently and thus the thermal quenching issue has garnered increasing attention, especially in high-power applications [19–22].

However, the potential application of QDs-WLEDs is severely suppressed by the thermal quenching challenge under high working temperature [23–35]. Cooling these luminescent nanoparticles is a tough target since they are not only self-heating due to the Stokes loss but also embedded in low-thermal-conductivity polymer matrix. In this chapter, we propose several solutions to release the severe thermal situation of QDs in WLEDs package.

7.1 Thermally Conductive QDs Composite

One possible solution to reduce the QDs' working temperature is to incorporate high-thermal-conductivity fillers into silicone gel, thus reinforcing the thermal conductivity of phosphor/QDs gel, and consequently enhancing the heat dissipation from silicone gel to ambient. To achieve this goal, the reinforcing fillers should be highly thermal-conductive and optically transparent without distinct light absorption effect. However, most of the reinforcing fillers such as graphene, carbon nanotube, and metals are light-absorbing materials that will lead to serious optical energy loss [24–26]. For example, Poostforush and Azizi mixed crystallized anodic aluminum oxide with epoxy resin at 39 vol% filler loading and the obtained composite had a through-plane thermal conductivity of $1.13\ \mathrm{W \cdot m^{-1} \cdot K^{-1}}$, while the transparency of the polymer matrix decreased by 23% [36]. Patel et al. utilized alumina powders to enhance the thermal conductivity of polymethylmethacrylate (PMMA) to $0.233\ \mathrm{W \cdot m^{-1} \cdot K^{-1}}$, but the transmittance decreased by 10% at a filling weight fraction of only 0.5% [37]. Therefore, there is a trade-off between transparency and thermal conductivity for these fillers. Even at low filling fractions, these fillers failed to achieve high thermal conductivity and high optical efficiency simultaneously.

Thermal Management for Opto-electronics Packaging and Applications, First Edition. Xiaobing Luo, Run Hu, and Bin Xie.

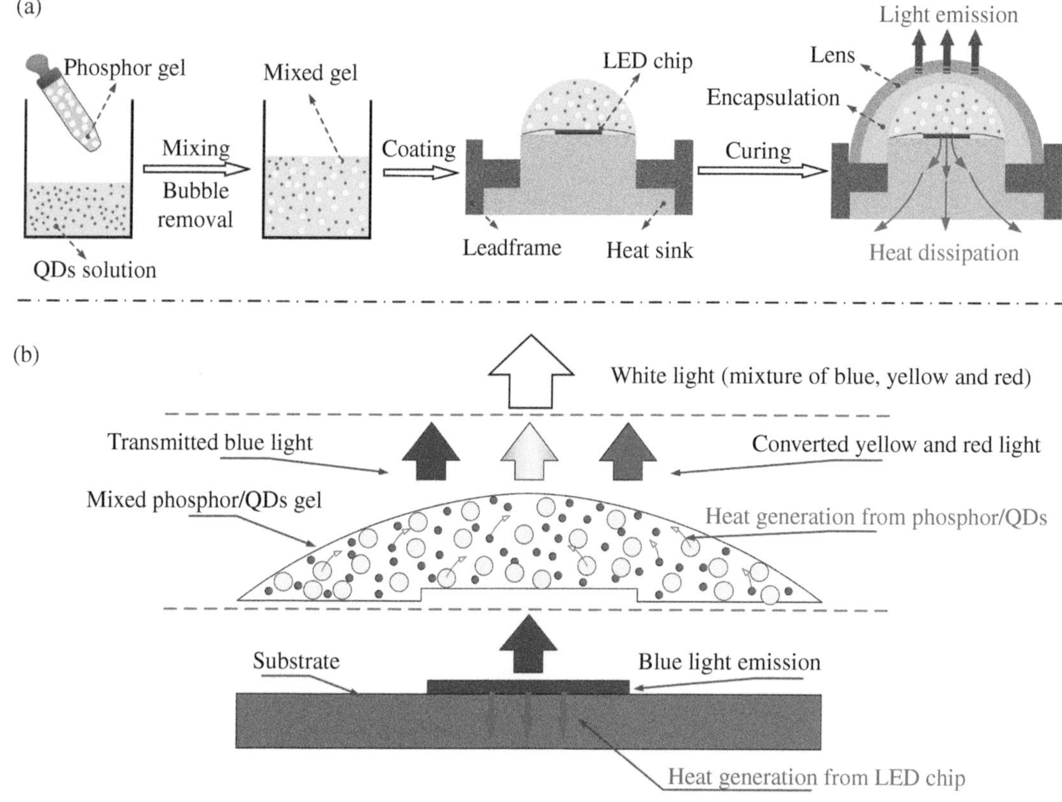

Figure 7.1 (a) Schematic fabrication process of QDs-WLEDs. (b) Schematic of the light emission and heat generation processes of QDs-WLEDs.

In recent years, a new reinforcing filler, hexagonal boron nitride (hBN), has attracted much attention due to its high thermal conductivity (~600 $W \cdot m^{-1} \cdot K^{-1}$) and negligible light absorption effect [38–40]. Thus, the hBN platelets seem promising as suitable filler candidate to reinforce the thermal conductivity of phosphor/QDs gel. Moreover, it is not enough to remove the heat from QDs by simply mixing hBN platelets with phosphor/QDs gel. An effective physical/chemical interaction between QDs and hBN platelets should be built so that the heat generation of QDs can be dissipated more effectively. In this section, we propose to prepare the highly thermal-conductive and luminescent QDs/hBN composites by coating QDs on hBN surface through electrostatic interaction to build a strong heat conduction from QDs to hBN.

Figure 7.2 depicts the schematic fabrication process. hBN platelets with an average diameter of 12 μm were provided by Momentive. First, aqueous hBN solutions were prepared at a concentration of $1\,g \cdot l^{-1}$ by sonication-assisted hydrolysis [41] under a bath sonicator for 8 hours. These hydroxyl-functional hBN platelets present negative surface charge due to the existence of hydroxyl groups. The resulting slurry was centrifuged at 6000 rpm for 5 minutes, and the hBN platelets were collected and dried in a vacuum oven under 80 °C for 12 hours. Highly luminescent red-emissive CdSe/ZnS core/shell QDs were provided by Poly OptoElectronics Co., Ltd., with some customization. It has been demonstrated that adding red QDs with peak wavelength of 626 nm into conventional phosphor-converted WLEDs could improve their color rendering index (CRI) significantly [9, 42]. Therefore, in this work, the QDs' peak wavelength was controlled to be around 626 nm by tuning the molar concentration of core and shell precursors. The QDs were precoated with cationic surface ligand (containing Cd^{2+} cation) before utilization [43]. Then, 1 g of the as-prepared hBN platelets was suspended in 10 ml of chloroform. Under magnetic stirring, 0.5 ml of QDs-chloroform solution (containing 5 mg QDs) was added dropwise. The zeta-potential measurements determined that the mean zeta-potential of QDs solution was positive (26.3 mV), while that of the hydroxyl-functional hBN was negative (−31.6 mV). Thus, the QDs nanoparticles attached to the surface of the platelets through the electrostatic interaction between the positively charged nanoparticles and the negatively charged platelets [44]. The suspension was incubated for 1 hour to coat the platelets with all the QDs nanoparticles. After that, the suspension was maintained at 60 °C until all the solvent evaporated. The final QDs/hBN composites were obtained by a further drying process in vacuum oven at 50 °C.

Figure 7.3(a) and (b) displays photographs of QDs solution, hBN platelets, and QDs/hBN composites under daylight and ultraviolet (UV) light irradiation. It can be seen that under daylight, the hBN platelets show pure white color with less light

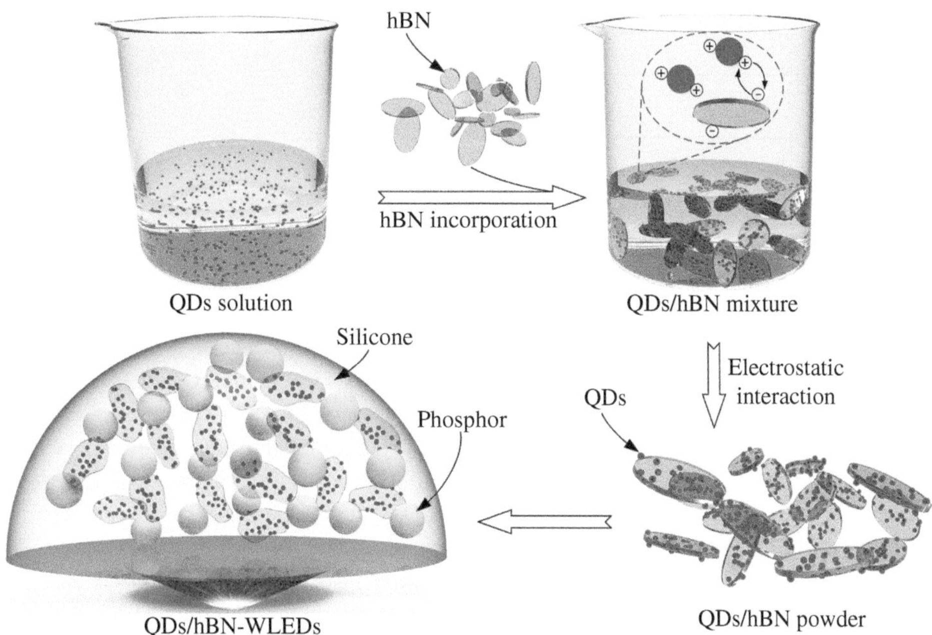

Figure 7.2 Schematic showing the preparation of QDs/hBN composites and the fabrication of QDs/hBN-WLEDs.

Figure 7.3 Photographs of the QDs solution, hBN, and QDs/hBN composites under (a) daylight and (b) UV light irradiation. (c) PL spectra of the QDs solution and QDs/hBN composites. (d) TRPL decay curves of the QDs and QDs/hBN composites.

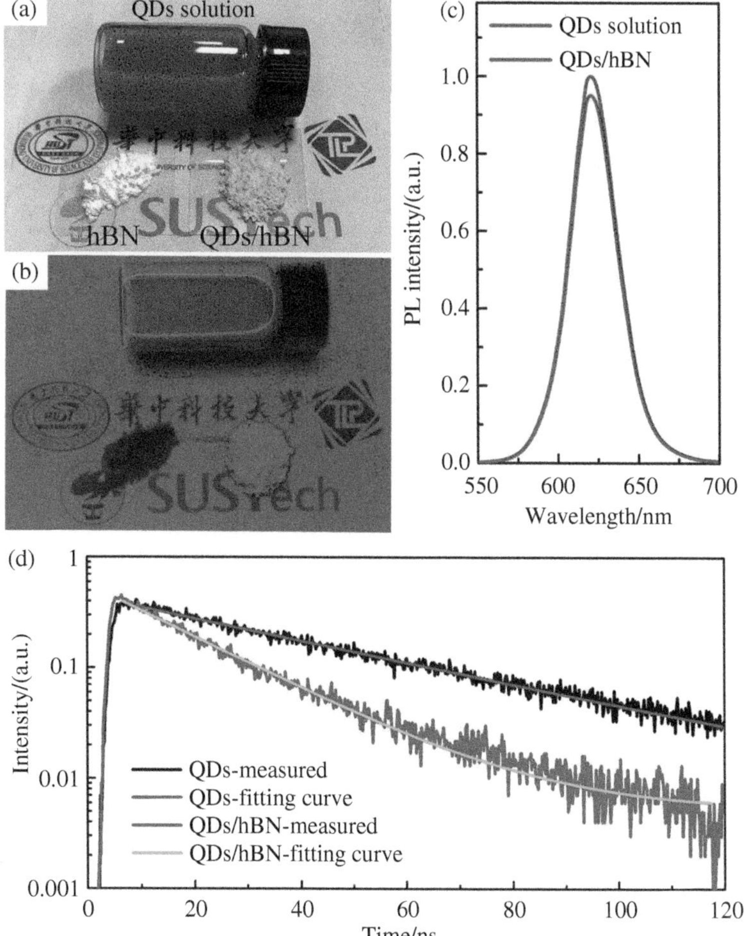

Table 7.1 Values for TRPL characteristics of CdSe/ZnS QDs and QDs/hBN composites.

Sample	τ_1	A_1
CdSe/ZnS QDs	41.9	357.127
QDs/hBN composites	18.1	398.347

absorption, and the QDs/hBN composites are yellowish powder. Under UV light, both the QDs solution and QDs/hBN composites show bright red emission, indicating that the incorporation of hBN platelets did not introduce significant efficiency drop toward QDs. Figure 7.3(c) provides the photoluminescence (PL) spectra of the QDs solution and QDs/hBN platelets. The peak wavelength and full-width-at-half-maximum (FWHM) of the QDs solution were measured as 625 and 34.5 nm, respectively. After incorporation with hBN platelets, QDs' peak wavelength and FWHM changed to 626 and 35 nm, respectively, which mean that the incorporation process has less effect on QDs spectral properties. The slight red shift of peak wavelength is attributed to the quantum states' overlapping effect among closed QDs, which is because the physical separations among QDs become smaller during the incorporation process [10]. The PL quantum yield (PL QY) decreased from 82% to 74% after incorporating QDs onto the hBN surface. To further investigate the underlying origins of QY decrease, the time-resolved PL (TRPL) spectra of both CdSe/ZnS QDs and QDs/hBN samples were measured in air, and the decay curves are depicted in Figure 7.3(d). These decay curves can be well fitted by an exponential function $I(t) = k_{\text{Fit}}e^{-t/\tau_1}$, where $I(t)$ is the PL initial intensity at time t. The fitting parameters k_{Fit} and PL lifetime are listed in Table 7.1. The PL lifetime of CdSe/ZnS QDs is calculated as 41.9 ns and that of QDs/hBN composites is 18.1 ns. The reduction of PL QY can be explained by the energy transfer process taking place from QDs donors to hBN acceptors or adjacent QDs acceptors because energy transfer process is an additional nonradiative de-excitation path.

Figure 7.4(a) shows the high-resolution transmission electron microscope (HRTEM) images of CdSe/ZnS core/shell QDs in different resolutions, respectively. The measured nanoparticles demonstrate uniform size distribution and clear lattice fringes that suggest monodispersed QDs with good crystallinity. The average size of QDs was measured as 6.8 nm. Figure 7.4(b) and (c) displays the scanning electron microscope (SEM) images of hBN platelets and QDs/hBN composites, respectively. The SEM characterization reveals that the diameter of hBN platelets ranges from 6 to 20 μm, and the size distribution of platelets changes very little during the QDs incorporation process. Further, the energy dispersive spectroscopy (EDS) analysis of QDs/hBN composites was conducted. The results reveal the presence of 32.1 at% boron, 36.2 at% nitrogen, 0.14 at% cadmium, and 0.21 at% zinc. Moreover, the elemental mapping by EDS shows a uniform distribution of QDs, and no large agglomerate of nanoparticles is observed in Figure 7.4(d). To further observe the distribution of QDs on hBN platelets, scanning TEM (STEM) image of QDs/hBN composites was measured and displayed in Figure 7.4(e). The clear lattice fringes of both hBN and QDs demonstrate the desired QDs-on-hBN structure. Figure 7.4(f) shows the high-angle-annular-dark-field STEM (HAADF-STEM) image of QDs/hBN composites, from which we can directly observe the uniform distribution of QDs on hBN platelets without agglomeration. Figure 7.4(g) shows the EDS mapping images of boron and cadmium collected from the same location of Figure 7.4(f).

To investigate the effectiveness of QDs-on-hBN bonding structure, 0.5 g of QDs/hBN composites were redispersed in 3 ml of chloroform and allowed to stand for 1 hour. Then, the supernatant chloroform was collected separately from the QDs/hBN composites. As shown in Figure 7.5, under UV light irradiation, the QDs/hBN composites show bright red emission while the supernatant chloroform is colorless, indicating that the effective bonding of QDs and hBN platelets was built.

To study the optical properties of QDs/hBN composites, both QDs- and QDs/hBN-WLEDs samples were fabricated, and their correlated color temperature (CCT) were controlled to around 4300 K (neutral white color) by fine-tuning the gel volume. In the fabrication of QDs/hBN-WLEDs, blue InGaN LED chip with peak wavelength of 455 nm and yellow-greenish YAG:Ce phosphor with peak wavelength of 538 nm were used. For the preparation of QDs/hBN-WLEDs, 0.1 g of phosphor, 0.05 g of QDs/hBN composites, and 1 g of silicone gel (Dow corning OE 6550) were mixed and stirred for 20 minutes to be dispersed uniformly. Bubbles introduced during the stirring process were removed by applying alternating cycles of vacuum. Then, the mixed gel was coated onto the surface of the LED chip, followed by a curing process in vacuum oven at 150 °C for 1 hour. For comparative purposes, the QDs-WLEDs without hBN platelets were also prepared. First, 70 μl of QDs-chloroform solution and 0.12 g of phosphor were mixed with 1 g of silicone. Then, the mixture was heated at 60 °C to remove the chloroform completely. Finally, the mixture was coated onto the LED chip and cured. It is noted that these

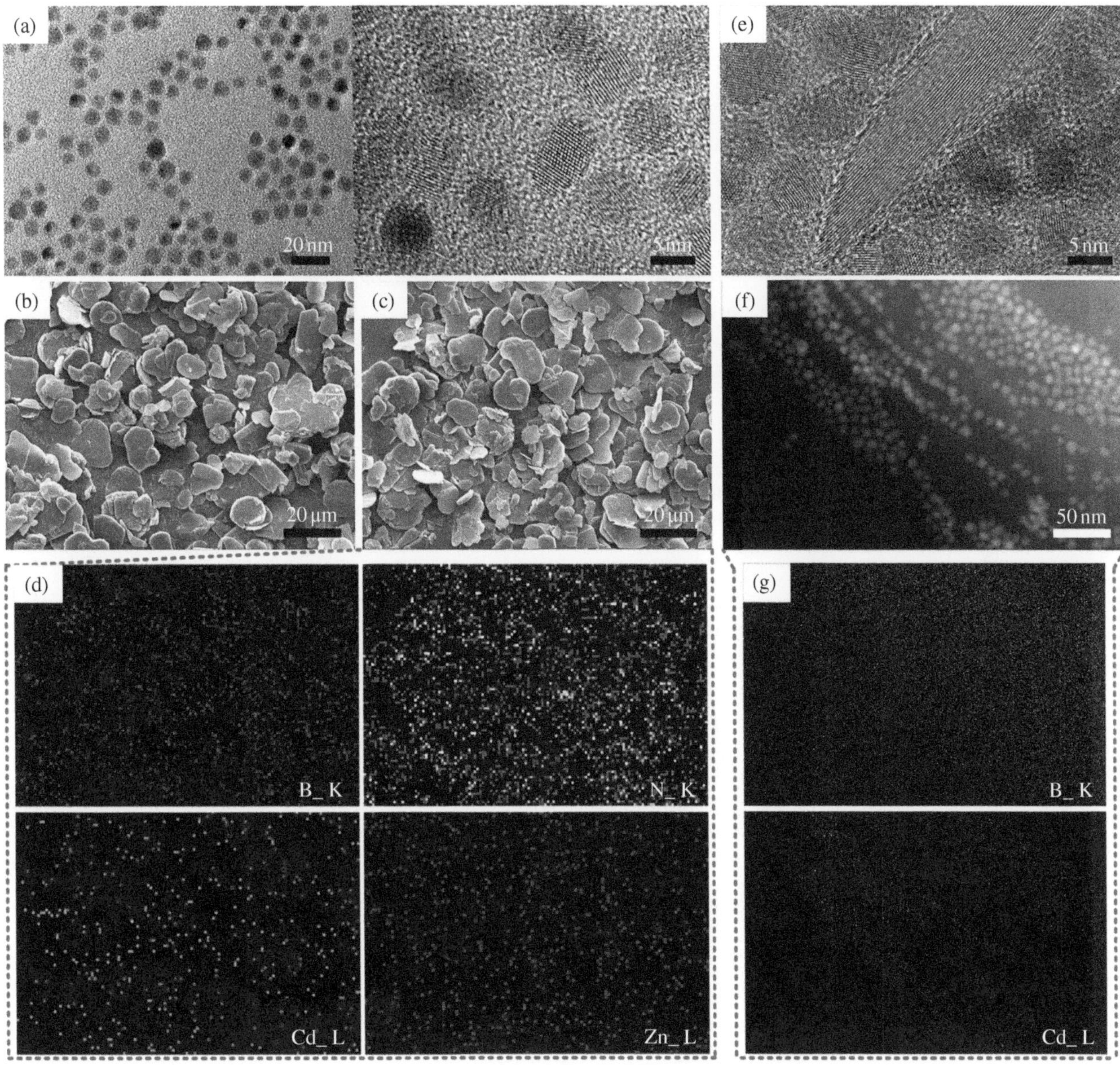

Figure 7.4 (a) HRTEM images of the QDs solution. (b) SEM image of the hBN platelets. (c) SEM image of the QDs/hBN composites. (d) Elemental mapping images of the QDs/hBN composites collected from the same location as (c). (e) STEM image of the QDs/hBN composites. (f) HAADF-STEM image of the QDs/hBN composites. (g) Elemental mapping images of the QDs/hBN composites collected from the same location as (f).

WLEDs were fabricated to have similar spectral power distribution (SPD), so that their thermal performances can be evaluated under the same optical conditions. The existence of hBN changes the light transmission/scattering situation. Therefore, the QDs' weight ratio for QDs/hBN-WLEDs and QDs-WLEDs to achieve the same SPD is different.

Figure 7.6(a) provides the spectra and optical properties of these WLEDs samples under driving current of 20 mA. It is seen that under similar spectral distribution, both WLEDs demonstrate high CRI of Ra > 93 and R9 > 90, which is the rendering index for deep red color. Meanwhile, the QDs-WLEDs present a high LE of 111.7 lm/W, and the LE of QDs/hBN-WLEDs is 108.5 lm/W. Figure 7.6(b) shows the CRI and LE variation with increasing driving currents from 20 to 300 mA. The results suggest that the CRIs of both WLEDs remain stable against increasing driving current, indicating their stable spectral distribution under different lighting conditions. The LEs of both WLEDs show a similar decreasing trend under increasing driving current. This is mainly attributed to the efficiency drop of LED chip itself. Therefore, the utilization

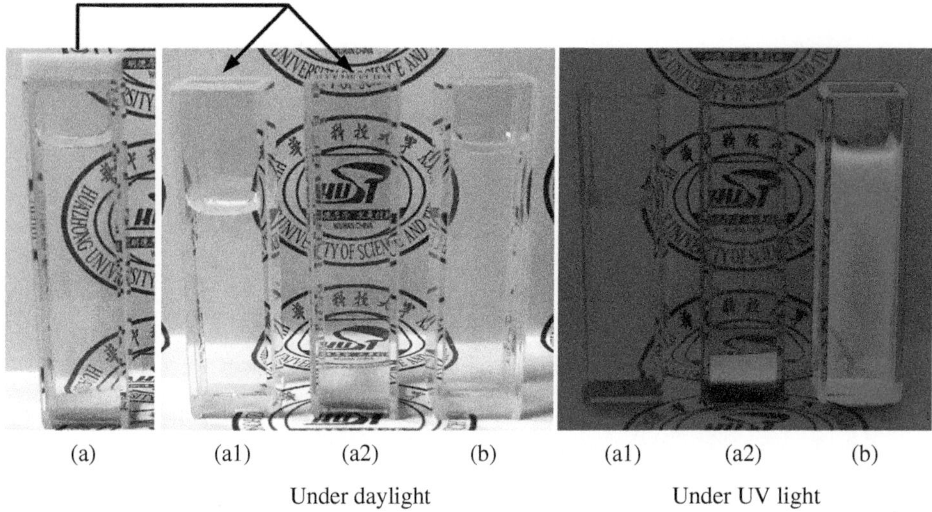

Figure 7.5 (a) QDs/hBN composites redispersed in chloroform and made to stand for 1 hour. (a1) The supernatant chloroform extracted from (a). (a2) The precipitation in (a). (b) QDs dispersed in chloroform.

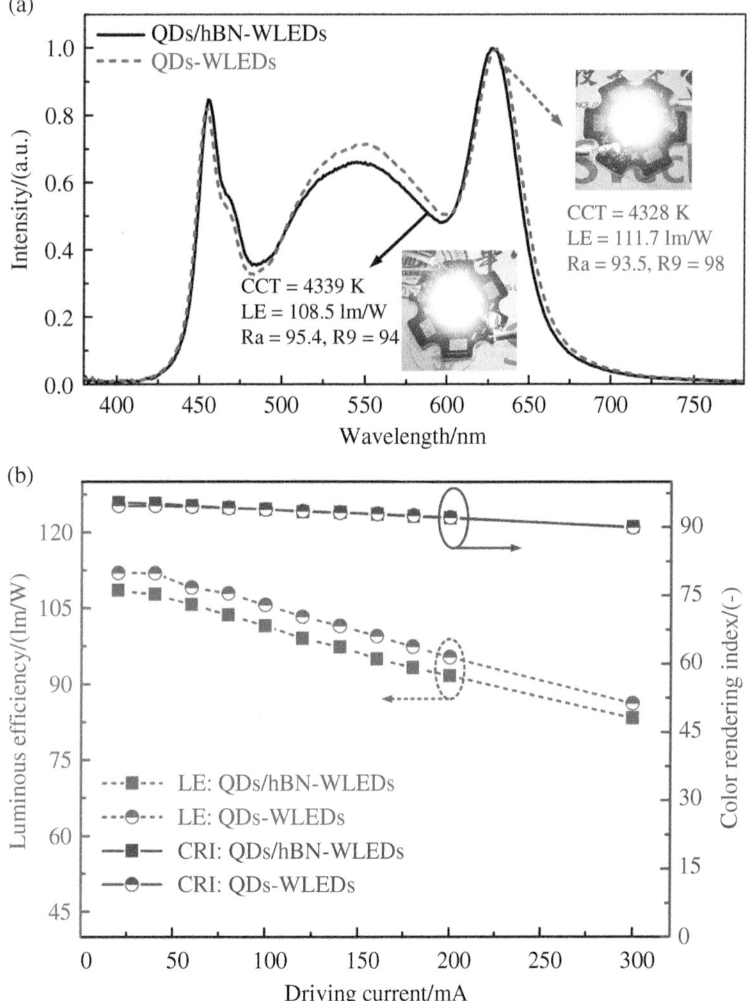

Figure 7.6 (a) Spectral distribution and optical properties of the QDs- and QDs/hBN-WLEDs under driving current of 20 mA. (b) LE and CRI of the QDs- and QDs/hBN-WLEDs under different driving currents from 20 to 300 mA.

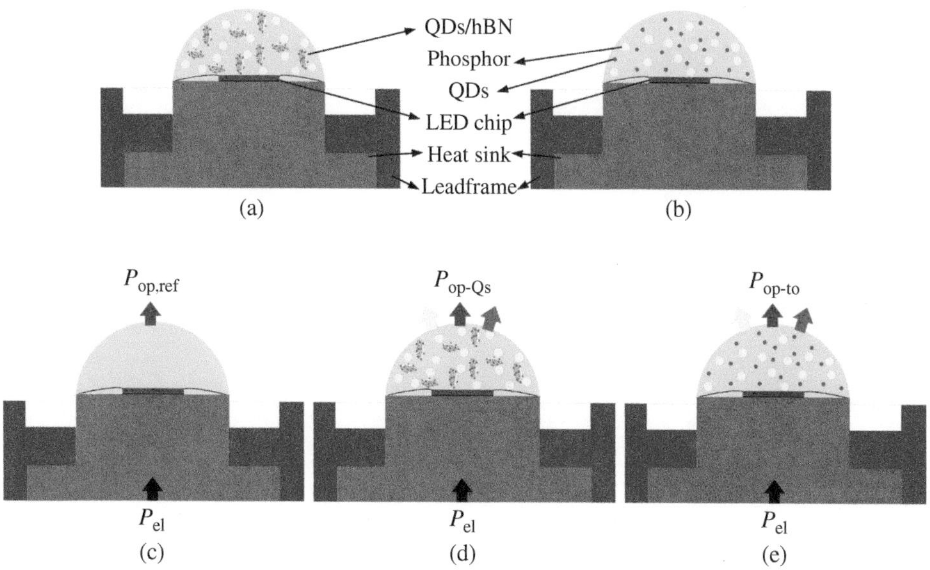

Figure 7.7 Structure schematics of (a) QDs/hBN-WLEDs and (b) QDs-WLEDs. (c)–(e) The heat power measuring process of LED chip, QDs/hBN-phosphor layer, and QDs-phosphor layer, respectively.

of QDs/hBN does not cause evident LE drop of WLEDs. Moreover, the QDs/hBN-WLEDs present the same optical performance and current stability as the QDs-WLEDs.

To further investigate the thermal differences between the QDs/hBN-WLEDs and QDs-WLEDs, the steady-state temperature fields of QDs/hBN- and QDs-WLEDs were examined by finite element thermal simulation. First, the heat generated by the LED chip, QDs/hBN-phosphor silicone layer, and QDs-phosphor silicone layer was measured by an integrating sphere system (ATA-1000, EVERFINE Inc.). The heat generation was calculated according to the optical energy loss within the corresponding layer. Figure 7.7 illustrates the measuring process and the structure schematics of QDs/hBN-WLEDs (Figure 7.7(a)) and QDs-WLEDs (Figure 7.7(b)). Briefly, the energy loss of LED chip is calculated by the difference in input electrical power and output optical power of LED module with only silicone gel (Figure 7.7(c)); the energy loss of QDs/hBN-phosphor silicone layer is calculated by the difference in output optical power of LED module with only silicone gel and output optical power of LED module with QDs/hBN-phosphor gel (Figure 7.7(d)); the energy loss of QDs-phosphor silicone layer is calculated by the difference in output optical power of LED module with only silicone gel and output optical power of LED module with QDs-phosphor gel (Figure 7.7(e)). The heat power of each layer is calculated as follows [25]:

$$P_{\text{heat-chip}} = P_{\text{el}} - P_{\text{op,ref}} \tag{7.1}$$

$$P_{\text{QDs/hBN-phosphor}} = P_{\text{op,ref}} - P_{\text{op-Qs}} \tag{7.2}$$

$$P_{\text{QDs-phosphor}} = P_{\text{op,ref}} - P_{\text{op-to}} \tag{7.3}$$

where P_{el} is the input electrical power of WLEDs package, $P_{\text{op-Qs}}$ is the optical power from QDs/hBN-WLEDs, and $P_{\text{op-to}}$ is the optical power from QDs-WLEDs.

Figure 7.8 illustrates the measured heat power of the LED chip, QDs/hBN-phosphor layer, and QDs-phosphor layer. The results reveal that the heat power of the LED chip is much larger than that of others, especially under large driving current. This is attributed to different kinds of nonradiative recombination and other causes of photon annihilation occurring within the LED chip [45]. The heat generation in QDs/hBN-phosphor is close to that in QDs-phosphor, and their heat power increases with the increase in driving current. For instance, under 300 mA illumination, the heat powers of QDs/hBN-phosphor layer and QDs-phosphor layer are 228.6 and 218.3 mW, respectively, while the heat power of the LED chip is nearly twice that of QDs/hBN-phosphor layer.

Then, the thermal conductivity κ of QDs/hBN-phosphor gel and QDs-phosphor gel was measured. The thermal conductivities of phosphor-QDs gel and phosphor-QDs/hBN gel were calculated by $\kappa = \alpha \cdot C_p \cdot \rho$, where α, C_p, and ρ are the thermal diffusivity, heat capacity, and density of the gel, respectively. α was measured by laser flash method using a Netzsch LFA 457. The geometry of the tested sample is a cylinder with a diameter of 10 mm and thickness of around 1 mm. Before the test,

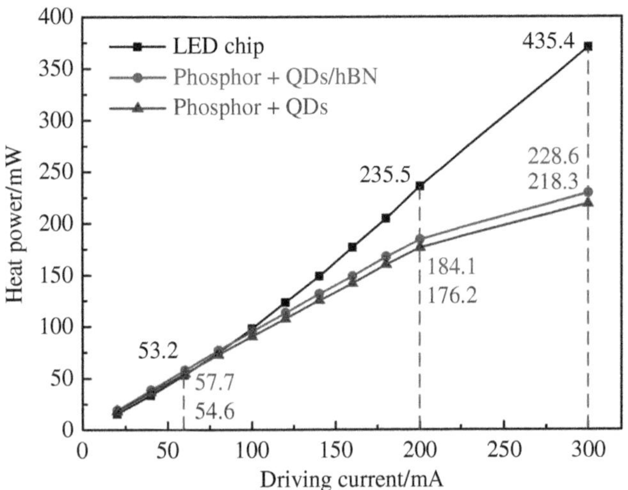

Figure 7.8 Measured heat generation in LED chip, QDs/hBN-phosphor layer, and QDs-phosphor layer under different driving currents.

a thin graphite film is applied on the composite surfaces to increase the energy absorption and emittance of the surfaces. During the test, heat propagates from the bottom to the top surface of the sample. C_p was determined by Perkin Elmer Diamond differential scanning calorimetry, and ρ was calculated from the weight fractions of the composite.

The α, C_p, and ρ of QDs-phosphor gel at 25 °C were measured as 0.098 $mm^2 \cdot s^{-1}$, 1.504 $J \cdot g^{-1} \cdot K^{-1}$, and 1.22 $g \cdot cm^{-3}$, respectively, and those of QDs/hBN-phosphor gel were measured as 0.136 $mm^2 \cdot s^{-1}$, 1.568 $J \cdot g^{-1} \cdot K^{-1}$, and 1.26 $g \cdot cm^{-3}$, respectively. Consequently, the κ of QDs-phosphor gel and QDs/hBN-phosphor gel were calculated as 0.18 and 0.27 $W \cdot m^{-1} \cdot K^{-1}$, respectively. The higher κ of QDs/hBN-phosphor gel is because of the formation of efficient thermal pathways made by the hBN platelets. In this case, the thermal conductivity was enhanced by 50% with a weight fraction of 4.3% (volume fraction of about 2%) of hBN platelets. The temperature-dependent κ data from 25 to 125 °C were also measured and displayed in Figure 7.9. It was found that the κ of both gels decreased slightly with increasing testing temperature. For instance, the κ of QDs-phosphor gel and QDs/hBN-phosphor gel were calculated as 0.173 and 0.25 $W \cdot m^{-1} \cdot K^{-1}$, respectively, at 125 °C. Despite this slight drop, the thermal conductivity was still enhanced effectively by 45%. Additionally, we also measured the thermal conductivity of QDs/hBN-phosphor gel with a higher weight fraction of 8.6% hBN platelets, and the κ was enhanced significantly by 88.9% (from 0.18 to 0.34 $W \cdot m^{-1} \cdot K^{-1}$), while the LE decreased from 111.7 to 97 lm/W due to the overload of hBN platelets, which interrupt the light output from QDs/hBN-phosphor gel. Therefore, it is validated that the utilization of QDs/hBN is an effective way to reinforce the thermal conductivity of phosphor gel, and the percentage of hBN platelets should be optimized to balance the heat dissipation and light output.

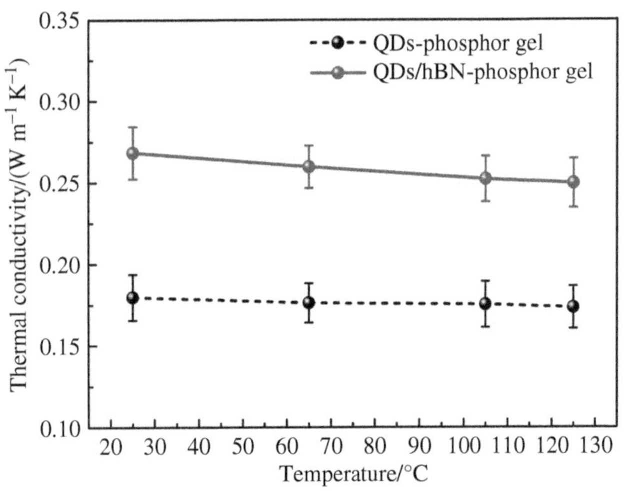

Figure 7.9 Temperature-dependent thermal conductivity of QDs-phosphor gel and QDs/hBN-phosphor gel.

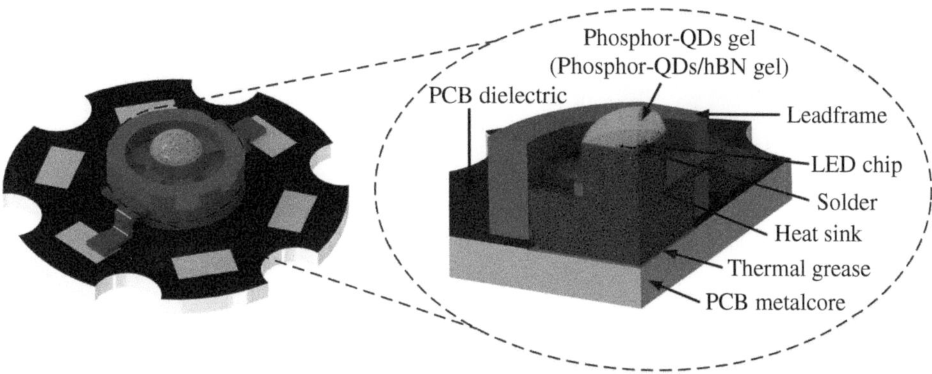

Figure 7.10 Physical model of the as-fabricated QDs/hBN- and QDs-WLEDs.

Table 7.2 Thickness and thermal conductivity used for thermal simulation.

Component	Thickness (mm)	κ ($W \cdot m^{-1} \cdot K^{-1}$)
PCB metal-core	0.98	170
Thermal grease	0.05	5
Heat sink	6	170
Solder	0.05	5
LED chip	0.1	65.6
Lead frame	6	0.36
QDs-phosphor gel	1.5	0.18
QDs/hBN-phosphor gel	1.5	0.27
PCB dielectric	0.02	0.2

Before conducting the thermal simulations, the corresponding physical models of QDs/hBN- and QDs-WLEDs were built, as shown in Figure 7.10. The WLEDs were mounted onto a metal-core printed circuit board (MCPCB) by thermal grease for electrical connection and heat dissipation. In the thermal simulation, only a quarter of the WLEDs model was simulated due to its symmetry. The boundary conditions of the model were set as follows: the ambient temperature was fixed at 25 °C; natural convection occurred at the bottom surface of the printed circuit board (PCB) with a heat transfer coefficient of $10\,W \cdot m^{-2} \cdot K^{-1}$, and other surfaces are cooled by natural convection with a heat transfer coefficient of $8\,W \cdot m^{-2} \cdot K^{-1}$. All the boundary conditions are similar to those in [44]. The thickness and thermal conductivity of each component used for thermal simulation are listed in Table 7.2.

Figure 7.11 gives the simulated steady-state temperature fields of the two WLEDs under driving currents of 60, 200, and 300 mA. It is seen that the highest temperature of all these WLEDs is located in the top surface of silicone gel. This is mainly attributed to the relatively low thermal conductivity of silicone gel that is incapable of dissipating the heat quickly [46, 47]. Under the same driving current, the highest temperature in QDs/hBN-WLEDs is lower than that in QDs-WLEDs, and the temperature reduction is more apparent at larger driving current. For instance, the temperature reduction is 5, 16.3, and 20.1 °C at 60, 200, and 300 mA, respectively. The simulation results clearly show the effectiveness of QDs/hBN on the reinforcement of heat dissipation and temperature reduction.

To further validate our simulation results, the surface temperature distribution of the two WLEDs was measured. The QDs/hBN-WLEDs and QDs-WLEDs were connected by a series circuit so that their driving currents are the same, as displayed in Figure 7.12(a). The surface emissivities of QDs/hBN-phosphor gel and QDs-phosphor gel are set as 0.96 [48]. The distance between the camera lens and WLEDs is 0.3 m. Figure 7.12(b)–(d) provides the measured steady-state temperature fields at 60, 200, and 300 mA, respectively. It is seen that the simulated temperature distributions between these two WLEDs are in good accordance with the measured results, and the maximum relative deviation of the highest temperature between

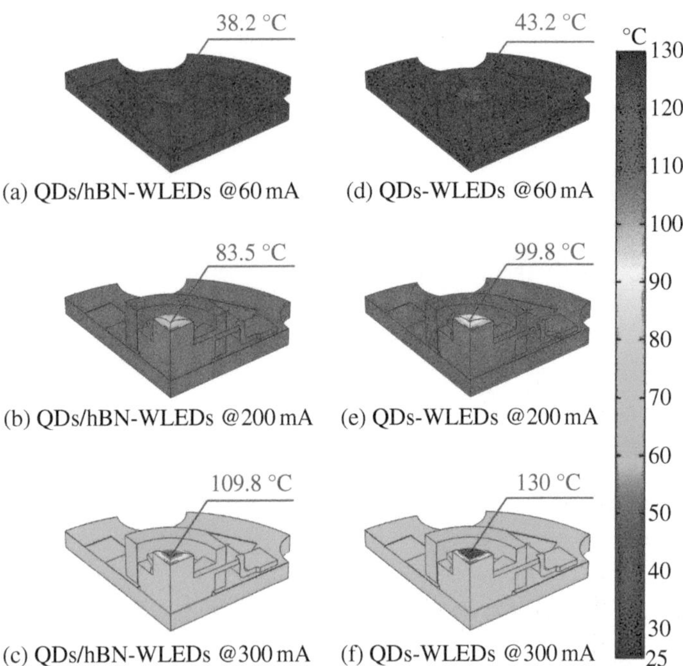

Figure 7.11 Simulated steady-state temperature fields of the two WLEDs under driving currents of 60, 200, and 300 mA.

experimental and simulated results is 12.8% (QDs-WLEDs at 60 mA), which is within the acceptable level. The measured results demonstrate that 22.7 °C of temperature reduction could be reached at 300 mA by the use of QDs/hBN composites. Thus, both simulated and experimental results show that the use of QDs/hBN can significantly enhance the heat dissipation and temperature reduction of WLEDs. As a result, the thermal quenching of QDs should also be reduced by the use of QDs/hBN. Therefore, QDs-LEDs consisting of blue LED chip and QDs-silicone gel, and QDs/hBN-LEDs consisting of blue LED chip and QDs/hBN silicone gel were fabricated for long-term aging test under a driving current of 200 mA. Their color coordinates were controlled at around (0.42, 0.17) for comparison. The QDs' working temperatures of QDs-LEDs and QDs/hBN-LEDs were measured as 125.8 and 102.5 °C (without MCPCB), respectively. The PL intensity of QDs was monitored every 24 hours. Figure 7.13 shows the decay curves of these samples. It is seen that after 144 hours of aging, 44% of the original PL intensity is still preserved for the QDs/hBN-LEDs, while the QDs-LEDs only retain 18% of the original PL intensity. From the illumination pattern, we can clearly observe the degradation of QDs-LEDs, which lost most of the red emission at the end of the aging process, and its Commission Internationale de L'Eclairage (CIE) color coordinates shifted to (0.2021, 0.0599), while the color coordinates of QDs/hBN-LEDs only shifted to (0.2837, 0.1012). Therefore, thermal-induced quenching of QDs can also be released by the QDs/hBN composites.

7.2 Heat Transfer Reinforcement Structures

7.2.1 Directional Heat Conducting QDs-Polymer

In Section 7.1, we designed thermally conductive composites by coating QDs onto hBN platelets, which reduced the working temperature of QDs from 127.2 to 104.5 °C. However, the greatest advantage of hBN platelets, also known as hBN sheets (hBNS), in enhancing the thermal conductivity of the luminescent layer has not been taken into account. The hBNS have a high aspect ratio and highly anisotropic thermal property ($20\ \mathrm{W \cdot m^{-1} \cdot K^{-1}}$ through plane, $585\ \mathrm{W \cdot m^{-1} \cdot K^{-1}}$ in plane for the monoisotopic ^{10}B hBNS) [49, 50]. Composite materials with anisotropic thermal conductivity have been produced by directionally arranging hBNS in polymers through various ways, such as magnetic field driving and ice templates shaping [38, 51–55]. However, the application of such materials in WLEDs is a big challenge and has not been researched, because the electoral conductivities and dark colors of the indispensable auxiliary materials such as ferroferric oxide particles and graphene oxide (GO) sheets are unacceptable in the luminescent layer of WLEDs [38, 52].

In this section, the ice templates method is modified by an acceptable material, sodium carboxymethylcellulose (SCMC), an industrial binder with white color and insulation properties, to substitute the traditional GO to connect hBNS, and thus the vertical thermal-conductive QDs-WLEDs (Ver-WLEDs) were manufactured by directionally arranging hBNS in the QDs-phosphor layer. Due to the vertical thermal conduction channels made by hBNS/SCMC, heat generated from QDs and phosphors can quickly diffusive to the heat sink, avoiding the severe heat aggregation and high working temperature in the luminous layer. The corresponding methodology, experiments, and discussion in detail are provided next.

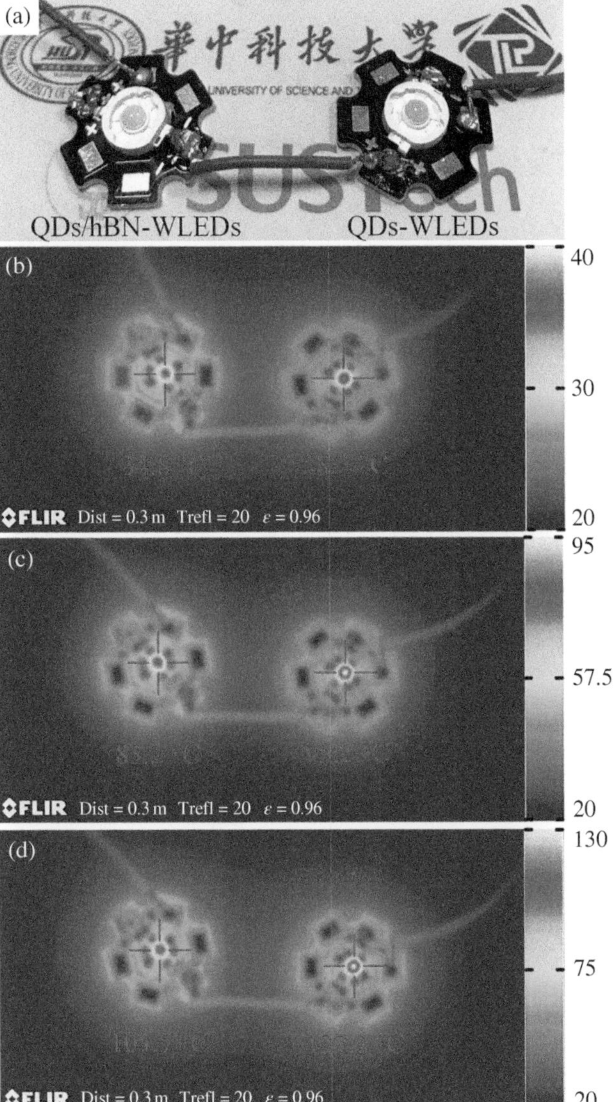

Figure 7.12 (a) Photograph of the QDs/hBN-WLEDs and QDs-WLEDs in a series circuit. (b)–(d) The measured temperature fields at 60, 200, and 300 mA, respectively.

Wong et al. used the growth of ice templates to shape vertical hBNS/GO templates in which GO plays an essential role in linking hBNS [52]. Figure 7.14 shows the schematic of ice template assembling progress of composite filled with vertically arranging hBNS, including the three main steps. The first step is to freeze hBNS/SCMC solution at a vertical temperature gradient, with liquid nitrogen as a cold source located at the bottom. A few minutes are needed for the solution to freeze to ice, and ice templates grow from the bottom to top surfaces as well as hBNS/SCMC templates. During the process of vertical crystal growth [56], the ice forms many separated vertical ice templates, while the SCMC is squeezed into the space between the ice templates, thus forming the vertical SCMC templates. Due to the unique lamellar structure, during the crystallization process of the vertical ice templates, hBNS will be deflected under the action of torque. Until the surface is parallel to the vertical ice templates, the torque will disappear, balance will be restored, and finally hBNS will be vertically arranged in the SCMC templates. In the second freeze-drying step, the ice with vertical hBNS/SCMC templates is set in vacuum freeze dryer for 12 hours before the ice templates are sublimated away and the hBNS/SCMC skeleton is left. The final polymer infiltration step is to fill the vacancy in the hBNS/SCMC skeleton with luminescent composite polymer by vacuuming for 2 hours.

The traditional QDs-WLEDs include blue-emitting chips, yellow-emitting phosphors (YAG:Ce), and red-emitting QDs (CdSe/CdS). It is noted that phosphors and QDs are added in various steps. The intervals between the mentioned hBNS/SCMC templates are smaller than the average diameter of yellow-emitting phosphors and larger than that of red-emitting QDs. This means QDs can permeate into the skeleton like the polymer but phosphors can't. For this reason, in the preparation process shown in Figure 7.15(a), phosphors are added in the freeze-drying solution and then evenly distributed through the skeleton. QDs are selected to add in the silicone polymer for uniformly filling the skeleton because the hydrophobic QDs will aggregate in the freeze-drying solution. Applying above manufacture craft, a novel anisotropic thermal-conductive luminescent composite with vertically arranging hBNS can be prepared.

Based on the preparation flow in Figure 7.15(a), the thermal conduction mechanism schematic of the anisotropic thermal-conductive luminescent composite is shown in Figure 7.15(b), including the physical structure, heat source distribution, and heat conduction channels. The luminescent composite is made up of two kinds of alternate vertical arranging templates: ① the silicone templates embedded by QDs and ② the hBNS/SCMC templates inlaid by phosphors. The silicone and hBNS/SCMC templates feature low and high thermal conductivities, respectively. These two kinds of alternative vertical templates

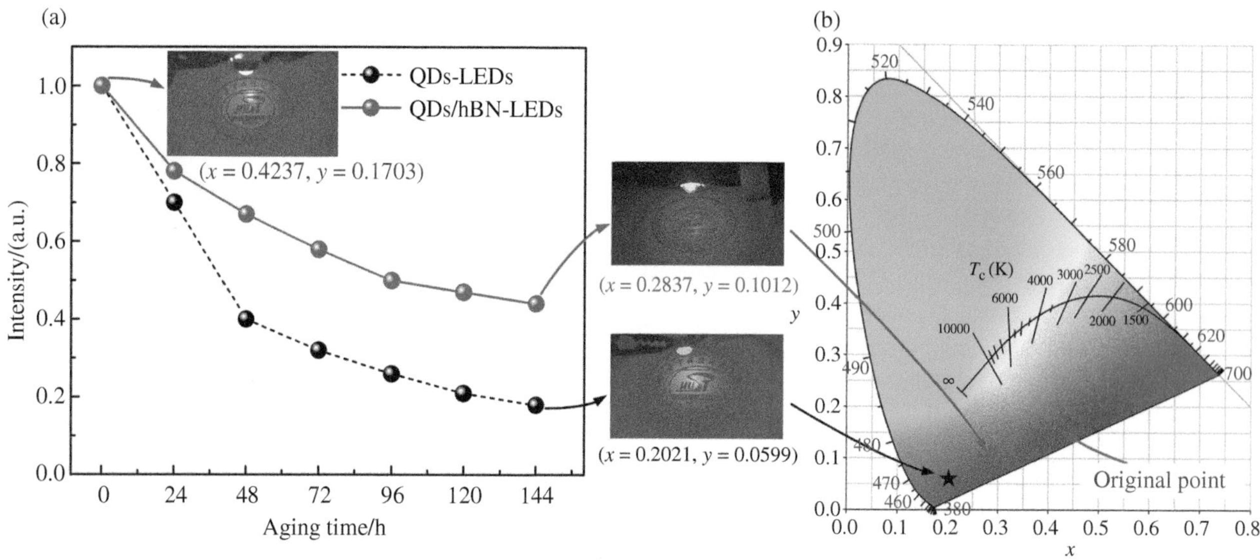

Figure 7.13 (a) Normalized QDs' intensity decay curves of QDs-LEDs and QDs/hBN-LEDs under driving current of 200 mA, ambient temperature. (b) CIE color coordinates of these LEDs before and after 144 hours of aging. Insets: corresponding illumination patterns under different aging time.

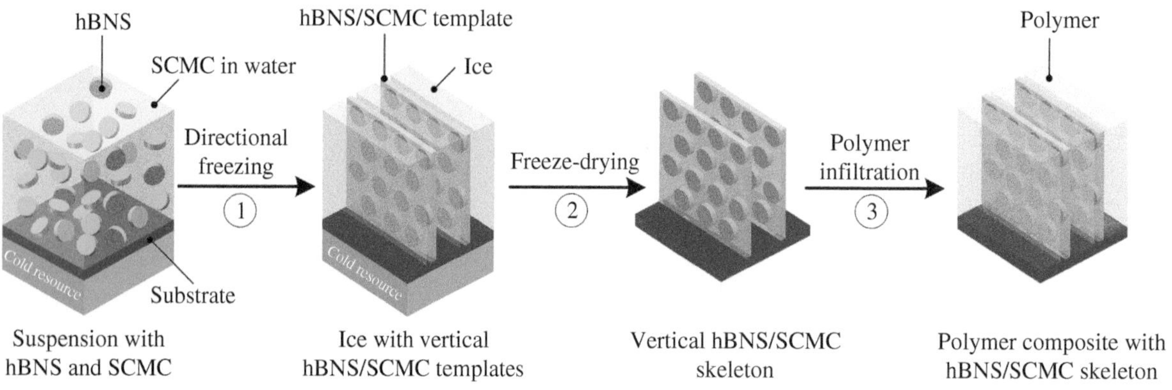

Figure 7.14 Schematic of the ice template assembly method to generate composite with vertically arranging hBNS.

endow the luminous composite with a high vertical thermal conductivity and a low horizontal thermal conductivity. In addition, the main heat sources are QDs, phosphors, and hBNS/SCMC. The high-thermal-conductive vertical hBNS/SCMC templates are not only the main heat-diffusive channels of phosphors but also beneficial to cool the adjacent QDs/silicone templates.

Based on the manufacturing flow in Figure 7.15(a), the luminous layer with vertical hBNS was coated on the surface of the chips of LED. A Ver-WLEDs was packaged and shown in Figure 7.16(a). The hBNS with 12 μm of average diameter and SCMC were purchased from Momentive Company and Aladdin Industrials Corporation, respectively. The luminescent layer was separated from the device and displayed in various directions in Figure 7.16(b)–(e), revealing the uniform distribution of phosphors/QDs and the full filling of silicone. For comparison, the isotropic thermal-conductive QDs-WLEDs with isotropic arranging hBNS (Iso-WLEDs) and the common QDs-WLEDs without hBNS (Com-WLEDs) were also prepared. To make a fair comparison of LE, we adjusted the concentrations of phosphors and QDs in the abovementioned three kinds of QDs-WLEDs with similar relative spectral distribution, as shown in Table 7.3.

The morphologies of the SCMC, hBNS/SCMC, and hBNS/SCMC/phosphor skeleton were captured by SEM. The thermal diffusion coefficient of the luminescent film, α, was measured by a laser flash method (Netzsch LFA 457). The thermal conductivity κ is calculated by the formula, $\kappa = \alpha \cdot C_p \cdot \rho$, where ρ and C_p are the density and specific heat of the film,

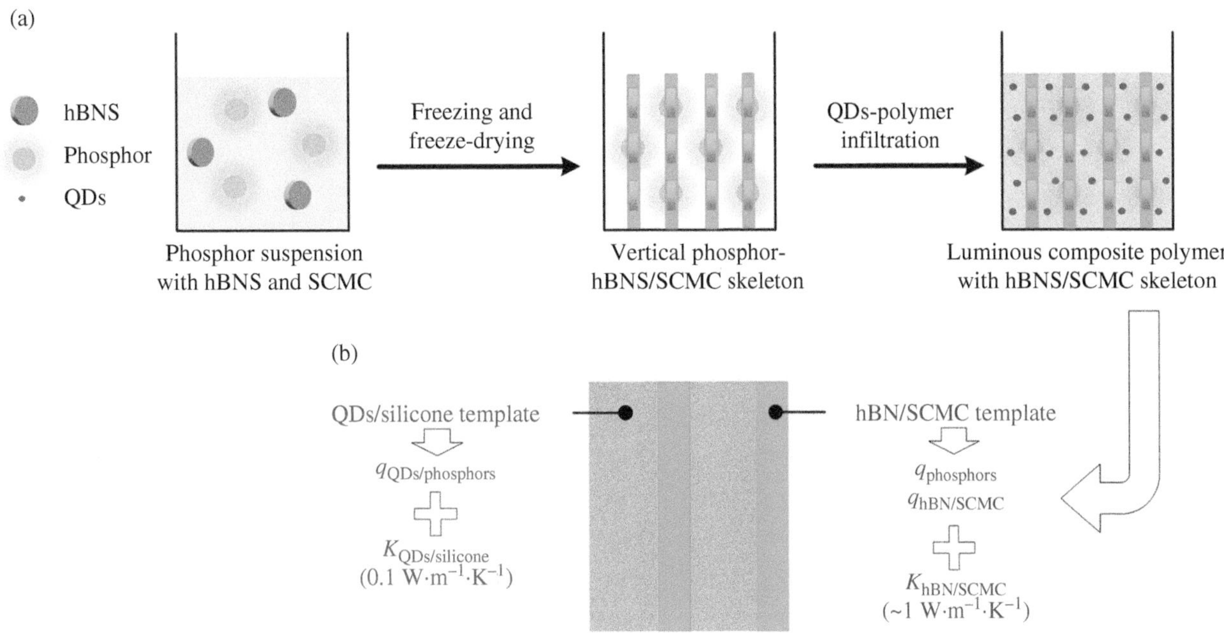

Figure 7.15 Schematics of (a) the preparation flow and (b) the thermal conduction mechanism of luminescent composite polymer with vertical hBNS/SCMC templates.

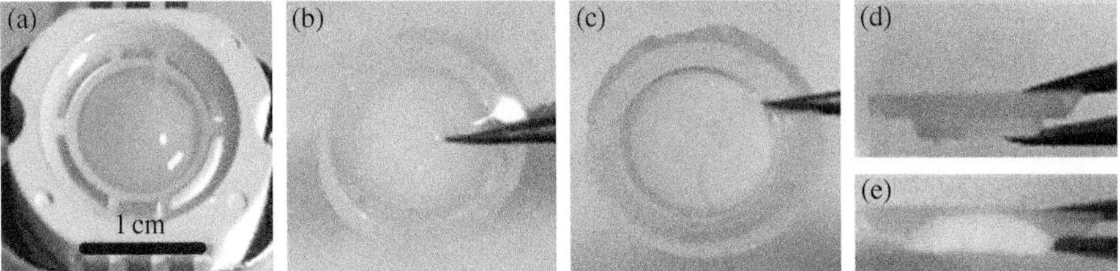

Figure 7.16 (a) Ver-WLEDs and the corresponding luminescent layer's (b) top surface, (c) lower surface, (d) lateral view, and (e) sectional view. Zhou et al. [57]/with permission of American Chemical Society.

Table 7.3 Different densities of phosphors and QDs in three kinds of device.

	Weight fractions (per gram of silicone)			
Samples	**Phosphors**	**QDs**	**hBNS**	**SCMC**
Ver-WLEDs	2%	0.01%	2%	1.5%
Iso-WLEDs	2%	0.008%	2%	0
Com-WLEDs	3%	0.01%	0	0

respectively. Furthermore, the temperature and spectra tests of the WLEDs were carried out by infrared photography (FLIR SC620) and integrating sphere (Everfine, ATA-1000), respectively.

Figure 7.17(a) and (b) show the morphologies of SCMC-based skeleton observed in the direction parallel to the SCMC templates' surface, which clearly show the thin vertical SCMC templates. The hBNS and spherical phosphors are shown in Figure 7.17(c) and (d), respectively. Seen in the perpendicular direction, as shown in Figure 7.17(e) and (f), the hBNS are embedded in the SCMC templates, almost parallel to the SCMC templates. In the same observing direction, the phosphors are also embedded in the SCMC templates as shown in Figure 7.17(g) and (h). It is noted that a high-volume ratio of SCMC

Figure 7.17 SEM results of (a, b) SCMC skeleton in a direction parallel to the templates' surface, (c) hBNS, (d) phosphor, (e, f) hBNS/SCMC, and (g, h) hBNS/SCMC/phosphor skeleton in a direction perpendicular to the templates' surface. Zhou et al. [57]/with permission of American Chemical Society.

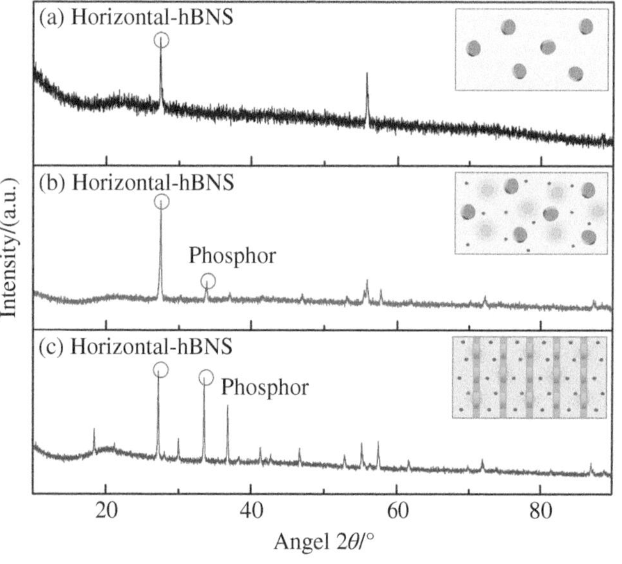

Figure 7.18 XRD results of (a) silicone film with hBNS, (b) phosphor/QDs-silicone film with hBNS and SCMC, and (c) phosphor/QDs-silicone film with vertical hBNS/SCMC templates.

to hBNS insures the skeleton does not collapse during the filling of polymer. Meanwhile, the additions of SCMC and hBNS are limited by their negative influence on the optical performance. For balancing the structure stability and the luminous property of the luminescent composite, the concentrations of SCMC and hBNS were experimentally selected as weight fractions of 1.5% and 2%, respectively. To evaluate the performance of vertical hBNS in the composite, the X-ray diffraction (XRD) results of the silicone film with hBNS, the phosphor/QDs film with hBNS and SCMC, and the phosphor/QDs film with vertical hBNS/SCMC templates are shown in Figure 7.18. Based on the previous work [38], the maximum peak, in Figure 7.18(a), is related to the proportion of horizontal hBNS in the film. Comparing the marked ratios of the peak of horizontal hBNS to phosphor in Figure 7.18(b) and (c), the film with vertical hBNS/SCMC templates has a relatively much lower peak of horizontal hBNS than the film with isotropically arranged hBNS and SCMC, which indicates the considerably large deflection angle of hBNS resulting in vertical hBNS.

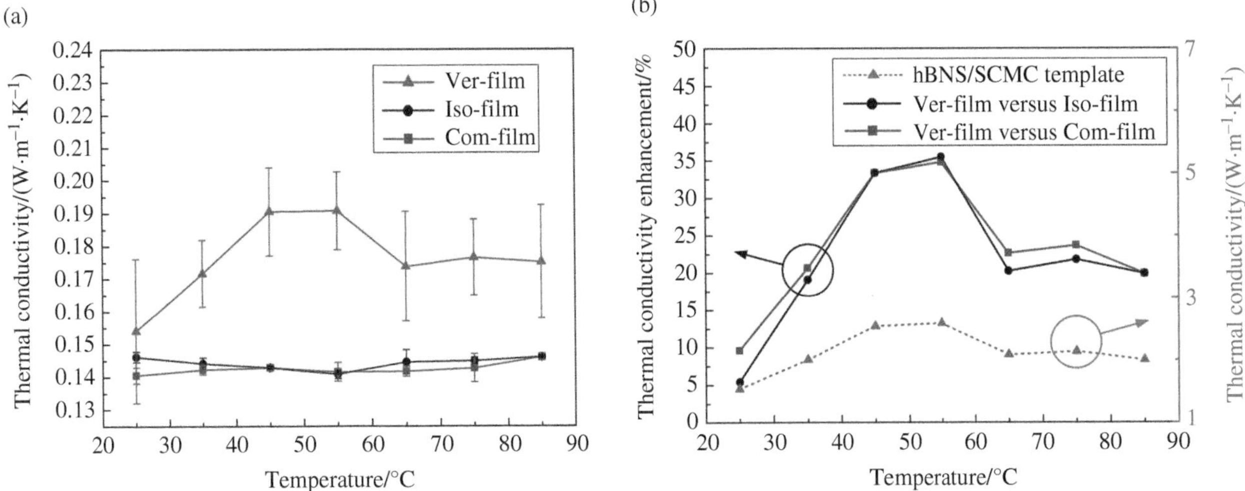

Figure 7.19 (a) Thermal conductivity of three different films: Ver-, Iso-, and Com-films. (b) Thermal conductivity enhancements of Ver-film compared to Iso- and Com-films, respectively, and the efficient thermal conductivity of hBNS/SCMC templates in the Ver-film.

Luminescent films containing the vertical hBNS/SCMC templates (Ver-film), with isotropic distributed hBNS (Iso-film) and without hBNS (Com-film), were made. Their thermal conductivities were tested. As shown in Figure 7.19(a), the much higher thermal conductivities of Ver-film at varied temperatures are due to the vertical orientation of high-thermal-conductive hBNS. There are bigger error bars in the thermal conductivities of Ver-film compared to the other two kinds of films, which mainly result from the inhomogeneous interval between the hBNS/SCMC templates. As shown in Figure 7.19(b), the average thermal conductivities of Ver-film are 25% larger than those of Iso- and Com-films. They are also much higher than the thermal enhancement performance of the vertical hBNS-based polymer in magnetic field method, which is about 6%, in the same volume ratio. The higher thermal conductivity of Ver-film at 45 and 55 °C is mainly caused by the corresponding variation of the specific heat of SCMC. Moreover, we inferred the efficient thermal conductivity of hBNS/SCMC template by efficient medium theory, $\kappa_y = \kappa_1 f + \kappa_2(1-f)$, where κ_y, κ_1 and κ_2 are the thermal conductivities of the composite films in the y-direction, silicone, and hBNS/SCMC templates, respectively; f is the volume fraction of silicone. Based on the tested thermal conductivities and the schematic of thermal mechanism in Figure 7.15(b), the hBNS/SCMC templates have an efficient average thermal conductivity more than 2 $W \cdot m^{-1} \cdot K^{-1}$, as the red dashed line shown in Figure 7.19(b), near 20-fold the lateral silicone templates. Though hBNS possess fairly high thermal conductivity through the surface, the low thermal conductivity of SCMC about 0.1 $W \cdot m^{-1} \cdot K^{-1}$ greatly limits the enhancement of the efficient thermal conductivity of hBNS/SCMC templates.

Based on the three kinds of luminescent films mentioned earlier, the Ver-WLEDs, Iso-WLEDs, and Com-WLEDs were packaged. They were set in a series circuit and driven by currents from 100 to 1000 mA. Their maximal working temperature (MWT) variations with currents are shown in Figure 7.20. Under the same currents, the largest and smallest MWTs were owned by the Iso- and Ver-WLEDs, respectively. Specifically, when driven by 1000 mA current, the MWTs of Com-, Iso-, and Ver-WLEDs were 137, 152, and 116 °C, respectively. In other words, compared to Com-WLEDs, the MWT of Ver-WLEDs decreased by 15%, but that of Iso-WLEDs increased by 12%. The opposite results imply one big difference between the thermal diffusion mechanisms of Ver- and Iso-WLEDs. Compared to Com-WLEDs, the decrease in usage of luminescent material of Iso-WLEDs and Ver-WLEDs, in Table 7.3, is because of the blue light loss in luminous layer caused by hBNS and hBNS/SCMC, which means more heat loss generated in Ver- and Iso-WLEDs, respectively. However, the isotropic thermal-conductive luminescent layer in Iso-WLEDs failed to diffuse all the increased heat in time, resulting in an unexpected temperature rise. On the contrary, owing to the high thermal conductivity of the vertical hBNS/SCMC templates and the special heat source distribution, which have been shown in Figure 7.15(b), the vertical thermal-conductive luminescent layer in Ver-WLEDs rapidly conducted the increased heat to the heat sink, leading to a lower MWT. It's noted that, in previous work, the hBNS-added WLEDs have smaller MWT than common WLEDs owing to QDs being absorbed on the surfaces of hBNS for heat diffusion [27]. But, in this work, QDs and hBNS are separated in the luminous layer of Iso-WLEDs. Despite the enhancement of thermal conductivity of the luminous layer of Iso-WLEDs, the heat loss caused by hBNS still leads to a higher MWT than that of Com-WLEDs.

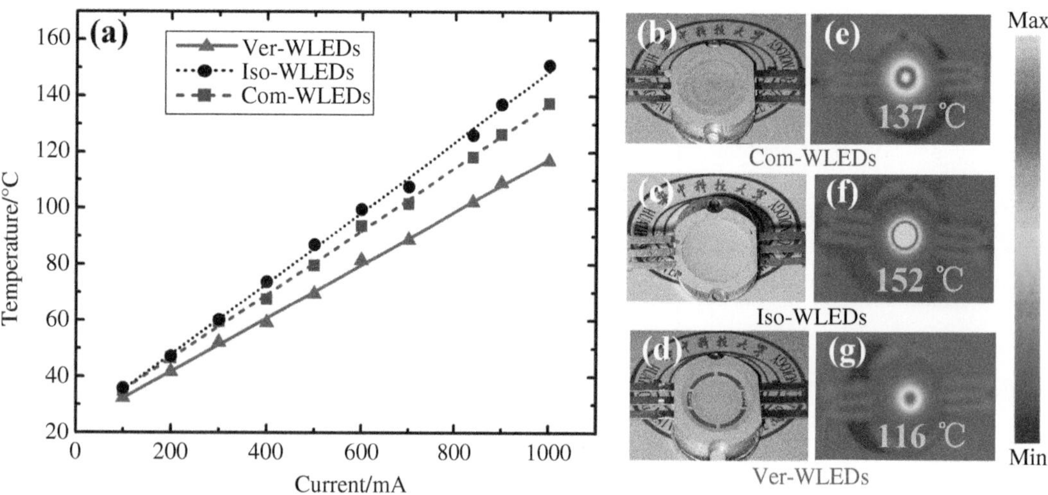

Figure 7.20 (a) Maximal working temperature with currents. (b–d) Photos and (e–g) infrared images under working current of 1000 mA of three kinds of QDs-WLEDs. (b–g) Zhou et al. [57]/with permission of American Chemical Society.

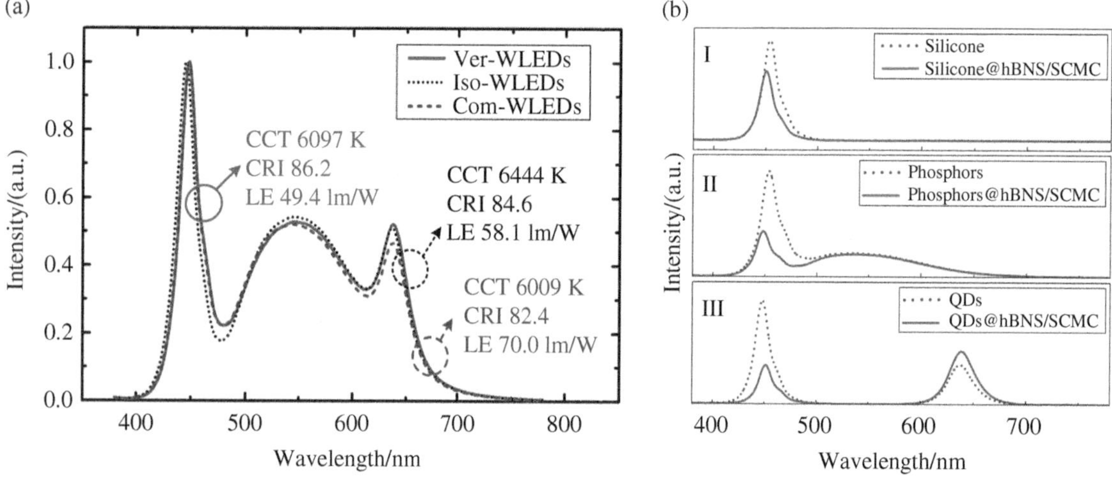

Figure 7.21 (a) Relative luminescent spectra of three kinds of QDs-WLEDs. (b) Luminescent spectra effects of hBNS/SCMC skeleton on silicone, phosphors-, and QDs-added silicone.

The relative luminescent spectra of the three kinds of QDs-WLEDs were tested under 1000 mA direct current, as shown in Figure 7.21(a). To ensure the consistency of relative spectra, all the samples generated similar high-quality light with CCT of about 6000 K and CRI of about 85. But the LE of each varied: the LE of Ver-WLEDs (49.4 $\mathrm{lm \cdot W^{-1}}$) was 15% and 29% less than that of Iso-WLEDs (58.1 $\mathrm{lm \cdot W^{-1}}$) and Com-WLEDs (70.0 $\mathrm{lm \cdot W^{-1}}$), respectively. This is mainly because of the inevitable light absorption effects of SCMC and hBNS, guided trapping effects of SCMC, and scattering effects of hBNS. Figure 7.21(b) shows the effect of hBNS/SCMC skeleton on the spectra of the luminescent layers with only phosphors or QDs. Although the yellow light from phosphors remained the same in phosphors/hBNS/SCMC case and the red light from QDs increased in QDs/hBNS/SCMC case, the light converting efficiencies of both are reduced, owing to the blue light absorption and loss caused by hBNS/SCMC. In addition, the blue light reduction of phosphors and QDs is larger than that of silicone, because phosphors and QDs enhanced the scattering of blue light and thus caused more light loss.

7.2.2 Thermally Conductive Composites Annular Fins

Besides the directional arrangement method mentioned earlier, there is another strategy that rearranges the distributions of hBN platelets (hBNPs) to enhance the heat dissipation of QDs. In QDs-WLEDs, the heat generated in the silicone gel is mainly dissipated by the metallic heat sink attached to the bottom of the gel. For the nanoscale "heat source" QDs which

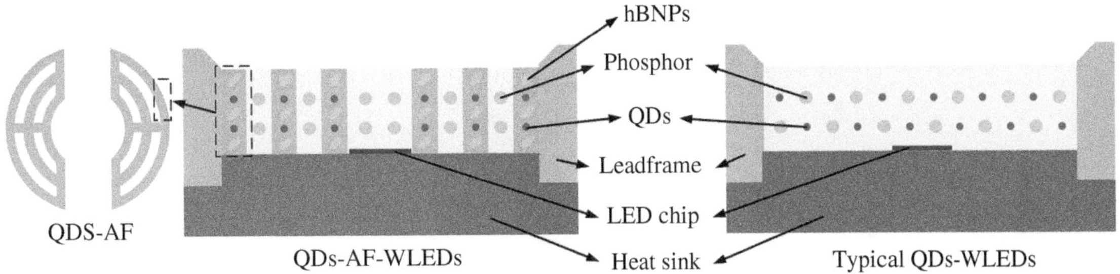

Figure 7.22 Schematic of QDs-AF-WLEDs and typical QDs-WLEDs.

are embedded in the low-thermal-conductivity gel, the thermal resistance between them and the heat sink is enormous, which hinders heat dissipation. In this section, QDs/hBNPs/silicone annular fins (QDs-AF) are proposed to establish rapid heat dissipation pathways between QDs and the heat sink. Figure 7.22 shows the schematic of QDs-AF-WLEDs and typical QDs-WLEDs. With the incorporation of hBNPs into the QDs/silicone, the QDs are more likely to conduct heat to the high thermal-conductive hBN platelets, thus their heat can be dissipated through the QDs–hBNPs–silicone route and even QDs–hBNPs–hBNPs route. Therefore, the overall thermal dissipation performance is improved.

We designed three initial structures of QDs-AF and utilized thermal simulation to attain the optimized structure parameters. The space between the two pieces is left for the LED chip and its bonding wire. There are different numbers of annular fins of the initial structures: three fins with two interspaces (3F), four fins with three interspaces (4F), and five fins with four interspaces (5F). The fin thickness ranges from 0.18 to 0.38 mm. The thicknesses of fins and interspaces are correlative. Table 7.4 lists their thickness and thermal conductivity used for simulation. The thermal conductivity of the QDs-AF in the simulation was set as $0.274\ \mathrm{W \cdot m^{-1} \cdot K^{-1}}$, which corresponds to the proper mass concentration of hBNS (15 wt%) in QDs-AF with acceptable light scattering effect and obvious thermal performance increase. The thickness of phosphor gel is equal to the interspace width, ranging from 0.705 to 0.0125 mm with the increase in fin thickness.

The settings of the boundary conditions are as follows: ambient temperature is constant at 25 °C; convection heat transfer coefficient of the bottom surface of sink is $245\ \mathrm{W \cdot m^{-2} \cdot K^{-1}}$; and natural convection heat transfer coefficient of the other surfaces is $5\ \mathrm{W \cdot m^{-2} \cdot K^{-1}}$.

Figure 7.23(a) shows the simulative maximum working temperatures of QDs-AF-WLEDs with different initial structures and fin thicknesses. It is noted that with the same number of fins, the interspace width decreases as the fin thickness increases. The increased fin thickness resulted in an increase in the contact area between the QDs-AF and the heat sink, enhancing heat conduction from QDs-AF to the heat sink. However, when the fin is too thick, the interspace width becomes too thin, thus leading to high heat generation density in the phosphor gel, which results in a considerable temperature rise. Therefore, to achieve the lowest working temperature, there is an optimal fin thickness of the QDs-AF. It can be found from the simulation results of the lowest maximum temperatures and related fin thicknesses: 125.4 °C and 0.36 mm for three-fins (3F) QDs-AF-WLEDs, 125.4 °C and 0.28 mm for 4F one, and 125.4 °C and 0.23 mm for 5F one. Figure 7.23(b) and (c) exhibit three optimal structures with different numbers of fins and temperature distribution of the corresponding WLEDs, whose emitting layers' temperature distribution curves are shown in Figure 7.23(d).

We chose four-fins QDs-AF with a fin thickness of 0.28 mm to fabricate QDs-AF-WLEDs. Figure 7.24(a) shows the schematics of the components in QDs-AF-WLEDs. The fabrication process of QDs-AF and the corresponding WLEDs are illustrated, respectively, in Figure 7.24(b) and (c). For QDs-AF (containing hBNS of 15 wt%), red-emissive QDs (Poly

Table 7.4 Thickness and thermal conductivity parameters for simulation.

Component	κ ($\mathrm{W \cdot m^{-1} \cdot K^{-1}}$)	Height (mm)	Heat power (W)	Thickness (mm)
Heat sink	130	20	0	—
Lead frame	1	30	0	—
Blue LED chip	63	0.2	4.36	—
QDs-AF	0.274	1	0.22	0.18–0.38
Phosphor gel	0.182	1	0.26	0.0125–0.705

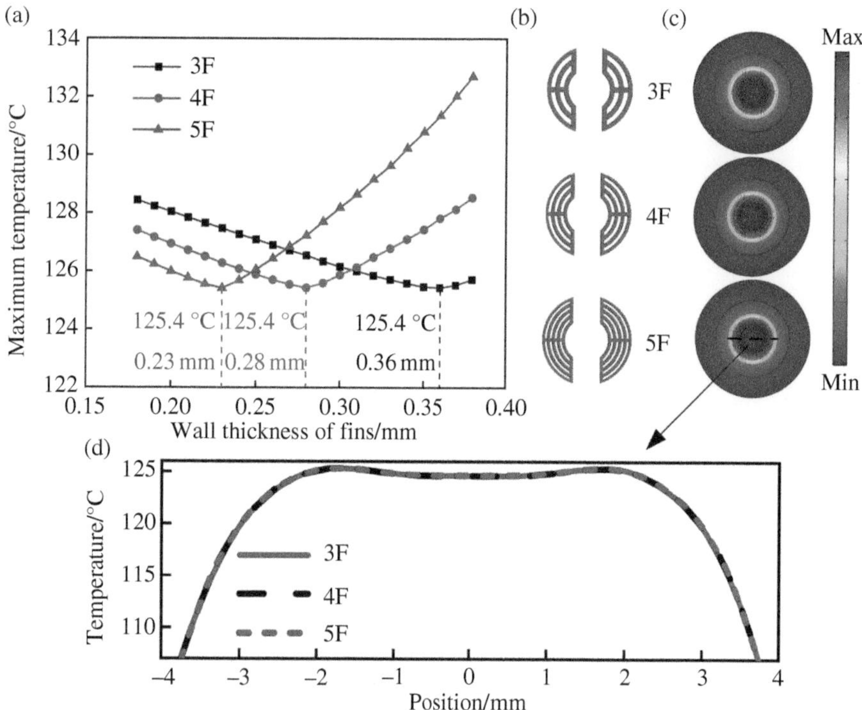

Figure 7.23 (a) Maximum temperatures corresponding to fin thicknesses of different initial structures. (b) Three optimal structures. (c) Temperature distributions of three optimal structures. (d) Temperature distribution curves of the emitting layer.

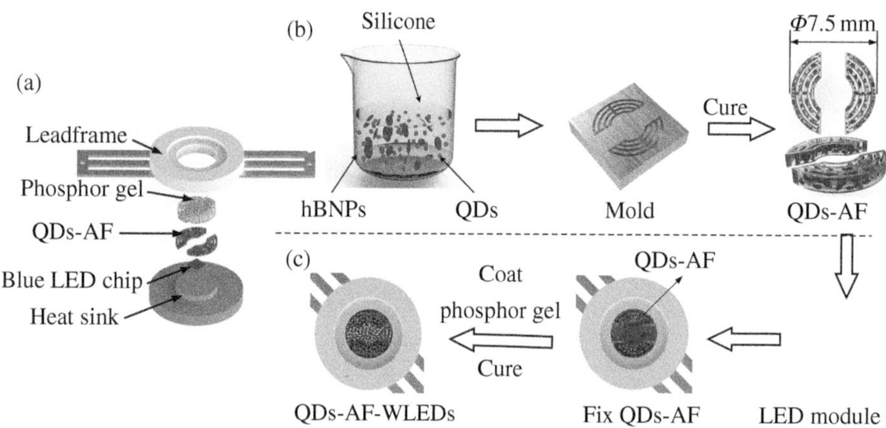

Figure 7.24 (a) Schematics of the components in QDs-AF-WLEDs. (b) Fabrication of QDs-AF. (c) Fabrication of QDs-AF-WLEDs.

OptoElectronics Ltd), hBNS (Momentive) with an average diameter of 19 μm, and silicone (Dow corning 184) were mixed, vacuumed and poured into the preprocessed mold of the optimized QDs-AF followed by a thermal curing process. Then, QDs-AF was coated on LED model and filled with phosphor gel to fabricate the QDs-AF-WLEDs. For QDs-WLEDs, QDs, yellow-emissive phosphor (Intematix), and silicone were mixed and coated on LED module. After fabrication, optical performance and surface temperature distribution of these WLEDs were measured by an EVERFINE ATA-1000 Auto-temperatured LED Opto-electronic Analyzer and a FLIR SC620 infrared thermal imager, respectively.

The fabricated four-fins QDs-AF under daylight and UV light irradiation are shown in the insets in Figure 7.25(a), respectively. The packaged QDs-WLEDs and QDs-AF-WLEDs are connected in series to be under the same driving currents, as shown in Figure 7.25(b). Figure 7.25(a) shows the maximum surface working temperature of QDs-WLEDs and QDs-AF-WLEDs under increasing driving currents. Compared with QDs-WLEDs, QDs-AF-WLEDs showed much lower temperatures under the same driving currents. In detail, the maximum surface working temperatures of QDs-AF-WLEDs were

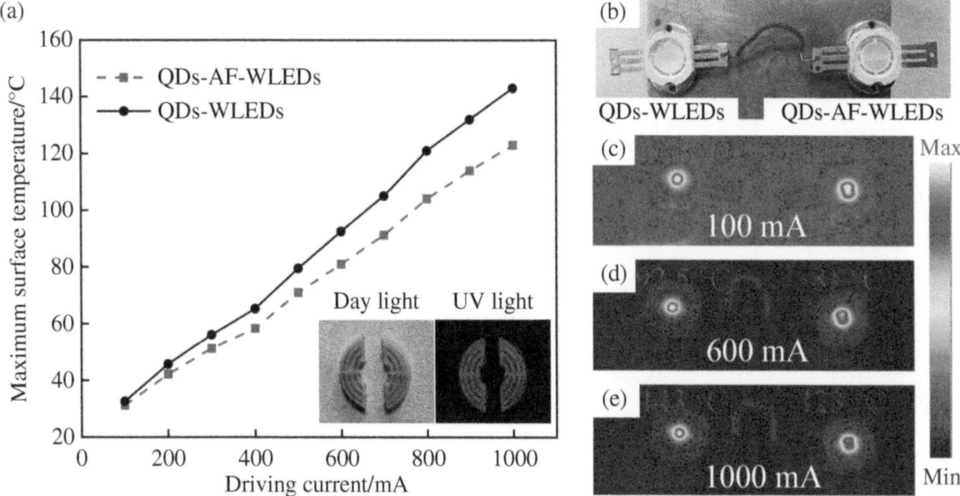

Figure 7.25 (a) Maximum surface temperatures of QDs-WLEDs and QDs-AF-WLEDs with increasing driving currents. Insets are the QDs-AF under daylight and UV light. (b) QDs-WLEDs and QDs-AF-WLEDs in series and their steady-state temperature distributions at (c) 100 mA, (d) 600 mA, and (e) 1000 mA.

lower than that of QDs-WLEDs by 1.4, 11.5, and 20 °C at 100, 600, and 1000 mA, respectively, as shown in Figure 7.25(c) and (d). The temperature reductions increased with increasing driving current, implying the application potential of QDs-AF in higher-power circumstances.

The optical performances of QDs-AF-WLEDs and QDs-WLEDs were also measured. As shown in Figure 7.26(a), the two WLEDs exhibited similar spectra under driving current of 20 mA. Furthermore, Figure 7.26(b)–(d) gives the CRI, LE, and CCT curves under increasing driving currents. Under a driving current of 20 mA, QDs-AF-WLEDs and QDs-WLEDs, respectively, showed a high CRI of 90.3 and 91, high LE of 124.1 and 131.8 $lm \cdot W^{-1}$, CCT of 5536 and 5315 K, and similar CIE coordinates of (0.3318, 0.3646) and (0.3377, 0.3695), as shown in the inset of Figure 7.26(d). The scattering effect of hBNPs causes light loss resulting in the lower LE of QDs-AF-WLEDs. The red light scattering by hBNPs increases the reabsorption of red light by QDs, which results in the higher CCT. In addition, QDs-AF-WLEDs showed similar optimal performance to QDs-WLEDs under increasing driving currents. These similarities proved that the package structure in QDs-AF-WLEDs had a negligible effect on their optical performance.

7.2.3 Packaging Structure Optimization for Temperature Reduction

As we have depicted at the beginning of this chapter, the low thermal conductivity of polymer composites is one of the main reasons that causes high working temperature of QDs. Therefore, in addition to solutions that introduce thermally conductive fillers into QDs-polymer composites, the working temperature of QDs can also be reduced by optimizing the packaging structure of QDs-WLEDs. Since the heat generation of QDs-polymer composites is mainly dissipated from LED chip to lead frame, and finally to the heat sink, the QDs' heat generation can be dissipated more quickly if they are packaged in a location closer to the LED chip. Thus, an obvious solution is to directly coat QDs nanoparticles onto the top surface of LED chip. However, the QDs nanoparticles easily self-aggregate if they are directly in contact with each other, consequently causing PL efficiency decrease. Therefore, the QDs should be chemically/physically protected before they can be packaged onto the LED chip. Several solutions have been proposed to solve this problem, such as QDs' surface chemistry modification [58, 59], incorporating QDs into mesoporous microspheres [60], and coating silica barrier layer on QDs' surface [61, 62]. Among these methods, coating silica layer onto QDs' surface to form the QDs-silica-coated nanoparticles (QSNs) is an optimal option that can simultaneously solve the QDs' self-aggregation problem and chemical incompatibility issue between QDs and polymer. Thus, in this chapter, we will introduce a new packaging structure for QDs-WLEDs, namely, the QSNs-on-chip WLEDs, to reduce the QDs' working temperature.

Figure 7.27 shows the schematic of QSNs-on-chip WLEDs (type I) and conventional QSNs/phosphor-mixed WLEDs (type II). Red-emissive CdSe/ZnS core/shell QDs were synthesized, and QSNs were prepared by a microemulsion reaction. For the fabrication of QSNs-on-chip WLEDs packaging, QSNs were directly coated onto the LED chip; then, yellow-emissive YAG:Ce phosphor silicone gel was coated. Both thermal simulations and thermal infrared temperature measurements

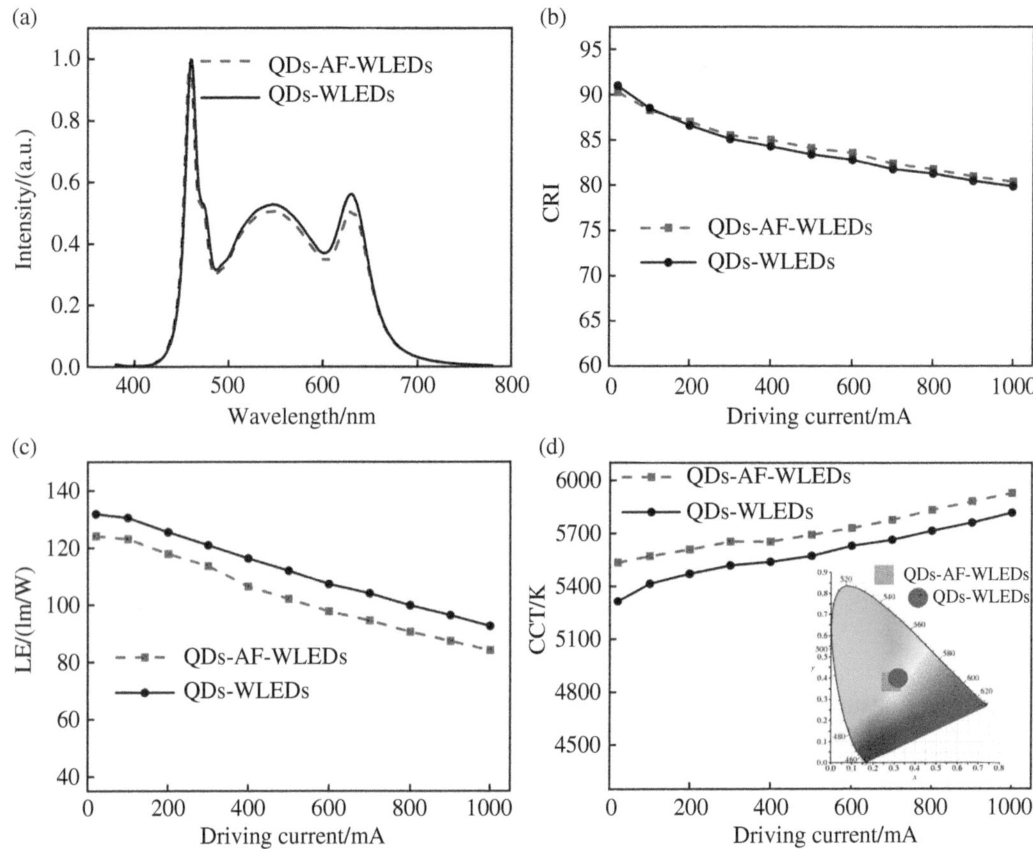

Figure 7.26 (a) Spectra of QDs-AF-WLEDs and QDs-WLEDs under driving current of 20 mA. (b) CRI, (c) LE, and (d) CCT of the WLEDs varies with driving currents. Inset is the CIE coordinates of the WLEDs at 20 mA.

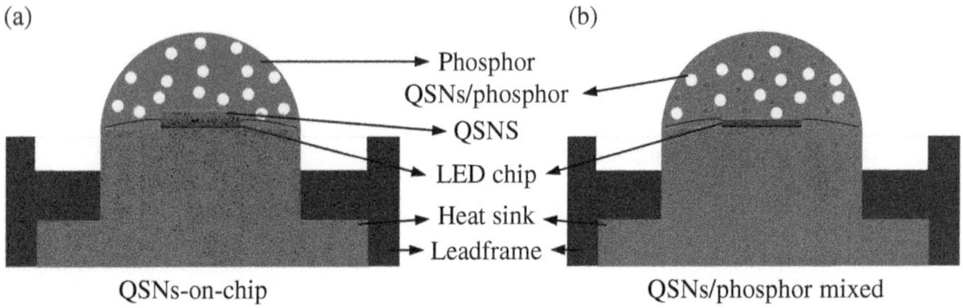

Figure 7.27 Schematic of the (a) QSNs-on-chip and (b) mixed-type WLEDs.

were conducted to analyze the optical and thermal performances of this newly proposed WLEDs packaging. Following are the details about the packaging procedures.

First is the preparation of QSNs. Red-emissive CdSe/ZnS core/shell QDs were prepared by our proposed tri-*n*-octylphosphine (TOP)-assisted successive ionic layer adsorption and reaction (SILAR) method [63]. This synthesis route enables high QYs of QDs at high coverage of the shell.

To coat the QDs with silica shells, a microemulsion reaction was applied [64]. Figure 7.28 shows the schematic of the silica coating process. Briefly, 25 ml of cyclohexane as a solvent and 3 g of IGEPAL CO-520 (Sigma-Aldrich) as a surfactant were mixed at room temperature. Into this mixture, 0.5 mg of QDs dispersed in 0.5 ml of toluene was introduced, and then 0.5 ml (9 mmol) of tetraethyl orthosilicate (TEOS) was added. The reaction was initiated by adding 0.5 ml of ammonia at the rate of $0.2\ \text{ml} \cdot \text{min}^{-1}$ and then it was allowed to proceed for 40 hours. After completion of the silica growth, the solution was precipitated by adding methanol and centrifuged to isolate the QSNs from the microemulsion. The resulting QSNs was washed sequentially with cyclohexane and *n*-hexane and finally dispersed in methanol.

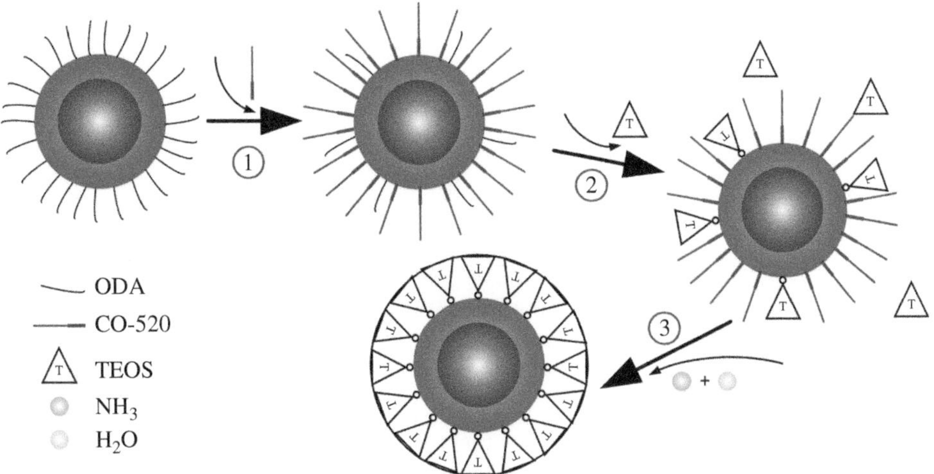

Figure 7.28 Schematic of the silica coating process.

Second is the fabrication of WLEDs. In the fabrication of QSNs-on-chip WLEDs, QSNs-methanol solution was first dropped onto the InGaN LED chip (455 nm); then, the module was placed on the hotplate at 70 °C for 10 minutes. After all the methanol had evaporated, YAG:Ce phosphor (538 nm, Intematix Co.) was mixed homogenously with the silicone gel (Dow Corning OE 6550, A:B = 1:1). Bubbles during the mixing process were removed by applying alternating cycles of vacuum. Then, the phosphor gel was coated onto the LED module to cover the LED chip and QSNs. Finally, the phosphor gel was cured by an annealing step at 150 °C for 30 minutes.

For comparison, the mixed-type WLEDs were also fabricated. First, QSNs-methanol solution was mixed homogenously with phosphor silicone gel. Then, the methanol solvent was removed by applying 60 °C heating and vacuum alternately. After all the methanol had evaporated, the resultant was coated onto the LED chip, followed by an annealing step at 150 °C for 30 minutes. For convenience, the QSNs-on-chip WLEDs were labeled as type I and those of the mixed type were labeled as type II.

After that, the temperature field of these two different WLEDs was simulated and measured to validate their heat dissipation capability. Figure 7.29 shows the schematic of the heat power measuring process. First, the optical powers of the two types of WLEDs were measured by an integrating sphere (ATA-1000, EVERFINE Inc.), and then, the heat powers of LED

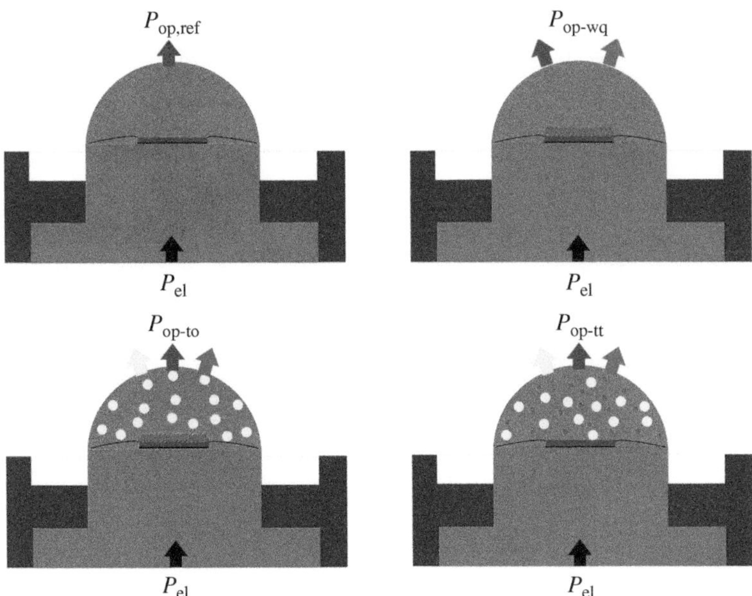

Figure 7.29 Schematic showing the heat generation measuring processes.

chip ($P_{\text{heat-chip}}$), QSNs (P_{QSNs}), phosphor silicone gel (P_{Phosphor}), and QSNs/phosphor silicone gel ($P_{\text{QSNs/phosphor}}$) were calculated according to the optical losses of the corresponding layer [42]:

$$P_{\text{heat-chip}} = P_{\text{el}} - P_{\text{op,ref}} \tag{7.4}$$

$$P_{\text{QSNs}} = P_{\text{op,ref}} - P_{\text{op-wq}} \tag{7.5}$$

$$P_{\text{phosphor}} = P_{\text{op-wq}} - P_{\text{op-to}} \tag{7.6}$$

$$P_{\text{QSNs/phosphor}} = P_{\text{el}} - P_{\text{op-tt}} \tag{7.7}$$

where P_{el} is the input electric power, $P_{\text{op,ref}}$ is the optical power from WLEDs with only silicone, $P_{\text{op-wq}}$ is the optical power from WLEDs with QSNs and silicone, $P_{\text{op-to}}$ is the optical power from the type I WLEDs, and $P_{\text{op-tt}}$ is the optical power from the type II WLEDs. Herein, the optical loss in lead frame is negligible due to its high reflectance and low absorption, and the scattering coefficient of QDs is assumed zero due to their small size [65]. Besides, it is assumed that the entire optical power loss is converted into heat power within each layer [27, 66].

Finite element models (FEM) of the two WLEDs were built to conduct thermal simulations, as depicted in Figure 7.30. The WLEDs were mounted onto an MCPCB for electrical connection and heat dissipation. Thickness and thermal conductivity of each component used for simulation are listed in Table 7.5.

The thermal conductivity of QSNs was calculated by [67]

$$\kappa_{\text{QSNs}} = \frac{\kappa_{\text{silica}} M^2}{(M - \delta_r)\ln(1 + M) + \delta_r M} \tag{7.8}$$

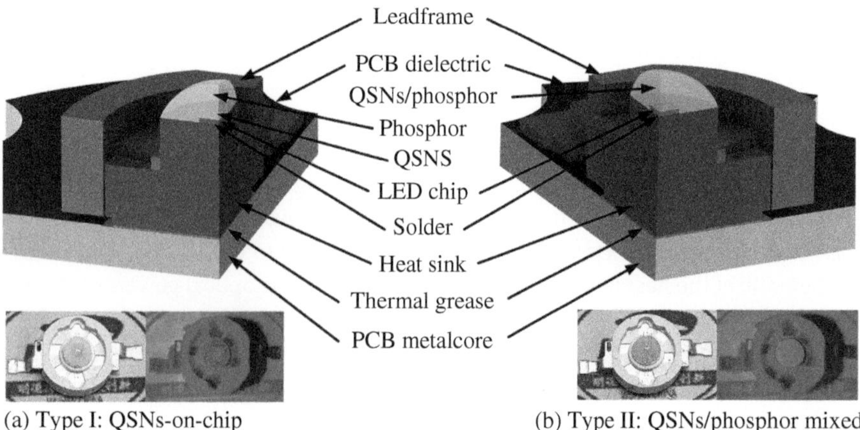

Figure 7.30 FEM setup of (a) QSNs-on-chip and (b) mixed-type WLEDs. The insets show the corresponding photographs of the WLEDs under daylight and UV light.

Table 7.5 Thickness and thermal conductivity of each component.

Component	Thickness (mm)	κ (W·m^{-1}·K^{-1})
PCB metal-core	0.98	170
Thermal grease	0.05	5
Heat sink	6	170
Solder	0.05	5
LED chip	0.1	65.6
QSNs	0.0001	11.3
Phosphor	3	0.18
QSNs/phosphor	3	0.18
PCB dielectric	0.02	0.2
Lead frame	6	0.36

with $M = \kappa_p(1+\delta_r) - 1$, where $\kappa_p = \kappa_{QDs}/\kappa_{silica}$ is the reduced thermal conductivity of the nanoparticle and $\delta_r = \delta/r_{QDs}$ is the ratio of the thickness of silica to the original QDs radius. Due to the symmetry, only a quarter of the WLEDs model was utilized to simulate the temperature field. The boundary conditions of the FEM model were set as follows: the ambient temperature was fixed as 25 °C; natural convection occurred at the bottom surface of the MCPCB with a heat transfer coefficient of $10\,W \cdot m^{-2} \cdot K^{-1}$; and other surfaces were cooled by natural convection with a heat transfer coefficient of $8\,W \cdot m^{-2} \cdot K^{-1}$. All the boundary conditions were similar to those in [44].

Figure 7.31 shows the HRTEM images of the as-prepared CdSe/ZnS QDs and the QSNs and the absorption/emission spectra of the core and core/shell QDs. From Figure 7.31(a), the QDs' average size is measured as 6.4 nm with a uniform size distribution. Benefited from the TOP-assisted SILAR method, in which the surface lattice imperfections are avoided by redissolution of the surface ions and lattice rearrangement during the whole ZnS shell formation process, the absolute PLQY was enhanced efficiently from 45% (core) to 69% (core–shell), as depicted in Figure 7.31(c). Figure 7.31(b) shows transmission electron microscope (TEM) images of the as-prepared QSNs, which demonstrate good dispersity and uniform size distribution. The average particle size is measured as 32 nm with a silica coating thickness of 12.8 nm. This coating thickness can provide effective protection for QDs and retain the QDs' PL intensity.

Figure 7.32(a) illustrates the EL spectra of the as-fabricated WLEDs under 20 mA illumination. With similar SPD, these WLEDs present a CCT of around 4250 K (natural white) and high CRI of Ra > 92, R9 > 90. Note that the QSNs-on-chip WLEDs achieved a high LE of 144.3 lm/W, which is 16.7% higher than that of the mixed type. This is mainly attributed to the separated structure which reduces the reabsorption losses between QSNs and phosphor particles. Figure 7.32(b) shows the current-dependent heat power in each component of QDs-WLEDs. It is seen that the heat generation in LED chip is much higher than those in QDs and phosphor. Besides, the heat generation of QSNs/phosphor in the mixed type is slightly higher than QSNs plus phosphor in the separated type. Therefore, the measured heat generation confirms the reduction of reabsorption losses in our proposed QSNs-on-chip structure.

To investigate the temperature difference between these two types of WLEDs, the measured heat generations were loaded onto the FEM, and thermal simulations were conducted. Figure 7.33 shows the simulated temperature fields of these WLEDs under driving current. The result shows that the highest temperature of all the WLEDs is located at the top of phosphor/QSNs silicone gel. This is mainly due to the low thermal conductivity of silicone gel, and the heat generated by phosphor/QSNs cannot be quickly dissipated from WLEDs to the surrounding air. The highest temperature in type I is lower than that in type II, and this difference is more apparent at larger driving current. For instance, the temperature difference is

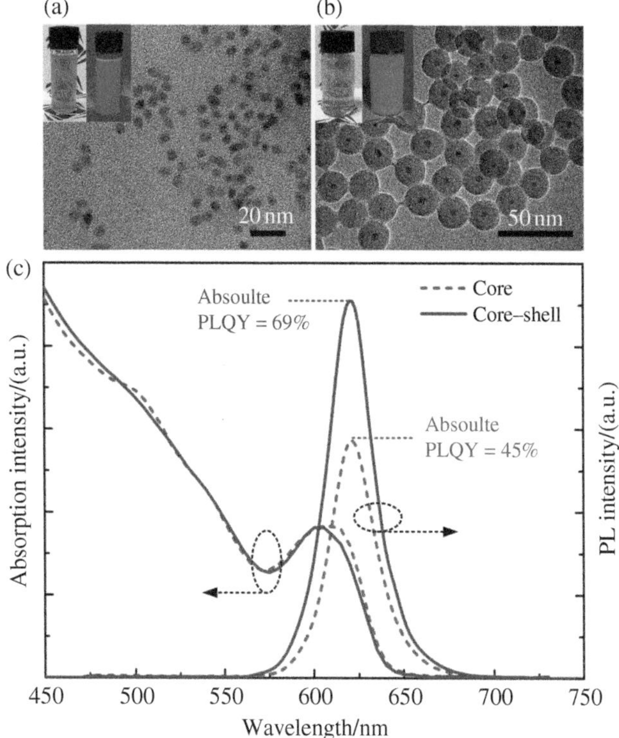

Figure 7.31 HRTEM images of the CdSe/ZnS QDs (a) and the QSNs (b). Insets show the corresponding photographs under daylight and UV light. (c) Absorption and PL spectra of the CdSe core QDs and CdSe/ZnS core/shell QDs.

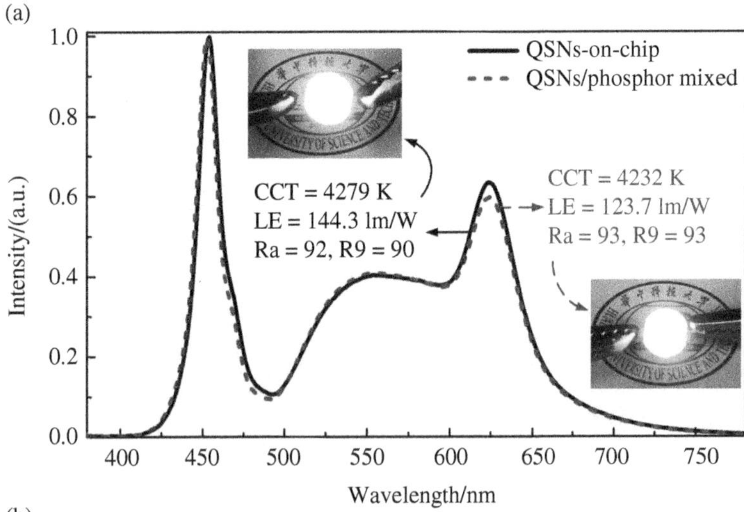

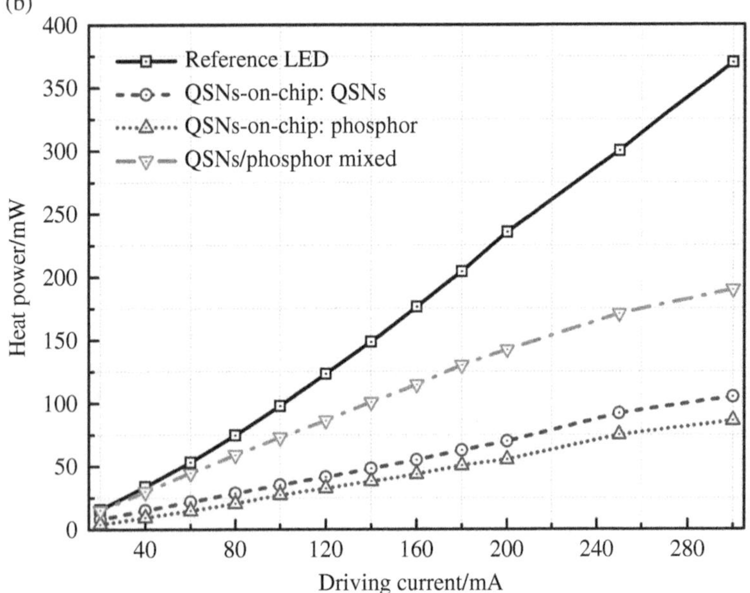

Figure 7.32 (a) EL spectra of the as-fabricated WLEDs under driving current of 20 mA; insets show their illuminated photographs. (b) Heat generation of each component under different driving currents.

11.5, 21.3, and 30.3 °C under driving currents of 80, 200, and 300 mA, respectively. Therefore, benefiting from the proposed QSNs-on-chip configuration, the QDs' working temperature can be significantly reduced.

To validate these simulation results, the corresponding temperature fields of the two WLEDs were measured by an infrared thermal imager (FLIR SC620). The emissivity of silicone gel was set as 0.96 [48], and the distance between WLEDs and the camera lens was fixed as 0.3 m. All the temperature fields were obtained after the WLEDs were lighted up for 30 minutes. Figure 7.34 shows the measured steady-state temperature fields. It is seen that the simulated temperature distributions between these WLEDs agree well with the experimental results, and the maximum relative deviation of the highest temperature between the experimental and simulation results is 5.9% (76.4 and 80.9 °C, respectively, at 300 mA). Therefore, it was confirmed by the experimental results that the highest temperature in QSNs-on-chip type is lower than that in the conventional mixed type.

7.3 3D-Interconnected Thermal Conduction of QDs

It has been validated in the earlier sections that incorporated hBN platelets successfully enhanced the heat dissipation. But, when further increasing the hBN filling fraction, the luminous efficacy of QDs-WLEDs decreased because of the increased scattering effect and reabsorption energy loss. Therefore, due to the trade-off between light efficiency and heat dissipation performance, the hBN filling load must be restricted to a relatively low range, but the cooling capability is also limited. In

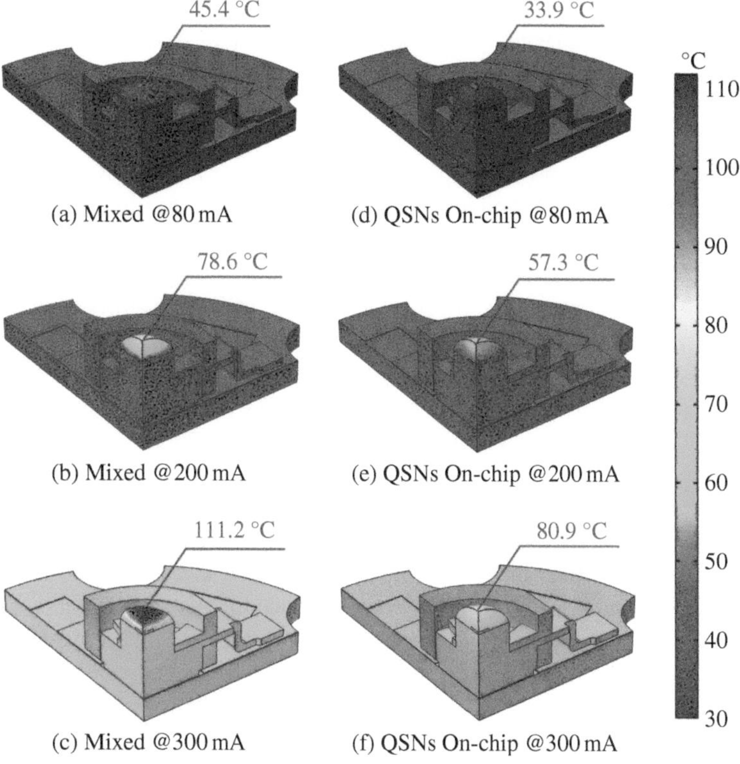

Figure 7.33 Simulated steady-state temperature fields of two WLEDs under driving currents of 80, 200, and 300 mA.

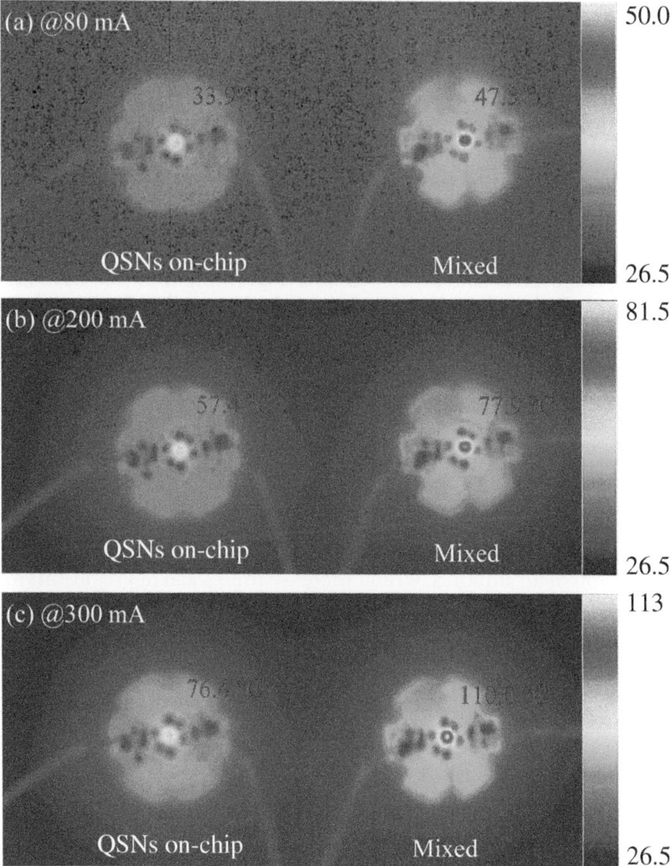

Figure 7.34 Temperature fields of two WLEDs under driving currents of (a) 80 mA, (b) 200 mA, and (c) 300 mA.

this section, we report an effective cooling strategy for QDs in WLEDs through establishing a three-dimensional (3D) hBN network in the QDs composites. Owing to the lateral expulsion of massive microscale air bubbles, 3D-interconnected thermal dissipation pathways were established under extremely low filler fraction. With this efficient 3D/hBN network, the heat generated by QDs and phosphor could be dissipated quickly, and the light energy could be extracted out efficiently through the porous 3D/hBN structure.

Figure 7.35 shows the as-proposed reinforced air-bubbles-assembly method for constructing *in situ* 3D/hBN aerogel in WLEDs. Typically, 0.5 ml alkyl polyglucoside (APG) was uniformly dissolved in 25 ml deionized water to form a colorless mixture. Then, 0.5 g curdlan, 0.3 g gelrite, 2.5 g hBN platelets, and 1.5 g phosphor powder were added sequentially, followed by a 30-minute magnetic stirring under 1500 rpm. During this process, the hBN and phosphor were dispersed uniformly, and curdlan/gelrite were fully dissolved to form a luminescent slurry. Besides, owing to the stirring of APG, massive air bubbles were introduced into the slurry, leading to an obvious volume expansion of the mixture. Next, the slurry was moved to a water bath of 90 °C and stirred at 1000 rpm for another 3–5 minutes, and then immediately poured into a LED module or other molds. In this heating–cooling process, the hBN platelets and phosphor particles were cross-linked by the curdlan/gelrite molecules, forming a thermally conductive network in the slurry. After cooling down to room temperature, the mixture formed a jelly-like hydrogel due to the solidification of curdlan/gelrite chains. Subsequently, the hydrogel was freeze-dried (Freeze dryer, SCIENTZ-12N, China) at low temperature (−55 °C) and pressure (20 Pa) for 12 hours to remove the water molecules and obtain the 3D/hBN-luminescent aerogel. The 3D/hBN-WLEDs is prepared as follows. The silicone and curing agent were first uniformly mixed at room temperature by a weight ratio of 10:1. Then, 25 μl QDs-chloroform solution (50 mg/ml) was added into 2 g silicone gel, followed by vacuum treatment to remove the chloroform and bubbles. Next, the QDs-silicone gel was infiltrated into the as-prepared 3D/hBN-luminescent aerogel of WLEDs under vacuum atmosphere to fill up the voids and obtain the 3D/hBN-composites. Finally, the composites were cured at 90 °C for 1 hour to obtain the 3D/hBN-WLEDs. As referential samples, the WLEDs with randomly distributed hBN platelets (namely, the R/hBN-WLEDs with R/hBN-composites) and traditional WLEDs without hBN platelets (T-WLEDs with T-composites) were also fabricated. For the fabrication of R/hBN-WLEDs, phosphor powder, QDs solution, and the same weight fraction of hBN platelets were directly mixed and vacuum-treated to prepare the luminescent composites. Then, the composites were poured into the cavity of the blue LED module, followed by a thermal curing process. For the fabrication of T-WLEDs, the procedure is similar to that of R/hBN-WLEDs, except for the addition of hBN platelets.

The aim of this work is to build a highly conductive thermal dissipation network inside the QDs composites without deteriorating the light output performances. In this regard, efficient interconnections between thermally conductive fillers and expedite escaping channels for photons should be established simultaneously. The air-bubbles-assembly method has been proposed and validated to construct ultralight, yet thermally conductive aerogel, like graphene aerogel [28, 68, 69] and hBN aerogel [70]. But, graphene aerogel is not practical since the GO gelling agent is strongly absorptive for light. Moreover, according to the bubbles-templated method of preparing hBN aerogel proposed by Li et al. [70], we failed to build a stable hBN aerogel structure under <5 wt% hBN filling fraction, because the weak shearing force provided by curdlan molecules is insufficient to support the whole network, and the hydrogel would either collapse during the air-dry process or disintegrate during the freeze-dry process.

After persistent trials, we found that gelrite, as another type of bacterial polysaccharide [71], could be used with curdlan to form a reinforced 3D/hBN aerogel under extremely low hBN filling load (~2.5 wt%). Therefore, we proposed a reinforced air-bubbles-assembly method, using both curdlan and gelrite as gelling agents, to prepare the 3D/hBN-luminescent composites and 3D/hBN-WLEDs. As shown in Figure 7.35, the hBN platelets, phosphor, curdlan, and gelrite powder were first dispersed in aqueous solution. Then, the mixture was intensively stirred to dissolve the curdlan and gelrite, and massive air bubbles were generated via the foaming function of APG. In this process, the hBN platelets were excluded by the lateral expulsion of air bubbles, which increase their local aggregations and force them to interconnect with each other. After transferring to a water bath and stirring for several minutes, the slurry was quickly cooled down to room temperature. During this process, the slurry was irreversibly thermal-gelled due to the interconnection between curdlan and gelrite molecules, forming a luminescent hydrogel.

Figure 7.36 displays the photographs and microstructures of the as-prepared 3D/hBN-luminescent hydrogel, 3D/hBN-luminescent aerogel, and 3D/hBN-composites. From Figure 7.36(a) and (b), it is seen that after the freeze-drying process, the 3D/hBN-luminescent aerogel keeps the original shape of the hydrogel, which means that the geometry of the aerogel could be conveniently controlled by simply designing the shape of the mold. Figure 7.36(c)–(f) shows the microstructures of the 3D/hBN-luminescent aerogels under different magnification ratios. It is seen from Figure 7.36(c) and (e) that the massive open pores with a diameter of 100–400 μm were left after the freeze-drying process, forming a 3D-interconnected hBN/phosphor

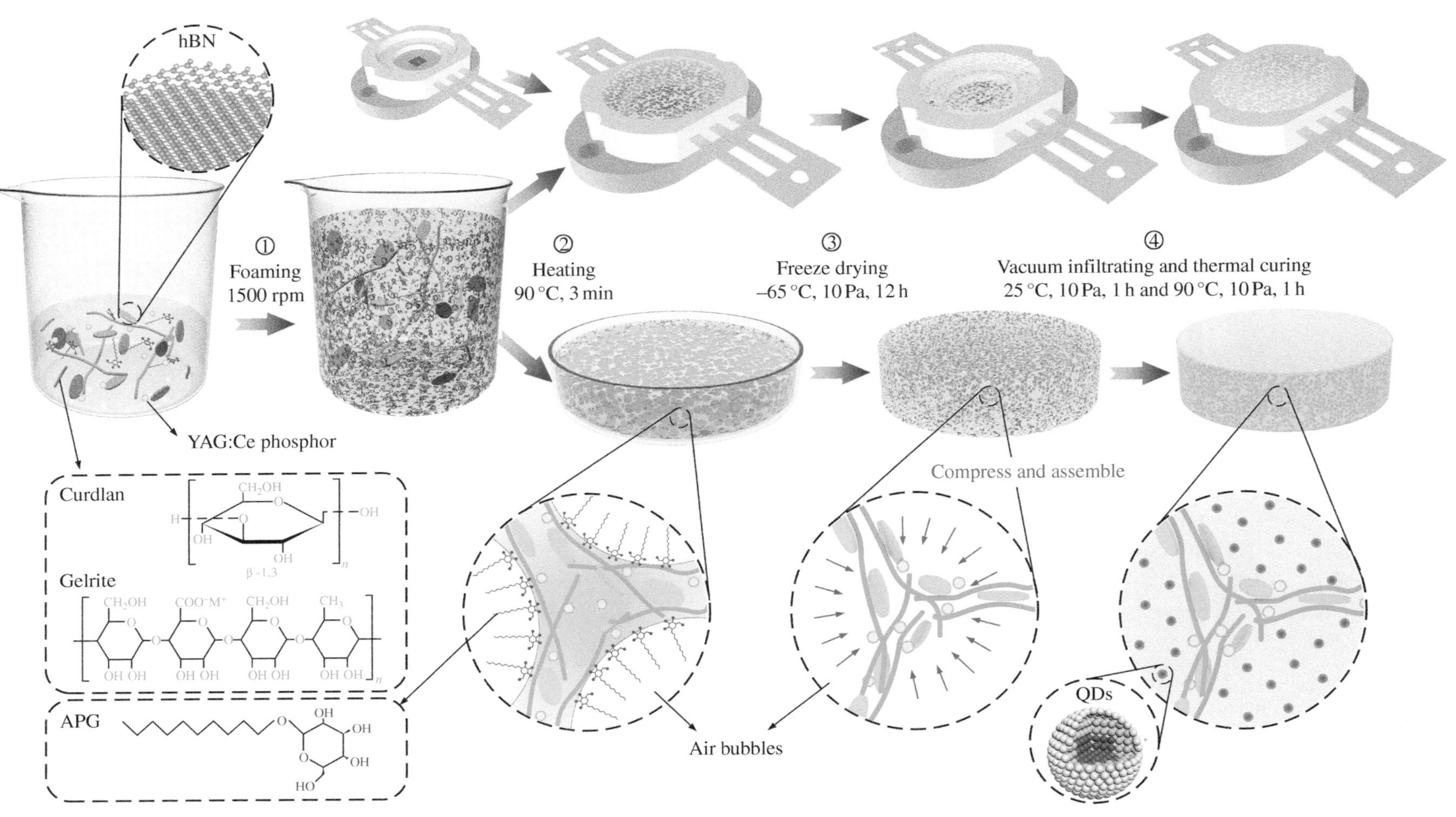

Figure 7.35 Schematic of the air-bubbles-assembly method for preparing the 3D/hBN-luminescent aerogel and the corresponding 3D/hBN-WLEDs.

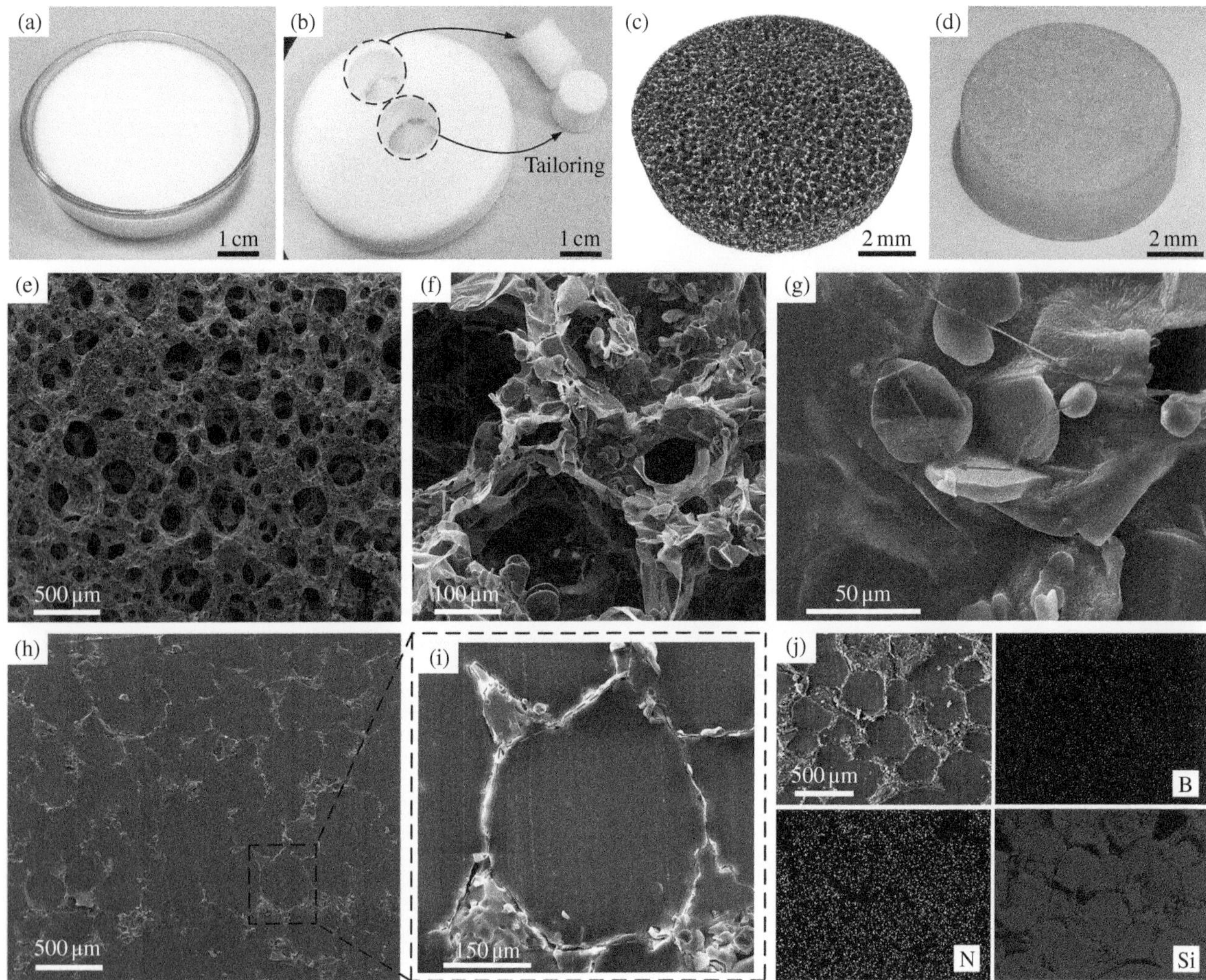

Figure 7.36 (a) Photograph of the 3D/hBN-luminescent hydrogels. (b) Photograph of the 3D/hBN-luminescent aerogels. (c) X-ray tomography (XRT) image of the 3D/hBN-luminescent aerogels. (d) Photograph of the 3D/hBN-composites. (e–g) SEM images of the 3D/hBN-luminescent aerogels under different magnifications. (h) SEM images of the 3D/hBN-composites. (i) Enlarged view of a local region in (h). (j) SEM image and the corresponding EDS mapping images of the 3D/hBN-composites. Xie et al. [72]/with permission of Elsevier.

network. Figure 7.36(f) and (g) show the detailed distribution of hBN platelets on the skeleton, from which the continuously connected hBN chains could be observed. Therefore, when the thermal phonons are transferred to the skeleton, they could be dissipated out quickly through the highly thermal-conductive hBN chains, as schematically illustrated in Figure 7.36(g). Moreover, the abundant open pores in the 3D/hBN-luminescent aerogel enable light rays to transmit through the whole aerogel. After QDs-silicone infiltration, all the voids in the aerogel were filled by the gel, and the as-prepared 3D/hBN-composites show a fully solid state, as displayed in Figure 7.36(h). Figure 7.36(i) displays a typical heat dissipation network of the composites, in which the hBN chains connected with each other to form an interconnected spatial network, and the silicone has fully infiltrated into the remaining space. From the EDS mapping images shown in Figure 7.36(j), the distribution of B, N, and Si elements is clearly illustrated, which confirms the existence of hBN network and the efficient filling of silicone. It should be noted that due to the low concentration of QDs and phosphor, their distributions are not indicated by the EDS images.

To investigate the effect of 3D hBN network on the optical response of QDs and phosphor, we measured the UV–VIS absorption spectra and PL spectra of the different composites, as shown in Figure 7.37(a) and (b). It is seen that QDs show multiple absorption peaks in silicone, of which the highest is located at 280 nm, and the phosphor shows an absorption peak around 300 nm. In the T-composites and 3D/hBN-composites, the absorption peak contributed by QDs is still apparent without evident shift. Moreover, the QDs in the different types of composites demonstrate a stable PL spectrum with peak

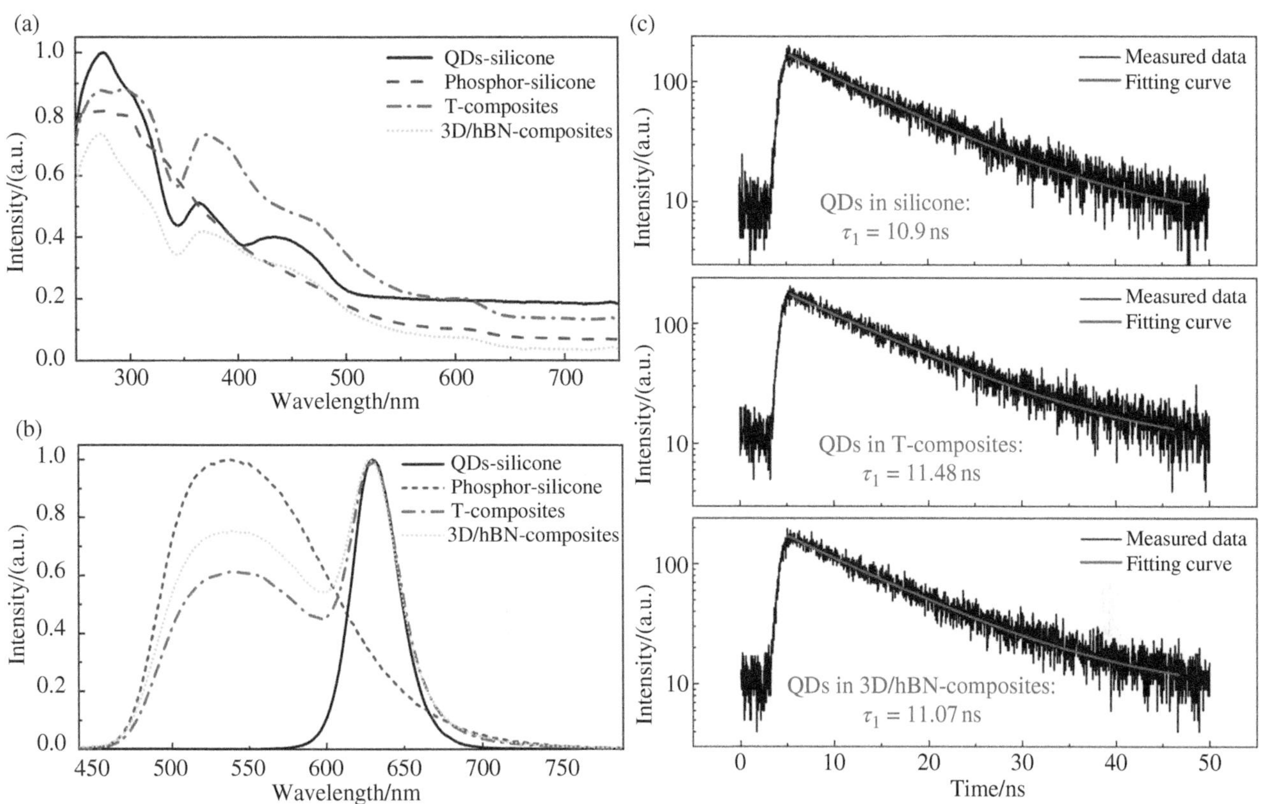

Figure 7.37 (a) UV–vis absorption spectra and (b) photoluminescent spectra of QDs-silicone, phosphor-silicone, T-composites, and 3D/hBN-composites. (c) TRPL decay curves of QDs in silicone, T-composites, and 3D/hBN-composites.

emission wavelength and FWHM of 630 and 34.5 nm, respectively. To further investigate the underlying mechanism of QDs' stability in these different composites, the PL lifetime of QDs in these composites was measured, as shown in Figure 7.37(c). These TRPL decay curves can be well fitted by an exponential function $I(t) = k_{\mathrm{Fit}}e^{-t/\tau_1}$, where $I(t)$ is the PL initial intensity at time t. The PL lifetime of QDs in silicone is calculated as 10.9 ns, and those in T-composites and 3D/hBN-composites are 11.48 and 11.07 ns, respectively. The stable PL lifetime of QDs in these composites indicates that energy transfer from QDs donors to hBN acceptors was low. Therefore, the existence of 3D hBN network has a negligible effect on QDs' optical properties.

Figure 7.38(a) shows the measured thermal conductivities of T-composites, R/hBN-composites, and 3D/hBN-composites with different hBN filling fractions (2.74 and 4.5 wt%). It is seen that the thermal conductivity of T-composites is quite low ($0.158\ \mathrm{W\cdot m^{-1}\cdot K^{-1}}$) due to the intrinsic weak phonon transport ability of the unordered polymer chains [73]. More specifically, phonon is the thermal energy carrier in the polymer. In the ordered region of the polymer, the atoms are in crystallized connect with each other closely and vibrate slightly near equilibrium position, so that the phonon transfers rapidly in the direction of molecule chains. However, in the unordered region of the polymer, the crystallinity is not very high due to the random entanglement of the polymer chains, thus the phonons are unable to transport quickly in the unordered region. Besides, the unordered polymer chains also introduce extra boundary scattering toward phonon, resulting in low phonon transport ability. After randomly filled with hBN platelets, the thermal conductivity of the R/hBN-composites shows a limited enhancement, for example, from 0.158 to $0.186\ \mathrm{W\cdot m^{-1}\cdot K^{-1}}$ under hBN filling load of 2.74 wt%. This is mainly because of the relatively low hBN volume fraction which is incapable of forming continuous thermal dissipation channels inside the composites. But, under the same hBN filling fraction, the thermal conductivity of 3D/hBN-composites was significantly reinforced from 0.158 to $0.317\ \mathrm{W\cdot m^{-1}\cdot K^{-1}}$, which is 100% enhanced compared to the T-composites. When the hBN loading is increased to 4.5 wt%, the R/hBN-composites still show a moderate enhancement of thermal conductivity ($0.216\ \mathrm{W\cdot m^{-1}\cdot K^{-1}}$), while the thermal conductivity of the 3D/hBN-composites continues to increase remarkably ($0.374\ \mathrm{W\cdot m^{-1}\cdot K^{-1}}$).

To reveal the underlying heat transfer mechanisms of these three composites, FEM was utilized to simulate the temperature distribution of these composites. The corresponding physical models and boundary conditions are displayed in Figure 7.39. The blue dashed circles indicate the previous location of the air bubbles. The right chart shows the thermal

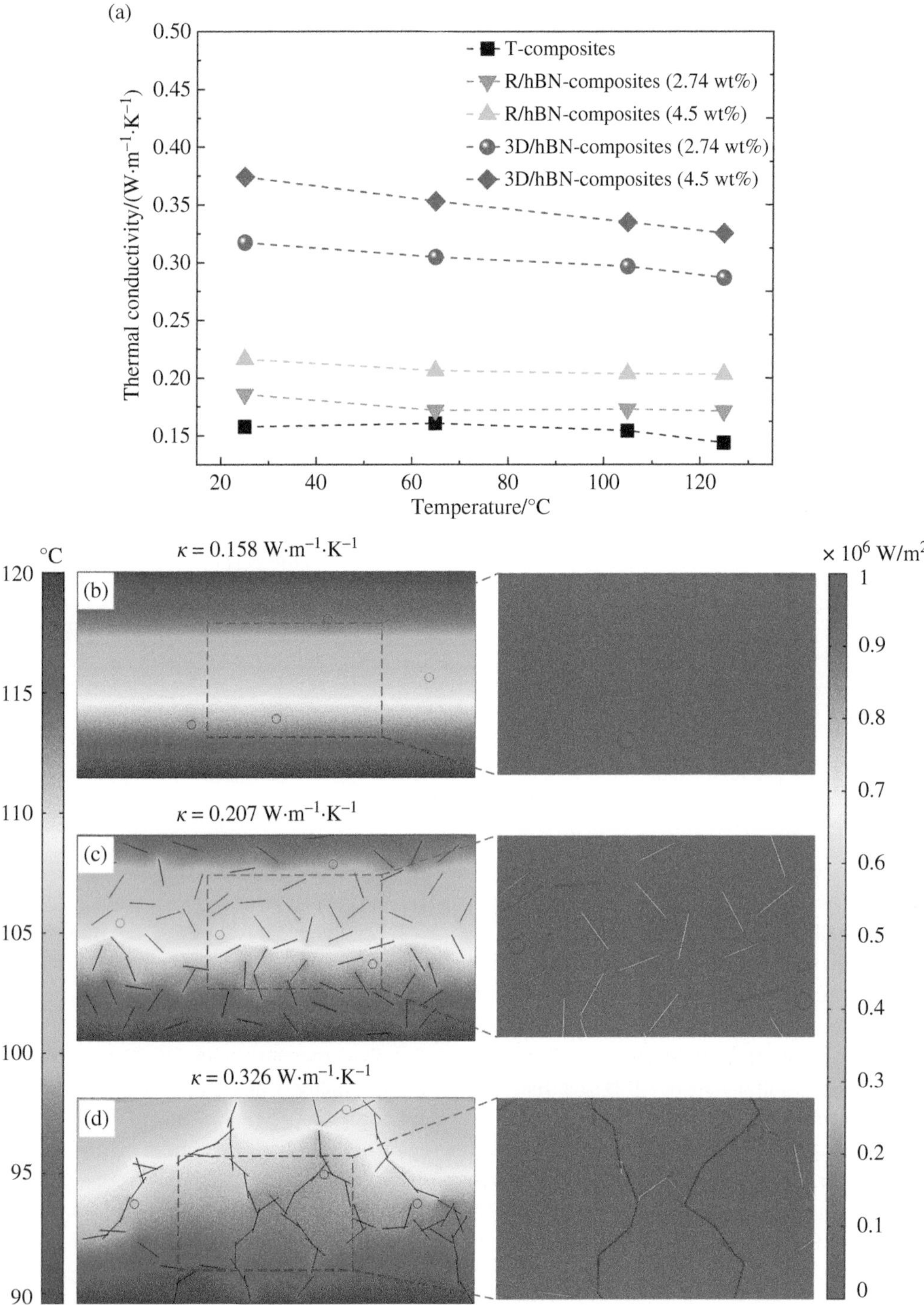

Figure 7.38 (a) Measured thermal conductivities of T-composites, R/hBN-composites, and 3D/hBN-composites under different ambient temperature. (b–d) Simulated temperature and heat flux distributions of the three composites.

resistance network of the composites, where T_b, T_t, and T_a represent the bottom temperature, top temperature, and ambient temperature, respectively. h is the conventional heat transfer coefficient at the top surface. κ represents the thermal conductivity of the composites, and δ is the thickness of the model. Figure 7.38(b)–(d) shows the simulated temperature and heat flux distributions of the three composites. Concretely, from the FEM results in Figure 7.38(d), it is seen that due to the existence of air bubbles, the hBN platelets have higher probability to contact with each other, thus forming an interconnected microstructure. Consequently, the heat flux in these interconnected channels is significantly higher than in the other

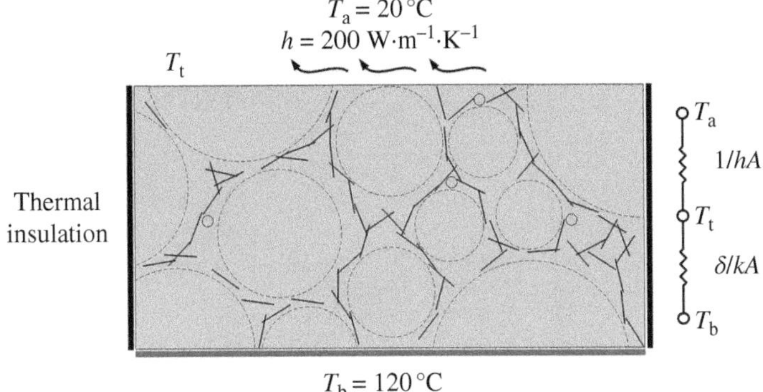

Figure 7.39 Thermal boundary conditions of the 3D/hBN-composites.

region, which confirms the formation of thermal dissipation channels. It is seen that the temperature of the top surface in 3D/hBN-composites is obviously higher than that of others, and the T-composites possess the lowest temperature of the top surface. Therefore, the heat transfer process from the bottom surface to the top surface is smoother in the 3D/hBN-composites than in the other two composites.

From the corresponding heat flux distributions of these composites, we can see that the heat flux density in hBN platelets is significantly higher than that in silicone, which means that the heat transfer in the composites mainly relies on hBN. Moreover, the heat flux was severely confined in the isolated hBN platelets of the R/hBN-composites. Therefore, although the phonons transport quickly inside the hBN platelets, they cannot be dissipated rapidly because the hBN platelets are still surrounded by low-thermal-conductivity silicone. But, in the 3D/hBN-composites, continuous hBN chains are formed, thus, the thermal phonons can transport quickly throughout the hBN chains. In other words, the interconnected hBN chains provide an efficient escape pathway for the phonons. Consequently, the heat flux density in these continuous hBN chains is an order of magnitude higher than that in isolated hBN platelets. This phenomenon supports the effectiveness of the established 3D-interconnected hBN network in heat transfer.

To study the optical performances of the as-prepared 3D/hBN-composites, the corresponding WLEDs (3D/hBN-WLEDs) were fabricated by embedding 3D/hBN-composites into the blue LED module. As reference samples, the WLEDs based on T-composites and R/hBN-composites were also prepared; their CCT were controlled to around 5500 K (neutral white color) by tuning the dosage of QDs/phosphor, and the hBN filling load in the R/hBN-WLEDs and 3D/hBN-WLEDs was controlled to 2.74 wt%. Figure 7.40(a) displays the SPD of these three WLEDs under the same driving current of 500 mA. It is seen that under similar spectrum, these WLEDs demonstrate high efficiency (LE > 83 $lm \cdot W^{-1}$) as well as high CRI (CRI > 92), indicating their superior ability to render the true color of objects. Meanwhile, their LE performances were quite close (within deviation of 4%), which means that this low hBN filling load contributes negligible effect on the light output efficiency of WLEDs. To further investigate the light transmittance and conversion process in the three composites, the transmittance of blue light and the conversion ratio of yellow and red lights in these WLEDs were calculated according to the measured data, as shown in Figure 7.41. The blue light power is calculated from the wavelength range of 380–490 nm, and the converted light power is calculated from the wavelength range of 490–780 nm. The transmittance of blue light in different composites is defined as the ratio of blue light power from WLEDs to the blue light power from bare LEDs. The conversion ratio of converted light in different composites is defined as the ratio of converted blue and yellow light power to the blue light power from bare LEDs. It is seen that about 15% of the total blue light energy was transmitted from the composites, and about 40% is converted into yellow and red light by phosphor and QDs, respectively. Thus, about 45% of the total blue light is converted into heat power. Moreover, the transmittance and conversion ratio among these WLEDs vary insignificantly (less than 2%), which also indicate the negligible influence of 3D/hBN network on the optical properties of WLEDs. Figure 7.40(b) displays the photographs of these WLEDs with and without illumination, which clearly demonstrate the comparable appearance and optical performance of 3D/hBN-WLEDs. Figure 7.40(c) shows their LE and CRI variation under different driving currents from 100 to 800 mA. From the results, we can see that the CRI of these WLEDs remain stable with the increasing driving current, suggesting their consistent color rendering ability under various lighting conditions. Meanwhile, these WLEDs demonstrate similar trends of LE drop under increasing driving current. This is mainly because of the efficiency drop of the LED chip itself under crowded electron-hole pairs [74]. Figure 7.40(d) indicates the location of these WLEDs in the CIE 1931 diagram. It is seen that they are quite close to the central region of the Planck blackbody curve, indicating a neutral, high-quality white light illumination. Additionally, we also measured the optical properties of the 3D/hBN-WLEDs, which contained 4.5 wt% of hBN platelets. The results indicate that the LE was decreased to 76.6 $lm \cdot W^{-1}$ due to the overload of hBN

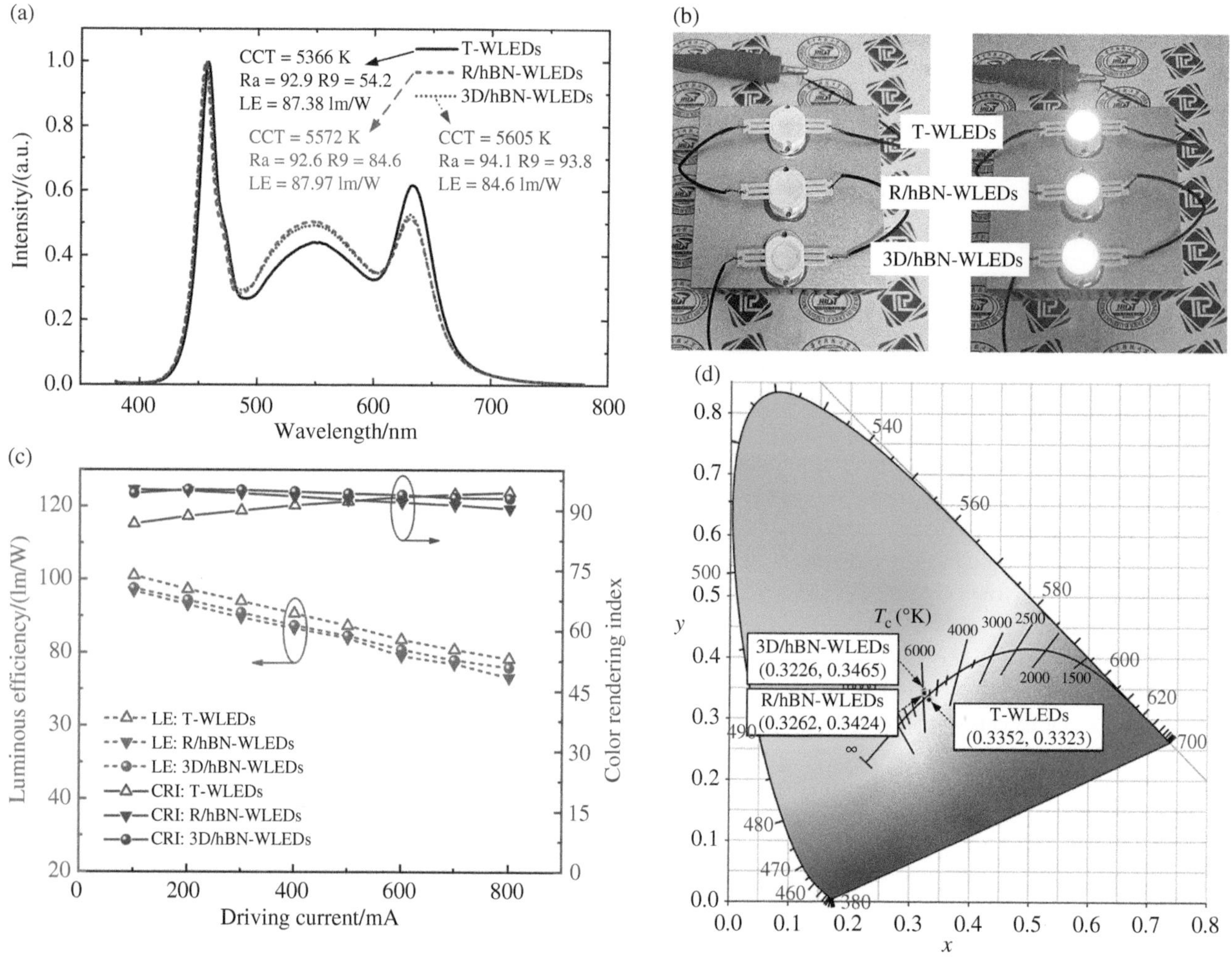

Figure 7.40 (a) Spectral power distributions and optical properties of the three kinds of WLEDs under a driving current of 500 mA. (b) Photographs of these WLEDs with and without illumination. Xie et al. [72]/with permission of Elsevier. (c) LE and CRI of the three kinds of WLEDs under different driving currents from 100 to 800 mA. (d) The coordinates of these WLEDs in the CIE 1931 diagram.

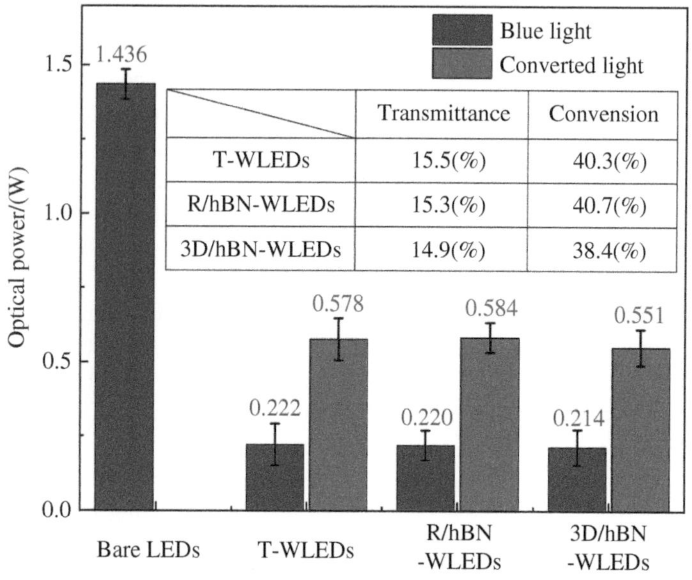

	Transmittance	Convension
T-WLEDs	15.5(%)	40.3(%)
R/hBN-WLEDs	15.3(%)	40.7(%)
3D/hBN-WLEDs	14.9(%)	38.4(%)

Figure 7.41 Calculated optical power data of the transmitted blue light and the converted yellow and red lights in different LEDs. The inserted table shows the transmittance of blue light and the conversion ratio of yellow and red lights.

platelets which obstruct the light output from the composites. Based on these results, it is validated that the construction of 3D/hBN aerogel in WLEDs does not cause evident efficiency drop under an optimized hBN filling load.

To study the thermal performances of the as-prepared 3D/hBN-WLEDs, the heat powers generated in the LED chip, T-composites, R/hBN-composites, and 3D/hBN-composites were measured and calculated according to the energy conservation law [42, 75]. The measured heat power is illustrated in Figure 7.42(a). From the results, we can see that the LED chip generates more heat than those luminescent composites, especially under higher driving current. For instance, the heat power of LED chip is 2.18 W under 700 mA, which is over two times as those in the composites (less than 1 W). In addition, the heat generation in T-composites, R/hBN-composites, and 3D/hBN-composites is quite close, which also proves that the introduction of 3D/hBN aerogel does not cause extra heat generation to the composites.

Based on the measured heat power results, the steady-state temperature distributions of these three WLEDs were simulated. The physical model is illustrated in Figure 7.42(b), and the results are displayed in Figure 7.42(c) and (d). It is seen that, thanks to the robust heat conduction ability of Al-alloy fins, most of the heat was dissipated to ambient air. But, due to the low thermal conductivity of the T-composites, the heat generated by QDs/phosphor was confined and accumulated inside the composites, resulting in a high working temperature of 102.3 °C under 500 mA. After incorporating R/hBN into the WLEDs, the highest working temperature was reduced to 87.6 °C. By constructing 3D/hBN network inside the WLEDs, the highest working temperature was further decreased to 70.2 °C. Furthermore, under a higher driving current of 700 mA, the situation of QDs in the T-WLEDs turns worse with a highest temperature of 142.7 °C, while the QDs in the 3D/hBN-WLEDs stay cool with a highest temperature of 89.1 °C. Therefore, the results clearly demonstrate the effectiveness of 3D/hBN network in the temperature reduction of WLEDs, especially under high-power condition.

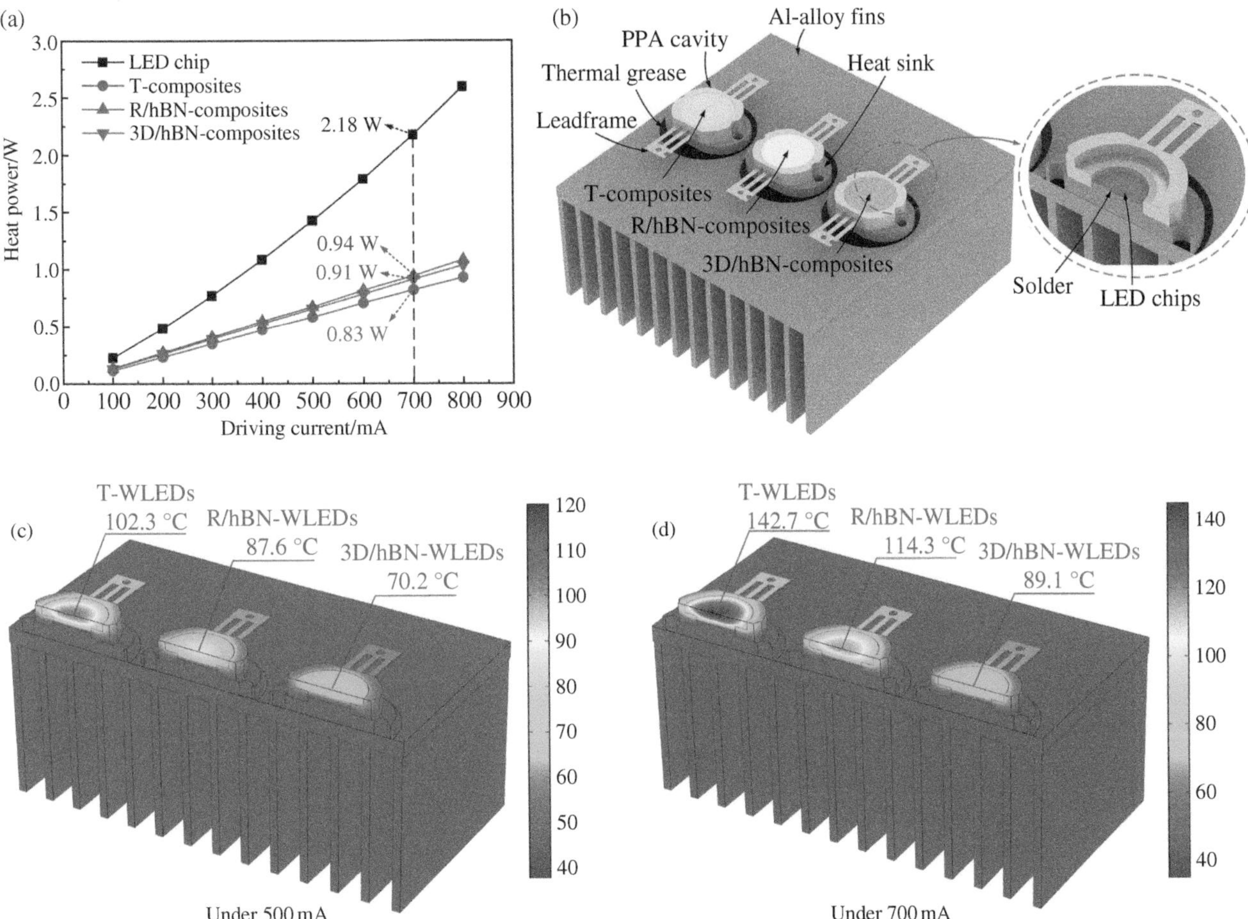

Figure 7.42 (a) Measured heat generation in LED chip, T-composites, R/hBN-composites, and 3D/hBN-composites under different driving currents from 100 to 800 mA. (b) 3D physical model for the thermal simulation. Simulated temperature distribution of these WLEDs under (c) 500 mA and (d) 700 mA, respectively.

To validate the simulation results, we also measured the surface temperature distributions of these WLEDs under 500 and 700 mA by an infrared thermal imager, as displayed in Figure 7.43(a) and (b), respectively. These WLEDs were mounted onto Al-alloy fins by thermal grease, and they were connected in series by electrical wires to make sure that they were in the same working condition, as shown in Figure 7.43(c). The surface emissivity of the composites was set as 0.96, and the testing distance between the camera and these WLEDs was controlled as 0.3 m. Their surface temperature distributions were recorded after the temperature variation of each WLEDs was less than 1% within 3 minutes, which took about 3 minutes since the WLEDs were lighted up. From the measured temperature distributions, it is seen that the experimental results have verified the FEM simulation results, and the maximum error of highest working temperature is less than 5%. Under 700 mA, the measured highest working temperature of WLEDs is reduced by 57.3 °C by the proposed 3D/hBN aerogel. Besides, it can be observed from Figure 7.43(d) that the temperature reduction is more apparent under higher driving current. Moreover, under the same working condition, the measured surface temperature of bare LEDs (without luminescent composites) in Figure 7.43(d) is significantly lower than those WLEDs with luminescent composites. This is mainly due to the good heat dissipation environment of LED chips that are directly in contact with the metal heat sink and Al-alloy fins.

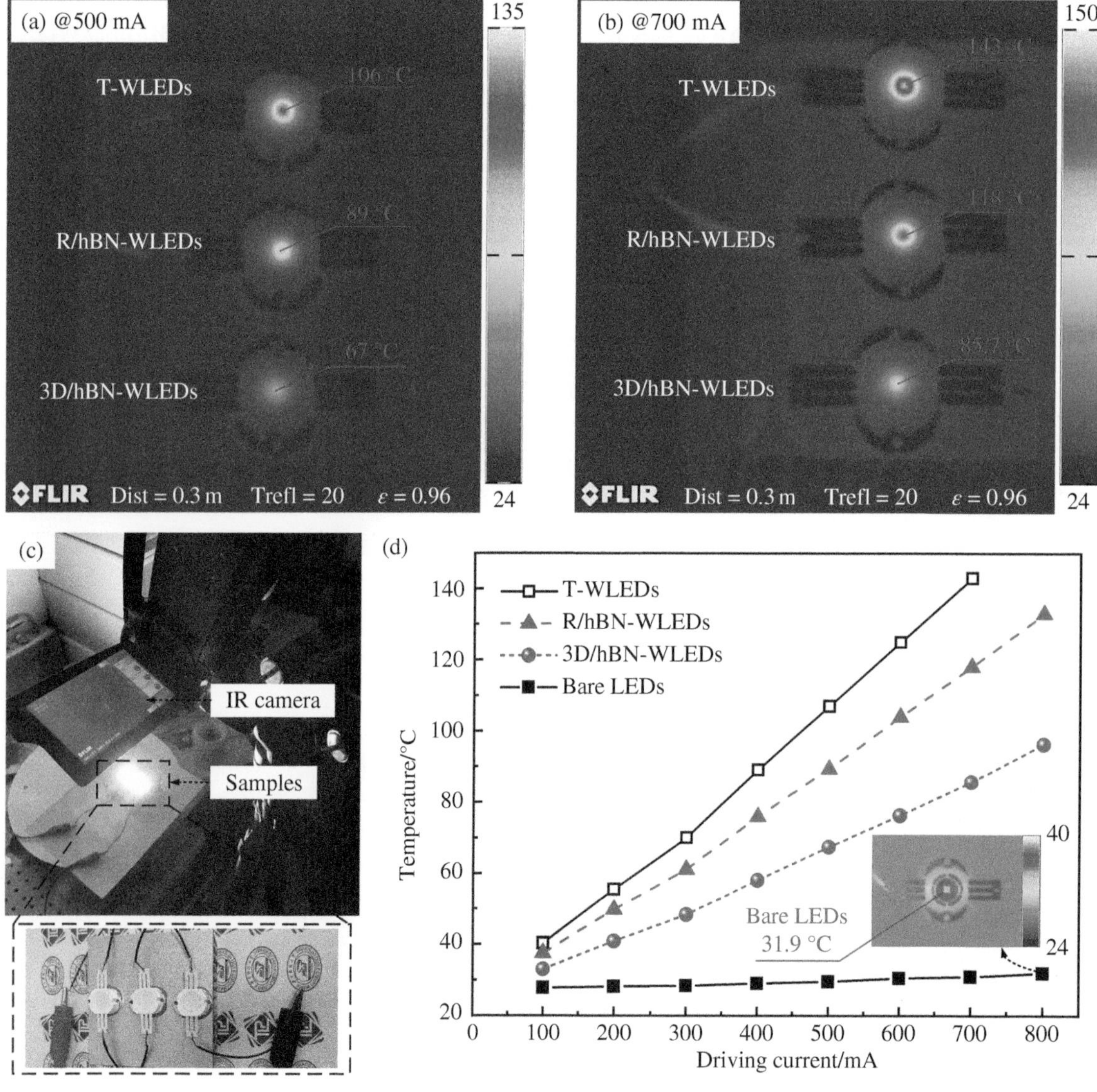

Figure 7.43 Measured surface temperature distributions of the three WLEDs under driving currents of (a) 500 mA and (b) 700 mA, respectively. (c) Temperature measurement setup. The lower figure shows the zoomed photograph of these WLEDs in the current-off status. Xie et al. [72]/with permission of Elsevier. (d) Measured highest surface temperature of the three WLEDs under different driving currents from 100 to 800 mA. The inset shows the surface temperature of bare LEDs under driving current of 800 mA.

Therefore, both the simulated and measured results have demonstrated that the heat dissipation bottleneck in WLEDs is the low-thermal-conductivity luminescent composites, and the utilization of 3D/hBN aerogel is effective in reinforcing the thermal dissipation of luminescent composites, thus solving the thermal dissipation challenge of WLEDs. Because of the decreased working temperature, the long-term stability of QDs is supposed to be enhanced. Therefore, QDs-LEDs samples which consist of LED chip and QDs-silicone (labeled as T-LEDs) and QDs-LEDs samples which consist of LED chip and 3D/hBN-QDs-silicone (labeled as 3D/hBN-LEDs) were fabricated to directly validate the benefit of 3D/hBN network toward QDs stability. Their color coordinates were controlled at around (0.34, 0.13) by slightly tuning the QDs dosage. Then, they were mounted onto an Al-alloy substrate for long-term aging test under a driving current of 200 mA, and the PL intensity of QDs was monitored every 24 hours. Before aging, the highest working temperature of QDs in the T-LEDs and 3D/hBN-LEDs was measured as 100.3 and 82.9 °C, respectively. Figure 7.44 shows the relative PL decay curves of these samples. It is

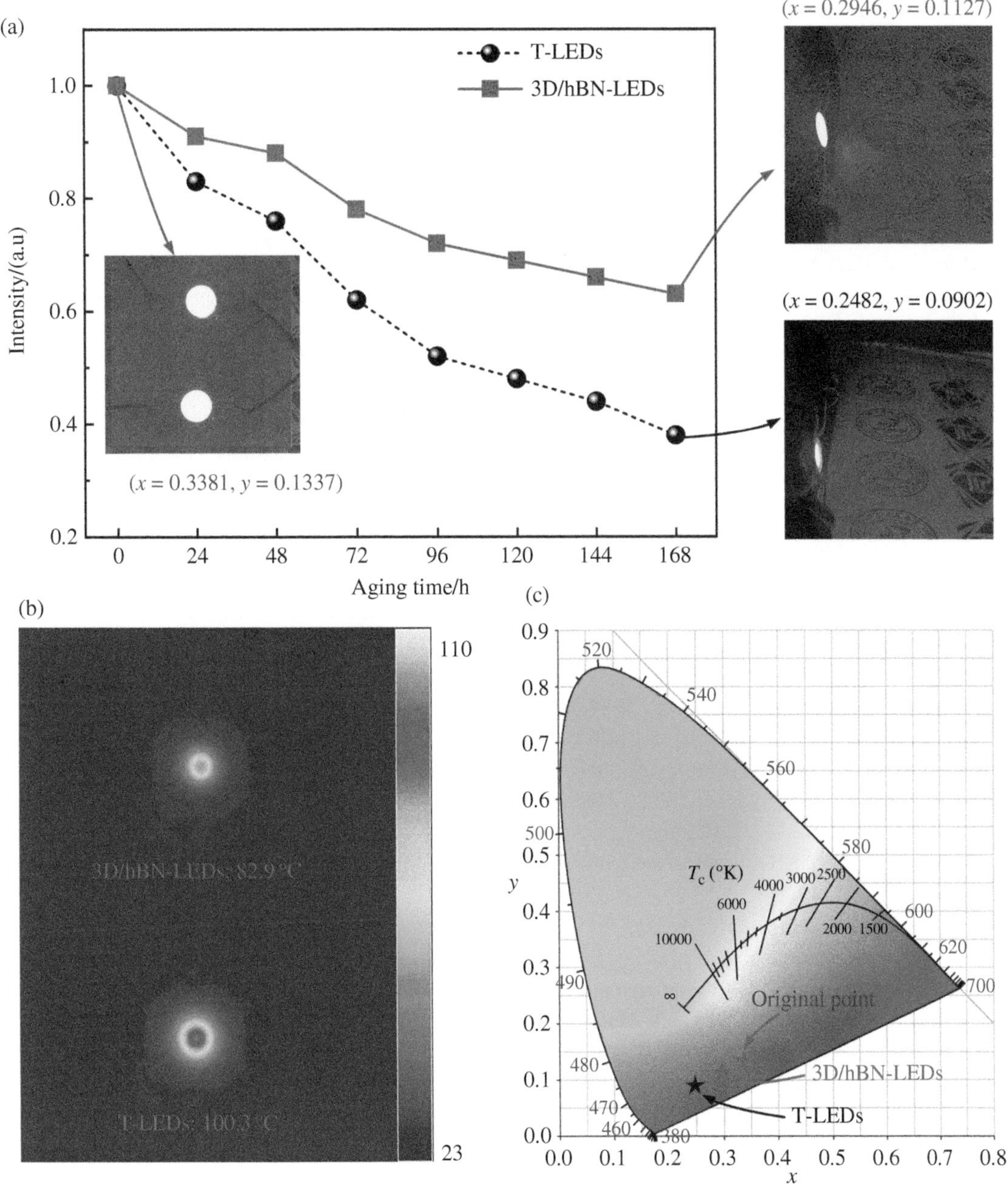

Figure 7.44 (a) PL decay curves of T-LEDs and 3D/hBN-LEDs. Insets show the illumination pattern of these LEDs under different aging stages. (b) Surface temperature distribution of the LEDs before aging, under 200 mA. (c) The color coordinates shift of the LEDs after aging for 168 hours.

seen that after 168 hours, 63% of the original PL intensity is still retained in the 3D/hBN-LEDs, while that in the T-LEDs only retains 38%. In addition, it is clearly seen from their illumination pattern that the T-LEDs lost a majority of red emission after 168 hours, and their color coordinates shifted to (0.2482, 0.0902), while the 3D/hBN-LEDs only shifted to (0.2946, 0.1127). The aging test has directly proved that the QDs stability can be enhanced by reinforcing their thermal dissipation through the 3D/hBN network.

7.4 Chapter Summary

In this chapter, we proposed several packaging-inside thermal management strategies for QDs-WLEDs in three routes: ① fabricate thermally conductive QDs composites by combining QDs with fillers with hBN; ② design packaging-inside heat transfer reinforcement structure by arranging hBN or optimizing packaging structure; and ③ establish 3D heat dissipation pathways by constructing 3D-interconnected hBN networks under extremely low filler fractions. The proposed strategies have shown marvelous ability in thermal management and maintained the devices' optical performance simultaneously, which have paved a new door for cooling QDs and other photoluminescent nanoparticles in high-power applications.

References

1 Kwak, J., Bae, W.K., Lee, D. et al. (2012). Bright and efficient full-color colloidal quantum dot light-emitting diodes using an inverted device structure. *Nano Lett.* 12 (5): 2362–2366.

2 Cho, K.-S., Lee, E.K., Joo, W.-J. et al. (2009). High-performance crosslinked colloidal quantum-dot light-emitting diodes. *Nat. Photon.* 3 (6): 341–345.

3 Zhang, Y., Xie, C., Su, H. et al. (2011). Employing heavy metal-free colloidal quantum dots in solution-processed white light-emitting diodes. *Nano Lett.* 11 (2): 329–332.

4 Pimputkar, S., Speck, J.S., DenBaars, S.P. et al. (2009). Prospects for LED lighting. *Nat. Photon.* 3 (4): 179–181.

5 Luo, X., Hu, R., Liu, S. et al. (2016). Heat and fluid flow in high-power LED packaging and applications. *Prog. Energy Combust. Sci.* 56: 1–32.

6 Jang, H.S. and Jang, H.S. (2007). White light emission from blue and near ultraviolet light-emitting diodes precoated with a Sr_3SiO_5: Ce^{3+}, Li^+ phosphor. *Opt. Lett.* 32 (23): 3444–3446.

7 Zhu, R., Luo, Z., Chen, H. et al. (2015). Realizing Rec 2020 color gamut with quantum dot displays. *Opt. Express* 23 (18): 23680.

8 Chen, H., He, J., and Wu, S.-T. (2017). Recent advances on quantum-dot-enhanced liquid-crystal displays. *IEEE J. Sel. Top. Quant.* 23 (5): 1–11.

9 Xie, B., Zhang, J., Chen, W. et al. (2017). Realization of wide circadian variability by quantum dots-luminescent mesoporous silica-based white light-emitting diodes. *Nanotechnology* 28 (42): 425204.

10 Chen, W., Wang, K., Hao, J. et al. (2016). High efficiency and color rendering quantum dots white light emitting diodes optimized by luminescent microspheres incorporating. *Nanophotonics* 5 (4): 565–572.

11 Chung, W., Yu, H.J., Park, S.H. et al. (2011). YAG and CdSe/ZnSe nanoparticles hybrid phosphor for white LED with high color rendering index. *Mater. Chem. Phys.* 126 (1–2): 162–166.

12 Dai, X., Zhang, Z., Jin, Y. et al. (2014). Solution-processed, high-performance light-emitting diodes based on quantum dots. *Nature* 515 (7525): 96–99.

13 Zhang, F., Zhong, H., Chen, C. et al. (2015). Brightly luminescent and color-tunable colloidal $CH_3NH_3PbX_3$ (X = Br, I, Cl) quantum dots: potential alternatives for display technology. *ACS Nano* 9 (4): 4533–4542.

14 Li, J., Xu, L., Wang, T. et al. (2016). 50-fold EQE improvement up to 6.27% of solution-processed all-inorganic perovskite $CsPbBr_3$ QLEDs via surface ligand density control. *Adv. Mater.* 29 (5): 1603885.

15 Jang, E., Jun, S., Jang, H. et al. (2010). White-light-emitting diodes with quantum dot color converters for display backlights. *Adv. Mater.* 22 (28): 3076–3080.

16 Wang, X., Yan, X., Li, W. et al. (2012). Doped quantum dots for white-light-emitting diodes without reabsorption of multiphase phosphors. *Adv. Mater.* 24 (20): 2742–2747.

17 Aboulaich, A., Michalska, M., Schneider, R. et al. (2013). Ce-doped YAG nanophosphor and red emitting $CuInS_2$/ZnS core/shell quantum dots for warm white light-emitting diode with high color rendering index. *ACS Appl. Mater. Interfaces* 6 (1): 252–258.

18 Jang, H.S., Yang, H., Kim, S.W. et al. (2008). White light-emitting diodes with excellent color rendering based on organically capped CdSe quantum dots and $Sr_3SiO_5:Ce^{3+}$, Li^+ phosphors. *Adv. Mater.* 20 (14): 2696–2702.

19 Zhao, Y., Riemersma, C., Pietra, F. et al. (2012). High-temperature luminescence quenching of colloidal quantum dots. *ACS Nano* 6 (10): 9058–9067.

20 Gao, F., Yang, W., Liu, X. et al. (2021). Highly stable and luminescent silica-coated perovskite quantum dots at nanoscale-particle level via nonpolar solvent synthesis. *Chem. Eng. J.* 407: 128001.

21 Yang, W., Gao, F., Qiu, Y. et al. (2019). $CsPbBr_3$-quantum-dots/polystyrene@silica hybrid microsphere structures with significantly improved stability for white LEDs. *Adv. Opt. Mater.* 7 (13): 1900546.

22 Yang, W., Fei, L., Gao, F. et al. (2020). Thermal polymerization synthesis of $CsPbBr_3$ perovskite-quantum-dots@copolymer composite: towards long-term stability and optical phosphor application. *Chem. Eng. J.* 387: 124180.

23 Xie, B., Cheng, Y., Hao, J. et al. (2018). White light-emitting diodes with enhanced efficiency and thermal stability optimized by quantum dots-silica nanoparticles. *IEEE Trans. Electron. Dev.* 65 (2): 605–609.

24 Lian, G., Tuan, C.-C., Li, L. et al. (2016). Vertically aligned and interconnected graphene networks for high thermal conductivity of epoxy composites with ultralow loading. *Chem. Mater.* 28 (17): 6096–6104.

25 Wang, M., Chen, H., Lin, W. et al. (2013). Crack-free and scalable transfer of carbon nanotube arrays into flexible and highly thermal conductive composite film. *ACS Appl. Mater. Interfaces* 6 (1): 539–544.

26 Goyal, V. and Balandin, A.A. (2012). Thermal properties of the hybrid graphene-metal nano-micro-composites: Applications in thermal interface materials. *Appl. Phys. Lett.* 100 (7): 073113.

27 Xie, B., Liu, H., Hu, R. et al. (2018). Targeting cooling for quantum dots in white QDs-LEDs by hexagonal boron nitride platelets with electrostatic bonding. *Adv. Funct. Mater.* 28 (30): 1801407.

28 Yang, H., Li, Z., Lu, B. et al. (2018). Reconstruction of inherent graphene oxide liquid crystals for large-scale fabrication of structure-intact graphene aerogel bulk toward practical applications. *ACS Nano* 12 (11): 11407–11416.

29 Camargo, P.H.C., Lee, Y.H., Jeong, U. et al. (2007). Cation exchange: a simple and versatile route to inorganic colloidal spheres with the same size but different compositions and properties. *Langmuir* 23 (6): 2985–2992.

30 Miszta, K., Dorfs, D., Genovese, A. et al. (2011). Cation exchange reactions in colloidal branched nanocrystals. *ACS Nano* 5 (9): 7176–7183.

31 Jo, J.-H., Kim, M.-S., Han, C.-Y. et al. (2018). Effective surface passivation of multi-shelled InP quantum dots through a simple complexing with titanium species. *Appl. Surf. Sci.* 428: 906–911.

32 Li, Z.-T., Chen, Y.-J., Li, J.-S. et al. (2020). Toward one-hundred-watt-level applications of quantum dot converters in high-power light-emitting diode system using water-cooling remote structure. *Appl. Therm. Eng.* 179: 115666.

33 Peng, Y., Mou, Y., Wang, T. et al. (2019). Effective heat dissipation of QD-based WLEDs by stacking QD film on heat-conducting phosphor-sapphire composite. *IEEE Trans. Electron. Dev.* 66 (6): 2637–2642.

34 Wang, J., Liu, D., Li, Q. et al. (2019). Lightweight, superelastic yet thermoconductive boron nitride nanocomposite aerogel for thermal energy regulation. *ACS Nano* 13 (7): 7860–7870.

35 Fang, H., Bai, S.-L., and Wong, C.P. (2016). "White graphene"-hexagonal boron nitride based polymeric composites and their application in thermal management. *Compos. Commun.* 2: 19–24.

36 Poostforush, M. and Azizi, H. (2014). Superior thermal conductivity of transparent polymer nanocomposites with a crystallized alumina membrane. *Express Polym. Lett.* 8 (4): 293–299.

37 Patel, T., Suin, S., Bhattacharya, D. et al. (2013). Transparent and thermally conductive polycarbonate (PC)/alumina (Al_2O_3) nanocomposites: Preparation and characterizations. *Polym. Plast. Technol.* 52 (15): 1557–1565.

38 Yuan, C., Duan, B., Li, L. et al. (2015). Thermal conductivity of polymer-based composites with magnetic aligned hexagonal boron nitride platelets. *ACS Appl. Mater. Interfaces* 7 (23): 13000–13006.

39 Song, W.L., Wang, P., Cao, L. et al. (2012). Polymer/boron nitride nanocomposite materials for superior thermal transport performance. *Angew. Chem. Int. Ed.* 51 (26): 6498–6501.

40 Lin, Z., Liu, Y., Raghavan, S. et al. (2013). Magnetic alignment of hexagonal boron nitride platelets in polymer matrix: toward high performance anisotropic polymer composites for electronic encapsulation. *ACS Appl. Mater. Interfaces* 5 (15): 7633–7640.

41 Lin, Y., Williams, T.V., and Connell, J.W. (2009). Soluble, exfoliated hexagonal boron nitride nanosheets. *J. Phys. Chem. Lett.* 1 (1): 277–283.

42 Xie, B., Chen, W., Hao, J. et al. (2016). Structural optimization for remote white light-emitting diodes with quantum dots and phosphor: packaging sequence matters. *Opt. Express* 24 (26): A1560.

43 Luther, J.M. and Pietryga, J.M. (2013). Stoichiometry control in quantum dots: a viable analog to impurity doping of bulk materials. *ACS Nano* 7 (3): 1845–1849.

44 Hu, R., Luo, X., and Zheng, H. (2012). Hotspot location shift in the high-power phosphor-converted white light-emitting diode packages. *Jpn. J. Appl. Phys.* 51 (9S2): 09mk05.
45 Xie, B., Hu, R., and Luo, X. (2016). Quantum dots-converted light-emitting diodes packaging for lighting and display: status and perspectives. *J. Electron. Packag.* 138 (2): 020803.
46 Liang, R., Dai, J., Ye, L. et al. (2017). Improvement of interface thermal resistance for surface-mounted ultraviolet light-emitting diodes using a graphene oxide silicone composite. *ACS Omega* 2 (8): 5005–5011.
47 Hu, R., Zheng, H., Hu, J. et al. (2013). Comprehensive study on the transmitted and reflected light through the phosphor layer in light-emitting diode packages. *J. Disp. Technol.* 9 (6): 447–452.
48 Orloff, L., De Ris, J., and Markxstein, G.H. Upward turbulent fire spread and burning of fuel surface. *Proc. Combust. Inst.* 15 (1): 183–192.
49 Zhang, K., Feng, Y., Wang, F. et al. (2017). Two dimensional hexagonal boron nitride (2D-hBN): synthesis, properties and applications. *J. Mater. Chem. C* 5 (46): 11992–12022.
50 Yuan, C., Li, J., Lindsay, L. et al. (2019). Modulating the thermal conductivity in hexagonal boron nitride via controlled boron isotope concentration. *Commun. Phys.* 2 (1): 43.
51 Yuan, J., Qian, X., Meng, Z. et al. (2019). Highly thermally conducting polymer-based films with magnetic field-assisted vertically aligned hexagonal boron nitride for flexible electronic encapsulation. *ACS Appl. Mater. Interfaces* 11 (19): 17915–17924.
52 Zeng, X., Yao, Y., Gong, Z. et al. (2015). Ice-templated assembly strategy to construct 3D boron nitride nanosheet networks in polymer composites for thermal conductivity improvement. *Small* 11 (46): 6205–6213.
53 Yao, Y., Sun, J., Zeng, X. et al. (2018). Construction of 3D skeleton for polymer composites achieving a high thermal conductivity. *Small* 14 (13): 1704044.
54 Bo, Z., Ying, C., Zhu, H. et al. (2019). Bifunctional sandwich structure of vertically-oriented graphenes and boron nitride nanosheets for thermal management of LEDs and Li-ion battery. *Appl. Therm. Eng.* 150: 1016–1027.
55 Han, J., Du, G., Gao, W. et al. (2019). An anisotropically high thermal conductive boron nitride/epoxy composite based on nacre-mimetic 3D network. *Adv. Funct. Mater.* 29 (13): 1900412.
56 Deville, S., Saiz, E., Nalla, R. et al. (2006). Freezing as a path to build complex composites. *Science* 311 (5760): 515–518.
57 Zhou, S., Xie, B., Ma, Y. et al. (2019). Effects of hexagonal boron nitrides sheets on the optothermal performances of quantum dots-converted white LEDs. *IEEE Trans. Electron Dev.* 66 (11): 4778–4783.
58 Tamborra, M., Striccoli, M., Comparelli, R. et al. (2004). Optical properties of hybrid composites based on highly luminescent CdS nanocrystals in polymer. *Nanotechnology* 15 (4): S240–S244.
59 Zhang, H., Cui, Z., Wang, Y. et al. (2003). From water-soluble CdTe nanocrystals to fluorescent nanocrystal-polymer transparent composites using polymerizable surfactants. *Adv. Mater.* 15 (10): 777–780.
60 Chen, W., Wang, K., Hao, J. et al. (2015). Highly efficient and stable luminescence from microbeans integrated with Cd-free quantum dots for white-light-emitting diodes. *Part. Part. Syst. Charact.* 32 (10): 922–927.
61 Zhao, B., Yao, Y., Gao, M. et al. (2015). Doped quantum dot@silica nanocomposites for white light-emitting diodes. *Nanoscale* 7 (41): 17231–17236.
62 Zhou, C., Shen, H., Wang, H. et al. (2012). Synthesis of silica protected photoluminescence QDs and their applications for transparent fluorescent films with enhanced photochemical stability. *Nanotechnology* 23 (42): 425601.
63 Hao, J.-J., Zhou, J., and Zhang, C.-Y. (2013). A tri-n-octylphosphine-assisted successive ionic layer adsorption and reaction method to synthesize multilayered core–shell CdSe–ZnS quantum dots with extremely high quantum yield. *Chem. Commun.* 49 (56): 6346–6348.
64 Li, H., Wu, K., Lim, J. et al. (2016). Doctor-blade deposition of quantum dots onto standard window glass for low-loss large-area luminescent solar concentrators. *Nat. Energy* 1 (12): 16157.
65 Şahin, D., Ilan, B., and Kelley, D.F. (2011). Monte-Carlo simulations of light propagation in luminescent solar concentrators based on semiconductor nanoparticles. *J. Appl. Phys.* 110 (3): 033108.
66 Ma, Y., Hu, R., Yu, X. et al. (2017). A modified bidirectional thermal resistance model for junction and phosphor temperature estimation in phosphor-converted light-emitting diodes. *Int. J. Heat Mass Transf.* 106: 1–6.
67 Xie, H., Fujii, M., and Zhang, X. (2005). Effect of interfacial nanolayer on the effective thermal conductivity of nanoparticle-fluid mixture. *Int. J. Heat Mass Transf.* 48 (14): 2926–2932.
68 Zhang, X., Zhang, T., Wang, Z. et al. (2018). Ultralight, superelastic, and fatigue-resistant graphene aerogel templated by graphene oxide liquid crystal stabilized air bubbles. *ACS Appl. Mater. Interfaces* 11 (1): 1303–1310.

69 Lv, L., Zhang, P., Cheng, H. et al. (2016). Solution-processed ultraelastic and strong air-bubbled graphene foams. *Small* 12 (24): 3229–3234.

70 Li, J., Li, F., Zhao, X. et al. (2020). Jelly-inspired construction of the three-dimensional interconnected BN network for lightweight, thermally conductive, and electrically insulating rubber composites. *ACS Appl. Electron. Mater.* 2 (6): 1661–1669.

71 Aslam, M., Imam, S.S., Aqil, M. et al. (2016). Levofloxacin loaded gelrite-cellulose polymer based sustained ocular drug delivery: formulation, optimization and biological study. *J. Polym. Eng.* 36 (8): 761–769.

72 Xie, B., Wang, Y., Liu, H. et al. (2022). Targeting cooling for quantum dots by 57.3°C with air-bubbles-assembled three-dimensional hexagonal boron nitride heat dissipation networks. *Chem. Eng. J.* 427: 130958.

73 Xu, Y., Kraemer, D., Song, B. et al. (2019). Nanostructured polymer films with metal-like thermal conductivity. *Nat. Commun.* 10 (1): 1771.

74 Zinovchuk, A.V., Malyutenko, O.Y., Malyutenko, V.K. et al. (2008). The effect of current crowding on the heat and light pattern in high-power AlGaAs light emitting diodes. *J. Appl. Phys.* 104 (3): 033115.

75 Xie, Y., Yang, D., Zhang, L. et al. (2019). Highly efficient and thermally stable QD-LEDs based on quantum dots-SiO_2-BN nanoplate assemblies. *ACS Appl. Mater. Interfaces* 12 (1): 1539–1548.

8

Thermal Management in Downhole Devices

Unlike conventional electronics, the electronic devices at high temperature need to be thermally insulated from the harsh environment, instead of dissipating the generated heat to environment. In this chapter, the downhole devices are taken as an example to describe the thermal management measures in high-temperature environments.

The downhole devices were widely used in the oil and gas industries to detect the viscosity, pressure, and temperature [1–4]. In general, the temperature and pressure in a downhole environment exceed 200 °C and 135 MPa, such that the standard electronics will quickly exceed their operational temperatures (125 °C) and suffer from the high failure risk [5, 6]. Ongoing operation in such environments can degrade the detection performance as well as cause severe accidents. The use of electronic components that can effectively operate at high temperatures would solve the temperature problem. However, the current options require expensive silicon-on-insulator (SOI) designs and die-attach materials [7–9]. High-temperature electronic components exist that can operate at temperatures as high as 300 °C. However, the number of commercial vendors and the variety of high-temperature electronic components and sensors that can withstand such high temperatures are very limited. Available high-temperature electronic components, which often require SOI technology at the semiconductor level, and high-temperature die-attach materials, solder, and printed circuit boards at the packaging level can be very expensive.

A thermal management system (TMS) for downhole electronics is currently more cost-effective and reliable than developing specialized electronic components. However, the harsh downhole conditions and the complex well-logging structures have impeded the implementation of current TMS designs. Figure 8.1 shows a well-logging schematic. The logging tool burrows into the high-temperature and high-pressure (HTHP) muds that are thousands of meters deep. Hence, unlike conventional electronics [10–12], a downhole system needs to be hermetically sealed and protected from high pressures. Furthermore, the electronic components inside the system must be shielded from the corrosive downhole fluids.

Various active cooling techniques have been proposed for thermal management, including thermoelectric cooling [5], vapor compression refrigeration [13], sorption cooling [14], convection cooling cycles [15], refrigerant circulation cooling [16], and thermoacoustic refrigeration [17]. These cooling techniques dissipate excess heat from the electronics into the surrounding downhole fluids. However, they generally require extra power, cooling liquids, and other moving components, which further complicate the system.

In addition to active solutions, passive cooling techniques have also been developed. Parrott et al. [18] proposed a passive cooling system that combined a Dewar flask with a heat sink. However, the overall thermal resistance between the electronics and the heat sink was quite large, resulting in a high-temperature difference and a reduced reliability period for the electronic components. Jakaboski [19] developed a closed coolant flow loop to provide a thermal path between the electronic components and the heat sink. However, the coolant pump and fluid expansion compensator complicated the system.

8.1 Experimental Analysis of Passive Thermal Management Systems

To tackle this issue, a passive thermal management system (PTMS) was proposed for downhole electronics in harsh thermal environments. A vacuum flask was employed to insulate the electronic components from the harsh external environment. The internal heat generation must also be transferred out of the components and stored elsewhere. Phase change materials (PCMs) are viable thermal storage materials due to their significant latent heat capacity [20–24]. The TMS developed here

Thermal Management for Opto-electronics Packaging and Applications, First Edition. Xiaobing Luo, Run Hu, and Bin Xie.

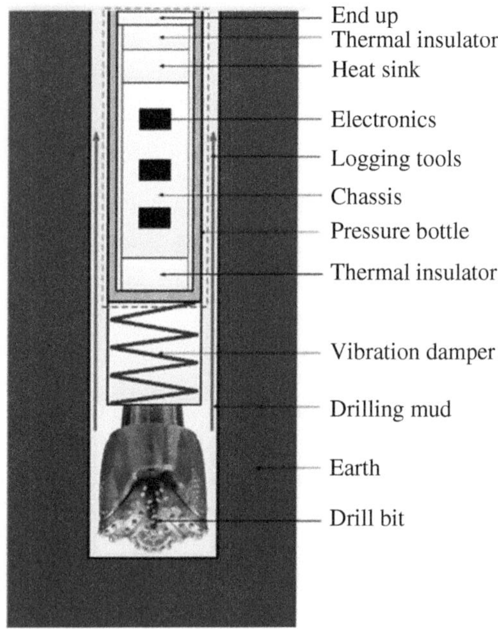

Figure 8.1 Schematic of well logging.

used eutectic salts as the PCM. Heat pipes were used to provide an efficient heat transfer path from the electronic components to the PCM to reduce the temperature difference between them and, thereby, extend the viable operating time.

8.1.1 Experimental Setup

The experimental setup was designed to simulate practical operating conditions. Figure 8.2(a) shows a schematic of the experimental setup. An oven created a high-temperature environment with temperatures from 5 to 300 °C, with an accuracy of ±1 °C. To ensure uniform heating, two Teflon blocks were used to prevent direct contact between the logging tools and the oven. K-type thermocouples detected the electronic component temperatures to evaluate the TMS characteristics.

Figure 8.2(b) shows a schematic of the original logging tool, which was comprised of a pressure bottle, electronic components, chassis, heat sink, and adiabatic plug. Two resistance heaters (40 mm × 40 mm × 3 mm) were used as the electronic components. Thermal grease (FL-658) was used as the thermal interface material (TIM) to attach the heaters to the chassis. The grease thermal conductivity was approximately $3\,\mathrm{W \cdot m^{-1} \cdot K^{-1}}$. Figure 8.2(c) shows a schematic of the logging tool with the TMS having a vacuum flask, heat pipes, and the PCM.

The vacuum flask was made of titanium alloy, which could maintain a pressure of 1000 MPa. The flask was built of two concentric stainless steel tubes that were permanently cold-welded at both ends. The annular space between the tubes was evacuated at high temperatures, providing an excellent barrier to heat transfer via conduction or convection. Both ends of the flask were sealed to reduce heat gain from the environment. The vacuum flask was 900 mm in length with inner and outer diameters of 73 and 90 mm, respectively.

The round heat pipe was made of copper with a length of 250 mm and a diameter of 5 mm. Water was used as the working fluid inside the pipe. The TMS had five heat pipes to connect the chassis to the PCM.

The PCM provided thermal storage for the electronic components and any heat gain into the vacuum flask. The low melting point and high latent heat of the PCM kept the electronic temperatures below their maximum temperature for a longer period. Therefore, the PCM selection is critical for the TMS characteristics. Eutectic salts, organic paraffin, and eutectic metal alloys are common PCMs with high latent heat and low melting points. The PCM properties were evaluated to improve the system design.

Differential scanning calorimeter (DSC) tests were conducted to obtain the thermal properties and phase transition characteristics of the PCM, with a temperature rise rate of $5\,^{\circ}\mathrm{C \cdot min^{-1}}$. Figure 8.3 shows the typical DSC curves. The PCM have different melting temperatures and different heat transfer rates with the latent heats calculated from the DSC test curves as:

$$L_{\mathrm{PCM}} = \frac{\int_{\mathrm{Onset}}^{\mathrm{End}} q\mathrm{d}T}{K_{\mathrm{VT}}m} \tag{8.1}$$

Where q is the heat flux density, T is the temperature, K_{VT} is the temperature rise rate, and m is the sample mass.

Onset represents the start of the temperature change and the end represents the end of the temperature change. The top of the curve represents the melting point. A TC3000 (Xiatech) was used to measure the thermal conductivities. Table 8.1 lists the properties of the different PCMs. The eutectic salts and the organic paraffin have higher latent heat than the eutectic metal alloy. However, the eutectic metal alloy has a much higher thermal conductivity. A good PCM should have a high thermal conductivity and a high latent heat. However, where the overall system mass is limited, a larger mass impedes the logging system. Although the eutectic metal alloy has a significant thermal conductivity, a large mass is still required for the logging tool. Based on latent heat, thermal conductivity, and mass considerations, the present study used the eutectic salts for the PCM.

The experiments were divided into three groups to highlight the effects of the vacuum flask, PCM, and heat pipes on the electronic component temperatures. Two K-type thermocouples were attached to the surfaces of the electrical heaters and

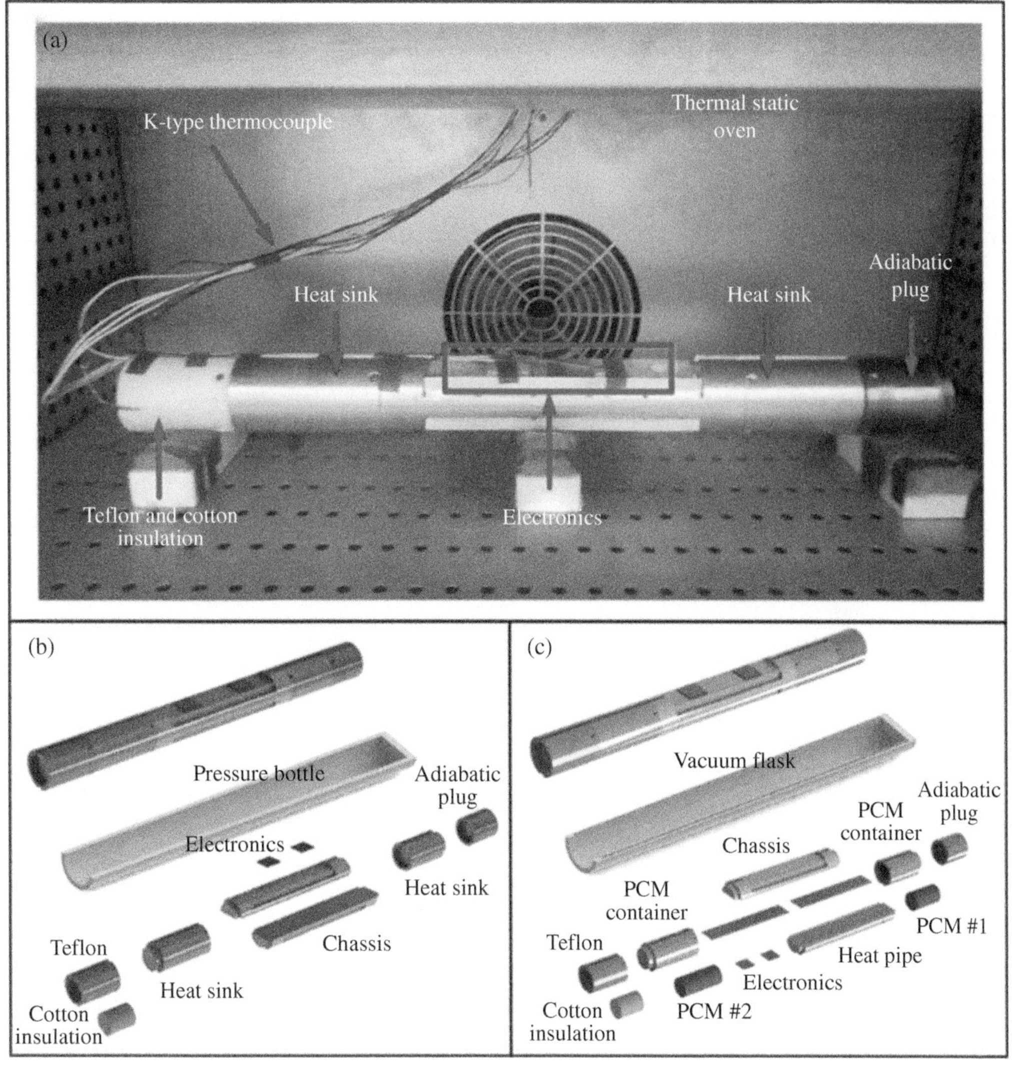

Figure 8.2 (a) Schematic of the experimental setup: (b) the original logging tool and (c) the logging tool with TMS.

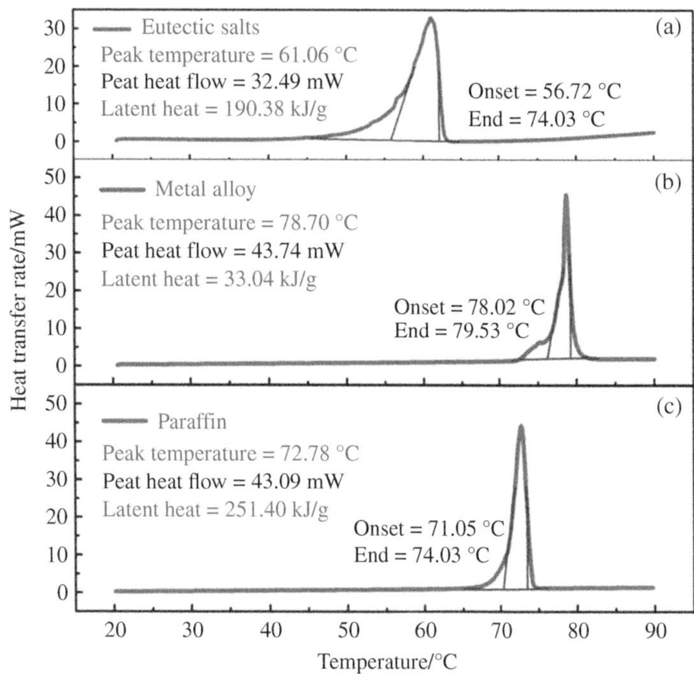

Figure 8.3 (a–c) DSC curve of different PCMs.

Table 8.1 Properties of PCMs.

PCMs	Melting point (°C)	Latent heat of fusion ($kJ \cdot kg^{-1}$)	Thermal conductivity ($W \cdot m^{-1} \cdot K^{-1}$)	Density ($kg \cdot m^{-3}$)	Specific heat solid ($J \cdot kg^{-1} \cdot K^{-1}$)	Specific heat liquid ($J \cdot kg^{-1} \cdot K^{-1}$)
Eutectic salts	61.1	190.4	0.50(s) 0.73(l)	1485	1930	1930
Organic paraffin	72.8	251.4	0.2	860	2000	2000
Metal alloys	78.7	33.04	18.8	9580	146	184

Table 8.2 Experimental setups of different groups.

Group	Vacuum flask	PCMs	Heat pipes	Boundary conditions
1	√	×	×	$P = 0\text{–}50\ W$
2	√	√	×	$P = 30\ W$
3	√	√	√	$P = 30\ W$

two more were inserted into the PCM. An additional thermocouple was fixed in the oven to measure the ambient temperature. The initial test temperature was room temperature. Table 8.2 lists the three experimental setups used to assess the thermal characteristics of the TMS.

8.1.2 Experimental Results

The Group 1 experiments assess the thermal insulation characteristics of the vacuum flask. Experiments were conducted with and without the vacuum flask using heating powers ranging from 0 to 50 W to replicate the actual working conditions. Figure 8.4 shows the electronic component temperatures at different heating powers. The temperatures exceed 125 °C after 0.38 hours without the vacuum flask. The vacuum flask keeps the temperatures below 125 °C for 6 hours for heating powers lower than 20 W. This indicates that the vacuum flask can prevent heat gain in a hot downhole environment. However, when the heating power exceeds 20 W, the operating time sharply decreases because the heat generated by the electronic components accumulates internally, which increases the component temperatures.

The Group 2 experiments analyze the thermal storage characteristics of the PCM. The PCM was 200 mm long and weighed 700 g. Figure 8.5 compares the electronic component temperatures with and without the PCM. The results show

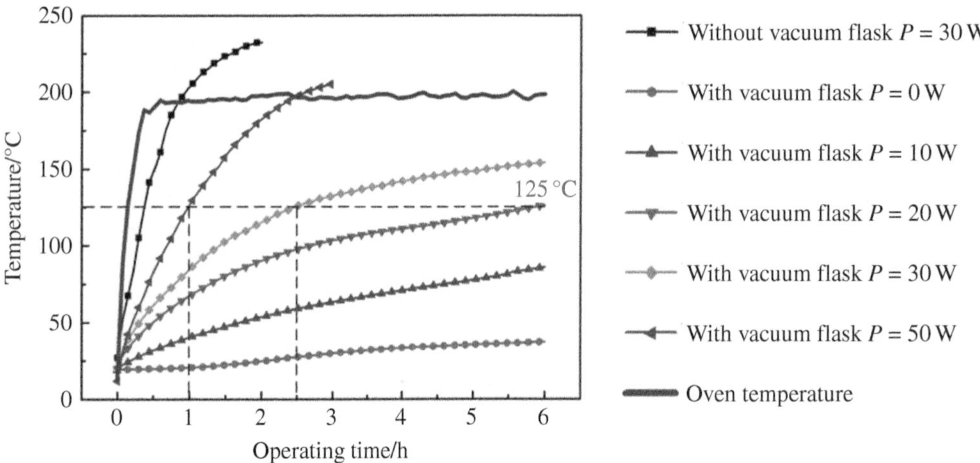

Figure 8.4 The electronics temperature at different heating power versus operating time.

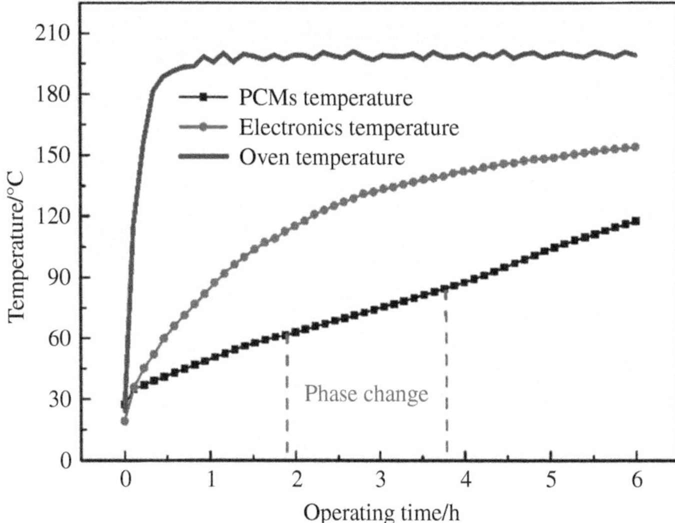

Figure 8.5 Comparison of the electronics temperature with and without PCMs.

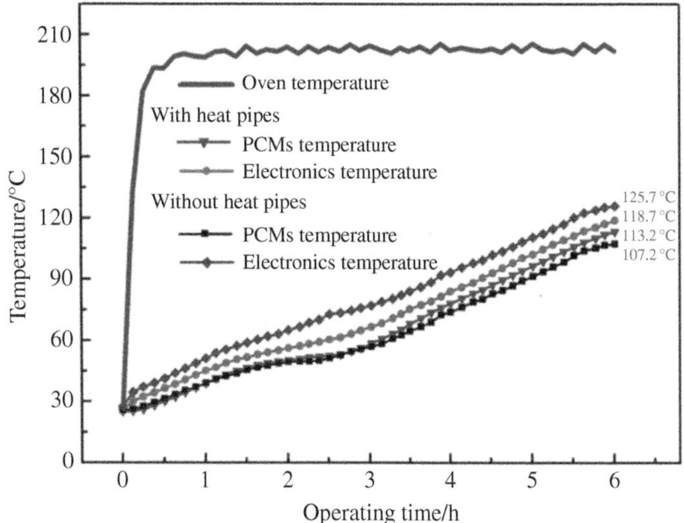

Figure 8.6 Comparison of the electronics temperature with and without heat pipes.

that the electronic component temperatures are reduced from 154 to 124 °C by the PCM. The lower electronic component temperatures indicate that the PCM introduces additional thermal storage for the components so that the component temperatures rise less as the energy is absorbed by the phase change process. The temperatures then remain stable at lower temperatures for an extended period, which improves component performance.

The Group 3 experiments investigated the effect of the heat pipes on the component temperatures. Figure 8.6 compares the component temperatures with and without the heat pipes. The temperature is 125.7 °C without a heat pipe. The temperature difference between the components and the PCM is 17.5 °C which shows that the thermal storage capacity of the PCM is not fully utilized. After the heat pipes are added, the temperature difference between the components and PCM decreases to 5.5 °C, with a component temperature of 118.7 °C and a reduction of 7.0 °C. This confirms that the heat pipes provide an efficient heat transfer path from the components to the PCM.

Figure 8.7 shows the thermal characteristics of the TMS. The total power of the electronic components is 30 W. The PCM mass is 700 g. The results show that the electronic component temperature is 119 °C after 6 hours, indicating that the TMS effectively maintains the component temperature below 125 °C for an operating time of 6 hours. This result completely satisfies the requirements for the downhole electronic components.

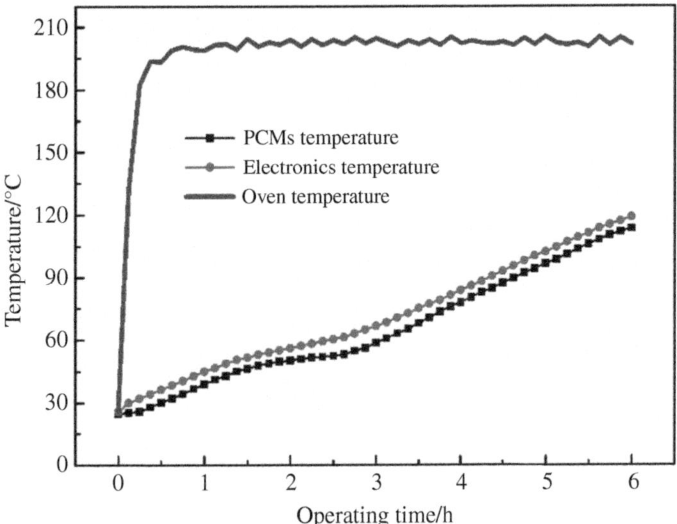

Figure 8.7 Thermal performance of the proposed TMS.

8.1.3 Finite-element Analysis

The system thermal characteristics were then analyzed using a finite-element analysis using a two-dimensional symmetrical model. Figure 8.8(a) and (b) show schematics of the physical model with and without the TMS. The boundary conditions were the same as those in the experiments. A heat transfer model was applied in the porous media to simulate the phase change process using a variable specific heat to represent the latent heat storage process to study the phase change thermal storage mechanism. The model included the radiation term in the equivalent thermal conductivity, with the predictions compared with experimental data to confirm the accuracy.

Figure 8.8(c) and (d) show the temperature fields for both cases. The logging tool internal temperature with the TMS is lower than without the TMS where the electronic component temperatures exceed 200 °C within an hour. The results agree with the experimental data. Without the thermal insulation, the ambient heat immediately transfers into the logging tool through conduction and radiation due to the significant temperature difference between the tool and the environment.

The heat flow inside the logging tool was also investigated. Figure 8.8(e) and (f) show the simulated heat flux vectors near the electronics at 10 minutes where the arrow size is proportional to the heat flux. Without the TMS, heat is transferred from the heat sink to the electronic components, thus accelerating the temperature increase in the components. With the TMS, the heat dissipates from the electronic components to the PCM, which reduces the component temperatures. Although heat flows into the components from the pressure bottle side, there is almost no heat flow from the vacuum flask side, which further confirms the thermal insulation characteristics of the vacuum flask.

A qualitative analysis was conducted to further understand the lower temperatures of the electronic components with the TMS. Heat is gained from the environment and from the heat generated by the electronic components. The sensible heat is given by:

$$\mathrm{SH} = mC_\mathrm{p}(T_\mathrm{e} - T_0) \tag{8.2}$$

The energy gained by the latent heat is given by:

$$\mathrm{LH} = m_\mathrm{PCM}\ L_\mathrm{PCM} \tag{8.3}$$

Where SH is the sensible heat storage, LH is the latent heat storage, m is the mass, C_p is the constant specific heat, T_0 is the initial temperature, T_e is the final temperature. The heat transfer is the difference between the thermal storage and the heat generated by the electronic components.

Figure 8.9 shows the heat output and the thermal storage for both cases. The TMS reduces the heat gain by 37.6%, which reduces the thermal storage. Per Equation (8.2), T_e decreases as SH decreases. The PCM absorbs 20.7% of the thermal storage, which results in a lower electronic component temperature.

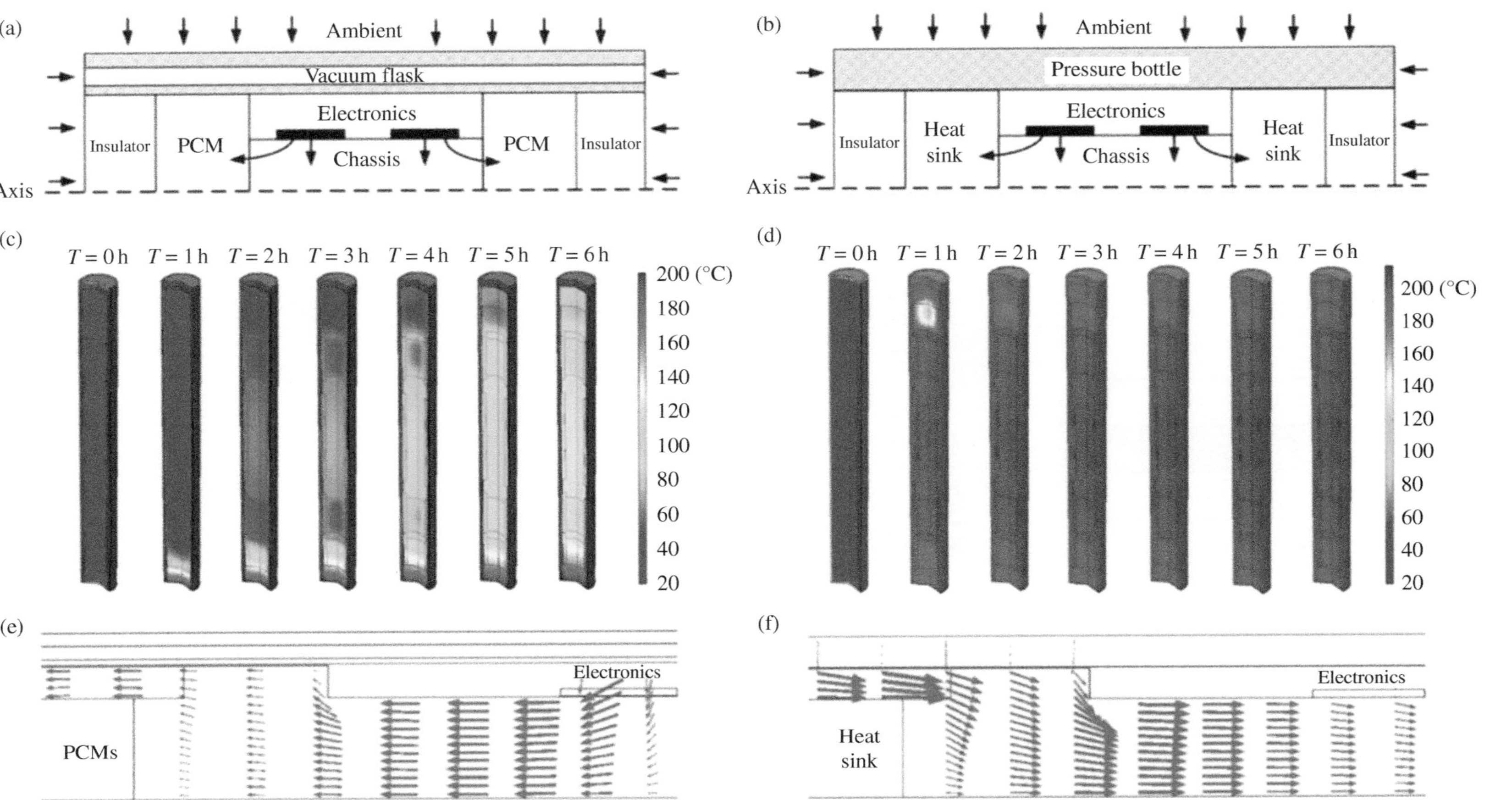

Figure 8.8 Schematic diagram of the physical model: (a) with the addition of TMS; (b) without the addition of TMS; (c) and (d) temperature field of the system versus operating time of the two cases, respectively; (e) and (f) simulated vectors of heat flux near the electronics of the two cases at 10 minutes, respectively.

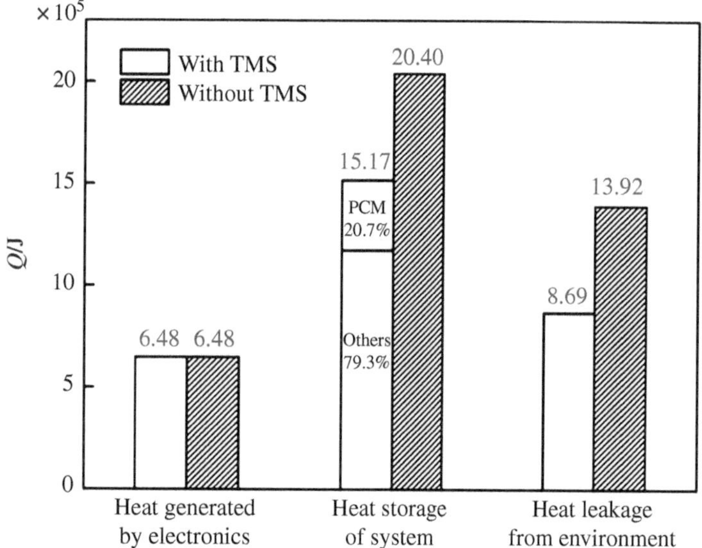

Figure 8.9 Heat output and heat storage of the system with and without TMS.

8.2 Thermal Modeling for Downhole Devices

Previous studies have shown that PTMSs have successfully protected electronics that operate in high ambient temperature environments for up to several hours. However, the percentage of the contribution from different heat transfer modes remains unclear, and the numerical models remain incomplete. To solve this issue, a 3D numerical model that couples all heat transfer modes was proposed for PTMSs of logging tool. Unlike previous simplistic models, the one proposed here considers solid heat conduction, natural air convection, thermal radiation, and the phase change heat storage processes.

In this section, a 3D numerical PTMS model for the logging tool was proposed. The proposed model considers solid heat conduction, natural air convection, thermal radiation, and phase change processes simultaneously (named as the SNTP numerical model). The SNTP numerical model was compared to the previous model with the same geometric model. The previous model only considers solid heat conduction and phase change processes (named as the SP numerical model). Additionally, an experiment was conducted to verify the accuracy of the SNTP numerical model. Finally, the transfer process was elucidated, and quantitative analyses of solid heat conduction, natural air convection, thermal radiation, and phase change heat storage were carried out through numerical simulation.

8.2.1 Thermal Modeling

Figure 8.10 shows the geometric model (i.e. diagram) for the PTMS of a typical logging tool. A vacuum bottle is utilized to prevent radial heat transfer from the high-temperature downhole environment. The insulators 1 and 2 are located at the end of the vacuum bottle to block the axial environmental heat transfer. It is worth noting that there is a vacuum layer on the closed side of the vacuum bottle, but not on the open side. The space between the vacuum bottle and the metal skeleton is filled with air. Heat source 1 (20 W) and heat source 2 (10 W) are connected to the metal skeleton through thermal silicone pads. PCMs are applied for heat storage at both ends of the skeleton. Forced convection heat exchange occurs between the high-temperature mud and the outer wall of the metal bottle during logging operation.

The SNTP numerical model is described in detail. Unlike previous simplistic models, the one proposed here considers solid heat conduction, natural air convection, thermal radiation, and phase change process inside the vacuum bottle. To simulate the transient heat transfer process efficiently and accurately, the following assumptions are made:

1) The complex heat transfer process of the vacuum layer is equated to a solid heat transfer process with very low thermal conductivity [25].
2) The effect of contact thermal resistance on heat transfer is ignored [26].
3) The change in the physical properties of materials with respect to temperature is ignored [27].
4) The electronics are treated as uniform heat sources.

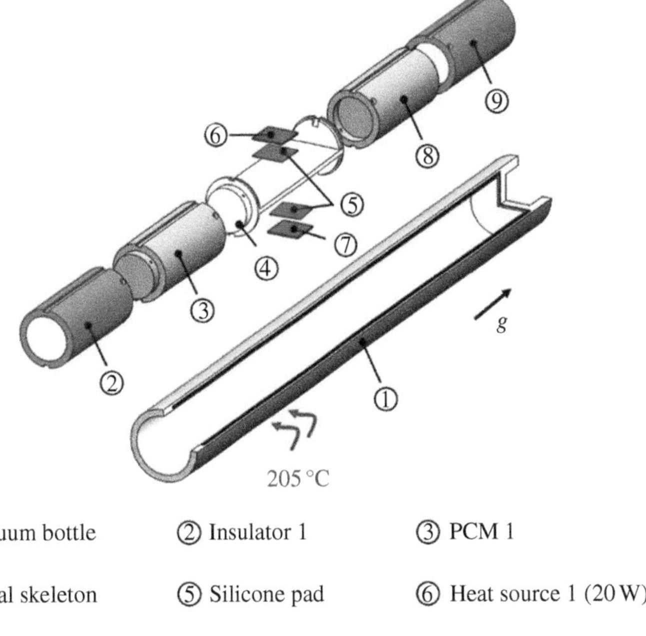

Figure 8.10 Geometric PTMS model of a typical logging tool.

The electronics inside the logging tool constantly generate heat during operation. Most of the heat is dissipated through the metal skeleton by solid heat conduction [28], which can be expressed as:

$$\rho C_\mathrm{p}\frac{\partial T}{\partial t} = \nabla\cdot(\kappa\nabla T) + q \tag{8.4}$$

Where ρ, C_p, κ, T and t are the density, specific heat, thermal conductivity, temperature of solid material and operation time, respectively. q means the heat power.

Since air exists between the skeleton and the inner wall of the vacuum bottle, natural convection occurs within the closed cavity due to the temperature difference and gravity. In this study, the Boussinesq approximation is adopted to calculate the natural air convection in the closed cavity [29–31]. It is assumed that the air density change has no effect on the flow field but affects the buoyancy force. The continuity equation can be simplified as:

$$\nabla\cdot\boldsymbol{u} = 0 \tag{8.5}$$

The conservation of momentum equation for natural air convection [32] can be expressed as:

$$\rho_0\left(\frac{\partial\boldsymbol{u}}{\partial t} + \boldsymbol{u}\cdot\nabla\boldsymbol{u}\right) = -\nabla p + \mu_\mathrm{air}\nabla^2\boldsymbol{u} + \rho_0\mathbf{g} - \rho_0\left(\frac{T-T_0}{T}\right)\mathbf{Fg} \tag{8.6}$$

where $\boldsymbol{u}$ is the air velocity, p is the air pressure, ρ_0 is the air initial density, μ_air is the air dynamic viscosity, T is the air temperature, and $\mathbf{Fg}$ is the gravity.

The equation of energy conservation for natural convection can be expressed as:

$$\rho_\mathrm{air}C_\mathrm{air}\frac{\partial T}{\partial t} + \rho_\mathrm{air}C_\mathrm{air}\boldsymbol{u}\cdot\nabla T = \nabla\cdot(\kappa_\mathrm{air}\nabla T) + q \tag{8.7}$$

Where ρ_air, C_air and κ_air are the density, specific heat capacity and thermal conductivity of air respectively.

Since the air inside the vacuum bottle can be regarded as a radiation transparent medium, the radiant heat transfer that occurs among the skeleton, the heat source surface, and the inner wall of the vacuum bottle can be calculated using the surface-to-surface radiation model. For surface i, the effective radiation can be expressed as:

$$J_i = (1-\varepsilon)G_i + \varepsilon\sigma T_i^{\,4} \tag{8.8}$$

Where, J_i refers to the effective radiation of the surface i, ε is the surface emissivity, T_i is the temperature of the surface i, σ is the Stefan–Boltzmann constant with a value of $5.67 \times 10^{-8}\ \mathrm{W \cdot m^{-2} \cdot K^{-4}}$, and G_i refers to the input radiation of the surface i, which can be expressed as:

$$G_i = \sum_{j=1}^{N} F_{ij} J_j \tag{8.9}$$

Where, J_i refers to the effective radiation of the surface i, and $k_{F_{ij}}$ is the radiation angle coefficient from surface i to surface j, which can be expressed as:

$$k_{F_{ij}} = \frac{1}{A_i} \int_{A_i} \int_{A_j} \frac{\cos\theta'_i \cos\theta'_j}{\pi d_{ij}^{\ 2}} dA_i dA_j \tag{8.10}$$

Where, θ'_i and θ'_j are the emission angle of surface i and the surface j, respectively. d_{ij} represents the distance between the two surfaces. A_i and A_j are the area of surface i and surface j.

From Equations (8.8) to (8.10), the radiation heat transfer of surface i can be expressed as:

$$q_i = A_i \sum_{j=1}^{N} F_{ij}(J_i - J_j) = \sum_{j=1}^{N} \frac{J_i - J_j}{(A_i F_{ij})^{-1}} \tag{8.11}$$

Since the phase change involves a nonlinear process, it is calculated by the equivalent heat capacity method [33]. The equivalent heat capacity of PCMs can be expressed as:

$$C_{\mathrm{eff}} = \begin{cases} C_{\text{PCM-S}} & (T < T_{\text{s-ini}}) \\ \dfrac{1}{\rho_{\mathrm{PCM}}}\left[(1-f_{\mathrm{PCM}}) \cdot \rho_{\text{PCM-S}} \cdot C_{\text{PCM-S}} + f_{\mathrm{PCM}} \cdot \rho_{\text{PCM-L}} \cdot C_{\text{PCM-L}}\right] + \dfrac{L_{\mathrm{PCM}}}{T_{\mathrm{l}} - T_{\text{s-ini}}} & (T_{\text{s-ini}} \le T \le T_{\mathrm{l}}) \\ C_{\text{PCM-L}} & (T_{\mathrm{l}} < T) \end{cases} \tag{8.12}$$

The equivalent density can be expressed as:

$$\rho_{\mathrm{PCM}} = (1-f_{\mathrm{PCM}}) \cdot \rho_{\text{PCM-S}} + f_{\mathrm{PCM}} \cdot \rho_{\text{PCM-L}} \tag{8.13}$$

The equivalent thermal conductivity can be expressed as:

$$\kappa_{\mathrm{PCM}} = (1-f_{\mathrm{PCM}}) \cdot \kappa_{\text{PCM-S}} + f_{\mathrm{PCM}} \cdot \kappa_{\text{PCM-L}} \tag{8.14}$$

f_{PCM} is a function of temperature. It can be expressed as:

$$f_{\mathrm{PCM}} = \begin{cases} 0 & (T < T_{\mathrm{s}}) \\ \dfrac{V_{\text{PCM-L}}}{V_{\text{PCM-L}} + V_{\text{PCM-S}}} & (T_{\mathrm{s}} \le T \le T_{\mathrm{l}}) \\ 1 & (T_{\mathrm{l}} < T) \end{cases} \tag{8.15}$$

Where, $V_{\text{PCM-S}}$, $V_{\text{PCM-L}}$ means the volume of the solid PCMs and liquid PCM, ρ_{PCM}, $\rho_{\text{PCM-S}}$, $\rho_{\text{PCM-L}}$ means the density of the PCM, solid PCM and liquid PCM, κ_{PCM}, $\kappa_{\text{PCM-S}}$, $\kappa_{\text{PCM-L}}$ means the thermal conductivity of the PCM, solid PCM and liquid PCM and C_{eff}, $C_{\text{PCM-S}}$, $C_{\text{PCM-L}}$ means the heat capacity of the PCM, solid PCMs and liquid PCM. $T_{\text{s-ini}}$ and T_{l} mean phase change onset temperature and phase change end temperature of PCM, respectively.

Convective heat transfer between the metal vacuum bottle and the high-temperature environment and the average convective heat transfer coefficient can be expressed as [34]

$$h_{\mathrm{L}} = \frac{1}{l_{\log}} \int_0^{L_{\log}} h(x) = \frac{3\kappa_{\mathrm{mud}}}{4} \left[\frac{u_{\mathrm{mud}}}{45(r_{\mathrm{h}} - r_{\log})^2 \alpha_{\mathrm{mud}} l_{\log}}\right]^{\frac{1}{3}} \left[(11r_{\mathrm{h}} - 5r_{\log})^{\frac{1}{3}} + \left(\frac{29r_{\mathrm{h}} - 5r_{\log}}{16}\right)^{\frac{1}{3}}\right] \tag{8.16}$$

Where r_{h} is the radius of the wellbore wall, $r_{\log}$ is the radius of the logging tool, u_{mud} is the velocity of the logging tool movement, $l_{\log}$ is the length of the logging tool and α_{mud} is thermal diffusivity of the mud.

In the simulation, COMSOL software was used to solve the numerical 3D model. The governing equations from the previous section were first added to the computational fluid dynamics (CFD) solver. Then, the 3D PTMS model for the logging tool was imported into the solver, and the unstructured tetrahedral mesh was divided. Subsequently, the materials and

Table 8.3 Materials and thermal properties of the logging tool.

Name	Material	Thermal conductivity ($W \cdot m^{-1} \cdot K^{-1}$)	Density ($kg \cdot m^{-3}$)	Heat capacity ($J \cdot kg^{-1} \cdot K^{-1}$)	Emissivity	Thermal expansion coefficient ($10^{-5} \cdot K^{-1}$)
Vacuum bottle	Inconel 718	14.7	8240	436	0.2	—
Vacuum layer	Composite	0.0002	100	1200	—	—
Skeleton	Aluminum alloy 6061	167	2710	896	0.16	—
Heat sources	Ceramic	30	3960	850	0.9	—
PCM	Wood alloy	19.8	9657.9	166.7(s) 184(l)	—	6.47
Insulator shell	PTFE	0.25	2200	1000	0.747	—
Insulator core	Aluminum silicate wool	0.035	400	794.2	—	—

thermal properties of each component were defined as shown in Table 8.3 Notably, the vacuum layer of the metal vacuum bottle was equivalent to a solid layer with a thermal conductivity of 0.0002 $W \cdot m^{-1} \cdot K^{-1}$ [25]. The phase change interval of the used PCM was 71.03–76.43 °C with a latent heat of 36.68 $kJ \cdot kg^{-1}$. The thermal expansion coefficient of the PCM was 6.47×10^{-5} K^{-1} [35]. The surface emissivity of the inner wall, skeleton, heat source, and insulator were 0.2, 0.16, 0.9, and 0.747, respectively [36–38]. Subsequently, the heating power of heat source 1 and heat source 2 were set to 20 and 10 W, respectively. The external surface of the vacuum bottle was set as the convective heat transfer boundary condition with an ambient temperature of 205 °C. The initial temperature of the logging tool was set to 20 °C. Based on the transient CFD solver, the heat transfer process of the logging tool was calculated from 0 to 360 minutes with a time step of 10 minutes.

To ensure the accuracy of the calculated results, a grid-independence analysis was conducted. Numerical models with grid numbers 50516, 77259, 148701, and 402411 were calculated. The calculated results are shown in Table 8.4. When considering the calculation error and computational resources, the numerical calculation results with grid number 148701 were finally selected for the subsequent analysis with a relative tolerance of 0.0001. In addition, to verify that the SNTP numerical model is more accurate than the SP numerical model, numerical simulations for the SP numerical model using the same geometric model were also performed.

8.2.2 Experimental Setup

To verify the accuracy of the simulated results, the experimental measurements of the SNTP numerical PTMS model were conducted. As shown in Figure 8.11(a), a skeleton prototype of the logging tool with the PTMS was fabricated on a scale of one to one with the simulated model. The two ceramic heating elements (40 mm × 40 mm × 2 mm, 10 W/5 V or 20 W/12 V, Zhengzhou Xindeng Electrothermal Ceramics Ltd.) were adhered to the skeleton by thermal silicone pads (LC120, 1 $W \cdot m^{-1} \cdot K^{-1}$), Shenzhen Liantengda Technology Ltd.). The contact surfaces are filled with TIM to reduce the contact thermal resistance. Subsequently, thermocouples (K type, 2 mm × 0.3 mm, Shenzhen Yibulan Electronics Ltd.) were positioned at several temperature measurement points, such as at the heat sources, insulators, and PCMs, as shown in Figure 8.11(a). The metal vacuum bottle (JP90/73 × 900, Xi'an Yufeng Electronics Company Ltd.) was utilized to contain the prototype with heat sources and thermocouples. Then, the metal vacuum bottle with the prototype was put into the oven (DHG-9205A, temperature range 10–300 °C, accuracy ±0.5 °C, Shanghai Hecheng Instrument Manufacturing Ltd.), as shown in Figure 8.11(b). In addition, adjustable direct current (DC) voltage regulators (MS-3010D, 0–30 V/10 A, Dongguan Meisheng Power Technology Ltd.) were implemented to supply power to the ceramic heating elements. A data acquisition

Table 8.4 Grid independent analysis.

Grid number	Temperature of heat source 1 (°C)	Grid number	Temperature of heat source 1 (°C)
50516	162.18	148701	159.15
77259	160.61	402411	159.34

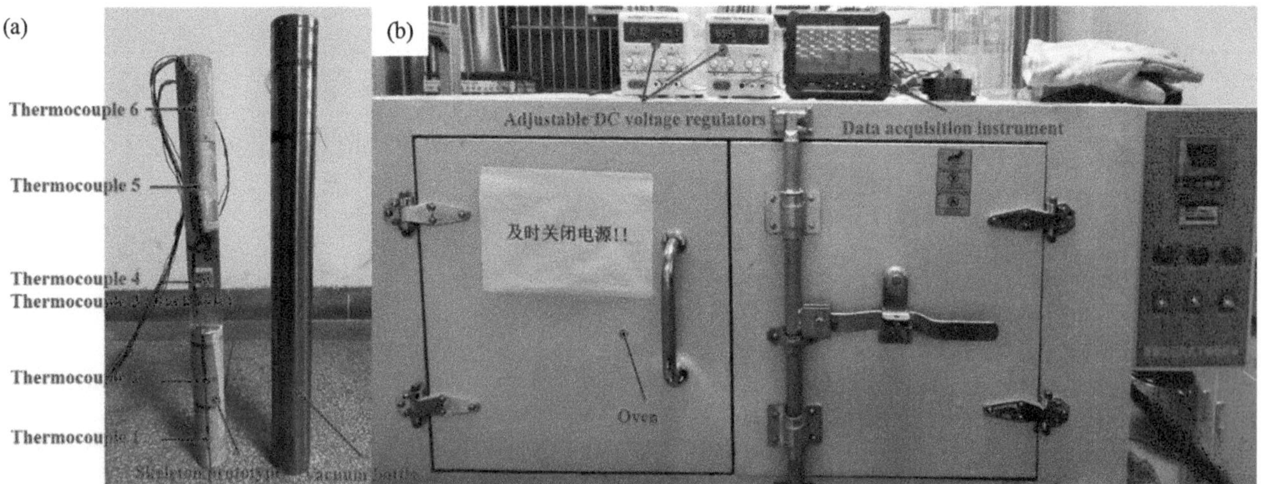

Figure 8.11 Experimental test: (a) skeleton prototype of the typical logging tool and the metal vacuum bottle; (b) high-temperature oven temperature test site.

instrument (MIK-R6000F, temperature measurement accuracy 0.2% FS ± 1D, sampling frequency 1 Hz, Hangzhou Mecon Automation Technology Ltd.) was used to collect and deal with temperature measurement signals. Finally, the oven temperature was controlled by a proportional–integral–derivative (PID) controller to maintain a temperature of 205 °C for 6 hours.

8.2.3 Experimental and Simulated Results

Figure 8.12 shows the comparison of the experimental temperatures and simulated temperatures of the PTMS. The temperatures of heat source 1, heat source 2, PCM 1, PCM 2, and insulator 2 are selected to compare the numerical and experimental results. The experimental temperatures show the same trends as the simulated temperature. During the whole process, the maximum absolute errors between the measured and simulated temperatures of heat source 1, heat source 2, PCM 1, PCM 2, and insulator 2 were 5.39, 5.41, 8.52, 9.43, and 3.78 °C, respectively. The maximum absolute errors were within 10 °C, and the average absolute error was only 3.02 °C, which demonstrated the accuracy of the simulated results.

The errors between the experiment and simulation mainly originated from the assumptions regarding the numerical simulations and the experiment setup. First, the contact thermal resistance was neglected in the simulation, which reduced

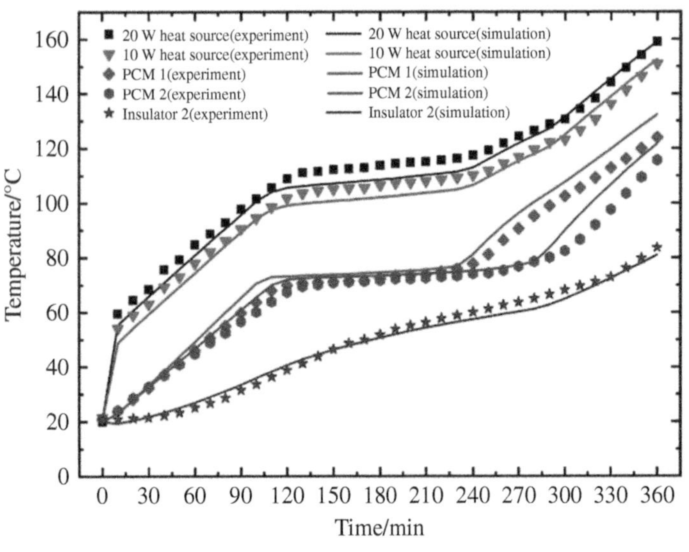

Figure 8.12 Comparison of the simulated and experimental temperatures of the PTMS.

the heat transfer thermal resistance between the heat sources and PCMs. Therefore, the simulated temperatures of the heat sources were generally lower than the experimental temperature during the 290 minutes operation, while the simulated temperature of the PCMs was higher than the experiment temperature. In addition, the contact thermal resistance delayed the onset of the phase transition of the PCMs in the experiment. Since the contact surfaces in the experiment were filled with TIM, the error was within the accepted range. Second, the error originated from the assumption that the change in physical properties of materials with temperature was ignored. When the temperature was higher, the material properties deviated from the defined properties at room temperature. Therefore, the heating rate of the heat sources and PCMs in the simulation was different from that in the experiment after 290 minutes. The change of material properties with temperature had a minor effect on the overall system. Finally, the experiment tests also introduced several errors. On the one hand, the experimental error derived from the temperature collection accuracy of the equipment, which was ±0.7 °C in this study. On the other hand, the ambient temperature in the experiment gradually rose from room temperature to 205 °C, while the simulation was directly set to the ambient convective heat transfer temperature of 205 °C. These differences had a small effect on the internal temperature rise due to the excellent insulation of the vacuum bottle and the insulator. Overall, the errors between the experimental results and the simulated results were within acceptable limits.

Figure 8.13 shows the heat source temperature time series for different numerical PTMS models. The SP numerical model only considers solid heat conduction and phase change. The temperature curves of the proposed numerical model were closer to the experimental results than those of the SP numerical model. The simulated results of the SP numerical model were not much different from that of the SNTP numerical model before 210 minutes. However, the temperature difference between them gradually increased due to the enhancement of thermal convection and radiation after 210 minutes. Overall, the heat source temperature difference between the simulated results of the SP numerical model and the experimental results reached a maximum absolute error of 8.61 °C, while the maximum error between the simulated results of the SNTP numerical model and the experimental test results was within 6 °C. The maximum percentage error between the experiment and simulation in the SNTP numerical model was 4.71%, compared to 6.59% in the SP numerical model. As a result, the SNTP numerical model accurately characterized the real heat transfer process in the logging tool. In addition, the computation times of the SP numerical model and the SNTP numerical model were 635 and 29,296 seconds, respectively.

Figure 8.14 shows the temperature distribution of the SNTP numerical model toward PTMS at different time intervals. The temperature of the PTMS increased with time due to the combined effect of operating within a high-temperature environment and self-generated heat. Since the opening of the vacuum bottle was in direct contact with the external high-temperature environment, insulator 1 displayed a large temperature gradient and the heat from the environment continuously penetrated into the interior of the vacuum bottle. However, insulator 2 was near the closed side of the vacuum bottle, which was less affected by the high-temperature environment, leading to its lower temperature. Due to the accumulation of the generated heat, the heat source temperature was generally higher than that of the metal skeleton and the nearby PCM. The maximum temperature of the heat sources reached 159.15 °C after 6 hours. In addition, the PCMs underwent a phase

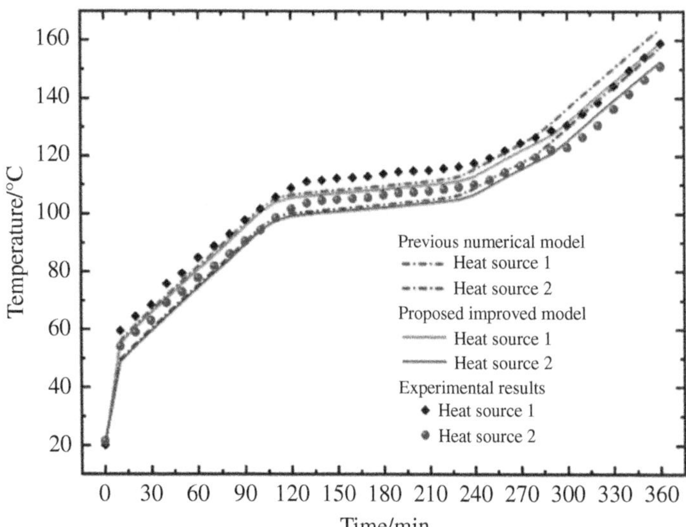

Figure 8.13 The heat source temperature time series for different numerical PTMS models.

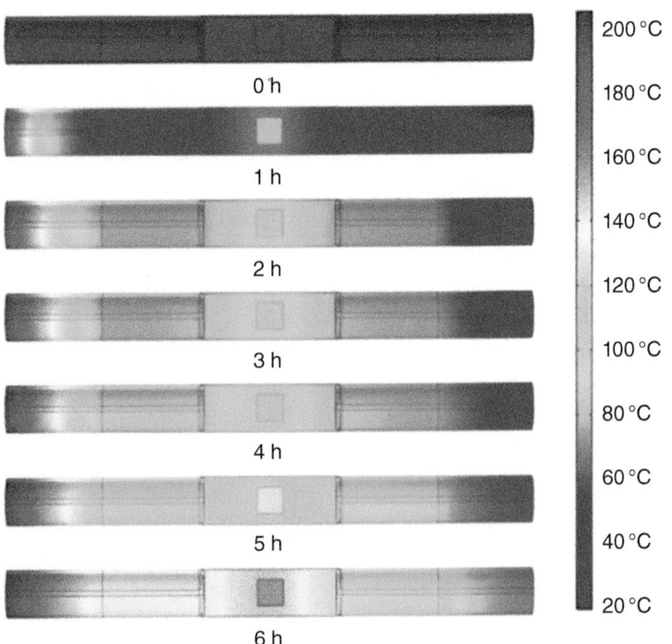

Figure 8.14 Temperature distribution of the PTMS for a typical logging tool.

change process during the 2–4 hours interval, and their temperature remained constant near the melting point. Therefore, the overall temperature field of PTMS remained almost the same during the 2–4 hours interval, reflecting the significant effect of the PCMs' latent heat on the temperature control.

Figure 8.15 shows the temperature curves of the logging tool over time. Influenced by the generated heat, the temperature of heat source 1 rose rapidly to approximately 50 °C in the first stage, resulting in a certain temperature difference with the PCMs. In the second stage, it rose at a uniform rate of 30.4 °C/hour due to the sensible heat of the metal skeleton and the PCMs. In the third stage, PCM 1 and PCM 2 finished the phase change process and absorbed excessive heat from the heat sources in the form of latent heat. The temperature of the heat sources remained almost constant between 100 and 280 minutes. In the fourth stage, the PCMs continued to absorb heat in the form of sensible heat, thus leading to the temperature rise of the heat sources at a rate of 25.97 °C/hour.

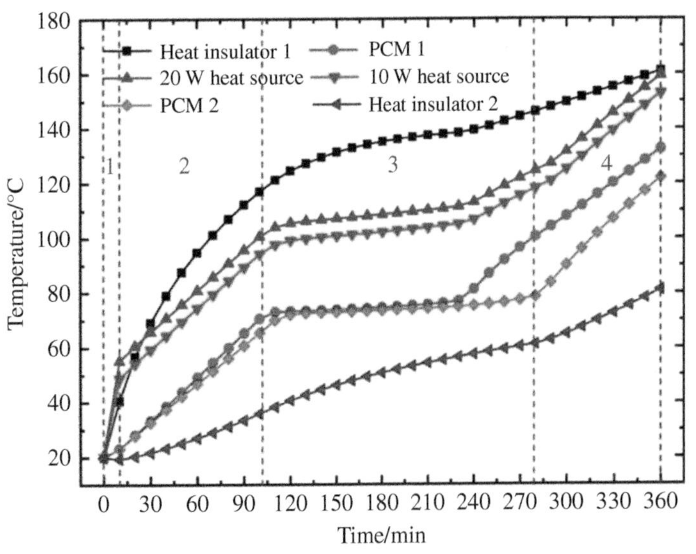

Figure 8.15 Temperature curves of the logging tool versus time.

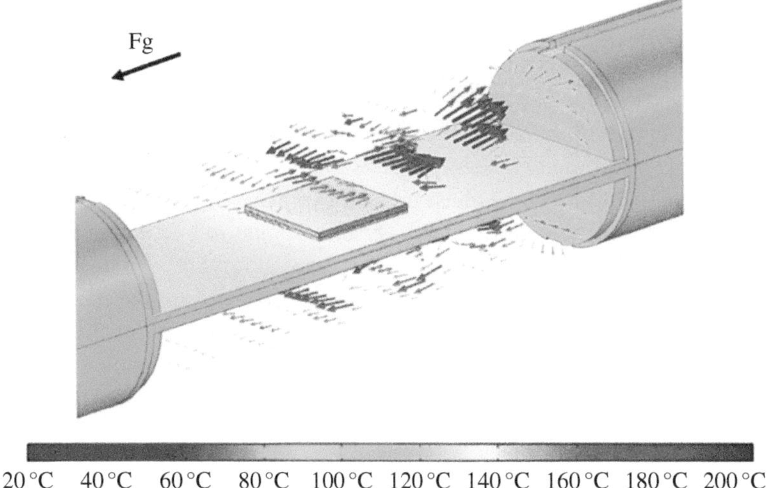

Figure 8.16 The airflow near the heat source in the vacuum bottle.

Figure 8.16 depicts the airflow near the heat source inside the vacuum bottle. The direction of the red arrows indicates the direction of the natural air convection. The size of the arrow indicates the magnitude of the air velocity. The colors of the heat sources and the metal skeleton denote the temperature. The temperature of the air near the heat sources is higher and has a lower density than the airflow inside the logging wool. The air moves upward due to the combined effect of the temperature gradient and gravity and is then dispersed in all directions after encountering the barrier imposed by PCM 1. In addition, the lower temperature of the inner vacuum bottle wall contributed to the cooling of the nearby air, which resulted in a downward return flow.

Figure 8.17 shows the velocity distribution of the air inside the vacuum bottle, where the arrows indicate the direction of the airflow and the colors reflect the magnitude of the velocity. In the beginning, there was no temperature difference inside

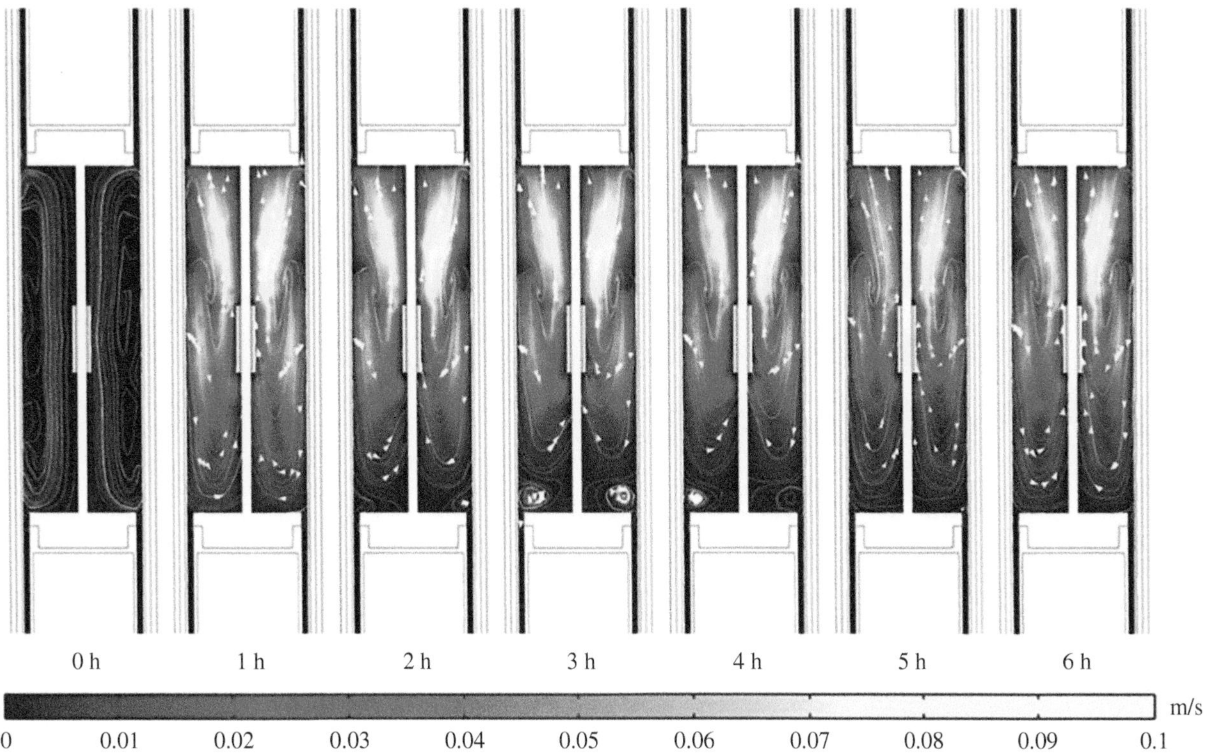

Figure 8.17 The velocity distribution of the air inside the vacuum bottle. Peng et al. [39]/with permission of Elsevier.

the vacuum bottle, and the air was stationary. The air flowed upward due to the change in density after 1 hour, and it descends due to the cooling of the inner wall of the vacuum bottle. Two vortices were clearly visible near the heat source. The flow field from 2 to 4 hours was approximately the same as that at 1 hour. The minor difference was that two new small reverse vortices appeared below the metal skeleton near the inner wall of the vacuum bottle. The temperature of PCM 2 was lower than that of the nearby vacuum bottle wall due to the phase change of PCM 2, and thus two small reverse vortices appeared on account of the temperature difference. The temperature of the metal skeleton further climbed again from 5 to 6 hours due to the end of the phase transition, which was higher than that of the nearby wall of the vacuum bottle. Hence, the small reverse vortices disappeared. Overall, the natural air convection velocity inside the vacuum bottle ranges from 0 to 0.1 m/s. The air between the PCMs and the vacuum bottle is nearly stationary. The natural convection of air inside the vacuum bottle accelerates the dissipation of the heat from the heat sources.

As the main heat storage pool of PTMS, PCMs are crucial for the temperature control of the heat source. Therefore, it is necessary to analyze the phase change process of PCMs. Figure 8.18(a) shows the phase change of the PCMs over time, where 0% and 100% indicate the solid PCM and the liquid PCM, respectively. 0%–100% indicate the solid–liquid mixing area of PCM, which reflects the phase change interface. The PCMs were solid from 0 to 1 hour, as shown in Figure 8.18(a). The phase change process started from the sides nearest the heat source at approximately 2 hours and then gradually advanced toward the two ends of the logging instruments until 5 hours, at which time, the phase change was complete. The generated heat from the heat sources was the main reason for the phase change process. The extreme ambient temperatures had less effect on internal phase changes due to the superior thermal insulation of the vacuum bottle and insulators. In addition, PCM 1 completed the phase change process approximately 1 hour earlier than PCM 2 due to the influence of the high-temperature environment.

Figure 8.18(b) displays the fraction of liquid PCMs curves as a percentage of volume over time. The x-axis is the time, and the y-axis is the percentage of the liquid PCM volume (0%–100%). Both PCM 1 and PCM 2 started the phase transition process at approximately 100 minutes. Since PCM 1 was located at the open side of the vacuum bottle, the phase transition of PCM 1 was completed at 240 minutes. The phase transition time of PCM 2 was relatively slower, and the phase transition was completed at 290 minutes. Since the phase transition process of PCM 1 was not synchronized with that of PCM 2, the total PCMs phase transition trend was significantly altered by the time the phase change of PCM 1 was completed.

The heat storage of the PCMs is further shown in Figure 8.19. When the PCMs were pure solid or pure liquid, the PCMs only accumulated heat in the form of sensible heat with a lower heat storage rate. When the PCMs were under the melt phase transition, their temperature was maintained near the melting point for a longer period of time, which showed a great heat storage effect and kept the temperature of other components from rising. Therefore, the PCM heat storage time series showed two inflection points, which represented the beginning and end of the phase change, and the slope of the curves

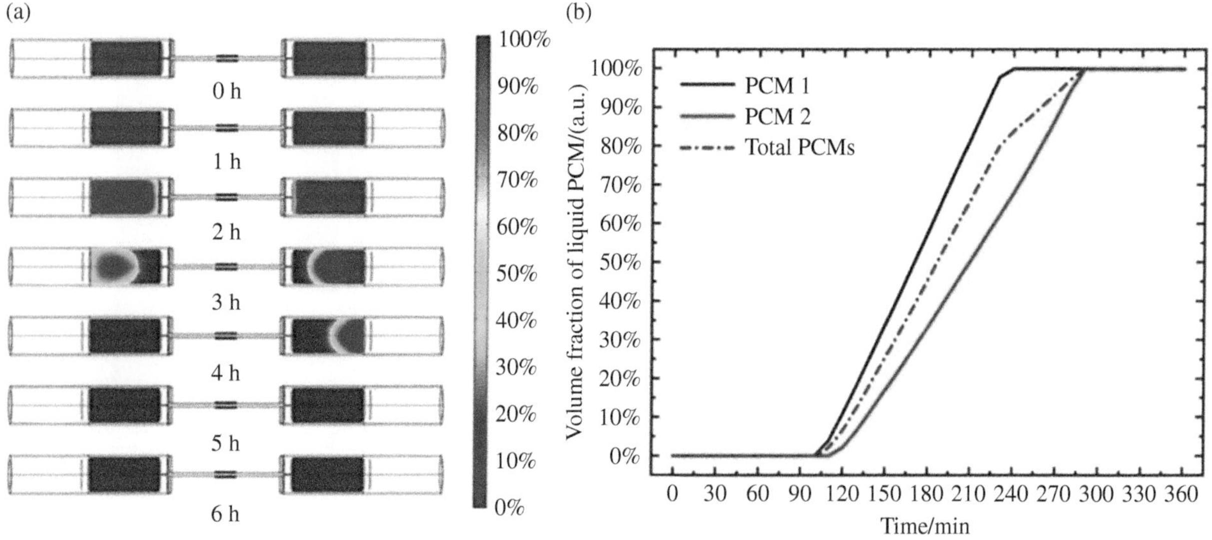

Figure 8.18 PCM phase transition: (a) phase transition of PCMs within the logging tool at 1 hour intervals; (b) liquid PCMs as a percentage of volume versus time.

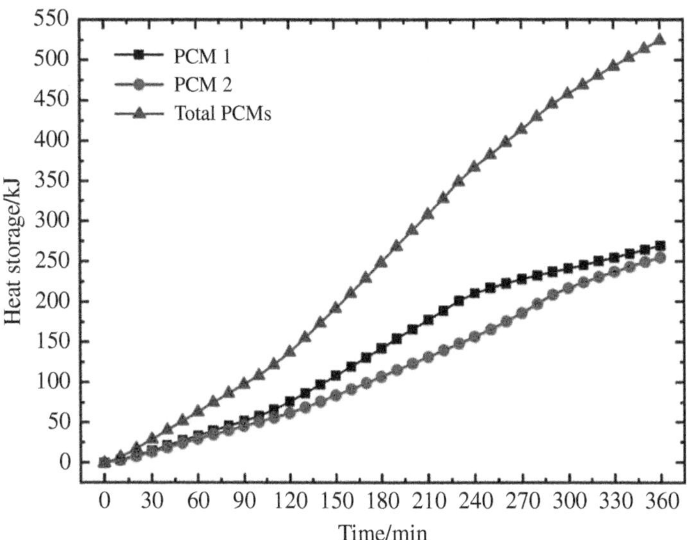

Figure 8.19 PCM heat storage time series.

between the two inflection points was significantly larger. The total heat storage of the PCMs reached 525.354 kJ (the absorbed heat of PCM 1 and PCM2 were 269.654 and 255.700 kJ, respectively). Among them, latent heat storage accounted for 58.1%, which played a significant role in the whole heat storage process of PCMs.

Figure 8.20(a) shows final the temperature distribution and heat flow of PTMS, where the color indicates the temperature and the red arrows indicate the direction and size of the heat flow. The great majority of heat from the heat sources was dissipated to the metal skeleton through the thermal silicone pad and then transferred to the PCMs along the skeleton to the

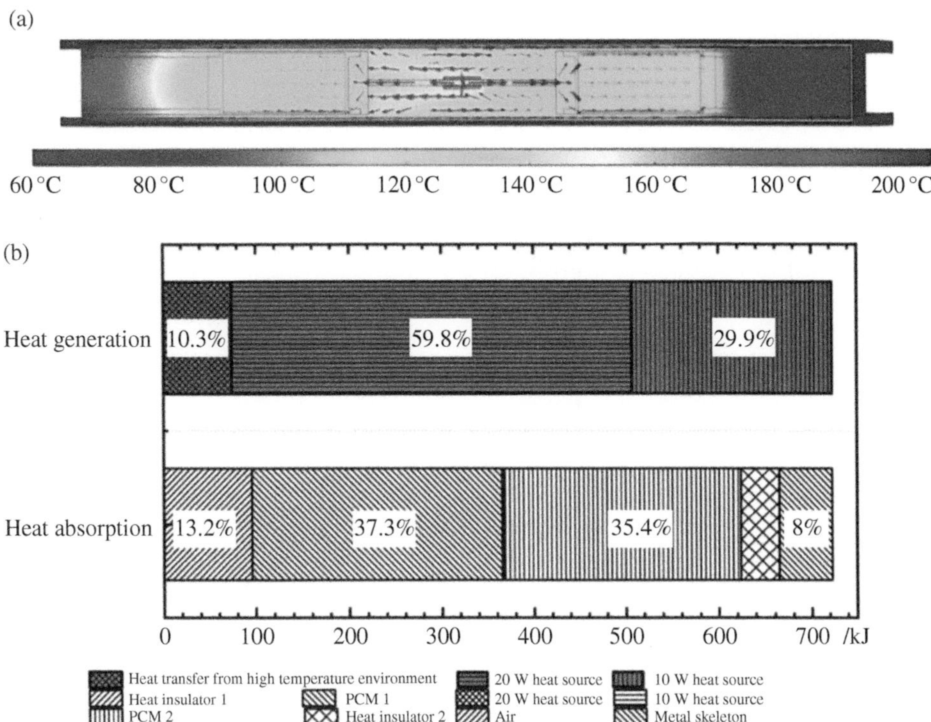

Figure 8.20 Heat flow distribution, heat generation, and absorption: (a) the final temperature and heat flow distribution of PTMS; (b) analysis of the heat generation and heat absorption of the PTMS.

left and right. At the junction of the metal skeleton and the PCMs, there was a significant heat diffusion process due to the change in cross-sectional area. After the heat flow entered the interior of the PCMs, it decreased, which visually reflected the heat storage process offered by the PCMs. Concerning the heat exchange by natural convection of air, the direction of heat flow was consistent with the velocity of air movement, which illustrated that the macroscopic movement of air could strengthen the heat exchange process. In addition, due to the outstanding thermal insulation of the vacuum layer, there was a "thermal bridge effect" at the opening of the vacuum bottle [38]. The heat from the external high-temperature environment could bypass the vacuum layer and seep into the interior of the vacuum bottle, which further intensified the environmental heat transfer at the opening of the vacuum bottle.

The heat generation and absorption capacities of the PTMS are further quantified, as shown in Figure 8.20(b). First, the heat generation was mainly divided into two parts: heat generated from the heat sources and heat transfer from the high-temperature environment. Among them, the heat transfer from the high-temperature environment reached 74.67 kJ, which only accounted for 10.3% of the total heat. The heat sources constantly generated a total of 648 kJ of heat during the 6 hours of operation, which accounted for 89.7% of the total heat. Since the heat cannot be dissipated to the high-temperature environment, all the heat was absorbed by each component of the PTMS, resulting in an increase in the overall enthalpy and temperature. PCM 1 and PCM 2 possessed abundant heat storage capacity, which accounted for 37.3% and 35.4% of the total heat storage, respectively, reflecting the essential role of PCMs in the PTMS. The PCMs were the most significant heat storage component of the logging tool when subjected to high ambient temperatures. The metal skeleton, insulator 1, and insulator 2 absorbed 8%, 13.2%, and 5.6% of the heat in the form of sensible heat, respectively. In addition, part of the heat was still trapped in heat source 1 and heat source 2, which was 1382.7 and 1320.4 J, respectively, and caused their temperature to rise.

The heat source temperature is a key indicator of the PTMS for the logging tool. The final temperature is tightly related to the heat dissipation capability. Therefore, it is necessary to focus on the heat dissipation path of the heat sources. Figure 8.21 displays the heat dissipation paths of heat source 1 and heat source 2. The majority of the generated heat from the heat sources was transferred to other components via heat conduction, natural air convection, and thermal radiation, and only 0.47% and 0.86% of the total heat were trapped within heat source 1 and heat source 2, respectively. Among the three forms of heat transfer, thermal conduction played a dominant role, where 94.0% of heat source 1 and 92.5% of heat source 2 were dissipated through the metal skeleton. Overall, the proportions of solid heat conduction, natural air convection, and thermal radiation heat exchange are 93.89%, 4.32%, and 1.79%, respectively, which reflects the absolute dominance of solid heat conduction in the whole heat exchange process.

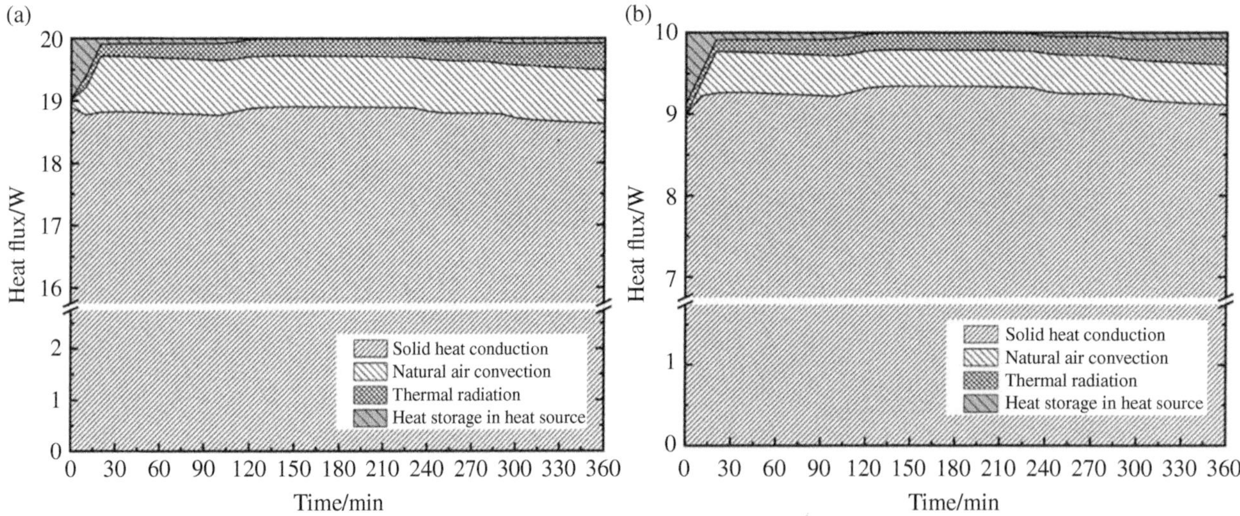

Figure 8.21 Heat dissipation path analysis: (a) heat source 1 and (b) heat source 2.

8.3 Phase-Change Materials Design

PCMs play a crucial role in TMSs. However, conventional organic PCMs such as paraffin have a low thermal conductivity, which results in a limited heat storage rate. To solve this issue, an ultrasonication method was proposed to improve the thermal conductivity of expanded graphite (EG)/paraffin composite phase change materials (CPCMs) at lower EG loading. Samples were prepared through thermal expansion, magnetic stirring, and ultrasonication for converting EG into nanosheets. Different sizes of graphene nanosheets (GNs) were prepared by controlling the ultrasonic time ranging from 0 to 30 minutes. The phase change characteristics and thermal conductivity of the CPCMs were investigated experimentally. The effect of particle size on the thermal conductivity of CPCMs was also studied.

8.3.1 Material Preparation

Figure 8.22 shows the process of preparing GNs/paraffin CPCMs including thermal expansion, magnetic stirring, and ultrasonication. The starting graphite powders (XFNANO, INC) were adopt to yield EG using a microwave oven with an overall power of 800 W for 40 seconds. Then the EG is mixed with liquid paraffin and fully stirred for 10 hours. After that, the mixed liquid composites were ultrasonically exfoliated in a water bath at 80 °C. The ultrasonic crushing system uses a cell disrupter (Xinzhi JY-92-IIN), which produces ultrasonic waves that can form small and energetic bubbles in the liquid medium. The bubbles burst instantaneously in the medium and release a huge amount of energy to exfoliate the EG. In the experiment, the ultrasonic power is 650 W and the ultrasonic frequency is 20–25 kHz. The ultrasonic procedure is set to be 2 seconds with an interval of 4 seconds. Finally, the prepared samples were dried under a vacuum at 80 °C to remove the bubbles.

8.3.2 Characteristics and Thermal Performance

Successful exfoliation requires to overcome the van der Waals force between the adjacent layers. Solvents with surface tension of about 40 $mJ \cdot m^{-2}$ are the best solvents for the dispersion of graphitic flakes [40]. If the interfacial tension between solid and liquid is high, the flakes tend to adhere to each other and the work of cohesion between them is high, hindering their dispersion in the liquid. In order to confirm that the paraffin is appropriate to exfoliate the EG, we measured the surface tension of the paraffin with drop shape analyzer (DSA25). The results showed that the surface tension of paraffin is 34 $mJ \cdot m^{-2}$, which indicate that paraffin can successfully exfoliate the EG into nanosheets.

DSC tests were conducted to obtain the thermal properties and phase change characteristics of the CPCMs. The test temperature ranges from 20 to 70 °C with a temperature rise rate of 5 °C/min. Figure 8.23(a) and (b) show the DSC curves of pure paraffin and GNs/paraffin CPCMs with a GNs loading of 0.5 wt%, respectively. T_1, T_2, and T_3 represent the melting point, phase change temperature, and end point, respectively. Thermal analyses of the samples with various GNs loadings were performed, and it was found that the DSC curves of all the test specimens are very similar, irrespective of the GNs loading.

Figure 8.24 shows the phase change characteristics of GNs/paraffin CPCMs including phase change temperature and latent heat. Figure 8.24(a) is the phase change temperatures of different samples. It can be seen that the phase change temperature of pure paraffin is 46.23 °C, while the phase change temperatures of other samples are within 46.23 ± 1 °C. Figure 8.24(b) is the comparison of the latent heat between the theoretical values and the experimental results. The theoretical latent heat could be calculated by Equation (8.17):

$$L_{comp} = (1 - \omega_{GN}) \cdot L_{para} \tag{8.17}$$

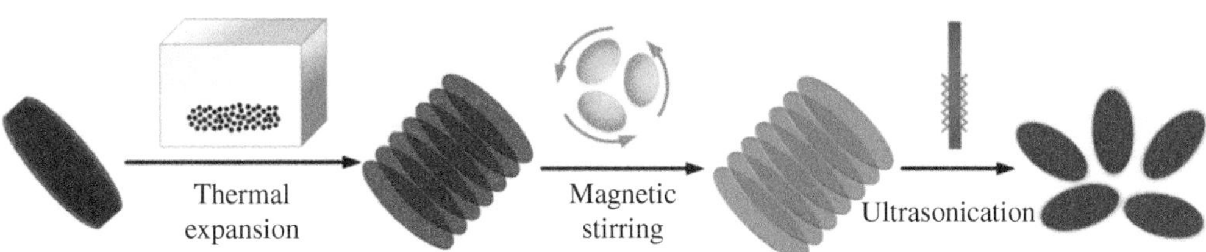

Figure 8.22 Process of preparing GNs/paraffin CPCMs.

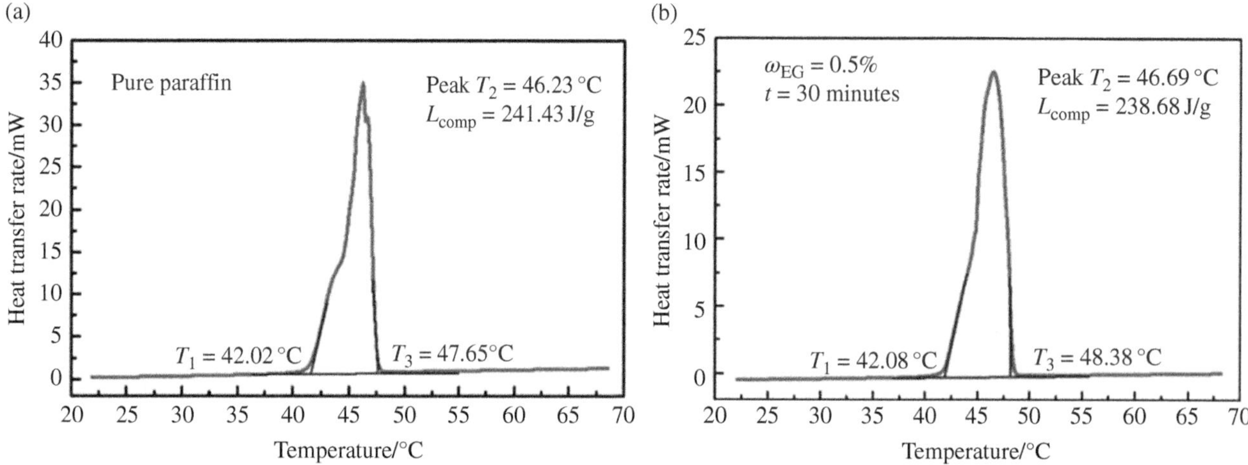

Figure 8.23 DSC curves of (a) pure paraffin and (b) GNs/paraffin CPCMs with a GNs loading of 0.5 wt%.

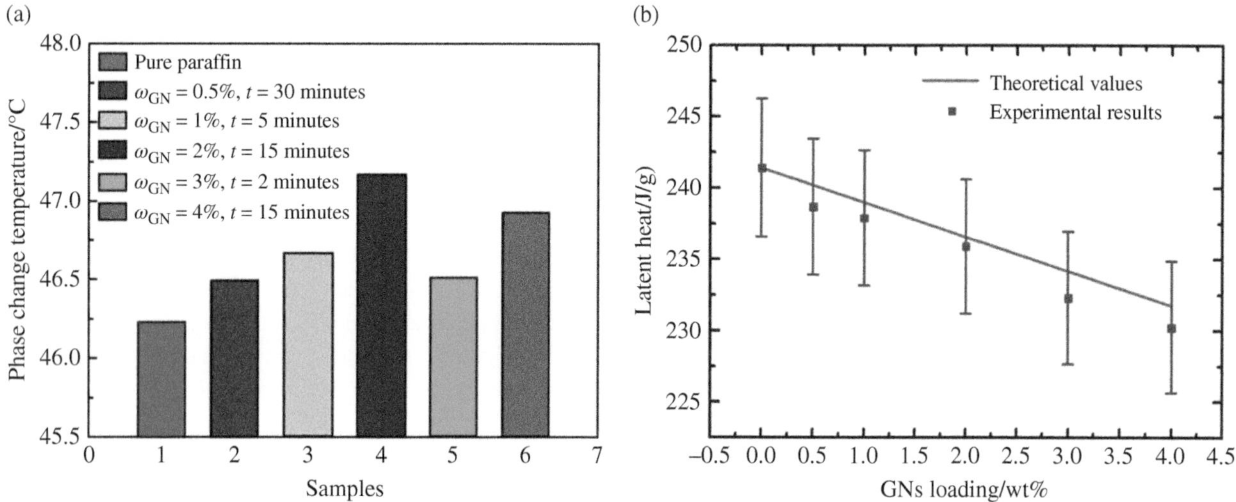

Figure 8.24 Phase change characteristics of GNs/paraffin CPCMs: (a) phase change temperature and (b) latent heat.

where L_{comp} and L_{para} are the latent heat of CPCMs and pure paraffin, respectively. ω_{GN} is the mass fraction of GNs. Obviously, the latent heat is inversely proportional to the GNs loading, although the thermal conductivity increased with GNs loading. We can find that the variations of the latent heat between theoretical values and experimental results are within 2%.

The thermal properties including melting point, phase change temperature, and latent heat are listed in Table 8.5. The average melting point of the GNs/paraffin CPCMs is 42.16 ± 0.37 °C, and the average phase change temperature of the GNs/paraffin CPCMs is 46.70 ± 0.77 °C, which are almost the same as that of the paraffin. According to the experimental results, we can find that the melting point and phase change temperature of the composites were similar to pure paraffin because no chemical reaction occurred between GNs and paraffin in the temperature range from 20 to 70 °C. As a consequence, the ultrasonic effect and the addition of GNs have negligible effect on the phase change characteristics of CPCMs.

The thermal enhancement effect of the ultrasonication method was investigated experimentally. Various samples with the mass fraction of GNs ranging from 0 to 4 wt% were prepared, and the ultrasound times were 0, 2, 5, 15, and 30 minutes, respectively. Figure 8.25 shows some samples prepared for thermal conductivity measurement. Both the length and width of the sample are 9.8 ± 0.2 mm, and the thickness is 1.58 ± 0.02 mm. It can be seen that the particle size of the GNs decreased with the ultrasonic time. The worm-like EG is dispersed in paraffin and cannot form a thermally conductive channel when

Table 8.5 Melting point, phase transition temperature, and latent heat of GNs/paraffin CPCMs.

GNs loading (wt%)	Ultrasound time (min)	Melting point (°C)	Phase-change temperature (°C)	Latent heat (kJ · kg^{-1})
Paraffin	0	42.02	46.23	241.44
0.5	30	42.08	46.49	238.68
1	5	42.10	46.25	237.90
2	15	42.11	47.17	236.00
3	2	42.53	46.67	232.32
4	15	41.99	46.93	230.26

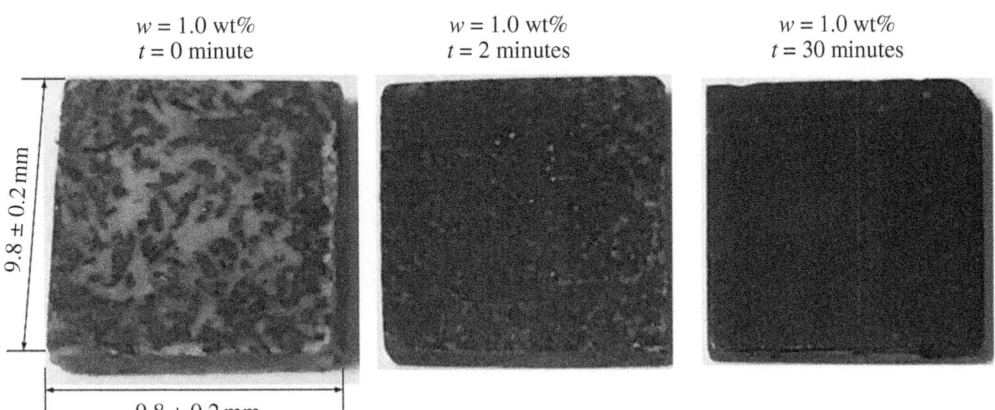

Figure 8.25 Samples prepared for thermal conductivity measurement by the laser flashing method. Shang et al. [41]/with permission of Elsevier.

the ultrasonic time is 0 minute. As the ultrasound time increases, the GNs were gradually uniformly distributed in paraffin. The thermal conductivities are measured by the laser flash method (NETZSCH-LFA467). All the samples have been measured repeatedly to reduce the deviation.

Figure 8.26 shows the thermal conductivities as a function of GNs loading at different ultrasonic times. As the GNs loading increased, the samples showed a better performance in thermal conductivity consistently. When the GNs loading is 1 wt%, the thermal conductivities of CPCMs increased by 63%, 61%, 47%, and 19% compared with the EG/paraffin CPCMs, respectively. Similarly, the thermal conductivities increased by 102%, 74%, 43%, and 26% while GNs loading is 3 wt%.

Figure 8.27 shows the comparison of the relative thermal conductivity enhancements with the GNs thermal-enhanced composites reported in the literature. The relative enhancement is defined as the ratio of the thermal conductivity between the composites and the base material. Results show that the relative enhancements of thermal conductivity obtained in the present study are at a higher level, showing a promising potential in thermal energy storage (TES) systems.

Furthermore, we could find from Figure 8.28 that the CPCMs with an ultrasonic time of 2 minutes possessed the highest thermal conductivity. As the particle size of the GNs decreased with the ultrasonic time, the inference could be drawn that the thermal conductivity decreased with the particle size, which was opposite to the previous researches [42–44]. Therefore, we speculate that there might be a peak value of thermal conductivity with the change of ultrasonic time. We further refined the time interval and explored the effect of ultrasonic time on the thermal conductivity of CPCMs. The samples with the ultrasonic time of 15 seconds, 30 seconds, 2 minutes, 5 minutes, 15 minutes, and 30 minutes have been prepared, respectively. Figure 8.28 shows the thermal conductivities of these samples as a function of ultrasonic time at different GNs loading. It shows an obvious peak value with respect to the ultrasonic time for all studied GNs loadings. For the CPCMs at GNs loading of 4 wt%, the thermal conductivity presents a peak value of 3.0 W · m^{-1} · K^{-1} at 2 minutes. And all the other measured samples present the same tendency. From the previous analysis, the thermal conductivity of the CPCMs is related to

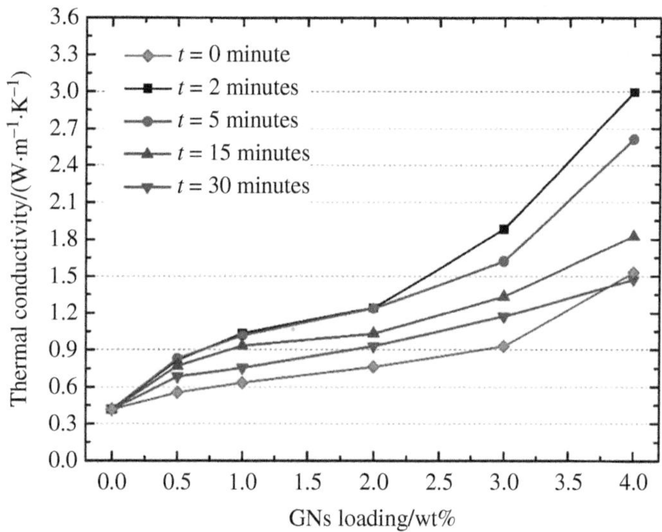

Figure 8.26 Thermal conductivity of CPCMs as a function of GNs loading with different ultrasonic times.

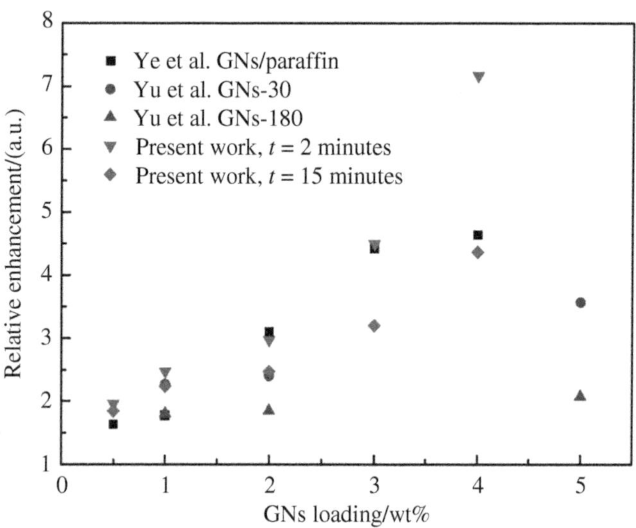

Figure 8.27 Comparison of the relative thermal conductivity enhancements with the GNs thermal-enhanced composites reported in the literature.

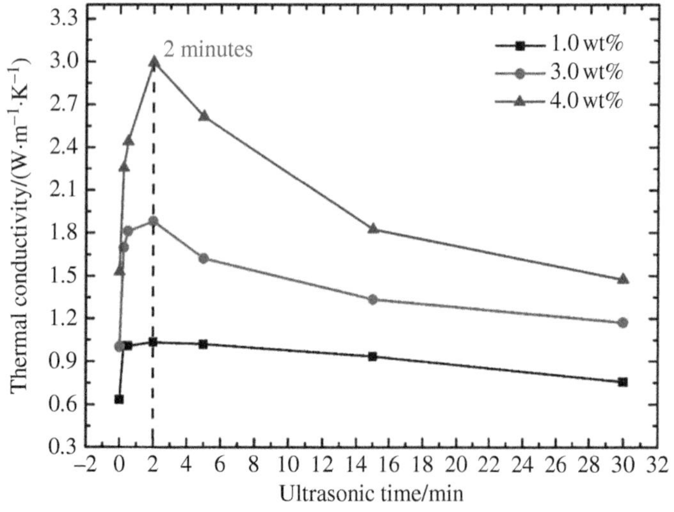

Figure 8.28 Thermal conductivities of CPCMs as a function of ultrasonic time at different GNs loading.

the aspect ratio of the particles [42–46]. When the EG was subjected to ultrasonic effect, the worm-like structure was broken and converted into GNs. The aspect ratio was significantly increased and heat transfer through the GNs and across the interfaces between the GNs and paraffin is more efficient, resulting in a thermal enhancement of CPCMs. As ultrasound time further increased, the in-plane size of GNs was also reduced. The decrease in in-plane size produced an increased number of interfaces, resulting in increased phonon boundary scattering at both the GNs and paraffin interfaces. The stronger boundary scattering led to an increase in interface thermal resistance and then decreased the intrinsic thermal conductivity of GNs. The negative influence of the interface thermal resistance surpassed the positive influence of the aspect ratio evolution. Therefore, the thermal conductivity gradually decreased with the ultrasonic time.

A scanning electron microscope (SEM) was used to demonstrate our statement. SEM was performed on the samples using the techniques described in [47]. Briefly, a brittle fracture was performed on each of the samples in order to avoid significant displacement of GNs within the paraffin. Frictional forces from traditional cutting techniques were found to result in extremely high local temperatures at the CPCMs interface, which melts the paraffin and alters the distribution of GNs. Figure 8.29 shows the SEM images of the samples with the ultrasound time of 2 and 30 minutes, respectively. It should be noted that the GNs are coated in paraffin, thus their exact geometries cannot be directly extrapolated from these images. It can be seen from Figure 8.29(a)–(c) that the particle sizes of GNs were significantly reduced after 2 minutes of ultrasonication, and an obvious sheet structure was observed in Figure 8.29(c), indicating that the ratio aspect dramatically increased after ultrasonic effect. Figure 8.29(d)–(f) proves that the actual diameter of the GNs decreases sharply with the extended ultrasound time. Individual nanosheets with dimensions less than 10 μm were obtained after 30 minutes of treatment of ultrasonication, while for GNs under 2 minutes of ultrasonication, the diameter can be as high as 50 μm. The narrow and thin sheets produced an increased number of interfaces, thus increasing the interface thermal resistance. As a consequence, the thermal conductivity of the GNs/paraffin composite is determined by the positive effect of aspect ratio and the negative influence of interface thermal resistance.

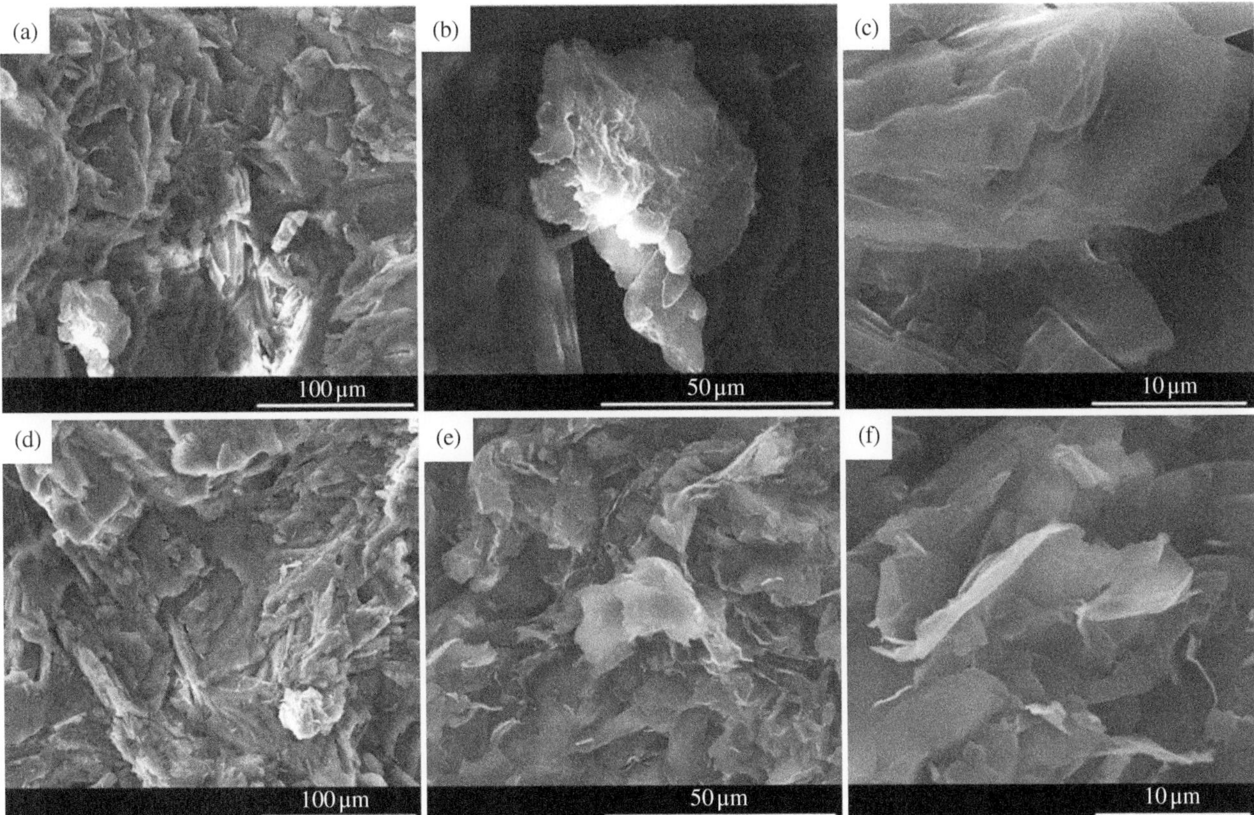

Figure 8.29 SEM images of the samples with different ultrasonic time: (a–c) ω_{GN} = 1 wt%, t = 2 minutes; (d–f) ω_{GN} = 1 wt%, t = 30 minutes.

8.4 Distributed PCM-Based Thermal Management Systems

In Section 8.1, a PTMS was proposed for the downhole electronics at temperature containing skeletons, circuit boards, PCMs, insulators as well as vacuum flask. Among them, a vacuum flask is utilized as a shell to protect the inside components from the outside environment with high pressure and high temperature. The thermal insulator is placed at the end of the vacuum flask to insulate the heat leakage from the opening. PCMs are adjacent to the thermal insulator and connected to the skeleton where plenty of circuit boards are mounted [48–50]. This typical structure is quite effective for the logging tool with short skeleton and few circuit boards. However, with the number of heat sources as well as the length of the skeleton increase, the thermal resistance between PCMs and heat sources becomes larger, which would deteriorate the temperature control effect of TMS severely. As a result, the TMS presented in Section 8.1 is only suitable for the logging tool with short skeleton and few circuit boards and unable to satisfy the temperature control demands of long-skeleton and multi-heat-source logging tool due to the significant thermal resistance.

8.4.1 System Design

In this section, a distributed thermal management system (DTMS) for long-skeleton, multi-circuit-board logging tool was proposed to reduce the thermal resistance between heat sources and PCMs. Figure 8.30(a) shows the structure of the logging tool with a centralized thermal management system (CTMS). It is based on a realistic instrument called the nuclear magnetic resonance logging tool. It can be seen that the vacuum flask is utilized as a shell to protect the inner electronic components from the high-temperature environment. However, the heat leakage still intrudes into the vacuum flask through the opening at both ends. Hence, two thermal insulators are placed at both ends of the vacuum flask, as a means to prevent the heat flow from entering the thermos. The skeleton is between two insulators where 14 pieces of heat sources are mounted. The heat power of the electronic components is listed in Table 8.6, and the total heat power is up to 90 W. It is noteworthy that the logging tool with CTMS only utilizes one piece of PCMs for centralized heat storage, which is adjacent to the insulator 1. As the distance between heat sources and PCMs becomes longer, the thermal resistance increases. As a result, the heat generated by the remote electronic components is hardly absorbed by the PCMs, thus the temperature control effect on the remote electronic components declines. To tackle this issue, a distributed PTMS is proposed for the

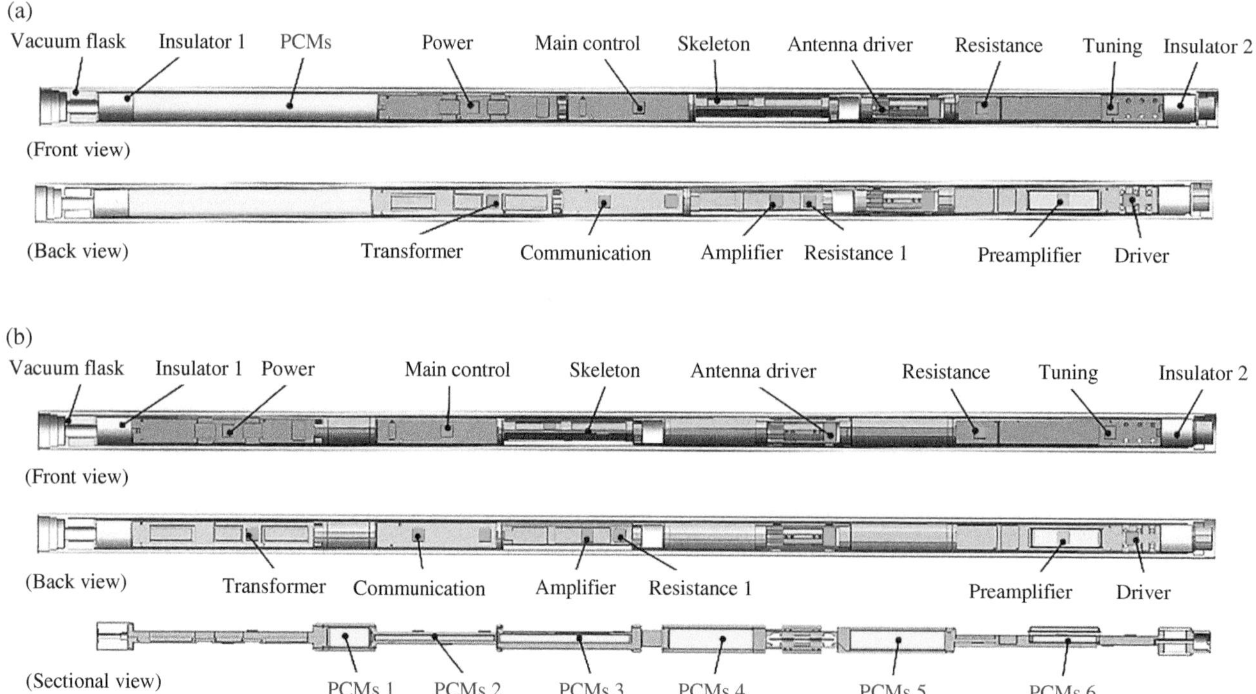

Figure 8.30 Structure of the logging tool with (a) CTMS and (b) DTMS.

Table 8.6 Materials and thermal properties of logging tool components.

Component	Material	Thermal conductivity ($W \cdot m^{-1} \cdot K^{-1}$)	Density ($kg \cdot m^{-3}$)	Heat capacity ($J \cdot kg^{-1} \cdot K^{-1}$)
Vacuum flask	TC-11	7.5	4480	550
Vacuum layer	Composite	0.0002	100	1200
Skeleton	Aluminum alloy	130	2810	960
Simulated heat sources	Ceramic	20	2145	750
PCMs	Paraffin	880(s)/770(l)	0.2	2000
PCMs container	304	7930	16.3	500
Heat sink	Copper	400	8700	385
Insulator 1	PTFE	0.25	2200	1000
Insulator 2	Aerogel blanket	0.025	250	502.3

long-skeleton and multi-heat-source logging tool, as shown in Figure 8.30(b). The PCMs are separated into six small parts and then distributed to the skeleton. Each heat source can store heat in the nearby PCMs, thus the thermal resistance between heat sources and PCMs decreases. In consequence, the quantity of heat storage in PCMs increases, and the temperature of the logging tool becomes lower and more uniform.

To make a fair comparison, the type of the PCMs and the total amount of PCMs in both TMSs are the same. Besides, the size of the skeleton in DTMS is equivalent to CTMS. Both TMSs are required to stay in the environment of 205 °C for 9 hours.

8.4.2 Simulated and Experimental Results

The temperature distribution of the logging tool was simulated by the commercial finite element software. First, the governing equations mentioned earlier were added to the CFD solver. Then the three-dimensional geometry model was imported, and the unstructured grids were generated. The computational domain includes the air domain, which exists in the gap between the inner side of the vacuum flask and the skeleton. Afterward, the materials of the components were defined, and the thermal properties are shown in Table 8.6. It is noteworthy that the melting point of PCMs (RT-70, RUBITHERM, Germany) is 70.84 °C and the latent heat reaches 258.6 $kJ \cdot kg^{-1}$, which are measured by the DSC test. Besides, the vacuum layer was assumed as a solid layer with an equivalent thermal conductivity of 0.0002 $W \cdot m^{-1} \cdot K^{-1}$ according to the parameters given by the manufacturer. Next, some essential boundary conditions were established. The initial temperature of all domains was set to 26 °C, and the environment temperature was set to 205 °C. Furthermore, the heat power of the heat sources was set according to Table 8.7. In the end, the case was calculated for 9 hours with a time step of 10 minutes.

To distinguish the differences between DTMS and CTMS, a temperature analysis is proposed. Figure 8.31(a) presents the temperature distribution of the logging tool with CTMS (left) and DTMS (right) at 9 hours. It is obvious that the maximum temperature of CTMS is higher than DTMS. And the temperature almost linearly increases with the distance to PCMs except antenna driver due to its highest heat power. On the contrary, DTMS displays a lower and more uniform temperature

Table 8.7 Heat sources of the logging tool.

Component	Power (W)	Component	Power (W)
Power	7	Communication	2 × 2
Main control	2	Amplifier	3 × 3
Antenna driver	24	Resistance 1	14
Resistance	15	Preamplifier	1
Tuning	4	Driver	3
Transformer	7		
Total power			90

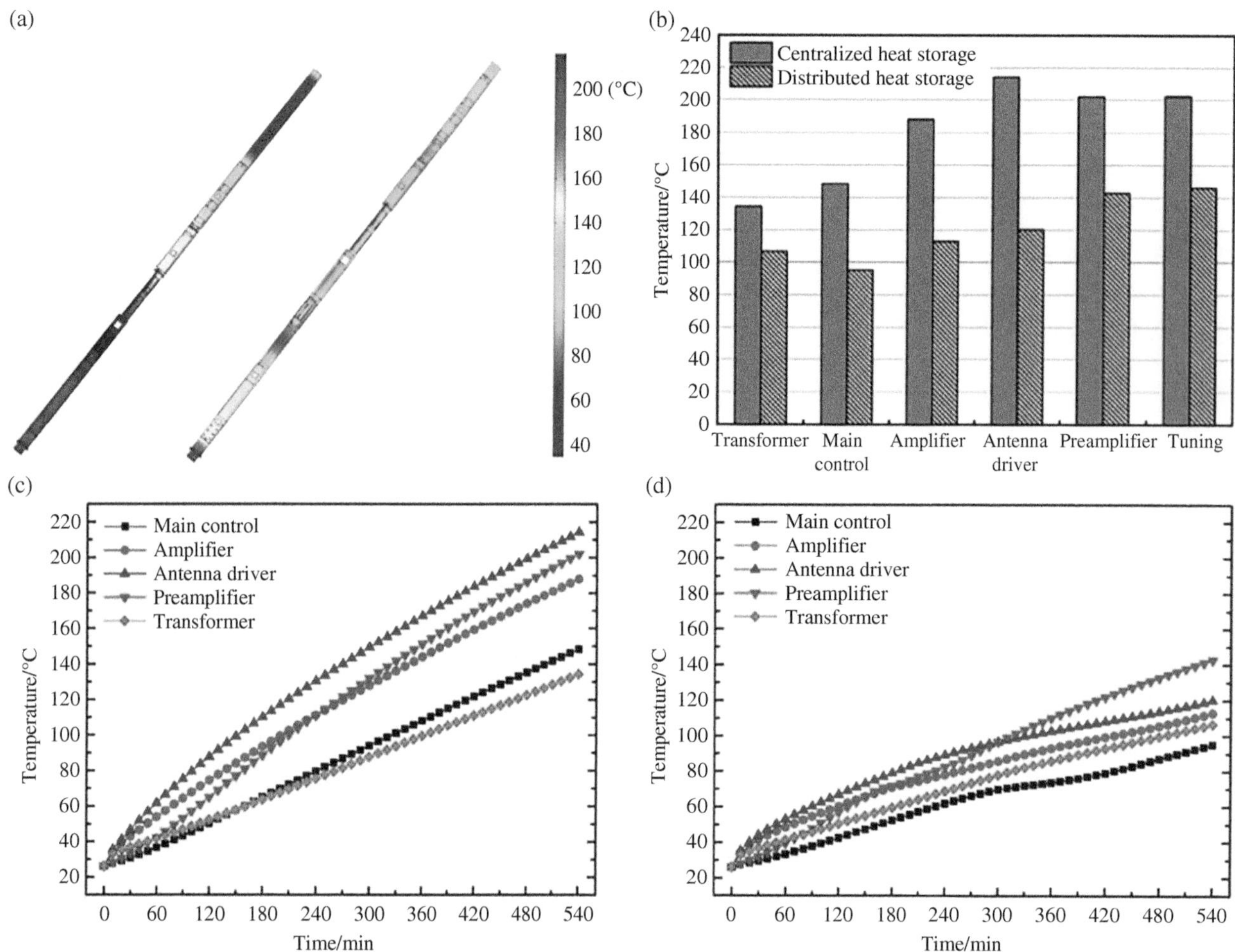

Figure 8.31 (a) Temperature contour of logging tool with CTMS (left) and DTMS (right) at 9 hours; (b) comparison of heat sources temperature between CTMS and DTMS; (c) curves of heat sources temperature versus time in CTMS; (d) curves of heat sources' temperature versus time in DTMS.

distribution on account of the reduction of thermal resistance between heat sources and PCMs. Figure 8.31(b) quantitatively compares the temperature of several typical heat sources. The bar chart shows a similar trend to Figure 8.31(a), and the temperature of the selected heat sources in DTMS is apparently lower than that in CTMS. The maximum temperature of CTMS reaches as high as 214 °C, far exceeding the operating temperature of the electronic components. By contrast, the maximum temperature sharply drops to 146 °C in the logging tool managed by DTMS, which is still within the allowable temperature range of the electronic components. The mean temperature drop of these five heat sources is up to 60.9 °C. According to the analysis above, it can be concluded that DTMS possesses a better temperature control performance than CTMS.

Figure 8.31(a) and (b) is merely focused on the temperature distribution of the logging tool at the end of the time, while Figure 8.31(c) and (d) display the time-varying process of temperature. As can be seen from the graphs below, the curves of the heat sources' temperature in CTMS rise at a roughly constant slope owing to the insignificant effects of PCMs. Similarly, the curves of temperature in DTMS lift sharply at first. However, when the temperature rises slightly above the PCMs melting point (70.84 °C), the PCMs maintain a constant temperature owing to the large latent heat, thus effectively restraining the rising rate of heat sources' temperature. As a result, the temperature of electronic components in DTMS is notably lower than the temperature in CTMS eventually.

The performances of PCMs in DTMS and CTMS during the heating-up process are investigated. Figure 8.32(a) presents the contour of PCMs' phase transition process in CTMS, where 0 represents the solid PCMs while 1 represents the PCMs that

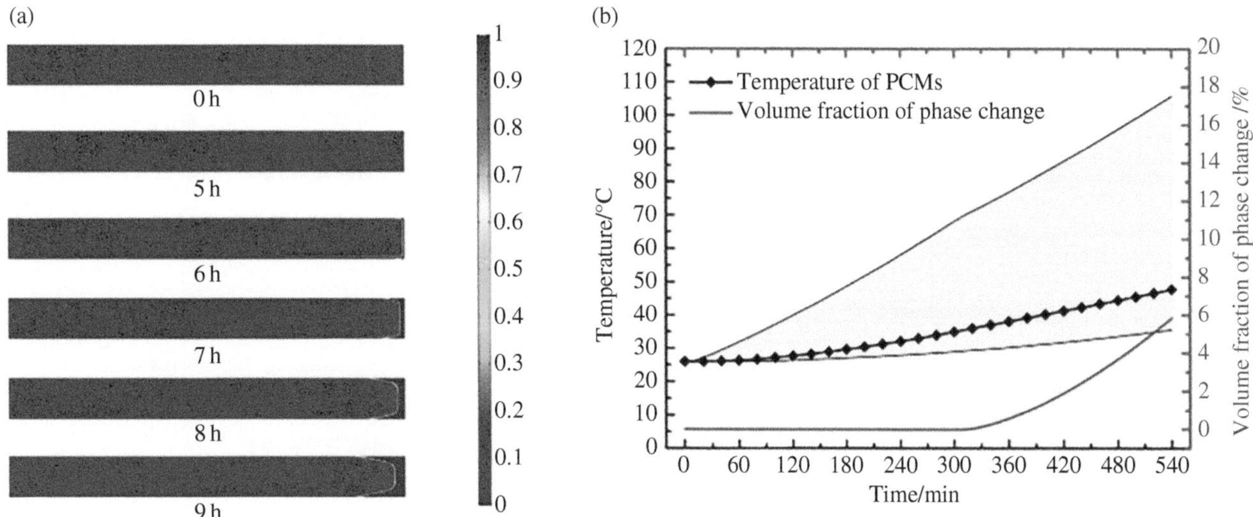

Figure 8.32 (a) Contour of PCMs' phase transition process in CTMS; (b) curves of PCMs' temperature and phase change percentage versus time in CTMS.

have been totally molten. As shown in Figure 8.32(a), the phase transition does not occur within the first 5 hours since the maximum temperature of PCMs does not reach the melting point. After 6 hours, the right side of PCMs begins to melt, whereas the center and the left side are still unmolten. It can be explained by two reasons. For one thing, PCMs are heated by the metal skeleton, thus the surfaces of PCMs reach the highest temperature and melt first. For another, a notable thermal resistance exists inside the PCMs leading to the nonuniform temperature distribution. At the end of 9 hours, only the left side of the PCMs is melting, which implies that a large amount of heat storage capacity of PCMs is waste.

Figure 8.32(b) presents a curve of PCMs' average body temperature along with a curve of phase change percentage versus time in CTMS. The yellow zone means the temperature range of the PCMs, and the edges of the yellow area imply the maximum temperature and minimum temperature of PCMs. At the end of 9 hours, the maximum temperature of PCMs is 106 °C, while the minimum temperature is 35 °C. The vast difference between maximum temperature and minimum temperature implies the significant thermal resistance inside the PCMs. Besides, the curve of phase change percentage in CTMS indicates that the phase change process begins at around 6 hours, and the final utilization rate of the latent heat reaches up to 6%.

The contour of PCMs' phase transition process in DTMS is shown in Figure 8.33. Six pieces of PCMs behave differently. For instance, owing to a massive amount of heat leakage from the opening of the vacuum flask, PCMs 6 melts at first and completes the phase transition process beyond other PCMs. At the end of the operation time, PCMs 2, PCMs 3, and PCMs 6 accomplish the phase transformation revealing that the latent heat of these PCMs is fully utilized. Though the rest of the PCMs have not melted totally, they still absorb a large amount of heat, as shown in Table 8.8. The total amount of PCMs heat storage in DTMS is 961.10 kJ, which is 3.5 times larger than that in CTMS.

Further analysis shows a quantitative description of the phase-change volume fraction of each PCM versus time in DTMS, as shown in Figure 8.34. The latent heat utilization rates of PCMs 1, PCMs 4, and PCMs 5 are 70%, 92.4%, and 55%, respectively. The rest of the PCMs reach up to 100% utilization rate of latent heat; hence, the total PCMs make 77.6% use of latent heat during the operation time. Nevertheless, the utilization rate of latent heat in DTMS is still far higher than that in CTMS. As a result, DTMS makes tremendous progress in temperature control performance.

To verify the simulation, the temperature experiment of the logging tool with DTMS was conducted. As shown in Figure 8.35(a), a prototype of the logging tool with DTMS was fabricated on a scale of one to one, and its size was Φ 91 mm × 3820 mm. It was assembled by several components including six pieces of PCMs. Among the PCMs modules, three pieces were the parts of the skeleton, while the others were embedded into the skeleton. Figure 8.35(b) shows all of the PCMs packaging modules.

After the fabrication of the prototype, two preparations were required for the logging tool temperature experiment. First, several pieces of ceramic heating elements (40 mm × 40 mm × 2 mm, heat power customization, Beijing youpu science and

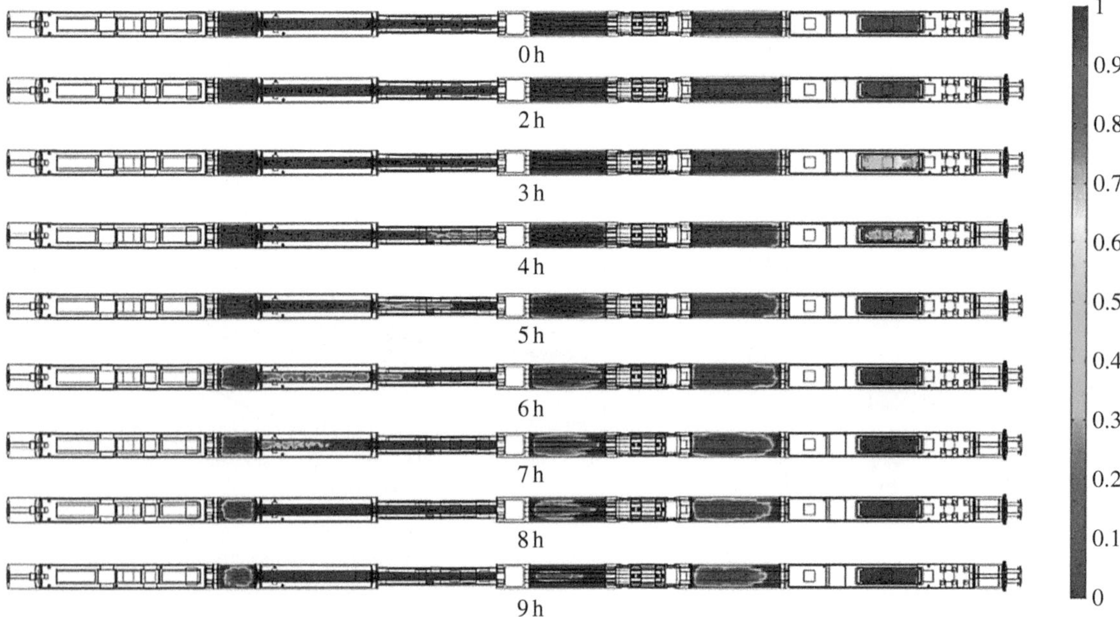

Figure 8.33 Contour of PCMs' phase transition process in DTMS.

Table 8.8 Heat storage of PCMs in DTMS and CTMS.

	PCMs 1	PCMs 2	PCMs 3	PCMs 4	PCMs 5	PCMs 6	Total
DTMS	152.60 kJ	78.02 kJ	74.46 kJ	369.33 kJ	204.36 kJ	82.32 kJ	961.10 kJ
CTMS	—	—	—	—	—	—	213.93 kJ
Heat storage difference							747.17 kJ

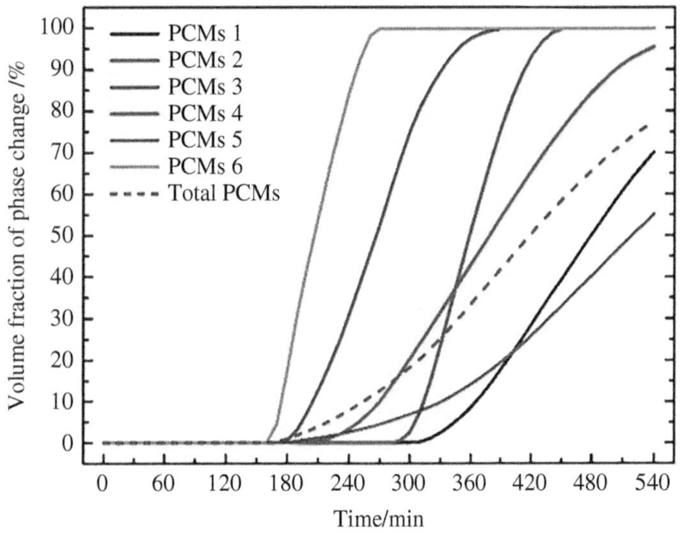

Figure 8.34 Curves of PCMs' phase change volume fraction versus time in DTMS.

Figure 8.35 (a) Prototype of the logging tool with DTMS; (b) PCMs packaging modules in DTMS; (c) simulated heat source and thermocouple; (d) prototype of DTMS with simulated heat sources and thermocouples; (e) experimental site; (f) prototype in the oven.

Technology Centre) were utilized as the simulated heat sources, and they were attached to the skeleton by TIMs ($2\,\text{W}\cdot\text{m}^{-1}\cdot\text{K}^{-1}$, XK-P20S20, GLPOLY). The layout of the simulated heat sources was the same as in Figure 8.30(b). Second, thermocouples (K-Type, 2 mm × 0.3 mm) were fixed on 13 temperature measuring points, including the outer surface of the vacuum flask, insulators, several simulated heat sources, and the PCMs modules. Figure 8.35(c) shows the simulated heat source and the thermocouple. The white plate implied the ceramic heating elements, and the pink pad was the thermal silicon pad. The blue line meant the thermocouple, which was stuck on the surface of the heating elements. Figure 8.35 (d) shows the full view of the prototype with simulated heat sources and thermocouples. The red–black power supply cord entered from one end, and the thermocouple wires entered from the other end.

Afterward, the prototype with simulated heat sources and thermocouples was stuffed in the vacuum flask (YJPST128x4031/92.5, Xi'an Yufeng Electronic Engineering Company, China). Then the prototype and the vacuum flask were lifted into the oven (operating range 51–260 °C, PTC1-40, Despatch), as shown in Figure 8.35(e) and (f). The temperature of the oven was set to 205 °C, and the experiment lasted for 9 hours. All of the temperature data was recorded by a data acquisition instrument (accuracy 0.2% FS ± 1D, MIK-6000F, Hangzhou Meacon Automation Technology Company) with a sampling period of 1 second.

Figure 8.36 shows the experimental temperature curves of the measuring points versus time. Since the number of the experimental curves is up to 13, it would not be easy to find out each curve if all the curves are displayed in one figure. Therefore, the experimental results are divided into three figures by category. From Figure 8.36(a), the environment temperature rises sharply at first and then maintains at 205 °C after 60 minutes. The temperature curves of several heat sources are also investigated here. And the temperature of them is below 150 °C during the heating-up process, which is still within the electronic operating temperature range. Among all of the heat sources, the temperature of the tuning board is the

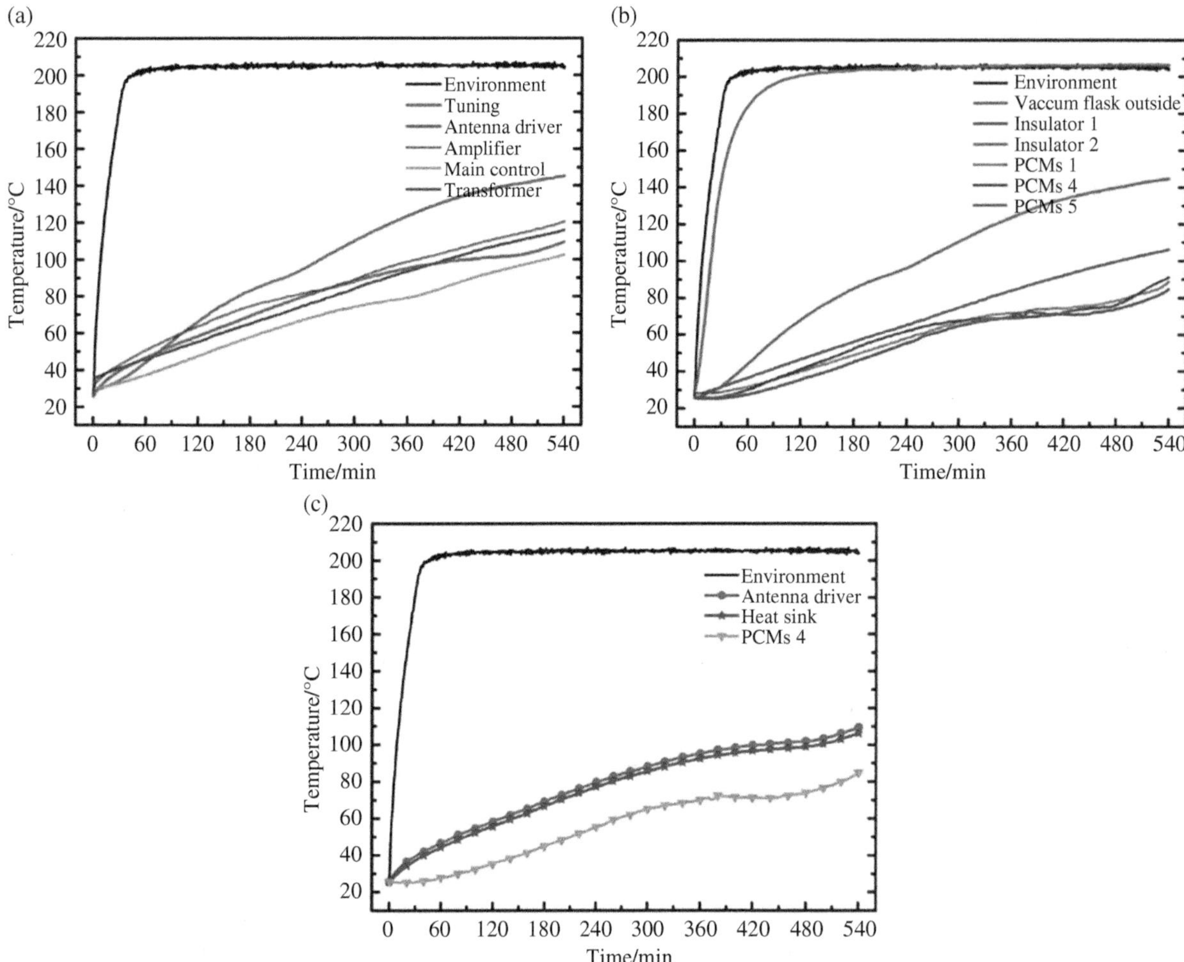

Figure 8.36 (a) Experimental temperature curves of heat sources versus time; (b) experimental temperature curves of insulators and PCMs; (c) experimental temperature curves of antenna driver under the control of PCMs 4.

highest owing to the great influence of the heat leakage from a high-temperature environment. Figure 8.36(b) shows the experimental temperature curves of insulators and PCMs. There are two apparent turning points and a flat zone in three curves of PCMs, indicating that the PCMs pass through the phase transition process during the experiment. Compared to insulator 1, the temperature of insulator 2 is much higher as a consequence of the severe heat leakage from the large opening of the vacuum flask. Figure 8.36(c) selects three representative curves to illustrate the temperature control effect of the PCMs on the heat sources. It is shown that PCMs 4 maintains the melting point from 360 to 480 minutes, and the temperature curves of the antenna driver and the heat sink are obviously restrained similar to the curve of PCMs 4. The result confirms the temperature control effect of the PCMs on the heat sources.

To verify the simulation, the temperature of five measuring points between the experiment and simulation is compared in Figure 8.37. The curves represent the temperature results of the simulation, while the scatter points represent the experimental results. Though the curves are unable to match the scatter points perfectly, the trend of the simulation results is similar to the experimental results. The curves and the scatter plots become mild during the phase change process, especially for PCMs 1. To quantitatively compare the experiment data and simulation data, several statistics are utilized to judge the agreement between simulation results and experimental results, including relative error (RE), root mean square error (RMSE), and correlation coefficient (r). Among them, RE is utilized to measure the temperature error between simulation and experiment at a specific moment, while RMSE is used to assess the temperature error through the whole heating-up

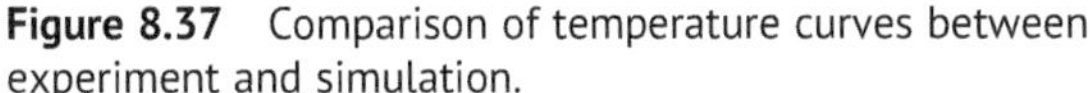

Figure 8.37 Comparison of temperature curves between experiment and simulation.

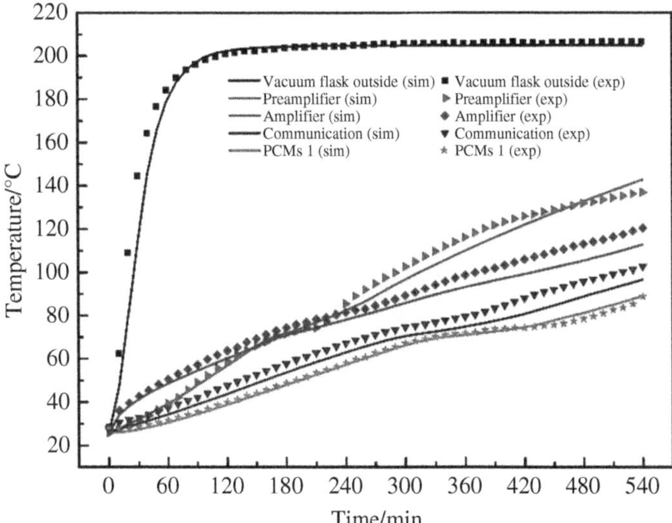

process. Besides, r is utilized to measure the strength of a linear relationship between the simulation data and experimental data. First, the simulation temperature and experimental temperature at the end of time are compared, the maximum RE of all the measuring points is lower than 10% and the average RE is below 5%. Then the agreement between the data sets of simulation and experiment is further investigated. The results show that RMSE between the two data sets is within 9 °C, and r between the simulation results and the experimental results is larger than 0.99, which implies that the trend of the simulation data is highly consistent with the experimental data. In summary, the analyses above indicate that the simulation of the logging tool is reasonable; hence, the discussions about the thermal performance of the logging tool with DTMS and CTMS are convincing.

Furthermore, the reasons behind the deviation between the experiment and simulation are studied. The deviation is mainly caused by the assumptions in the simulation. First, the neglect of the heat convection and radiation inside the flask attenuates the heat transfer between the vacuum flask and the skeleton, which would reduce the simulation temperature of the skeleton and the heat sources. Second, owing to the disregard of the contact thermal resistance, the thermal resistance between the adjacent components would decline, leading to the enhancement of the heat transfer between heat sources and PCMs. Hence, the temperature of PCMs would be lower in the simulation. Third, the vacuum layer is equivalent to a solid layer with low thermal conductivity, regardless of the radiation between the radiation shields inner the vacuum layer. Although an equivalent thermal conductivity is given by the manufacturer, it may still cause a deviation.

8.5 Thermal Optimization of High-Temperature Downhole Electronic Devices

The TMS mentioned in Section 8.4 can protect the electronics for several hours. However, these PTMSs of the logging tool in previous studies have not been optimized, failing to exert a better temperature control performance. In general, longer insulators are instrumental in reducing heat leakage, and more PCMs are beneficial for heat storage, both of which contribute to a lower temperature of heat sources. Due to the limited space in the logging tool, the total length of insulators and PCMs is fixed. Therefore, under the condition of a fixed total length of the logging tool, it is a challenging task to optimize the lengths of insulators and PCMs to obtain a better thermal management effect.

8.5.1 Optimization Method

In this section, the finite-element method and the Nelder–Mead method were combined to optimize the thermal design of the logging tool. The finite-element method was adopted to simulate the thermal performance of the current logging tool

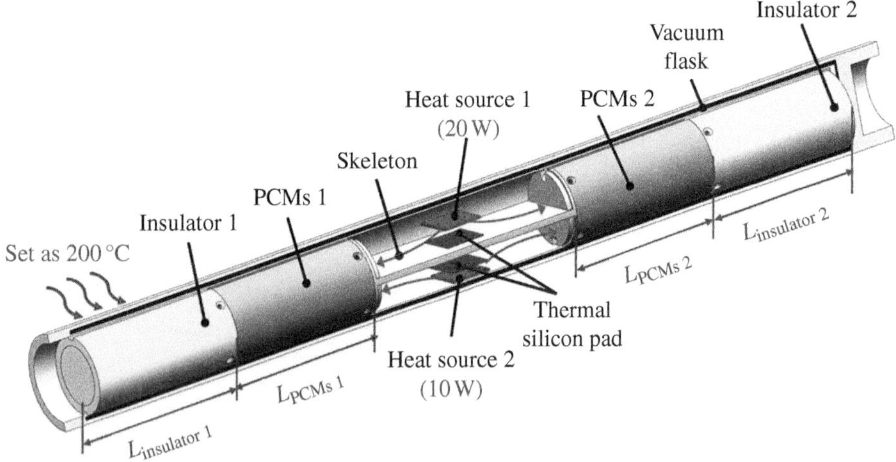

Figure 8.38 Diagram of the logging tool.

structure, and then the Nelder–Mead algorithm was used to optimize the design of the logging tool based on the simulation results. The above processes were repeated until the optimized thermal design was obtained. Afterward, the optimization process and the thermal performance of the logging tool before and after optimization were investigated, including temperature, phase change process, and the amount of heat generation/storage. Finally, an experiment was conducted to verify the effectiveness of thermal optimization on the logging tools.

Figure 8.38 displays a diagram of the logging tool composed of a vacuum flask, two insulators, two pieces of PCMs, a skeleton, and two heat sources. Since the air gap inside the vacuum flask is below 0.5 mm, it is reasonable to assume that the heat convection and radiation of the gap are ignored. Therefore, the simulation of the logging tool is simplified to a transient heat transfer problem and it is governed by differential heat conduction equation.

The square heat capacity curve method is utilized to describe the phase change process of PCMs, and its governing equations are shown in Equations (8.12)–(8.15). Considering the limited computing resources, the convection inside the liquid PCMs is neglected in our simulation. Actually, the natural convection of liquid PCMs will affect the heat transfer process, which is evident in the PCMs with low thermal conductivity [51, 52]. In our study, the thermal conductivity of Bendalloy PCMs reaches 19.83 $W \cdot m^{-1} \cdot K^{-1}$, about 100 times that of paraffin. Therefore, the natural convection effect of liquid PCMs does not occupy a dominant position in our cases. To clarify about the accuracy of the simplification, 2D axisymmetric simulations with convection and without convection are conducted, and the results show that the average temperature difference between the two cases is 1.64 °C, and the melting time in the convection case is shortened by 1.5% compared with the only heat conduction case. The results indicate that the neglect of convection inside the liquid PCMs is acceptable in our study.

The simulation and optimization flow chart are shown in Figure 8.39. The Nelder–Mead method, a classical gradient-free optimization algorithm, was utilized to combine the simulation and control the whole optimization process [53, 54]. First, a parametric three-dimensional geometry model of the logging tool was built in Solidworks and then imported into CFD solver automatically. Next, the thermal performance of the logging tool was simulated, and the maximum temperature of heat sources was obtained. Afterward, the optimization parameters of the logging tool were modified by the Nelder–Mead algorithm based on the simulated results. Iteratively calculated the above processes until the optimized solution was obtained. In our cases, the optimization parameters are $l_{\text{insulator 1}}$, $l_{\text{PCMs 1}}$, $l_{\text{PCMs 2}}$, and $l_{\text{insulator 2}}$, which represent the length of insulator 1, the length of PCMs 1, the length of PCMs 2, and the length of insulator 2, respectively. Generally, a uniform length distribution of PCMs and insulators is adopted as the initial thermal design of the logging tool, and, in our cases, the initial values of the four optimization parameters are set to 150 mm simultaneously. The optimization objective is to obtain the lowest heat-source temperature, which is expressed by the following equation:

$$\text{minimize}(T_{\text{heat source 1}} + T_{\text{heat source 2}})/2 \tag{8.18}$$

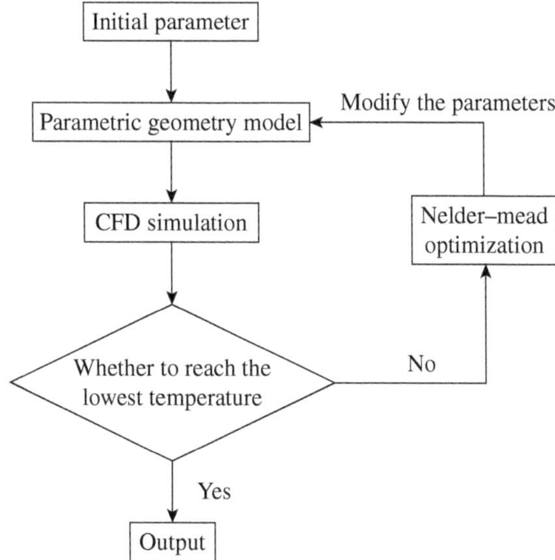

Figure 8.39 Simulation and optimization flow chart.

To maintain the size of the logging tool unchanged during the optimization process, the total length of four optimization parameters should be kept constant. Moreover, the value of each optimization parameter should be within a certain range to meet the assembly requirements. Therefore, the optimization constraints can be determined by the following equations:

$$\begin{cases} l_{\text{insulator 1}} + l_{\text{PCMs 1}} + l_{\text{insulator 2}} + l_{\text{PCMs 2}} = 600\ \text{mm} \\ 30\ \text{mm} \leq l_{\text{insulator 1}} \leq 300\ \text{mm} \\ 30\ \text{mm} \leq l_{\text{PCMs 1}} \leq 300\ \text{mm} \\ 30\ \text{mm} \leq l_{\text{PCMs 2}} \leq 300\ \text{mm} \\ 30\ \text{mm} \leq l_{\text{insulator 2}} \leq 300\ \text{mm} \end{cases} \tag{8.19}$$

Every optimization iteration needs to calculate the root mean square of the function value difference between each vertex in the simplex and the optimal vertex. The stop condition is that the root mean square is less than the pre-set value, which can be expressed by the following equation:

$$\left\{ \frac{1}{N_{\text{V}}} \sum_{i=1}^{k} [f(X_i) - f(X_1)]^2 \right\}^{1/2} \leq \xi \tag{8.20}$$

where N_v means the number of vertexes in the simplex and equals 5 in our cases. X_i represents the ith vertex, and X_l represents the optimal vertex. ξ means the pre-set value, which is equal to 0.5.

Some details about CFD simulation are shown in this section. First, the geometry model was imported into the solver and then automatically generated the tetrahedral mesh. Next, materials along with the thermal properties of each component were defined in our cases, as shown in Table 8.9. Among them, the vacuum layer was assumed as a solid layer with an equivalent low thermal conductivity of 0.0002 $\text{W} \cdot \text{m}^{-1} \cdot \text{K}^{-1}$, regardless of the influence of thermal radiation [27]. Low melting point alloys with a latent heat of 36.68 $\text{kJ} \cdot \text{kg}^{-1}$ were utilized for heat storage. Generally, the melting range determines the temperature control range of PCMs and has a strong impact on the equivalent square heat capacity model. To obtain the melting range of PCMs, a DSC test was conducted at a heating rate of 5 $°\text{C} \cdot \text{min}^{-1}$. According to the DSC curve, the temperature ranges between the onset temperature (T_{onset}) and the endset temperature (T_{endset}) is recognized as the actual melting temperature range, which is from 72.03 to 75.43 °C in our cases. Afterward, the boundary conditions were defined according to the actual logging situation. To simulate a high-temperature downhole environment, the outside surface temperature of the vacuum flask was set to 200 °C. Besides, the heat sources were set to generate heat uniformly, where heat source 1 was set to 20 W and heat source 1 was set to 10 W. The initial temperature of all domains was set to 20 °C, and the relative tolerance factor was set to 0.0001. In the end, the case was calculated for 6 hours with the time step of 10 minutes.

Table 8.9 Materials and thermal properties of logging tool components.

Name	Material	Thermal conductivity ($W \cdot m^{-1} \cdot K^{-1}$)	Density ($kg \cdot m^{-3}$)	Heat capacity ($J \cdot kg^{-1} \cdot K^{-1}$)
Vacuum flask	Inconel 718	14.7	8240	436
Vacuum layer	Composite	0.0002	200	1200
Skeleton	Aluminum alloy	167	2710	896
Heat source	Ceramics	30	3960	850
Thermal silicon pad	Silica gel	1	1810	923
Shell of insulator	PEEK	0.2	2710	880
Core of insulator	Felt	0.035	400	794.2
Shell of PCMs	Aluminum alloy	167	2710	896
PCMs	Bendalloy	19.83	9658	166.7(s)/184(l)

To eliminate the influence of mesh number, the mesh independence test was conducted. In consideration of the errors and the computing time, 146,317 mesh elements were chosen for further simulation. In the meantime, to avoid the step-jump problem, we investigated the effect of the different relative tolerances. By lowering the relative tolerance, the software automatically decreases the timestep size to maintain the desired relative tolerance, which is beneficial to increase the accuracy of the solution. Finally, the relative tolerance is 0.0001 in our cases.

8.5.2 Experimental Setup

To verify the thermal optimization effectiveness of the logging tool, the temperature experiment was carried out. Figure 8.40 (a) and (b) display an initial logging tool and an optimized logging tool, respectively, which were fabricated on a scale of one to one with the size of Φ 72 mm × 820 mm. In preparation for the experiment, two ceramic heating elements (40 mm × 40 mm × 2 mm, 20 W/12 V or 10 W/5 V, Shenzhen Yibolan Electronics Co., Ltd.) were first attached to the skeleton by thermal silicon pad ($1\ W \cdot m^{-1} \cdot K^{-1}$, XK-P10, GLPOLY). Afterward, thermocouples (K-Type, 2 mm × 0.3 mm) were affixed to eight temperature measuring points, including the heat sources, PCMs, insulators, outer surface of the vacuum flask, and the oven, as shown in Figure 8.40. Next, the prototypes with heat sources and thermocouples were stuffed in the vacuum flask (JP90x900/73, Xi'an Yufeng Electronic Engineering Company). And then the prototype and the vacuum flask were put in the oven (operating range 5–300 °C, Shanghai Hecheng Instrument Manufacturing Co., Ltd). The temperature of the oven was set to 200 °C, and the two heat sources were powered by two DC power supplies (0–30 V/0–10 A, MS-3010D, Dongguan Meisheng Power Technology Co., Ltd.), as shown in Figure 8.40(c) and (d). The experiment lasted for 6 hours. In the meantime, the temperature data were recorded every second by a data acquisition instrument (accuracy 0.2% FS ± 1D, MIK-6000F, Hangzhou Meacon Automation Technology Company).

8.5.3 Thermal Optimization Results

As shown in Figure 8.41(a), the lengths of PCMs and insulators change versus optimization iterations. In the beginning, the initial lengths of all the components are 150 mm. During the optimization process, the lengths of PCMs 1 and PCMs 2 show a fluctuating growth. On the contrary, the length of insulator 1 and insulator 2 decreases with the optimization iterations. Notably, the sum of all lengths remains constant, which implies that the size of the logging tool stays the same. After 30 optimization iterations, all the lengths change little, which indicates that the residuals of iteration are converging. The optimization process comes to an end at the 61st optimization iteration. The optimized lengths of PCMs 1, PCMs 2, insulator 1, and insulator 2 are 279.99, 200.03, 88.15, and 31.83 mm, respectively.

Figure 8.41(b) displays the maximum temperature of heat source 1, heat source 2, PCMs 1, and PCMs 2 versus optimization iterations. Under the control of Nelder–Mead optimization algorithm, the temperature of all the components decreases rapidly at first and then fluctuates around the minimum value after 30 optimization iterations. The results imply that the temperature trend of the logging tool is consistent with the optimization goal, and the optimized thermal design of the logging tool is obtained in the end.

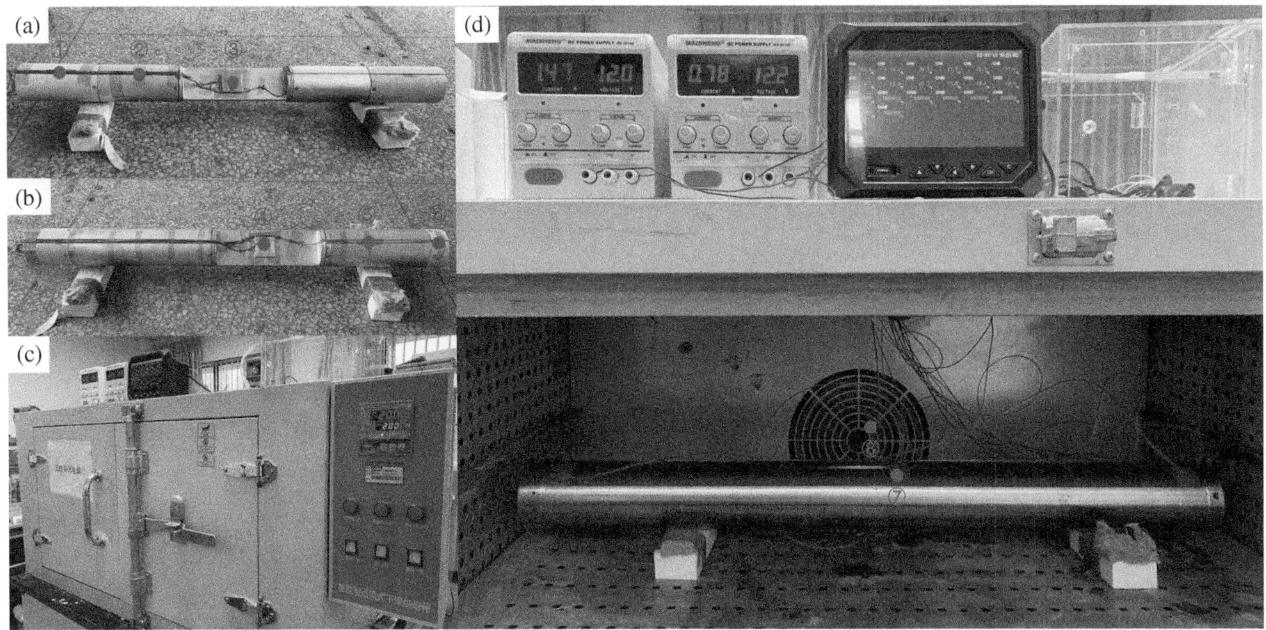

Figure 8.40 (a) Prototype of the initial logging tool; (b) prototype of the optimized logging tool; (c) temperature test site; (d) experimental site layout and data recording.

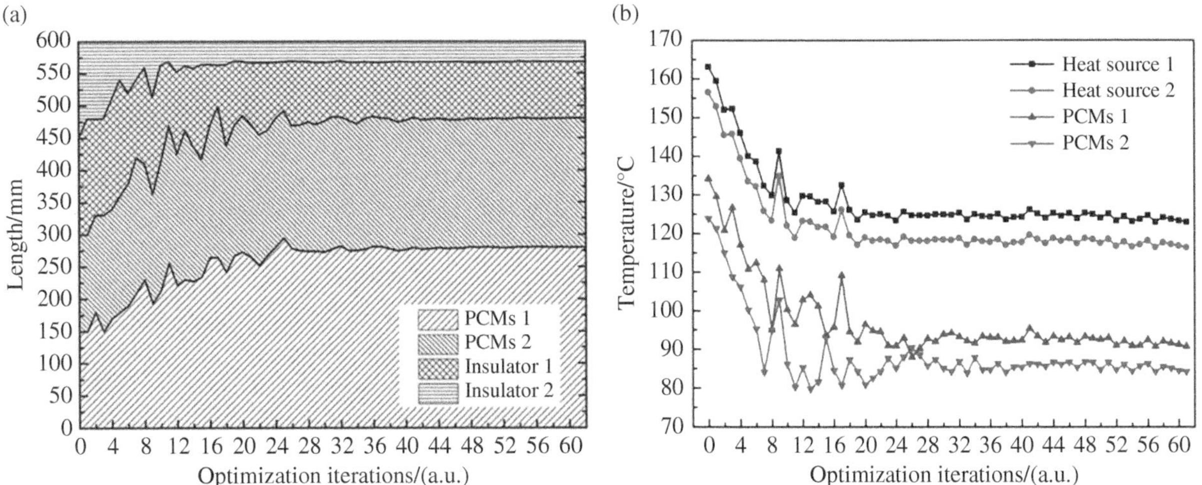

Figure 8.41 (a) The lengths of PCMs and insulators versus optimization iterations; (b) temperature versus optimization iterations.

To illustrate the effectiveness of the optimization, the temperature performances of the logging tool before and after optimization are compared in Figure 8.42. Figure 8.42(a) displays the temperature contour of the initial logging tool (left) and the optimized logging tool (right). As shown in the temperature contour before optimization, the maximum temperature appears at insulator 1 owing to the severe heat leakage from the opening. In contrast, insulator 2 possesses the lowest temperature, revealing that the heat leakage in this end is much milder. Therefore, it is wise to cut down the length of insulator 2 to increase the lengths of other components. Compared with the initial temperature contour, the length of insulator 2 is obviously cut down in the optimized thermal design, and the lengths of both PCMs increase for the enhancement of heat capacity. As a result, the temperature distribution in the optimized case is much lower and more uniform.

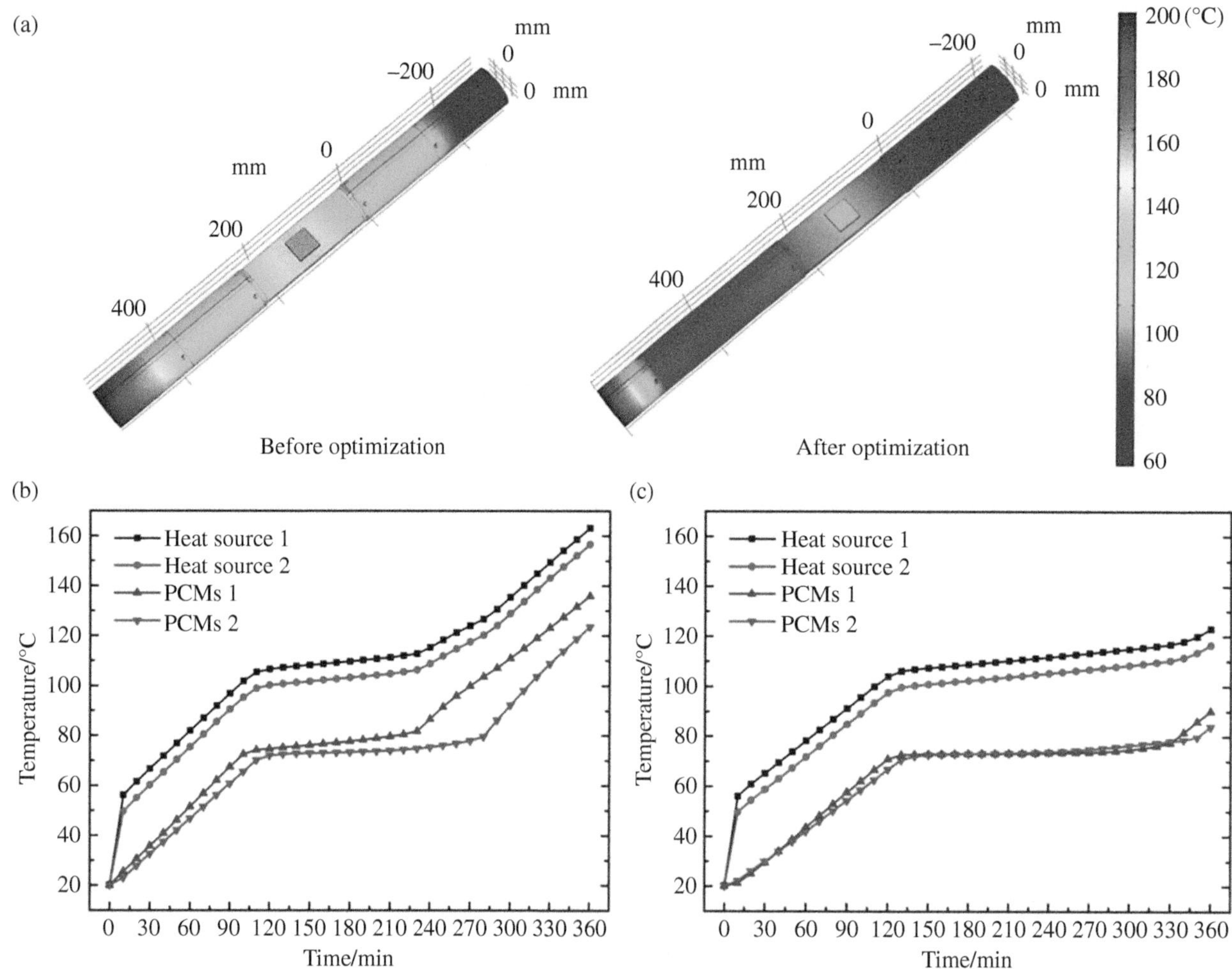

Figure 8.42 (a) Temperature contours before and after optimization; (b) temperature versus time before optimization; (c) temperature versus time after optimization.

Figure 8.42(b) and (c) show the temperature curves versus time before and after optimization, respectively. Owing to the higher heat power, the temperature of heat source 1 is the highest among the four components in both figures. All the curves experience three stages. In the first stage, the temperature curves of the four components rise rapidly at first, in the meanwhile, the heat absorbed by PCMs is transformed into their own sensible heat. In the second stage, the rising rate of temperature is restrained due to the phase change process of PCMs. After the phase change process, the temperature curves rise rapidly again. The comparison between Figure 8.42(b) and (c) show that the rising rate of temperature in the optimized logging tool is lower, and the phase change process extends from 180 to 240 minutes, which is prolonged by about 33%. As a consequence, at the end of the operation time, the temperature of all components after optimization is lower than that before optimization. The maximum temperature before optimization is 163.1 °C, and the maximum temperature after optimization is 123 °C, decreasing by 40.1 °C. The average temperature of all the components reduces by 41.4 °C in the optimized logging tool. In summary, the thermal management effect in the optimized logging tool is enhanced, and the temperature performance is much better than that in the initial case.

According to the former analysis, the phase change process of PCMs plays a significant role in the inhibition of temperature rise rate. In this section, the performances of PCMs before and after optimization are investigated during the heating-up process. Figure 8.43(a) presents the phase change contour of PCMs versus time in the initial logging tool, where 0 represents the solid PCMs and 1 represents the molten PCMs. As can be seen in Figure 8.43(a), the phase change process does not occur in the first 2 hours since the maximum temperature of PCMs does not reach the melting point. After 3 hours, PCMs begin to melt from the edge next to the heat sources. And then the phase change interface diffuses to both

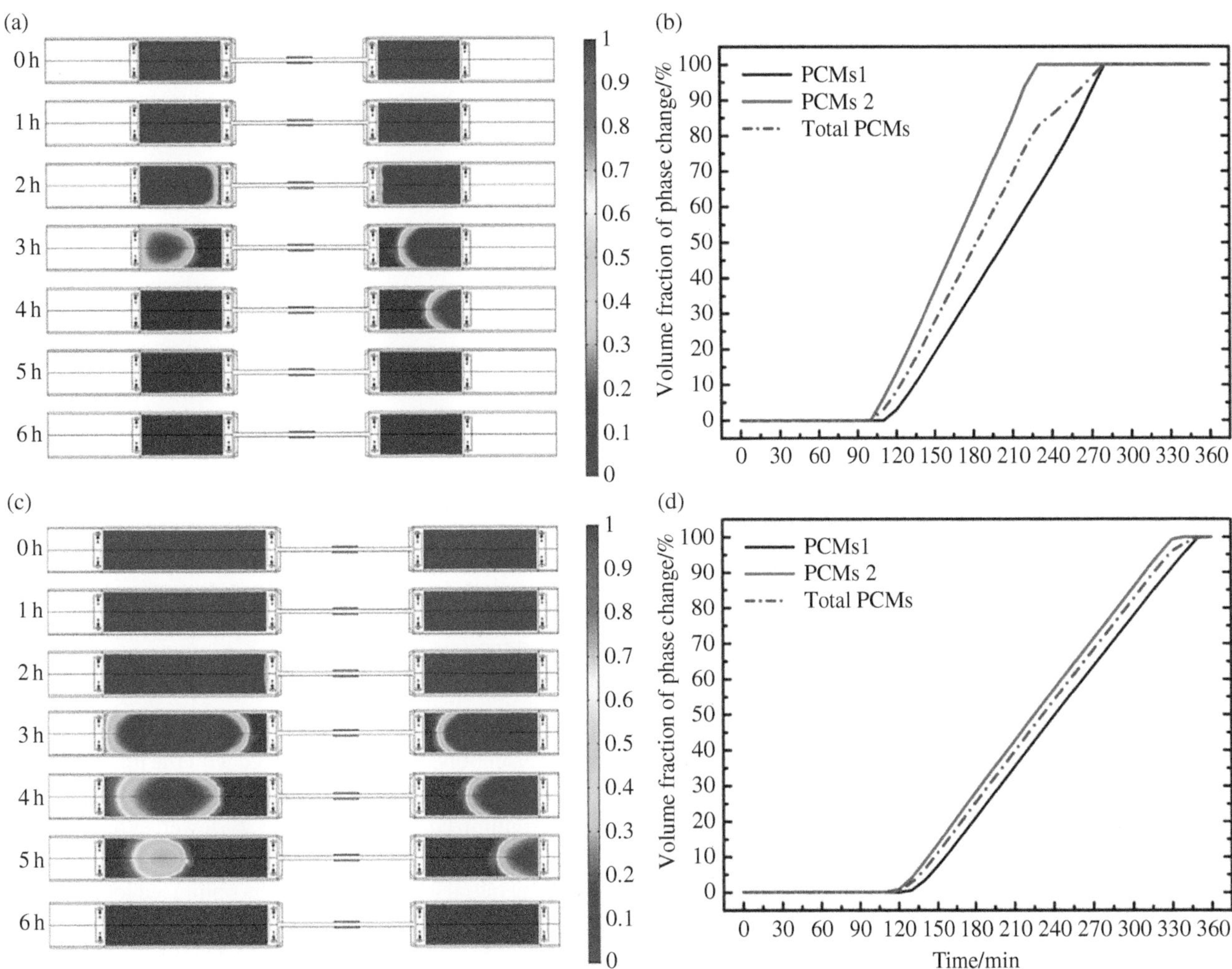

Figure 8.43 (a) Phase change contour of PCMs versus time before optimization; (b) curves of phase change volume fraction versus time before optimization; (c) phase change contour of PCMs versus time after optimization; (d) curves of phase change volume fraction versus time after optimization.

ends until the PCMs melt completely. Figure 8.43(b) displays the curves of phase change percentage versus time before optimization. The PCMs 1 starts melting at around 100 minutes and finishes the whole phase change process at 220 minutes. Besides, the PCMs 2 begins to melt at the same time as PCMs 1 and completes the phase transition at 280 minutes owing to less heat leakage. In the end, the phase-change volume fraction of the total PCMs reaches 100%.

In comparison with the initial logging tool, the phase change contour of PCMs versus time after optimization is presented in Figure 8.43(c). Similar to the phase change contour before optimization, the PCMs begin to melt from the edge next to the heat sources and complete the phase transition process in the end. But the phase transition process starts 20 minutes later than the initial case because the PCMs of the optimized case possess more sensible heat capacity. Besides, since the length of insulator 1 reduces in the optimized case, more heat leakage enters the PCMs 1, and thus the phase change interface of PCMs 1 diffuses from two ends to the center. As shown in Figure 8.43(d), the curves of PCMs 1 and PCMs 2 are almost the same, which simultaneously start to melt at 120 minutes and rise at a similar rate. The similar performances of PCMs 1 and PCMs 2 are beneficial to the more uniform heat storage, and thus further contribute to a well-distributed temperature of the logging tool.

To further explain the reason for temperature reduction in the optimized logging tool, the heat flow of all the components is investigated in this section. Figure 8.44 compares the amount of heat generation or heat storage before and after optimization. During the whole operation process, heat source 1 and heat source 2 generate 432 kJ heat and 216 kJ heat,

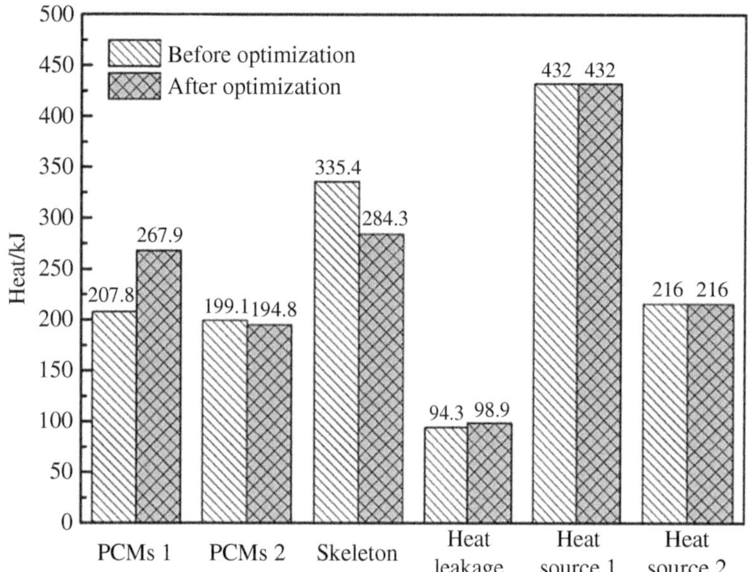

Figure 8.44 Comparison of heat generation/storage before and after optimization.

respectively, which are equal before and after optimization. Besides, the heat leakage from the high-temperature environment in the optimized logging tool is 4.9% higher on account of shorter insulators. Further, compared with the initial logging tool, the PCMs in the optimized logging tool store 13.7% more heat, and thus avoid heat accumulation of the heat sources and the skeleton. It is notable that the heat stored by the skeleton of the optimized logging tool is 15.2% lower; as a result, the skeleton and the heat sources present a lower temperature performance.

Figure 8.45(a) and (b) show the experimental temperature curves of the initial logging tool and the optimized logging tool, respectively. It can be seen that the environment of the two experiments is almost the same. The environmental temperature rises sharply at first and then fluctuates around 200 °C after 45 minutes in both experiments. The temperature of the outer surface of the vacuum flask follows environmental temperature closely and reaches 200 °C after 60 minutes. Due to the severe heat leakage from the opening, the temperature of insulator 1 is much higher than that of insulator 2. As for PCMs, the phase change process in the optimized logging tool starts at 134 minutes, which is about 20 minutes later than that in the initial logging tool. Further, the phase transition time of the optimized logging tool reaches 226 minutes, which is extended by 29.9%. Therefore, the temperature rise rate of heat sources in the optimized logging tool was restrained for a longer time. As a consequence, the maximum temperature of heat sources in the optimized logging tool is 130.6 °C, which is 30.1 °C lower than that in the initial logging tool.

The temperature of heat sources and PCMs between experiment and simulation is compared in Figure 8.45(c) and (d). The curves represent the experimental results, while the dotted line represents the simulated results. Though the curves are unable to match the dotted line perfectly, the trend of the simulated results is similar to the experimental results. As shown in Figure 8.45(c), the maximum error between simulated results and experimental results of the initial logging tool is 11.83 °C, and the average error is just 4.31 °C. Similarly, Figure 8.45(d) shows that the experimental results in the optimized logging tool match the simulated results with a maximum error of 13.2 °C and an average error of 4.98 °C. On the whole, the average error between simulated results and experimental results of both cases is 4.64 °C.

Further, the deviation between the experiment and simulation is mainly caused by two reasons. On the one hand, the neglect of the contact thermal resistance in the simulation enhances the heat transfer between heat sources and PCMs. Hence, the simulated temperature of heat sources is lower than the experimental temperature, and the simulated temperature of PCMs tends to be higher. On the other hand, the experimental ambient temperature rises from room temperature to 200 °C, different from the boundary conditions in the simulation, which would reduce the heat leakage from the high-temperature environment. Besides, neglecting the change of materials properties with temperature and the air convection inside the vacuum flask would also lead to errors. In summary, the temperature experiments verify the effectiveness of thermal optimization with a 30.1 °C decrease in the optimized logging tool. And the simulated results well match the

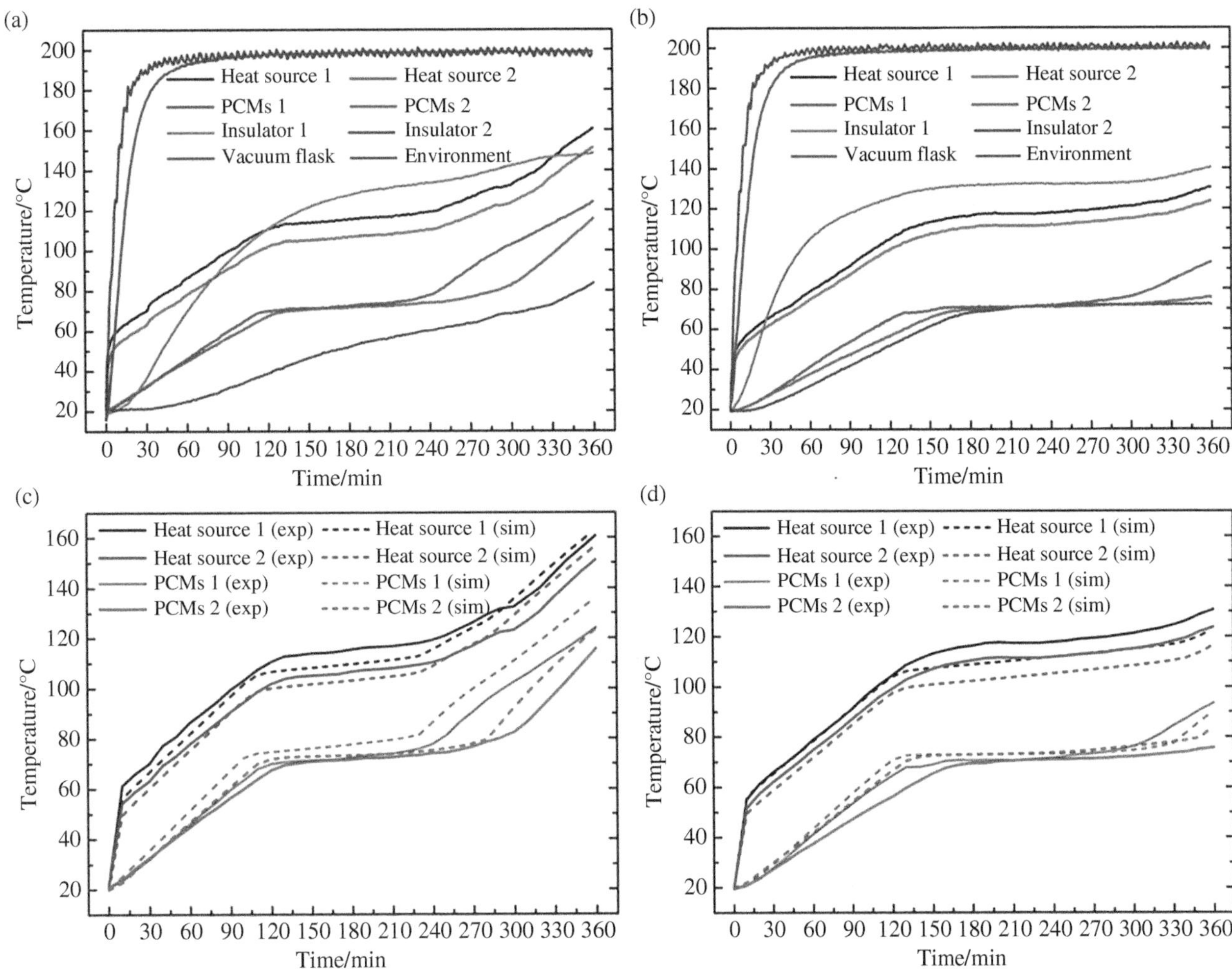

Figure 8.45 (a) Experimental temperature curves versus time in the initial logging tool; (b) experimental temperature curves versus time in the optimized logging tool; (c) comparison of temperature curves between experiment and simulation in the initial logging tool; (d) comparison of temperature curves between experiment and simulation in the optimized logging tool.

experimental results with an average error of 4.64 °C, revealing that the discussions about the thermal performance in former sections are convincing.

8.6 Chapter Summary

In Section 8.1, a passive thermal management system for downhole electronic devices was proposed in thermally harsh environment. The thermal performance of the proposed system was experimentally evaluated. The temperature of the electronic components can be maintained below 125 °C for 6-hour operating time, making the proposed TMS a promising candidate for further application in oil and gas industry. In Section 8.2, a numerical model that coupled multiple heat transfer modes was proposed to create a passive thermal management system for logging tools. The proposed model considers solid heat conduction, natural air convection, thermal radiation, and phase change processes simultaneously. The experimental results showed that the maximum absolute error between simulation and experiment was 8.61 °C, and the average absolute error was only 3.02 °C, thus demonstrating the accuracy of the numerical model. The proposed model can characterize the real heat transfer process accurately with a deviation of 4.71% from the experiment. In Section 8.3, an ultrasonication method was demonstrated to prepare thermal enhanced GNs/paraffin CPCMs. The thermal conductivity could be increase by 102.26% at 4 wt% GNs loading, with negligible effect on the phase change characteristics. In addition, a peak value of thermal conductivity existed while changing the particle size of GNs, instead of a monotonic trend presented by previous researches. In Section 8.4, A DTMS was proposed for the long-skeleton and multi-heat-source logging tool by dispersing the PCMs into the skeleton to decrease the thermal resistance between heat sources and PCMs. The maximum temperature dropped from 214 to 146 °C. Besides, the utilization rate of latent heat in DTMS was up to 77.6%. In Section 8.5, the finite element method and the Nelder–Mead algorithm was combined to optimize the thermal design of the logging tool. After 61 optimization iterations, we obtained the optimized thermal design of the logging tool. Compared with the initial logging tool, the phase change process is prolonged by about 33%, and thus the maximum temperature of heat sources in optimized case drops from 163.1 to 123 °C. Besides, the phase transition performance of two pieces of PCMs in the optimized logging tool are more consistent, which store 13.7% more heat in total, avoiding the heat accumulation of the skeleton and the heat-sources.

References

1 Boyes, J. (1981). The eyes of the oil industry. *Electron. Power* 27: 484–488.

2 Hyne, N.J. (2012). *Nontechnical Guide to Petroleum Geology, Exploration, Drilling, and Production*. PennWell Books.

3 Cheng, W.L., Nian, Y.L., Li, T.T. et al. (2014). A novel method for predicting spatial distribution of thermal properties and oil saturation of steam injection well from temperature logs. *Energy* 66: 898–906.

4 Zhao, Q.M., Fan, J.C., Yan, D. et al. (2012). Development of downhole tools for stratified testing in oil and gas open-hole wells. *Adv. Mater. Res.* 422: 614–618.

5 Werner, M.R. and Fahrner, W.R. (2001). Review on materials, microsensors, systems and devices for high-temperature and harsh-environment applications. *IEEE Trans. Ind. Electron.* 48: 249–257.

6 Sinha, A. and Joshi, Y.K. (2011). Downhole electronics cooling using a thermoelectric device and heat exchanger arrangement. *J. Electron. Packag.* 133: 041005.

7 Tian, B., Liu, H., Yang, N. et al. (2015). Note: High temperature pressure sensor for petroleum well based on silicon over insulator. *Rev. Sci. Instrum.* 86: 126103.

8 Ohme, B.W., Johnson, B.J., and Larson, M.R. (2012). *SOI CMOS for Extreme Temperature Application*. Honeywell Aerospace, Defense & Space, Honeywell International.

9 Evans, J.L., Lall, P., Knight, R. et al. (2008). System design issues for harsh environment electronics employing metal-backed laminate substrates. *IEEE Trans. Compon. Packag. Technol.* 31: 74–85.

10 Yuan, C., Li, L., Duan, B. et al. (2016). Locally reinforced polymer-based composites for efficient heat dissipation of local heat source. *Int. J. Therm. Sci.* 102: 202–209.

11 Yuan, C., Xie, B., Huang, M.Y. et al. (2016). Thermal conductivity enhancement of platelets aligned composites with volume fraction from 10% to 20%. *Int. J. Heat Mass Transf.* 94: 20–28.

12 Luo, X.B., Hu, R., Liu, S. et al. (2016). Heat and fluid flow in high-power LED packaging and applications. *Prog. Energy Combust. Sci.* 56: 1–32.

13 Flores, A.G. (1997). Apparatus and method for actively cooling instrumentation in a high temperature environment. US Patent 5,701,751, issued 10 May1996, filed 30 December 1997.

14 DiFoggio, R. (2002). Downhole sorption cooling of electronics in wireline logging and monitoring while drilling. US Patent 6,341,498, issued 8 January 1994, filed 29 January 2002.

15 Martin, F.S. and Bearden, J.L. (1996). Downhole motor cooling and protection system. US Patent 5,554,897, issued 22 April 1994, filed 10 September 1996.

16 De, K.S. (1983). Downhole tool cooling system. US Patent 4,407,136, issued 29 March 1982, filed 4 October 1983.

17 Bennett, G.A. (1991). Active cooling for downhole instrumentation: miniature thermoacoustic refrigerator. PhD thesis. The University of New Mexico, New Mexico.

18 Parrott, R.A., Song, H., and Chen, K. (2002). Cooling system for downhole tools. US Patent 6,336,408, issued 29 January 1999, filed 8 January 2002.

19 Jakaboski, J.C. (2004). Innovative thermal management of electronics used in oil-well logging, MS thesis. Georgia Institute of Technology, Atlanta, GA.

20 Kandasamy, R., Wang, X., and Mujumdar, A.S. (2007). Application of phase change materials in thermal management of electronics. *Appl. Therm. Eng.* 27: 2822–2832.

21 Wang, Z.C., Zhang, Z.Q., Li, J. et al. (2015). Paraffin and paraffin/aluminum foam composite phase change material heat storage experimental study based on thermal management of Li-ion battery. *Appl. Therm. Eng.* 78: 428–436.

22 Tomizawa, Y., Sasaki, K., Kuroda, A. et al. (2016). Experimental and numerical study on phase change material (PCM) for thermal management of mobile devices. *Appl. Therm. Eng.* 98: 320–329.

23 Zhao, J.T., Rao, Z.H., Liu, C.Z. et al. (2016). Experimental investigation on thermal performance of phase change material coupled with closed-loop oscillating heat pipe (PCM/CLOHP) used in thermal management. *Appl. Therm. Eng.* 93: 90–100.

24 Hu, J.Y., Hu, R., Zhu, Y.M. et al. (2016). Experimental investigation on composite phase-change material (CPCM)-based substrate. *Heat Transf. Eng.* 37: 351–358.

25 Peng, J.L., Lan, W., Wang, Y.J. et al. (2020). Thermal management of the high-power electronics in high temperature downhole environment. *Proceedings of IEEE 22nd Electronics Packaging Technology Conference (EPTC)*, 369–375.

26 Zhang, J.W., Lan, W., Deng, C. et al. (2021). Thermal optimization of high-temperature downhole electronic devices. *IEEE Trans. Compon. Packag. Manuf. Technol.* 11: 1816–1823.

27 Lan, W., Zhang, J.W., Peng, J.L. et al. (2020). Distributed thermal management system for downhole electronics at high temperature. *Appl. Therm. Eng.* 180: 115853.

28 Holman, J.P. (2009). *Heat Transfer*, 10e. New York: McGraw Hill Education.

29 Gray, D.D. and Giorgini, A. (1976). The validity of the Boussinesq approximation for liquids and gases. *Int. J. Heat Mass Transf.* 19 (5): 545–551.

30 Blackburn, H.M., Lopez, J.M., Singh, J. et al. (2021). On the Boussinesq approximation in arbitrarily accelerating frames of reference. *J. Fluid Mech.* 924 (1): 1–11.

31 Lyubimov, D.V., Lyubimova, T.P., Alexander, J.I.D. et al. (1998). On the Boussinesq approximation for fluid systems with deformable interfaces. *Adv. Space Res.* 22 (8): 1159–1168.

32 Mayeli, P. and Sheard, G.J. (2021). Buoyancy-driven flows beyond the Boussinesq approximation: a brief review. *Int. Commun. Heat Mass Transf.* 125: 105316.

33 Hu, Y., Guo, R., Heiselberg, P.K. et al. (2020). Modeling PCM phase change temperature and hysteresis in ventilation cooling and heating applications. *Energies* 13 (23): 6455.

34 Flores, A.G. (1996). *Active Cooling for Electronics in a Wireline Oil-exploration Tool.* Cambridge, MA: Massachusetts Institute of Technology.

35 Costa, S.C. and Kenisarin, M. (2022). A review of metallic materials for latent heat thermal energy storage: thermophysical properties, applications, and challenges. *Renew. Sust. Energ. Rev.* 154: 111812.

36 Fishenden, M. and Saunders, O.A. (1950). *An Introduction to Heat Transfer.* Oxford: Clarendon Press.

37 Dropkin, D. and Somerscales, E. (1965). Heat transfer by natural convection in liquids confined by two parallel plates which are inclined at various angles with respect to the horizontal. *J. Heat Transf.* 87 (1): 77–82.

38 Zhou, F. and Zheng, X. (2015). Heat transfer in tubing-casing annulus during production process of geothermal systems. *J. Earth Sci.* 26 (1): 116–123.

39 Peng, J., Lan, W., Wei, F. et al. (2023). A numerical model coupling multiple heat transfer modes to develop a passive thermal management system for logging tool. *Appl. Therm. Eng.* 223: 120011.

40 Ciesielski, A. and Samorì, P. (2014). Graphene via sonication assisted liquid-phase exfoliation. *Chem. Soc. Rev.* 43 (1): 381.

41 Shang, B., Wu, R., Hu, J. et al. (2018). Non-monotonously tuning thermal conductivity of graphite-nanosheets/ paraffin composite by ultrasonic exfoliation. *Int. J. Therm. Sci.* 131: 20–26.

42 Xiang, J. and Drzal, L.T. (2011). Investigation of exfoliated graphite nanoplatelets (x GnP) in improving thermal conductivity of paraffin wax-based phase change material. *Sol. Energ. Mater. Sol. Cell.* 95 (95): 1811–1818.

43 Debelak, B. and Lafdi, K. (2007). Use of exfoliated graphite filler to enhance polymer physical properties. *Carbon* 45 (9): 1727–1734.

44 Yu, A.P., Ramesh, P., Itkis, M.E. et al. (2007). Graphite nanoplatelet-epoxy composite thermal interface materials. *J. Phys. Chem. C* 111: 7565–7569.

45 Warzoha, R.J. and Fleischer, A.S. (2014). Effect of graphene layer thickness and mechanical compliance on interfacial heat flow and thermal conduction in solid-liquid phase change materials. *ACS Appl. Mater. Interfaces* 6 (15): 12868–12876.

46 Fang, X., Ding, Q., Li, L.Y. et al. (2015). Tunable thermal conduction character of graphite-nanosheets-enhanced composite phase change materials via cooling rate control. *Energ. Convers. Manage.* 103: 251–258.

47 Warzoha, R.J., Weigand, R.M., and Fleischer, A.S. (2014). Temperature-dependent thermal properties of a paraffin phase change material embedded with herringbone style graphite nanofibers. *Appl. Energy* 716–725.

48 Shang, B.F., Ma, Y.P., Hu, R. et al. (2017). Passive thermal management system for downhole electronics in harsh thermal environments. *Appl. Therm. Eng.* 118: 593–599.

49 Rafie, S. (2007). Thermal management of downhole oil & gas logging sensors for HTHP applications using nanoporous materials. *ASME 2007 2nd Energy Nanotechnology International Conference*, 1–6.

50 Ma, Y.P., Shang, B.F., Hu, R. et al. (2016). Thermal management of downhole electronics cooling in oil & gas well logging at high temperature. *17th International Conference on Electronic Packaging Technology*, 623–627.

51 Ezan, M.A., Yüksel, C., Alptekin, E. et al. (2018). Importance of natural convection on numerical modelling of the building integrated PVP/PCM systems. *Sol. Energy* 159: 616–627.

52 Samara, F., Groulx, D., and Biwole, P.H. (2012). Natural convection-driven melting of phase change material: comparison of two methods. *Excerpt from the Proceeding of the COMSOL Conference.*

53 Singer, S. and Nelder, J.A. (2009). Nelder-Mead algorithm. *Scholarpedia* 4 (7): 2928.

54 Nelder, J.A. and Mead, R. (1965). A simplex method for function minimization. *Comput. J.* 7 (4): 308–313.

9

Liquid Cooling for High-Heat-Flux Electronic Devices

Due to the increasing integration density, thermal management of high-power electronics, such as insulated gate bipolar transistors (IGBTs), high electron mobility transistors (HEMTs), light-emitting diodes (LEDs), and high-performance computing chips, has become one of the major challenges [1–6]. With the growing need for performance, several hundred watts of heat may be generated by millimeter-sized electronics. Liquid cooling is the mainstream method to control the temperature of high-power electronics within a proper range, owing to its large heat transfer efficiency and compactness. Various efficient liquid cooling technologies have been developed to dissipate the tremendous heat from small chips, including microchannel cooling [7–9], spray cooling [10–12], and jet impingement cooling [13–15]. Despite the fact that liquid cooling systems have been well developed over the years, there still remain some problems when they are applied to specific high-heat-flux electronic devices. Improving the overall cooling performance and maintaining uniform chip temperature are two core needs in the thermal management of high-flux electronic devices, which are always pursued by both industry and academics. This chapter will present several advances in liquid cooling technologies for high-power electronic devices, including double-nozzle spray cooling, direct body cooling, integrated piezoelectric jet cooling, and microchannel liquid cooling. These technologies offer improved heat dissipation, enhanced temperature uniformity, and reduced system volume, making them valuable for various high-power electronic applications.

9.1 Double-Nozzle Spray Cooling for High-Power LEDs

LEDs are widely used around the world due to their long lifetime and ecofriendliness. The past few decades have witnessed significant breakthroughs in the material, packaging, and design of LEDs [16, 17], which are developing in the trends of miniaturization and high-density packaging. The heating power of high-power LEDs can account for 60%–70% of the input electric power due to the nonradiative recombination with the multiple quantum well (MQW) layer and light confinement within the LEDs packaging, which will lead to a sharp temperature rise of the device [1]. As a result, luminous flux degradation, wavelength drift, lifetime reduction, and even irreversible damage will occur in LEDs [18]. Therefore, thermal management is critical for high-power LEDs.

The local heat-flux level of high-power LEDs can reach 500 W/cm^2 and the junction temperature should be controlled below 120 °C [19]. Limited by the packaging constraints, thermal characteristics, and materials factors, the forced air convection or passive liquid cooling methods [20–23] fail to meet the heat dissipation requirement of high-power LEDs, while active liquid cooling technology can, due to its better heat transfer performance [24, 25]. Active liquid cooling technologies, including microchannel liquid cooling, jet impingement liquid cooling, and spray cooling, have been widely studied [6, 26]. Microchannel liquid cooling and jet impingement liquid cooling have shown excellent thermal management capacity in low-power LEDs (0.05–1 W, mA or μA) but are not suitable for high-power LEDs (3 W or even tens of watts, several amperes) [27–35]. Spray cooling, regarded as the promising thermal management solution for ultra-high heat-flux (>1000 W/cm^2) electronic devices, has drawn tremendous attention, of which the studies focus on revealing the heat transfer mechanism, enhancing the heat dissipation capacity, and promoting the applications [12, 36, 37]. The flow parameters and nozzle configuration have been proven to have a significant influence on the heat transfer performance of spray cooling. Meanwhile, as an advanced thermal management method, spray cooling has been successfully applied to many fields, such as aerospace equipment [38], motor [39], and lithium-ion battery [40], but studies on spray cooling of LEDs are very limited.

Thermal Management for Opto-electronics Packaging and Applications, First Edition. Xiaobing Luo, Run Hu, and Bin Xie.

Only a few studies take advantage of single-nozzle spray cooling for the thermal management of LEDs and the effectiveness has been verified [41, 42]. However, high-power LED is a temperature-sensitive optical device that requires high heat transfer coefficient throughout the whole surface. A single spray may not provide sufficient surface coverage for LEDs' unique shape and packaging constraints while double spray can cover a larger surface area. In order to explore how to integrate laboratory spray cooling systems with industrial applications, the coupling effect of various parameters and the packaging process are worthy of in-depth analysis. Proper design of the thermal management system of high-power LEDs will greatly reduce energy consumption and effectively utilize low-grade thermal energy.

Herein, a double-nozzle spray cooling system was proposed for 300 W high-power LEDs. The effects of nozzle configuration and the spray parameters including flow rate and nozzle-to-surface distance on the system were investigated experimentally. At the same time, in order to accurately explore the temperature distribution and junction temperature of LED chips in spray cooling, a detailed three-dimensional (3D) numerical simulation was carried out.

9.1.1 Spray Cooling System

The experiments were conducted in an open system. Figure 9.1 shows the schematic diagram of the apparatus used in the spray cooling experimental setup. The whole experimental system consists of four parts, namely, the flow loop circuit system, the high-power LEDs testing framework, the data acquisition system, and the visualization system.

Deionized water was chosen as the working coolant stored in the reservoir (DC-2010, temperature control range: −20 to 100 °C, accuracy: ±0.1%, Zhulan). A gear pump (NP400, 24 V-400 W, Suofu) was used to drive the deionized water onto the heated surface through a solid cone pressure nozzle (B1/4TT-SS+TG-SS0.3, Spraying Systems Company). Under the experimental condition of double nozzle, there were two nozzles arranged side by side, parallelly, to the horizontal orientation, 25 mm apart. The flow rate of the nozzle and the pressure drop across the nozzle were measured by the flow meter (CX-P2-F, measuring range: 3–300 ml/min, accuracy: ±0.5%, Gnflowmeter) and pressure gauge (SIN-Y190, measuring range: 0–5 MPa, accuracy: ±1%, Sinomeasure), respectively.

In order to get the thermal management capacity of spray cooling and prevent the LEDs from short circuit, a LEDs testing framework was designed as shown in Figure 9.2(a). The high-power LEDs were assembly integrated with a copper spreader

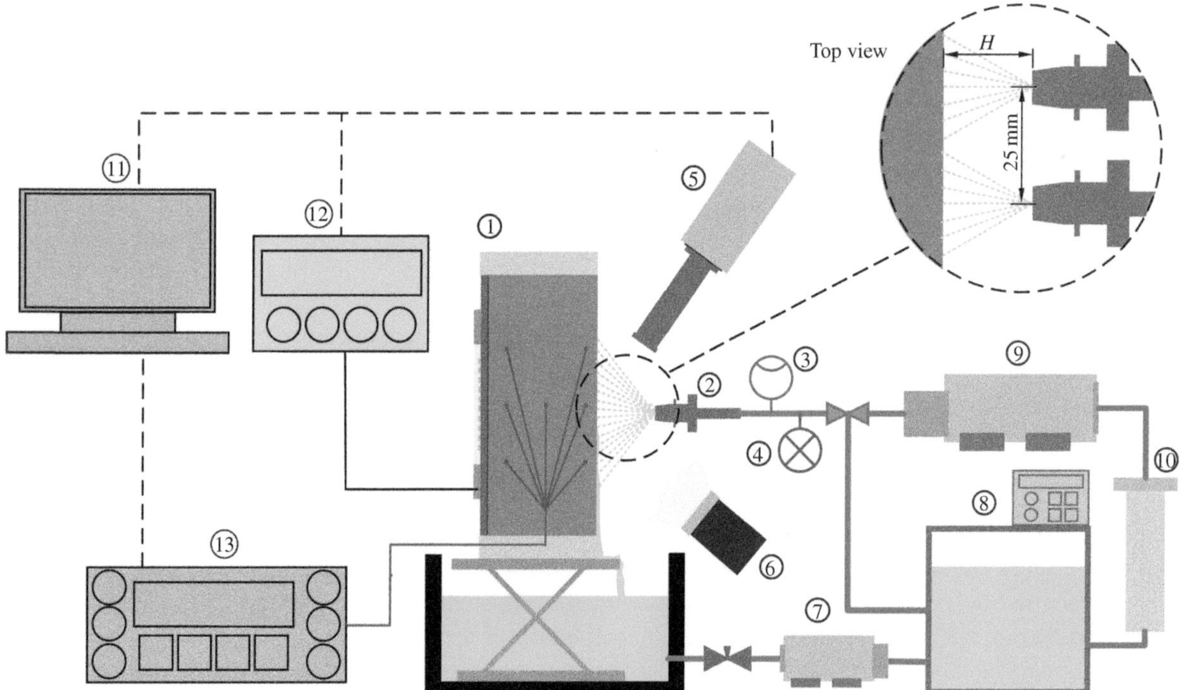

Figure 9.1 Schematic layout of the experimental setup: 1. the LEDs testing device, 2. nozzle, 3. pressure gauge, 4. flow rate gauge, 5. a high-speed camera, 6. light source, 7. diaphragm pump, 8. reservoir, 9. gear pump, 10. filter, 11. computer, 12. data acquisition system, 13. DC power.

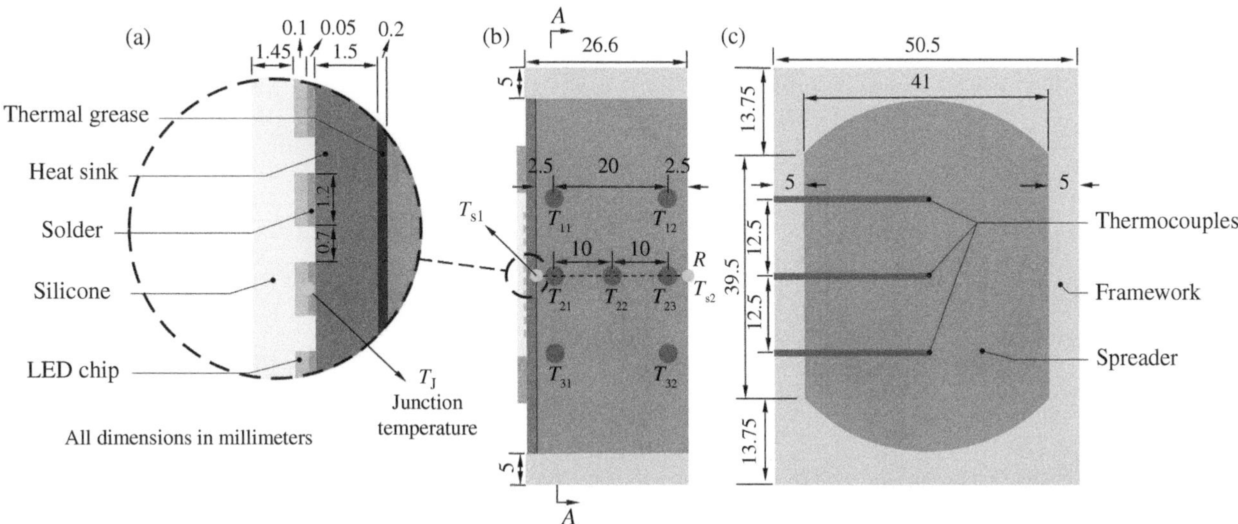

Figure 9.2 High-power LEDs assembly: (a) the detailed structure of LEDs, (b) the cross-sectional view with the location of thermocouples, (c) the view at *A*–*A* plane.

and protected by a framework made by polyether-ether-ketone (PEEK). Also, the excellent thermal insulation ability of PEEK minimized heat loss. Thermal grease (OT-201, Omega Engineering) was used to decrease the thermal contact resistance between the LEDs heat sink and the spreader. Furthermore, seven K-type thermocouples (TT-K-36, ETA) were embedded in the spreader to measure the heat flux and temperature distribution accurately, as shown in Figure 9.2(b). In order to clearly show the lateral distribution of the thermocouples, the cross-section view at *A*–*A* plane is shown in Figure 9.2(c). The insertion depth of the unshown four thermocouples is the same as the three thermocouples shown in Figure 9.2(c). A programmable direct current (DC) power supply (FTP032-300-16, Faith) was employed to provide stable input electric power for high-power LEDs. The inlet coolant temperature and ambient temperature were given by the thermocouples placed in the reservoir and spray chamber. A data acquisition instrument (Keithley 2700) was used to collect, record, and display the signals of all thermocouples.

A visualization system was built to observe and assess the performance of the spray cooling system. A high-speed camera (SA2 120K, Photron) was used to obtain the spray morphology. The Sauter mean diameter (SMD,D_s) and the droplet velocity were measured by using Phase Doppler Interferometer.

9.1.2 Data Analysis Method and Uncertainty Analysis

The one-dimensional (1D) Fourier's law of heat conduction can be applied and the heat flux is estimated at the middle of the spreader as

$$q = -\kappa \frac{\Delta T(y)}{\Delta y} \tag{9.1}$$

$\Delta T(y)/\Delta y$ is the temperature gradient fitted by three thermocouples T_{21}, T_{22}, and T_{23}.

The substrate temperature T_{sub} and the surface temperature T_{sur} are estimated from the temperatures of T_{21} and T_{23}, respectively:

$$T_{\text{sub}} = T_{21} - \frac{qy_1}{\kappa} \tag{9.2}$$

$$T_{\text{sur}} = T_{23} - \frac{qy_2}{\kappa} \tag{9.3}$$

The heat transfer coefficient h can be evaluated as:

$$h = \frac{q}{T_{\text{sur}} - T_{\text{in}}} \tag{9.4}$$

The uncertainties of the thermocouples and thermocouple position are about ±0.8 °C and ±0.1 mm, respectively. Based on the error-transfer functions on this experimental bench, the uncertainties of surface temperature, heat flux, and heat transfer coefficient can be calculated by

$$U_y = \sqrt{\sum_{i=1}^{n} \left(\frac{\partial f}{\partial x_i}\right)^2 U_{x_i}^2} \tag{9.5}$$

The results showed that the maximum uncertainties of heat flux, surface temperature, and heat transfer coefficient are 2.1%, ±3.2%, and ±2.2%, respectively.

9.1.3 Simulations for Junction Temperature Evaluation

The COMSOL Multiphysics® Software was employed in this study and it is a supplement to the experiments where the junction temperature is hard to obtain. The simulation model is shown in Figures 9.3 and 9.4. The thermal conductivity

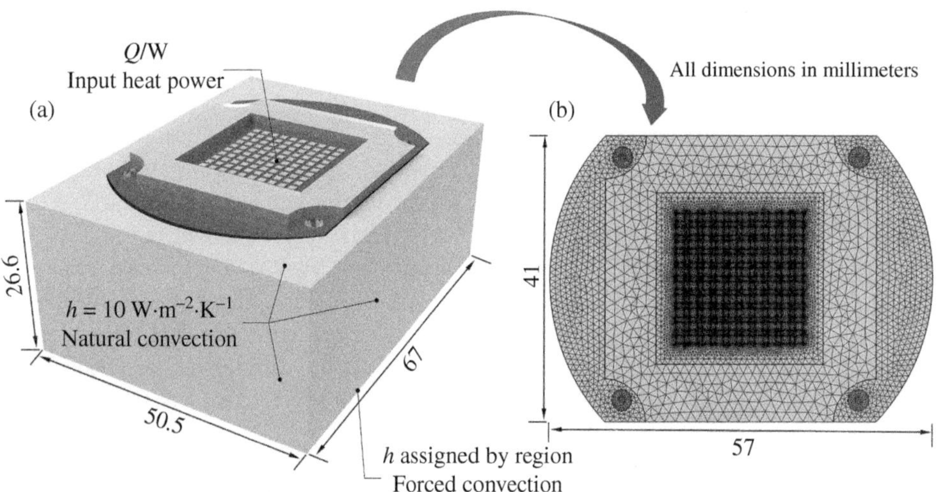

Figure 9.3 (a) Computational domain used for the simulation, (b) nature of meshing the LEDs module used for the simulation.

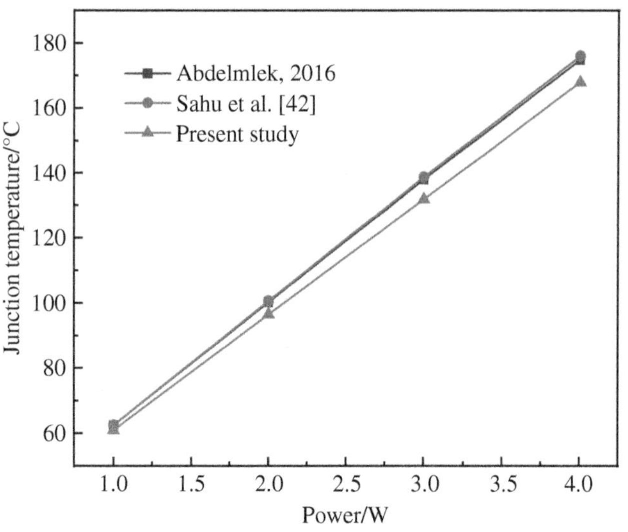

Figure 9.4 Validation plot of previous numerical studies for the simulation model.

Table 9.1 Thermal conductivity of materials used in the simulation.

	Materials	Thermal conductivity κ ($W \cdot m^{-1} \cdot K^{-1}$)
Spreader	Copper	398
Heat sink	Copper	398
Framework	Peek	0.29
Thermal grease	Silicone	16
Solder	H20E	29
LEDs chip	GaP, GaAlAs, GaAs, GaN	65.6
Encapsulation layer	PDMS	0.15

of the materials used in the simulation is shown in Table 9.1. The results of the numerical simulation were obtained under steady state. The governing equations are

$$\rho C_p \frac{\partial T}{\partial t} + \nabla \cdot q = Q \tag{9.6}$$

$$q = -\kappa \nabla T = h(T - T_{in}) \tag{9.7}$$

The boundary conditions and assumptions used in the simulation analysis are mentioned as follows:

1) A constant heat transfer coefficient can be applied to the target surface as a boundary condition in the simulation. Here, the heat transfer coefficient is the heat transfer coefficient of the target surface derived from the results of the experiments.
2) Heat dissipation of thermal radiation is ignored, and the heat is mainly lost through the surface of the copper block by convection heat transfer. The heat transfer coefficient of other surfaces exposed to environment is set to $10\ W \cdot m^{-2} \cdot K^{-1}$ due to the natural convection.
3) Blue LEDs module was used in this work. The light absorption of silica gel can be ignored.
4) The heat power of each chip is uniform, which is the total heat power divided by the volume of the chip.
5) The thermal properties (thermal conductivity, heat capacity, and density) of all materials are assumed to be uniform, isotropic, and independent of temperature.
6) The material contact through the interface is considered perfect (no air gap and surface roughness).

In order to validate the proposed simulation model, the numerical models of Ben Abdelmlek et al. [43] and Sahu et al. [42] are compared with our model, as shown in Figure 9.4. The maximum difference in junction temperature is about 4.8%. Grid independence test is conducted under different input electric powers, and three different grid sizes are chosen, namely finer, extra fine, and extremely fine. The total numbers of grids are 1,786,541, 3,538,358, and 9,072,035, respectively. The maximum temperature deviation between the grid size of finer and extra fine is 0.01 °C while there is no temperature deviation between the grid size of extra fine and extremely fine. Based on the accuracy and the simulation times, the finer size is selected for the remainder of the study.

9.1.4 Characteristics of High-Power LEDs Module and Spray Droplets

The 300 W blue LEDs (SP100-I6BB-100, Shenzhen SkyBright Optoelectronics) were chosen to be tested. This LEDs module has 100 chips arranged in a square array with an area of 20 mm × 20 mm, and the maximum input electric power of each chip is 3 W. The maximum input electric power of the LEDs module can reach 300 W under good heat dissipation conditions. The optical performance was obtained by an integrating sphere system (ATA-1000, Everfine) and the relationship between the light efficiency of the LEDs and the input current at different temperatures was shown in Figure 9.5(a). The results show that as the temperature increases, the light efficiency will gradually decrease. The light power accounts

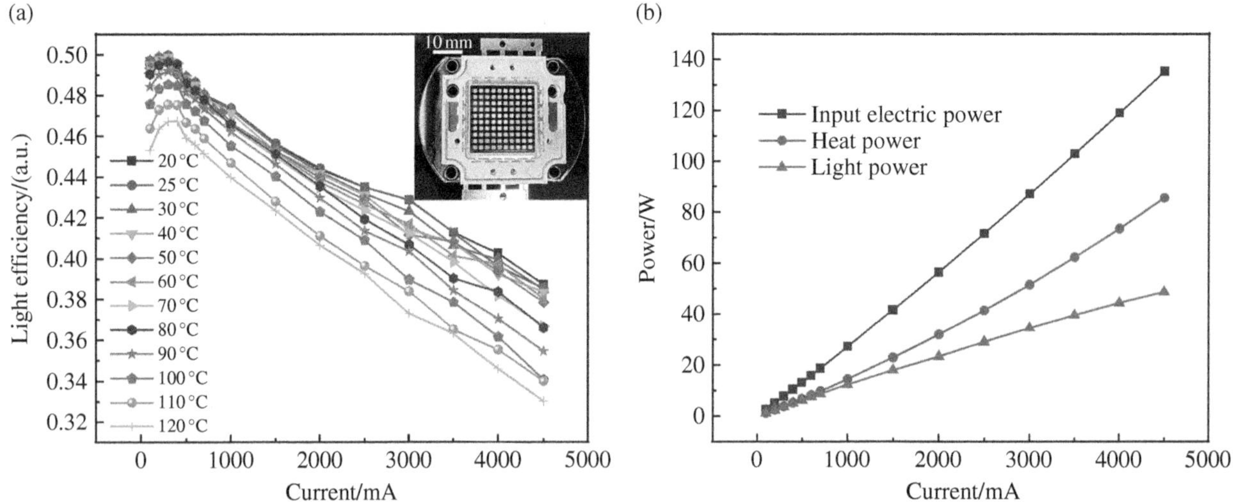

Figure 9.5 (a) The relationship between light efficiency and driven current under various temperatures; (b) the input electric power, light power, and heat power of the LEDs module under various driven currents.

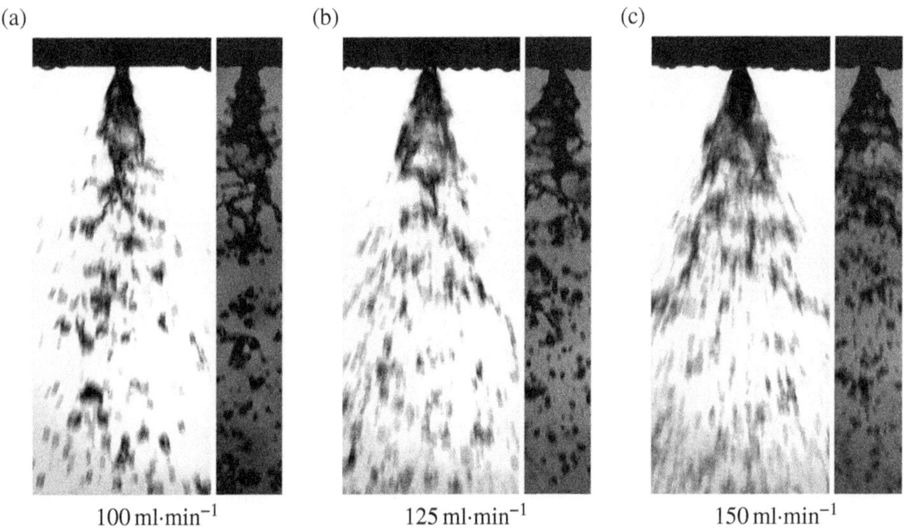

Figure 9.6 Photographic details of spray morphology of different flow rates: (a) flow rate of 100 ml/min, (b) flow rate of 125 ml/min, (c) flow rate of 150 ml/min.

for about 30%–40% of the input electric power, while the heat power accounts for about 60%–70% of the input electric power as shown in Figure 9.5(b).

Spray cooling removes the heat from the target surface by employing the droplet clusters ejected from the nozzle. The characteristics of the droplet and spray significantly influence the performance of the spray cooling. The spray morphology of different flow rates is displayed in Figure 9.6 and each set of image includes two pictures. The picture on the left is the overall morphology of the spray, and the picture on the right is the morphology near the center line of the nozzle outlet. The value of different parameters obtained by our characterization is shown in Table 9.2.

9.1.5 Results and Discussion

As shown in Figure 9.2, seven thermocouples are embedded in the copper block to evaluate the thermal management capacity of spray cooling such as heat transfer coefficient (h), substrate temperature (T_{sub}), surface temperature (T_{sur}), and

Table 9.2 Characterization of nozzle.

Nozzle-to-surface distance (mm)	Flow rate (ml/min)	Sauter mean diameter (μm)	Velocity at the left line (m/s)	Pressure drop across nozzle (MPa)	Spray angle (°)
10	100	264.31	3.35	0.04	34.0
20	100	255.41	2.98	0.04	34.0
30	100	259.65	1.47	0.04	34.0
10	125	217.59	3.74	0.06	44.7
10	150	188.24	3.94	0.09	57.6
30	125	99.50	1.76	0.06	44.7
30	150	101.70	1.65	0.09	57.6

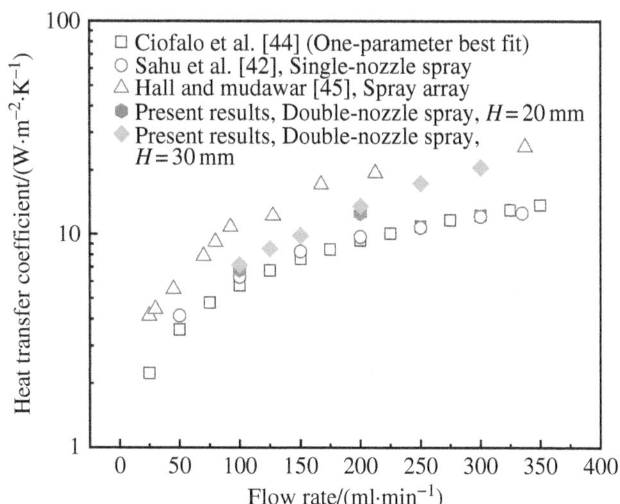

Figure 9.7 Heat transfer coefficient h at stagnation point (R) in comparison with the previous studies. Adapted from Sahu et al. [42], Ciofalo et al. [44], and Hall and Mudawar [45].

temperature distribution. In this study, the effects of flow rate (Q_f) and nozzle-to-surface distance (H) on the spray cooling system are investigated. The performances of the single- and double-nozzle spray cooling system are compared as three different expressions are used to identify different experimental conditions, namely "Double-nozzle-1," "Double-nozzle-2," and "Single-nozzle." The flow rate of "Double-nozzle-1" and "Single-nozzle" refers to the flow rate of each nozzle while the flow rate of "Double-nozzle-2" refers to the total flow rate of two nozzles.

To verify the rationality of the experimental data and the reliability of the experimental apparatus, the h obtained in the present study is compared with the reported value in the published literature [44, 45] as shown in Figure 9.7. The experiments are conducted with the input electric power of 300 W for comparison based on H of 20 and 30 mm. The results are in good agreement with the previous experiments, no matter they being experiments of single-phase spray cooling [44, 45] or experiments of the thermal management of LED [42]. Moreover, the present study of double-nozzle spray cooling shows better heat dissipation capacity than single-nozzle spray cooling [42, 44] while it is weaker than spray array [45].

9.1.5.1 Effect of Nozzle Configuration and Flow Rate

There is a consensus that the flow rate (Q_f) is one of the most important hydrodynamic parameters in spray cooling. The flow rate is controlled at 100, 125, and 150 ml/min in this study and two different nozzle configurations are investigated, namely double nozzle and single nozzle.

Figure 9.8 gives the thermo-hydraulic evaluation based on different nozzle configurations and flow rates. The h increases with the Q_f but remains similar around different input electric powers under each nozzle configuration as shown in

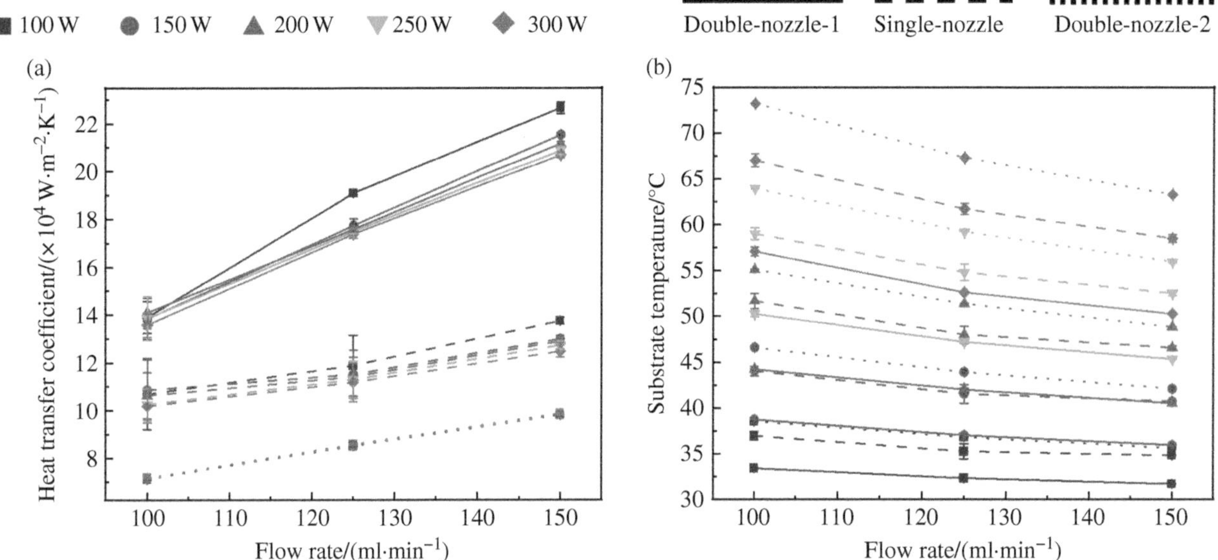

Figure 9.8 Thermohydraulic evaluation of different nozzle configurations: (a) heat transfer coefficient h versus flow rate Q, (b) substrate temperature T_{s1} versus flow rate Q.

Figure 9.8(a). The h of Double-nozzle-1 is larger than Single-nozzle and Double-nozzle-2 over the entire input electric power range. At the Q_f of 150 ml/min, there is a 64.9% growth in h when the experimental condition changes from Single-nozzle to Double-nozzle-1 and a 113.8% growth in h when it changes from Double-nozzle-2 to Double-nozzle-1. Figure 9.8(b) shows that T_{sub} decreases with the increase of Q_f. The T_{sub} of Double-nozzle-1 is lower than the T_{sub} of Double-nozzle-2 and Single-nozzle. At the maximum input electric power of 300 W, the highest h reaches 20.7 $kW \cdot m^{-2} \cdot K^{-1}$ and the lowest T_{sub} reaches 50.3 °C.

Double-nozzle-1 has a larger total flow rate than Double-nozzle-2 and Single-nozzle. Larger Q_f refers that more coolant can reach the target surface and the droplet velocity is faster, which can be proved by the characterization of the nozzle. The faster droplet velocity makes it easier to break the boundary layer and provides timely fluid replenishment to form new liquid membrane for stronger heat dissipation. However, although Double-nozzle-2 has the same Q_f as Single-nozzle, the Q_f of each nozzle of Double-nozzle-2 is half of Single-nozzle. The lower flow rate will lead to smaller droplet velocity, larger droplet size, and smaller spray covering area according to the characterization of the nozzle. Smaller droplet velocity and larger droplet size will lead to a thicker liquid membrane and a slower rate of liquid removal which are unfavorable for heat dissipation. Furthermore, as for double nozzle, the smaller spray covering area of each nozzle will lead to a separated state. There will be flow interactions and blockage between neighboring sprays, which lead to worse heat dissipation capacity. Therefore, under the same total flow rate, the performance of Double-nozzle-2 is not as good as that of Single-nozzle.

Compared to Single-nozzle, Double-nozzle-1 can give larger h in a greater area. Figure 9.9 compares the temperature at T_{11}, T_{21}, and T_{31} under different experimental conditions. At the Q_f of 100 and 150 ml/min, the T_{sub} of Double-nozzle-1 is lower than Double-nozzle-2 and Single-nozzle. The temperature of T_{21} is higher than T_{11} and T_{31} which are similar to each other. LED is a temperature-sensitive optical device, and if any chip burns out, the module will be invalid. Therefore, the smaller the temperature difference between the thermocouples in the same height, namely, the better the temperature uniformity, the better the thermal management capacity.

In order to visualize the temperature nonuniformity ΔT_{uni}, the standard deviation of the temperature variation has been calculated as

$$\Delta T_{uni} = S = \sqrt{\frac{\sum_{i=1}^{3} \left(T_{i1} - \overline{T}\right)^2}{2}} \tag{9.8}$$

Figure 9.10 shows ΔT_{uni} versus input electric power under different nozzle configurations and Q_f. As for Double-nozzle-1 and Double-nozzle-2, the ΔT_{uni} are similar among different Q_f. However, the larger the Q_f, the larger the ΔT_{uni} for

Single-nozzle. Combined with the characterization of nozzle, it can be found that the increase of Q_f leads to the increase of spray angle. As a result, the spray impact area will increase at the same H. The mean volumetric flux $\left(\overline{Q}''\right)$, which is defined by dividing the total volume flow rate of the spray by the portion of the surface directly impacted by the spray, is 6.2×10^{-3}, 4.4×10^{-3}, and 2.9×10^{-3} $m^3 \cdot s^{-1} \cdot m^{-2}$, for 100, 125, and 150 ml/min, respectively. More specifically, the decrease of $\overline{Q}''$ will lower the heat dissipation capacity of the central area under a large Q_f, resulting in higher temperature and larger temperature nonuniformity.

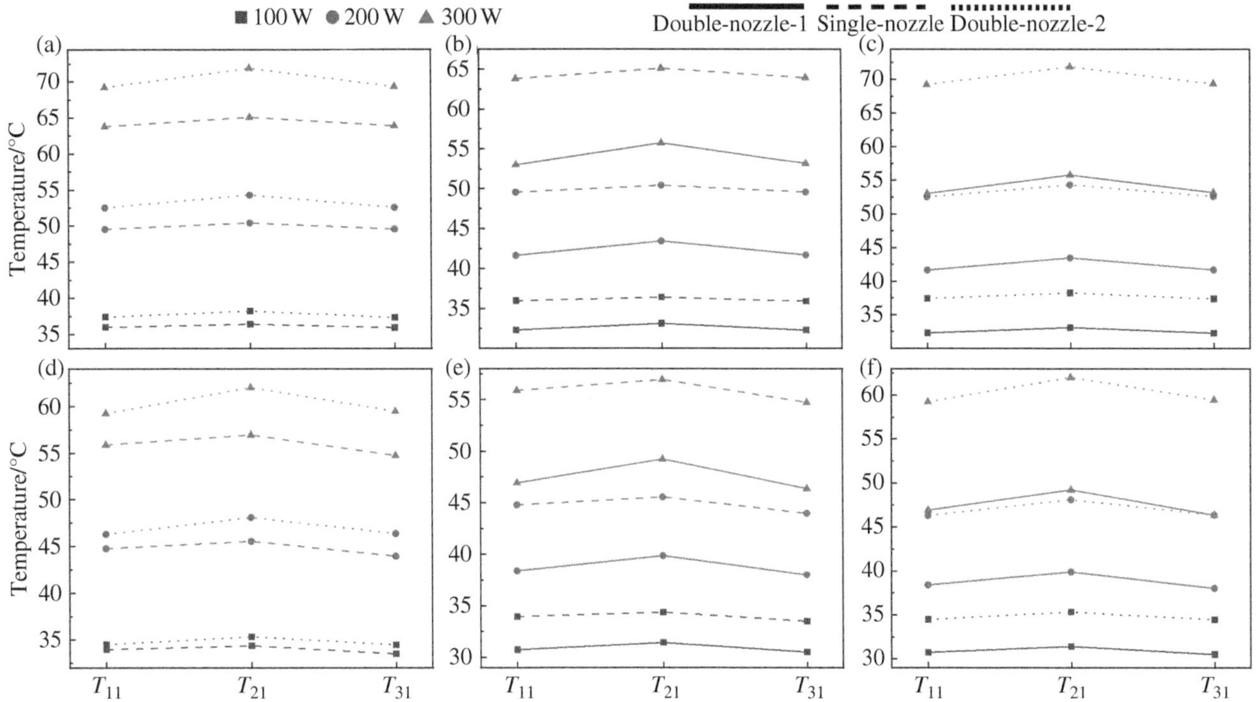

Figure 9.9 Temperature of T_{11}, T_{21}, and T_{31} with different flow rates and input electric power: (a) comparison of Single-nozzle and Double-nozzle-2 at Q of 100 ml/min, (b) comparison of Single-nozzle and Double-nozzle-1 at Q of 100 ml/min, (c) comparison of Double-nozzle-1 and Double-nozzle-2 at Q of 100 ml/min, (d) comparison of Single-nozzle and Double-nozzle-2 at Q of 150 ml/min, (e) comparison of Single-nozzle and Double-nozzle-1 at Q of 150 ml/min, (f) comparison of Double-nozzle-1 and Double-nozzle-2 at Q of 150 ml/min.

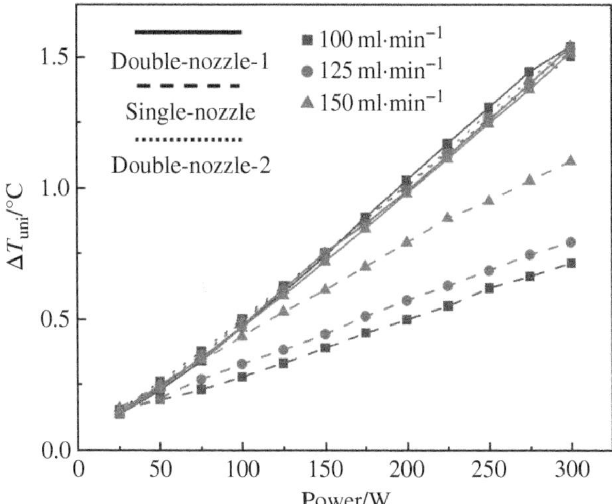

Figure 9.10 Temperature nonuniformity versus input electric power with different nozzle configurations and flow rates.

In addition, the ΔT_{uni} of double-nozzle configuration is larger than single-nozzle configuration. The reason is not duplicating for Double-nozzle-1 and Double-nozzle-2. The temperature distribution of the LEDs module of Double-nozzle-1 is lower than Single-nozzle, and that of Single-nozzle is lower than Double-nozzle-2. As for lower temperature distribution, it is more difficult to achieve a uniform temperature distribution inside the module. Meanwhile, the lack of heat dissipation capacity can also lead to a nonuniform temperature distribution, just like Double-nozzle-2. Therefore, the judgment criterion of temperature nonuniformity is worthy of further discussion.

9.1.5.2 Effect of Nozzle-to-Surface Distance

The packaging constraints and the trend toward miniaturization of electronic devices both limit the volume of high-power LEDs. Exploring the effect of nozzle-to-surface distance (H) on spray cooling of high-power LEDs can provide guidance for the design of the thermal management system. The H is chosen as 10, 20, and 30 mm. All experiments are conducted based on Double-nozzle-1 and Q_f of 100 ml/min.

The thermal evaluation is plotted in Figure 9.11. The h increases with the increase of the H and it remains similar between different input electric powers. The T_{sub} goes down with the increase of the H. In short, the farther the H, the better the heat dissipation capacity. As shown in Table 9.2, the droplet velocity is lower at the farther H. Nevertheless, the surface area directly impacted by the spray is larger at the farther H. The heat transfer mechanism is different in the direct impact area and the nondirect impact area. In the direct impact area, high-speed droplets impact the surface, break the boundary layer, take away the heat, and replenish the liquid membrane timely. In the nondirect impact area, forced convection of coolant is the main mechanism to carry away the heat load even though it is inefficient. Hence, the size of the direct impact area is a more important factor than droplet velocity in spray cooling.

The temperature nonuniformity evaluation based on different H is shown in Figure 9.12. The temperature of T_{21} is higher than T_{11} and T_{31}, which are similar to each other as shown in Figure 9.12(a). Figure 9.12(b) gives the ΔT_{uni} versus input electric power under different H. The ΔT_{uni} is similar among H of 10 and 20 mm, while the ΔT_{uni} is smaller at H of 30 mm. The reason why ΔT_{uni} goes up slightly when H is 10 and 20 mm is that the lower the H, the larger the $\overline{Q}''$ and the droplet velocity, while the direct impact area is smaller. As a result, the huge difference in heat dissipation capacity between direct impact area and nondirect impact area will cause a larger ΔT_{uni}. But, in general, as shown in Figure 9.12(b), ΔT_{uni} is not sensitive to the changes of H.

As mentioned earlier, the junction temperature (T_J) is the critical indicator of LEDs module, which is incapable of being measured by the thermocouple directly. The simulation is employed to obtain the T_J of the center chip column.

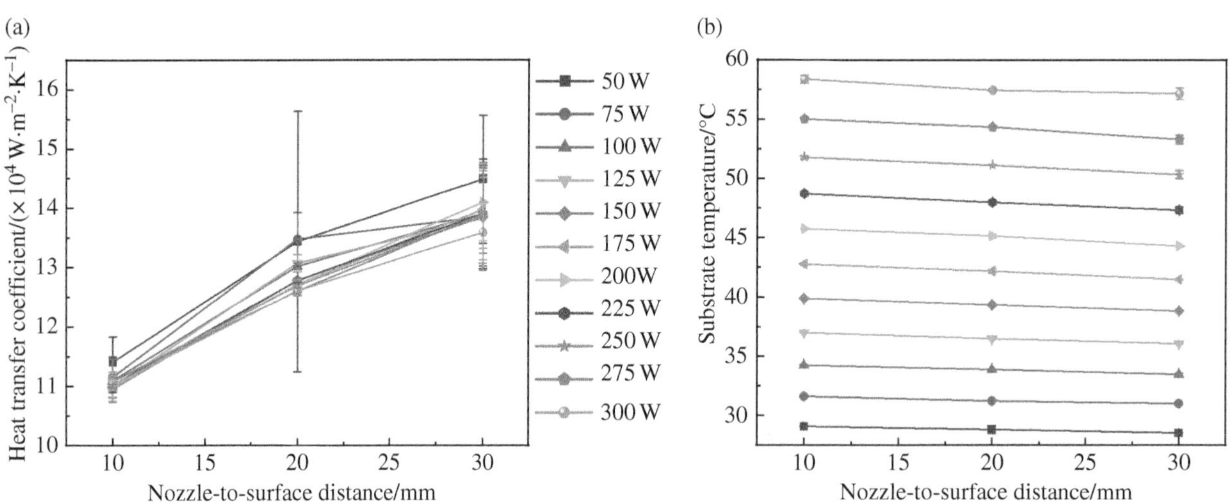

Figure 9.11 Thermal evaluation of Double-nozzle-1 at the flow rate of 100 ml/min: (a) heat transfer coefficient h versus nozzle-to-surface distance H, (b) substrate temperature T_{s1} versus nozzle-to-surface distance H.

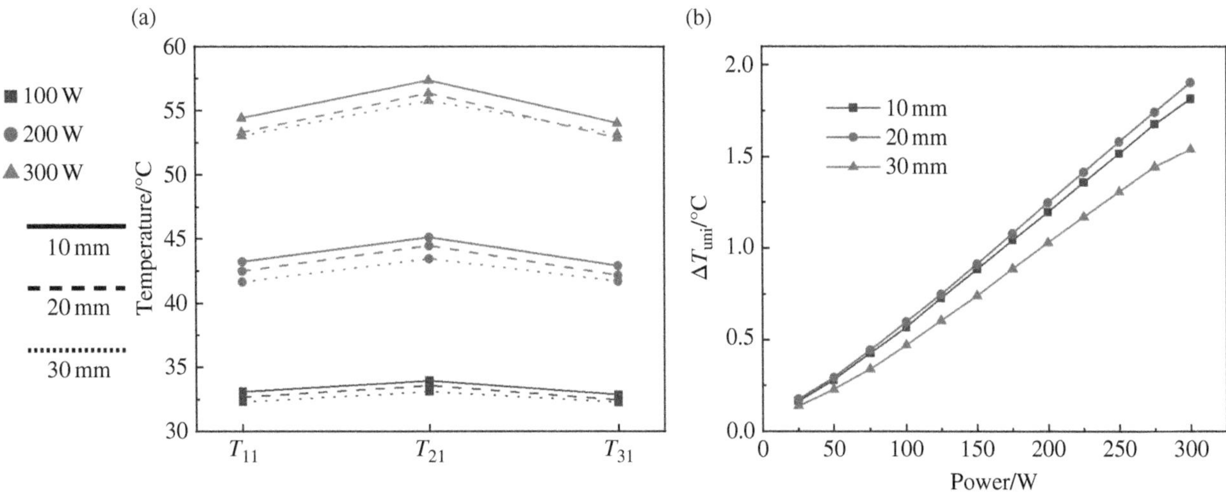

Figure 9.12 Temperature nonuniformity evaluation of Double-nozzle-1 at Q of 100 ml/min: (a) temperature of T_{11}, T_{21}, and T_{31} with different nozzle-to-surface distance H, (b) temperature nonuniformity versus input electric power with different nozzle-to-surface distance H.

9.1.5.3 Validation Study

It can be drawn from the previous analysis that the heat dissipation capacity of spray cooling is different in the direct impact area and the nondirect impact area. In the simulation, we adopt the method of assigning heat transfer coefficient by region to make the simulation more accurate. The surface of the copper block exposed to the environment is divided into the direct impact area and the nondirect impact area of which the sizes are determined by the spray angle, H, and D. The nozzle configuration of single nozzle and double nozzle are both validated at Q_f of 150 ml/min among the whole range of input electric power. Figure 9.13 expresses the comparison of steady-state temperature between the experiment and the simulation at different thermocouple positions. The data of experiment and simulation are in good agreement with each other.

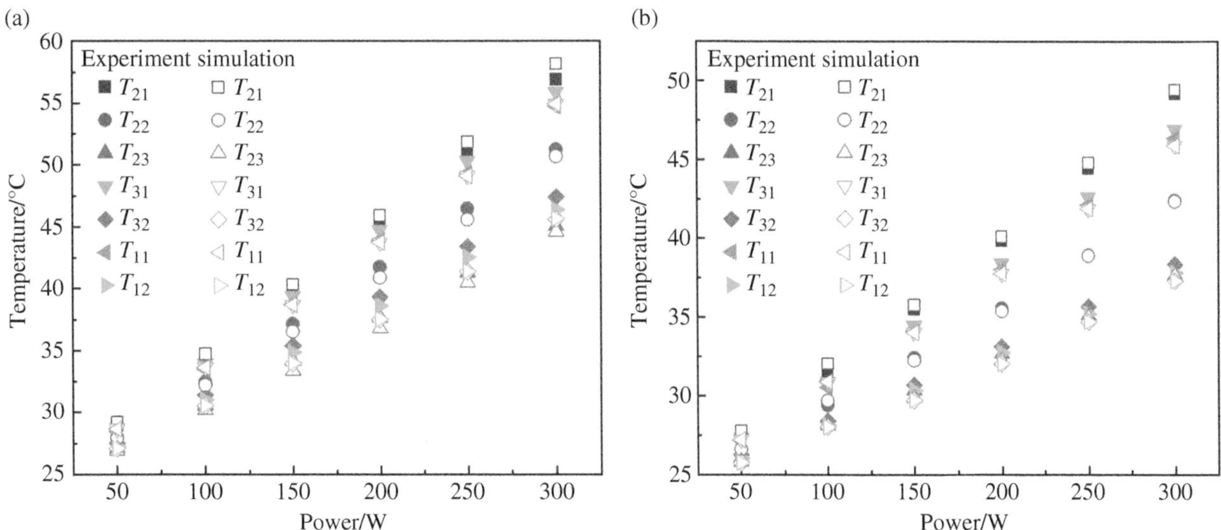

Figure 9.13 Steady-state temperature at different positions for various input electric powers at flow rate of 150 ml/min: (a) Single-nozzle, (b) Double-nozzle-1.

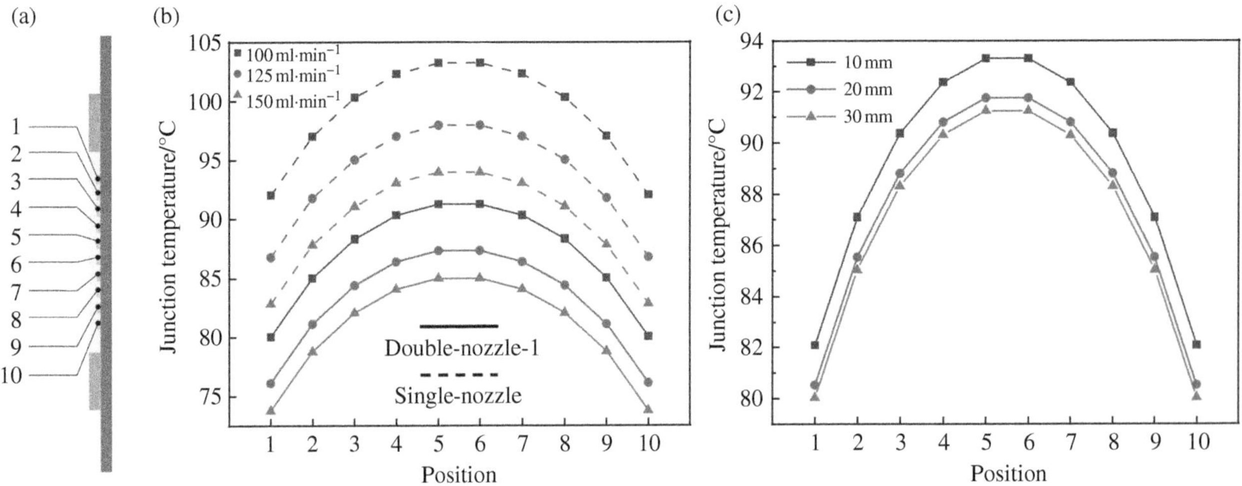

Figure 9.14 Junction temperature T_J of high-power LEDs under the input electric power of 300 W: (a) cross-sectional view of LEDs module at the center chip column, (b) comparison of Double-nozzle-1 and Single-nozzle at different flow rates, (c) comparison of different nozzle-to-surface distance H of Double-nozzle-1.

9.1.5.4 Estimation of Junction Temperature

The chips are distributed in a square array in the LEDs module. Thus, there will be heat accumulation in the center of LEDs module and the T_J of chip column at the center is chosen for analysis when the input electric power is 300 W. Figure 9.14(a) is the diagram of different chip locations. The T_J distribution of Double-nozzle-1 and Single-nozzle at different Q_f is shown in Figure 9.14(b). Figure 9.14(c) compares T_J of different H of Double-nozzle-1. The T_J distribution of Double-nozzle-1 is lower than that of Single-nozzle, and Figure 9.14(b) and (c) prove that the T_J is symmetrically distributed around the LEDs center. As mentioned earlier, the temperature sensitivity of the LEDs module causes it to be completely damaged due to local overheating. Hence, the temperature of the center chip is analyzed. The result shows that when Q_f is 150 ml/min, Double-nozzle-1 achieves the lowest T_J of the center chip, which is 85 °C. Under the same condition, the value of T_J is 94 °C of Single-nozzle, which is 10% higher than that of Double-nozzle-1. Additionally, the farther the H, the lower the T_J. The decrease of T_J resulting from increasing the Q_f and nozzle numbers is greater than that by increasing the H. When H changes from 10 to 20 mm, the highest T_J decrease is by 1.7% while it is only 0.5% when H changes from 20 to 30 mm. The results show the consistency of the effect of H on h, ΔT_{uni}, and T_J. In other words, the effect of changing H on the heat transfer capacity of spray cooling is very limited. However, this also implies that adjusting H to fit the packaging constraint has greater application potential.

9.2 Direct Body Liquid Cooling

Most liquid cooling heat sinks are attached to the electronic device by the thermal interface material (TIM) to reduce the contact thermal resistance, which is called indirect liquid cooling [4, 46, 47]. However, the large thermal resistance caused by the low thermal conductivity of the TIM is the major obstacle for thermal management of high-power electronics. Although a lot of researches have been done to increase the thermal conductivity of the TIM [24, 48], it is still far from meeting the requirements of engineering applications. Therefore, direct liquid cooling, which eliminates the contact thermal resistance by making the coolant directly touch the electronics, is a better option for heat dissipation of electronics with high heat flux.

Jet impingement and spray cooling are commonly used direct liquid cooling techniques. Spray cooling was reported to have the highest heat dissipation capability owing to the boiling of the micron liquid droplets [11, 36]. However, the system complexity, critical heat flux, and instability caused by the vapor and bubbles limit its application. Jet impingement liquid cooling with array has also been studied by many researchers in the past decades [13, 49, 50]. The heat transfer and flow mechanism of the free, submerged, and confined impinging liquid jets have been fully investigated by theory, numerical simulation, and experiments. When the jet flow is ejected to a surface, very thin hydrodynamic and thermal boundary layers form in the impingement region, resulting in extremely high heat transfer coefficients within the stagnation zone. Garimella and coworker experimentally investigated the influence of fluid thermo-physical properties on heat transfer from confined and submerged impinging jets [51], and the generalized correlations for stagnation-point Nusselt number and area-averaged Nusselt number were obtained. Confined jet impingement direct liquid cooling is often used to cool the electronics because of its compact structure. Bandhauer et al. developed a jet impingement direct liquid cooling device for high-performance integrated circuits (ICs), and the average heat transfer coefficient of 13,100 $W \cdot m^{-2} \cdot K^{-1}$ was achieved using a nozzle diameter of 300 μm [52]. However, the heat sink device was made of epoxy by using the 3D printing method, which is unreliable at high temperature, and the lifetime is short. Jorg et al. presented an approach of direct single-jet impingement liquid cooling of a typical metal-oxide semiconductor field-effect transistor (MOSFET) power module [53]. Heat transfer coefficient of up to 12,000 $W \cdot m^{-2} \cdot K^{-1}$ was achieved using only 10.8 cm^2 assembly space for the cooling device. Besides, the hybrid microchannel jet impingement cooling method was proposed and studied to enhance the heat transfer of jet impingement cooling [54–57].

However, almost all the previous studies about the jet impingement direct liquid cooling were focused on cooling the top surface of the electronic device or the chip. The existence of the thickness of the chip may cause large temperature rise inside the chip under a large heat load. The thickness of the electronic components or chips could be as large as several millimeters, especially for the high-power electronic packaging unit [58]. In addition, there are various shapes and sizes of the electronics, which might cause the ratio of the top surface area to the overall surface area to be very small. In such cases, traditional direct liquid cooling on the top surface would not be sufficient for temperature control of the high-power electronics with a certain thickness. For example, the size of the simulated high-power chip in Figure 9.15 is 10 mm × 50 mm × 4 mm. The calculated surface temperature distribution of the high-power chip (2500 W) with top surface cooling and with the five-surface cooling is shown in Figure 9.15(a) and (b), respectively. The only surface that is not cooled in Figure 9.15(b) is used for electric connection. To achieve the same maximum temperature, the heat transfer coefficient h of the top surface cooling should only be 2.5 times as large as that of the five-surface cooling case. The cooling surface area of the case in Figure 9.15(a) is 9.8 cm^2, which is only 1.96 times (less than 2.5) as large as the cooling surface area of the case in Figure 9.15(b). Therefore, the heat dissipation improvement of the body cooling in Figure 9.15(b) is due not only to the increase in the cooling surface area, but also to the 3D thermal conduction of the bulk high-power chip. When all the available surfaces are cooled, the thermal resistance between the coolant and the heat source is smaller than in the case when only one surface is cooled.

Therefore, direct liquid cooling on all the available surfaces of the electronic component is required to decrease its temperature rise, taking full advantage of the heat dissipation area. This cooling method is called body cooling, which is also

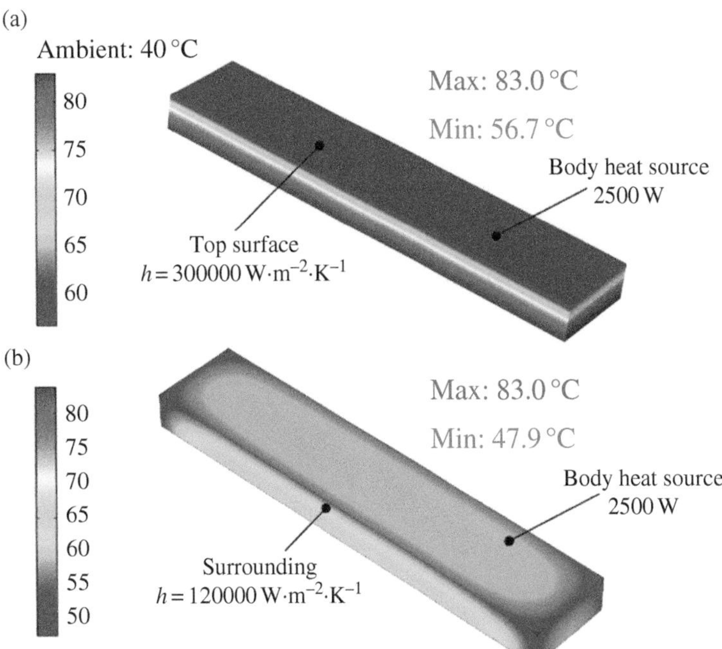

Figure 9.15 Calculated surface temperature with cooling: (a) only one surface and (b) five surfaces.

known as the side and end cooling [59]. However, there are few studies about body cooling jet impingement. In addition, the thermal model of the direct liquid body cooling method has not been developed, which is not convenient for the structure design and optimization of the body cooling methods.

A jet impingement body cooling (JIBC) device and a channel/jet impingement hybrid body cooling (HBC) device were developed using the traditional mechanical machining method with micronozzles. The thermal model for the JIBC, HBC, and traditional jet impingement surface cooling (JISC) was established. The thermal model is helpful to guide the structure design of the JIBC, HBC and JISC. The heat dissipation capabilities of the three cooling methods were compared by theory, numerical simulation, and experiments.

The developed JIBC and HBC structures are shown in Figure 9.16(a) and (b). The traditional JISC structure is shown in Figure 9.16(c). In the JIBC case, the jet impingement nozzles are set opposite to all the surfaces of the heated chip to let the coolant directly impinge the exposed surfaces of the heated chip. In the HBC case, however, the nozzles are only set opposite to the top surface of the heated chip, so that the coolant can only be ejected to the top surface. After being ejected to the top surface, the coolant flows through the side surfaces of the heated chip to cool these surfaces and then flows out of the device. Thus, the HBC combines jet impingement cooling for the top surface of the chip and channel cooling for the side surfaces. If some microstructures are machined at the side surfaces of the chip or the corresponding opposite walls, the channel cooling may turn to the mini/microchannel cooling. In the JISC case, the nozzles are also set opposite to the top surface of the heated chip. After being ejected to the top surface, the coolant directly flows out of the device without touching the side surfaces.

9.2.1 Calculation of Surface Heat Transfer Coefficient

During the engineering design, the edge length of the chip probably may not be divided evenly by the designed jet-to-jet spacing S. S is the square unit cell dimension of the jet impingement array footprint [31]. For this reason, as is shown in Figure 9.17(a), some part of the array footprint may be missing on the chip surface, and some extra part may exist. Therefore, the area-averaged convective heat transfer coefficient cannot be simply calculated by using correlations for the area-averaged Nusselt number in literatures. The chip surface must be discretized to obtain the local convective heat transfer coefficient $h(x, y)$. The single-phase heat transfer profile of the jet impingement is verified as a bell-shaped local heat transfer coefficient distribution with a maximum value at the stagnation point and a monotonic decrease in the outward radial direction [60]. The local convective heat transfer coefficient can be obtained by

$$h'(r) = \left[C_1 - C_2 \exp\left(\frac{-r^2}{2}\right)\right]^{-1} \tag{9.9}$$

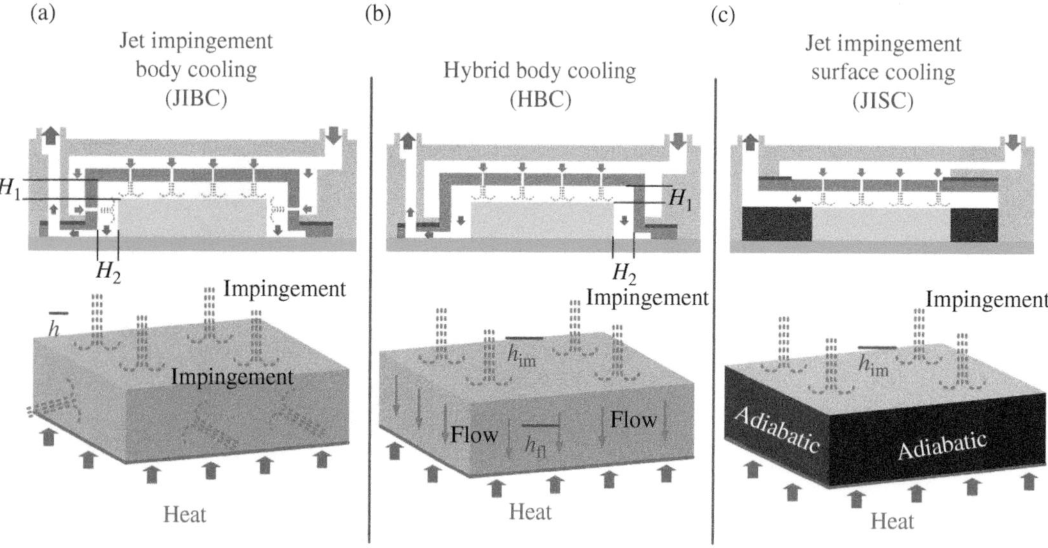

Figure 9.16 Schematic diagram of (a) the jet impingement body cooling, (b) the hybrid body cooling, and (c) the traditional jet impingement surface cooling.

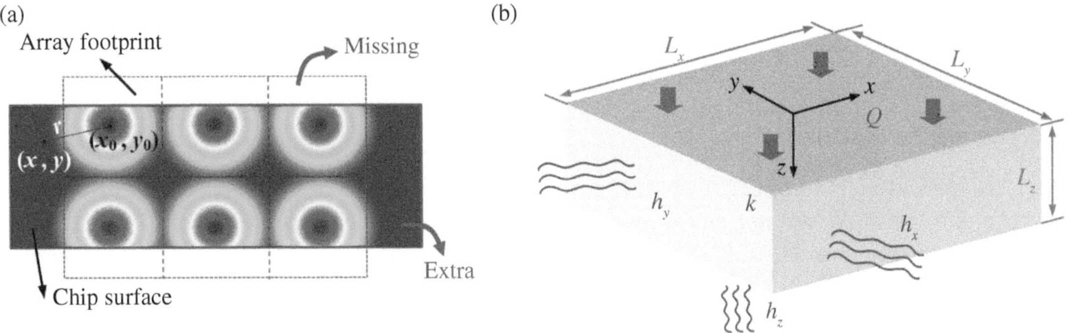

Figure 9.17 Schematic diagram of (a) the distribution of the convective heat transfer coefficient on the surfaces and (b) the chip with body cooling.

where C_1 and C_2 can be calculated by the stagnation Nusselt number and the area-averaged Nusselt number of the array footprint. These two numbers were calculated by using the empirical correlations verified by experiments previously. The stagnation Nusselt number is given by [51]

$$\mathrm{Nu}_0 = 1.409\,\mathrm{Re}^{0.497}\mathrm{Pr}^{0.444}(l_\mathrm{n}/D_\mathrm{n})^{-0.058}(2r_\mathrm{eq}/D_\mathrm{n})^{-0.272} \tag{9.10}$$

where $r_\mathrm{eq} = S/\sqrt{\pi}$ and l_n is the nozzle length. D_n is the nozzle diameter. The area-averaged Nusselt number of the array footprint is given by [61]

$$\begin{aligned}\overline{\mathrm{Nu}_i} &= 0.5\,\mathrm{Re}^{0.667}\mathrm{Pr}^{0.42}(L/D_\mathrm{n})^{-0.058}\left[1+\left(\frac{(H/D_\mathrm{n}\sqrt{\pi}/2)}{0.6(S/D_\mathrm{n})}\right)^6\right]^{-0.05}\\ &\times\left(\frac{\sqrt{\pi}}{S/D_\mathrm{n}}\right)\frac{1-1.1\sqrt{\pi}/(S/D_\mathrm{n})}{1+0.1(H/D_\mathrm{n}-6)\sqrt{\pi}/(S/D_\mathrm{n})}\end{aligned} \tag{9.11}$$

where H is the jet-to-target spacing, which is shown in Figure 9.16 as H_1 and H_2. The calculation of C_1 and C_2 and more details about calculating the $h'(r)$ may be found in Ref. [55]. Thus, the local convective heat transfer coefficient $h(x, y)$ can be calculated by

$$h(x,y) = h'\left(\sqrt{(x-x_0)^2+(y-y_0)^2}\right) \tag{9.12}$$

where (x_0, y_0) is the coordinate of the nozzle closest to the point (x, y). And the area-averaged surface convective heat transfer coefficient of the chip surface is calculated as

$$\overline{h} = \left[\frac{1}{A}\iint\frac{1}{h(x,y)}\mathrm{d}x\mathrm{d}y\right]^{-1} \tag{9.13}$$

where A is the surface area. The jet impingement heat transfer coefficient $\overline{h_\mathrm{im}}$ of the HBC and JISC can be calculated by the method mentioned earlier, with Re calculated by using V'_D.

Assume that the channel flow of the coolant through the side surfaces of the chip is uniform. Thus, the downward flow velocity can be calculated by

$$V_\mathrm{side} = \frac{Q_\mathrm{v}}{2(l_x+l_y)H_2} \tag{9.14}$$

where Q_v is the total volume flow rate of the coolant and l_x and l_y are edge lengths of the top surface of the chip. The channel flow is usually laminar flow because of the small equivalent diameter. In this work, the laminar channel flow is regarded as the fluid flowing through the plate. Therefore, the heat transfer coefficient of the side surfaces in the HBC is given by

$$\overline{h_\mathrm{fl}} = 0.664\frac{\kappa_\mathrm{f}}{l_z}\,\mathrm{Re}_\mathrm{fl}^{0.5}\mathrm{Pr}^{1/3} \tag{9.15}$$

where κ_f is the thermal conductivity of the coolant, l_z is the thickness of the chip, and $\text{Re}_{\text{fl}} = \rho V_{\text{side}} l_z/\mu$. ρ and μ are the density and viscosity of the coolant, respectively.

9.2.2 Body Cooling Thermal Conductive Model

As shown in Figure 9.17(b), when the uniform heat flux P_{thermal} is added to one surface of the chip, if the area-averaged convective heat transfer coefficients of the side and end walls of the chip are known, the temperature field $T(x, y, z)$ of the chip can be obtained. The temperature field $T(x, y, z)$ should be obtained according to the steady-state thermal conduction equation in three dimensions, given as:

$$\nabla^2 T = \frac{\partial^2 T}{\partial x^2} + \frac{\partial^2 T}{\partial y^2} + \frac{\partial^2 T}{\partial z^2} = 0 \tag{9.16}$$

The boundary conditions are prescribed as follows:

$$\begin{aligned}
& x = \frac{L_x}{2}, \quad \frac{\partial T}{\partial x} = 0 \\
& x = 0 \text{ or } L_x, \quad \frac{\partial T}{\partial x} + \frac{h_y}{\kappa}(T - T_f) = 0 \\
& y = \frac{L_y}{2}, \quad \frac{\partial T}{\partial y} = 0 \\
& y = 0 \text{ or } L_y, \quad \frac{\partial T}{\partial y} + \frac{h_x}{\kappa}(T - T_f) = 0 \\
& z = 0, \quad \frac{\partial T}{\partial z} + \frac{Q}{L_x L_y \kappa} = 0 \\
& z = L_z, \quad \frac{\partial T}{\partial z} + \frac{h_z}{\kappa}(T - T_f) = 0
\end{aligned} \tag{9.17}$$

where κ is the thermal conductivity of the chip, h_z is the area-averaged heat transfer coefficient of the top surface of the chip ($\bar{h}$ and $\overline{h_{\text{im}}}$), and h_x and h_y are the corresponding area-averaged heat transfer coefficient of the side surfaces. According to the method of separation of variables, the temperature field is given as

$$\begin{aligned} T(x, y, z) - T_f = & \sum_{m=1}^{\infty} \sum_{n=1}^{\infty} \cos(\lambda_{xm} x) \cos(\lambda_{ym} y) \\ & \times [A_{mn} \cosh(\beta_{mn} z) + B_{mn} \sinh(\beta_{mn} z)] \end{aligned} \tag{9.18}$$

where T_f is the fluid temperature, which is assumed to be the inlet fluid temperature of the heat sink, and λ_{xm}, λ_{yn}, and β_{mn} are the eigenvalues and are obtained from the iterative solution of the following equations:

$$\lambda_{xm} \sin\left(\frac{\lambda_{xm} L_x}{2}\right) = \frac{h_y}{\kappa} \cos\left(\frac{\lambda_{xm} L_x}{2}\right) \tag{9.19}$$

$$\lambda_{yn} \sin\left(\frac{\lambda_{yn} L_y}{2}\right) = \frac{h_x}{\kappa} \cos\left(\frac{\lambda_{yn} L_y}{2}\right) \tag{9.20}$$

$$\beta_{mn} = \sqrt{\lambda_{xm}^2 + \lambda_{yn}^2} \tag{9.21}$$

and A_{mn} and B_{mn} are obtained from the following equations:

$$A_{mn} = -B_{mn} \frac{k\beta_{mn} + h_z \tanh(\beta_{mn} L_z)}{h_z + k\beta_{mn} \tanh(\beta_{mn} L_z)} \tag{9.22}$$

$$B_{mn} = \frac{-4Q \sin(\lambda_{xm} L_x/2) \sin(\lambda_{yn} L_y/2)}{\kappa L_x L_y \beta_{mn} [\sin(\lambda_{xm} L_x)/2 + \lambda_{xm} L_x/2] [\sin(\lambda_{yn} L_y)/2 + \lambda_{yn} L_y/2]} \tag{9.23}$$

For more details about calculating the temperature field, refer to Ref. [31].

9.2.3 Experiment

The heat sink device of the three cooling cases was designed, manufactured, and tested by experiments. The structure of the JIBC device is shown in Figure 9.18(a). Rather than using the 3D printing method to manufacture the whole device with epoxy in one piece, the device is separated to be composed of three components: the shell, the gasket, and the nozzle substrate, which were all machined by traditional mechanical computer numerical control (CNC) machining method with aluminum alloy. As shown in Figure 9.18(b), nozzles with diameter of 300 μm were machined by electro-sparking method on the side walls of the nozzle substrate. The viton rubber gasket was set between the gap of the nozzle substrate and the shell to prevent leakage and make sure that all the coolant flows through the nozzles. The gasket and nozzle substrate were fixed to the shell by screws. In the JIBC case, nozzles were machined on the top wall and side walls (left, front, right, and back walls). However, in the HBC case, the jet impingement nozzles were not machined on the side walls, but only machined on the top wall of the nozzle substrate. As is shown in Figure 9.16(c), the nozzle substrate of the JISC case was machined as a plate with jet impingement nozzles for the top surface of the chip, which is similar to the traditional structures in previous studies. The device has a long life and is reliable even in high-temperature environment.

The nozzle arrangement for the JIBC case is shown in Figure 9.18(c). The nozzle arrangements of the top wall in the HBC case and the JISC case are the same as in the JIBC case. More details about the nozzle arrangement of the three cases are listed in Table 9.3, in which N is the nozzle number. The size of the heated chip is 10 mm × 50 mm × 4 mm and the chip was made of copper.

The test facility in this work is schematically shown in Figure 9.19 DI-water is circulated through the flow loop driven by a hydrodynamically levitated centrifugal micropump [62]. The flow rate can be set by tuning the rotation speed of the micropump and by the valve. The volume flow rate is measured by a turbine flow meter (YF-S401) with ±2% accuracy. The air-cooled heat exchanger with 16 copper pipes was used to cool the coolant and remove the heat to the ambient. A 40 μm filter was positioned upstream of the heat sink to prevent blocking caused by impurities in the coolant. A water tank was assembled upstream of the filter and a thermocouple was put inside the water tank to measure the inlet coolant temperature. The inlet fluid temperature was controlled at 40 °C stably by tuning the rotating speed of the fans in the air-cooled heat exchanger in all the experiments. Seven heating rods were inserted in the copper block to provide the heating power. In order to prevent heat loss, thermal insulation cotton was used to wrap the copper block up. The maximum heating loss

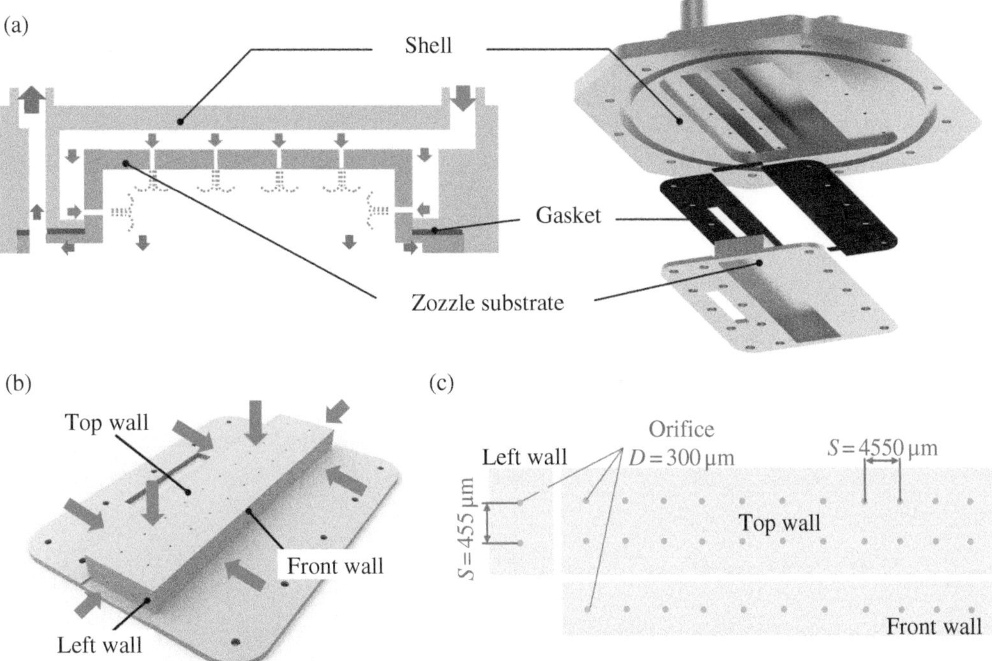

Figure 9.18 Schematic diagram of (a) the structure of the JIBC device, (b) the structure of the nozzle substrate, and (c) the nozzle arrangement on the nozzle substrate walls.

Table 9.3 Structure parameters of the nozzle substrate.

		JIBC	HBC	JISC
D (μm)		300	300	300
S (μm)		4550	4550	4550
Top	N	2 × 11	2 × 11	2 × 11
	H (μm)	400	400	400
Left	N	2 × 1	—	—
	H_2 (μm)	400	400	—
Front	N	1 × 11	—	—
	H_2 (μm)	400	400	—

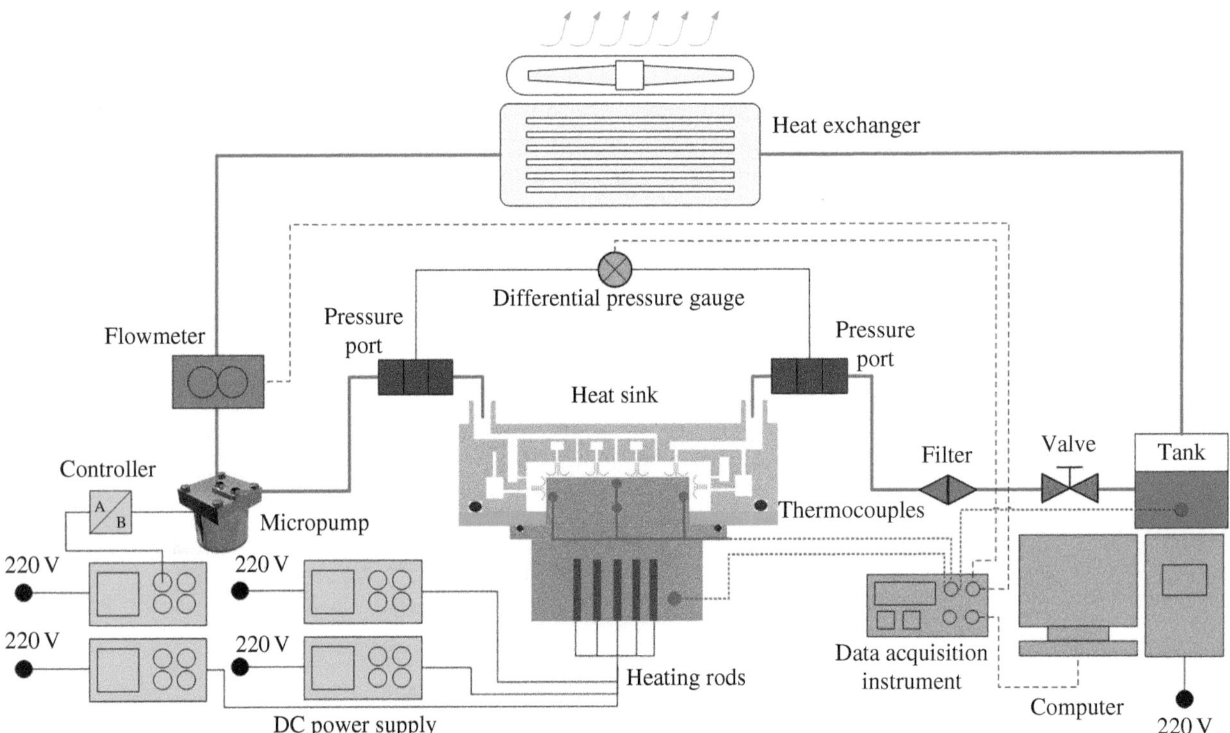

Figure 9.19 The system diagram of the test facility.

was briefly calculated using the Newton's law of cooling $q_{loss} = h_n \times A_c \times (T_c - T_a)$ with the maximum heating power (800 W), where h_n is the natural convective heat transfer coefficient, A_c is the surface area of the copper block, T_c is the cotton temperature, and T_a is the ambient temperature. In maximum, h_n was set to be $10\,W \cdot m^{-2} \cdot K^{-1}$. By measuring the cotton and ambient temperatures, the heat loss was calculated to be 15.97 W with heating power of 800 W, meaning that the heat loss was no more than 2%.

9.2.4 Numerical Simulation

Numerical simulations were done to investigate the heat transfer and fluid flow mechanism by using the commercial multi-physics software COMSOL MULTIPHYSICS 5.3a. The Heat Transfer and computational fluid dynamics (CFD) packages were coupled in this study. To reduce the computational costs, only the dominant fluid domain was considered in the simulation, as shown in Figure 9.20, rather than using the whole fluid domain of the heat sink device. To characterize the heat

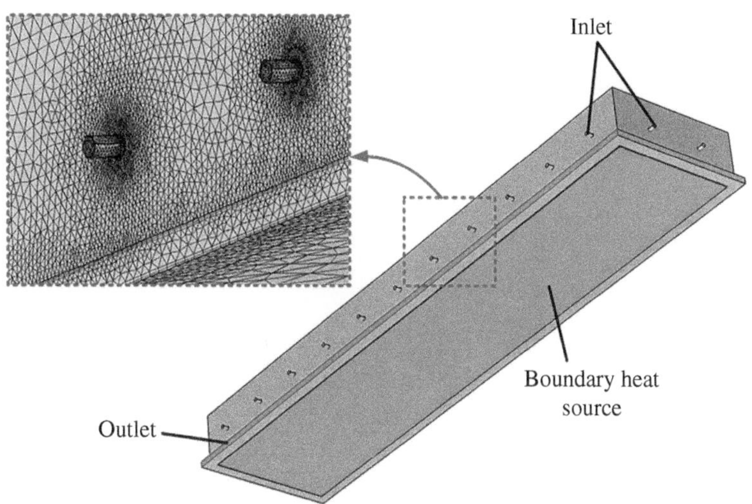

Figure 9.20 Simulation model of the JIBC.

Table 9.4 The physical property parameters of materials used in the numerical simulation.

Material	κ (W·m^{-1}·K^{-1})	ρ (kg/m^3)	μ (mPa·s)	C_p (J·kg^{-1}·K^{-1})	$T_{fluid,in}$ (°C)
Water	0.635	992.2	0.6533	4174	40
Pure copper	380	—	—	—	—

transfer and fluid flow process more accurately, grids were meshed finer in the jet impingement zone. The nozzle orifices were set as the fluid inlet. The inlet fluid temperature was set to be 40 °C. Boundary heat source was added on the bottom surface of the chip. The material of the chip was set as copper and the coolant was the DI-water. The physical property parameters of the materials used in the numerical simulation are listed in Table 9.4. The grid independence test was done. The maximum temperature of the chip of the JIBC case was calculated with the grid numbers 313,135, 702,516, 1,922,799, and 2,289,566. The discrepancies were less than 5.5%. Taking the computational cost into consideration, the mesh parameters with grid number 702,516 were used in all the simulations, meaning that the "Normal" grid was selected in the COMSOL.

9.2.5 Performance of the Developed JIBC Device

As the chip is cooled by all the exposed surfaces, the temperature field inside the chip is highly nonuniform. Therefore, the average temperature of the chip is not easy to be characterized. Thus, the maximum temperature of the chip was used to analyze the cooling performance in this work. In the experiment, seven separated thermocouples were put at the bottom surface region of the chip to measure the maximum temperature. As the inlet fluid temperature was controlled at 40 °C in the model, simulation, and the experiment, the maximum temperature rise $\Delta T_{max} = T_{max} - T_f$ was used in the following results, where T_{max} is the maximum temperature. Experimental results of the maximum temperature rise versus the heating power of the JIBC, HBC, and JISC cases are shown in Figure 9.21 It can be seen that the maximum temperature rise changes linearly with the heating power, indicating that the total thermal resistance is independent of the heat load, which is the general result that is obtained using the single-phase liquid cooling. When the volume flow rate increases from 1000 to 1500 ml/min, the maximum temperature decreases, and the difference becomes larger with increasing heat load. According to the results shown in Figure 9.21(a), the temperature difference between the experiment and model of the JIBC case is less than 8.0%. Similarly, results in Figure 9.21(b) and (c) indicate that the temperature differences between the experiment and model of the HBC and JISC cases are less than 18.1% and 6.2%, respectively. Thus, results of the experiment, simulation, and model are in good agreement. The measured maximum temperature of the HBC is slightly higher than that obtained by simulation and model, because the flow velocity of the side-wall channel cooling is mainly dominated by the channel width

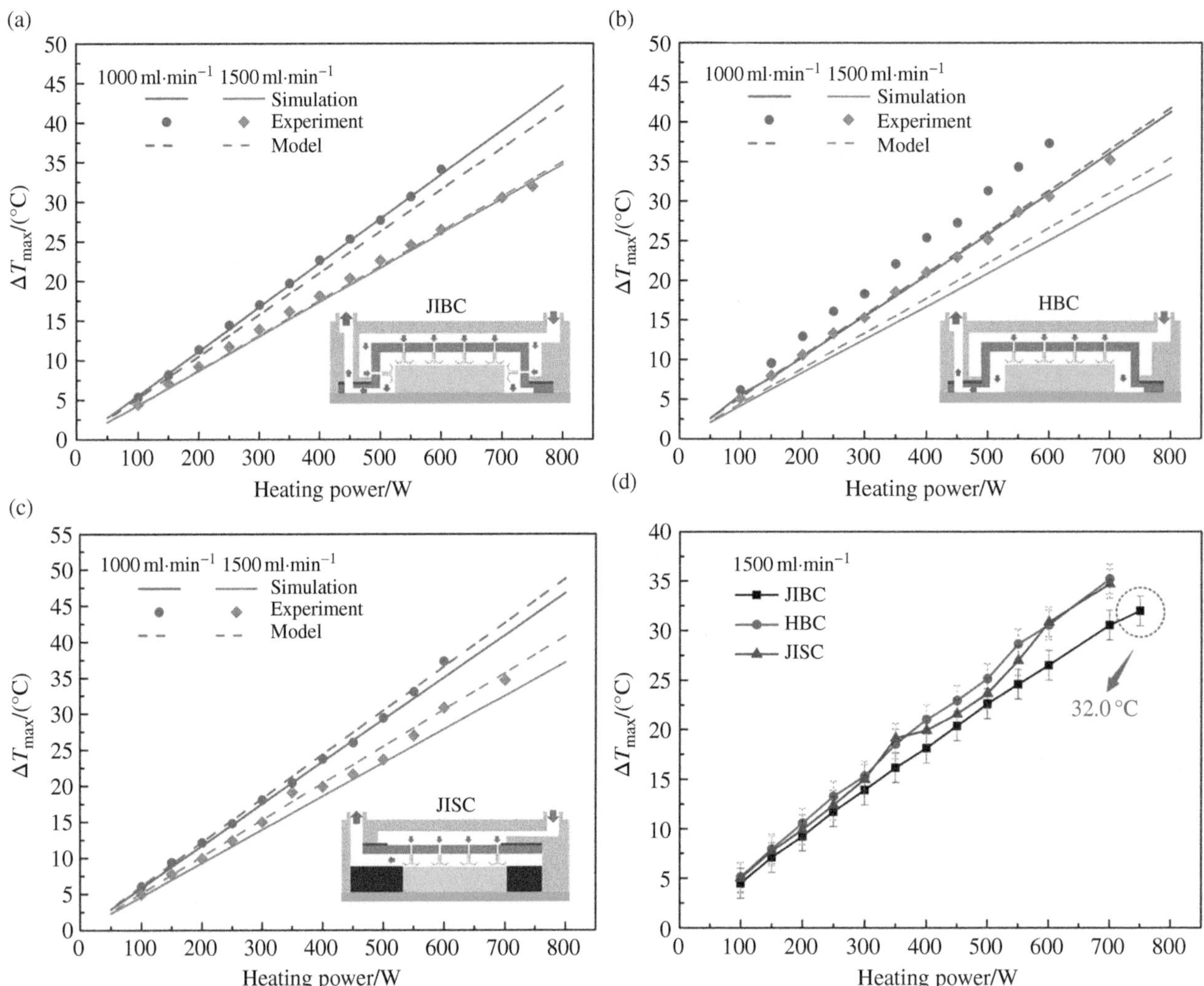

Figure 9.21 The maximum temperature rise versus the heating power of (a) the JIBC case, (b) the HBC case, (c) the JISC case, and (d) the three different cooling cases obtained by experiments with the volume flow rate of 1500 ml/min.

H_2, where machining and assembly errors may exist. From the experimental results shown in Figure 9.21(d), it can be observed that the JIBC has the best cooling performance. When the heating power is as large as 750 W, the maximum temperature rise of the chip is only 32.0 °C with the flow rate of 1500 ml/min by using the JIBC.

The range of the jet Reynolds number is from 1344 to 4030 in this work. The jet flow is transition flow with Reynolds number from 1344 to 3000, and turbulent flow with Reynolds number from 3000 to 4030. Therefore, the jet flow was assumed to be turbulent in the COMSOL simulation. Experimental results of the maximum temperature rise versus the volume flow rate of the JIBC, HBC, and JISC cases with the heat power of 500 W are shown in Figure 9.8. The maximum temperature rise decreases with increasing volume flow rate. The trend of results indicates that the temperature would not decrease much when the flow rate is very large, because the convective heat transfer coefficient would be large enough so that the thermal conductive resistance of the chip dominates.

According to the results shown in Figure 9.22(a), the temperature difference between the experiment and model of the JIBC case is less than 10.8%. Results in Figure 9.22(b) and (c) indicate that the temperature differences between the experiment and model of the HBC and JISC cases are less than 21.0% and 19.3%, respectively. The difference between experiment and model is mainly due to the assumptions and the accuracy of the adopted empirical heat transfer correlations. Even so, the model still fits well with the simulation. Within the margin of errors, the developed thermal model is accurate enough for the heat transfer study, quick design, and structure optimization of the JIBC, HBC, and JISC. Figure 9.22(d) shows the

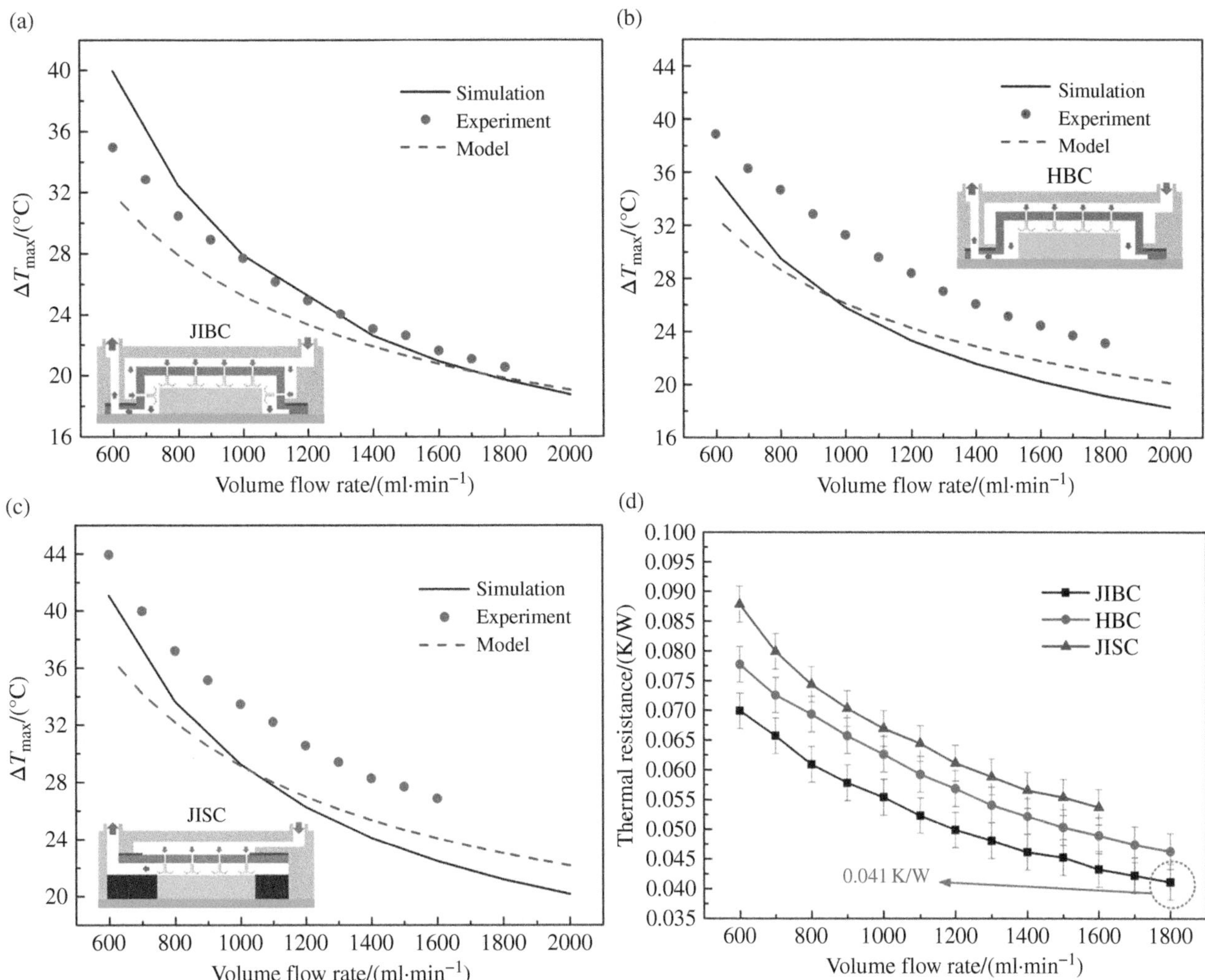

Figure 9.22 The maximum temperature rise versus the volume flow rate of (a) the JIBC case, (b) the HBC case, and (c) the JISC case and (d) maximum thermal resistance versus the volume flow rate of the three different cooling cases obtained by experiments with the heating power of 500 W.

experimental results of the maximum total thermal resistance, which is given as $R_{total} = (T_{max} - T_f)/ P_{thermal}$. The decreasing trends of the three devices are the same. The minimum total thermal resistance of 0.041 K/W was experimentally obtained by using the JIBC with the volume flow rate of 1800 ml/min. On comparison, the thermal resistance of the commercial TIM with high thermal conductivity of $3\ W \cdot m^{-1} \cdot K^{-1}$ is as large as 0.033 K/W, considering the thickness of 50 μm and the surface size of 10 mm × 50 mm. Therefore, the extremely low total thermal resistance was obtained.

In Figure 9.23(b), we can see that the heat dissipation capabilities for the top surface of the HBC and JISC are almost the same. However, the HBC can provide extra cooling for the chip by channel cooling to the side surfaces, which is not available in the JISC. That's the reason why the HBC has lower total thermal resistance than the JISC shown in Figure 9.22(d), although the side cooling h is only nearly 1/4 of the top cooling h of the HBC. Figure 9.23(a) shows the simulated and modeled jet impinging heat transfer coefficient of the top, left, and front surfaces in the JIBC case. The model-calculated h fits well with the simulated results. Although the nozzle parameters of all the nozzle substrate walls are set as the same, the jet impinging heat transfer coefficients for the top, left, and front surfaces are all different. This is mainly due to the missing and extension of the array footprint as described in Section 9.2.1 and Figure 9.17(a).

Owing to the chip size and the nozzle arrangement, the JIBC provides the maximum heat transfer coefficient for the front and back surfaces, and the minimum heat transfer coefficient for the top surface. In the JIBC case, the diameters of all the

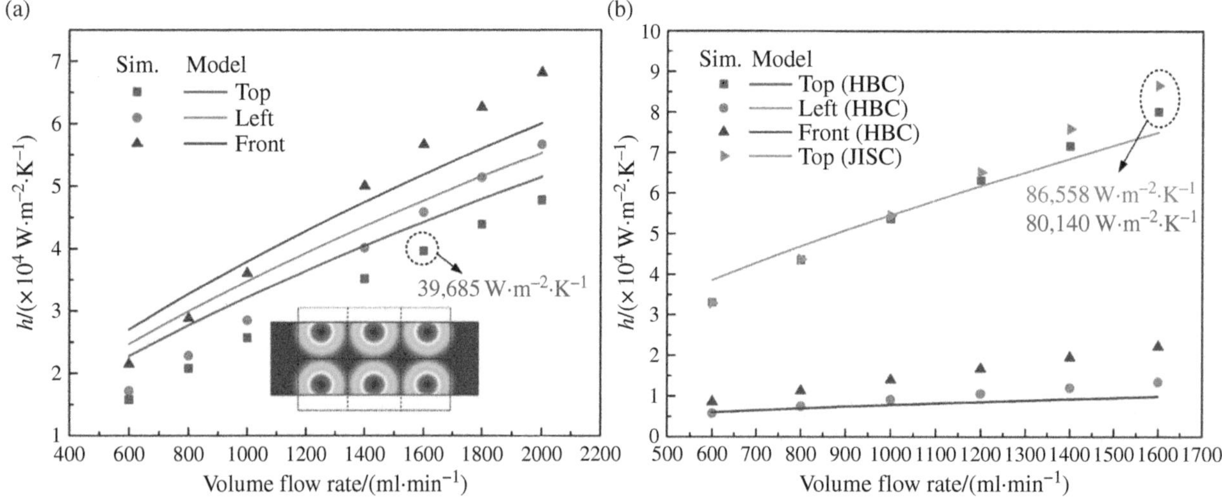

Figure 9.23 The convective heat transfer coefficient of the chip surfaces for (a) the JIBC and (b) the HBC and the JISC.

nozzles were set to be the same, so that the flow rate at each nozzle was assumed to be equal. The fluid flow velocity at the nozzles can be calculated as

$$V_{\mathrm{D}} = \frac{Q_{\mathrm{v}}}{(N_{\mathrm{top}} + 2N_{\mathrm{front}} + 2N_{\mathrm{right}})A_{\mathrm{D}}} \tag{9.24}$$

where N_{top}, N_{front}, and N_{right} are the nozzle number at the top, front, and right surface, respectively. $A_{\mathrm{D}} = \pi D^2/4$ is the nozzle area. The flow velocity at the nozzles for the HBC and JISC is given by

$$V'_{\mathrm{D}} = \frac{Q_{\mathrm{v}}}{N_{\mathrm{top}}A_{\mathrm{D}}} \tag{9.25}$$

The nozzle Reynolds number Re can be calculated according to the nozzle flow velocity. In this work, the nozzle number of the JIBC is 48, and the nozzle numbers of the HBC and JISC are both 22. Therefore, the nozzle flow velocity of the HBC and JISC is more than two times that of the JIBC, leading to much larger heat transfer coefficient for the top surface. Although the top surface heat transfer coefficient of the HBC and JISC is more than two times that of the JIBC, the total thermal resistance of the JIBC is still lower than all the other cooling devices. The reason is that the heat transfer coefficients for the side surfaces of the JIBC are nearly two times larger than that of the HBC, which confirmed that the side cooling is quite important.

The heat flux ratio of the surface is calculated by $\eta = P_{\mathrm{thermal,sur}}/P_{\mathrm{thermal}}$, where $P_{\mathrm{thermal,sur}}$ represents the heat dissipated from the surface. Obviously, the heat flux ratio of the top surface in the JISC case is 100%. As is shown in Figure 9.24, nearly 75% of the heat is dissipated from the top surface in the HBC case, while only less than 45% of the heat is dissipated from the top surface in the JIBC case. Much more heat is dissipated from the side surfaces of the JIBC because of the larger h of the side surfaces. Although the sum of the areas of the front and back surfaces is only 80% of the area of the top surface, the dissipated heat from the front and back surfaces is larger than that from the top surface in the JIBC, which is mainly owing to the larger side cooling h than the top surface.

Figure 9.25 shows the simulated surface temperature field of the JIBC, HBC, and JISC. The temperature field clearly reflects the arrangement of the jet impingement nozzles. We can see that the temperature at the edges and corners of the chip is much lower in the JIBC and HBC case due to the body cooing effect. The same result does not exist in the JISC case without body cooling. The nozzle flow velocity of the JIBC is only 4.91 m/s, while those of the HBC and JISC cases are both 10.72 m/s. The larger nozzle flow velocity results in greater heat dissipation capability, leading to the lower top surface temperature of the HBC than the JIBC. Although the top surface heat dissipation capability of the HBC and the JISC are equal, the top surface temperature of the JISC is much higher, because of the larger heat flux. Different from the other two cases, the temperature field of the HBC shows obvious asymmetry. This result indicates that the flow in the HBC is more sensitive to the pressure difference between the inlet and the outlet. Figure 9.26(a) shows the simulated streamline of the

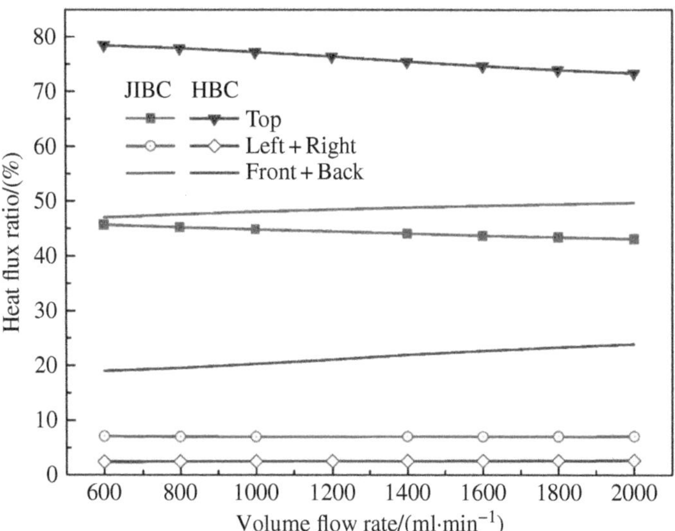

Figure 9.24 Heat flux ratio of different surfaces.

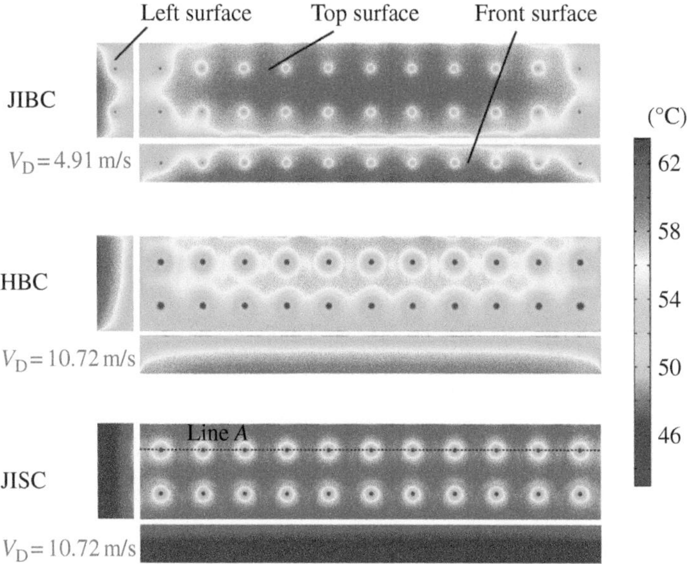

Figure 9.25 The simulated temperature contour of the chip surfaces.

cross section at line A of the JISC. The high-speed jet flow impinges on the hot surface and then turns to the flow in the axial direction. As is shown in Figure 9.26(b), the thickness of the boundary layer gradually increases in the radial flow region and the temperature of the fluid rises due to the absorption of heat. Owing to the confinement of the nozzle substrate, the adjacent jet flow interacts with each other and forms the vortexes. The fluid temperature in the vortex region is nearly 10 °C higher than the nozzle region, which might result in preheating of the inlet fluid. As is shown in Figure 9.26(c), the temperature at the center of the impingement region is very low. Thanks to the extremely high-stagnation heat transfer coefficient, the minimum temperature is only 44.4 °C, 4.4 °C higher than the inlet fluid temperature. However, the local heat transfer coefficient sharply drops from $2.2 \times 10^5\ \mathrm{W \cdot m^{-2} \cdot K^{-1}}$ to only $5.1 \times 10^4\ \mathrm{W \cdot m^{-2} \cdot K^{-1}}$, resulting in large temperature difference on the surface. The temperature difference is as large as 20 °C. Therefore, the thermal model is very important for the optimal design to decrease the temperature difference.

The aforementioned results show that the JIBC device has the best cooling performance with the aforementioned nozzle arrangement and chip structure. However, the convective heat transfer coefficient of the top surface of the JIBC is much lower than that of the HBC. Thus, if the nozzle arrangement and the chip size have changed, the HBC might have better cooling performance. Figure 9.27(a) shows the changes of the total thermal resistance of the three cases with the nozzle

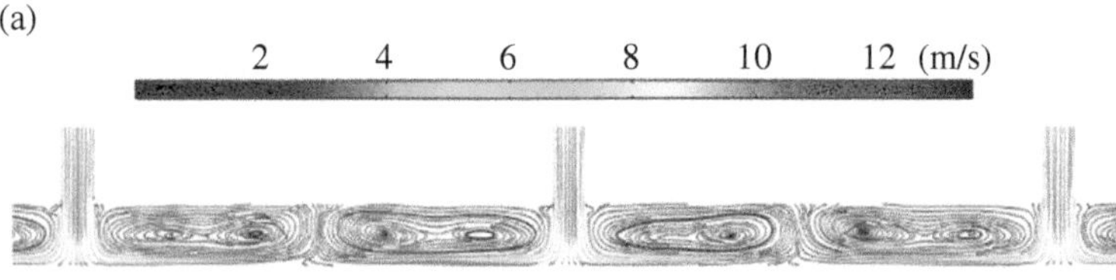

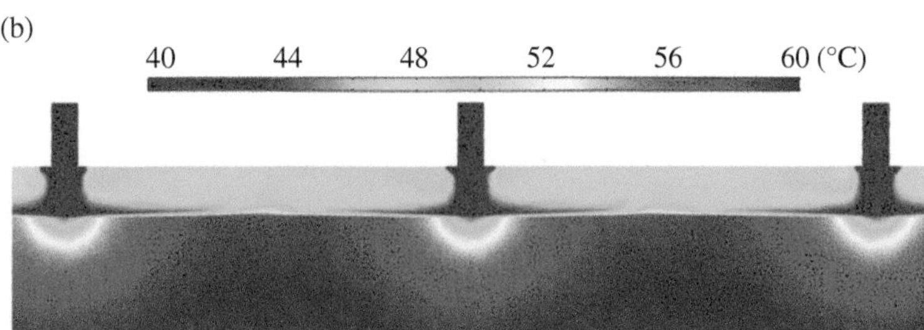

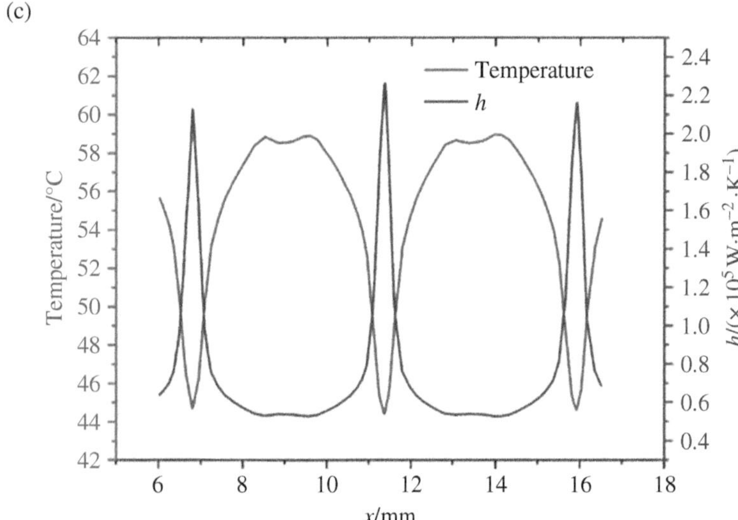

Figure 9.26 (a) The streamline of the cross section at line A of the JISC, (b) the temperature distribution of the cross section at line A of the JISC, and (c) the local temperature and local convective heat transfer coefficient on line A of the JISC.

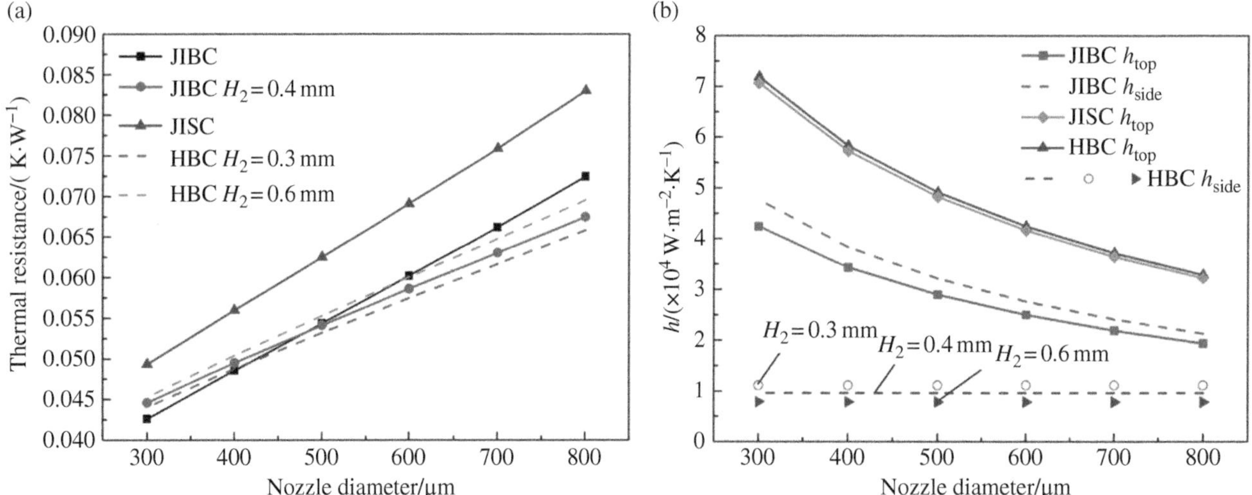

Figure 9.27 (a) The total thermal resistance versus the nozzle diameter and (b) the surface h versus nozzle diameter.

diameter. When the nozzle diameter increases, the thermal resistance of the three cases increases linearly. When the nozzle diameter is larger than 500 μm, the thermal resistance of the HBC becomes lower than that of the HBC, which means that the cooling performance of the HBC is better than the JIBC when the nozzle diameter is larger than 500 μm. When the nozzle diameter increases, the jet impinging convective heat transfer coefficient decreases, but the channel cooling convective heat transfer coefficient does not change, as is shown in Figure 9.27(b). This is due to the fact that the flow velocity of the channel cooling is only determined by the flow rate and the cross-section area of the channel, and it is independent of the nozzle arrangement. Therefore, when the H_2 increases from 0.4 to 0.6 mm, the cross-section area of the channel increases and the flow velocity decreases. As a result, the heat transfer coefficient h of the channel cooling decreases, leading to increasing thermal resistance of the HBC. Thus, the critical nozzle diameter increases to 600 μm. On the contrary, when the H_2 is reduced to 0.3 mm, the critical nozzle diameter would decrease to 400 μm. Therefore, the cooling performance of the JIBC is not always better than the HBC when changing the nozzle arrangement, H_2, or the chip size. The developed thermal model is essential for choosing which of the three cooling methods is the best under the given working condition.

9.3 Integrated Piezoelectric Pump Cooling

With the advantages of high thermal performance, efficiency, and flexibility, liquid cooling systems have been widely used in electric cars [63, 64] and data centers [65, 66]. However, thermal management systems utilizing liquid cooling technologies usually have large volume, which restricts their applications in small spaces, such as aerospace and portable electronics. A compact liquid cooling system is thus of great importance.

Liquid cooling system consists of the following functional components: pump, heat sink, and heat exchanger. Substantial research efforts have focused on the miniaturization of these components [67–69]. Duan et al. developed a high-performance mechanical micropump for liquid cooling systems with an external dimension of ϕ43 mm × 65 mm [70]. Yueh et al. applied a miniature piezoelectric micropump (30 mm × 15 mm × 3.8 mm) to an active fluidic cooling system [71]. Jung et al. developed a silicon microchannel cold plate with a low profile for power electronics [72]. To reduce the size and weight of the thermal management system, Kwon et al. developed a 1 cm^3 heat exchanger by electrical discharge machining [73]. These works successfully decrease the size of the individual components and save some space for the mount of the cooling system. However, it is still not enough for those extremely room-limited applications.

Since the sizes of single components have almost been reduced to the limit, some new schemes of components integration are developed to further decrease the total volume of liquid cooling systems [47, 74–76]. Among these schemes, integrating heat sink with micropump is a feasible solution, where piezoelectric micropump is employed as a substitute of conventional micropump due to its unique advantages in integration [77, 78]. For example, Ma et al. developed a mini-channel cold plate integrated with a piezoelectric micropump for laptops, which dissipated a heat load of 60 W [79]. Tang et al. experimentally investigated the thermal performance of an integrated heat sink with a piezoelectric pump, which achieved a convective heat transfer coefficient of 4800 $W \cdot m^{-2} \cdot K^{-1}$ [76]. These previous studies demonstrate the feasibility of pump-integrated liquid cooling system. However, there is still considerable room for the enhancement of the heat dissipation capability of the cooling system. For instance, the heat sink can be designed according to the driving characteristics of the piezoelectric micropumps. Piezoelectric micropumps show a relatively high output pressure but a low flow rate due to the driving characteristic of the piezoelectric actuator [77, 80]. Therefore, it's greatly significant to develop an integrated cooling system, which can make full use of the high pressure to achieve high thermal performance under a low flow rate.

An effective way to provide high heat removal ability and reduce the size of the liquid cooling system was proposed by integrating jet impingement cooling with piezoelectric micropump. The jet impingement cooling system ejects high-speed coolants onto the cooling surface through small nozzles and produces a high convective heat transfer coefficient with a relatively low flow rate. A compact jet array impingement cooling system with integrated piezoelectric micropump (JAICIPM) was designed, fabricated, and tested. The thermal performance of the JAICIPM was investigated by experiments and simulations.

9.3.1 Design and Fabrication of JAICIPM

Figure 9.28(a) shows the schematic diagram of the JAICIPM. The JAICIPM is mainly composed of a piezoelectric actuator and a 3D printed base with internal flow channels and nozzles. The piezoelectric actuator is fixed onto the base by the cover and screws. All the interfaces are sealed by O-rings. Two external check valves, designed as in our previous study [81], are

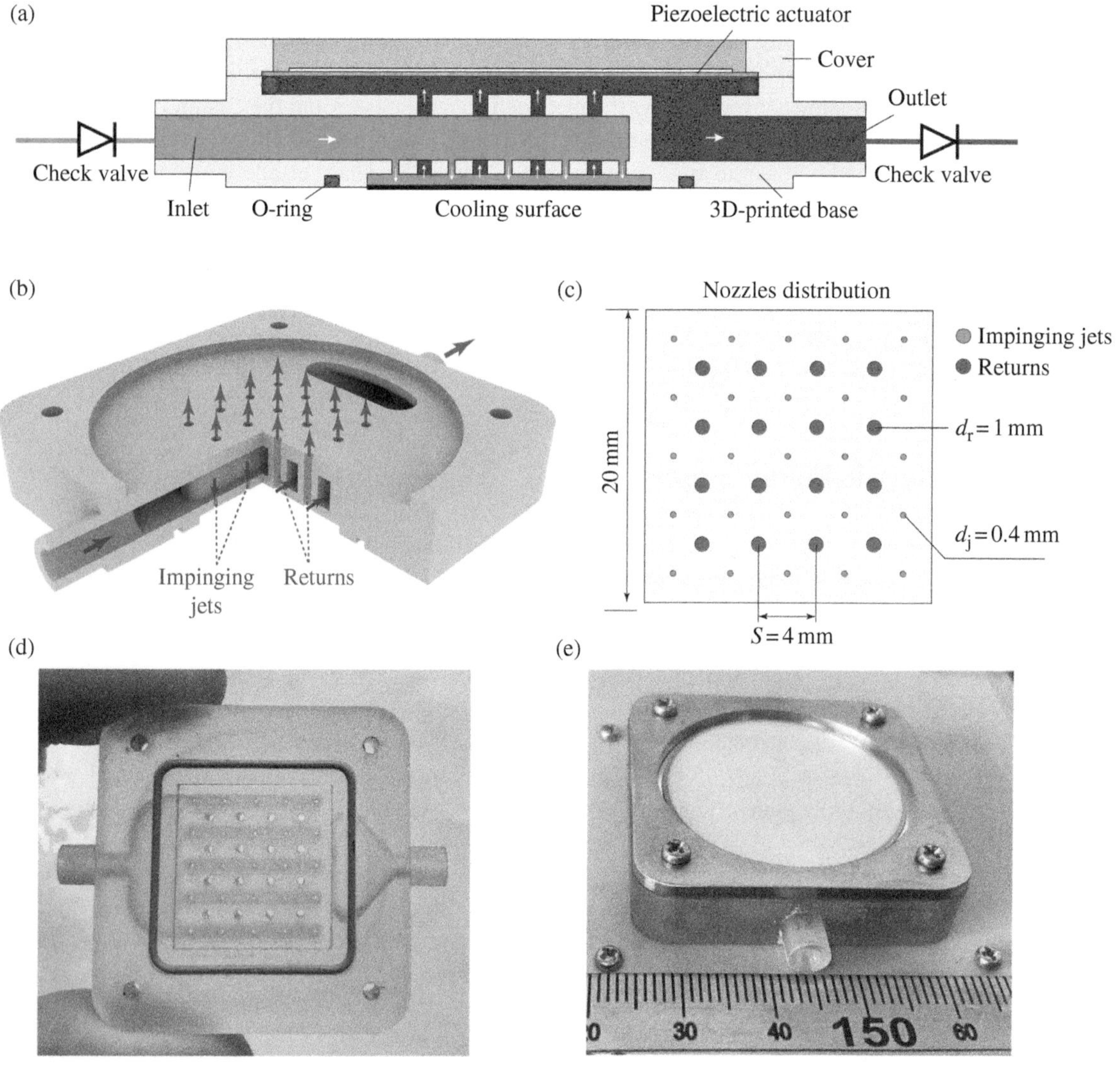

Figure 9.28 (a) Schematic diagram of the structure of the JAICIPM, (b) the flow path of the coolant, (c) the arrangement of the impinging jets and returns, (d) the picture of the 3D printed base, (e) the picture of the assembled JAICIPM.

connected to the inlet and outlet to control the direction of fluid flow. Figure 9.28(b) shows the flow path of the coolant. Jet array impingement cooling with distributed returns is designed to maintain uniform cooling. The coolant flows into the JAICIPM from the inlet and is then distributed to the jet nozzles through the channels. The coolant is directly ejected onto the cooling surface and flows out through the return nozzles. The arrangement of the impinging jets and returns is shown in Figure 9.28(c). The area of the jet impinging is 20 mm × 20 mm. The distance between two adjacent nozzles S is 4 mm. Thus, there are 25 impingement nozzles and 16 return nozzles. The diameters of the jet nozzles d_j and return nozzles d_r were designed as 0.4 and 1.0 mm, respectively. Previous studies indicate that optimal heat transfer could be obtained when H/d_j is in the range of 2–4 for jet array impingement [13]. The height of the jet cavity H is designed as 0.8 mm. The base was fabricated by stereolithography (SLA) 3D printing. The 3D printing technology enables the complicated array jet impingement structures in a small size without assembly and sealing. Figure 9.28(d) and (e) show the pictures of the 3D printed base and the assembled JAICIPM. The dimensions of the JAICIPM were examined by a microscope. The diameter of the nozzles is measured to be $d_j = 425 \pm 28$ μm and $d_r = 1023 \pm 26$ μm, respectively. The external dimension of JAICIPM is 40 mm × 40 mm × 10 mm (16 cm^3) after assembly. The detailed parameters of the components are shown in Table 9.5.

Table 9.5 Material and size of the components.

Component	Material	Dimensions (mm)
Cover	Stainless steel	$40 \times 40 \times 2.5$
Piezoelectric actuator	Brass plate and PZT ceramic	$\phi 35 \times 0.53$
3D printed base	Photocurable resin	$40 \times 40 \times 7.5$

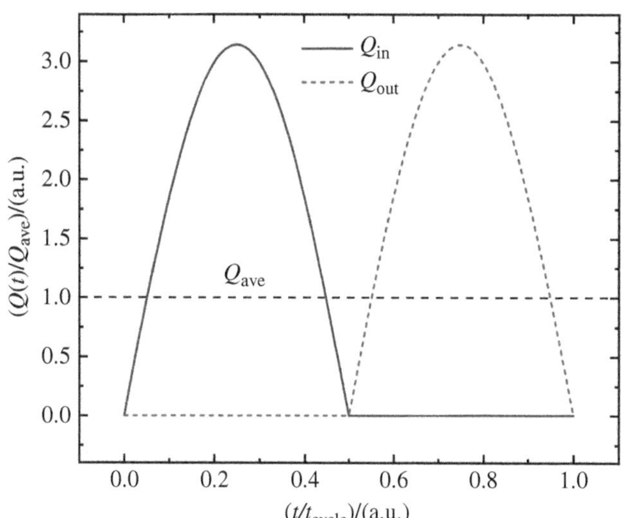

Figure 9.29 The ideal transient flow rate of the JAICIPM.

With the vibration of the piezoelectric actuator under alternating current (AC) voltage, the coolant is sucked into and discharged from the chamber of the JAICIPM periodically. Due to the actions of the check valves, the flow of the coolant is unidirectional. As the volume of the chamber increases, the coolant is sucked into the chamber through the inlet and nozzles and ejected onto the cooling surface to extract the generated heat. As the volume decreases, the coolant is discharged to take the heat away. Then, electronics can be cooled continuously in the repeated process. The piezoelectric actuator is driven by a sine waveform in this study. Therefore, the ideal transient flow rate of the inlet and the outlet is shown as Figure 9.29. The flow rate varies from 0 to π times the average flow rate in a half cycle and remains 0 in another half cycle. The flow rate of the jet impingement is equal to the flow rate of the inlet. The changing flow rate leads to a changing thermal performance in the cycle. Although the jet impinging is intermittent, the cooling performance of the JAICIPM could remain stable because of the thermal capacity and a high driving frequency.

9.3.2 Numerical Simulation

3D finite volume method simulations were conducted to evaluate the thermal performance of the JAICIPM. Due to the driving characteristics of the piezoelectric micropump, the velocity of the jet impinging changes at different periods. Therefore, transient simulations are necessary to investigate the transient thermal performance. Steady simulations were performed before the transient simulations to initialize the cases. RNG k-epsilon turbulence model with enhanced wall treatment was chosen to simulate the flow field according to previous studies [54, 82]. Pressure-based solver with SIMPLE method was selected, and the convergence criteria is 1×10^{-6} for the energy term and 1×10^{-3} for the others. The boundary conditions setup is shown in Figure 9.30(a). The bottom surface of the chip was set to a constant heat flux of 50 W/cm^2. The inlet boundary was set to velocity inlet with an user defined file (UDF), and the outlet was set to pressure outlet. The other walls were set adiabatic. The inlet temperature and the reference temperature were set to 298.15 K. The simulated chip was set as copper and the coolant was the deionized water. The physical property parameters of the materials used in the simulation are listed in Table 9.6. As the viscosity of water varies greatly with the temperature, the viscosity was set to piecewise interpolation. The detailed calculation of viscosity can be found in the supplementary material. Except viscosity, the

(a)

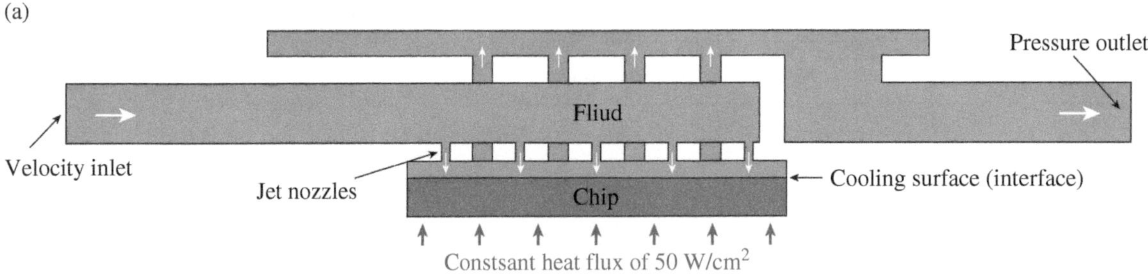

(b)

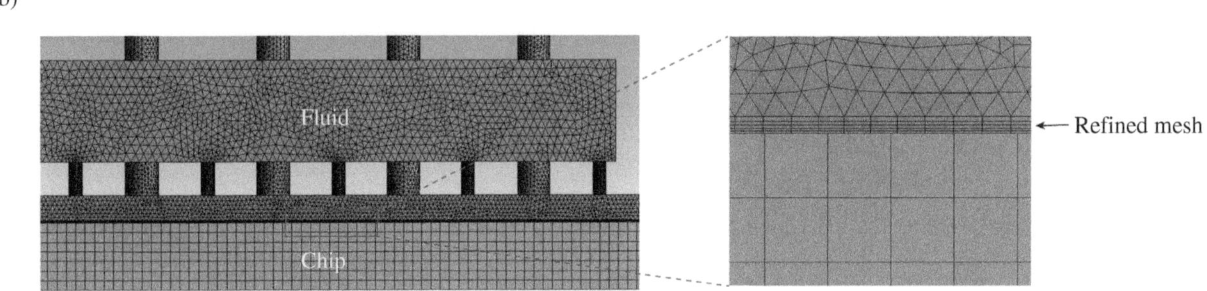

Figure 9.30 Setup of the numerical simulation: (a) the boundary conditions, (b) the mesh.

Table 9.6 The physical property parameters of materials used in the numerical simulation.

Material	κ (W · m^{-1} · K^{-1})	ρ (kg/m^3)	μ (mPa · s)	C_p (J · kg^{-1} · K^{-1})	T_{in} (°C)
Water	0.61	992.0	Temperature-varied	4179	25
Copper	380	8960	—	394	—

thermophysical properties of water are relatively stable with the change of temperature. Therefore, the other thermophysical properties were based on the inlet temperature. The time step of the transient simulation was set to $1/20\ t_{cycle}$, which is 1.11 ms when the driving frequency is 45 Hz. The number of time steps is set to 200 to simulate 10 cycles totally and the maximum iterations for each time step is 150 to insure convergence.

The mesh of the numerical simulation is shown in Figure 9.30(b). To meet the $y+$ requirements of the enhanced wall treatment, the grid height of the first layer was set to 0.01 mm, and the growth factor was set to 1.2. The grid independence was done by changing the size of the grids with steady simulations at a flow rate of 250 ml/min and a heat load of 200 W. The pressure drop and the average surface temperature rise of the JAICIPM were simulated with 2,193,030, 4,928,429, and 6,871,256 grids. As shown in Table 9.7, the discrepancies of pressure drop and average surface temperature rise are no more than 1%. Considering the calculation costs and accuracy, the grids of 4,928,429 were chosen in this study.

Table 9.7 Verification of grid independence.

Number of grids	Pressure drop (kPa)	Average wall temperature rise (K)
2,193,030	2.59	34.0
4,928,429	2.59	34.1
6,871,256	2.59	34.1

9.3.3 Experiment

In order to evaluate the thermal performance of the JAICIPM, a test facility was constructed, which is depicted in Figure 9.31. To simulate the high-power chip, four heating rods are inserted into the copper block (pure copper, 380 $W \cdot m^{-1} \cdot K^{-1}$), which can generate a heat load of 100–400 W. The area of the cooling surface is 4 cm^2 (20 mm × 20 mm), and the heat flux varies from 25 to 100 W/cm^2 correspondingly. A framework made by PEEK (0.29 $W \cdot m^{-1} \cdot K^{-1}$) is utilized to support the components and provide heat insulation. The JAICIPM is fixed onto the framework by screws. As for the driving of the JAICIPM, the power signal generator (Aigtek, ATG2031, China) generates a sinusoidal signal to drive the piezoelectric actuator and the power consumption is recorded by a power meter (HIOKI, PW3335, Japan). The voltage of the driving signal is set as 100–300 Vpp (35–106 VAC). Then, the coolant can be circulated with the JAICIPM and pipes, bringing the generated heat from the chip to the water tank. The water tank is equipped with a radiator which maintains the water inside at 25–30 °C. Three K-type thermocouples (T-K-36, ETA, China) are inserted into the copper clock to evaluate the thermal performance of the jet impingement cooling. And two thermocouples are used to measure the temperature of the coolant at the inlet and the outlet. All these temperatures are recorded by a data acquisition system (2700, Keithley, USA). The detailed dimensions of the heat source are shown in Figure 9.32. The distance between the thermocouple 1 and the upper surface is 2 mm. And the

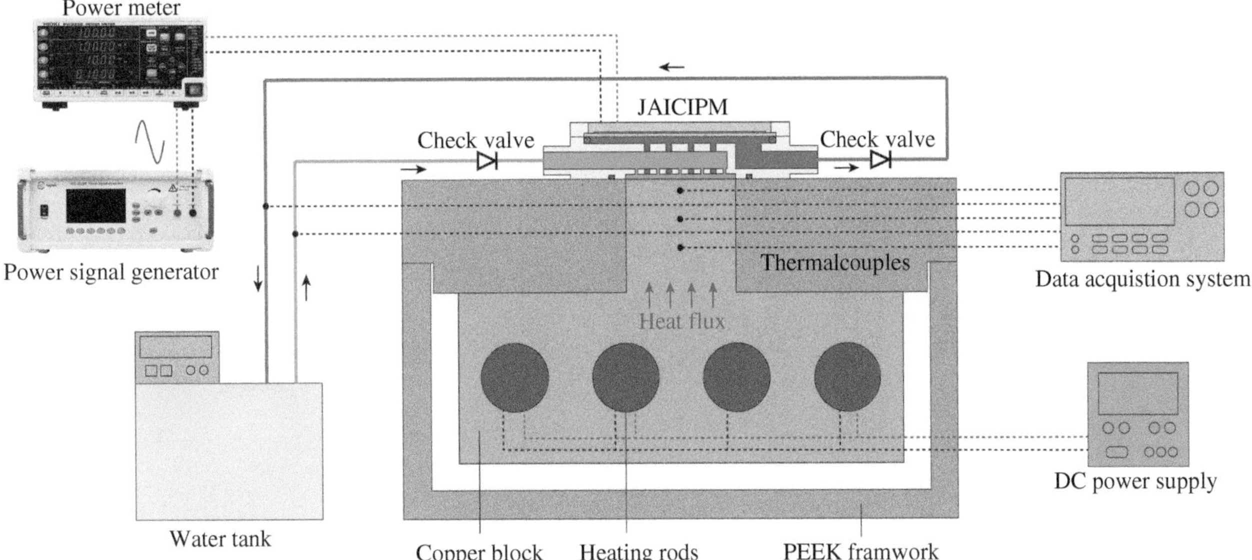

Figure 9.31 The schematic of the test facility.

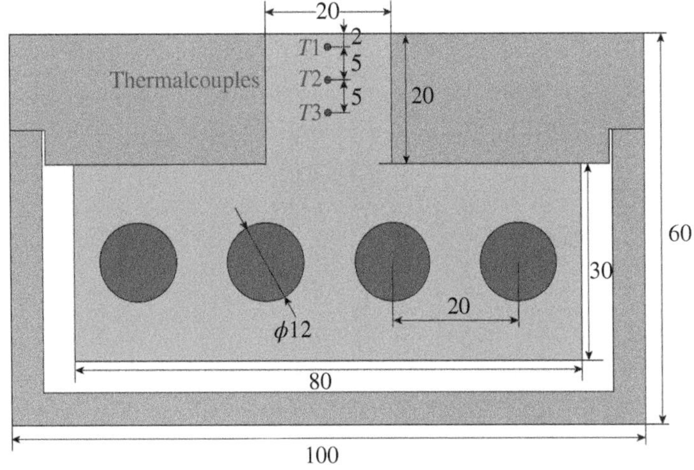

Figure 9.32 The dimensions of the heat source.

thermocouples were arranged with a spacing of 5 mm. Because of the thermal diffusion process, the temperature measurement didn't fluctuate with time in the experiments although the flow and surface temperature varied sinusoidally. Only average thermal performance could be obtained by the experiments.

According to the 1D Fourier's law of heat conduction, the effective heat flux of the cooling surface can be calculated with the temperature gradient of the copper block:

$$q = -\kappa \frac{dT}{dy} \tag{9.26}$$

where κ is the thermal conductivity of the copper block and y is the vertical distance from the cooling surface.

The temperature of the cooling surface T_s, the temperature rise ΔT, and the heat transfer coefficient h can be calculated as:

$$T_s = T_1 + y_1 \frac{dT}{dy} \tag{9.27}$$

$$\Delta T = T_s - T_{in} \tag{9.28}$$

$$h = \frac{q}{\Delta T} \tag{9.29}$$

where T_{in} is the inlet coolant temperature and y_1 is the distance between the surface and the thermocouple 1.

Then, the thermal resistance of the JAICIPM can be calculated as:

$$R = \frac{1}{hA} \tag{9.30}$$

As the coolant remains single phase in the experiment, the volumetric flow rate of the JAICIPM can be calculated by the conservation of energy:

$$Q_v = \frac{qA}{\rho C_p (T_{out} - T_{in})} \tag{9.31}$$

where C_p is the thermal capacity of water, T_{in} is the inlet temperature, and T_{out} is the outlet temperature.

As for the calculations of thermal performance with numerical results, T_s is the area average surface temperature, which is calculated as

$$T_s = \frac{\int T_{loc} dA}{A} \tag{9.32}$$

where T_{loc} is the local temperature of the surface.

The transient heat transfer coefficient is calculated as

$$h(t) = \frac{q}{(T_s(t) - T_{in})} \tag{9.33}$$

The impingement velocity is calculated as

$$V_j = \frac{Q_v}{N \times \frac{1}{4}\pi d_j^2} \tag{9.34}$$

The Reynolds and Nusselt numbers are defined as

$$\text{Re} = \frac{\rho_w V_j d_j}{\mu} \tag{9.35}$$

$$\text{Nu} = \frac{h d_j}{\kappa_w} \tag{9.36}$$

The absolute error of the K-type thermocouple is ±1.1 °C and the dimensional error of copper block is ±0.1 mm. According to the error-transfer methodology, the uncertainty of heat flux and average heat transfer coefficient can be calculated as

$$\frac{U_q}{q} = \sqrt{\left(\frac{U_{(T_3-T_1)}}{T_3 - T_1}\right)^2 + \left(\frac{U_{(y_3-y_1)}}{y_3 - y_1}\right)^2} \tag{9.37}$$

$$\frac{U_h}{h} = \sqrt{\left(\frac{U_q}{q}\right)^2 + \left(\frac{U_{(T_s - T_{in})}}{T_s - T_{in}}\right)^2} \tag{9.38}$$

The uncertainty of heat flux and average heat transfer coefficient were estimated to be ±6.7% and ±7.8% at the heat flux of 50 W/cm^2 and the flow rate of 250 ml/min.

9.3.4 Results and Discussion

As a pumping device, the flow rate and power consumption of the JAICIPM were experimentally tested at different driving voltages and frequencies. As shown in Figure 9.33(a), the flow rate increases with the increase of the driving voltage because of larger vibrating amplitude of piezoelectric vibrator, but the power consumption grows rapidly too. The flow rate achieves 138.6 ml/min at 100 Vpp with a power consumption of 23 mW. The maximum flow rate achieves 306 ml/min at 300 Vpp with a power consumption of 232 mW. The average jet velocity is proportional to the flow rate, ranging from 0.73 to 1.62 m/s in the experiments. As for the output pressure of the JAICIPM, it increases linearly with the driving voltage and achieves the maximum pressure of 32 kPa at 300 Vpp. As shown in Figure 9.33(b), the flow rate increases first and then decreases with the increase of driving frequency. The maximum flow rate was obtained at 45 Hz. The power consumption at different frequencies shares the same trend with the flow rate below 70 Hz but continues to rise at a higher frequency. It can be concluded that changing the driving voltage is more energy-efficient than changing the driving frequency when adjusting the flow rate of the JAICIPM.

The thermal performance of the JAICIPM was tested at different heat loads and driving voltages. The driving frequency is set to 45 Hz to get the best performance. The inlet temperature was maintained within 25–30 °C, and surface temperature rise was controlled below 45 °C in the experiments. As shown in Figure 9.34(a), the heat transfer coefficient increases with the increase of the driving voltage and heat load. When the driving voltage is 100 Vpp, the maximum heat load achieves 200 W, which means that 200 W heat load can be dissipated with a power consumption of 23 mW. And the heat transfer coefficient (HTC) rises from 8843 to 11,090 $W \cdot m^{-2} \cdot K^{-1}$ with the increase of heat load. When the driving voltage is 200 Vpp, the maximum heat load achieves 301 W, and the HTC rises to 17,241 $W \cdot m^{-2} \cdot K^{-1}$. The highest heat transfer coefficient of 20,572 $W \cdot m^{-2} \cdot K^{-1}$ was achieved at 300 Vpp and 366 W. As mentioned earlier, a higher flow rate can be obtained with a higher driving voltage, which increases the jet velocity. Meanwhile, the temperature of the water near the cooling surface continues to rise with the increase of heat load, reducing the viscosity of the water. And the higher jet velocity and smaller viscosity increase the local Reynolds number. Therefore, a thinner boundary layer can be obtained, resulting in a higher heat transfer coefficient. The calculated thermal resistance at different heat loads and driving voltages is shown in Figure 9.34(b). When the driving voltage is 100 Vpp, the thermal resistance varies from 0.225 to 0.283 K/W with the change of heat load. With the increase of driving voltage, the thermal resistance decreases obviously. When increasing the driving

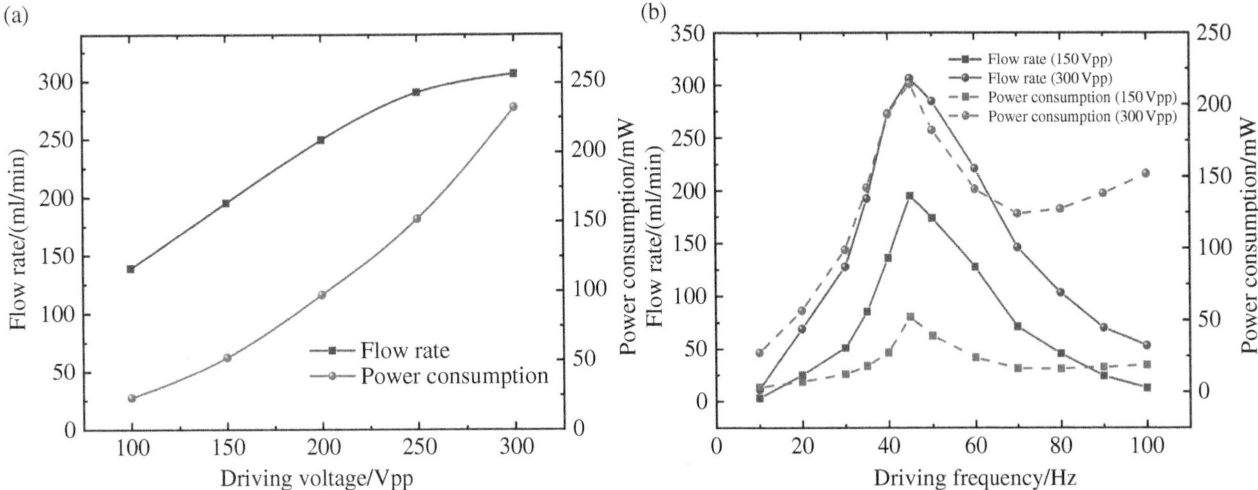

Figure 9.33 Flow rate and power consumption of the JAICIPM: (a) at different driving voltages, (b) at different driving frequencies.

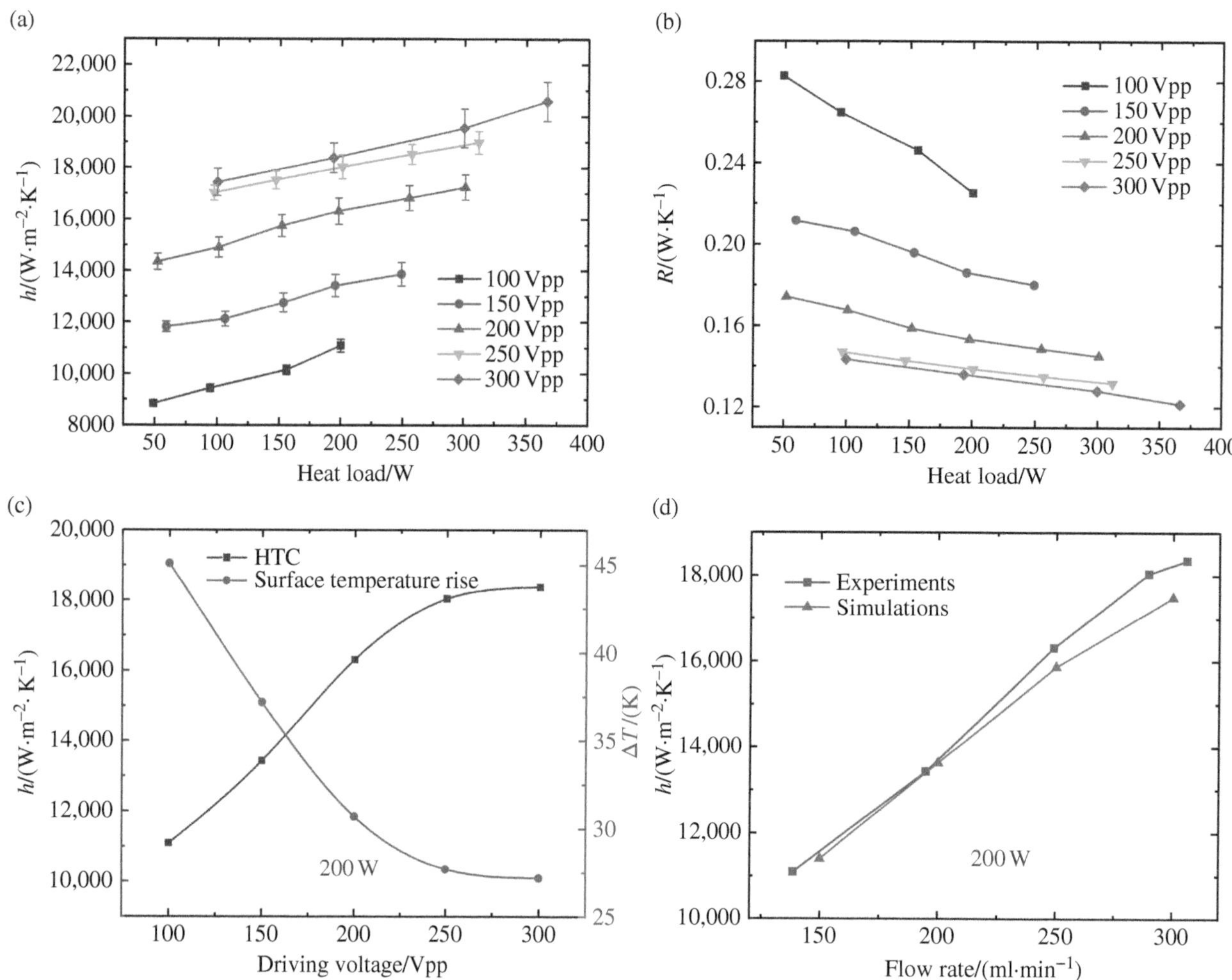

Figure 9.34 (a–d) Thermal performance of the JAICIPM.

voltage to 200 Vpp, the thermal resistance could drop to 0.145 K/W. The lowest thermal resistance of 0.122 K/W is obtained at 300 Vpp and 366 W.

To get a clear view of the influence of the driving voltage, the thermal performance of the JAICIPM at 200 W is shown in Figure 9.34(c). As the driving voltage increases, the HTC increases and the surface temperature decreases. The curve of the HTC is similar to the flow rate at different driving voltages, which increases rapidly first and then slows down. The surface temperature rise is 45 °C when the driving voltage is 100 Vpp. When the driving voltage increases to 300 Vpp, the wall temperature rise decreases to 27.2 °C.

Due to the unsteady jet impingement, the heat transfer coefficient and the surface temperature change through the cycles. Numerical simulations were performed to investigate the transient thermal performance. For the convenience of comparison, the heat load was set to 200 W and the flow rate was set to 150–300 ml/min. The transient flow rate change was achieved with an UDF by changing the inlet velocity. The results from simulations and experiments were compared with the average heat transfer coefficient. As shown in Figure 9.34(d), the simulations agree well with the experiments with a difference within 5%.

The Nusselt number Nu as a function of Reynolds number Re is shown in Figure 9.35. The Nusselt number and Reynolds number were calculated based on the jet diameter. In this study, the Reynolds number ranged from 335 to 1054, and the Nusselt number ranged from 5.8 to 12.7. Nonlinear fitting of these two parameters was performed, and the fitting result is

$$\mathrm{Nu} = 0.075\,\mathrm{Re}^{0.75} \tag{9.39}$$

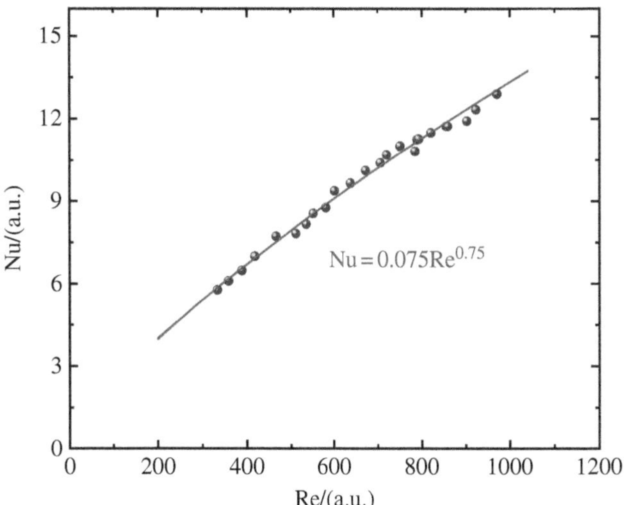

Figure 9.35 Nusselt number versus Reynolds number.

Table 9.8 Performance of the liquid cooling systems driven by piezoelectric micropumps.

Author	Type	Pump size (mm)	Q_{max} (ml/min)	P_{max} (kPa)	Heat load (W)
Yueh et al. [71]	Nonintegrated	30 × 15 × 3.8	7	50	8
Lu et al. [83]	Nonintegrated	270 × 55 × 28	251.1	60.2	12.8
Ma et al. [84]	Integrated	—	141	6.4	60
Tang et al. [76]	Integrated	50 × 45 × 12	210	—	80
This work	Integrated	40 × 40 × 10	306	32	366

As shown in Table 9.8, the performance of the JAICIPM is compared with the reported liquid cooling systems driven by piezoelectric micropumps. The nonintegrated system means that the piezoelectric micropump and heat sink are separated, while the integrated system means that the piezoelectric micropump and heat sink are integrated. Results show the JAICIPM has a much better heat dissipation ability than other studies. At the same time, the JAICIPM remains small and consumes little electricity.

To investigate the transient heat transfer process of the JAICIPM, the case of 250 ml/min was analyzed in detail. The numerically calculated transient heat transfer coefficient and surface temperature rise are shown in Figure 9.36. At 0–0.5 t_{cycle}, the coolant is ejected onto the cooling surface. When $t = 0$, the flow rate is zero and the surface temperature is high. With the increase of flow rate, the heat transfer coefficient increases to 27,969 $W \cdot m^{-2} \cdot K^{-1}$ at 0.3 t_{cycle}. At this period, the surface temperature increases slightly and then decreases significantly. The jet impinging is finished at 0.5 t_{cycle}, and the HTC continues to decrease to 8547 $W \cdot m^{-2} \cdot K^{-1}$ at 1 t_{cycle}. The surface temperature rise varies from 31.05 to 31.97 K in the cycle, and the time-averaged surface temperature rise is 31.52 K. Therefore, the average heat transfer coefficient is calculated as 15,863 $W \cdot m^{-2} \cdot K^{-1}$. The variation of surface temperature rise is within 1 K, which indicates the stable cooling performance of the JAICIPM.

The temperature field of the cooling surface at different times is shown in Figure 9.37. The surface temperature distribution keeps changing at different times. When $t = 0$, the surface temperature is high in all areas. When $t = 0.25\ t_{cycle}$, the temperature field shows a typical array jet cooling pattern and the surface temperature in the jet area is much lower. Because of the arrangement of the nozzles, few coolants flow through the corners. And the flow rate distribution of jet nozzles is a little different due to the hydraulic resistance of the channels, which affects the temperature distribution too. Therefore, the highest temperature appears at the corner near the outlet. As time goes by, the jet velocity reduces and the temperature rises in the jet area. The temperature difference is within 6 K through the whole cycle. The cooling performance is uniform due to the jet array design.

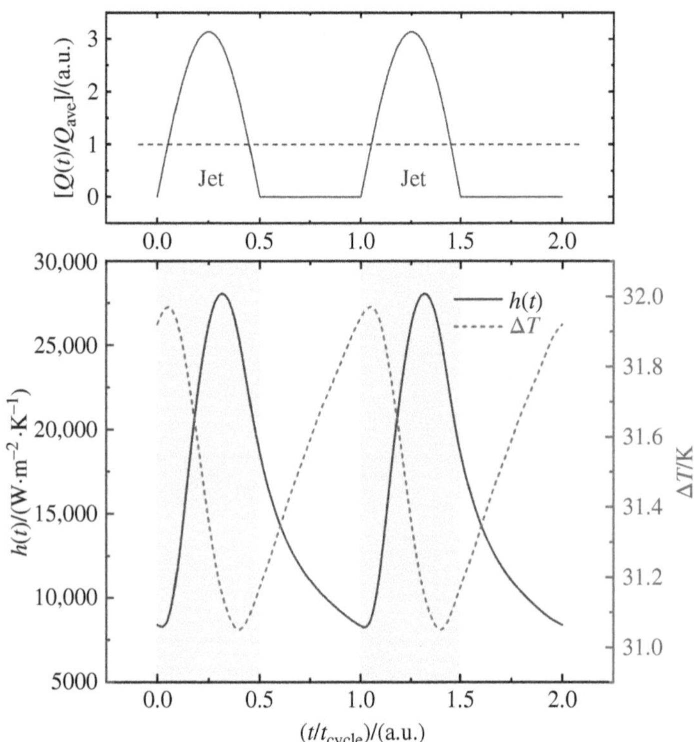

Figure 9.36 Transient HTC and surface temperature rise.

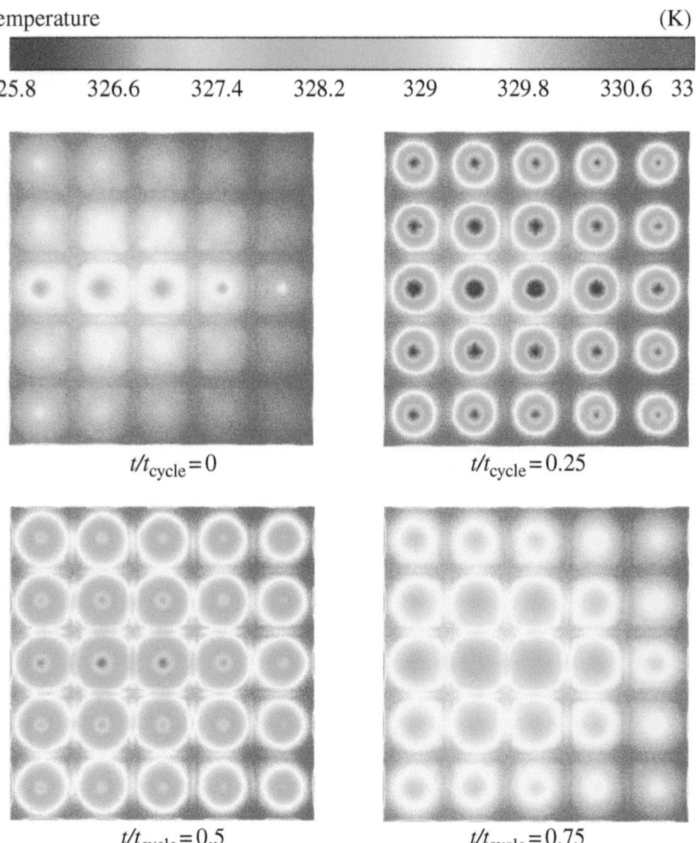

Figure 9.37 Temperature distribution of the surface at different times.

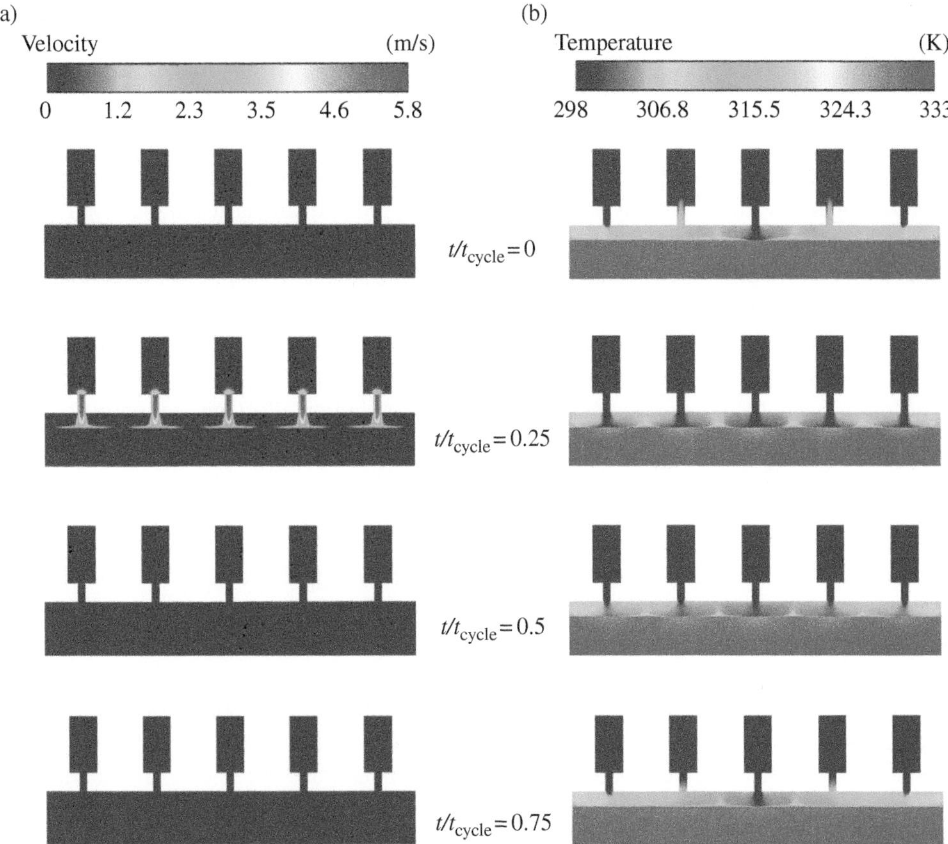

Figure 9.38 Fields of the JAICIPM at different times: (a) velocity field, (b) temperature field.

To explain the process further, the velocity field and the temperature field were also studied. Figure 9.38 shows the velocity field and the temperature field of the JAICIPM at different times. When $t = 0$, the jet velocity remains zero, and the temperature of the chip and the water nearby is high. When $t = 0.25\ t_{cycle}$, high-speed water is ejected onto the surface and then spreads, cooling down the surface and bringing in the low-temperature water. The average jet velocity achieves the maximum of 4.2 m/s and a thin thermal boundary is formed. Although the jet ends at 0.5 t_{cycle}, the temperature of the water near the surface remains relatively low. At 0.5–1 t_{cycle}, the heat of the chip is mainly dissipated by the heat conduction between the chip and the water nearby. It can be seen that the temperature of the water nearby continues to rise from 0.5 t_{cycle}.

9.4 Microchannel Cooling for Uniform Chip Temperature Control

Semiconductor chips are recognized as the heart of advanced technologies such as communications, LEDs, high-performance computers, and mobile devices [1, 2, 6]. Driven by Moore's law, the number of internal transistors of the latest core series has reached 580 million [85], resulting in high-heat-flux density and high chip temperature. Starting with the seminal work of Tuckerman and Pease [86], single-phase microchannel liquid cooling, which has heat dissipation capability of heat flux up to 790 W/cm^2, has been widely investigated and applied to cool the chips [87, 88]. In the past decades, most researches on microchannel liquid cooling focused on maintaining the maximum junction temperature under a critical value, regardless of the temperature uniformity [89–91]. However, due to the different thermal expansion coefficient (CTE), large temperature gradient increases the thermal stress in the chip and even causes interfacial delamination, resulting in performance degradation and low reliability [92]. On the other hand, single-phase microchannel liquid cooling usually has uniform heat dissipation capability, which means that when hotspots are kept under critical temperature, the other places (backgrounds) along the straight channels may be subcooled, resulting as a waste of energy [6]. For example, it is investigated that roughly 33% of the total electricity consumed is allocated to thermal management systems of server electronics [93]. Temperature gradient is caused not only by nonuniform heat source, but also by temperature rise of the flowing fluid in microchannel, which is the inherent disadvantage of single-phase liquid cooling.

Recently, extensive researches have been done to solve the temperature uniformity problem including two-phase flow boiling and single-phase microchannel liquid cooling with various local channel densities. Although two-phase flow boiling presents much larger heat dissipation capability, the instability and critical heat flux problem are hard to handle and still far from industrial application [94–96]. Luo et al. designed a tree-like microchannel to deal with temperature uniformity of multiple heat sources [92] and the temperature difference can be controlled within 1.3 °C. Lorenzini et al. used microgaps with variable pin fin clustering in the microchannel and had a great cooling effect on the hotspot with high heat flux [97]. Sharma et al. used the manifold microchannel with alternately fine and coarse channels to cool the electronics with multiple heat sources [46]. Although locally finer channels can provide larger local heat transfer capability owing to the larger heat transfer area, the flow rate may decrease due to the larger flow resistance and maldistribution effect, on the contrary, resulting in higher surface temperature. So, the structure of the microchannel, such as how wide the local channel is, has a significant influence on the surface temperature that cannot be simply described by experience. Therefore, an accurate model and optimal method are urgently required for the microchannel structural design.

A 1D thermal resistance model has been developed to predict the microchannel junction temperature. Most researches only considered the average surface temperature [98, 99], regardless of the temperature gradient. Mao et al. developed a compact thermal model for microchannel substrate with high-temperature uniformity subjected to multiple heat sources [100]. Sharma et al. proposed a hotspot-targeted semi-empirical design method, considering the temperature rise caused by heated fluid along the flowing direction [101]. However, the 3D thermal conduction and heat spreading effect are not taken into account in the 1D model, thus temperature uniformity in the 1D model is not satisfied and far from the industry requirement. Abdoli et al. [102] developed a conjugate heat transfer model to study the multifloor microchannel based on the 3D steady thermal conduction and the quasi-1D thermofluid model. However, they only studied the microchannel with uniform heat source and uniform distribution of the channel width. A bilayer compact microchannel thermal model has proved effective in achieving uniform chip temperature, with nonuniform heat sources that considered the 3D thermal conduction from the heat source to microchannel and the 1D thermal resistance model of the microchannel with fluid temperature rise.

9.4.1 Bilayer Compact Thermal Model

Figure 9.39(a) shows the structure of a typical plate-fin microchannel with nonuniform heat source. Cold fluid enters the microchannel and flows along the channels. Heat generated by the nonuniform heat source is conducted and diffused in the solid layer and then absorbed by fluid. The heated fluid flows out of the microchannel and carries the heat away. Based on the heat transfer process, the compact thermal model is composed of two parts. One is the 3D thermal conduction model that describes the heat conduction in the solid layer [103]. The other is the 1D thermal resistance model considering fluid temperature rise [101] that describes the heat transfer to fluid in the so-called convection layer.

9.4.2 Heat Transfer in the Solid Layer

Heat conduction process in the solid layer is described by solving the 3D thermal conduction partial differential equation. Figure 9.39(b) shows some of the required parameters of the model and the heat transfer process in the solid layer. The length and width of the microchannel are a and b, respectively. The thickness of the solid layer is t. The top surface is the chip heat source with nonuniform heat flux, and is described as $P_{\text{thermal,sur}}(x, y)$. The bottom surface is cooled by convection layer with an effective heat transfer coefficient of $h(x, y)$. The thermal conductivity of the microchannel material is κ. The governing energy equation for the temperature field $T(x, y, z)$ is given by

$$\frac{\partial^2 T}{\partial x^2} + \frac{\partial^2 T}{\partial y^2} + \frac{\partial^2 T}{\partial z^2} = 0 \tag{9.40}$$

with the boundary conditions described as

$$\left.\frac{\partial T}{\partial x}\right|_{x=0,a} = 0 \tag{9.41}$$

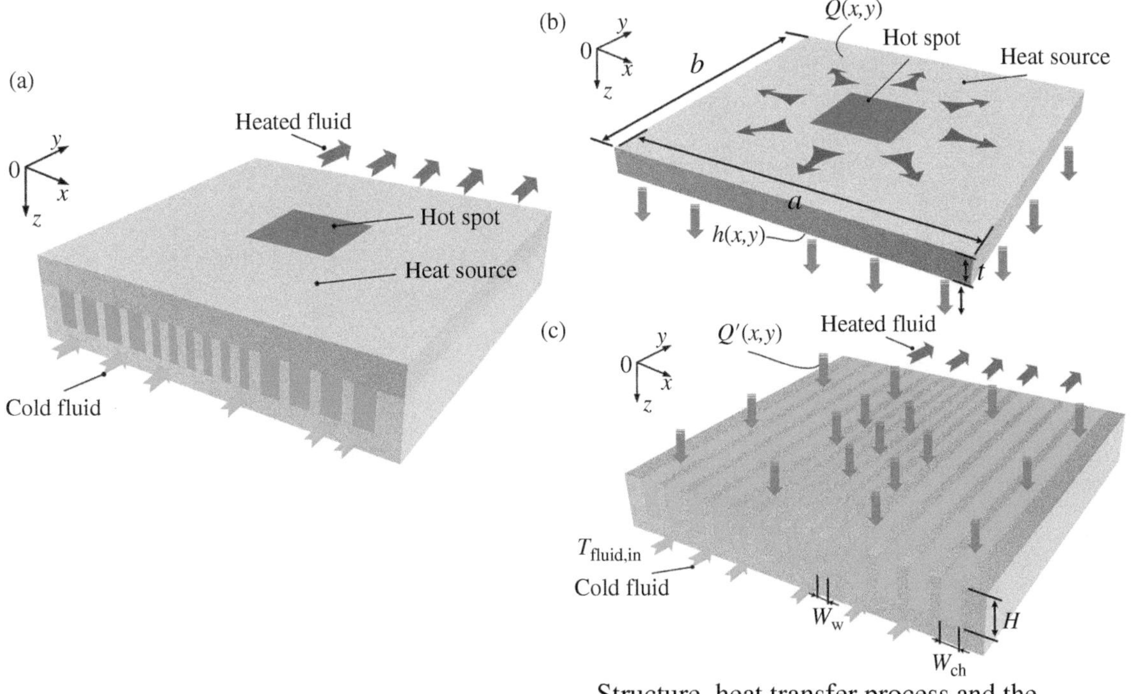

Figure 9.39 Schematic diagram of (a) a typical plate-fin microchannel with nonuniform heat source; (b) structure and the heat transfer process of the solid layer; (c) structure, heat transfer process, and the fluid flow of the convection layer.

$$\left.\frac{\partial T}{\partial y}\right|_{y=0,b} = 0 \tag{9.42}$$

$$\kappa \left.\frac{\partial T}{\partial z}\right|_{z=0} + Q(x,y) = 0 \tag{9.43}$$

$$\kappa \left.\frac{\partial T}{\partial z}\right|_{z=t} + h(x,y)(T(x,y,t) - T_{\text{in}}) = 0 \tag{9.44}$$

where T_{in} is the inlet fluid temperature as shown in Figure 9.39(c). The effective heat transfer coefficient $h(x, y)$ describes the heat transfer from the bottom surface of the solid layer to the heat sink with temperature T_{in}, which is the cold source of the microchannel. To solve these equations, temperature field can be written as the two-variable Fourier series and derived as the following equation:

$$T(x,y,z) = \left(A_{00} - \frac{U_{00}}{\kappa_z} z\right) + \sum_{n=0}^{N} \sum_{m=0}^{M} \begin{pmatrix} A_{nm}(\exp(\lambda_{nm} z) + \exp(-\lambda_{nm} z)) \\ + \dfrac{U_{nm}}{\lambda_{nm}\kappa} \exp(-\lambda_{nm} z) \end{pmatrix} \cos\left(\frac{n\pi x}{a}\right) \cos\left(\frac{m\pi x}{b}\right) \tag{9.45}$$

where

$$U_{00} = \frac{1}{ab} \int_0^b \int_0^a Q(x,y) \mathrm{d}x\mathrm{d}y \tag{9.46}$$

$$U_{nm} = \frac{\varepsilon_{nm}}{ab}\int_0^b\int_0^a Q(x,y)\cos\left(\frac{n\pi x}{a}\right)\cos\left(\frac{m\pi y}{b}\right)\mathrm{d}x\mathrm{d}y \tag{9.47}$$

$$\varepsilon_{nm} = \begin{cases} 2 & n = 0 \text{ or } m = 0 \\ 4 & n \neq 0, m \neq 0 \end{cases} \tag{9.48}$$

$$\lambda_{nm} = \sqrt{\left(\frac{n\pi}{a}\right)^2 + \left(\frac{m\pi}{b}\right)^2} \tag{9.49}$$

and the parameters A_{00} and A_{nm} satisfy the following group of equations:

$$\begin{aligned} &A_{00}\int_0^b\int_0^a h(x,y)\mathrm{d}x\mathrm{d}y \\ &+ \sum_{n=0}^{N}\sum_{m=0}^{M} A_{nm}(\exp(\lambda_{nm}t) + \exp(-\lambda_{nm}t))\int_0^b\int_0^a h(x,y)\cos\left(\frac{n\pi x}{a}\right)\cos\left(\frac{m\pi y}{b}\right)\mathrm{d}x\mathrm{d}y \\ &= U_{00}ab + U_{00}\frac{t}{\kappa}\int_0^b\int_0^a h(x,y)\mathrm{d}x\mathrm{d}y \\ &- \sum_{n=0}^{N}\sum_{m=0}^{M}\frac{U_{nm}}{\kappa\lambda_{nm}}\exp(-\lambda_{nm}t)\int_0^b\int_0^a h(x,y)\cos\left(\frac{n\pi x}{a}\right)\cos\left(\frac{m\pi y}{b}\right)\mathrm{d}x\mathrm{d}y \end{aligned} \tag{9.50}$$

$$A_{00}d_{ij} + A_{ij}e_{ij} + \sum_{n=0}^{N}\sum_{m=0}^{M} A_{nm}f_{ijnm} = U_{00}g_{ij} + U_{ij}l_{ij} + \sum_{n=0}^{N}\sum_{m=0}^{M} U_{nm}S_{ijnm} \tag{9.51}$$

where

$$d_{ij} = \int_0^a\int_0^b h(x,y)\cos\left(\frac{i\pi x}{a}\right)\cos\left(\frac{j\pi y}{b}\right)\mathrm{d}x\mathrm{d}y \tag{9.52}$$

$$e_{ij} = \kappa\lambda_{ij}\left(\exp\left(\lambda_{ij}t\right) - \exp\left(-\lambda_{ij}t\right)\right)N_{x,i}N_{y,j} \tag{9.53}$$

$$\begin{aligned} f_{ijnm} = &(\exp(\lambda_{nm}t) + \exp(-\lambda_{nm}t))\int_0^a\int_0^b h(x,y) \\ &\times \cos\left(\frac{n\pi x}{a}\right)\cos\left(\frac{m\pi x}{b}\right)\cos\left(\frac{i\pi x}{a}\right)\cos\left(\frac{j\pi x}{a}\right)\mathrm{d}x\mathrm{d}y \end{aligned} \tag{9.54}$$

$$g_{ij} = \frac{t}{\kappa}f_{ij} \tag{9.55}$$

$$l_{ij} = \exp(-\lambda_{nm}t)N_{x,i}N_{y,j} \tag{9.56}$$

$$S_{ijnm} = \frac{-\exp(-\lambda_{nm}t)}{\kappa\lambda_{nm}\left(\exp(\lambda_{nm}t) + \exp(-\lambda_{nm}t)\right)}f_{ijnm} \tag{9.57}$$

$$N_{x,i} = \begin{cases} a & i = 0 \\ a/2 & i \neq 0 \end{cases}, \quad N_{y,j} = \begin{cases} b & j = 0 \\ b/2 & j \neq 0 \end{cases} \tag{9.58}$$

where $i = 0,1,2,\ldots,N$ and $j = 0,1,2,\ldots,M$, while i and j cannot be zero at the same time. The upper limits of Fourier series are set to be $n = N$ and $m = M$ for computation. It should be noted that, in all equations in this section, both n and m cannot be zero at the same time. According to Equations (9.50) and (9.51), there is a set of $(N+1)\times(M+1)$ linear equations in $(N+1)\times(M+1)$ variables, with unknown parameters A_{00},A_{01},A_{02},...,A_{NM}. This set of equations can be solved by matrix inversion using the Matlab. Then the temperature field may be obtained. For details about the derivation of the earlier equations, refer to these works [63–65].

The heat transfer coefficient of the bottom surface is related to the total thermal resistance $R_{total}(x,y)$ of the convection layer according to energy conservation, satisfying

$$h(x,y)(T(x,y,t) - T_{in}) = \frac{T(x,y,t) - T_{in}}{R_{total}(x,y)} \tag{9.59}$$

9.4.3 Heat Transfer in the Convection Layer

Figure 9.39(c) shows the structure of the convection layer, in which $P_{thermal}'(x,y)$ is the input heat flux from the solid layer. wid_{ch}, wid_w, and H_{ch} are the channel width, channel wall width, and channel height, respectively. In general, the heat flux, channel width, and channel wall may all be nonuniformly distributed. Therefore, the computational domain must be discretized (Figure 9.40(a)) such that each cell of the discretized domain encompasses an area of nearly constant quantity $F(i,j)$. More specifically, $F(i,j)$ may represent $P_{thermal}'(i,j)$, $R_{total}(i,j)$, $wid_{ch}(i,j)$, $wid_w(i,j)$, etc. It should be noted that it is assumed that the heat transfer is 1D from the bottom surface of the solid layer to the fluid in the relevant channel of the same discrete cell. And the heat transfer between the channels in the y-direction is ignored.

Figure 9.40(a) shows the flow direction, and the average coolant temperature in the (i,j)th cell is calculated according to energy conservation and is given by

$$T_{fluid}(i,j) = T_{fluid,in} + \sum_{k=1}^{i-1} \frac{P_{thermal}'(k,j)\Delta x_k \Delta y_j}{\rho Q_{N_x}(j) C_p} + \frac{P_{thermal}'(i,j)\Delta x_i \Delta y_j}{2\rho Q_{N_y}(j) C_p} \tag{9.60}$$

where $Q_{N_y}(j)$ represents the volume flow rate in the jth line and c_p is the fluid heat capacity. The number of discrete mesh in x-direction and y-direction is N_x and N_y respectively. So the total input volume flow rate should be $Q_{N_y,tot} = \sum_{j=1}^{V} Q_{N_y}(j)$. Figure 9.40(b) shows that, in analogy with the Ohm's law in circuitry, volume flow rate of each line satisfies the following equation:

$$Q_{N_y}(j) = \frac{\Delta P}{\sum_{k=1}^{S} R_f(i,j)} \tag{9.61}$$

where ΔP represents the total pressure loss of the microchannel. $R_f(i,j)$ represents the flow resistance in the (i,j)th cell and is described as follows:

$$R_f(i,j) = \frac{2\mu_f (f\mathrm{Re})_{(i,j)} (wid_{ch}(i,j) + wid_w(i,j))\Delta x_i}{D_{ch}^2(i,j) wid_{ch}(i,j) H_{ch} \Delta y_j} \tag{9.62}$$

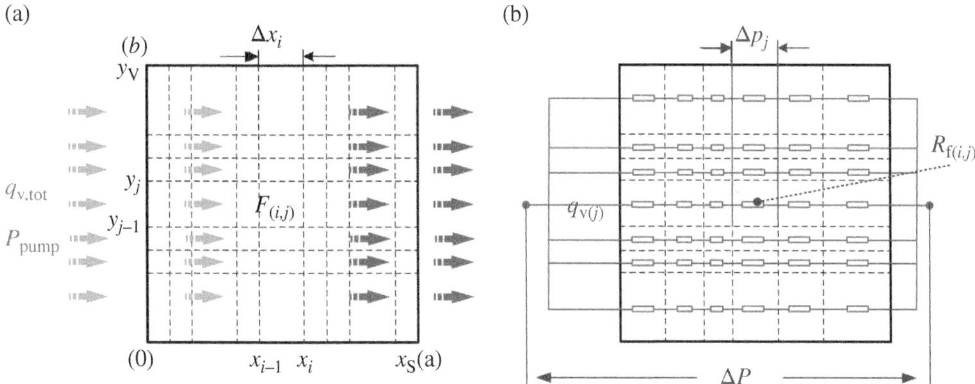

Figure 9.40 Schematic diagram of (a) the discretization of the computational domain, (b) the flow resistance network of the discrete convection layer.

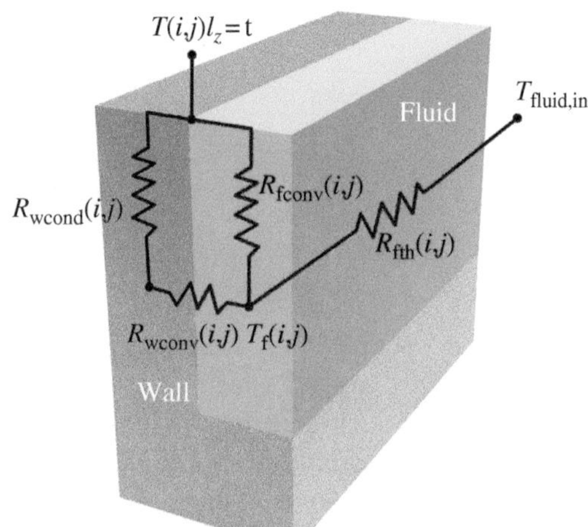

Figure 9.41 Thermal resistance model of the convection layer.

where $D_{\text{ch}}(i,j) = 2\text{wid}_{\text{ch}}(i,j)H_{\text{ch}}/(\text{wid}_{\text{ch}} + H_{\text{ch}})$ represents the local equivalent hydraulic diameter. In the calculation, at least one input quantity (Q_{N_y},tot, ΔP, and input pumping power P_{pump}) should be known. In this work, we care more about the energy consumption, so the pumping power was set as the known constant input parameter and it satisfies the relationship $P_{\text{pump}} = \Delta P \cdot Q_{N_y}$,tot. It can be seen from Equations (9.60) and (9.61) that the fluid temperature and the flow resistance are only dependent on the microchannel structure and the input heat flux.

Figure 9.41 shows the thermal resistance model of the convection layer. According to the thermal resistant network, the thermal resistances can be calculated by

$$R_{\text{tot}}(i,j) = R_{\text{fth}}(i,j) + \left(\frac{1}{R_{\text{fconv}}(i,j)} + \frac{1}{R_{\text{wcond}}(i,j) + R_{\text{wconv}}(i,j)}\right)^{-1} \tag{9.63}$$

where R_{fth}, R_{fconv}, R_{wcond}, and R_{wconv} represent the effective fluid thermal resistance, fluid convection thermal resistance, conduction thermal resistance to the wall, and wall conduction thermal resistance, respectively. The resistances are given by

$$R_{\text{fth}}(i,j) = \frac{T_{\text{fluid}}(i,j) - T_{\text{in}}}{P_{\text{thermal}}(i,j)} \tag{9.64}$$

$$R_{\text{fconv}}(i,j) = \frac{\text{wid}_{\text{ch}}(i,j) + \text{wid}_{\text{w}}(i,j)}{h_{\text{f}}(i,j)\text{wid}_{\text{ch}}(i,j)} \tag{9.65}$$

$$R_{\text{wcond}}(i,j) + R_{\text{wconv}}(i,j) = \frac{\text{wid}_{\text{ch}}(i,j) + \text{wid}_{\text{w}}(i,j)}{2h_{\text{f}}(i,j)H_{\text{ch}}\gamma(i,j)} \tag{9.66}$$

where γ is the fin efficiency [101]. h_{f} is the convection heat transfer coefficient of the flowing fluid, which is given by $h_{\text{f}}(i,j) = \text{Nu}_{\text{D}}(i,j)\kappa_{\text{f}}/D_{\text{ch}}(i,j)$. κ_{f} is the thermal conductivity of the fluid. Flow in the microchannel is usually laminar [66] because of the small hydraulic diameter. It is assumed to be fully developed and the relations between Nu and fRe in Equation (9.62) are adopted from Shah and London [104].

9.4.4 Heat Flux Iteration

Resulting from the heat spreading in the solid layer, heat flux distribution $Q'(x, y)$ is different from $Q(x, y)$ and is unknown. Solving the 3D heat conduction partial differential equation requires the effective heat transfer coefficient h of the bottom surface to be known. However, h cannot be obtained unless $Q'(x, y)$ is known, according to Section 9.4.1. Therefore, solving $Q'(x, y)$ is the key for the bilayer compact thermal model. Combining the 3D heat conduction model in the solid layer with the 1D thermal resistance model, the bottom surface temperature field of the solid layer can be solved by $T(x, y, t) = f(Q'(x, y))$ and h is obtained by $h = g(Q'(x, y))$. In addition, at the interface of the two layers, the heat flux of the bottom surface of the solid layer and $Q'(x, y)$ must be equal. Thus, we have

$$\begin{aligned} Q'(x,y) &= h[T(x,y,t) - T_{\text{in}}] \\ &= g(Q'(x,y))[f(Q'(x,y)) - T_{\text{in}}] \end{aligned} \tag{9.67}$$

Equation (9.67) shows that $Q'(x, y)$ is the fixed point matrix and we can solve it by iteration. Figure 9.42 shows the flow chart of the iteration process. The initiating value should be $Q(x, y)$. ε represents the error and we set it to be 0.01 for computation.

9.4.5 Genetic Algorithm Optimization

When the pumping power is set to be constant, we believe that the optimum microchannel structure is to make the junction surface temperature uniform. By regulating the local channel width and wall width, the local flow rate and heat transfer

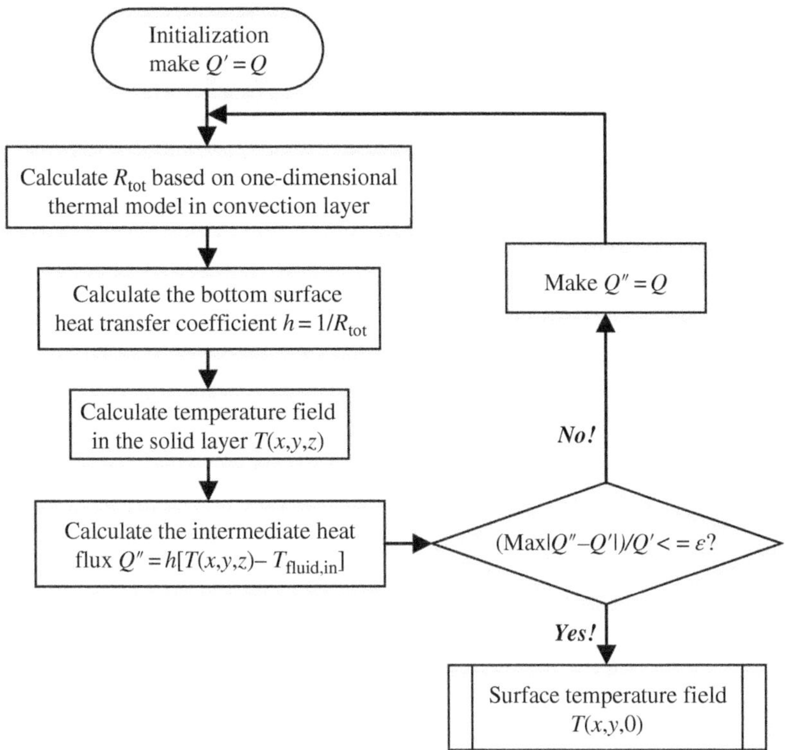

Figure 9.42 Flow chart of the heat flux iteration process of the bilayer compact thermal model.

coefficient can be changed, and eventually the best temperature uniformity is reached. The genetic algorithm (GA) is invoked to do the optimization. It is a semi-stochastic global search method, based on an analogy with Darwin's laws of natural selection. The GA generally consists of the population initialization, selection, crossover, and mutation processes [105]. The selection rule in this work is set as the standard deviation of the temperature field which is given by

$$\begin{aligned}&\min\left(\sqrt{\frac{\sum(T-\overline{T})^2}{N_{Tx}N_{Ty}-1}}\right)\\&=\min f(w_{ch}(1,1),w_{ch}(1,2),\ldots,w_{ch}(S,V),w_w(1,1),w_w(1,2),\ldots,w_w(S,V))\end{aligned} \tag{9.68}$$

where $\overline{T}$ is the average junction surface temperature. N_{Tx} and N_{Ty} are the *x*- and *y*-direction discrete numbers of the surface temperature gradient, respectively. The GA in this work is aimed at finding the chrome that has the minimum standard deviation. And confinements of variables are as follows:

$$w_{ch,min} \le w_{ch}(i,j) \le w_{ch,max} \qquad i = 1,2,\ldots,S \; j = 1,2,\ldots,V \tag{9.69}$$

$$w_{w,min} \le w_w(i,j) \le w_{w,max} \qquad i = 1,2,\ldots,S \; j = 1,2,\ldots,V \tag{9.70}$$

9.4.6 Validation

The bilayer compact thermal model was coded using the Matlab. Since the fabrication of the microchannel on the substrate in chip scale is quite difficult and expensive, the proposed thermal model was verified by COMSOL simulation. As the flow and heat transfer in the plate-fin microchannel are not that complex, the coupled CFD and Heat Transfer package in COMSOL MULTIPHYSICS is accurate enough for the simulation [106]. In the experiment, the microchannel was made by copper with uniform channel width. The channel width and wall width are both 300 μm. Ceramic heater was used to simulate the chip heat source. The size of the cooling surface of the microchannel was set to be equal to the size of the heater, which is

20 mm × 20 mm. The thickness of the heater is 2 mm. The ceramic heater was attached to the microchannel by the TIM (Dow Corning TC-5121). The whole experimental setup was wrapped by thermal insulation cotton to avoid heat loss. As shown in Figure 9.43, the temperature at the center of the ceramic heater was measured by a K-type thermocouple. The coolant was water and the inlet fluid temperature was set to be 24.7 °C in the simulation, which is the measured temperature in the experiment. The thermal conductivities of the ceramic heater and the microchannel were set to be 3 and 380 $W \cdot m^{-1} \cdot K^{-1}$ in the COMSOL simulation, respectively. The experiments and simulations about the cooling situation under the input heat power of 50 and 80 W with various coolant flow rates were done. Figure 9.43 shows that the simulation results of the temperature at the center of the ceramic heater are slightly smaller than the experiments since the thermal contact resistance was not considered in the simulation. Regardless of this, the simulation results are in good agreement with experiments. Therefore, the COMSOL simulation method was validated and it was able to be used to verify the thermal model.

The calculated chip microchannel structure parameters and material physical properties in this work are listed in Table 9.9. The material of the chip microchannel is silicon, and water is selected as the cooling fluid. Considering the fabrication complexity, the channel wall width is set to be constant and uniform.

In order to verify the bilayer compact thermal model, three cases with various channel distribution and heat source distribution are compared. Case A represents the simplest condition that the heat source and channel distribution are both uniform. Case B represents microchannel with uniform channel distribution and nonuniform heat source. Case C represents microchannel with nonuniform channel distribution and heat source. The specific parameters of cases A, B, and C are listed in Table 9.10. For simplicity, the nonuniform heat source in this study is composed of a central hotspot with higher heat flux and a background with lower heat flux. $a_{hs} \times b_{hs}$ in Table 9.10 represents the dimension of the hotspot. However, it should be noted that the bilayer compact thermal model is also suitable for the heat source with multiple hotspots, because heat flux in this model is described by a function $Q(x, y)$.

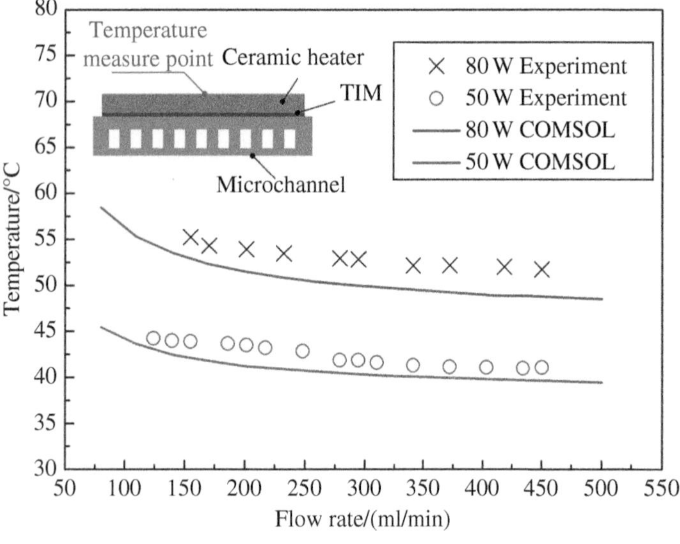

Figure 9.43 Validation of the numerical scheme by experiments.

Table 9.9 Microchannel structure and physical property parameters.

	$a \times b$ (mm) 12 × 16	t (μm) 225		wid$_w$ (μm) 30	H_{ch} (μm) 300
Material	**κ ($W \cdot m^{-1} \cdot K^{-1}$)**	**ρ (kg/m^3)**	**μ (Pa · s)**	**C_p ($J \cdot kg^{-1} \cdot K^{-1}$)**	**$T_{fluid,in}$ (°C)**
Water	0.599	998.2	0.001	4180	20
Silicon	150	—	—	—	—

Table 9.10 Heat source and channel width distribution for cases A, B, and C.

	Q (HS) (W/cm^2)	Q (BG) (W/cm^2)	$a_{hs} \times b_{hs}$ (mm)	wid$_{ch}$ (HS) (μm)	wid$_{ch}$ (BG) (μm)
Case A	100	100	—	80	80
Case B	200	66.6667	6 × 8	80	80
Case C	200	66.6667	6 × 8	50	80

Table 9.11 Number of elements, calculated maximum temperature, and pressure difference for the grid independence test.

	Grids	Coarser	Coarse	Normal	Fine	Finer	Extra fine
Case B	$N_{element}$	979,801	1,180,054	1,394,957	1,993,502	2,704,988	4,136,868
	T_{max} (°C)	71.31	71.28	71.2	71.52	71.06	71.42
	ΔP (Pa)	10,972.41	10,972.41	10,978.24	10,964.55	11,028.65	10,930.65
Case C	$N_{element}$	1,587,939	1,963,106	2,664,902	3,119,260	4,719,531	9,237,535
	T_{max} (°C)	95.03	95.13	94.28	94.28	94.15	94.14
	ΔP (Pa)	12,579.93	12,579.93	12,498.2	12,493.75	12,485.31	12,463.45

The grid independence test was done based on the representative cases in this work, which are cases B and C. Six different grids were used to do the numerical simulation for cases B and C using the default mesh parameters in the COMSOL, from "coarser" to "extra fine." The number of elements for this grid independence test is listed in Table 9.11. The number of elements of case C is relatively larger than that of case B under the same type of default mesh parameters, because the microchannel width is much thinner at the hotspot region in case C. Both the calculated maximum temperature of the heat source surface and the pressure difference between the inlet and outlet do not change much from the "coarser" grid to the "extra fine" grid, indicating that the numerical results are independent of the number of grid elements. The difference between the maximum temperatures and the difference between the pressure differences are within 1.05% and 0.93%, respectively. With the comprehensive consideration of calculating cost and quality, we chose the default mesh parameter of "coarse" to do the rest of the numerical simulations in this work. The average elements' qualities for cases B and C with the default "coarse" grid are 0.38 and 0.41, respectively.

Figure 9.44 shows the surface temperature comparison between results from bilayer model, COMSOL simulation, and 1D model [101]. Results show that, for case A, the 1D model and the bilayer compact thermal model can both predict the surface temperature accurately. However, for cases B and C, the bilayer compact thermal model can still accurately predict the temperature contour, but the 1D model shows large discrepancy from the COMSOL simulation. This is because the 1D model is unable to describe the hotspot heat spreading from high temperature to low temperature in the solid layer. The Matlab code for the bilayer compact thermal model only costs a few seconds, rather than several hours required for the COMSOL simulation.

The assignment of heat sources in cases A, B. and C makes the average heat flux to be equal, that is, 100 W/cm^2, and the total thermal power to be 192 W. And the input cooling energy, namely, the pumping power, is set to be constant as 0.015 W. Figure 9.45 shows the calculated surface temperature for cases A, B, and C, in which Line D represents the line $y = 8$ mm on the top surface and Line E represents the line $x = 9$ mm on the top surface. We can see that along the flow direction (x-direction), calculated results of bilayer model and 1D model both fit well with that of the COMSOL simulation. However, the calculated results of 1D model along y-direction are not accurate. The maximum temperature discrepancy in Figure 9.45 between the bilayer thermal compact model and the COMSOL simulation is only 4.8%. The surface temperature rises along the flow direction because of the heat absorption of fluids. This is one of the main reasons that cause the nonuniform temperature. Results in Figure 9.45 show that, even when the average heat flux is kept the same, surface temperatures of cases B and C are much larger than that of case A, indicating that the existence of hotspots with locally high heat flux will cause higher temperature. In addition, surface temperature of case C with locally fine channel is much higher than that of cases

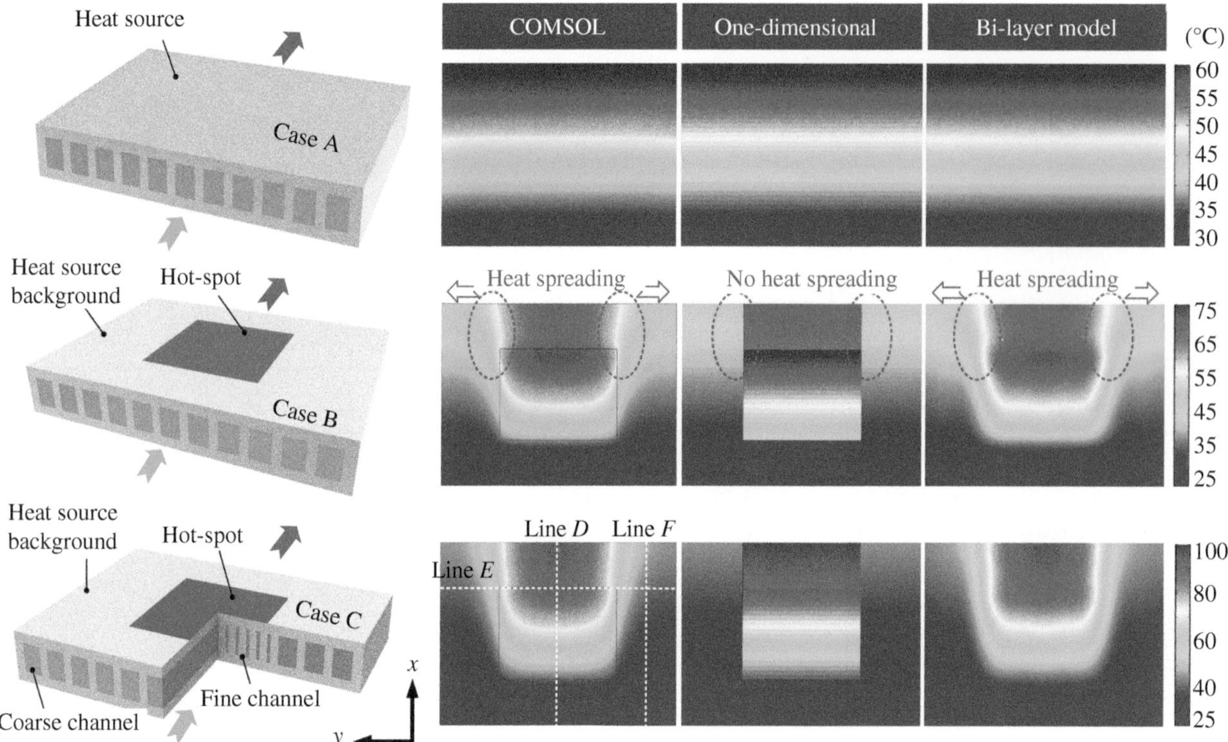

Figure 9.44 Calculated junction surface temperature contour by COMSOL, 1D thermal resistance model, and our bilayer compact thermal model for cases A, B, and C.

A and B. This means that locally refined channels might not help decrease the temperature. On the contrary, it may weaken the cooling effect.

We can see clearly in Figure 9.46 that this phenomenon is mainly caused by higher fluid temperature. It shows that the fluid temperature of case C is much larger than that of case B, which is caused by smaller flow rate in case C. As the fine channels in case C give rise to larger flow resistances, limited by the constant pumping power, the flow rate in case C should be the smallest among the three cases. What's more, the flow rate at the hotspot region is much smaller because of the flow rate distribution. On the other hand, using the 1D theory, the surface temperature can be described as $T_{sur} = QR_{cond} + QR_{conv} + T_{fluid}$, where R_{cond} is the conduction thermal resistance of the solid layer and R_{conv} is the convection thermal resistance of the convection layer, which is shown as the second term in Equation (9.63). The solid layer structure and heat source of cases B and C are nearly the same. The R_{conv} of case C is smaller than that of case B, owing to the larger heat transfer area of fine channels. Thus, the temperature difference ΔT_B is larger than ΔT_C. Therefore, the junction surface temperature gradient may be regulated by modifying the local channel structure, which is responsible for the change of the local R_{conv} and flow rate.

Figure 9.47(a) shows the simulated pressure distribution for the case C at $z = 0.375$ cross section. The pressure distribution in the hotspot region is obviously different. It can be seen from Figure 9.47(d) that the calculated pressure using the bilayer model fits well with the COMSOL simulation, and the flow resistance in the hotspot region is larger. Line F represents the line $y = 2$ in the cross section. Figure 9.47(b) shows the simulated flow velocity. The maximum velocity is 0.85 m/s, so that the Reynolds number is less than 107.2, indicating that it is laminar flow in the microchannel. The right side of Figure 9.47(b) is the upstream channels of the hotspot region and the left side the channels at the background region. Because of the flow resistance difference, flow rate in the latter is apparently larger than that in the former. Figure 9.47(c) shows the channel structure at the interface of the fine and coarse channels. The distance between the fine and coarse channels is set to be 50 μm for fabrication simplicity. It can be seen that the fluid flows into the corresponding downstream channels without obvious vortex generated at the interface region.

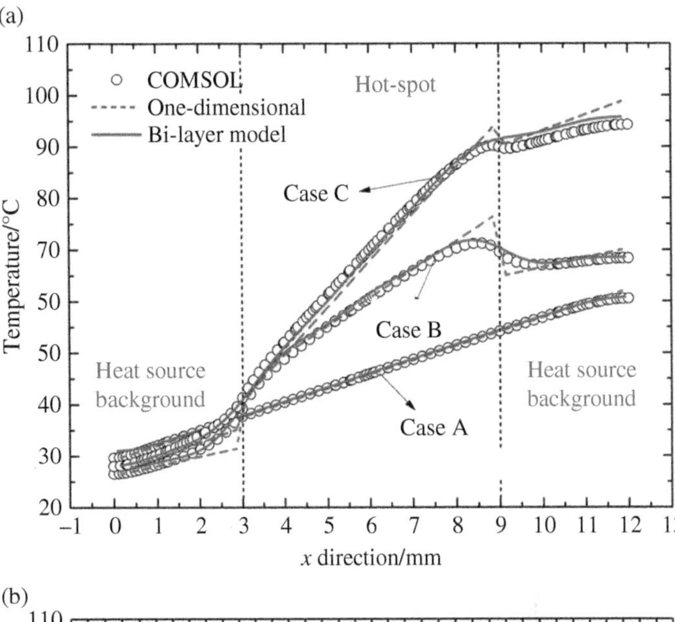

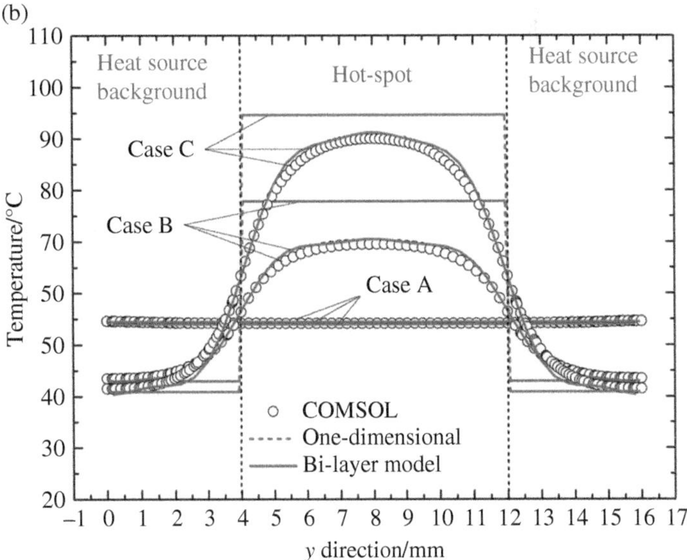

Figure 9.45 Calculated surface temperature for cases A, B, and C of (a) line D (y = 8 mm) and (b) line E (x = 9 mm).

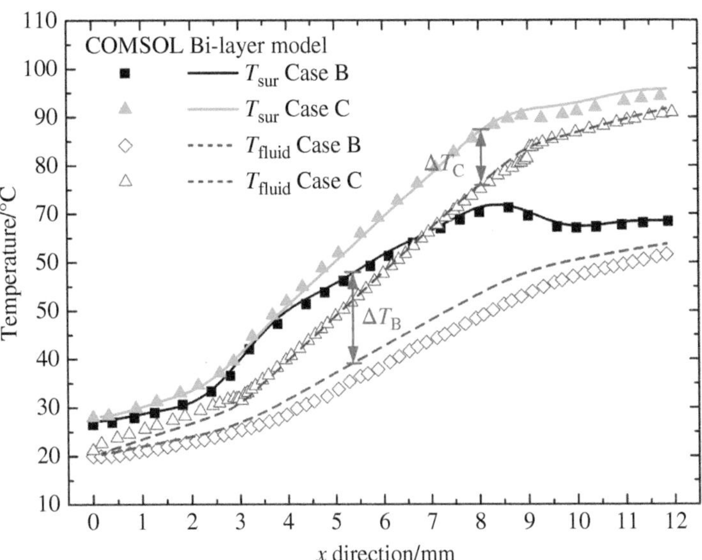

Figure 9.46 Calculated surface temperature and fluid temperature along line D for cases B and C.

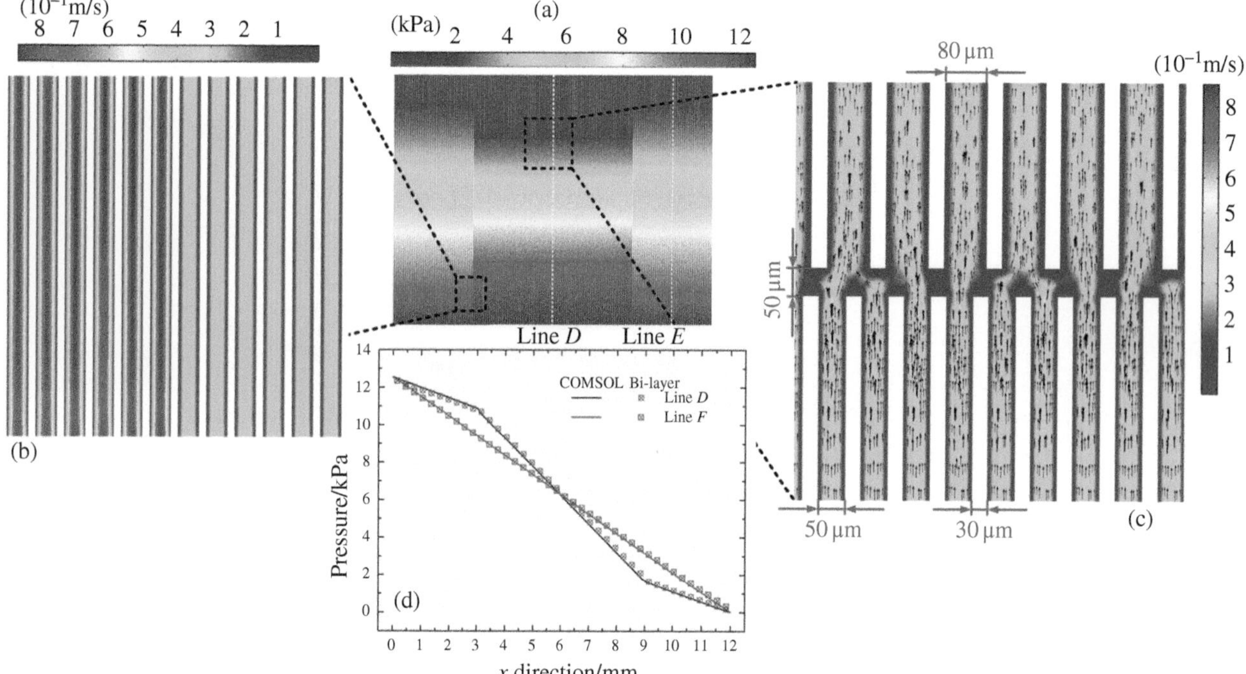

Figure 9.47 (a) The pressure contour of the $z = 0.375$ mm cross section, (b) the flow velocity contour of the $z = 0.375$ mm cross section, (c) the streamline of the $z = 0.375$ mm cross section, and (d) the pressure distribution along line D and line F.

The most important improvement of the bilayer thermal compact model is that it uses the 3D heat conduction model to describe the heat spreading of the nonuniform heat source. Thus, we verified this proposed model through changing the thickness of the solid layer t. Results in Figure 9.48(a)–(c) show that the bilayer model predicts the temperature accurately along the flow direction with t varying from 100 μm to 2 mm. But the 1D results show huge discrepancies. Figure 9.48(d)–(f) shows the same conclusion for the calculated temperature along the line E. When the solid layer thickness increases, the heat spreading of heat sources is more sufficient, so the 1D model can't properly predict the surface temperature gradient in this case.

In spite of the larger thermal conduction resistance induced by increasing the solid layer thickness, the temperature uniformity is better and the maximum temperature is reduced. Therefore, when locally high heat flux exists, the heat spreading may be more important than heat conduction resistance. Therefore, the nonuniform microchannel design, to obtain a nonuniform local heat transfer coefficient, which results in the heat spreading effect, is very helpful for temperature uniformity and hotspot cooling.

Figure 9.49(a) shows that when the channel width is increased for case B, the maximum temperature first decreases to 69.72 °C, then increases to over 100 °C, regardless of the monotonically increasing flow rate. Figure 9.49(b) shows the maximum and minimum temperatures vary with channel width at hotspot region with background channel width of 80 and 150 μm, respectively. The best hotspot channel width for lowest maximum temperature and best uniformity is 90 μm for background channel width of 80 μm and 100 μm for background channel width of 150 μm. The flow rate and heat transfer coefficient corresponding to the channel width together determine the surface temperature gradient. So channel width distribution optimization is required.

Considering a maximum feasible aspect ratio of 10 for microchannels and the feasibility of the deep reactive ion etching technology, the silicon microchannel height H is assumed to be 300 μm and the channel width ranges from 30 to 300 μm. The channel wall width is set as a constant variable w_w. The inlet fluid temperature is 20 °C and the input pumping power is limited as 0.015 W. The nonuniform heat source is set the same as case B.

Using the GA described in Section 9.4.3, we obtained the microchannel structure with optimized channel distribution, which is shown in Figure 9.50(a). The overall input volume flow rate of the optimized microchannel is 95.1 ml/min. The channel wall width is optimized as 32 μm. Figure 9.50(b) shows the optimized w_{ch} distribution for the microchannel. It can

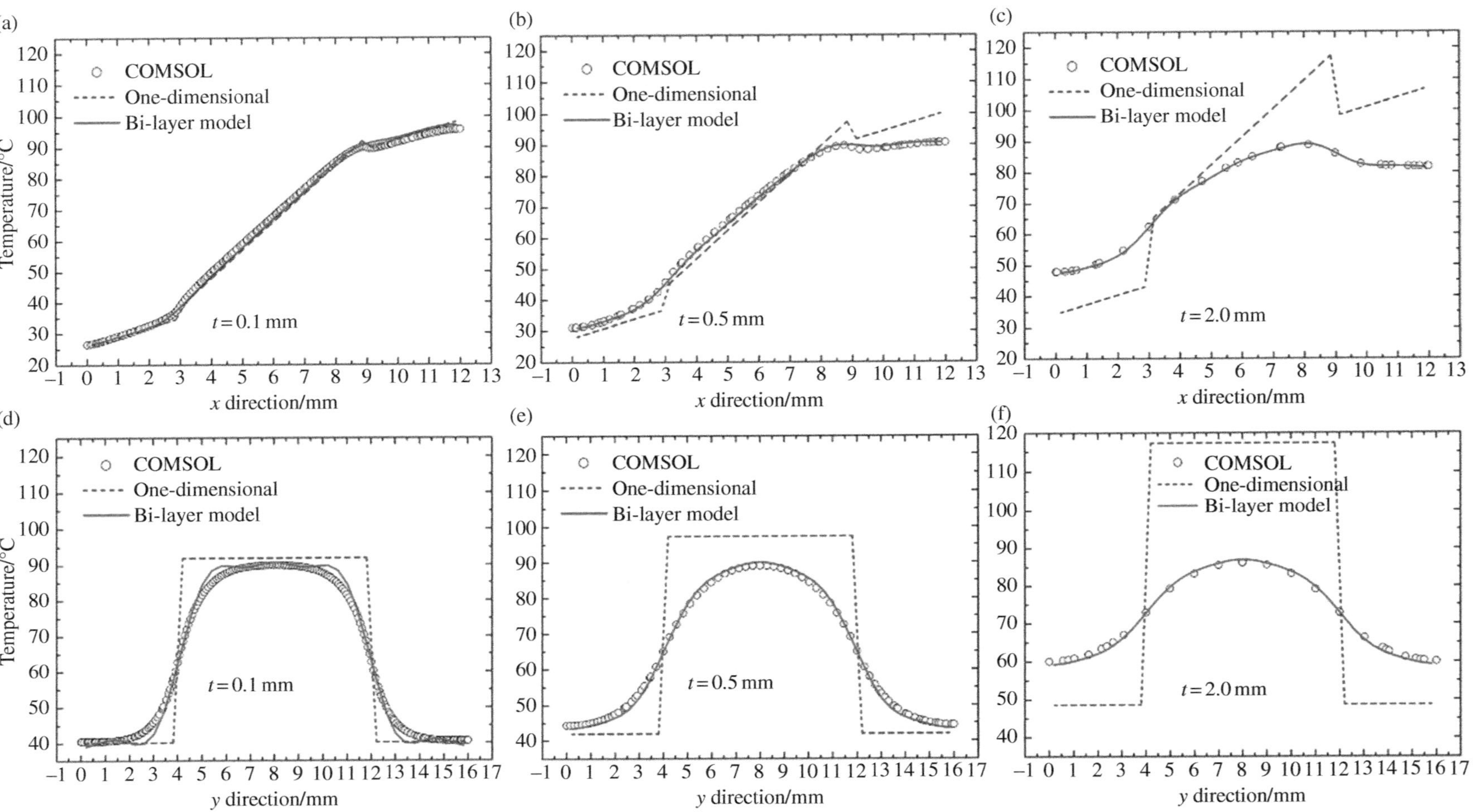

Figure 9.48 Calculated surface temperature for case C of (a) line D (y = 8 mm) with t = 0.1 mm, (b) line D (y = 8 mm) with t = 0.5 mm, (c) line D (y = 8 mm) with t = 2 mm, (d) line E (x = 9 mm) with t = 0.1 mm, (e) line E (x = 9 mm) with t = 0.5 mm, and (f) line E (x = 9 mm) with t = 2 mm.

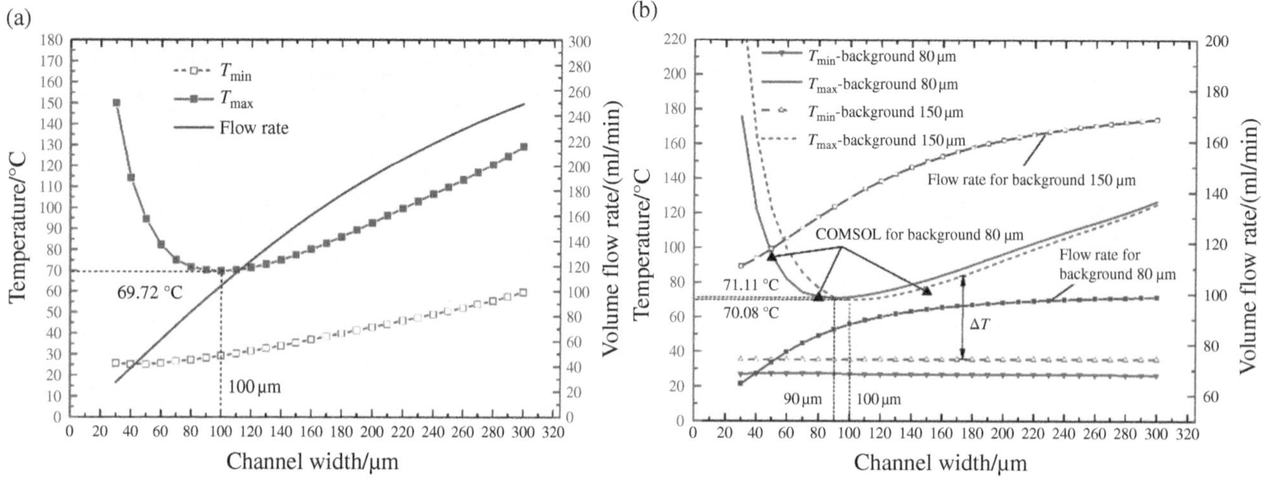

Figure 9.49 Calculated maximum and minimum surface temperatures with various channel widths for (a) uniform channel like case B and (b) nonuniform channel like case C.

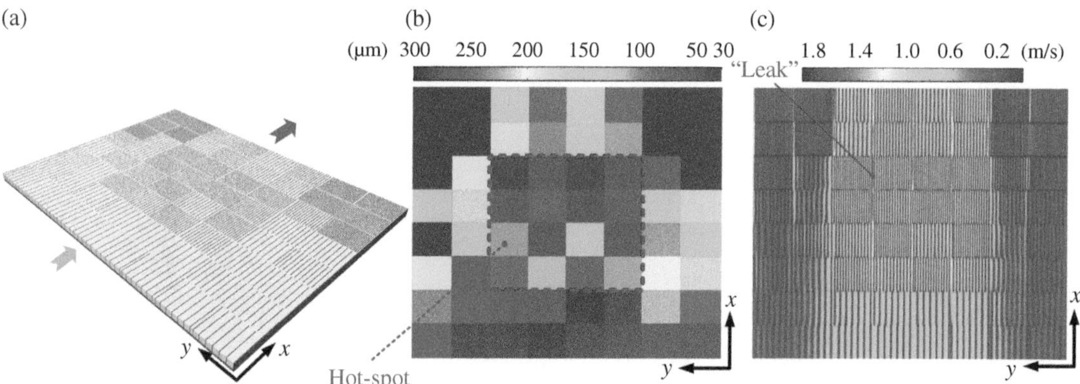

Figure 9.50 (a) Structure of the optimized microchannel, (b) the optimized w_{ch} distribution for the microchannel, and (c) the simulated fluid velocity distribution of the optimized microchannel at the z = 0.375 mm cross section.

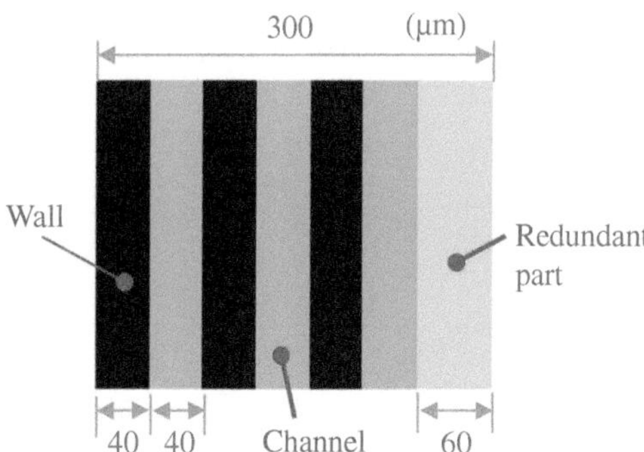

Figure 9.51 Schematic diagram of the structure setup principle of the microchannel in the simulation of this work.

be observed that the w_{ch} gradually decreases along the flow direction in the background region, so that the temperature gradient along the flow direction caused by fluid temperature is as small as possible, because large w_{ch} causes large convective thermal resistance, which makes up the low upstream fluid temperature. In addition, the w_{ch} of hotspot is small to get large convective thermal resistance. The w_{ch} distribution makes a large aspect ratio of flow rate in the hotspot region, to provide enough heat dissipation capability in this region (Figure 9.50(c)). Figure 9.50(c) shows the fluid velocity distribution of the optimized microchannel. The discrete cell cannot be divided exactly by the calculated local w_{ch} and w_w. The redundant part in each discrete cell makes a channel with a larger w_{ch}. For example, in Figure 9.51, assuming that the width of the discrete cell is 300 μm, and the channel width and the wall width are both 40 μm, there should be a redundant part with the width of 60 μm that is not able to be

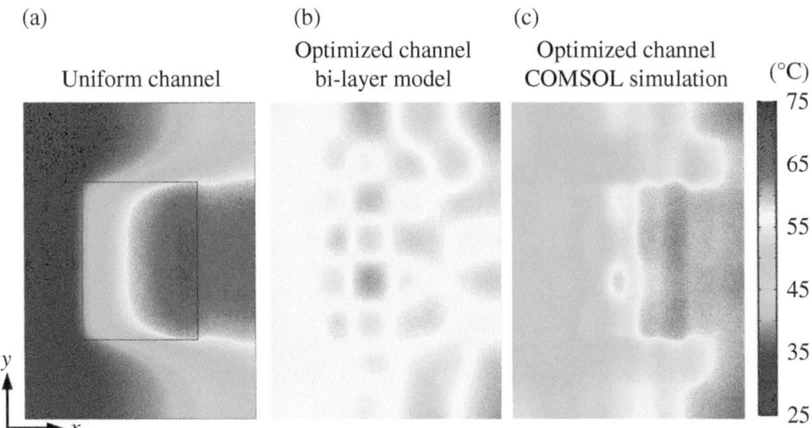

Figure 9.52 (a) Simulated junction surface temperature for microchannel with uniform channels, (b) the bilayer compact thermal model calculated junction surface temperature for optimized microchannel, and (c) the simulated junction surface temperature for optimized microchannel.

assigned to a pair of channel and wall. Therefore, we should simply add the redundant part to the nearest channel and make the channel width to be 100 μm.

It should be noted that, in reality, this trouble may commonly exist in the fabrication of the microchannel. This strategy is only used in this work for simplicity of the simulation and it may not be the best choice in industry. However, no matter what strategy is applied, some of the channel widths must be enlarged. Wider channel has a smaller flow resistance, resulting in larger flow rate and larger flow velocity. Therefore, a certain part of the fluid in each discrete cell may "leak" through the wide channel (Figure 9.50(c)). This is one of the main errors between the simulation results and the model (Figure 9.52), which result in the slightly different temperature distributions between Figure 9.52(b) and (c). Therefore, the presented methodology cannot be used to predict the real temperature distribution of the fabricated microchannel accurately because of the structure difference between the real microchannel and the presented model. However, it is totally sufficient for guiding the design of the microchannels, especially the microchannels embedded in chips. And the structure considering the redundant part can be optimized further.

Figure 9.52 shows the comparison of junction surface temperature between the optimized microchannel and the uniform microchannel in case B. Results show that the temperature uniformity of the optimized microchannel is apparently better than that of the uniform microchannel (Figure 9.52(a)). In addition, the maximum temperature of the optimized microchannel is lower than the nonoptimized one. The model-calculated maximum surface temperature is 63 °C and the simulated maximum temperature is 65 °C. The model-calculated surface temperature difference of optimized microchannel is only 13 °C, which is much smaller than that of the uniform microchannel, 45 °C.

9.5 Chapter Summary

In Section 9.1, a high-performance spray cooling system for co-designing nozzle configuration and electronics for thermal management is presented, and it is demonstrated on high-power LEDs. Co-design of nozzle configurations and electronics for spray cooling system can provide the optimal thermal management solution for electronics, which may aid in solving the critical challenges in spray cooling applications in electronics, as well as enabling energy-efficient cooling. In Section 9.2, a JIBC device and a channel/jet impingement HBC device with micronozzles using the traditional mechanical machining method were introduced. The body cooling devices provide high efficient cooling to the top surface and the side surfaces of the chip. The thermal model for the developed cooling methods was established to predict the chip temperature and characterize the heat transfer mechanism. The results reveal that the critical nozzle diameter exists, determining which one of the two developed body cooling devices has the better cooling performance. The proposed thermal model is found to be necessary for the quick and optimal design of the body cooling device for high-power and high-heat-flux electronics. In Section 9.3, we introduced a compact jet array impingement cooling system driven by integrated piezoelectric micropump.

The JAICIPM provides a high thermal performance and reduces the size of the thermal management system by integrating jet impingement cooling with piezoelectric micropump. The JAICIPM achieved a high thermal performance with a small size and may become a promising option for cooling electronics in compact systems. In Section 9.4, we introduced a bilayer compact thermal model by considering 3D heat spreading and thermal conduction for the junction temperature gradient prediction in microchannels with nonuniform heat source. The present model and design strategy can be used to diminish the hotspot and achieve uniform temperature in high-power chips for extensive applications like ICs, IGBTs, and HEMTs.

References

1 Luo, X.B., Hu, R., Liu, S. et al. (2016). Heat and fluid flow in high-power LED packaging and applications. *Prog. Energ. Combust.* 56: 1–32.

2 Chu, R.C., Simons, R.E., Ellsworth, M.J. et al. (2004). Review of cooling technologies for computer products. *IEEE Trans. Device Mater. Reliab.* 4 (4): 568–585.

3 Murshed, S.M.S. and de Castro, C.A.N. (2017). A critical review of traditional and emerging techniques and fluids for electronics cooling. *Renew. Sust. Energ. Rev.* 78: 821–833.

4 Laloya, E., Lucia, O., Sarnago, H. et al. (2016). Heat management in power converters: from state of the art to future ultrahigh efficiency systems. *IEEE Trans. Power Electron.* 31 (11): 7896–7908.

5 Ma, Y., Lan, W., Xie, B. et al. (2018). An optical-thermal model for laser-excited remote phosphor with thermal quenching. *Int. J. Heat Mass Transf.* 116: 694–702.

6 Kheirabadi, A.C. and Groulx, D. (2016). Cooling of server electronics: a design review of existing technology. *Appl. Therm. Eng.* 105: 622–638.

7 Hou, F., Zhang, H., Huang, D. et al. (2020). Microchannel thermal management system with two-phase flow for power electronics over 500 W/cm^2 heat dissipation. *IEEE Trans. Power Electron.* 35 (10): 10592–10600.

8 Zhang, X.-D., Yang, X.-H., Zhou, Y.-X. et al. (2019). Experimental investigation of galinstan based minichannel cooling for high heat flux and large heat power thermal management. *Energy Convers. Manag.* 185: 248–258.

9 Shi, M., Yu, X., Tan, Y. et al. (2022). Thermal performance of insulated gate bipolar transistor module using microchannel cooling base plate. *Appl. Therm. Eng.* 201: 117718.

10 Wang, Y., Zhou, N.Y., Yang, Z. et al. (2016). Experimental investigation of aircraft spray cooling system with different heating surfaces and different additives. *Appl. Therm. Eng.* 103: 510–521.

11 Xu, R., Wang, G., and Jiang, P. (2022). Spray cooling on enhanced surfaces: a review of the progress and mechanisms. *J. Electron. Packag.* 144 (1): 010802.

12 Liang, G. and Mudawar, I. (2017). Review of spray cooling - part 1: single-phase and nucleate boiling regimes, and critical heat flux. *Int. J. Heat Mass Transf.* 115: 1174–1205.

13 Ekkad, S.V. and Singh, P. (2021). A modern review on jet impingement heat transfer methods. *J. Heat Transf. Trans. ASME* 143 (6): 064001.

14 Wu, R., Hong, T., Cheng, Q. et al. (2019). Thermal modeling and comparative analysis of jet impingement liquid cooling for high power electronics. *Int. J. Heat Mass Transf.* 137: 42–51.

15 de Brun, C., Jenkins, R., Lupton, T.L. et al. (2017). Confined jet array impingement boiling. *Exp. Thermal Fluid Sci.* 86: 224–234.

16 Xie, B., Liu, H., Hu, R. et al. (2018). Targeting cooling for quantum dots in white QDs-LEDs by hexagonal boron nitride platelets with electrostatic bonding. *Adv. Funct. Mater.* 28 (30): 1801407.

17 Yu, X., Xiang, L., Zhou, S. et al. (2021). Effect of refractive index of packaging materials on the light extraction efficiency of COB-LEDs with millilens array. *Appl. Opt.* 60 (2): 306–311.

18 Hamidnia, M., Luo, Y., and Wang, X.D. (2018). Application of micro/nano technology for thermal management of high power LED packaging - a review. *Appl. Therm. Eng.* 145: 637–651.

19 Khandekar, S., Sahu, G., Muralidhar, K. et al. (2021). Cooling of high-power LEDs by liquid sprays: challenges and prospects. *Appl. Therm. Eng.* 184: 115640.

20 Tamdogan, E. and Arik, M. (2015). Natural convection immersion cooling with enhanced optical performance of light-emitting diode systems. *J. Electron. Packag.* 137 (4): 041006.

21 Kumar, P., Sahu, G., Chatterjee, D. et al. (2022). Copper wick based loop heat pipe for thermal management of a high-power LED module. *Appl. Therm. Eng.* 211: 118459.

22 Luo, X., Hu, R., Guo, T. et al. (2010). Low thermal resistance LED light source with vapor chamber coupled fin heat sink. *Proceedings of the 60th Electronic Components and Technology Conference*, Las Vegas, NV (1–4 June 2010). pp. 1347–1352.

23 Xiao, C., Liao, H., Wang, Y. et al. (2017). A novel automated heat-pipe cooling device for high-power LEDs. *Appl. Therm. Eng.* 111: 1320–1329.

24 Zhang, X., Xie, B., Zhou, S. et al. (2022). Radially oriented functional thermal materials prepared by flow field-driven self-assembly strategy. *Nano Energy* 104: 107986.

25 Mudawar, I. (2013). Recent advances in high-flux, two-phase thermal management. *J. Therm. Sci. Eng. Appl.* 5 (2): 021012.

26 Agostini, B., Fabbri, M., Park, J.E. et al. (2007). State of the art of high heat flux cooling technologies. *Heat Transf. Eng.* 28 (4): 258–281.

27 Seo, J.-H. and Lee, M.-Y. (2018). Illuminance and heat transfer characteristics of high power LED cooling system with heat sink filled with ferrofluid. *Appl. Therm. Eng.* 143: 438–449.

28 Chung, Y.-C., Chung, H.-H., Lee, Y.-H. et al. (2020). Heat dissipation and electrical conduction of an LED by using a microfluidic channel with a graphene solution. *Appl. Therm. Eng.* 175: 115383.

29 Lin, X., Mo, S., Mo, B. et al. (2020). Thermal management of high-power LED based on thermoelectric cooler and nanofluid-cooled microchannel heat sink. *Appl. Therm. Eng.* 172: 115165.

30 Jeong, M.W., Jeon, S.W., Lee, S.H. et al. (2015). Effective heat dissipation and geometric optimization in an LED module with aluminum nitride (AlN) insulation plate. *Appl. Therm. Eng.* 76: 212–219.

31 Ramos-Alvarado, B., Feng, B., and Peterson, G.P. (2013). Comparison and optimization of single-phase liquid cooling devices for the heat dissipation of high-power LED arrays. *Appl. Therm. Eng.* 59 (1–2): 648–659.

32 Deng, X., Luo, Z., Xia, Z. et al. (2017). Active-passive combined and closed-loop control for the thermal management of high-power LED based on a dual synthetic jet actuator. *Energy Convers. Manag.* 132: 207–212.

33 Luo, X., Chen, W., Sun, R. et al. (2008). Experimental and numerical investigation of a microjet-based cooling system for high power LEDs. *Heat Transf. Eng.* 29 (9): 774–781.

34 Gatapova, E.Y., Sahu, G., Khandekar, S. et al. (2021). Thermal management of high-power LED module with single-phase liquid jet array. *Appl. Therm. Eng.* 184: 116270.

35 Liu, S., Yang, J., Gan, Z. et al. (2008). Structural optimization of a microjet based cooling system for high power LEDs. *Int. J. Therm. Sci.* 47 (8): 1086–1095.

36 Kim, J. (2007). Spray cooling heat transfer: the state of the art. *Int. J. Heat Fluid Flow* 28 (4): 753–767.

37 Xiang, L., Yu, X., Hong, T. et al. (2023). Performance of spray cooling with vertical surface orientation: an experimental investigation. *Appl. Therm. Eng.* 219: 119434.

38 Wang, J.-X., Guo, W., Xiong, K. et al. (2020). Review of aerospace-oriented spray cooling technology. *Prog. Aerosp. Sci.* 116: 100635.

39 Dong, H., Ruan, L., Wang, Y. et al. (2021). Performance of air/spray cooling system for large-capacity and high-power-density motors. *Appl. Therm. Eng.* 192: 116925.

40 Lei, S., Shi, Y., and Chen, G. (2020). A lithium-ion battery-thermal-management design based on phase-change-material thermal storage and spray cooling. *Appl. Therm. Eng.* 168: 114792.

41 Hsieh, S.-S., Hsu, Y.-F., and Wang, M.-L. (2014). A microspray-based cooling system for high powered LEDs. *Energy Convers. Manag.* 78: 338–346.

42 Sahu, G., Khandekar, S., and Muralidhar, K. (2022). Thermal characterization of spray impingement heat transfer over a high-power LED module. *Therm. Sci. Eng. Prog.* 32: 101332.

43 Ben Abdelmlek, K., Araoud, Z., Ghnay, R. et al. (2016). Effect of thermal conduction path deficiency on thermal properties of LEDs package. *Appl. Therm. Eng.* 102: 251–260.

44 Ciofalo, M., Caronia, A., Di Liberto, M. et al. (2007). The Nukiyama curve in water spray cooling: its derivation from temperature-time histories and its dependence on the quantities that characterize drop impact. *Int. J. Heat Mass Transf.* 50 (25–26): 4948–4966.

45 Hall, D.D. and Mudawar, I. (1995). Experimental and numerical study of quenching complex-shaped metallic alloys with multiple, overlapping sprays. *Int. J. Heat Mass Transf.* 38 (7): 1201–1216.

46 Sharma, C.S., Tiwari, M.K., Zimmermann, S. et al. (2015). Energy efficient hotspot-targeted embedded liquid cooling of electronics. *Appl. Energy* 138: 414–422.

47 Zhang, L.-Y., Zhang, Y.-F., Chen, J.-Q. et al. (2015). Fluid flow and heat transfer characteristics of liquid cooling microchannels in LTCC multilayered packaging substrate. *Int. J. Heat Mass Transf.* 84: 339–345.

48 Yuan, C., Xie, B., Huang, M. et al. (2016). Thermal conductivity enhancement of platelets aligned composites with volume fraction from 10% to 20%. *Int. J. Heat Mass Transf.* 94: 20–28.

49 Cui, F.L., Hong, F.J., and Cheng, P. (2018). Comparison of normal and distributed jet array impingement boiling of HFE-7000 on smooth and pin-fin surfaces. *Int. J. Heat Mass Transf.* 126: 1287–1298.

50 Devahdhanush, V.S. and Mudawar, I. (2021). Review of critical heat flux (CHF) in jet impingement boiling. *Int. J. Heat Mass Transf.* 169: 120893.

51 Li, C.Y. and Garimella, S.V. (2001). Prandtl-number effects and generalized correlations for confined and submerged jet impingement. *Int. J. Heat Mass Transf.* 44 (18): 3471–3480.

52 Bandhauer, T.M., Hobby, D.R., Jacobsen, C. et al. (2016). Thermal performance of micro-jet impingement device with parallel flow, jet-adjacent fluid removal. *Proceedings of the 16th ASME International Conference on Nanochannels, Microchannels, and Minichannels (ICNMM 2018)*, Dubrovnik, Croatia (10–13 June 2018).

53 Jorg, J., Taraborrelli, S., Sarriegui, G. et al. (2018). Direct single impinging jet cooling of a MOSFET power electronic module. *IEEE Trans. Power Electron.* 33 (5): 4224–4237.

54 Cui, H.C., Xie, J.H., Zhao, R.Z. et al. (2022). Thermal-hydraulic performance analysis of a hybrid micro pin-fin, jet impingement heat sink with non-uniform heat flow. *Appl. Therm. Eng.* 208: 118201.

55 Kim, C.-B., Leng, C., Wang, X.-D. et al. (2015). Effects of slot-jet length on the cooling performance of hybrid microchannel/slot-jet module. *Int. J. Heat Mass Transf.* 89: 838–845.

56 Han, Y., Lau, B.L., Zhang, X. et al. (2014). Thermal management of hotspots with a microjet-based hybrid heat sink for GaN-on-Si devices. *IEEE Trans. Compon. Packag. Manuf. Technol.* 4 (9): 1441–1450.

57 Sung, M.K. and Mudawar, I. (2008). Single-phase hybrid micro-channel/micro-jet impingement cooling. *Int. J. Heat Mass Transf.* 51 (17–18): 4342–4352.

58 Qian, C., Gheitaghy, A.M., Fan, J. et al. (2018). Thermal management on IGBT power electronic devices and modules. *IEEE Access* 6: 12868–12884.

59 Yovanovich, M.M. (2003). Thermal resistances of circular source on finite circular cylinder with side and end cooling. *J. Electron. Packag.* 125 (2): 169–177.

60 Chang, C., Kojasoy, G., Landis, F.P. et al. (1995). Confined single- and multiple-jet impingement heat transfer—I. Turbulent submerged liquid jets. *Int. J. Heat Mass Transf.* 38: 833–842.

61 Robinson, A.J., Kempers, R., Colenbrander, J. et al. (2018). A single phase hybrid micro heat sink using impinging micro-jet arrays and microchannels. *Appl. Therm. Eng.* 136: 408–418.

62 Wu, R., Duan, B., Liu, F. et al. (2017). Design of a hydro-dynamically levitated centrifugal micro-pump tor the active liquid cooling system. *2017 18th International Conference on Electronic Packaging Technology (ICEPT)*. pp. 402–406.

63 Wang, X., Li, B., Gerada, D. et al. (2022). A critical review on thermal management technologies for motors in electric cars. *Appl. Therm. Eng.* 201: 117758.

64 Yang, X.-H., Tan, S.-C., and Liu, J. (2016). Thermal management of Li-ion battery with liquid metal. *Energy Convers. Manag.* 117: 577–585.

65 Khalaj, A.H. and Halgamuge, S.K. (2017). A review on efficient thermal management of air- and liquid-cooled data centers: from chip to the cooling system. *Appl. Energy* 205: 1165–1188.

66 Zimmermann, S., Meijer, I., Tiwari, M.K. et al. (2012). Aquasar: a hot water cooled data center with direct energy reuse. *Energy* 43 (1): 237–245.

67 Jiang, L.-J., Jiang, S.-L., Cheng, W.-L. et al. (2019). Experimental study on heat transfer performance of a novel compact spray cooling module. *Appl. Therm. Eng.* 154: 150–156.

68 Tang, G., Han, Y., Lau, B.L. et al. (2016). Development of a compact and efficient liquid cooling system with silicon microcooler for high-power microelectronic devices. *IEEE Trans. Compon. Packag. Manuf. Technol.* 6 (5): 729–739.

69 Ning, J., Wang, X., Sun, Y. et al. (2022). Experimental and numerical investigation of additively manufactured novel compact plate-fin heat exchanger. *Int. J. Heat Mass Transf.* 190: 122818.

70 Duan, B., Guo, T., Luo, M. et al. (2014). A mechanical micropump for electronic cooling. *Proceedings of the 14th InterSociety Conference on Thermal and Thermomechanical Phenomena in Electronic Systems (ITherm)*, Orlando, FL (27–30 May 2014). pp. 1038–1042.

71 Yueh, W., Wan, Z., Xiao, H. et al. (2017). Active fluidic cooling on energy constrained system-on-chip systems. *IEEE Trans. Compon. Packag. Manuf. Technol.* 7 (11): 1813–1822.

72 Jung, K.W., Kharangate, C.R., Lee, H. et al. (2019). Embedded cooling with 3D manifold for vehicle power electronics application: single-phase thermal-fluid performance. *Int. J. Heat Mass Transf.* 130: 1108–1119.

73 Kwon, B., Maniscalco, N.I., Jacobi, A.M. et al. (2018). High power density air-cooled microchannel heat exchanger. *Int. J. Heat Mass Transf.* 118: 1276–1283.

74 van Erp, R., Soleimanzadeh, R., Nela, L. et al. (2020). Co-designing electronics with microfluidics for more sustainable cooling. *Nature* 585 (7824): 211–216.

75 Peng, Y.-H., Wang, D.-H., Li, X.-Y. et al. (2022). Cooling chip on PCB by embedded active microchannel heat sink. *Int. J. Heat Mass Transf.* 196: 123251.

76 Tang, Y., Jia, M., Ding, X. et al. (2019). Experimental investigation on thermal management performance of an integrated heat sink with a piezoelectric micropump. *Appl. Therm. Eng.* 161.

77 Li, H., Liu, J., Li, K. et al. (2021). A review of recent studies on piezoelectric pumps and their applications. *Mech. Syst. Signal Process.* 151: 107393.

78 Nisar, A., Aftulpurkar, N., Mahaisavariya, B. et al. (2008). MEMS-based micropumps in drug delivery and biomedical applications. *Sensor Actuator B Chem* 130 (2): 917–942.

79 Ma, H.-K., Chen, B.-R., Gao, J.-J. et al. (2009). Development of an OAPCP-micropump liquid cooling system in a laptop. *Int. Commun. Heat Mass Transf.* 36 (3): 225–232.

80 Laser, D.J. and Santiago, J.G. (2004). A review of micropumps. *J. Micromech. Microeng.* 14 (6): R35–R64.

81 Fan, Y., Zhao, W., Zhang, X. et al. (2022). Development of a piezoelectric pump with unfixed valve. *J. Micromech. Microeng.* 32 (5): 055004.

82 Cui, H.C., Lai, X.T., Wu, J.F. et al. (2021). Overall numerical simulation and experimental study of a hybrid oblique-rib and submerged jet impingement/microchannel heat sink. *Int. J. Heat Mass Transf.* 167: 120839.

83 Lu, S., Yu, M., Qian, C. et al. (2020). A quintuple-bimorph tenfold-chamber piezoelectric pump used in water-cooling system of electronic chip. *IEEE Access* 8: 186691–186698.

84 Ma, H.K., Hou, B.R., Lin, C.Y. et al. (2008). The improved performance of one-side actuating diaphragm micropump for a liquid cooling system. *Int. Commun. Heat Mass Transf.* 35 (8): 957–966.

85 Sun, B. and Liu, H.-L. (2017). Flow and heat transfer characteristics of nanofluids in a liquid-cooled CPU heat radiator. *Appl. Therm. Eng.* 115: 435–443.

86 Tuckerman, D.B. and Pease, R.F.W. (1981). High-performance heat sinking for VLSI. *IEEE Electron Device Lett.* 2 (5): 126–129.

87 Gong, L., Zhao, J., and Huang, S. (2015). Numerical study on layout of micro-channel heat sink for thermal management of electronic devices. *Appl. Therm. Eng.* 88: 480–490.

88 Mu, Y.-T., Chen, L., He, Y.-L. et al. (2015). Numerical study on temperature uniformity in a novel mini-channel heat sink with different flow field configurations. *Int. J. Heat Mass Transf.* 85: 147–157.

89 Chein, R. and Chuang, J. (2007). Experimental microchannel heat sink performance studies using nanofluids. *Int. J. Therm. Sci.* 46: 57–66.

90 Koo, J.-M., Im, S., Jiang, L. et al. (2005). Integrated microchannel cooling for three-dimensional electronic circuit architectures. *J. Heat Transf.* 127: 49–58.

91 Kandlikar, S.G. (2012). History, advances, and challenges in liquid flow and flow boiling heat transfer in microchannels: a critical review. *J. Heat Transf. Trans. ASME* 134 (3): 034001.

92 Luo, X. and Mao, Z. (2012). Thermal modeling and design for microchannel cold plate with high temperature uniformity subjected to multiple heat sources. *Int. Commun. Heat Mass Transf.* 39: 781–785.

93 Garimella, S.V., Yeh, L.T., and Persoons, T. (2012). Thermal management challenges in telecommunication systems and data centers. *IEEE Trans. Compon. Packag. Manuf. Technol.* 2: 1307–1316.

94 Thome, J.R. (2006). State-of-the-art overview of boiling and two-phase flows in microchannels. *Heat Transf. Eng.* 27: 19–14.

95 Cheng, L. and Xia, G. (2017). Fundamental issues, mechanisms and models of flow boiling heat transfer in microscale channels. *Int. J. Heat Mass Transf.* 108: 97–127.

96 Zhang, T., Wen, J.T., Julius, A.A. et al. (2011). Stability analysis and maldistribution control of two-phase flow in parallel evaporating channels. *Int. J. Heat Mass Transf.* 54: 5298–5305.

97 Lorenzini, D., Green, C.E., Sarvey, T.E. et al. (2016). Embedded single phase microfluidic thermal management for non-uniform heating and hotspots using microgaps with variable pin fin clustering. *Int. J. Heat Mass Transf.* 103: 1359–1370.

98 Brask, A., Kutter, J.P., and Bruus, H. (2005). Long-term stable electroosmotic pump with ion exchange membranes. *Lab Chip* 5 (7): 730–738.

99 Yang, X., Tan, S.-C., Ding, Y. et al. (2017). Flow and thermal modeling and optimization of micro/mini-channel heat sink. *Appl. Therm. Eng.* 117: 289–296.

100 Mao, Z., Luo, X., and Liu, S. (2011). Compact thermal model for microchannel substrate with high temperature uniformity subjected to multiple heat sources. *IEEE 61st Electronic Components and Technology Conference (ECTC)*, Lake Buena Vista, FL, USA. 1662–1672.

101 Sharma, C.S., Tiwari, M.K., and Poulikakos, D. (2016). A simplified approach to hotspot alleviation in microprocessors. *Appl. Therm. Eng.* 93: 1314–1323.

102 Abdoli, A. and Dulikravich, G.S. (2013). Optimized multi-floor throughflow micro heat exchangers. *Int. J. Therm. Sci.* 78: 111–123.

103 Luhar, S., Sarkar, D., and Jain, A. (2017). Steady state and transient analytical modeling of non-uniform convective cooling of a microprocessor chip due to jet impingement. *Int. J. Heat Mass Transf.* 110: 768–777.

104 Shah, R.K. and London, A.L. (1978). *Laminar Flow Forced Convection in Ducts*. Academic Press, New York.

105 Hu, R., Cheng, T., Li, L. et al. (2014). Phosphor distribution optimization to decrease the junction temperature in white pc-LEDs by genetic algorithm. 77: 891–896.

106 Lee, H., Agonafer, D., Won, Y. et al. (2016). Thermal modeling of extreme heat flux microchannel coolers for GaN-on-SiC semiconductor devices. *J. Electron. Packag* 138: 010907.

Index

Thermal Management for Opto-electronics Packaging and Applications, First Edition. Xiaobing Luo, Run Hu, and Bin Xie.

d

e

i

q

t

u

v

w

x

y

z